College Algebra

Concepts and Models

Fifth Edition

Ron Larson
Pennsylvania State University
The Behrend College

Robert Hostetler
Pennsylvania State University
The Behrend College

Anne V. Hodgkins
Phoenix College

HOUGHTON MIFFLIN COMPANY   Boston   New York

Vice President and Publisher: Jack Shira
Associate Sponsoring Editor: Cathy Cantin
Development Manager: Maureen Ross
Associate Editor: Jennifer King
Editorial Assistant: Elizabeth Kassab
Supervising Editor: Karen Carter
Senior Project Editor: Patty Bergin
Editorial Assistant: Julia Keller
Production Technology Supervisor: Gary Crespo
Senior Marketing Manager: Danielle Potvin
Marketing Coordinator: Nicole Mollica
Senior Manufacturing Coordinator: Marie Barnes

We have included examples and exercises that use real-life data as well as technology output from a variety of software. This would not have been possible without the help of many people and organizations. Our wholehearted thanks goes to all for their time and effort.

Trademark acknowledgments: TI and CBR are registered trademarks of Texas Instruments, Inc.

Custom Publishing Editor: Dan Luciano
Custom Publishing Production Manager: Tina Kozik
Project Coordinator: Andrea Wagner

Cover Art: Photodisc

This book contains select works from existing Houghton Mifflin Company resources and was produced by Houghton Mifflin Custom Publishing for collegiate use. As such, those adopting and/or contributing to this work are responsible for editorial content, accuracy, continuity and completeness.

Printed in the United States of America.

ISBN: 0-618-65878-5
N-04989

2 3 4 5 6 7 8 9 – DS – 07 06 05

Houghton Mifflin
Custom Publishing

222 Berkeley Street • Boston, MA 02116

Address all correspondence and order information to the above address.

CONTENTS

A Word from the Authors (Preface)   vii

Highlights of Features   x

CHAPTER R1   Fundamental Concepts of Algebra

R1.1   Real Numbers: Order and Absolute Value   R2

R1.2   The Basic Rules of Algebra   R10

R1.3   Integer Exponents   R20

R1.4   Radicals and Rational Exponents   R29

Mid-Chapter Quiz   R39

R1.5   Polynomials and Special Products   R40

R1.6   Factoring   R48

R1.7   Fractional Expressions   R55

Make a Decision Project: Musical Notes and Frequencies   R62

Summary R63   **Review Exercises** R64   **Chapter Test** R67

CHAPTER R2   Equations and Inequalities

R2.1   Linear Equations   R70

R2.2   Mathematical Modeling   R79

R2.3   Quadratic Equations   R93

R2.4   The Quadratic Formula   R104

Mid-Chapter Quiz   R114

R2.5   Other Types of Equations   R115

R2.6   Linear Inequalities   R126

R2.7   Other Types of Inequalities   R138

Make a Decision Project: Salaries in Professional Baseball   R148

Summary R149   **Review Exercises** R150   **Chapter Test** R154

Cumulative Test: Chapters R1–R2   R155

CHAPTER 1 Functions and Graphs

1.1   Graphs of Equations   2

1.2   Lines in the Plane   16

1.3   Linear Modeling and Direct Variation   28

1.4   Functions   40

Mid-Chapter Quiz   53

1.5   Graphs of Functions   54

1.6   Transformations of Functions   65

1.7   The Algebra of Functions   74

1.8   Inverse Functions   84

Make a Decision Project: Demographics   94

Summary   95   **Review Exercises**   96   **Chapter Test**   101

CHAPTER 2 Polynomial and Rational Functions

2.1   Quadratic Functions and Models   104

2.2   Polynomial Functions of Higher Degree   116

2.3   Polynomial Division   126

2.4   Real Zeros of Polynomial Functions   136

Mid-Chapter Quiz   150

2.5   Complex Numbers   151

2.6   The Fundamental Theorem of Algebra   161

2.7   Rational Functions   169

Make a Decision Project: Renewable Energy   181

Summary   182   **Review Exercises**   183   **Chapter Test**   187

CHAPTER 3 Exponential and Logarithmic Functions

3.1 Exponential Functions 190

3.2 Logarithmic Functions 202

3.3 Properties of Logarithms 212

Mid-Chapter Quiz 220

3.4 Solving Exponential and Logarithmic Equations 221

3.5 Exponential and Logarithmic Models 231

Make a Decision Project: Newton's Law of Cooling 244

Summary 245 **Review Exercises** 246 **Chapter Test** 250

Cumulative Test: Chapters 1–3 251

CHAPTER 4 Systems of Equations and Inequalities

4.1 Solving Systems Using Substitution 254

4.2 Solving Systems Using Elimination 264

4.3 Linear Systems in Three or More Variables 276

Mid-Chapter Quiz 289

4.4 Systems of Inequalities 290

4.5 Linear Programming 300

Make a Decision Project: Newspapers 310

Summary 311 **Review Exercises** 312 **Chapter Test** 317

CHAPTER 5 Matrices and Determinants

5.1 Matrices and Linear Systems 320

5.2 Operations with Matrices 334

5.3 The Inverse of a Square Matrix 349

Mid-Chapter Quiz 359

5.4 The Determinant of a Square Matrix 360

5.5 Applications of Matrices and Determinants 370

Make a Decision Project: Market Share 379

Summary 380 **Review Exercises** 381 **Chapter Test** 385

CHAPTER 6   Sequences, Series, and Probability

6.1   Sequences and Summation Notation   388

6.2   Arithmetic Sequences and Partial Sums   398

6.3   Geometric Sequences and Series   407

6.4   The Binomial Theorem   417

Mid-Chapter Quiz   424

6.5   Counting Principles   425

6.6   Probability   435

6.7   Mathematical Induction   447

Make a Decision Project: The Multiplier Effect   458

Summary 459   **Review Exercises** 460   **Chapter Test** 464

Cumulative Test: Chapters 4–6 465

APPENDICES

Appendix A

An Introduction to Graphing Utilities   A1

Appendix B

Conic Sections

B.1   Conic Sections   A8

B.2   Conic Sections and Translations   A20

Appendix C

Further Concepts in Statistics (available on the web/CD only)**

C.1   Data and Linear Modeling

C.2   Measures of Central Tendency and Dispersion

Answers to Warm Ups, Odd-Numbered Exercises, Quizzes, and Tests A29

Index of Applications A99

Index A104

**For Appendix C and other resources, please visit *math.college.hmco.com/students*.

A Word from the Authors

Welcome to *College Algebra: Concepts and Models,* Fifth Edition. In this revision, we continue to focus on developing students' conceptual understanding of college algebra while offering opportunities for them to hone their problem-solving skills.

We have found that many college algebra students grasp theoretical concepts more easily when they work with them in the context of a real-life situation. Throughout the Fifth Edition, students have many opportunities to collect, analyze, and model real data. We updated all real-data applications or replaced them with new applications that use current data. Applications involving inflation, interest rates, demographics, and technology have been updated to represent current trends. We have enriched many applications throughout the text with a *Make a Decision* feature, encouraging students to think beyond the mathematical answer and make decisions about that answer within its real-life context.

As always, there are many opportunities in the Fifth Edition for students to use a graphing utility in the problem-solving process—to visualize and explore theoretical concepts, to analyze real data, and to verify alternative solution methods. To accommodate a variety of teaching and learning styles, the use of graphing technology is always optional.

New Features

- A new *Make a Decision* feature has been incorporated throughout the text. It appears in each Chapter Opener application, each section's examples and exercises, each chapter's review exercises, and each end-of-chapter project. *Make a Decision* provides students with the next level of problem solving; students are asked to think about whether an answer fits within the context of the real-life situation, to interpret answers in the real-life context, to choose an appropriate model for a data set, or to decide whether a current model will continue to be accurate in the future. The student must examine all data and determine the appropriate answer.

- To further emphasize the connection between mathematics and real life, each chapter now contains two *In the News* features. After reading an excerpt from a current magazine, newspaper, or Internet article, students must answer questions that connect the article and the algebra learned in that section.

- After each numbered example, a *Checkpoint* now directs the student to a matched problem in the exercise set. Students should try the *Checkpoint* exercise before continuing with the lesson.

- A section reference has been added to each set of *Warm Up* exercises that directs the student to the proper section(s) to review if they struggle with the *Warm Up* exercises.

Changes in the Table of Contents

- Two review chapters are now included at the beginning of the text. The first is Chapter R1 *Fundamental Concepts of Algebra,* which was Appendix A in the Fourth Edition. The concepts presented in this chapter have been expanded and more exercises have been added to each exercise set. The second review chapter is Chapter R2 *Equations and Inequalities,* which was Chapter 1 in the Fourth Edition. These review chapters are targeted for students who do not have a strong background in Intermediate Algebra.

- Chapter 1 is made up of sections from Chapters 2 and 3 in the Fourth Edition. This chapter is retitled as *Functions and Graphs.*

- Appendix C *Further Concepts in Statistics* is now available only on the website that accompanies this text.

- The *Introduction to Graphing Utilities* has been moved from the beginning of the text to Appendix A (pages A1–A7), and a description of the *maximum* and *minimum* features of a graphing utility has been added. Margin notes, provided at point-of-use, guide students to this appendix where they can learn how to use specific graphing utility features. These notes also direct students to the *Graphing Technology Guide* on the textbook website for keystroke support that is available for several models of graphing utilities.

Additional Pedagogical Features

- Each section begins with a list of learning *Objectives*, the algebra skills presented in the section. This conceptual outline functions as a useful tool for reference and review for the students and class planning for the instructor.

- Approximately halfway through the chapter, students will find a *Mid-Chapter Quiz*. This study tool provides additional practice for the first half of the lessons in each chapter. The answers to all Mid-Chapter Quiz exercises are provided in the back of the book.

- At the end of each chapter, students will find the Chapter Summary, Review Exercises, and Chapter Test study tools. The *Chapter Summary* lists the objectives of each chapter keyed to appropriate Review Exercises. The *Review Exercises* at the end of each chapter offer students an additional opportunity for practice and review. The answers to odd-numbered Review Exercises are provided in the back of the book. The *Chapter Test* gives the students an opportunity to assess their knowledge of the chapter. The answers to all Chapter Test exercises are provided in the back of the book.

- The *Cumulative Tests* appearing after every few chapters provide an opportunity for students to assess their retention of the material covered as the course progresses. The answers to all Cumulative Test exercises are provided in the back of the book.

- The *Make a Decision Projects* are extensions of the applications presented in the Chapter Opener. Students are guided through a set of multi-part exercises using modeling, graphing, and critical thinking skills to analyze the data.

For more information about the pedagogical features of the text, see pp. x–xi.

We hope you will enjoy using the Fifth Edition in your college algebra class. We think that its straightforward, readable style, engaging applications, and study tools will appeal to your students and contribute to their success in this course.

Ron Larson

Robert Hostetler

Anne Hodgkins

Highlights of Features

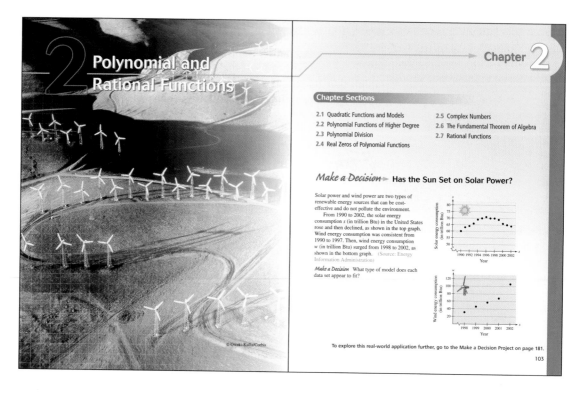

Polynomial and Rational Functions

Chapter 2

Chapter Sections

2.1 Quadratic Functions and Models
2.2 Polynomial Functions of Higher Degree
2.3 Polynomial Division
2.4 Real Zeros of Polynomial Functions

2.5 Complex Numbers
2.6 The Fundamental Theorem of Algebra
2.7 Rational Functions

Make a Decision **Has the Sun Set on Solar Power?**

Solar power and wind power are two types of renewable energy sources that can be cost-effective and do not pollute the environment.
From 1990 to 2002, the solar energy consumption *s* (in trillion Btu) in the United States rose and then declined, as shown in the top graph. Wind energy consumption was consistent from 1990 to 1997. Then, wind energy consumption *w* (in trillion Btu) surged from 1998 to 2002, as shown in the bottom graph. (Source: Energy Information Administration)

Make a Decision What type of model does each data set appear to fit?

To explore this real-world application further, go to the Make a Decision Project on page 181.

103

© Owaki-Kulla/Corbis

Make a Decision Chapter Opener and Chapter Project

Chapter Openers introduce the Make a Decision feature, which encourages students to draw connections between the mathematical concepts of algebra and the real world. An application is introduced in the opener, and then developed as a full project found at the end of the chapter. This *Make a Decision Project* guides students through multi-part exercises using modeling, graphing, and critical thinking skills to analyze real data.

Chapter Project   181

Make a Decision **Project: Renewable Energy**

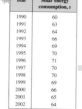

Year	Solar energy consumption, s
1990	60
1991	63
1992	64
1993	66
1994	69
1995	70
1996	71
1997	70
1998	70
1999	69
2000	66
2001	65
2002	64

Year	Wind energy consumption, w
1998	31
1999	46
2000	57
2001	68
2002	106

Solar energy uses photovoltaic cells that convert sunlight directly into electricity. These photovoltaic cells once were used only in space and in solar powered watches and calculators. The energy that wind contains is converted to electricity as air flows past a wind turbine rotor. One-third of the United States has enough wind to generate electricity.
The tables at the left show the solar energy consumption s (in trillion Btu) in the United States from 1990 to 2002 and the wind energy consumption w (in trillion Btu) in the United States from 1998 to 2002. Use this information to investigate the following questions. (Source: Energy Information Administration)

1. **Inspecting Data** Study the tables at the left. What pattern do you see in each set of data?

2. *Make a Decision* Use a graphing utility to create a scatter plot of each data set. How do the scatter plots compare to your answer to Question 1? Do the points appear to resemble a quadratic pattern or a cubic pattern? Do you expect the model to have a negative or a positive leading coefficient?

3. **Fitting a Model to Data** The standard form of a quadratic equation is $y = a(x - h)^2 + k$, where (h, k) is the vertex. Use the vertex (h, k) and any other ordered pair (x, y) to determine a value for a in the quadratic equation for solar energy consumption. Write the quadratic model in the general form $y = ax^2 + bx + c$.

4. **Fitting a Model to Data** Use the *regression* feature of a graphing utility to find another quadratic model for the solar energy consumption data. Write the quadratic model in the general form $y = ax^2 + bx + c$. Then find a cubic model for the wind energy consumption data.

5. **Finding the Vertex of a Model** Algebraically find the vertex of the quadratic model for solar energy consumption found in Question 4. Then use the *trace* feature or the *maximum* feature of a graphing utility to find the vertex. Does the vertex of the model found in Question 4 agree with the actual data given in the table?

6. **Comparing Actual Data with Models** Use the *table* feature of a graphing utility to evaluate the solar energy consumption values that both models give for each year. How do the two models' values compare to the actual values from the table?

7. *Make a Decision* Of the models that you found in Questions 3 and 4, select the one that appears to best fit the actual data. Use a graphing utility to graph the equation with the original data.

8. **Graphing Models** Use a graphing utility to graph the models you found in Question 4 in the same viewing window. Use the *trace* feature of the graphing utility to approximate the year in which wind energy consumption surpassed solar energy consumption.

9. **Research** Use your school's library, the Internet, or some other reference source to research the advantages and disadvantages of using renewable energy.

Examples and Other Features of the Text

Worked-out examples will enhance understanding of the concepts. Each is followed by a *Checkpoint*, connecting the example to a similar odd-numbered exercise. A variety of features such as *Discovery*, *Technology*, and *Study Tips*, provide explorations, alternative approaches, hints, and gentle reminders. Most sections end with *Discussing the Concept*, which can serve as a group project or help to motivate class discussion.

Exercise Sets

Each Exercise Set begins with ten *Warm Up* exercises, which offer a review of the algebra students have just learned, or will need to review in order to successfully complete the following exercises. Each set of exercises contains *Make a Decision* exercises for which students will often use a graphing calculator to model real-life data and choose the model that best fits the problem, or decide which answer fits within the context of a problem based on the given information. The *In the News* feature further connects algebra to the real world by presenting an excerpt from a magazine, Internet, or newspaper article, followed by related exercises.

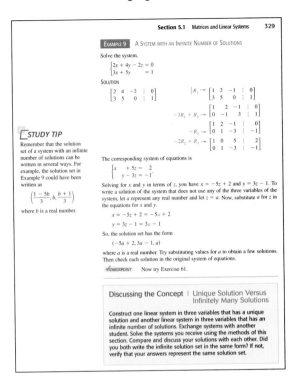

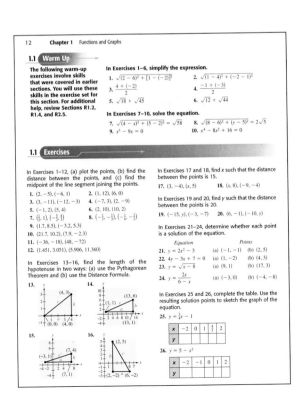

Acknowledgments

We would like to thank our colleagues who have helped us develop this project throughout this and previous editions. Their encouragement, criticisms, and suggestions have been invaluable to us.

Reviewers

Rosalie Abraham, Florida Community College at Jacksonville; Judith A. Ahrens, Pellisippi State Technical Community College; Sandra Beken, Horry-Georgetown Technical College; Diane Benjamin, University of Wisconsin—Platteville; Dona Boccio, Queensborough Community College, New York; Kent Craghead, Colby Community College; Carol Edwards, St. Louis Community College at Florissant Valley; Thomas L. Fitzkee, Francis Marion University, South Carolina; Michael Frantz, University of La Verne, California; Nick Geller, Collin County Community College; Carolyn H. Goldberg, Niagara County Community College; Carl Hughes, Fayetteville State University; Buddy A. Johns, Wichita State University; Annie Jones, John C. Calhoun State Community College; Steven Z. Kahn, Anne Arundel Community College; Claire Krukenberg, Eastern Illinois University; John Kubicek, Southwest Missouri State University; Charles G. Laws, Cleveland State Community College, Tennessee; John A. Lewallen, Southeastern Louisiana University; Gael Mericle, Mankato State University; Michael Montano, Riverside Community College; Sue Neal, Wichita State University, Kansas; Terrie L. Nichols, Cuyamaca College; Mark Omodt, Anoka-Ramsey Community College; G. Bryan Stewart, Tarrant County Junior College; Jacqueline Stone, University of Maryland; David Surowski, Kansas State University; Pamela K. Trim, Southwest Tennessee Community College; Jamie Whitehead, Texarkana College

In addition, we would like to thank the staff of Larson Texts, Inc., who assisted in preparing the manuscript, rendering the art package, and typesetting and proofreading the pages and supplements.

A special note of appreciation goes to all the instructors and students who have used previous editions of the text.

On a personal level, we would like to thank our families, especially Deanna Gilbert Larson, Eloise Hostetler, and Jay N. Torok, for their love, patience, and support. Also, special thanks goes to R. Scott O'Neil.

If you have suggestions for improving the text, please feel free to write to us. Over the past 20 years, we have received many useful comments from both instructors and students.

Ron Larson
Robert Hostetler
Anne V. Hodgkins

Supplements

The integrated learning system for *College Algebra: Concepts and Models,* Fifth Edition, addresses the changing needs of today's instructors and students, offering dynamic teaching tools for instructors and interactive learning resources for students in print, CD-ROM, and online formats.

Resources

Eduspace® is an on-line learning environment that combines algorithmic tutorials, homework capabilities, and testing. Text-specific content, organized by section, is available to help students understand the mathematics covered in this text.

For the Student

Study and Solutions Guide by Anne V. Hodgkins (Phoenix College) and Dianna L. Zook (Purdue University and Indiana University at Fort Wayne) contains the complete worked-out solutions to selected problems.

Student Website (math.college.hmco.com/students)
Contains the *Student Success Organizer* (a study and note-taking guide keyed to the text by section), Digital Lessons, downloadable graphing calculator programs, and other resources. The *Graphing Technology Guide* is now available on the web.

HM mathSpace® Student CD-ROM
Contains algorithmically generated exercises with step-by-step solutions for extra practice as well as many additional resources.

SMARTHINKING.com On-line Tutoring

Instructional Videotapes/DVD by Dana Mosely complement the textbook topic coverage should a student struggle with the algebra concepts or miss a class.

Instructional Videotapes for Graphing Calculators by Dana Mosely

For the Instructor

Instructor's Annotated Edition

HM ClassPrep with HM Testing CD-ROM (Windows, Macintosh)
ClassPrep contains a Complete Solutions Guide, IAE version of the Student Success Organizer, Test Item File, Digital Figures (ppts), and other resources. HM Testing offers a computerized test generator with algorithmically generated test items.

Instructor Website (math.college.hmco.com/instructors) offers pdfs of the Complete Solutions Guide, IAE version of the Student Success Organizer, Test Item File, and Digital Figures and Lessons (ppts). Appendix C is also posted on the website.

R1 Fundamental Concepts of Algebra

© Reuters/CORBIS

Chapter Sections

R1.1 Real Numbers: Order and Absolute Value

R1.2 The Basic Rules of Algebra

R1.3 Integer Exponents

R1.4 Radicals and Rational Exponents

R1.5 Polynomials and Special Products

R1.6 Factoring

R1.7 Fractional Expressions

Make a Decision ➤ Is There Math in Middle C?

Musical notes are characterized by their frequencies and amplitudes. High and low frequencies correspond to high and low notes.

The musical note A-440 (the first A above middle C) has a frequency of 440 vibrations per second. This note is often the first to be tuned on a musical instrument. The frequency F of any note can be found by the model

$$F = 440 \cdot \sqrt[12]{2^n}$$

where n represents the number of notes above or below A-440. The table at the right lists the frequency of eight notes.

Make a Decision Two notes are one or more octaves apart if the ratio of the higher frequency to the lower frequency is an integer. Which notes in the table are one or more octaves apart?

Distance from A-440, n	Frequency, F
−24	110
−18	$220/\sqrt{2}$
−12	220
−6	$440/\sqrt{2}$
0	440
6	$440\sqrt{2}$
12	880
18	$880\sqrt{2}$

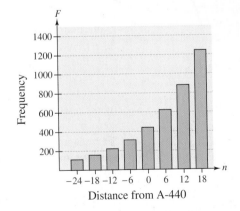

Distance from A-440

To explore this real-world application further, go to the Make a Decision Project on page R62.

R1.1   Real Numbers: Order and Absolute Value

Objectives

- Classify real numbers as natural numbers, integers, rational numbers, or irrational numbers.
- Order real numbers.
- Give a verbal description of numbers represented by an inequality.
- Use inequality notation to describe a set of real numbers.
- Interpret absolute value notation.
- Find the distance between two numbers on the real number line.
- Use absolute value to solve an application problem.

Real Numbers

The formal term that is used in mathematics to refer to a collection of objects is the word **set.** For instance, the set

$$\{1, 2, 3\}$$

contains the three numbers 1, 2, and 3. Note that a pair of braces { } is used to enclose the members of the set. In this text, a *pair* of braces will always indicate the members of a set. Parentheses ( ) and brackets [ ] are used to represent other ideas.

The set of numbers that is used in arithmetic is the set of **real numbers.** The term *real* distinguishes real numbers from *imaginary* or *complex* numbers.

A set A is called a **subset** of a set B if every member of A is also a member of B. Here are two examples.

- $\{1, 2, 3\}$ is a subset of $\{1, 2, 3, 4\}$.   • $\{0, 4\}$ is a subset of $\{0, 1, 2, 3, 4\}$.

One of the most commonly used subsets of real numbers is the set of **natural numbers** or **positive integers**

$$\{1, 2, 3, 4, . . .\}. \qquad \text{Set of positive integers}$$

Note that the three dots indicate that the pattern continues. For instance, the set also contains the numbers 5, 6, 7, and so on.

Positive integers can be used to describe many quantities that you encounter in everyday life—for instance, you might be taking four classes this term, or you might be paying $700 a month for rent. But even in everyday life, positive integers cannot describe some concepts accurately. For instance, you could have a zero balance in your checking account, or the temperature could be $-10°$ (10 degrees below zero). To describe such quantities, you need to expand the set of positive integers to include **zero** and the **negative integers.** The expanded set is called the set of **integers,** which can be written as follows.

$$\underbrace{\{. . . , -3, -2, -1,}_{\text{Negative integers}} \overset{\text{Zero}}{\;0,\;} \underbrace{1, 2, 3, . . .\}}_{\text{Positive integers}}$$

The set of integers is a subset of the set of real numbers. This means that every integer is a real number.

Even with the set of integers, there are still many quantities in everyday life that you cannot describe accurately. The costs of many items are not in whole dollar amounts, but in parts of dollars, such as $1.19 or $39.98. You might work $8\frac{1}{2}$ hours, or you might miss the first *half* of a movie. To describe such quantities, the set of integers is expanded to include **fractions.** The expanded set is called the set of **rational numbers.** Formally, a real number is called **rational** if it can be written as the ratio p/q of two integers, where $q \neq 0$. (The symbol $\neq$ means **not equal to.**) For instance,

$$2 = \frac{2}{1}, \quad 0.333\ldots = \frac{1}{3}, \quad 0.125 = \frac{1}{8}, \quad \text{and} \quad 1.126126\ldots = \frac{125}{111}$$

are rational numbers. Real numbers that cannot be written as the ratio of two integers are called **irrational.** For instance, the numbers

$$\sqrt{2} = 1.4142135\ldots \quad \text{and} \quad \pi = 3.1415926\ldots$$

are irrational numbers. The decimal representation of a rational number is either *terminating* or *repeating*. For instance, the decimal representation of

$$\tfrac{1}{4} = 0.25 \qquad\qquad\qquad \text{Terminating decimal}$$

is terminating, and the decimal representation of

$$\tfrac{4}{11} = 0.363636\ldots = 0.\overline{36} \qquad \text{Repeating decimal}$$

is repeating. (The line over "36" indicates which digits repeat.)

The decimal representation of an irrational number neither terminates nor repeats. When you perform calculations using decimal representations of nonterminating decimals, you usually use a decimal approximation that has been **rounded** to a certain number of decimal places. For instance, rounded to four decimal places, the decimal approximations of $\frac{2}{3}$ and π are

$$\frac{2}{3} \approx 0.6667 \qquad \text{and} \qquad \pi \approx 3.1416.$$

The symbol $\approx$ means **approximately equal to.**

The Venn diagram in Figure R1.1 shows the relationships between the real numbers and several commonly used subsets of the real numbers.

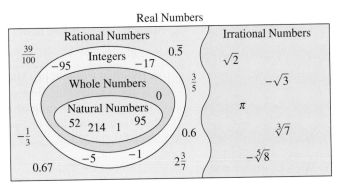

Figure R1.1

The Real Number Line and Ordering

The picture that is used to represent the real numbers is the **real number line.** It consists of a horizontal line with a point (the **origin**) labeled as 0 (zero). Points to the left of zero are associated with **negative numbers,** and points to the right of zero are associated with **positive numbers,** as shown in Figure R1.2. The real number zero is neither positive nor negative. So, when you want to talk about real numbers that might be positive *or* zero, you can use the term **nonnegative real numbers.**

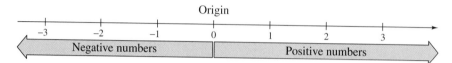

Figure R1.2   The Real Number Line

Each point on the real number line corresponds to exactly one real number, and each real number corresponds to exactly one point on the real number line, as shown in Figure R1.3. The number associated with a point on the real number line is the **coordinate** of the point.

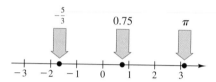

Figure R1.3   Every real number corresponds to a point on the real number line.

The real number line provides you with a way of comparing any two real numbers. For instance, if you choose any two (different) numbers on the real number line, one of the numbers must be to the left of the other number. The number to the left is **less than** the number to the right, and the number to the right is **greater than** the number to the left.

Definition of Order on the Real Number Line

If the real number a lies to the left of the real number b on the real number line, a is **less than** b, which is denoted by

$a < b$

as shown in Figure R1.4. This relationship can also be described by saying that b is **greater than** a and writing $b > a$. The inequality $a \leq b$ means that a is **less than or equal to** b, and the inequality $b \geq a$ means that b is **greater than or equal to** a.

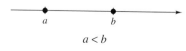

$a < b$

Figure R1.4 a is to the left of b.

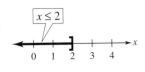

(a)

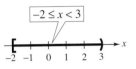

(b)

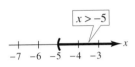

(c)

Figure R1.5

The symbols $<$, $>$, $\leq$, and $\geq$ are called **inequality symbols.** Inequalities are useful in denoting subsets of real numbers, as shown in Examples 1 and 2.

EXAMPLE 1   INTERPRETING INEQUALITIES

a. The inequality $x \leq 2$ denotes all real numbers that are less than or equal to 2, as shown in Figure R1.5(a).

b. The inequality $-2 \leq x < 3$ means that $x \geq -2$ *and* $x < 3$. This **double inequality** denotes all real numbers between -2 and 3, including -2 but *not* including 3, as shown in Figure R1.5(b).

c. The inequality $x > -5$ denotes all real numbers that are greater than -5, as shown in Figure R1.5(c).

✓CHECKPOINT   Now try Exercise 21.

In Figure R1.5, notice that a bracket is used to *include* the endpoint of an interval and a parenthesis is used to *exclude* the endpoint.

EXAMPLE 2   INEQUALITIES AND SETS OF REAL NUMBERS

a. "c is nonnegative" means that c is greater than or equal to zero, which you can write as $c \geq 0$.

b. "b is at most 5" can be written as $b \leq 5$.

c. "d is negative" can be written as $d < 0$, and "d is greater than -3" can be written as $-3 < d$. Combining these two inequalities produces $-3 < d < 0$.

d. "x is positive" can be written as $0 < x$, and "x is not more than 6" can be written as $x \leq 6$. Combining these two inequalities produces $0 < x \leq 6$.

✓CHECKPOINT   Now try Exercise 31.

The following property of real numbers is called the **Law of Trichotomy.** As the "tri" in its name suggests, this law tells you that for any two real numbers a and b, precisely one of *three* relationships is possible.

$$a < b, \quad a = b, \quad \text{or} \quad a > b \qquad \text{Law of Trichotomy}$$

Absolute Value and Distance

The **absolute value** of a real number is its *magnitude,* or its value disregarding its sign. For instance, the absolute value of -3, written $|-3|$, has the value of 3.

STUDY TIP

Be sure you see from the definition that the absolute value of a real number is never negative. For instance, if $a = -5$, then $|-5| = -(-5) = 5$.

Definition of Absolute Value

Let a be a real number. The **absolute value** of a, denoted by $|a|$, is

$$|a| = \begin{cases} a, & \text{if } a \geq 0 \\ -a, & \text{if } a < 0 \end{cases}.$$

The absolute value of any real number is either positive or zero. Moreover, 0 is the only real number whose absolute value is zero. That is, $|0| = 0$.

EXAMPLE 3   FINDING ABSOLUTE VALUE

a. $|-7| = 7$

b. $\left|\frac{1}{2}\right| = \frac{1}{2}$

c. $|-4.8| = 4.8$

d. $-|-9| = -(9) = -9$

✓*CHECKPOINT*   Now try Exercise 37.

EXAMPLE 4   COMPARING REAL NUMBERS

Place the correct symbol ($<$, $>$, or $=$) between the real numbers.

a. $|-4|$ $|4|$ **b.** $|-5|$ 3 **c.** $-|-1|$ $|-1|$

SOLUTION

a. $|-4| = |4|$, because both are equal to 4.

b. $|-5| > 3$, because $|-5| = 5$ and 5 is greater than 3.

c. $-|-1| < |-1|$, because $-|-1| = -1$ and $|-1| = 1$.

✓*CHECKPOINT*   Now try Exercise 47.

Properties of Absolute Value

Let a and b be real numbers. Then the following properties are true.

1. $|a| \geq 0$ **2.** $|-a| = |a|$

3. $|ab| = |a|\,|b|$ **4.** $\left|\dfrac{a}{b}\right| = \dfrac{|a|}{|b|}, \, b \neq 0$

Absolute value can be used to define the distance between two numbers on the real number line. To see how this is done, consider the numbers -3 and 4, as shown in Figure R1.6. To find the distance between these two numbers, subtract *either* number from the other and then take the absolute value of the difference.

$$(\text{Distance between } -3 \text{ and } 4) = |-3 - 4| = |-7| = 7$$

Figure R1.6   The distance between -3 and 4 is 7.

Distance Between Two Numbers

Let a and b be real numbers. The **distance between a and b** is given by

$$\text{Distance} = |b - a| = |a - b|.$$

EXAMPLE 5	FINDING THE DISTANCE BETWEEN TWO NUMBERS

a. The distance between 2 and 7 is

$$\text{Distance} = |2 - 7| = |-5| = 5.$$

b. The distance between 0 and -4 is

$$\text{Distance} = |0 - (-4)| = |4| = 4.$$

c. The statement "the distance between x and 2 is at least 3" can be written as

$$|x - 2| \geq 3.$$

✓CHECKPOINT   Now try Exercise 59.

Application

EXAMPLE 6	BUDGET VARIANCE	

Make a Decision As an accountant for a contracting company, one of your jobs is to determine whether the monthly expenses of the other departments vary "too much" from their budgets. According to your company, the absolute value of the difference between actual and budgeted expenses must be less than or equal to $500 *and* less than or equal to 5% of the budgeted expenses. By letting a represent the actual expenses and b the budgeted expenses, you can translate these requirements as follows.

$$|a - b| \leq 500 \qquad \text{and} \qquad |a - b| \leq 0.05b$$

For travel, the budgeted expenses were $12,500 and the actual expenses were $12,872.56. For office supplies, the budgeted expenses were $750 and the actual expenses were $704.15. For wages, the budgeted expenses were $84,600 and the actual expenses were $85,143.95. Are these amounts within budget restrictions?

© Chuck Savage/CORBIS

Math plays an important part in keeping your personal finances in order as well as a company's expenses and budget.

SOLUTION

One way to determine whether these three expenses are within budget restrictions is to create a table, as follows.

	Budgeted Expenses, b	Actual Expenses, a	$|a - b|$	$0.05b$
Travel	$12,500	$12,872.56	$372.56	$625.00
Office supplies	$750	$704.15	$45.85	$37.50
Wages	$84,600	$85,143.95	$543.95	$4230.00

From this table, you can see that the expenses for travel pass both tests, so the travel expenses are within budget restrictions. The expenses for office supplies pass the first test but fail the second test, so they are *not* within budget restrictions. The expenses for wages fail the first test and pass the second test, so they are *not* within budget restrictions.

✓CHECKPOINT   Now try Exercise 73.

R1.1 Exercises

In Exercises 1–6, determine which numbers in the set are (a) natural numbers, (b) integers, (c) rational numbers, and (d) irrational numbers.

1. $\left\{-9, -\frac{7}{2}, 5, \frac{2}{3}, \sqrt{2}, 0.1\right\}$

2. $\left\{\sqrt{5}, -7, -\frac{7}{3}, 0, 3.12, \frac{5}{4}\right\}$

3. $\left\{12, -13, 1, \sqrt{4}, \sqrt{6}, \frac{3}{2}\right\}$

4. $\left\{3, -1, \frac{1}{3}, \frac{6}{3}, -\frac{1}{2}\sqrt{2}, -7.5\right\}$

5. $\left\{\frac{8}{2}, -\frac{8}{3}, \sqrt{10}, -4, 9, 14.2\right\}$

6. $\left\{25, -17, \frac{12}{5}, \sqrt{9}, \sqrt{8}, -\sqrt{8}\right\}$

In Exercises 7–10, use a calculator to find the decimal form of the rational number. If the number is a nonterminating decimal, write the repeating pattern.

7. $\dfrac{5}{8}$

8. $\dfrac{1}{3}$

9. $\dfrac{6}{11}$

10. $\dfrac{41}{333}$

In Exercises 11 and 12, approximate the numbers and place the correct symbol (< or >) between them.

11.

12.

In Exercises 13–18, plot the two real numbers on the real number line and place the appropriate inequality sign (< or >) between them.

13. $\frac{3}{2}, 7$

14. $-4, -8$

15. $-3.5, 1$

16. $1, \frac{16}{3}$

17. $\frac{5}{6}, \frac{2}{3}$

18. $-\frac{8}{7}, -\frac{3}{7}$

In Exercises 19 and 20, use a calculator to order the numbers from least to greatest.

19. $\dfrac{7071}{5000}, \dfrac{584}{413}, \sqrt{2}, \dfrac{47}{33}, \dfrac{127}{90}$

20. $\dfrac{26}{15}, \sqrt{3}, 1.73\overline{20}, \dfrac{381}{220}, \sqrt{10} - \sqrt{2}$

In Exercises 21–30, give a verbal description of the subset of real numbers that is represented by the inequality, and sketch the subset on the real number line.

21. $x < 0$

22. $x < 2$

23. $x \le 5$

24. $x \ge -2$

25. $x > 3$

26. $x \ge 4$

27. $-2 < x < 2$

28. $0 \le x \le 5$

29. $-1 \le x < 0$

30. $0 < x \le 6$

In Exercises 31–36, use inequality notation to describe the subset of real numbers.

31. x is negative.

32. y is greater than 5 and less than or equal to 12.

33. The person's age A is at least 35.

34. The yield Y is no more than 42 bushels per acre.

35. The annual rate of inflation r is expected to be at least 3.5%, but no more than 6%.

36. The price p of unleaded gasoline is not expected to go below \$1.30 per gallon during the coming year.

In Exercises 37–46, write the expression without using absolute value signs.

37. $|-10|$

38. $|0|$

39. $-3 - |-3|$

40. $|-1| - |-2|$

41. $-3|-3|$

42. $-5|-5|$

43. $\dfrac{-5}{|-5|}$

44. $\dfrac{|-4|}{-4}$

45. $|3 - \pi|$

46. $|4 - \pi|$

In Exercises 47–52, place the correct symbol (<, >, or =) between the two real numbers.

47. $|-7|$ ___ $|7|$

48. -5 ___ $-|5|$

49. $|-3|$ ___ $-|-3|$

50. $-|-6|$ ___ $|-6|$

51. $-|-2|$ ___ $-|2|$

52. $-(-2)$ ___ -2

In Exercises 53–62, find the distance between a and b.

53.

$a = -1$ $b = 3$

-1   0   1   2   3

54. $a = -\frac{5}{2}$ $b = 0$

55. $a = -4, b = -\frac{3}{2}$ **56.** $a = \frac{1}{4}, b = \frac{11}{4}$

57. $a = -\frac{7}{2}, b = 0$ **58.** $a = \frac{3}{4}, b = \frac{9}{4}$

59. $a = 126, b = 75$ **60.** $a = -126, b = -75$

61. $a = \frac{16}{5}, b = \frac{112}{75}$ **62.** $a = 9.34, b = -5.65$

In Exercises 63–68, use absolute value notation to describe the sentence.

63. The distance between z and $\frac{3}{2}$ is greater than 1.

64. The distance between x and 5 is no more than 3.

65. The distance between x and -10 is at least 6.

66. The distance between z and 0 is less than 8.

67. y is at least six units from 0.

68. y is at most two units from a.

69. Travel   While traveling on the Pennsylvania Turnpike, you pass milepost 57 near Pittsburgh, then milepost 236 near Gettysburg. How far do you travel during that time period?

70. Travel   While traveling on the Pennsylvania Turnpike, you pass milepost 326 near Valley Forge, then milepost 351 near Philadelphia. How far do you travel during that time period?

71. Temperature   The temperature in Bismarck, North Dakota was $60°$ at noon, then $23°$ at midnight. What was the change in temperature over the 12-hour period?

72. Temperature   The temperature in Chicago, Illinois was $48°$ last night at midnight, then $82°$ today at noon. What was the change in temperature over the 12-hour period?

Make a Decision: **Budget Variance**   In Exercises 73–78, the accounting department of a logistics company is checking to see whether the actual expenses of a department differ from the budgeted expenses by more than $500 or more than 5%. Complete the missing parts of the table. Then determine whether each actual expense passes the "budget variance test."

Budgeted Expense, b	Actual Expense, a	$|a - b|$	0.05b
73. $25,000	$24,872.12		

Budgeted Expense, b	Actual Expense, a	$|a - b|$	0.05b
74. $112,700	$113,356.52		
75. $9400	$9972.59		
76. $7500	$7104.68		
77. $37,640	$36,968.25		
78. $2575	$2613.15		

Make a Decision: **Quality Control**   In Exercises 79–84, the quality control inspector for a tire factory is testing the rim diameter of various tires. A tire is rejected if its rim diameter varies too much from its expected measure. Each diameter should not differ by more than 0.03 inch or by more than 0.2% of the expected diameter measure. Complete the missing parts of the table. Then determine whether each tire is passed or rejected according to the inspector's guidelines.

Expected Diameter, b	Actual Diameter, a	$|a - b|$	0.002b
79. 15 in.	14.98 in.		
80. 16 in.	15.969 in.		
81. 16.5 in.	16.545 in.		
82. 17 in.	17.033 in.		
83. 18 in.	18.035 in.		
84. 19 in.	18.974 in.		

85. Think About It   Consider $|u + v|$ and $|u| + |v|$.

(a) Are the values of the expressions always equal? If not, under what conditions are they unequal?

(b) If the two expressions are not equal for certain values of u and v, is one of the expressions always greater than the other? Explain.

86. Think About It   Is there a difference between saying that a real number is positive and saying that a real number is nonnegative? Explain.

87. Describe the differences among the sets of natural numbers, integers, rational numbers, and irrational numbers.

88. *Make a Decision*   You may hear it said that to take the absolute value of a real number you simply remove any negative sign and make the number positive. Can it ever be true that $|a| = -a$ for a real number a? Explain.

R1.2   The Basic Rules of Algebra

Objectives

- Identify the terms of an algebraic expression.
- Evaluate an expression.
- Identify rules of algebra.
- Perform operations on real numbers.
- Use a calculator to evaluate an expression.

Algebraic Expressions

One of the basic characteristics of algebra is the use of letters (or combinations of letters) to represent numbers. The letters used to represent numbers are called **variables,** and combinations of letters and numbers are called **algebraic expressions.** Here are a few examples of algebraic expressions.

$$5x, \qquad 2x - 3, \qquad \frac{4}{x^2 + 2}, \qquad 7x + y$$

Algebraic Expression

A collection of letters (called **variables**) and real numbers (called **constants**) that are combined using the operations of addition, subtraction, multiplication, and division is an **algebraic expression.** (Other operations can also be used to form an algebraic expression.)

The **terms** of an algebraic expression are those parts that are separated by addition. For example, the algebraic expression $x^2 - 5x + 8$ has three terms: x^2, $-5x$, and 8. Note that $-5x$, rather than $5x$, is a term, because

$$x^2 - 5x + 8 = x^2 + (-5x) + 8.$$

The terms x^2 and $-5x$ are the **variable terms** of the expression, and 8 is the **constant term** of the expression. The numerical factor of a variable term is the **coefficient** of the variable term. For instance, the coefficient of the variable term $-5x$ is -5, and the coefficient of the variable term x^2 is 1.

STUDY TIP

When adding, subtracting, multiplying, or dividing more than two numbers, it is important to use symbols of grouping (such as parentheses) to indicate the order of operations.

EXAMPLE 1   IDENTIFYING THE TERMS OF AN ALGEBRAIC EXPRESSION

Algebraic Expression	Terms
a. $4x - 3$	$4x, -3$
b. $2x + 4y - 5$	$2x, 4y, -5$

✓CHECKPOINT   Now try Exercise 1.

EXAMPLE 2 SYMBOLS OF GROUPING

a. $7 - 3(4 - 2) = 7 - 3(2) = 7 - 6 = 1$

b. $(4 - 5) - (3 - 6) = (-1) - (-3) = -1 + 3 = 2$

✓CHECKPOINT Now try Exercise 7.

The **Substitution Principle** states, "If $a = b$, then a can be replaced by b in any expression involving a." You use this principle to **evaluate** an algebraic expression by substituting numerical values for each of the variables in the expression. In the first evaluation shown below, 3 is substituted for x in the expression $-3x + 5$.

Expression	Value of Variable	Substitution	Value of Expression
$-3x + 5$	$x = 3$	$-3(3) + 5$	$-9 + 5 = -4$
$3x^2 + 2x - 1$	$x = -1$	$3(-1)^2 + 2(-1) - 1$	$3 - 2 - 1 = 0$
$-2x(x + 4)$	$x = -2$	$-2(-2)(-2 + 4)$	$-2(-2)(2) = 8$
$\dfrac{1}{x - 2}$	$x = 2$	$\dfrac{1}{2 - 2}$	Undefined

EXAMPLE 3 EVALUATING ALGEBRAIC EXPRESSIONS

Evaluate each algebraic expression when $x = -2$ and $y = 3$.

a. $4y - 2x$ **b.** $5 + x^2$ **c.** $5 - x^2$

SOLUTION

a. When $x = -2$ and $y = 3$, the expression $4y - 2x$ has a value of
$$4(3) - 2(-2) = 12 + 4 = 16.$$

b. When $x = -2$, the expression $5 + x^2$ has a value of
$$5 + (-2)^2 = 5 + 4 = 9.$$

c. When $x = -2$, the expression $5 - x^2$ has a value of
$$5 - (-2)^2 = 5 - 4 = 1.$$

✓CHECKPOINT Now try Exercise 11.

TECHNOLOGY

To evaluate the expression $3 + 4x$ for the values 2 and 5, use the *last entry* feature of a graphing utility.

1. Evaluate $3 + 4 \cdot 2$.

2. Press [2ND] [ENTER] (recalls previous expression to the home screen).

3. Cursor to 2, replace 2 with 5, and press [ENTER].

For specific keystrokes for the *last entry* feature, go to the text website at *college.hmco.com*.

Basic Rules of Algebra

The four basic arithmetic operations are **addition, multiplication, subtraction,** and **division,** denoted by the symbols $+$, $\times$ or $\cdot$, $-$, and $\div$, respectively. Of these, addition and multiplication are considered to be the two primary arithmetic operations. Subtraction and division are defined as the inverse operations of addition and multiplication, as follows.

Subtraction: Add the opposite

$$a - b = a + (-b)$$

Division: Multiply by the reciprocal

If $b \neq 0$, then $a \div b = a\left(\dfrac{1}{b}\right) = \dfrac{a}{b}$.

In these definitions, $-b$ is called the **additive inverse** (or opposite) of b, and $1/b$ is called the **multiplicative inverse** (or reciprocal) of b. In place of $a \div b$, you can use the fraction symbol a/b. In this fractional form, a is called the **numerator** of the fraction and b is called the **denominator.**

Be sure you see that the **basic rules of algebra,** listed below, are true for variables and algebraic expressions as well as for real numbers.

STUDY TIP

Because subtraction is defined as "adding the opposite," the Distributive Property is also true for subtraction. For instance, the "subtraction form" of $a(b + c) = ab + ac$ is $a(b - c) = a[b + (-c)] = ab + a(-c) = ab - ac$.

Basic Rules of Algebra

Let a, b, and c be real numbers, variables, or algebraic expressions.

Property	*Example*
Commutative Property of Addition	
$a + b = b + a$	$4x + x^2 = x^2 + 4x$
Commutative Property of Multiplication	
$ab = ba$	$(4 - x)x^2 = x^2(4 - x)$
Associative Property of Addition	
$(a + b) + c = a + (b + c)$	$(-x + 5) + 2x^2 =$ $-x + (5 + 2x^2)$
Associative Property of Multiplication	
$(ab)c = a(bc)$	$(2x \cdot 3y)(8) = (2x)(3y \cdot 8)$
Distributive Property	
$a(b + c) = ab + ac$	$3x(5 + 2x) = 3x \cdot 5 + 3x \cdot 2x$
$(a + b)c = ac + bc$	$(y + 8)y = y \cdot y + 8 \cdot y$
Additive Identity Property	
$a + 0 = a$	$5y^2 + 0 = 5y^2$
Multiplicative Identity Property	
$a \cdot 1 = a$	$(4x^2)(1) = 4x^2$
Additive Inverse Property	
$a + (-a) = 0$	$5x^3 + (-5x^3) = 0$
Multiplicative Inverse Property	
$a \cdot \dfrac{1}{a} = 1, \quad a \neq 0$	$(x^2 + 4)\left(\dfrac{1}{x^2 + 4}\right) = 1$

| EXAMPLE 4 | IDENTIFYING THE BASIC RULES OF ALGEBRA |

Identify the rule of algebra illustrated by each statement.

a. $(4x^2)5 = 5(4x^2)$

b. $(2y^3 + y) - (2y^3 + y) = 0$

c. $(4 + x^2) + 3x^2 = 4 + (x^2 + 3x^2)$

d. $(x - 5)7 + (x - 5)x = (x - 5)(7 + x)$

e. $2x \cdot \dfrac{1}{2x} = 1, \quad x \neq 0$

SOLUTION

a. This equation illustrates the Commutative Property of Multiplication.

b. This equation illustrates the Additive Inverse Property.

c. This equation illustrates the Associative Property of Addition. In other words, to form the sum $4 + x^2 + 3x^2$, it doesn't matter whether 4 and x^2 are added first or x^2 and $3x^2$ are added first.

d. This equation illustrates the Distributive Property in reverse order.

$$ab + ac = a(b + c) \qquad \text{Distributive Property}$$
$$(x - 5)7 + (x - 5)x = (x - 5)(7 + x)$$

e. This equation illustrates the Multiplicative Inverse Property. Note that it is important that x be a nonzero number. If x were allowed to be zero, you would be in trouble because the reciprocal of zero is undefined.

✓**CHECKPOINT**   Now try Exercise 19.

The following three lists summarize the basic properties of negation, zero, and fractions. When you encounter such lists, you should not only *memorize* a verbal description of each property, but you should also try to gain an *intuitive feeling* for the validity of each.

Properties of Negation

Let a and b be real numbers, variables, or algebraic expressions.

Property	*Example*
1. $(-1)a = -a$	$(-1)7 = -7$
2. $-(-a) = a$	$-(-6) = 6$
3. $(-a)b = -(ab) = a(-b)$	$(-5)3 = -(5 \cdot 3) = 5(-3)$
4. $(-a)(-b) = ab$	$(-2)(-6) = 12$
5. $-(a + b) = (-a) + (-b)$	$-(3 + 8) = (-3) + (-8)$

Be sure you see the difference between the opposite of a number and a negative number. If a is negative, then its opposite, $-a$, is positive. For instance, if $a = -5$, then $-a = -(-5) = 5$.

Properties of Zero

Let a and b be real numbers, variables, or algebraic expressions. Then the following properties are true.

1. $a + 0 = a$ and $a - 0 = a$

2. $a \cdot 0 = 0$

3. $\dfrac{0}{a} = 0, \quad a \neq 0$

4. $\dfrac{a}{0}$ is undefined.

5. Zero-Factor Property: If $ab = 0$, then $a = 0$ or $b = 0$.

The "or" in the Zero-Factor Property includes the possibility that both factors are zero. This is called an **inclusive or,** and it is the way the word "or" is always used in mathematics.

Properties of Fractions

Let a, b, c, and d be real numbers, variables, or algebraic expressions such that $b \neq 0$ and $d \neq 0$. Then the following properties are true.

1. *Equivalent fractions:* $\dfrac{a}{b} = \dfrac{c}{d}$ if and only if $ad = bc$.

2. *Rules of signs:* $-\dfrac{a}{b} = \dfrac{-a}{b} = \dfrac{a}{-b}$ and $\dfrac{-a}{-b} = \dfrac{a}{b}$

3. *Generate equivalent fractions:* $\dfrac{a}{b} = \dfrac{ac}{bc}, \quad c \neq 0$

4. *Add or subtract with like denominators:* $\dfrac{a}{b} \pm \dfrac{c}{b} = \dfrac{a \pm c}{b}$

5. *Add or subtract with unlike denominators:* $\dfrac{a}{b} \pm \dfrac{c}{d} = \dfrac{ad \pm bc}{bd}$

6. *Multiply fractions:* $\dfrac{a}{b} \cdot \dfrac{c}{d} = \dfrac{ac}{bd}$

7. *Divide fractions:* $\dfrac{a}{b} \div \dfrac{c}{d} = \dfrac{a}{b} \cdot \dfrac{d}{c} = \dfrac{ad}{bc}, \quad c \neq 0$

In Property 1 (equivalent fractions) the phrase "if and only if" implies two statements. One statement is: If $a/b = c/d$, then $ad = bc$. The other statement is: If $ad = bc$, where $b \neq 0$ and $d \neq 0$, then $a/b = c/d$.

EXAMPLE 5   PROPERTIES OF ZERO AND PROPERTIES OF FRACTIONS

a. $x - \dfrac{0}{5} = x - 0 = x$ Properties 3 and 1 of zero

b. $\dfrac{x}{5} = \dfrac{3 \cdot x}{3 \cdot 5} = \dfrac{3x}{15}$ Generate equivalent fractions.

c. $\dfrac{x}{3} + \dfrac{2x}{5} = \dfrac{x \cdot 5 + 3 \cdot 2x}{15} = \dfrac{11x}{15}$ Add fractions with unlike denominators.

d. $\dfrac{7}{x} \div \dfrac{3}{2} = \dfrac{7}{x} \cdot \dfrac{2}{3} = \dfrac{14}{3x}$ Divide fractions.

✓*CHECKPOINT*   Now try Exercise 27.

If a, b, and c are integers such that $ab = c$, then a and b are **factors** or **divisors** of c. For example, 2 and 3 are factors of 6 because $2 \cdot 3 = 6$. A **prime number** is a positive integer that has exactly two factors: itself and 1. For example, 2, 3, 5, 7, and 11 are prime numbers, whereas 1, 4, 6, 8, 9, and 10 are not. The numbers 4, 6, 8, 9, and 10 are **composite** because they can be written as the products of two or more prime numbers. The number 1 is neither prime nor composite. The **Fundamental Theorem of Arithmetic** states that every positive integer greater than 1 can be written as the product of prime numbers in precisely one way (disregarding order). For instance, the *prime factorization* of 24 is

$$24 = 2 \cdot 2 \cdot 2 \cdot 3.$$

When adding or subtracting fractions with unlike denominators, you can use Property 4 of fractions by rewriting both fractions so that they have the same denominator. This is called the **least common denominator** method.

EXAMPLE 6   ADDING AND SUBTRACTING FRACTIONS

Evaluate $\dfrac{2}{15} - \dfrac{5}{9} + \dfrac{4}{5}$.

SOLUTION

By prime factoring the denominators ($15 = 3 \cdot 5, 9 = 3 \cdot 3$, and $5 = 5$) you can see that the least common denominator is $3 \cdot 3 \cdot 5 = 45$. It follows that

$$\frac{2}{15} - \frac{5}{9} + \frac{4}{5} = \frac{2 \cdot 3}{15 \cdot 3} - \frac{5 \cdot 5}{9 \cdot 5} + \frac{4 \cdot 9}{5 \cdot 9}$$

$$= \frac{6 - 25 + 36}{45}$$

$$= \frac{17}{45}.$$

✓*CHECKPOINT*   Now try Exercise 41.

Equations

An **equation** is a statement of equality between two expressions. So, the statement

$$a + b = c + d$$

means that the expressions $a + b$ and $c + d$ represent the same number. For instance, because $1 + 4$ and $3 + 2$ both represent the number 5, you can write $1 + 4 = 3 + 2$. Three important properties of equality follow.

Properties of Equality

Let a, b, and c be real numbers, variables, or algebraic expressions.

1. Reflexive: $a = a$

2. Symmetric: If $a = b$, then $b = a$.

3. Transitive: If $a = b$ and $b = c$, then $a = c$.

In algebra, you often rewrite expressions by making substitutions that are permitted under the Substitution Principle. Two important consequences of the Substitution Principle are the following rules.

1. If $a = b$, then $a + c = b + c$.   **2.** If $a = b$, then $ac = bc$.

The first rule allows you to add the same number to each side of an equation. The second allows you to multiply each side of an equation by the same number. The converses of these two rules are also true and are listed below.

1. If $a + c = b + c$, then $a = b$.   **2.** If $ac = bc$ and $c \neq 0$, then $a = b$.

So, you can also subtract the same number from each side of an equation as well as divide each side of an equation by the same nonzero number.

Calculators and Rounding

The specific keystrokes listed here correspond to a standard scientific calculator and a graphing calculator. The keystrokes listed here may not be the same as those for your graphing calculator. Consult your user's guide for specific keystrokes. Some comparisons between graphing calculator keys and scientific calculator keys follow.

1. The key marked $\boxed{\text{ENTER}}$ is similar to $\boxed{=}$.

2. The key marked $\boxed{(\text{--})}$ is similar to $\boxed{+/-}$.

3. The key marked $\boxed{\wedge}$ is similar to $\boxed{y^x}$.

4. The key marked $\boxed{x^{-1}}$ is similar to $\boxed{1/x}$.

For example, you can evaluate 13^3 on a graphing calculator or a scientific calculator as follows.

      Graphing Calculator   *Scientific Calculator*

     13 $\boxed{\wedge}$ 3 $\boxed{\text{ENTER}}$ 13 $\boxed{y^x}$ 3 $\boxed{=}$

<table>
<tr><td>**EXAMPLE 7**</td><td>USING A CALCULATOR</td></tr>
</table>

Scientific Calculator

Expression	*Keystrokes*	*Display*
a. $7 - (5 \cdot 3)$	7 ⊟ 5 ⊠ 3 ⊜	-8
b. $-12^2 - 100$	12 [x²] [+/−] ⊟ 100 ⊜	-244
c. $24 \div 2^3$	24 ⊡ 2 [yˣ] 3 ⊜	3
d. $3(10 - 4^2) \div 2$	3 ⊠ ⦅ 10 ⊟ 4 [x²] ⦆ ⊡ 2 ⊜	-9
e. 37% of 40	.37 ⊠ 40 ⊜	14.8

Graphing Calculator

Expression	*Keystrokes*	*Display*
a. $7 - (5 \cdot 3)$	7 ⊟ 5 ⊠ 3 (ENTER)	-8
b. $-12^2 - 100$	[(−)] 12 [x²] ⊟ 100 (ENTER)	-244
c. $24 \div 2^3$	24 ⊡ 2 [^] 3 (ENTER)	3
d. $3(10 - 4^2) \div 2$	3 ⦅ 10 ⊟ 4 [x²] ⦆ ⊡ 2 (ENTER)	-9
e. 37% of 40	.37 ⊠ 40 (ENTER)	14.8

✓CHECKPOINT   Now try Exercise 63.

TECHNOLOGY

Be sure you see the difference between the change sign key [+/−] or [(−)] and the subtraction key ⊟, as used in Example 7(b).

For all their usefulness, calculators do have a problem representing some numbers because they are limited to a finite number of digits. For instance, what does your calculator display when you compute $2 \div 3$? Some calculators simply truncate (drop) the digits that exceed their display range and display .66666666. Others round the number and display .66666667. Although the second display is more accurate, *both* of these decimal representations of $\frac{2}{3}$ contain rounding errors. When rounding decimals in this text, use the following rule: *round up on 5 or greater, round down on 4 or less.*

<table>
<tr><td>**EXAMPLE 8**</td><td>ROUNDING DECIMAL NUMBERS</td></tr>
</table>

Number	*Rounded to Three Decimal Places*	
a. $\sqrt{2} = 1.4142135\ldots$	1.414	Round down.
b. $\pi = 3.1415926\ldots$	3.142	Round up.
c. $\frac{7}{9} = 0.7777777\ldots$	0.778	Round up.

✓CHECKPOINT   Now try Exercise 47.

One way to minimize error due to rounding is to leave numbers in your calculator until your calculations are complete. If you want to save a number for future use, store it in your calculator's memory.

R1.2 Warm Up

The following warm-up exercises involve skills that were covered in earlier sections. You will use these skills in the exercise set for this section. For additional help, review Section R1.1.

In Exercises 1–4, place the correct inequality symbol ($<$ or $>$) between the two numbers.

1. $-4 \quad -2$ 2. $0 \quad -3$ 3. $\sqrt{3} \quad 1.73$ 4. $-\pi \quad -3$

In Exercises 5–8, find the distance between the two numbers.

5. $4, 6$ 6. $-2, 2$ 7. $0, -5$ 8. $-1, 3$

In Exercises 9 and 10, evaluate the expression.

9. $|-7| + |7|$ 10. $-|8 - 10|$

R1.2 Exercises

In Exercises 1–6, identify the terms of the algebraic expression.

1. $7x + 4$
2. $-5 + 3x$
3. $x^2 - 4x + 8$
4. $4x^3 + x - 5$
5. $2x^2 - 9x + 13$
6. $3x^4 + 2x^3 - 1$

In Exercises 7–10, simplify the expression.

7. $(8 - 17) + 3$
8. $-3(5 - 2)$
9. $(4 - 7)(-2)$
10. $(-5)(-8)$

In Exercises 11–18, evaluate the expression for each value of x. (If not possible, state the reason.)

Expression	Values	
11. $4x - 6$	(a) $x = -1$	(b) $x = 0$
12. $5 - 3x$	(a) $x = -3$	(b) $x = 2$
13. $x^2 - 3x + 4$	(a) $x = -2$	(b) $x = 2$
14. $-x^2 + 3x - 4$	(a) $x = -1$	(b) $x = 1$
15. $-x^3 + 4x$	(a) $x = 0$	(b) $x = 2$
16. $x^3 - 2x - 1$	(a) $x = 0$	(b) $x = -1$
17. $\dfrac{x}{x + 3}$	(a) $x = -3$	(b) $x = 3$
18. $\dfrac{x + 1}{x - 1}$	(a) $x = 1$	(b) $x = -1$

In Exercises 19–34, identify the rule(s) of algebra illustrated by the statement.

19. $3 + 4 = 4 + 3$
20. $x + 9 = 9 + x$
21. $-15 + 15 = 0$
22. $(x + 2) - (x + 2) = 0$

23. $2(x + 3) = 2x + 6$
24. $2\left(\frac{1}{2}\right) = 1$
25. $\dfrac{1}{h + 6}(h + 6) = 1, \quad h \neq -6$
26. $(5 + 11) \cdot 6 = 5 \cdot 6 + 11 \cdot 6$
27. $h + 0 = h$
28. $(z - 2) + 0 = z - 2$
29. $57 \cdot 1 = 57$
30. $1 \cdot (1 + x) = 1 + x$
31. $6 + (7 + 8) = (6 + 7) + 8$
32. $x + (y + 10) = (x + y) + 10$
33. $x(3y) = (x \cdot 3)y = (3x)y$
34. $\frac{1}{7}(7 \cdot 12) = \left(\frac{1}{7} \cdot 7\right)12 = 1 \cdot 12 = 12$

In Exercises 35–38, write the prime factorization of the integer.

35. 36
36. 30
37. 200
38. 120

In Exercises 39–46, perform the indicated operation(s). (Write fractional answers in simplest form.)

39. $2\left(\dfrac{77}{-11}\right)$
40. $\dfrac{27 - 35}{4}$
41. $\frac{5}{8} + \frac{1}{4} - \frac{5}{6}$
42. $\frac{10}{11} + \frac{6}{33} - \frac{13}{66}$
43. $\frac{2}{5} \cdot \frac{7}{8}$
44. $\left(-\frac{2}{3}\right) \cdot \frac{5}{8} \cdot \frac{3}{4}$
45. $\frac{2}{3} \div 8$
46. $\left(\frac{3}{5} \div 3\right) - \left(6 \cdot \frac{4}{8}\right)$

In Exercises 47–50, use a calculator to evaluate the expression. (Round to two decimal places.)

47. $3\left(-\frac{5}{12} + \frac{3}{8}\right)$
48. $2\left(-7 + \frac{1}{6}\right)$

49. $\dfrac{11.46 - 5.37}{3.91}$ **50.** $\dfrac{-8.31 + 4.83}{7.65}$

In Exercises 51–54, use a calculator to solve the percent problem.

51. 33% of 57   **52.** 29% of 725

53. 121% of 34   **54.** 153% of 12

In Exercises 55 and 56, find the percent that corresponds to the unlabeled portion of the circle graph.

55.

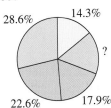

56.

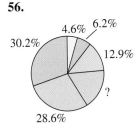

IN THE NEWS

Almost Half of Weekday Meals Now Prepared in Fewer Than Thirty Minutes, According to New Study

New York (PRWEB), February 23, 2004—About 44% of weekday meals in the U.S. are prepared in 30 minutes or less, and most consumers would like to cut that time even further, according to the U.S. Market for Ready Meals and Side Dishes, a new report by market research publisher Packaged Facts. With the number of two-working-parent households, single-parent households, and generation X and Y consumers who don't know how to cook steadily increasing, it's no wonder that quickly prepared, easily consumed, portable meal solutions are in demand. . . .

(Source: PR Web, "Almost Half of Weekday Meals Now Prepared in Fewer Than Thirty Minutes, According to New Study," news release, February 24, 2004.)

57. How many of the five weekday meals are prepared in 30 minutes or less?

58. Write a paragraph describing the types of weekday meals you prepare. List the times it takes to prepare your meals. How does your meal preparation compare to that described in the article?

59. *Make a Decision*: **Federal Government Expenses**
The circle graph shows the types of expenses for the human resources department of the federal government in 2002. (Source: Office of Management and Budget)

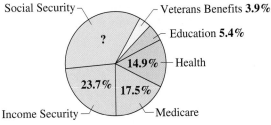

(a) What percent of the human resources budget was spent on Social Security?

(b) The total human resources expenses in 2002 were $1,317,800,000,000. Find the expense for each of the categories in the circle graph. (Round your answers to the nearest billion dollars.)

(c) Find the difference between the amounts spent on Social Security and health care. Do you think the difference between these two categories will be the same in the future? Explain.

60. College Enrollment The percent of students 24 years old or older at a college is 44.7%. The total enrollment for the college is 13,385. How many students are under 24 years old?

61. College Enrollment The percent of students who attend evening classes at a college is 44.1%. The total enrollment for the college is 11,875. How many students attend evening classes?

62. Calculator Keystrokes Write the algebraic expression that corresponds to each set of keystrokes.

(a) 5 ⊠ ⟮ 2.7 ⊟ 9.4 ⟯ ⊟   Scientific

    5 ⟮ 2.7 ⊟ 9.4 ⟯ ENTER   Graphing

(b) 2 ⊠ ⟮ 4 +/− ⊞ 2 ⟯ ⊟   Scientific

    2 ⟮ (−) 4 ⊞ 2 ⟯ ENTER   Graphing

63. Calculator Keystrokes Write the keystrokes used to evaluate each algebraic expression. (Base your description on either a scientific or a graphing calculator.)

(a) $5(18 - 2^3) \div 10$

(b) $-6^2 - [7 + (-2)^3]$

Objectives

- **Use properties of exponents.**
- **Use a calculator to raise a number to a power.**
- **Use interest formulas to solve an application problem.**

Properties of Exponents

Repeated multiplication of a real number by itself can be written in **exponential form.** Here are some examples.

Repeated Multiplication	Exponential Form
$7 \cdot 7$	7^2
$a \cdot a \cdot a \cdot a \cdot a$	a^5
$(-4)(-4)(-4)$	$(-4)^3$
$(2x)(2x)(2x)(2x)$	$(2x)^4$

STUDY TIP

It is important to recognize the difference between exponential forms such as $(-2)^4$ and -2^4. In $(-2)^4$, the parentheses indicate that the exponent applies the negative sign as well as the 2, but in $-2^4 = -(2^4)$, the exponent applies only to the 2. Similarly, in $(5x)^3$, the parentheses indicate that the exponent applies to the 5 as well as the x, whereas in $5x^3 = 5(x^3)$, the exponent applies only to the x.

Exponential Notation

Let a be a real number, a variable, or an algebraic expression, and let n be a positive integer. Then

$$a^n = \underbrace{a \cdot a \cdot a \cdots a}_{n \text{ factors}}$$

where n is the **exponent** and a is the **base.** The expression a^n is read as "a to the nth **power**" or simply "a to the nth."

When multiplying exponential expressions with the same base, *add* exponents.

$$a^m \cdot a^n = a^{m+n} \qquad \text{Add exponents when multiplying.}$$

For instance, to multiply 2^2 and 2^3, you can write

$$2^2 \cdot 2^3 = \overbrace{(2 \cdot 2)}^{\substack{\text{Two}\\\text{factors}}} \cdot \overbrace{(2 \cdot 2 \cdot 2)}^{\substack{\text{Three}\\\text{factors}}} = \overbrace{2 \cdot 2 \cdot 2 \cdot 2 \cdot 2}^{\substack{\text{Five}\\\text{factors}}}$$

$$= 2^{2+3} = 2^5.$$

On the other hand, when dividing exponential expressions, *subtract* exponents. That is,

$$\frac{a^m}{a^n} = a^{m-n}, \qquad a \neq 0. \qquad \text{Subtract exponents when dividing.}$$

These and other properties of exponents are summarized in the list on the following page.

Properties of Exponents

Let a and b be real numbers, variables, or algebraic expressions, and let m and n be integers. (Assume all denominators and bases are nonzero.)

Property	*Example*	
1. $a^m a^n = a^{m+n}$	$3^2 \cdot 3^4 = 3^{2+4} = 3^6$	Product of Powers
2. $\dfrac{a^m}{a^n} = a^{m-n}$	$\dfrac{x^7}{x^4} = x^{7-4} = x^3$	Quotient of Powers
3. $(ab)^m = a^m b^m$	$(5x)^3 = 5^3 x^3 = 125x^3$	Power of a Product
4. $\left(\dfrac{a}{b}\right)^m = \dfrac{a^m}{b^m}$	$\left(\dfrac{2}{x}\right)^3 = \dfrac{2^3}{x^3} = \dfrac{8}{x^3}$	Power of a Quotient
5. $(a^m)^n = a^{mn}$	$(y^3)^{-4} = y^{3(-4)} = y^{-12}$	Power of a Power
6. $a^{-n} = \dfrac{1}{a^n}$	$y^{-4} = \dfrac{1}{y^4}$	Definition of negative exponent
7. $a^0 = 1, \quad a \neq 0$	$(x^2 + 1)^0 = 1$	Definition of zero exponent
8. $\left(\dfrac{a}{b}\right)^{-n} = \left(\dfrac{b}{a}\right)^n, \quad a \neq 0, b \neq 0$	$\left(\dfrac{3}{2}\right)^{-3} = \left(\dfrac{2}{3}\right)^3$	
9. $|a^2| = |a|^2 = a^2$	$|(-2)^2| = |-2|^2 = (-2)^2 = 4$	

Be sure you see that these properties of exponents apply for *all* integers m and n, not just positive ones. For instance, by the Quotient of Powers Property,

$$\frac{3^4}{3^{-5}} = 3^{4-(-5)} = 3^{4+5} = 3^9.$$

DISCOVERY

Using your calculator, find the values of 10^3, 10^2, 10^1, 10^0, 10^{-1}, and 10^{-2}. What do you observe? Explain how to use the fact that $10^2 = 100$ and $10^1 = 10$ to find 10^0 and 10^{-1}.

EXAMPLE 1   USING PROPERTIES OF EXPONENTS

a. $3^4 \cdot 3^{-1} = 3^{4-1} = 3^3 = 27$

b. $\dfrac{5^6}{5^4} = 5^{6-4} = 5^2 = 25$

c. $5\left(\dfrac{2}{5}\right)^3 = 5 \cdot \dfrac{2^3}{5^3} = 5 \cdot 5^{-3} \cdot 2^3 = 5^{-2} \cdot 2^3 = \dfrac{2^3}{5^2} = \dfrac{8}{25}$

d. $(-5 \cdot 2^3)^2 = (-5)^2 \cdot (2^3)^2 = 25 \cdot 2^6 = 25 \cdot 64 = 1600$

e. $(-3ab^4)(4ab^{-3}) = -3(4)(a)(a)(b^4)(b^{-3}) = -12a^2 b$

f. $3a(-4a^2)^0 = 3a(1) = 3a, \quad a \neq 0$

g. $\left(\dfrac{5x^3}{y}\right)^2 = \dfrac{5^2(x^3)^2}{y^2} = \dfrac{25x^6}{y^2}$

✓**CHECKPOINT**   Now try Exercise 1.

The next example shows how expressions involving negative exponents can be rewritten using positive exponents.

EXAMPLE 2 REWRITING WITH POSITIVE EXPONENTS

Rewrite each expression with positive exponents.

a. x^{-1} **b.** $\dfrac{1}{3x^{-2}}$ **c.** $\dfrac{12a^3b^{-4}}{4a^{-2}b}$ **d.** $\left(\dfrac{3x^2}{y}\right)^{-2}$

SOLUTION

a. $x^{-1} = \dfrac{1}{x}$ Definition of negative exponent

b. $\dfrac{1}{3x^{-2}} = \dfrac{1(x^2)}{3} = \dfrac{x^2}{3}$ The exponent -2 does not apply to 3.

c. $\dfrac{12a^3b^{-4}}{4a^{-2}b} = \dfrac{12a^3 \cdot a^2}{4b \cdot b^4}$ Definition of negative exponent

$\qquad\quad = \dfrac{3a^5}{b^5}$ Product of Powers Property

d. $\left(\dfrac{3x^2}{y}\right)^{-2} = \dfrac{3^{-2}(x^2)^{-2}}{y^{-2}}$ Power of a Quotient and Power of a Product Properties

$\qquad\quad = \dfrac{3^{-2}x^{-4}}{y^{-2}}$ Power of a Power Property

$\qquad\quad = \dfrac{y^2}{3^2x^4}$ Definition of negative exponent

$\qquad\quad = \dfrac{y^2}{9x^4}$ Simplify.

✓CHECKPOINT   Now try Exercise 51.

EXAMPLE 3 RATIO OF VOLUME TO SURFACE AREA

The volume V and surface area S of a sphere are given by

$$V = \frac{4}{3}\pi r^3 \qquad \text{and} \qquad S = 4\pi r^2$$

where r is the radius of the sphere. A spherical weather balloon has a radius of 2 feet, as shown in Figure R1.7. Find the quotient of the volume and the surface area.

SOLUTION

Form the quotient V/S and simplify, as follows.

$$\frac{V}{S} = \frac{\frac{4}{3}\pi r^3}{4\pi r^2} = \frac{\frac{4}{3}\pi 2^3}{4\pi 2^2} = \frac{1}{3}(2) = \frac{2}{3}$$

✓CHECKPOINT   Now try Exercise 25.

$r = 2$ ft

Figure R1.7

Scientific Notation

Exponents provide an efficient way of writing and computing with very large (or very small) numbers. For instance, a drop of water contains more than 33 billion billion molecules—that is, 33 followed by 18 zeros.

33,000,000,000,000,000,000

It is convenient to write such numbers in **scientific notation.** This notation has the form $c \times 10^n$, where $1 \leq c < 10$ and n is an integer. So, the number of molecules in a drop of water can be written in scientific notation as

$$3.3 \times 10{,}000{,}000{,}000{,}000{,}000{,}000 = 3.3 \times 10^{19}.$$

The *positive* exponent 19 indicates that the number is *large* (10 or more) and that the decimal point has been moved 19 places. A *negative* exponent in scientific notation indicates that the number is *small* (less than 1). For instance, the mass (in grams) of one electron is approximately

$$9.0 \times 10^{-28} = 0.0000000000000000000000000009.$$

28 decimal places

EXAMPLE 4   CONVERTING TO SCIENTIFIC NOTATION

a. $0.0000572 = 5.72 \times 10^{-5}$ Number is less than 1.

b. $149{,}400{,}000 = 1.494 \times 10^8$ Number is greater than 10.

c. $32.675 = 3.2675 \times 10^1$ Number is greater than 10.

✓CHECKPOINT   Now try Exercise 53.

EXAMPLE 5   CONVERTING TO DECIMAL NOTATION

a. $3.125 \times 10^2 = 312.5$ Number is greater than 10.

b. $3.73 \times 10^{-6} = 0.00000373$ Number is less than 1.

c. $7.91 \times 10^5 = 791{,}000$ Number is greater than 10.

✓CHECKPOINT   Now try Exercise 57.

Most calculators automatically switch to scientific notation when they are showing large (or small) numbers that exceed the display range. Try multiplying $86{,}500{,}000 \times 6000$. If your calculator follows standard conventions, its display should be

5.19   11   or   5.19E11 .

This means that $c = 5.19$ and the exponent of 10 is $n = 11$, which implies that the number is 5.19×10^{11}. To *enter* numbers in scientific notation, your calculator should have an exponential entry key labeled [EXP] or [EE].

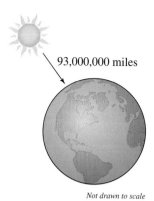

93,000,000 miles

Not drawn to scale

Figure R1.8

EXAMPLE 6   THE SPEED OF LIGHT

The distance between Earth and the sun is approximately 93 million miles, as shown in Figure R1.8. How long does it take for light to travel from the sun to Earth? Use the fact that light travels at a rate of approximately 186,000 miles per second.

SOLUTION

Using the formula Distance = (rate)(time), you find the time as follows.

$$\text{Time} = \frac{\text{distance}}{\text{rate}} = \frac{93 \text{ million miles}}{186,000 \text{ miles per second}}$$

$$= \frac{9.3 \times 10^7 \text{ miles}}{1.86 \times 10^5 \text{ miles/second}}$$

$$= 5 \times 10^2 \text{ seconds}$$

$$\approx 8.33 \text{ minutes}$$

Note that to convert 500 seconds to 8.33 minutes, you divide by 60, because there are 60 seconds in a minute.

✓CHECKPOINT   Now try Exercise 63.

Applications

One of the most useful features of a calculator is its ability to evaluate exponential expressions. Consult your user's guide for specific keystrokes.

EXAMPLE 7   USING A CALCULATOR TO RAISE A NUMBER TO A POWER

Scientific Calculator

Expression	Keystrokes	Display
a. $13^4 + 5$	13 [yˣ] 4 [+] 5 [=]	28566
b. $3^{-2} + 4^{-1}$	3 [yˣ] 2 [+/−] [+] 4 [yˣ] 1 [+/−] [=]	0.361111111
c. $\dfrac{3^5 + 1}{3^5 - 1}$	[(] 3 [yˣ] 5 [+] 1 [)] [÷] [(] 3 [yˣ] 5 [−] 1 [)] [=]	1.008264463

TECHNOLOGY

Make sure you include parentheses as needed when entering expressions in your calculator. Notice the use of parentheses in Example 7(c).

Graphing Calculator

Expression	Keystrokes	Display
a. $13^4 + 5$	13 [^] 4 [+] 5 [ENTER]	28566
b. $3^{-2} + 4^{-1}$	3 [^] [(−)] 2 [+] 4 [^] [(−)] 1 [ENTER]	.3611111111
c. $\dfrac{3^5 + 1}{3^5 - 1}$	[(] 3 [^] 5 [+] 1 [)] [÷] [(] 3 [^] 5 [−] 1 [)] [ENTER]	1.008264463

✓CHECKPOINT   Now try Exercise 69.

The formulas shown below can be used to find the balance in a savings account.

Balance in an Account

The balance A in an account that earns an annual interest rate of r (in decimal form) for t years is given by one of the following.

$$A = P(1 + rt) \qquad \text{Simple interest}$$

$$A = P\left(1 + \frac{r}{n}\right)^{nt} \qquad \text{Compound interest}$$

In both formulas, P is the principal (or the initial deposit). In the formula for compound interest, n is the number of compoundings *per year*. Make sure you convert all units of time t to years. For instance, 6 months $= \frac{1}{2}$ year. So, $t = \frac{1}{2}$.

EXAMPLE 8   FINDING THE BALANCE IN AN ACCOUNT

Make a Decision You are trying to decide how to invest $5000 for 10 years. Which savings plan will earn more money?

a. 4% simple annual interest

b. 3.5% interest compounded quarterly

SOLUTION

a. The balance after 10 years is

$$A = P(1 + rt)$$

$$= 5000[1 + 0.04(10)]$$

$$= \$7000$$

b. The balance after 10 years is

$$A = P\left(1 + \frac{r}{n}\right)^{nt}$$

$$= 5000\left(1 + \frac{0.035}{4}\right)^{(4)(10)}$$

$$= \$7084.54$$

Savings plan (a) will earn $7000 - 5000 = \$2000$ and savings plan (b) will earn $7084.54 - 5000 = \$2084.54$. So, plan (b) will earn more money.

✓CHECKPOINT   Now try Exercise 71.

In addition to finding the balance in an account, the compound interest formula can also be used to determine the rate of inflation. To apply the formula, you must know the cost of an item for two different years, as demonstrated in Example 9.

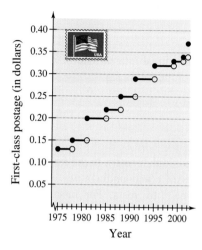

First-class postage (in dollars)

0.40
0.35
0.30
0.25
0.20
0.15
0.10
0.05

1975 1980 1985 1990 1995 2000
Year

Figure R1.9

EXAMPLE 9   FINDING THE RATE OF INFLATION

In 1976, the cost of a first-class postage stamp was $0.13. By 2002, the cost had risen to $0.37, as shown in Figure R1.9. Find the average annual rate of inflation for first-class postage over this 26-year period.   (Source: U.S. Postal Service)

SOLUTION

To find the average annual rate of inflation, use the formula for compound interest with *annual* compounding. So, you need to find the value of r that will make the following equation true.

$$A = P\left(1 + \frac{r}{n}\right)^{nt}$$

$$0.37 = 0.13(1 + r)^{26}$$

You can begin by guessing that the average annual rate of inflation was 5%. Entering $r = 0.05$ in the formula, you find that $0.13(1 + 0.05)^{26} \approx 0.4622$. Because this result is more than 0.37, try some smaller values of r. Finally, you can discover that

$$0.13(1 + 0.041)^{26} \approx 0.37.$$

So, the average annual rate of inflation for first-class postage from 1976 to 2002 was about 4.1%.

✓CHECKPOINT   Now try Exercise 73.

Discussing the Concept | Compound Interest

Suppose that you have $5000 to invest for 5 years in an account that pays compound interest according to the formula

$$A = 5000\left(1 + \frac{r}{n}\right)^{5n}.$$

Which of the following combinations would produce the highest balance?

	Annual Interest Rate	*Type of Compounding*
a.	5.60%	Daily ($n = 365$)
b.	5.65%	Weekly
c.	5.70%	Monthly
d.	5.75%	Quarterly

Choose one annual interest rate and construct a bar graph or table showing the balance that could be obtained with each type of compounding. Discuss how the balance is affected by the type of compounding.

R1.3 Warm Up

The following warm-up exercises involve skills that were covered in earlier sections. You will use these skills in the exercise set for this section. For additional help, review Section R1.2.

In Exercises 1–10, perform the indicated operation(s) and simplify.

1. $\left(\frac{2}{3}\right)\left(\frac{3}{2}\right)$

2. $\left(\frac{1}{4}\right)(5)(4)$

3. $3\left(\frac{2}{7}\right) + 11\left(\frac{2}{7}\right)$

4. $11\left(\frac{1}{4}\right) + \frac{5}{4}$

5. $\frac{1}{2} \div 2$

6. $\frac{1}{3} \div \frac{1}{3}$

7. $\frac{1}{7} + \frac{1}{3} - \frac{1}{21}$

8. $\frac{1}{3} + \frac{1}{2} - \frac{5}{6}$

9. $\frac{1}{12} - \frac{1}{3} + \frac{1}{8}$

10. $\left(\frac{1}{2} - \frac{1}{3}\right) \div \frac{1}{6}$

R1.3 Exercises

In Exercises 1–20, evaluate the expression. Write fractional answers in simplest form.

1. $2^2 \cdot 2^4$

2. $3 \cdot 3^5$

3. $\dfrac{2^6}{2^3}$

4. $\dfrac{4^5}{4^3}$

5. $(3^3)^2$

6. $(2^4)^2$

7. -3^4

8. $(-3)^4$

9. $8 \cdot 2^{-2} \cdot 4^{-1}$

10. $6 \cdot 2^{-3} \cdot 3^{-1}$

11. $\left(\frac{1}{2}\right)^{-3}$

12. $\left(\frac{2}{3}\right)^{-3}$

13. $5^{-1} - 2^{-1}$

14. $3^{-1} - 4^{-1}$

15. $(2^3 \cdot 3^2)^2$

16. $(-3 \cdot 4^2)^3$

17. $\left(-\frac{3}{5}\right)^3\left(\frac{5}{3}\right)^2$

18. $\left(\frac{-5}{4}\right)^3\left(\frac{4}{5}\right)^2$

19. 3^0

20. $(-2)^0$

In Exercises 21–24, evaluate the expression for the indicated value of x.

Expression	Value
21. $\dfrac{x^4}{2x^2}$	$x = -6$
22. $4x^{-3}$	$x = 2$
23. $7x^{-2}$	$x = 4$
24. $6x^0 - (6x)^0$	$x = -5$

In Exercises 25–42, simplify the expression.

25. $(-5z)^3$

26. $(-2w)^5$

27. $(8x^4)(2x^3)$

28. $5x^4(x^2)$

29. $10(x^2)^2$

30. $(5x^3)^3$

31. $(-z)^3(3z^4)$

32. $(6y^2)(2y^3)^3$

33. $\dfrac{25y^8}{10y^4}$

34. $\dfrac{3x^5}{x^3}$

35. $\left(\frac{4}{y}\right)^3\left(\frac{3}{y}\right)^4$

36. $\dfrac{15(x + 3)^3}{9(x + 3)^2}$

37. $\dfrac{7x^2}{x^3}$

38. $\dfrac{5z^5}{z^7}$

39. $\dfrac{x^2 \cdot x^n}{x^3 \cdot x^n}$

40. $\dfrac{x^n \cdot x^{2n}}{x^{3n}}$

41. $3^n \cdot 3^{2n}$

42. $2^m \cdot 2^{3m}$

In Exercises 43–52, rewrite the expression with positive exponents and simplify.

43. $(2x^5)^0, \quad x \neq 0$

44. $(x + 5)^0, \quad x \neq -5$

45. $(y + 2)^{-2}(y + 2)^{-1}$

46. $(x + y)^{-3}(x + y)$

47. $(4y^{-2})(8y^4)$

48. $(-2x^2)^3(4x^3)^{-1}$

49. $\left(\frac{x}{10}\right)^{-1}$

50. $\left(\frac{y}{5}\right)^{-2}$

51. $\left(\frac{x^{-3}y^4}{5}\right)^{-3}$

52. $\left(\frac{2z^2}{y}\right)^{-2}$

In Exercises 53–56, write the number in scientific notation.

53. Land Area of Earth: 57,300,000 square miles

54. Water Area of Earth: 139,500,000 square miles

55. Light Year: 9,461,000,000,000 kilometers

56. Thickness of a Soap Bubble: 0.0000001 meter

In Exercises 57–60, write the number in decimal notation.

57. Number of Air Sacs in Lungs: 3.5×10^8

58. Temperature of Core of Sun: 1.5×10^7 degrees Celsius

59. Charge of an Electron: 1.602×10^{-19} coulomb

60. Width of a Human Hair: 9.0×10^{-5} meter

IN THE NEWS

Scientists measure movement in split millionth of a second

VIENNA (AFP)—Austro-Hungarian physicist Ferenc Krauzs said scientists had developed a device that can measure the speed of atomic processes down to the smallest fraction of a second yet.

Describing the device as "the fastest stopwatch in the world", Krauzs said Thursday that it measures the movement of atomic particles in time units smaller than 100 attoseconds.

An attosecond is the name given to a quintillionth, or a millionth of a millionth of a millionth, of a second.

"This time is to a second what a minute is to the age of the universe," Krauzs explained. . . .

(Source: Agence France Presse, "Scientists measure movement in split of a millionth of a second," www.yahoo.com, February 26, 2004.)

61. Write the number of attoseconds in a second in scientific notation and in decimal notation.

62. How many seconds is 100 attoseconds?

In Exercises 63 and 64, evaluate each expression without using a calculator.

63. (a) $(1.2 \times 10^7)(5 \times 10^{-3})$ (b) $\dfrac{(6.0 \times 10^8)}{(3.0 \times 10^{-3})}$

64. (a) $(9.8 \times 10^{-2})(3 \times 10^7)$ (b) $\dfrac{9.0 \times 10^5}{4.5 \times 10^{-2}}$

In Exercises 65–68, use a calculator to evaluate each expression. Write your answer in scientific notation. (Round to three decimal places.)

65. (a) $0.000345(8,900,000,000)$

(b) $\dfrac{67,000,000 + 93,000,000}{0.0052}$

66. (a) $0.000045(9,200,000)$

(b) $\dfrac{0.0000928 - 0.0000021}{0.0061}$

67. (a) $(9.3 \times 10^6)^3(6.1 \times 10^{-4})$ (b) $\dfrac{(2.414 \times 10^4)^6}{(1.68 \times 10^5)^5}$

68. (a) $(1.2 \times 10^2)^2(5.3 \times 10^{-5})$ (b) $\dfrac{(3.28 \times 10^{-6})^{10}}{(5.34 \times 10^{-3})^{22}}$

69. Calculator Keystrokes Write the algebraic expression that corresponds to each set of keystrokes.

(a) ⟮ 5.1 ⊟ 3.6 ⟯ y^x 5 ⊟   Scientific

⟮ 5.1 ⊟ 3.6 ⟯ ⌃ 5 ⟮ENTER⟯   Graphing

(b) 1 ⊞ 3 ⊠ 2 ⊟ y^x 2 ⟮+/-⟯ ⊟   Scientific

1 ⊞ 3 ⊠ 2 ⟮ENTER⟯ ⌃ ⟮(-)⟯ 2 ⟮ENTER⟯   Graphing

70. *Make a Decision* Because $-2^3 = -8$ and $(-2)^3 = -8$, a student concludes that $-a^n = (-a)^n$, where n is an integer. Do you agree? Can you find an example where $-a^n \neq (-a)^n$?

71. *Make a Decision*: **Balance in an Account** You deposit $10,000 in an account with an annual interest rate of 6.5% for 10 years. Determine the balance in the account when the interest is compounded (a) daily ($n = 365$), (b) weekly, (c) monthly, and (d) quarterly. How is the balance affected by the type of compounding?

72. *Make a Decision*: **Balance in an Account** You deposit $2000 in an account with an annual interest rate of 7.75% for 15 years. Determine the balance in the account when the interest is compounded (a) daily ($n = 365$), (b) weekly, (c) monthly, and (d) quarterly. How is the balance affected by the type of compounding?

73. College Tuition The bar graph shows the average tuition at private four-year colleges in the United States from 1996 to 2002. Find the average annual rate of inflation for tuition over this seven-year period. (Source: U.S. National Center for Education Statistics)

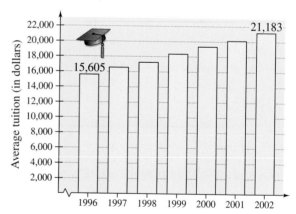

74. Minimum Wage The minimum wage in the United States in 1983 was $3.35. By 2003, the minimum wage had risen to $5.15. Find the average annual rate of inflation for the minimum wage over this 20-year period. (Source: U.S. Department of Labor)

R1.4 Radicals and Rational Exponents ⟵

Objectives

- **Simplify a radical.**
- **Rationalize a denominator.**
- **Use properties of rational exponents.**
- **Combine radicals.**
- **Use a calculator to evaluate a radical.**
- **Use a radical expression to solve an application problem.**

Radicals and Properties of Radicals

A **square root** of a number is defined as one of its two equal factors. For example, 5 is a square root of 25 because 5 is one of the two equal factors of 25. In a similar way, a **cube root** of a number is one of its three equal factors. Here are some examples.

Number	*Equal Factors*	*Root*
$25 = (-5)^2$	$(-5)(-5)$	-5 (square root)
$-64 = (-4)^3$	$(-4)(-4)(-4)$	-4 (cube root)
$81 = 3^4$	$3 \cdot 3 \cdot 3 \cdot 3$	3 (fourth root)

Definition of *n*th Root of a Number

Let a and b be real numbers and let n be a positive integer. If

$$a = b^n$$

then b is an ***n*th root of a.** If $n = 2$, the root is a **square root,** and if $n = 3$, the root is a **cube root.**

From this definition, you can see that some numbers have more than one *n*th root. For example, both 5 and -5 are square roots of 25. The following definition distinguishes between these two roots.

Principal *n*th Root of a Number

Let a be a real number that has at least one real *n*th root. The **principal *n*th root of a** is the *n*th root that has the same sign as a, and it is denoted by the **radical symbol**

$$\sqrt[n]{a}.\qquad \text{Principal } n\text{th root}$$

The positive integer n is the **index** (the plural of index is *indexes* or *indices*) of the radical, and the number a is the **radicand.** If $n = 2$, omit the index and write $\sqrt{a}$ rather than $\sqrt[2]{a}$.

| **EXAMPLE 1** | EVALUATING EXPRESSIONS INVOLVING RADICALS |

a. The principal square root of 121 is $\sqrt{121} = 11$ because $11^2 = 121$.

b. The principal cube root of $\frac{125}{64}$ is $\sqrt[3]{\frac{125}{64}} = \frac{5}{4}$ because $\left(\frac{5}{4}\right)^3 = \frac{5^3}{4^3} = \frac{125}{64}$.

c. The principal fifth root of -32 is $\sqrt[5]{-32} = -2$ because $(-2)^5 = -32$.

d. $-\sqrt{49} = -7$ because $7^2 = 49$.

e. $\sqrt[4]{-81}$ is not a real number because there is no real number that can be raised to the fourth power to produce -81.

✓CHECKPOINT   Now try Exercise 15.

From Example 1, you can make the following generalizations about nth roots of a real number.

1. If a is a positive real number and n is a positive *even* integer, then a has exactly two real nth roots, which are denoted by $\sqrt[n]{a}$ and $-\sqrt[n]{a}$.

2. If a is any real number and n is an *odd* integer, then a has only one (real) nth root. It is the principal nth root and is denoted by $\sqrt[n]{a}$.

3. If a is negative and n is an *even* integer, then a has no (real) nth root.

Integers such as 1, 4, 9, 16, 49, and 81 are called **perfect squares** because they have integer square roots. Similarly, integers such as 1, 8, 27, 64, and 125 are called **perfect cubes** because they have integer cube roots.

Properties of Radicals

Let a and b be real numbers, variables, or algebraic expressions such that the indicated roots are real numbers, and let m and n be positive integers. Then the following properties are true.

Property	*Example*
1. $\sqrt[n]{a^m} = \left(\sqrt[n]{a}\right)^m$	$\sqrt[3]{8^2} = \left(\sqrt[3]{8}\right)^2 = (2)^2 = 4$
2. $\sqrt[n]{a} \cdot \sqrt[n]{b} = \sqrt[n]{ab}$	$\sqrt{5} \cdot \sqrt{7} = \sqrt{5 \cdot 7} = \sqrt{35}$
3. $\dfrac{\sqrt[n]{a}}{\sqrt[n]{b}} = \sqrt[n]{\dfrac{a}{b}}, \quad b \neq 0$	$\dfrac{\sqrt[4]{27}}{\sqrt[4]{9}} = \sqrt[4]{\dfrac{27}{9}} = \sqrt[4]{3}$
4. $\sqrt[m]{\sqrt[n]{a}} = \sqrt[mn]{a}$	$\sqrt[3]{\sqrt{10}} = \sqrt[6]{10}$
5. $\left(\sqrt[n]{a}\right)^n = a$	$\left(\sqrt{3}\right)^2 = 3$
6. For n even, $\sqrt[n]{a^n} = |a|$.	$\sqrt{(-12)^2} = |-12| = 12$
For n odd, $\sqrt[n]{a^n} = a$.	$\sqrt[3]{(-12)^3} = -12$

A common special case of Property 6 is

$$\sqrt{a^2} = |a|.$$

Simplifying Radicals

An expression involving radicals is in **simplest form** when the following conditions are satisfied.

1. All possible factors have been removed from the radical.

2. All fractions have radical-free denominators (accomplished by a process called *rationalizing the denominator*).

3. The index of the radical has been reduced as far as possible.

To simplify a radical, factor the radicand into factors whose powers are multiples of the index. The roots of these factors are written outside the radical, and the "leftover" factors make up the new radicand.

EXAMPLE 2   SIMPLIFYING EVEN ROOTS

a. $\sqrt[4]{48} = \sqrt[4]{16 \cdot 3}$ Find largest fourth-power factor.

$\phantom{\sqrt[4]{48}} = \sqrt[4]{2^4 \cdot 3}$ Rewrite.

$\phantom{\sqrt[4]{48}} = 2\sqrt[4]{3}$ Find fourth root.

b. $\sqrt{75x^3} = \sqrt{25x^2 \cdot 3x}$ Find largest square factor.

$\phantom{\sqrt{75x^3}} = \sqrt{(5x)^2 \cdot 3x}$ Rewrite.

$\phantom{\sqrt{75x^3}} = 5x\sqrt{3x}, \quad x \geq 0$ Find root of perfect square.

c. $\sqrt[4]{(5x)^4} = |5x| = 5|x|$

✓*CHECKPOINT*   Now try Exercise 33.

In Example 2(c), note that the absolute value symbol is included in the answer because $\sqrt[4]{x^4} = |x|$.

EXAMPLE 3   SIMPLIFYING ODD ROOTS

a. $\sqrt[3]{24} = \sqrt[3]{8 \cdot 3}$ Find largest cube factor.

$\phantom{\sqrt[3]{24}} = \sqrt[3]{2^3 \cdot 3}$ Rewrite.

$\phantom{\sqrt[3]{24}} = 2\sqrt[3]{3}$ Find root of perfect cube.

b. $\sqrt[5]{32a^{11}} = \sqrt[5]{32a^{10} \cdot a}$ Find largest fifth-power factor.

$\phantom{\sqrt[5]{32a^{11}}} = \sqrt[5]{(2a^2)^5 \cdot a}$ Rewrite.

$\phantom{\sqrt[5]{32a^{11}}} = 2a^2 \sqrt[5]{a}$ Find fifth root.

c. $\sqrt[3]{-40x^6} = \sqrt[3]{(-8x^6) \cdot 5}$ Find largest cube factor.

$\phantom{\sqrt[3]{-40x^6}} = \sqrt[3]{(-2x^2)^3 \cdot 5}$ Rewrite.

$\phantom{\sqrt[3]{-40x^6}} = -2x^2 \sqrt[3]{5}$ Find root of perfect cube.

✓*CHECKPOINT*   Now try Exercise 35.

To **rationalize the denominator** of a fraction, multiply the numerator and denominator by an appropriate factor. To find the "appropriate factor," make use of the form $a + b\sqrt{m}$ and its **conjugate** $a - b\sqrt{m}$. The product of this conjugate pair has no radical. For instance,

$$\left(4 + \sqrt{3}\right)\left(4 - \sqrt{3}\right) = 4^2 - \left(\sqrt{3}\right)^2 = 16 - 3 = 13.$$

Therefore, to rationalize a denominator of the form $a - b\sqrt{m}$ (or $a + b\sqrt{m}$), you can multiply the numerator and denominator by the conjugate factor. If $a = 0$, the rationalizing factor of $\sqrt{m}$ is itself, $\sqrt{m}$.

EXAMPLE 4   RATIONALIZING SINGLE-TERM DENOMINATORS

a. To rationalize the denominator of the following fraction, multiply *both* the numerator and the denominator by $\sqrt{3}$ to obtain

$$\frac{5}{2\sqrt{3}} = \frac{5}{2\sqrt{3}} \cdot \frac{\sqrt{3}}{\sqrt{3}} = \frac{5\sqrt{3}}{2\sqrt{3^2}} = \frac{5\sqrt{3}}{2(3)} = \frac{5\sqrt{3}}{6}.$$

b. To rationalize the denominator of the following fraction, multiply *both* the numerator and the denominator by $\sqrt[3]{5^2}$. Note how this eliminates the radical from the denominator by producing a perfect *cube* in the radicand.

$$\frac{2}{\sqrt[3]{5}} = \frac{2}{\sqrt[3]{5}} \cdot \frac{\sqrt[3]{5^2}}{\sqrt[3]{5^2}} = \frac{2\sqrt[3]{5^2}}{\sqrt[3]{5^3}} = \frac{2\sqrt[3]{25}}{5}$$

✓CHECKPOINT   Now try Exercise 39.

EXAMPLE 5   RATIONALIZING A DENOMINATOR WITH TWO TERMS

$$\frac{2}{3 + \sqrt{7}} = \frac{2}{3 + \sqrt{7}} \cdot \frac{3 - \sqrt{7}}{3 - \sqrt{7}}$$ Multiply numerator and denominator by conjugate.

$$= \frac{2\left(3 - \sqrt{7}\right)}{3^2 - \left(\sqrt{7}\right)^2}$$ Multiply fractions.

$$= \frac{2\left(3 - \sqrt{7}\right)}{9 - 7}$$ Simplify.

$$= \frac{2\left(3 - \sqrt{7}\right)}{2}$$ Divide out like factors.

$$= 3 - \sqrt{7}$$ Simplify.

✓CHECKPOINT   Now try Exercise 41.

Don't confuse an expression such as $\sqrt{2} + \sqrt{7}$ with $\sqrt{2 + 7}$. In general, $\sqrt{x + y} \neq \sqrt{x} + \sqrt{y}$.

Rational Exponents

The definition on the following page shows how radicals are used to define **rational exponents.** Until now, work with exponents has been restricted to integer exponents.

STUDY TIP

The numerator of a rational exponent denotes the *power* to which the base is raised, and the denominator denotes the *index* or the *root* to be taken. Moreover, it doesn't matter in which order the two operations are performed, provided the *n*th root exists. Here is an example.

$$8^{2/3} = \left(\sqrt[3]{8}\right)^2 = 2^2 = 4$$

$$8^{2/3} = \sqrt[3]{8^2} = \sqrt[3]{64} = 4$$

Definition of Rational Exponents

If a is a real number and n is a positive integer such that the principal nth root of a exists, then $a^{1/n}$ is defined to be $a^{1/n} = \sqrt[n]{a}$.

If m is a positive integer that has no common factor with n, then $a^{m/n} = (a^{1/n})^m = \left(\sqrt[n]{a}\right)^m$ and $a^{m/n} = (a^m)^{1/n} = \sqrt[n]{a^m}$.

The properties of exponents that were listed in Section R1.3 also apply to rational exponents (provided the roots indicated by the denominators exist). Some of those properties are relisted here, with different examples.

Properties of Exponents

Let r and s be rational numbers, and let a and b be real numbers, variables, or algebraic expressions. If the roots indicated by the rational exponents exist, then the following properties are true.

Property	*Example*
1. $a^r a^s = a^{r+s}$	$4^{1/2}(4^{1/3}) = 4^{5/6}$
2. $\dfrac{a^r}{a^s} = a^{r-s}, \quad a \neq 0$	$\dfrac{x^2}{x^{1/2}} = x^{2-(1/2)} = x^{3/2}, x \neq 0$
3. $(ab)^r = a^r b^r$	$(2x)^{1/2} = 2^{1/2}(x^{1/2})$
4. $\left(\dfrac{a}{b}\right)^r = \dfrac{a^r}{b^r}, \quad b \neq 0$	$\left(\dfrac{x}{3}\right)^{1/3} = \dfrac{x^{1/3}}{3^{1/3}}$
5. $(a^r)^s = a^{rs}$	$(x^3)^{1/3} = x$
6. $a^{-r} = \dfrac{1}{a^r}, \quad a \neq 0$	$4^{-1/2} = \dfrac{1}{4^{1/2}} = \dfrac{1}{2}$
7. $\left(\dfrac{a}{b}\right)^{-r} = \left(\dfrac{b}{a}\right)^r, \quad a \neq 0, \quad b \neq 0$	$\left(\dfrac{x}{4}\right)^{-1/2} = \left(\dfrac{4}{x}\right)^{1/2} = \dfrac{2}{x^{1/2}}$

STUDY TIP

Rational exponents can be tricky, and you must remember that the expression $b^{m/n}$ is not defined unless $\sqrt[n]{b}$ is a real number. This restriction produces some unusual-looking results. For instance, the number $(-8)^{1/6}$ is not defined because $\sqrt[6]{-8}$ is not a real number. And yet, $(-8)^{1/3}$ is defined because $\sqrt[3]{-8} = -2$.

Rational exponents are particularly useful for evaluating roots of numbers on a calculator, for reducing the index of a radical, and for simplifying (and factoring) algebraic expressions. Examples 6 and 7 demonstrate some of these uses.

EXAMPLE 6   SIMPLIFYING WITH RATIONAL EXPONENTS

a. $(27)^{1/3} = \sqrt[3]{27} = 3$

b. $(-32)^{-4/5} = \left(\sqrt[5]{-32}\right)^{-4} = (-2)^{-4} = \dfrac{1}{(-2)^4} = \dfrac{1}{16}$

c. $(-5x^{2/3})(3x^{-1/3}) = -15x^{(2/3)-(1/3)} = -15x^{1/3}, \; x \neq 0$

✓CHECKPOINT   Now try Exercise 45.

| EXAMPLE 7 | REDUCING THE INDEX OF A RADICAL |

a. $\sqrt[6]{a^4} = a^{4/6} = a^{2/3} = \sqrt[3]{a^2}$

b. $\sqrt[3]{\sqrt{125}} = (125^{1/2})^{1/3}$ Rewrite with rational exponents.

$= (125)^{1/6}$ Multiply exponents.

$= (5^3)^{1/6}$ Rewrite base as perfect cube.

$= 5^{3/6}$ Multiply exponents.

$= 5^{1/2}$ Reduce exponent.

$= \sqrt{5}$ Rewrite as radical.

✓*CHECKPOINT*   Now try Exercise 53.

Radical expressions can be combined (added or subtracted) if they are **like radicals**—that is, if they have the same index and radicand. For instance, $2\sqrt{3x}$ and $\frac{1}{2}\sqrt{3x}$ are like radicals, but $\sqrt[3]{3x}$ and $2\sqrt{3x}$ are not like radicals.

| EXAMPLE 8 | SIMPLIFYING AND COMBINING LIKE RADICALS |

a. $2\sqrt{48} + 3\sqrt{27} = 2\sqrt{16 \cdot 3} + 3\sqrt{9 \cdot 3}$ Find square factors.

$= 8\sqrt{3} + 9\sqrt{3}$ Find square roots.

$= 17\sqrt{3}$ Combine like terms.

b. $\sqrt[3]{16x} - \sqrt[3]{54x} = \sqrt[3]{8 \cdot 2x} - \sqrt[3]{27 \cdot 2x}$ Find cube factors.

$= 2\sqrt[3]{2x} - 3\sqrt[3]{2x}$ Find cube roots.

$= -\sqrt[3]{2x}$ Combine like terms.

✓*CHECKPOINT*   Now try Exercise 61.

Radicals and Calculators

To evaluate square roots using a calculator, use the square root key $\boxed{\sqrt{\ }}$. For cube roots, use the cube root key $\boxed{\sqrt[3]{\ }}$. For other roots, first convert the radical to exponential form and then use the exponential key $\boxed{\wedge}$ or $\boxed{y^x}$.

| EXAMPLE 9 | EVALUATING A CUBE ROOT WITH A CALCULATOR |

To evaluate $\sqrt[3]{25}$, first rewrite it as $25^{1/3}$. Then enter the following keystrokes.

25 $\boxed{y^x}$ $\boxed{(}$ 1 $\boxed{\div}$ 3 $\boxed{)}$ $\boxed{=}$ Scientific

25 $\boxed{\wedge}$ $\boxed{(}$ 1 $\boxed{\div}$ 3 $\boxed{)}$ $\boxed{\text{ENTER}}$ Graphing

For either of these keystroke sequences, the calculator display should read 2.924017738. So,

$\sqrt[3]{25} \approx 2.924.$

✓*CHECKPOINT*   Now try Exercise 63.

EXAMPLE 10   EVALUATING RADICALS WITH A CALCULATOR

Evaluate the following radicals. Round to three decimal places.

a. $\sqrt[3]{-4}$ **b.** $(1.4)^{-2/5}$

SOLUTION

a. Because

$$\sqrt[3]{-4} = \sqrt[3]{(-1)(4)} = \sqrt[3]{-1} \cdot \sqrt[3]{4} = -\sqrt[3]{4}$$

you can attach the negative sign of the radicand as follows.

$$4 \boxed{y^x} \boxed{(} 1 \boxed{\div} 3 \boxed{)} \boxed{=} \boxed{+/-}$$ Scientific

$$\boxed{(-)} 4 \boxed{\wedge} \boxed{(} 1 \boxed{\div} 3 \boxed{)} \boxed{\text{ENTER}}$$ Graphing

The calculator display is -1.587401052, which implies that

$$\sqrt[3]{-4} \approx -1.587.$$

b. Using the following keystroke sequence

$$1.4 \boxed{y^x} \boxed{(} 2 \boxed{\div} 5 \boxed{+/-} \boxed{)} \boxed{=}$$ Scientific

$$1.4 \boxed{\wedge} \boxed{(} \boxed{(-)} 2 \boxed{\div} 5 \boxed{)} \boxed{\text{ENTER}}$$ Graphing

you obtain a calculator display of 0.874075175. So,

$$(1.4)^{-2/5} \approx 0.874.$$

✔CHECKPOINT   Now try Exercise 67.

Application

EXAMPLE 11   ESCAPE VELOCITY

Make a Decision   A rocket, launched vertically from Earth, has an initial velocity of 10,000 meters per second. All of the fuel is used for launching. The *escape velocity,* or the minimum initial velocity necessary for the rocket to travel forever into space, is

$$\sqrt{\frac{2(6.67 \times 10^{-11})(5.98 \times 10^{24})}{6.37 \times 10^6}}$$ meters per second.

Will the rocket travel forever into space?

SOLUTION

The escape velocity is

$$\sqrt{\frac{2(6.67 \times 10^{-11})(5.98 \times 10^{24})}{6.37 \times 10^6}} \approx 11{,}190.7 \text{ meters per second.}$$

The initial velocity of 10,000 meters per second is less than the escape velocity of 11,190.7 meters per second. The rocket will not travel forever into space.

✔CHECKPOINT   Now try Exercise 81.

© Mark M. Lawrence/CORBIS

R1.4 Warm Up

The following warm-up exercises involve skills that were covered in earlier sections. You will use these skills in the exercise set for this section. For additional help, review Section R1.3.

In Exercises 1–10, simplify the expression.

1. $\left(\frac{1}{3}\right)\left(\frac{2}{3}\right)^2$

2. $3(-4)^2$

3. $(-2x)^3$

4. $(-2x^3)(-3x^4)$

5. $(7x^5)(4x)$

6. $(5x^4)(25x^2)^{-1}$

7. $\dfrac{12z^6}{4z^2}$

8. $\left(\dfrac{2x}{5}\right)^2\left(\dfrac{2x}{5}\right)^{-4}$

9. $\left(\dfrac{3y^2}{x}\right)^0$, $x \neq 0$, $y \neq 0$

10. $[(x+2)^2(x+2)^3]^2$

R1.4 Exercises

In Exercises 1–12, fill in the missing form.

Radical Form	Rational Exponent Form
1. $\sqrt{9} = 3$	
2. $\sqrt[3]{64} = 4$	
3.	$32^{1/5} = 2$
4.	$-(144^{1/2}) = -12$
5.	$196^{1/2} = 14$
6.	$614.125^{1/3} = 8.5$
7. $\sqrt[3]{-216} = -6$	
8. $\sqrt[5]{-243} = -3$	
9. $\left(\sqrt[4]{81}\right)^3 = 27$	
10. $\sqrt[4]{81^3} = 27$	
11.	$27^{2/3} = 9$
12.	$16^{5/4} = 32$

In Exercises 13–30, evaluate the expression.

13. $\sqrt{9}$

14. $\sqrt[3]{64}$

15. $-\sqrt[3]{-27}$

16. $\sqrt[3]{0}$

17. $\dfrac{4}{\sqrt{64}}$

18. $\dfrac{\sqrt[4]{81}}{3}$

19. $\left(\sqrt[3]{-125}\right)^3$

20. $\sqrt[4]{562^4}$

21. $16^{1/2}$

22. $27^{1/3}$

23. $36^{3/2}$

24. $16^{3/2}$

25. $\sqrt{2} \cdot \sqrt{3}$

26. $\sqrt{2} \cdot \sqrt{5}$

27. $\left(\frac{16}{81}\right)^{-3/4}$

28. $\left(\frac{9}{4}\right)^{-1/2}$

29. $\left(-\frac{1}{64}\right)^{-1/3}$

30. $\left(-\frac{125}{27}\right)^{-1/3}$

In Exercises 31–36, simplify the expression.

31. $\sqrt[3]{16x^5}$

32. $\sqrt[4]{(3x^2)^4}$

33. $\sqrt{75x^2y^{-4}}$

34. $\sqrt[5]{96x^5}$

35. $\sqrt[5]{64y^{-5}}$

36. $\sqrt{8x^4y^3z^{-2}}$

In Exercises 37–44, rewrite the expression by rationalizing the denominator. Simplify your answer.

37. $\dfrac{1}{\sqrt{3}}$

38. $\dfrac{5}{\sqrt{10}}$

39. $\dfrac{8}{\sqrt[3]{2}}$

40. $\dfrac{5}{\sqrt[3]{(5x)^2}}$

41. $\dfrac{2x}{5 - \sqrt{3}}$

42. $\dfrac{5x}{\sqrt{14} - 2}$

43. $\dfrac{3}{\sqrt{5} + \sqrt{6}}$

44. $\dfrac{5}{2\sqrt{10} - 5}$

In Exercises 45–50, simplify the expression.

45. $5^{1/2} \cdot 5^{3/2}$

46. $4^{1/3} \cdot 4^{5/3}$

47. $\dfrac{2^{3/2}}{2}$

48. $\dfrac{5^{1/2}}{5}$

49. $\dfrac{x^2}{x^{1/2}}$

50. $\dfrac{x \cdot x^{1/2}}{x^{3/2}}$

In Exercises 51–56, use rational exponents to reduce the index of the radical.

51. $\sqrt{\sqrt{32}}$

52. $\sqrt{\sqrt{x^4}}$

53. $\sqrt[4]{3^2}$

54. $\sqrt[4]{(3x^2)^4}$

55. $\sqrt[6]{x^2}$

56. $\sqrt[6]{(x+2)^4}$

In Exercises 57–62, simplify the expression.

57. $5\sqrt{x} - 3\sqrt{x}$

58. $3\sqrt{x+1} + 10\sqrt{x+1}$

59. $5\sqrt{50} + 3\sqrt{8}$

60. $2\sqrt{27} - \sqrt{75}$

61. $2\sqrt{4y} - 2\sqrt{9y}$

62. $2\sqrt{80} + \sqrt{125}$

In Exercises 63–70, use a calculator to approximate the number. (Round to three decimal places.)

63. $\sqrt[3]{45}$

64. $\sqrt{57}$

65. $5.7^{2/5}$

66. $36.8^{1.2}$

67. $0.26^{-0.8}$

68. $3.75^{-1/2}$

69. $\dfrac{3 - \sqrt{5}}{2}$

70. $\dfrac{-4 + \sqrt{12}}{4}$

In Exercises 71–76, complete the statement with <, =, or >.

71. $\sqrt{5} + \sqrt{3}$ ____ $\sqrt{5+3}$

72. $\sqrt{3} - \sqrt{2}$ ____ $\sqrt{3-2}$

73. 5 ____ $\sqrt{3^2 + 2^2}$

74. 5 ____ $\sqrt{3^2 + 4^2}$

75. $\sqrt{3} \cdot \sqrt[4]{3}$ ____ $\sqrt[8]{3}$

76. $\sqrt{\dfrac{3}{11}}$ ____ $\dfrac{\sqrt{3}}{\sqrt{11}}$

77. Geometry Find the dimensions of a cube that has a volume of 13,824 cubic inches (see figure).

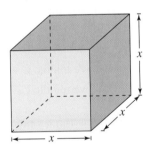

78. Geometry Find the dimensions of a square classroom that has 800 square feet of floor space (see figure).

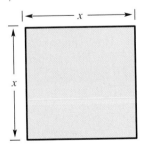

Declining Balances Depreciation In Exercises 79 and 80, find the annual depreciation rate r. To find the annual depreciation rate by the declining balances method, use the formula

$$r = 1 - \left(\frac{S}{C}\right)^{1/n}$$

where n is the useful life of the item (in years), S is the salvage value (in dollars), and C is the original cost (in dollars).

79. A truck whose original cost is $75,000 is depreciated over an eight-year period, as shown in the bar graph.

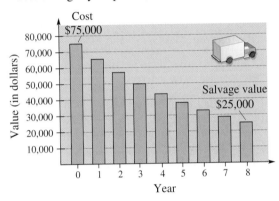

80. A printing press whose original cost is $125,000 is depreciated over a 10-year period, as shown in the bar graph.

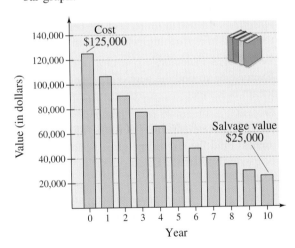

81. *Make a Decision*: **Escape Velocity** The escape velocity (in meters per second) for a rocket launched vertically from the moon is

$$\sqrt{\frac{2(6.67 \times 10^{-11})(7.36 \times 10^{22})}{1.74 \times 10^6}}.$$

If all the fuel is consumed during launching, will a rocket launched vertically from the moon with an initial velocity of 2000 meters per second travel forever into space? Explain your reasoning. (*Hint:* The initial velocity must reach or exceed the escape velocity for a rocket to travel forever in space.)

82. *Make a Decision*: **Escape Velocity** The escape velocity (in meters per second) for a rocket launched vertically from Mars is

$$\sqrt{\frac{2(6.67 \times 10^{-11})(6.42 \times 10^{23})}{3.37 \times 10^6}}.$$

If all the fuel is consumed during launching, will a rocket launched vertically from Mars with an initial velocity of 6000 meters per second travel forever into space? Explain your reasoning. (*Hint:* The initial velocity must reach or exceed the escape velocity for a rocket to travel forever in space.)

83. **Period of a Pendulum** The period T (in seconds) of a pendulum is given by

$$T = 2\pi\sqrt{\frac{L}{32}}$$

where L is the length (in feet) of the pendulum. Find the period of a pendulum whose length is 2 feet.

84. **Period of a Pendulum** Use the formula given in Exercise 83 to find the period of a pendulum whose length is 1.5 feet.

85. **Erosion** A stream of water moving at the rate of v feet per second can carry particles of size $0.03\sqrt{v}$ inches. Find the size of the largest particle that can be carried by a stream flowing at the rate of $\frac{3}{4}$ foot per second.

86. **Erosion** A stream of water moving at the rate of v feet per second can carry particles of size $0.03\sqrt{v}$ inches. Find the size of the largest particle that can be carried by a stream flowing at the rate of $\frac{5}{9}$ foot per second.

Notes on a Musical Scale In Exercises 87–90, find the frequency of the indicated note on a piano (see figure). The musical note A above middle C has a frequency of 440 vibrations per second. If we denote this frequency by F_1, then the frequency of the next higher note is given by $F_2 = F_1 \cdot 2^{1/12}$. Similarly, the frequency of the next note is given by $F_3 = F_2 \cdot 2^{1/12}$.

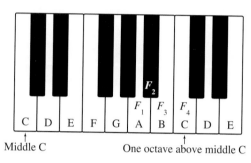

87. Find the frequency of the musical note B above middle C.

88. Find the frequency of the musical note C that is one octave above middle C.

89. *Make a Decision* Which note would you expect to have a higher frequency? Explain your reasoning.

(a) Musical note E above middle C

(b) Musical note D above middle C

90. *Make a Decision* Which note would you expect to have a higher frequency?

(a) Musical note D above middle C

(b) Musical note G above middle C

91. **Calculator Experiment** Enter any positive real number in your calculator and repeatedly take the square root. What real number does the display appear to be approaching?

92. **Calculator Experiment** Square the real number $2/\sqrt{5}$ and note that the radical is eliminated from the denominator. Is this equivalent to rationalizing the denominator? Why or why not?

93. **Think About It** How can you show that $a^0 = 1$, $a \neq 0$? (*Hint:* Use the property of exponents $a^m/a^n = a^{m-n}$.)

94. Explain why $\sqrt{4x^2} \neq 2x$.

95. Explain why $\sqrt{2} + \sqrt{3} \neq \sqrt{5}$.

Mid-Chapter Quiz

Take this quiz as you would take a quiz in class. After you are done, check your work against the answers given in the back of the book.

In Exercises 1 and 2, place the correct symbol ($<$, $>$, or $=$) between the two real numbers.

1. $-|-7|$ ▒▒▒ $|-7|$

2. $-(-3)$ ▒▒▒ $|-3|$

In Exercises 3 and 4, use inequality notation to describe the subset of real numbers.

3. x is positive or x is equal to zero.

4. The apartment occupancy rate r will be at least 96.5% during the coming year.

5. Describe the subset of real numbers that is represented by the inequality $0 \leq x < 3$, and sketch the subset on the real number line.

6. Identify the terms of the algebraic expression $3x^2 - 7x + 2$.

In Exercises 7–10, perform the indicated operation(s). Write fractional answers in simplest form.

7. $-4 - (-7)$

8. $\dfrac{28 - 4}{(-6)}$

9. $\dfrac{2}{3} \cdot \dfrac{5}{4} \cdot \dfrac{3}{7}$

10. $\dfrac{11}{15} \div \dfrac{3}{5}$

In Exercises 11–13, rewrite the expression with positive exponents and simplify.

11. $(-x)^3(2x^4)$

12. $\dfrac{5y^7}{15y^3}$

13. $\left(\dfrac{x^{-2}y^2}{3}\right)^{-3}$

14. You deposit \$1000 in an account with an annual interest rate of 8%, compounded monthly. Find the balance in the account after 10 years.

In Exercises 15 and 16, evaluate the expression.

15. $\dfrac{-\sqrt[4]{81}}{3}$

16. $\left(\sqrt[3]{-64}\right)^3$

In Exercises 17–19, simplify the expression.

17. $3^{1/2} \cdot 3^{3/2}$

18. $\sqrt[3]{81} - 4\sqrt[3]{3}$

19. $\sqrt[10]{12^5}$

20. Find the dimensions of a cube that has a volume of 17,576 cubic centimeters (see figure).

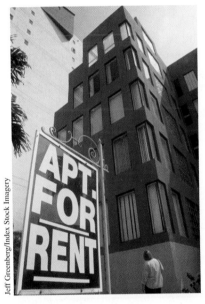

When occupancy rates are not maximized, renters can sometimes negotiate for a lower rent. But when the market is overwhelmed by renters, rates are driven up.

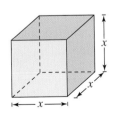

Figure for 20

R1.5 Polynomials and Special Products

Objectives

- Write a polynomial in standard form.
- Add, subtract, and multiply polynomials.
- Use special products to multiply polynomials.
- Use polynomials to solve an application problem.

Polynomials

One of the simplest and most common types of algebraic expressions is a **polynomial.** Here are some examples.

$$2x + 5, \qquad 3x^4 - 7x^2 + 2x + 4, \qquad 5x^2y^2 - xy + 3$$

The first two are *polynomials in x* and the third is a *polynomial in x and y.* The terms of a polynomial in x have the form ax^k, where a is the **coefficient** and k is the **degree** of the term. Because a polynomial is defined as an algebraic sum, the coefficients take on the signs between the terms. For instance, the polynomial

$$2x^3 - 5x^2 + 1 = 2x^3 + (-5)x^2 + (0)x + 1$$

has coefficients 2, -5, 0, and 1.

Definition of a Polynomial in x

Let $a_n, \ldots, a_2, a_1, a_0$ be real numbers and let n be a *nonnegative integer.* A **polynomial in x** is an expression of the form

$$a_n x^n + \cdots + a_2 x^2 + a_1 x + a_0$$

where $a_n \neq 0$. The polynomial is of **degree** n, and the number a_n is the **leading coefficient.** The number a_0 is the **constant term.** The constant term is considered to have a degree of zero.

Note in the definition of a polynomial in x that the polynomial is written in descending powers of x. This is called the **standard form** of a polynomial.

EXAMPLE 1   REWRITING A POLYNOMIAL IN STANDARD FORM

Polynomial	*Standard Form*	*Degree*
a. $4x^2 - 5x^3 - 2 + 3x$	$-5x^3 + 4x^2 + 3x - 2$	3
b. $4 - 9x^2$	$-9x^2 + 4$	2
c. 8	$8 \ (8 = 8x^0)$	0

✓CHECKPOINT   Now try Exercise 7.

Polynomials with one, two, and three terms are called **monomials, binomials,** and **trinomials,** respectively.

A polynomial that has all zero coefficients is called the **zero polynomial,** denoted by 0. This particular polynomial is not considered to have a degree.

EXAMPLE 2 IDENTIFYING A POLYNOMIAL AND ITS DEGREE

a. $-2x^3 + x^2 + 3x - 2$ is a polynomial of degree 3.

b. $\sqrt{x^2 - 3x}$ is not a polynomial because the radical sign indicates a noninteger power of x.

c. $x^2 + 5x^{-1}$ is not a polynomial because of the negative exponent.

✓CHECKPOINT   Now try Exercise 9.

For polynomials in more than one variable, the *degree of a term* is the sum of the powers of the variables in the term. The *degree of the polynomial* is the highest degree of all its terms. For instance, the polynomial

$$5x^3y - x^2y^2 + 2xy - 5$$

has two terms of degree 4, one term of degree 2, and one term of degree 0. The degree of the polynomial is 4.

Operations with Polynomials

You can **add** and **subtract** polynomials in much the same way that you add and subtract real numbers—you simply add or subtract the *like terms* (terms having the same variables to the same powers) by adding their coefficients. For instance, $-3x^2$ and $5x^2$ are like terms and their sum is given by

$$-3x^2 + 5x^2 = (-3 + 5)x^2 = 2x^2.$$

EXAMPLE 3 SUMS AND DIFFERENCES OF POLYNOMIALS

a. $(5x^3 - 7x^2 - 3) + (x^3 + 2x^2 - x + 8)$

$= (5x^3 + x^3) + (2x^2 - 7x^2) - x + (8 - 3)$ Group like terms.

$= 6x^3 - 5x^2 - x + 5$ Combine like terms.

b. $(7x^4 - x^2 - 4x + 2) - (3x^4 - 4x^2 + 3x)$

$= 7x^4 - x^2 - 4x + 2 - 3x^4 + 4x^2 - 3x$ Distribute sign.

$= (7x^4 - 3x^4) + (4x^2 - x^2) + (-4x - 3x) + 2$ Group like terms.

$= 4x^4 + 3x^2 - 7x + 2$ Combine like terms.

✓CHECKPOINT   Now try Exercise 17.

A common mistake is to fail to change the sign of *each* term inside parentheses preceded by a minus sign. For instance, note the following.

$-(3x^4 - 4x^2 + 3x) = -3x^4 + 4x^2 - 3x$ Correct

~~$-(3x^4 - 4x^2 + 3x) = -3x^4 - 4x^2 + 3x$~~   Common mistake

To find the **product** of two polynomials, you can use the left and right Distributive Properties. For example, if you treat $(5x + 7)$ as a single quantity, you can multiply $(3x - 2)$ by $(5x + 7)$ as follows.

$$(3x - 2)(5x + 7) = 3x(5x + 7) - 2(5x + 7)$$

$$= (3x)(5x) + (3x)(7) - (2)(5x) - (2)(7)$$

$$= 15x^2 + 21x - 10x - 14$$

Product of	Product of	Product of	Product of
First terms	Outer terms	Inner terms	Last terms

$$= 15x^2 + 11x - 14$$

With practice you should be able to multiply two binomials without writing all of the steps above. In fact, the four products in the boxes above suggest that you can put the product of two binomials in the FOIL form in just one step. This is called the **FOIL Method.**

EXAMPLE 4   USING THE FOIL METHOD

Use the FOIL Method to find the product of $2x - 4$ and $x + 5$.

SOLUTION

$$\overset{\text{F}\quad\text{O}\quad\text{I}\quad\text{L}}{(2x - 4)(x + 5) = 2x^2 + 10x - 4x - 20}$$

$$= 2x^2 + 6x - 20$$

✓CHECKPOINT   Now try Exercise 31.

When multiplying two polynomials, be sure to multiply *each* term of one polynomial by *each* term of the other. The following vertical pattern is a convenient way to multiply two polynomials.

EXAMPLE 5   USING A VERTICAL FORMAT TO MULTIPLY POLYNOMIALS

Multiply $(x^2 - 2x + 2)$ by $(x^2 + 2x + 2)$.

SOLUTION

$$
\begin{array}{ll}
x^2 - 2x + 2 & \text{Standard form} \\
x^2 + 2x + 2 & \text{Standard form} \\
\hline
x^4 - 2x^3 + 2x^2 & x^2(x^2 - 2x + 2) \\
\quad\quad 2x^3 - 4x^2 + 4x & 2x(x^2 - 2x + 2) \\
\quad\quad\quad\quad 2x^2 - 4x + 4 & 2(x^2 - 2x + 2) \\
\hline
x^4 + 0x^3 + 0x^2 + 0x + 4 = x^4 + 4 & \text{Combine like terms.}
\end{array}
$$

So, $(x^2 - 2x + 2)(x^2 + 2x + 2) = x^4 + 4$.

✓CHECKPOINT   Now try Exercise 51.

Special Products

Special Products

Let u and v be real numbers, variables, or algebraic expressions.

Special Product   *Example*

Sum and Difference of Two Terms

$(u + v)(u - v) = u^2 - v^2$ $(x + 4)(x - 4) = x^2 - 16$

Square of a Binomial

$(u + v)^2 = u^2 + 2uv + v^2$ $(x + 3)^2 = x^2 + 6x + 9$

$(u - v)^2 = u^2 - 2uv + v^2$ $(3x - 2)^2 = 9x^2 - 12x + 4$

Cube of a Binomial

$(u + v)^3 = u^3 + 3u^2v + 3uv^2 + v^3$ $(x + 2)^3 = x^3 + 6x^2 + 12x + 8$

$(u - v)^3 = u^3 - 3u^2v + 3uv^2 - v^3$ $(x - 1)^3 = x^3 - 3x^2 + 3x - 1$

EXAMPLE 6   SUM AND DIFFERENCE OF TWO TERMS

$(5x + 9)(5x - 9) = (5x)^2 - 9^2 = 25x^2 - 81$

✔*CHECKPOINT*   Now try Exercise 33.

EXAMPLE 7   SQUARE OF A BINOMIAL

$(6x - 5)^2 = (6x)^2 - 2(6x)(5) + 5^2$

$= 36x^2 - 60x + 25$

✔*CHECKPOINT*   Now try Exercise 35.

EXAMPLE 8   CUBE OF A BINOMIAL

$(3x + 2)^3 = (3x)^3 + 3(3x)^2(2) + 3(3x)(2)^2 + 2^3$

$= 27x^3 + 54x^2 + 36x + 8$

✔*CHECKPOINT*   Now try Exercise 41.

EXAMPLE 9   THE PRODUCT OF TWO TRINOMIALS

$(x + y - 2)(x + y + 2) = [(x + y) - 2][(x + y) + 2]$

$= (x + y)^2 - 2^2$

$= x^2 + 2xy + y^2 - 4$

✔*CHECKPOINT*   Now try Exercise 47.

Applications

Many families set up savings accounts to help pay their children's college expenses.

EXAMPLE 10   A SAVINGS PLAN

Make a Decision At the same time each year for 5 consecutive years, you deposit money in an account that pays 7% interest, compounded annually. You deposit $1500 the first year, $1800 the second year, $2400 the third year, $2600 the fourth year, and $3000 the fifth year. Is there enough money in the account to pay a $12,000 college tuition bill?

SOLUTION

Using the formula for compound interest, for *each* deposit you have

$$\text{Balance} = P\left(1 + \frac{r}{n}\right)^{nt} = P(1 + 0.07)^t = P(1.07)^t.$$

For the first deposit, $P = 1500$ and $t = 4$. For the second deposit, $P = 1800$ and $t = 3$, and so on. The balances for the five deposits are as follows.

Date	Deposit	Time in Account	Balance in Account
First Year	$1500	4 years	$1500(1.07)^4$
Second Year	$1800	3 years	$1800(1.07)^3$
Third Year	$2400	2 years	$2400(1.07)^2$
Fourth Year	$2600	1 year	$2600(1.07)$
Fifth Year	$3000	0 years	3000

By adding these five balances, you can find the total balance in the account to be

$$1500(1.07)^4 + 1800(1.07)^3 + 2400(1.07)^2 + 2600(1.07) + 3000.$$

Note that this expression is in polynomial form. By evaluating the expression, you can find the balance to be $12,701.03, as shown in Figure R1.10.

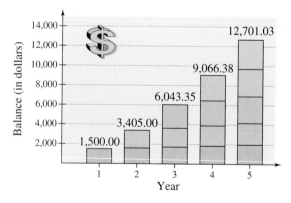

Figure R1.10

After the fifth deposit, there is enough money in the account to pay the college tuition bill.

✓CHECKPOINT   Now try Exercise 57.

EXAMPLE 11 GEOMETRY: VOLUME OF A BOX

An open box is made by cutting squares from the corners of a piece of metal that measures 16 inches by 20 inches and turning up the sides, as shown in Figure R1.11. If the edge of each cut-out square measures x inches, what is the volume of the box? Find the volume when $x = 1$, $x = 2$, and $x = 3$ inches.

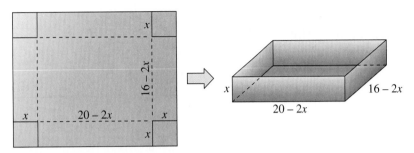

Figure R1.11

SOLUTION

Verbal Model: Volume = Length · Width · Height

Labels: Height = x (inches)
Width = $16 - 2x$ (inches)
Length = $20 - 2x$ (inches)

Equation: Volume = $(20 - 2x)(16 - 2x)(x)$

$= (320 - 72x + 4x^2)(x)$

$= 320x - 72x^2 + 4x^3$

When $x = 1$ inch, the volume of the box is

Volume = $320(1) - 72(1)^2 + 4(1)^3 = 252$ cubic inches.

When $x = 2$ inches, the volume of the box is

Volume = $320(2) - 72(2)^2 + 4(2)^3 = 384$ cubic inches.

When $x = 3$ inches, the volume of the box is

Volume = $320(3) - 72(3)^2 + 4(3)^3 = 420$ cubic inches.

✓CHECKPOINT Now try Exercise 61.

Discussing the Concept | Communicating Mathematically

Describe how you would show that $\sqrt{a^2 + b^2} \neq a + b$. Give an algebraic argument and a numerical example.

R1.5 Warm Up

The following warm-up exercises involve skills that were covered in earlier sections. You will use these skills in the exercise set for this section. For additional help, review Sections R1.3 and R1.4.

In Exercises 1–10, perform the indicated operation(s).

1. $(7x^2)(6x)$

2. $(10z^3)(-2z^{-1})$

3. $(-3x^2)^3$

4. $-3(x^2)^3$

5. $\dfrac{27z^5}{12z^2}$

6. $\sqrt{24} \cdot \sqrt{2}$

7. $\left(\dfrac{2x}{3}\right)^{-2}$

8. $16^{3/4}$

9. $\dfrac{4}{\sqrt{8}}$

10. $\sqrt[3]{-27x^3}$

R1.5 Exercises

In Exercises 1–6, find the degree and leading coefficient of the polynomial.

1. $2x^2 - x + 1$

2. $-3x^4 + 2x^2 - 5$

3. $x^5 - 1$

4. 3

5. $3x^5 - 6x^4 + x - 2$

6. $-3x$

In Exercises 7–12, determine whether the algebraic expression is a polynomial. If it is, write the polynomial in standard form and state its degree.

7. $2x - 3x^3 + 8$

8. $2x^3 + x - 3x^{-1}$

9. $\dfrac{3x + 4}{x}$

10. $\dfrac{2x^2 + 5x - 3}{3}$

11. $w^2 - w^4 + 2w^3$

12. $\sqrt{y^2 - y^4}$

In Exercises 13–16, evaluate the polynomial for each value of x.

13. $4x + 3$
   (a) $x = -1$ (b) $x = 0$
   (c) $x = 1$ (d) $x = 2$

14. $x^2 - 6$
   (a) $x = -2$ (b) $x = -1$
   (c) $x = 0$ (d) $x = 1$

15. $-2x^2 + 3x + 4$
   (a) $x = -2$ (b) $x = -1$
   (c) $x = 0$ (d) $x = 1$

16. $x^3 - 4x^2 + x$
   (a) $x = -1$ (b) $x = 0$
   (c) $x = 1$ (d) $x = 2$

In Exercises 17–28, perform the indicated operation(s) and write the resulting polynomial in standard form.

7. $(6x + 5) - (8x + 15)$

18. $(3x^2 + 1) - (2x^2 - 2x + 3)$

19. $-(x^3 + 5) + (3x^3 - 4x)$

20. $-(5x^2 - 1) + (-3x^2 + 5)$

21. $(15x^2 - 6) - (-8x^3 - 14x^2 - 17)$

22. $(15x^4 - 18x - 19) - (13x^4 - 5x + 15)$

23. $3x(x^2 - 2x + 1)$

24. $z^2(2z^2 + 3z + 1)$

25. $-4x(3 - x^3)$

26. $-5y(2y - y^2)$

27. $(-2x)(-3x)(5x + 2)$

28. $(1 - x^3)(4x)$

In Exercises 29–54, find the product.

29. $(x + 3)(x + 4)$

30. $(x - 5)(x + 10)$

31. $(3x - 5)(2x + 1)$

32. $(7x - 2)(4x - 3)$

33. $(x + 5)(x - 5)$

34. $(3x + 2)(3x - 2)$

35. $(x + 6)^2$

36. $(3x - 2)^2$

37. $(2x - 5y)^2$

38. $(5 - 8x)^2$

39. $[(x - 3) + y]^2$

40. $[(x + 1) - y]^2$

41. $(x + 1)^3$

42. $(x - 2)^3$

43. $(2x - y)^3$

44. $(3x + 2y)^3$

45. $(3y^2 - 1)(3y^2 + 1)$

46. $(3x^2 - 4y^2)(3x^2 + 4y^2)$

47. $(m - 3 + n)(m - 3 - n)$

48. $(x + y + 1)(x + y - 1)$

49. $(\sqrt{x} + \sqrt{y})(\sqrt{x} - \sqrt{y})$

50. $(5 + \sqrt{x})(5 - \sqrt{x})$

51. $(x^2 - x + 1)(x^2 + x + 1)$

52. $(x^2 + 3x - 2)(x^2 - 3x - 2)$

53. $5x(x + 1) - 3x(x + 1)$

54. $(2x - 1)(x + 3) + 3(x + 3)$

55. Compound Interest After 2 years, an investment of $500 compounded annually at an interest rate of r will yield an amount of $500(1 + r)^2$. Write this polynomial in standard form.

56. Compound Interest After 3 years, an investment of $1200 compounded annually at an interest rate of r will yield an amount of $1200(1 + r)^3$. Write this polynomial in standard form.

57. *Make a Decision*: Savings Plan At the same time each year for 5 consecutive years, you deposit money in an account that pays 6% interest, compounded annually. The deposits are $1200, $1400, $800, $2300, and $2900, respectively. Is there enough money in the account to pay a $10,000 college tuition bill?

58. *Make a Decision*: Savings Plan You have an investment that pays an annual dividend. At the same time each year for 6 consecutive years, you reinvest this dividend in an account that pays 6.5% interest, compounded annually. For the dividends shown in the table, find the balance in the account at the time of the sixth deposit. (Note: The first deposit will earn 5 years of interest, and the sixth deposit will earn no interest.) Is there enough money in the account for a $6500 down payment on a car?

$	Year	Dividend
	1	$840
	2	$930
	3	$760
	4	$1020
	5	$980
	6	$1130

59. Federal Pell Grants The total amount (in millions of dollars) of Federal Pell Grants from 1995 to 2002 can be represented by the polynomial

$$120.07x^2 - 1222.1x + 8775$$

where x represents the year, with $x = 5$ corresponding to 1995. Evaluate the polynomial when $x = 11$ and $x = 12$. Then describe your results in everyday terms. (Source: U.S. Department of Education)

60. Federal Pell Grants The average Federal Pell Grant (in dollars) awarded from 1995 to 2002 can be represented by the polynomial $7.06x^2 + 10.7x + 1278$, where x represents the year, with $x = 5$ corresponding to 1995. Evaluate the polynomial when $x = 11$ and $x = 12$. Then describe your results in everyday terms. (Source: U.S. Department of Education)

61. *Make a Decision*: Geometry A box has a length of $(54 - 2x)$ inches, a width of $(38 - 2x)$ inches, and a height of x inches. Find the volume when $x = 5$, $x = 7$, and $x = 9$ inches. Which of the x-values gives a maximum volume?

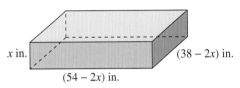

62. *Make a Decision*: Geometry A box has a length of $(45 - 2x)$ inches, a width of $(45 - 2x)$ inches, and a height of x inches. Find the volume when $x = 6$, $x = 8$, and $x = 10$ inches. Which of the x-values gives a maximum volume?

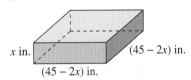

63. Geometry Find the area of the shaded region in the figure. Write your answer as a polynomial in standard form.

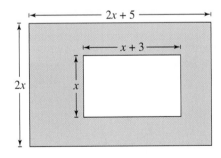

64. Geometry Find a polynomial that represents the total number of square feet in the floor plan.

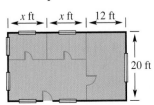

R1.6 Factoring

Objectives

- **Factor a polynomial by removing common factors.**
- **Factor a polynomial in a special form.**
- **Factor a trinomial as the product of two binomials.**
- **Factor a polynomial by grouping.**

Common Factors

The process of writing a polynomial as a product is called **factoring.** It is an important tool for solving equations and reducing fractional expressions.

A polynomial that cannot be factored using integer coefficients is called **prime** or **irreducible over the integers.** For instance, the polynomial $x^2 - 3$ is irreducible over the integers. [Over the *real numbers*, this polynomial can be factored as

$$x^2 - 3 = \left(x + \sqrt{3}\right)\left(x - \sqrt{3}\right).]$$

A polynomial is **completely factored** when each of its factors is prime. For instance,

$$x^3 - x^2 + 4x - 4 = (x - 1)(x^2 + 4) \qquad \text{Completely factored}$$

is completely factored, but

$$x^3 - x^2 - 4x + 4 = (x - 1)(x^2 - 4) \qquad \text{Not completely factored}$$

is not completely factored. Its complete factorization is

$$x^3 - x^2 - 4x + 4 = (x - 1)(x + 2)(x - 2).$$

The simplest type of factoring involves a polynomial that can be written as the product of a monomial and another polynomial. To factor such a polynomial, you can use the Distributive Property in the *reverse* direction.

$$ab + ac = a(b + c) \qquad \qquad a \text{ is a common factor.}$$

EXAMPLE 1   REMOVING COMMON FACTORS

Factor each expression.

a. $6x^3 - 4x$ **b.** $(x - 2)(2x) + (x - 2)(3)$

SOLUTION

a. Each term of this polynomial has $2x$ as a common factor.

$$6x^3 - 4x = 2x(3x^2) - 2x(2) = 2x(3x^2 - 2)$$

b. The binomial factor $(x - 2)$ is common to both terms.

$$(x - 2)(2x) + (x - 2)(3) = (x - 2)(2x + 3)$$

✓CHECKPOINT   Now try Exercise 5.

Factoring Special Polynomial Forms

Factoring Special Polynomial Forms

Factored Form	*Example*

Difference of Two Squares

$$u^2 - v^2 = (u + v)(u - v) \qquad 9x^2 - 4 = (3x + 2)(3x - 2)$$

Perfect Square Trinomial

$$u^2 + 2uv + v^2 = (u + v)^2 \qquad x^2 + 6x + 9 = (x + 3)^2$$

$$u^2 - 2uv + v^2 = (u - v)^2 \qquad x^2 - 6x + 9 = (x - 3)^2$$

Sum or Difference of Two Cubes

$$u^3 + v^3 = (u + v)(u^2 - uv + v^2) \quad x^3 + 8 = (x + 2)(x^2 - 2x + 4)$$

$$u^3 - v^3 = (u - v)(u^2 + uv + v^2) \quad 27x^3 - 1 = (3x - 1)(9x^2 + 3x + 1)$$

STUDY TIP

In Example 2, note that the first step in factoring a polynomial is to check for common factors. Once the common factor is removed, it is often possible to recognize patterns that were not obvious at first glance.

EXAMPLE 2   REMOVING A COMMON FACTOR FIRST

Factor $3 - 12x^2$.

SOLUTION

$$
\begin{aligned}
3 - 12x^2 &= 3(1 - 4x^2) & \text{3 is a common factor.} \\
&= 3[1^2 - (2x)^2] & \text{Difference of two squares} \\
&= 3(1 + 2x)(1 - 2x) & \text{Completely factored}
\end{aligned}
$$

✓**CHECKPOINT**   Now try Exercise 47.

EXAMPLE 3   FACTORING THE DIFFERENCE OF TWO SQUARES

a.
$$
\begin{aligned}
(x + 2)^2 - y^2 &= [(x + 2) + y][(x + 2) - y] \\
&= (x + 2 + y)(x + 2 - y) \\
&= (x + y + 2)(x - y + 2)
\end{aligned}
$$

b. You can factor $16x^4 - 81$ by applying the difference of two squares formula twice.

$$
\begin{aligned}
16x^4 - 81 &= (4x^2)^2 - 9^2 \\
&= (4x^2 + 9)(4x^2 - 9) & \text{First application} \\
&= (4x^2 + 9)[(2x)^2 - 3^2] \\
&= (4x^2 + 9)(2x + 3)(2x - 3) & \text{Second application}
\end{aligned}
$$

✓**CHECKPOINT**   Now try Exercise 7.

A perfect square trinomial is the square of a binomial, and it has the following form. Note that the first and last terms of a perfect square trinomial are squares and the middle term is twice the product of u and v.

$$u^2 + 2uv + v^2 = (u + v)^2 \qquad \text{or} \qquad u^2 - 2uv + v^2 = (u - v)^2$$

Same sign   Same sign

EXAMPLE 4   FACTORING PERFECT SQUARE TRINOMIALS

a. $16x^2 + 8x + 1 = (4x)^2 + 2(4x)(1) + 1^2 = (4x + 1)^2$

b. $x^2 - 10x + 25 = x^2 - 2(x)(5) + 5^2 = (x - 5)^2$

✓*CHECKPOINT*   Now try Exercise 13.

The next two formulas show that sums and differences of cubes factor easily. Pay special attention to the signs of the terms.

Like signs   Like signs

$$u^3 + v^3 = (u + v)(u^2 - uv + v^2) \qquad u^3 - v^3 = (u - v)(u^2 + uv + v^2)$$

Unlike signs   Unlike signs

EXAMPLE 5   FACTORING THE SUM AND DIFFERENCE OF CUBES

Factor each expression.

a. $x^3 - 27$ **b.** $3x^3 + 192$

SOLUTION

a. $x^3 - 27 = x^3 - 3^3$ Rewrite 27 as 3^3.

$\qquad = (x - 3)(x^2 + 3x + 9)$ Factor.

b. $3x^3 + 192 = 3(x^3 + 64)$ 3 is a common factor.

$\qquad = 3(x^3 + 4^3)$ Rewrite 64 as 4^3.

$\qquad = 3(x + 4)(x^2 - 4x + 16)$ Factor.

✓*CHECKPOINT*   Now try Exercise 19.

Trinomials with Binomial Factors

To factor a trinomial of the form $ax^2 + bx + c$, use the following pattern.

Factors of a

$$ax^2 + bx + c = (\quad x + \quad)(\quad x + \quad)$$

Factors of c

The goal is to find a combination of factors of a and c such that the outer and inner products add up to the middle term bx. For instance, for the trinomial

$$6x^2 + 17x + 5$$

you can write

$$
\begin{array}{cccc}
\text{F} & \text{O} & \text{I} & \text{L} \\
\downarrow & \downarrow & \downarrow & \downarrow
\end{array}
$$

$$(2x + 5)(3x + 1) = 6x^2 + 2x + 15x + 5$$

$$
\begin{array}{c}
\text{O} + \text{I} \\
\downarrow
\end{array}
$$

$$= 6x^2 + 17x + 5.$$

Note that the outer (O) and inner (I) products add up to $17x$.

EXAMPLE 6   FACTORING A TRINOMIAL: LEADING COEFFICIENT IS 1

Factor the trinomial $x^2 - 7x + 12$.

SOLUTION

For this trinomial, you have $a = 1$, $b = -7$, and $c = 12$. Because b is negative and c is positive, both factors of 12 must be negative. That is, $12 = (-2)(-6)$, $12 = (-1)(-12)$, or $12 = (-3)(-4)$. So, the possible factorizations of $x^2 - 7x + 12$ are

$$(x - 2)(x - 6), \qquad (x - 1)(x - 12), \qquad \text{and} \qquad (x - 3)(x - 4).$$

Testing the middle term, you can find the correct factorization to be

$$x^2 - 7x + 12 = (x - 3)(x - 4).$$

✓CHECKPOINT   Now try Exercise 25.

EXAMPLE 7   FACTORING A TRINOMIAL: LEADING COEFFICIENT IS NOT 1

Factor the trinomial $2x^2 + x - 15$.

SOLUTION

For this trinomial, you have $a = 2$ and $c = -15$, which means that the factors of -15 must have unlike signs. The eight possible factorizations are as follows.

$$
\begin{array}{ll}
(2x - 1)(x + 15) & (2x + 1)(x - 15) \\
(2x - 3)(x + 5) & (2x + 3)(x - 5) \\
(2x - 5)(x + 3) & (2x + 5)(x - 3) \\
(2x - 15)(x + 1) & (2x + 15)(x - 1)
\end{array}
$$

Testing the middle term, you can find the correct factorization to be

$$2x^2 + x - 15 = (2x - 5)(x + 3).$$

✓CHECKPOINT   Now try Exercise 33.

Factoring by Grouping

Sometimes polynomials with more than three terms can be **factored by grouping.**

EXAMPLE 8 FACTORING BY GROUPING

$$x^3 - 2x^2 - 3x + 6 = (x^3 - 2x^2) - (3x - 6) \qquad \text{Group terms.}$$
$$= x^2(x - 2) - 3(x - 2) \qquad \text{Factor groups.}$$
$$= (x - 2)(x^2 - 3) \qquad \text{Distributive Property}$$

✓CHECKPOINT   Now try Exercise 39.

When factoring by grouping, sometimes several different groupings will work. For instance, another way to factor the polynomial in Example 8 is to group the terms as follows.

$$x^3 - 2x^2 - 3x + 6 = (x^3 - 3x) - (2x^2 - 6)$$
$$= x(x^2 - 3) - 2(x^2 - 3)$$
$$= (x^2 - 3)(x - 2)$$

As you can see, you obtain the same result as in Example 8.

Factoring a trinomial can involve quite a bit of trial and error, which can be lessened by using factoring by grouping. The key to this method of factoring is knowing how to rewrite the middle term. In general, to factor a trinomial $ax^2 + bx + c$ by grouping, choose factors of the product ac that add up to b and use these factors to rewrite the middle term. This technique is illustrated in Example 9.

EXAMPLE 9 FACTORING A TRINOMIAL BY GROUPING

Use factoring by grouping to factor $2x^2 + 5x - 3$.

SOLUTION

In the trinomial $2x^2 + 5x - 3$, $a = 2$ and $c = -3$, which implies that the product ac is -6. Now, -6 factors as $(6)(-1)$, and $6 - 1 = 5 = b$. So, you can rewrite the middle term as $5x = 6x - x$. This produces the following.

$$2x^2 + 5x - 3 = 2x^2 + 6x - x - 3 \qquad \text{Rewrite middle term.}$$
$$= (2x^2 + 6x) - (x + 3) \qquad \text{Group terms.}$$
$$= 2x(x + 3) - (x + 3) \qquad \text{Factor groups.}$$
$$= (x + 3)(2x - 1) \qquad \text{Distributive Property}$$

So, the trinomial factors as $2x^2 + 5x - 3 = (x + 3)(2x - 1)$.

✓CHECKPOINT   Now try Exercise 75.

R1.6 Warm Up

The following warm-up exercises involve skills that were covered in earlier sections. You will use these skills in the exercise set for this section. For additional help, review Section R1.5.

In Exercises 1–10, find the product.

1. $3x(5x - 2)$

2. $-2y(y + 1)$

3. $(2x + 3)^2$

4. $(3x - 8)^2$

5. $(2x - 3)(x + 8)$

6. $(4 - 5z)(1 + z)$

7. $(2y + 1)(2y - 1)$

8. $(x + a)(x - a)$

9. $(x + 4)^3$

10. $(2x - 3)^3$

R1.6 Exercises

In Exercises 1–6, factor out the common factor.

1. $3x + 6$

2. $6y - 30$

3. $4x^3 - 8x$

4. $4x^3 - 6x^2 + 12x$

5. $(x - 1)^2 + 6(x - 1)$

6. $3x(x + 2) - 4(x + 2)$

In Exercises 7–12, factor the difference of two squares.

7. $x^2 - 36$

8. $x^2 - \frac{1}{9}$

9. $16x^2 - 9y^2$

10. $49 - 9y^2$

11. $(x - 1)^2 - 4$

12. $25 - (z + 5)^2$

In Exercises 13–18, factor the perfect square trinomial.

13. $x^2 - 4x + 4$

14. $x^2 + 10x + 25$

15. $4y^2 + 12y + 9$

16. $9x^2 - 12x + 4$

17. $25y^2 - 10y + 1$

18. $z^2 + z + \frac{1}{4}$

In Exercises 19–24, factor the sum or difference of cubes.

19. $x^3 - 8$

20. $x^3 - 27$

21. $y^3 + 125$

22. $z^3 + 64$

23. $8t^3 - 1$

24. $27x^3 + 8$

In Exercises 25–38, factor the trinomial.

25. $x^2 + x - 2$

26. $x^2 + 5x + 6$

27. $w^2 - 5w + 6$

28. $z^2 - z - 6$

29. $y^2 + y - 20$

30. $z^2 - 5z - 24$

31. $x^2 - 30x + 200$

32. $x^2 - 13x + 42$

33. $3x^2 - 5x + 2$

34. $2x^2 - x - 1$

35. $9x^2 - 3x - 2$

36. $12y^2 + 7y + 1$

37. $5x^2 + 26x + 5$

38. $5u^2 + 13u - 6$

In Exercises 39–44, factor by grouping.

39. $x^3 - x^2 + 2x - 2$

40. $x^3 + 5x^2 - 5x - 25$

41. $2x^3 - x^2 - 6x + 3$

42. $5x^3 - 10x^2 + 3x - 6$

43. $6 + 2y - 3y^3 - y^4$

44. $z^5 + 2z^3 + z^2 + 2$

In Exercises 45–68, completely factor the expression.

45. $4x^2 - 8x$

46. $12x^2 - 48$

47. $y^3 - 9y$

48. $3x^2 - 48$

49. $x^3 - 4x^2$

50. $6y^2 - 54$

51. $x^2 - 2x + 1$

52. $16 + 6x - x^2$

53. $1 - 4x + 4x^2$

54. $9x^2 - 6x + 1$

55. $2y^3 - 7y^2 - 15y$

56. $9x^2 + 10x + 1$

57. $-2x^2 - 4x + 2x^3$

58. $13x + 6 + 5x^2$

59. $3x^3 + x^2 + 15x + 5$

60. $5 - x + 5x^2 - x^3$

61. $x^4 - 4x^3 + x^2 - 4x$

62. $3u - 2u^2 + 6 - u^3$

63. $25 - (x + 5)^2$

64. $(t - 1)^2 - 49$

65. $(x^2 + 1)^2 - 4x^2$

66. $(x^2 + 8)^2 - 36x^2$

67. $2t^3 - 16$

68. $5x^3 + 40$

Geometric Modeling In Exercises 69–72, make a "geometric factoring model" to represent the given factorization. For instance, a factoring model for $2x^2 + 5x + 2 = (2x + 1)(x + 2)$ is shown below.

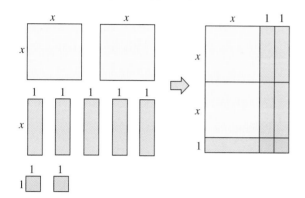

69. $3x^2 + 7x + 2 = (3x + 1)(x + 2)$

70. $x^2 + 4x + 3 = (x + 3)(x + 1)$

71. $2x^2 + 7x + 3 = (2x + 1)(x + 3)$

72. $x^2 + 3x + 2 = (x + 2)(x + 1)$

73. Geometry The room shown in the figure has a floor space of $2x^2 + 3x + 1$ square feet. If the width of the room is $(x + 1)$ feet, what is the length?

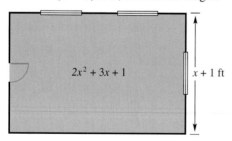

$2x^2 + 3x + 1$ $x + 1$ ft

74. Geometry The room shown in the figure has a floor space of $3x^2 + 8x + 4$ square feet. If the width of the room is $(x + 2)$ feet, what is the length?

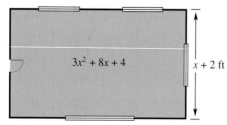

$3x^2 + 8x + 4$ $x + 2$ ft

75. *Make a Decision* Factor each trinomial. Tell whether you used factoring by grouping or factoring by trial and error.

(a) $3x^2 - 2x - 8$ (b) $x^2 + 11x + 30$

76. Find all integers b such that $x^2 + bx + 12$ can be factored. Describe how you found the values for b.

77. Find all integers $c > 0$ such that $x^2 + 10x + c$ can be factored. Describe how you found the values for c.

78. Think About It A student tells you that

$$x^3 - 8 = (x - 2)^3.$$

Explain how you would help the student realize that $x^3 - 8 = (x - 2)(x^2 + 2x + 4)$.

79. Think About It Describe two different ways to factor $2x^2 - 7x - 15$.

80. Geometric Modeling The figure shows two cubes: a large cube whose volume is a^3 and a smaller cube whose volume is b^3. If the smaller cube is removed from the larger, the remaining solid has a volume of $a^3 - b^3$ and is composed of three rectangular boxes, labeled Box 1, Box 2, and Box 3. Find the volume of each box and describe how these results are related to the factoring formula

$$a^3 - b^3 = (a - b)(a^2 + ab + b^2).$$

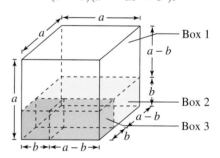

81. Geometric Modeling The figure shows two squares: a large square whose area is a^2 and a smaller square whose area is b^2. If the smaller square is removed from the larger square, the remaining figure has an area of $a^2 - b^2$. Use the geometric model to illustrate the factoring formula

$$a^2 - b^2 = (a - b)(a + b).$$

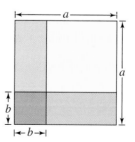

R1.7 Fractional Expressions

Objectives

- **Find the domain of an algebraic expression.**
- **Simplify a rational expression.**
- **Perform operations with rational expressions.**
- **Simplify a complex fraction.**

Domain of an Expression

The set of all real numbers for which an algebraic expression is defined is called the **domain** of the expression. For instance, the domain of

$$\frac{1}{x}$$

is all real numbers other than $x = 0$. Two algebraic expressions are **equivalent** if they have the same domain and yield the same value for all numbers in their domain. For instance, the expressions

$$[(x + 1) + (x + 2)] \quad \text{and} \quad 2x + 3$$

are equivalent.

EXAMPLE 1   FINDING THE DOMAIN OF AN ALGEBRAIC EXPRESSION

a. The domain of the polynomial

$$2x^3 + 3x + 4$$

is the set of all real numbers. In fact, the domain of any polynomial is the set of all real numbers (unless the domain is specifically restricted).

b. The domain of the polynomial

$$x^2 + 5x + 2, \qquad x > 0$$

is the set of positive real numbers, because the polynomial is specifically restricted to that set.

c. The domain of the radical expression

$$\sqrt{x}$$

is the set of nonnegative real numbers, because the square root of a negative number is not a real number.

d. The domain of the expression

$$\frac{x + 2}{x - 3}$$

is the set of all real numbers except $x = 3$, because the value of $x = 3$ would produce an undefined division by zero.

✓*CHECKPOINT*   Now try Exercise 3.

STUDY TIP

For all algebraic expressions, exclude from the domain all values that create *division by zero* or *the square root of a negative number.*

Simplifying Rational Expressions

The quotient of two algebraic expressions is a **fractional expression.** Moreover, the quotient of two *polynomials* such as

$$\frac{1}{x}, \qquad \frac{2x-1}{x+1}, \qquad \text{or} \qquad \frac{x^2-1}{x^2+1}$$

is a **rational expression.** Recall that a fraction is in simplest form if its numerator and denominator have no factors in common aside from ± 1. To write a fraction in simplest form, divide out common factors.

$$\frac{a \cdot \cancel{c}}{b \cdot \cancel{c}} = \frac{a}{b}, \qquad b \neq 0, \qquad c \neq 0$$

The key to success in simplifying rational expressions lies in your ability to *factor* polynomials. For example,

$$\frac{18x^2-18}{6x-6} = \frac{3(6)(x+1)\cancel{(x-1)}}{6\cancel{(x-1)}} = 3(x+1), \quad x \neq 1.$$

Note that the original expression is undefined when $x = 1$ (because division by zero is undefined). So, in order to make sure that the simplified expression is *equivalent* to the original expression, we must restrict the domain of the simplified expression by excluding the value $x = 1$.

EXAMPLE 2   SIMPLIFYING A RATIONAL EXPRESSION

$$\frac{x^2+4x-12}{3x-6} = \frac{(x+6)\cancel{(x-2)}}{3\cancel{(x-2)}} \qquad \text{Factor completely.}$$

$$= \frac{x+6}{3}, \quad x \neq 2 \qquad \text{Divide out common factors.}$$

✓**CHECKPOINT**   Now try Exercise 21.

In Example 2, do not make the mistake of trying to simplify further by dividing out *terms.*

$$\frac{x+6}{3} = \frac{\cancel{x}+\cancel{6}^2}{\cancel{3}_1} = x+2$$

Remember that to simplify fractions, you divide out *factors*, not terms.

When simplifying rational expressions, be sure to factor each polynomial completely before concluding that the numerator and denominator have no factors in common. Moreover, changing the sign of a factor may allow further simplification, as demonstrated in part (b) of the next example.

EXAMPLE 3 SIMPLIFYING RATIONAL EXPRESSIONS

a. $\dfrac{x^3 - 4x}{x^2 + x - 2} = \dfrac{x(x + 2)(x - 2)}{(x + 2)(x - 1)}$ Factor completely.

$= \dfrac{x(x - 2)}{x - 1}, \quad x \neq -2$ Divide out common factors.

b. $\dfrac{12 + x - x^2}{2x^2 - 9x + 4} = \dfrac{(4 - x)(3 + x)}{(2x - 1)(x - 4)}$ Factor completely.

$= \dfrac{-(x - 4)(3 + x)}{(2x - 1)(x - 4)}$ $4 - x = -(x - 4)$

$= -\dfrac{3 + x}{2x - 1}, \quad x \neq 4$ Divide out common factors.

✓*CHECKPOINT*   Now try Exercise 23.

To multiply or divide rational expressions, use the properties of fractions (see Section R1.2). Recall that to divide fractions you invert the divisor and multiply.

EXAMPLE 4 MULTIPLYING RATIONAL EXPRESSIONS

$\dfrac{6x^2 - 6x}{x^2 + 2x - 3} \cdot \dfrac{x^2 + x - 6}{2x}$ Original product

$= \dfrac{6x(x - 1)(x + 3)(x - 2)}{(x - 1)(x + 3)(2x)}$ Factor and multiply.

$= \dfrac{3(2x)(x - 1)(x + 3)(x - 2)}{(x - 1)(x + 3)(2x)}$ Divide out common factors.

$= 3(x - 2), \quad x \neq -3, x \neq 0, x \neq 1$ Simplify.

✓*CHECKPOINT*   Now try Exercise 31.

EXAMPLE 5 DIVIDING RATIONAL EXPRESSIONS

$\dfrac{2x}{3x - 12} \div \dfrac{x^2 - 2x}{x^2 - 6x + 8} = \dfrac{2x}{3x - 12} \cdot \dfrac{x^2 - 6x + 8}{x^2 - 2x}$ Invert and multiply.

$= \dfrac{(2x)(x - 2)(x - 4)}{(3)(x - 4)(x)(x - 2)}$ Factor and multiply.

$= \dfrac{(2x)(x - 2)(x - 4)}{(3)(x - 4)(x)(x - 2)}$ Divide out common factors.

$= \dfrac{2}{3}, \quad x \neq 0, x \neq 2, x \neq 4$ Simplify.

✓*CHECKPOINT*   Now try Exercise 41.

To add or subtract rational expressions, use the least common denominator method or the following basic definition for adding two fractions.

$$\frac{a}{b} \pm \frac{c}{d} = \frac{ad \pm bc}{bd}, \quad b \neq 0, d \neq 0$$

This definition is efficient for adding or subtracting *two* fractions that have no common factors in their denominators.

EXAMPLE 6 ADDING RATIONAL EXPRESSIONS

$$\frac{x}{x-3} + \frac{2}{3x+4} = \frac{x(3x+4) + 2(x-3)}{(x-3)(3x+4)} \qquad \frac{a}{b} + \frac{c}{d} = \frac{ad+bc}{bd}$$

$$= \frac{3x^2 + 4x + 2x - 6}{(x-3)(3x+4)} \qquad \text{Distributive Property}$$

$$= \frac{3x^2 + 6x - 6}{(x-3)(3x+4)} \qquad \text{Combine like terms.}$$

$$= \frac{3(x^2 + 2x - 2)}{(x-3)(3x+4)} \qquad \text{Factor.}$$

✓*CHECKPOINT*   Now try Exercise 49.

For fractions with a repeated factor in their denominators, the LCD method works well. Recall that the least common denominator of two or more fractions consists of the product of all prime factors in the denominators, with each factor given the highest power of its occurrence in any denominator.

EXAMPLE 7 COMBINING RATIONAL EXPRESSIONS: THE LCD METHOD

Perform the given operations and simplify.

$$\frac{3}{x-1} - \frac{2}{x} + \frac{x+3}{x^2-1}$$

SOLUTION

Using the factored denominators $(x-1)$, x, and $(x+1)(x-1)$, you can see that the least common denominator is $x(x+1)(x-1)$.

$$\frac{3}{x-1} - \frac{2}{x} + \frac{x+3}{x^2-1}$$

$$= \frac{3(x)(x+1)}{x(x+1)(x-1)} - \frac{2(x+1)(x-1)}{x(x+1)(x-1)} + \frac{(x+3)(x)}{x(x+1)(x-1)}$$

$$= \frac{3(x)(x+1) - 2(x+1)(x-1) + (x+3)(x)}{x(x+1)(x-1)}$$

$$= \frac{3x^2 + 3x - 2x^2 + 2 + x^2 + 3x}{x(x+1)(x-1)}$$

$$= \frac{2x^2 + 6x + 2}{x(x+1)(x-1)} = \frac{2(x^2 + 3x + 1)}{x(x+1)(x-1)}$$

✓*CHECKPOINT*   Now try Exercise 53.

Complex Fractions

Fractional expressions with separate fractions in the numerator, denominator, or both are called **complex fractions.** Here are two examples.

$$\frac{\left(\dfrac{1}{x}\right)}{x^2 + 1} \quad \text{and} \quad \frac{\left(\dfrac{1}{x}\right)}{\left(\dfrac{1}{x^2 + 1}\right)}$$

A complex fraction can be simplified by combining the fractions in its numerator into a single fraction and then combining the fractions in its denominator into a single fraction. Then invert the denominator and multiply.

EXAMPLE 8   SIMPLIFYING A COMPLEX FRACTION

$$\frac{\left(\dfrac{2}{x} - 3\right)}{\left(1 - \dfrac{1}{x - 1}\right)} = \frac{\left[\dfrac{2 - 3(x)}{x}\right]}{\left[\dfrac{1(x - 1) - 1}{x - 1}\right]} \qquad \text{Combine fractions.}$$

$$= \frac{\left(\dfrac{2 - 3x}{x}\right)}{\left(\dfrac{x - 2}{x - 1}\right)} \qquad \text{Simplify.}$$

$$= \frac{2 - 3x}{x} \cdot \frac{x - 1}{x - 2} \qquad \text{Invert and multiply.}$$

$$= \frac{(2 - 3x)(x - 1)}{x(x - 2)}, \quad x \neq 1$$

✓CHECKPOINT   Now try Exercise 55.

Another way to simplify a complex fraction is to multiply its numerator and denominator by the LCD of all fractions in its numerator and denominator. This method is applied to the fraction in Example 8 as follows.

$$\frac{\left(\dfrac{2}{x} - 3\right)}{\left(1 - \dfrac{1}{x - 1}\right)} = \frac{\left(\dfrac{2}{x} - 3\right)}{\left(1 - \dfrac{1}{x - 1}\right)} \cdot \frac{x(x - 1)}{x(x - 1)} \qquad \text{LCD is } x(x - 1).$$

$$= \frac{\left(\dfrac{2 - 3x}{x}\right) \cdot x(x - 1)}{\left(\dfrac{x - 2}{x - 1}\right) \cdot x(x - 1)}$$

$$= \frac{(2 - 3x)(x - 1)}{x(x - 2)}, \quad x \neq 1$$

R1.7 (Warm Up

The following warm-up exercises involve skills that were covered in earlier sections. You will use these skills in the exercise set for this section. For additional help, review Section R1.6

In Exercises 1–10, completely factor the polynomial.

1. $5x^2 - 15x^3$

2. $16x^2 - 9$

3. $9x^2 - 6x + 1$

4. $9 + 12y + 4y^2$

5. $z^2 + 4z + 3$

6. $x^2 - 15x + 50$

7. $3 + 8x - 3x^2$

8. $3x^2 - 46x + 15$

9. $s^3 + s^2 - 4s - 4$

10. $y^3 + 64$

R1.7 (Exercises

In Exercises 1–8, find the domain of the expression.

1. $3x^2 - 4x + 7$

2. $6x^2 + 7x - 9,\ x > 0$

3. $\dfrac{1}{x - 2}$

4. $\dfrac{x + 1}{2x + 1}$

5. $\dfrac{x - 1}{x^2 - 4x}$

6. $\dfrac{3x + 1}{x^2 - 16}$

7. $\sqrt{x + 1}$

8. $\dfrac{1}{\sqrt{x + 1}}$

In Exercises 9–14, find the missing factor in the numerator such that the two fractions will be equivalent.

9. $\dfrac{5}{2x} = \dfrac{5(\quad)}{6x^2}$

10. $\dfrac{3}{4} = \dfrac{3(\quad)}{4(x + 1)}$

11. $\dfrac{x + 1}{x} = \dfrac{(x + 1)(\quad)}{x(x - 2)}$

12. $\dfrac{3y - 4}{y + 1} = \dfrac{(3y - 4)(\quad)}{y^2 - 1}$

13. $\dfrac{3x}{x - 3} = \dfrac{3x(\quad)}{x^2 - x - 6}$

14. $\dfrac{1 - z}{z^2} = \dfrac{(1 - z)(\quad)}{z^3 + z^2}$

In Exercises 15–30, write the rational expression in simplest form.

15. $\dfrac{15x^2}{10x}$

16. $\dfrac{18y^2}{60y^5}$

17. $\dfrac{2x}{4x + 4}$

18. $\dfrac{9x^2 + 9x}{2x + 2}$

19. $\dfrac{x - 5}{10 - 2x}$

20. $\dfrac{3 - x}{8x - 24}$

21. $\dfrac{x^2 - 25}{5 - x}$

22. $\dfrac{x^2 - 16}{4 - x}$

23. $\dfrac{x^3 + 5x^2 + 6x}{x^2 - 4}$

24. $\dfrac{x^2 + 8x - 20}{x^2 + 11x + 10}$

25. $\dfrac{y^2 - 7y + 12}{y^2 + 3y - 18}$

26. $\dfrac{x + 1}{x^2 - 3x - 4}$

27. $\dfrac{2 - x + 2x^2 - x^3}{x - 2}$

28. $\dfrac{x^2 - 9}{x^3 + x^2 - 9x - 9}$

29. $\dfrac{z^3 - 8}{z^2 + 2z + 4}$

30. $\dfrac{y^3 - 2y^2 - 3y}{y^3 + 1}$

In Exercises 31–54, perform the indicated operations and simplify.

31. $\dfrac{5}{x - 1} \cdot \dfrac{x - 1}{25(x - 2)}$

32. $\dfrac{x + 13}{x^3(3 - x)} \cdot \dfrac{x(x - 3)}{5}$

33. $\dfrac{(x - 9)(x + 7)}{x + 1} \cdot \dfrac{x}{9 - x}$

34. $\dfrac{(x + 5)(x - 3)}{x + 2} \cdot \dfrac{1}{(x + 5)(x + 2)}$

35. $\dfrac{r}{r - 1} \cdot \dfrac{r^2 - 1}{r^2}$

36. $\dfrac{4y - 16}{5y + 15} \cdot \dfrac{2y + 6}{4 - y}$

37. $\dfrac{t^2 - t - 6}{t^2 + 6t + 9} \cdot \dfrac{t + 3}{t^2 - 4}$

38. $\dfrac{y^3 - 8}{2y^3} \cdot \dfrac{4y}{y^2 - 5y + 6}$

39. $\dfrac{x^2 + x - 2}{x^3 + x^2} \cdot \dfrac{x}{x^2 + 3x + 2}$

40. $\dfrac{x^3 - 1}{x + 1} \cdot \dfrac{x^2 + 1}{x^2 - 1}$

41. $\dfrac{3(x+y)}{4} \div \dfrac{x+y}{2}$

42. $\dfrac{x+2}{5(x-3)} \div \dfrac{x-2}{5(x-3)}$

43. $\dfrac{\left[\dfrac{x^2}{(x+1)^2}\right]}{\left[\dfrac{x}{(x+1)^3}\right]}$

44. $\dfrac{\left(\dfrac{x^2-1}{x}\right)}{\left[\dfrac{(x-1)^2}{x}\right]}$

45. $\dfrac{5}{x-1} + \dfrac{x}{x-1}$

46. $\dfrac{2x-1}{x+3} + \dfrac{1-x}{x+3}$

47. $\dfrac{2x}{x-5} - \dfrac{5}{5-x}$

48. $\dfrac{3}{x-2} + \dfrac{5}{2-x}$

49. $6 - \dfrac{5}{x+3}$

50. $\dfrac{3}{x-1} - 5$

51. $\dfrac{2}{x^2-4} - \dfrac{1}{x^2-3x+2}$

52. $\dfrac{x}{x^2+x-2} - \dfrac{1}{x+2}$

53. $-\dfrac{1}{x} + \dfrac{2}{x^2+1} + \dfrac{1}{x^3+x}$

54. $\dfrac{2}{x+1} + \dfrac{2}{x-1} + \dfrac{1}{x^2-1}$

In Exercises 55–60, simplify the complex fraction.

55. $\dfrac{\left(\dfrac{x}{2}-1\right)}{(x-2)}$

56. $\dfrac{(x-3)}{\left(\dfrac{x}{4}-\dfrac{4}{x}\right)}$

57. $\dfrac{\left(\dfrac{1}{x}-\dfrac{1}{x+1}\right)}{\left(\dfrac{1}{x+1}\right)}$

58. $\dfrac{\left(\dfrac{5}{y}-\dfrac{6}{2y+1}\right)}{\left(\dfrac{5}{y}+4\right)}$

59. $\dfrac{\left(\sqrt{x}-\dfrac{1}{2\sqrt{x}}\right)}{\sqrt{x}}$

60. $\dfrac{\left(\dfrac{x+4}{x+5}-\dfrac{x}{x+1}\right)}{4}$

Monthly Payment In Exercises 61 and 62, use the formula for the approximate annual interest rate r of a monthly installment loan

$$r = \dfrac{\left[\dfrac{24(NM-P)}{N}\right]}{\left(P+\dfrac{NM}{12}\right)}$$

where N is the total number of payments, M is the monthly payment, and P is the amount financed.

61. (a) Approximate the annual interest rate r for a four-year car loan of $15,000 that has monthly payments of $400.

(b) Simplify the expression for the annual interest rate r, and then rework part (a).

62. (a) Approximate the annual interest rate r for a five-year car loan of $18,000 that has monthly payments of $400.

(b) Simplify the expression for the annual interest rate r, and then rework part (a).

63. *Make a Decision*: **Refrigeration** When food (at room temperature) is placed in a refrigerator, the time required for the food to cool depends on the amount of food, the air circulation in the refrigerator, the original temperature of the food, and the temperature of the refrigerator. One model for the temperature of food that starts at 75°F and is placed in a 40°F refrigerator is

$$T = 10\left(\dfrac{4t^2+16t+75}{t^2+4t+10}\right), \quad t \geq 0$$

where T is the temperature (in degrees Fahrenheit) and t is the time (in hours). Sketch a bar graph showing the temperature of the food when $t = 0, 1, 2, 3,$ 4, and 5 hours. According to the model, will the food reach a temperature of 40°F after 6 hours?

64. *Make a Decision*: **Oxygen Level** The mathematical model

$$O = \dfrac{t^2-t+1}{t^2+1}, \quad t \geq 0$$

gives the percentage of the normal level of oxygen in a pond, where t is the time in weeks after organic waste is dumped into the pond. Sketch a bar graph showing the oxygen level of the pond when $t = 0, 1, 2, 3, 4,$ and 5 weeks. What conclusions can you make from your bar graph?

Make a Decision ► Project: Musical Notes and Frequencies

The musical note A-440 has a frequency of 440 vibrations per second. The frequency of any note can be found by the model

$$F = 440 \cdot \sqrt[12]{2^n}$$

where n represents the number of black and white keys below or above A-440. The figure below shows a piano keyboard with its corresponding written notes. It also shows the approximate ranges of the human voice.

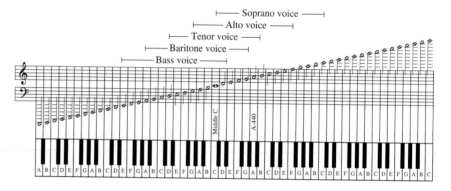

1. **Approximating Frequencies**   Approximate the lowest and highest frequencies of a soprano voice.

2. **Approximating Frequencies**   Approximate the lowest and highest frequencies of a tenor voice.

3. **Approximating Frequencies**   Approximate the lowest and highest frequencies of a bass voice.

4. **Approximating Frequencies**   From the results of Questions 1–3, approximate the range of the human voice.

5. *Make a Decision*   Notes that are an octave apart differ by 12 notes. So, you can represent the two notes by n and $n + 12$. What can you say about the frequencies of notes that are one octave apart? Two octaves apart?

6. **Finding Frequencies**   There are 88 keys on a piano—39 keys above A-440 and 48 keys below A-440. Find the highest and lowest frequencies on a piano.

7. **Harmonizing Notes**   Two notes are said to harmonize if the ratio of their frequencies is very close to an integer or a simple rational number. Find a note such that the frequency ratio of A-440 to that note is about 3 to 4. Find a note such that the frequency ratio of A-440 to that note is about 3 to 2.

8. *Make a Decision*   Make a table listing at least 20 notes, x and their frequencies, y. (Make sure to choose only positive values for x.) Enter the data into a graphing utility or computer program with power regression capabilities. Find a power regression equation to fit the data that you entered. Does the equation agree with the model given above? Explain.

Chapter Summary

After studying this chapter, you should have acquired the following skills. These skills are keyed to the Review Exercises that begin on page R64. Answers to odd-numbered Review Exercises are given in the back of the book.

R1.1 • Classify real numbers as natural numbers, integers, rational numbers, or irrational numbers.

Review Exercises 1, 2

• Order real numbers.

Review Exercises 3, 4

• Give a verbal description of numbers represented by an inequality.

Review Exercises 5, 6

• Use inequality notation to describe a set of real numbers.

Review Exercises 7, 8

• Interpret absolute value notation.

Review Exercises 9–12, 15, 16

• Find the distance between two numbers on the real number line.

Review Exercises 13, 14

R1.2 • Identify the terms of an algebraic expression.

Review Exercises 17, 18

• Evaluate an expression.

Review Exercises 19, 20

• Identify rules of algebra.

Review Exercises 21, 22

• Perform operations on real numbers.

Review Exercises 23–28

• Use a calculator to evaluate an expression.

Review Exercises 29, 30

R1.3 • Use properties of exponents.

Review Exercises 31–34

• Use scientific notation.

Review Exercises 35–38

• Use a calculator to raise a number to a power.

Review Exercises 39, 40

• Use interest formulas to solve an application problem.

Review Exercises 41, 42

R1.4 • Simplify a radical.

Review Exercises 45–48

• Rationalize a denominator.

Review Exercises 49, 50

• Use properties of rational exponents.

Review Exercises 55, 56

• Combine radicals.

Review Exercises 51–54

• Use a calculator to evaluate a radical.

Review Exercises 57, 58

R1.5 • Write a polynomial in standard form.

Review Exercises 59–66

• Add, subtract, and multiply polynomials.

Review Exercises 59–66

• Use special products to multiply polynomials.

Review Exercises 63–66

• Use polynomials to solve an application problem.

Review Exercises 67–70

R1.6 • Factor a polynomial by removing common factors.

Review Exercises 71–76

• Factor a polynomial in a special form.

Review Exercises 71–76

• Factor a trinomial as the product of two binomials.

Review Exercises 71–76

• Factor a polynomial by grouping.

Review Exercises 71–76

R1.7 • Find the domain of an algebraic expression.

Review Exercises 77–80

• Simplify a rational expression.

Review Exercises 81–86

• Perform operations with rational expressions.

Review Exercises 87–90

• Simplify a complex fraction.

Review Exercises 91–94

Review Exercises

In Exercises 1 and 2, determine which numbers in the set are (a) natural numbers, (b) integers, (c) rational numbers, and (d) irrational numbers.

1. $\left\{11, -14, -\frac{8}{9}, \frac{5}{2}, \sqrt{6}, 0.4\right\}$

2. $\left\{\sqrt{15}, -22, -\frac{10}{3}, 0, 5.2, \frac{3}{7}\right\}$

In Exercises 3 and 4, plot the two real numbers on the real number line and place the appropriate inequality sign ($<$ or $>$) between them.

3. $-4, -3$ **4.** $\frac{1}{5}, \frac{1}{6}$

In Exercises 5 and 6, give a verbal description of the subset of real numbers that is represented by the inequality, and sketch the subset on the real number line.

5. $x \leq -6$ **6.** $x \geq 2$

In Exercises 7 and 8, use inequality notation to describe the subset of real numbers.

7. x is nonnegative.

8. x is greater than 2 and less than or equal to 5.

In Exercises 9 and 10, write the expression without using absolute value signs.

9. $-|-14|$

10. $|-4 - 2|$

In Exercises 11 and 12, place the correct symbol ($<$, $>$, or $=$) between the two real numbers.

11. $|-12|$ ____ $-|12|$

12. $|9|$ ____ $|-9|$

In Exercises 13 and 14, find the distance between a and b.

13. $a = 48, \quad b = 45$

14. $a = 2, \quad b = -8$

In Exercises 15 and 16, use absolute value notation to describe the sentence.

15. The distance between x and 7 is at least 4.

16. The distance between x and 25 is no more than 10.

In Exercises 17 and 18, identify the terms of the algebraic expression.

17. $2x^2 - 3x + 4$ **18.** $3x^3 - 9x$

In Exercises 19 and 20, evaluate the expression for each value of x.

19. $-4x^2 - 6x$ (a) $x = -1$ (b) $x = 0$

20. $12 - 5x^2$ (a) $x = -2$ (b) $x = 3$

In Exercises 21 and 22, identify the rule(s) of algebra illustrated by the statement.

21. $5(x^2 + x) = 5x^2 + 5x$

22. $x + (2x + 3) = (x + 2x) + 3$

In Exercises 23–28, perform the indicated operations. Write fractional answers in simplest form.

23. $-3 - 2(4 - 5)$ **24.** $12(-3 + 5) - 20$

25. $\frac{1}{2} + \frac{1}{3} - \frac{1}{6}$ **26.** $\frac{5}{12} + \frac{3}{4}$

27. $5^2 \cdot 5^{-1}$ **28.** $(4^2)^2$

In Exercises 29 and 30, use a calculator to evaluate the expression. (Round to two decimal places.)

29. $4\left(\frac{1}{6} - \frac{1}{7}\right)$

30. $-2 + 3\left(\frac{1}{2} - \frac{1}{3}\right)$

In Exercises 31 and 32, evaluate the expression for the value of x.

31. $-2x^2, \quad x = -1$

32. $-\frac{(-x)^2}{6}, \quad x = -3$

In Exercises 33 and 34, simplify the expression.

33. $\frac{(4x)^2}{2x}$

34. $4(-3x)^3$

In Exercises 35 and 36, write the number in scientific notation.

35. Population of California: 35,116,000
(Source: U.S. Census Bureau)

36. Number of Meters in One Foot: 0.3048

In Exercises 37 and 38, write the number in decimal notation.

37. Distance Between Sun and Jupiter: 4.834×10^8 miles

38. Ratio of Day to Year: 2.74×10^{-3}

In Exercises 39 and 40, use a calculator to evaluate the expression. (Round to three decimal places.)

39. (a) $1800(1 + 0.08)^{24}$

(b) $0.0024(7,658,400)$

40. (a) $50,000\left(1 + \dfrac{0.075}{12}\right)^{48}$

(b) $\dfrac{28,000,000 + 34,000,000}{87,000,000}$

In Exercises 41 and 42, complete the table by finding the balance.

41. Balance in an Account You deposit $2000 in an account with an annual interest rate of 6%, compounded monthly.

Year	5	10	15	20	25
Balance					

42. Balance in an Account You deposit $10,000 in an account with an annual interest rate of 5.5%, compounded quarterly.

Year	5	10	15	20	25
Balance					

In Exercises 43 and 44, fill in the missing form.

Radical Form   *Rational Exponent Form*

43. $\sqrt{16} = 4$

44. $16^{1/4} = 2$

In Exercises 45 and 46, evaluate the expression.

45. $\sqrt{169}$ **46.** $\sqrt[3]{125}$

In Exercises 47 and 48, simplify by removing all possible factors from the radical.

47. $\sqrt{4x^4}$ **48.** $\sqrt[3]{\dfrac{2x^3}{27}}$

In Exercises 49 and 50, rewrite the expression by rationalizing the denominator. Simplify your answer.

49. $\dfrac{1}{2 - \sqrt{3}}$ **50.** $\dfrac{2}{3 - \sqrt{10}}$

In Exercises 51–54, simplify the expression.

51. $2\sqrt{x} - 5\sqrt{x}$

52. $\sqrt{72} + \sqrt{128}$

53. $\sqrt{5}\sqrt{2}$

54. $4^{1/3} \cdot 4^{5/3}$

In Exercises 55 and 56, use rational exponents to reduce the index of the radical.

55. $\sqrt[4]{5^2}$ **56.** $\sqrt[8]{x^4}$

In Exercises 57 and 58, use a calculator to approximate the number. (Round your answer to three decimal places.)

57. $\sqrt{127}$ **58.** $\sqrt[3]{52}$

In Exercises 59–66, perform the indicated operations and write the resulting polynomial in standard form.

59. $2(x - 3) - 4(2x - 8)$

60. $3(x^2 - 5x + 2) + 3x(2 - 4x)$

61. $x(x - 2) - 2(3x + 7)$

62. $2x(x + 1) + 3(x^2 - x)$

63. $(x + 1)(x - 2)$

64. $(x - 5)(x + 5)$

65. $(x - 1)(x^2 + 2)$

66. $(2x + 1)^2$

67. Home Prices The median sales price (in thousands of dollars) of a new one-family home in the northeastern United States from 1997 to 2002 can be represented by the polynomial

$15.07x + 79.9$

where x represents the year, with $x = 7$ corresponding to 1997. Evaluate the polynomial when $x = 12$. Then describe your result in everyday terms. (Source: U.S. Census Bureau and U.S. Department of Housing and Urban Development)

68. Home Prices   The median sales price (in thousands of dollars) of a new one-family home in the southern United States from 1997 to 2002 can be represented by the polynomial

$$6.57x + 83.9$$

where x represents the year, with $x = 7$ corresponding to 1997. Evaluate the polynomial when $x = 12$. Then describe your result in everyday terms. (Source: U.S. Census Bureau and U.S. Department of Housing and Urban Development)

69. Cell Phone Subscribers   The number of cell phone subscribers (in thousands) in the United States from 1995 to 2002 can be represented by the polynomial

$$16,069.7x - 53,216$$

where x represents the year, with $x = 5$ corresponding to 1995. Evaluate the polynomial when $x = 5$ and $x = 12$. Then describe your results in everyday terms.   (Source: Cellular Telecommunications & Internet Association)

70. Cell Sites   The number of cell sites in the United States from 1995 to 2002 can be represented by the polynomial

$$511.92x^2 + 8893.4x - 37,386$$

where x represents the year, with $x = 5$ corresponding to 1995. Evaluate the polynomial when $x = 5$ and $x = 12$. Then describe your results in everyday terms.   (Source: Cellular Telecommunications & Internet Association)

In Exercises 71–76, completely factor the expression.

71. $4x^2 - 36$ **72.** $x^2 - 4x - 5$

73. $-3x^2 - 6x + 3x^3$ **74.** $x^3 - 16x$

75. $x^3 - 4x^2 - 2x + 8$ **76.** $x^3 - 125$

In Exercises 77–80, find the domain of the expression.

77. $\dfrac{2x + 1}{x - 3}$ **78.** $\dfrac{x - 3}{x + 1}$

79. $2x^2 - 11x + 5$ **80.** $4\sqrt{2x}$

In Exercises 81 and 82, find the missing factor in the numerator so that the two fractions will be equivalent.

81. $\dfrac{4}{3x} = \dfrac{4()}{9x^2}$ **82.** $\dfrac{5}{7} = \dfrac{5()}{7(x + 2)}$

In Exercises 83–86, write the rational expression in simplest form.

83. $\dfrac{x^2 - 4}{2x + 4}$

84. $\dfrac{x^2 - 2x - 15}{x + 3}$

85. $\dfrac{2x^2 + 4x}{2x}$

86. $\dfrac{x^3 + 2x^2 - 3x}{x - 1}$

In Exercises 87–90, perform the operation and simplify.

87. $\dfrac{2x - 1}{x + 1} \cdot \dfrac{x^2 - 1}{2x^2 - 7x + 3}$

88. $\dfrac{x + 2}{x - 4} \div \dfrac{2x + 4}{8x}$

89. $\dfrac{x}{x - 1} + \dfrac{2x}{x - 2}$

90. $\dfrac{2}{x + 2} - \dfrac{3}{x - 2}$

In Exercises 91–94, simplify the complex fraction.

91. $\dfrac{\left(\dfrac{x^2 - 1}{x}\right)}{\left[\dfrac{(x - 1)^2}{x}\right]}$

92. $\dfrac{(x - 4)}{\left(\dfrac{x}{4} - \dfrac{4}{x}\right)}$

93. $\dfrac{\left(\dfrac{1}{2x - 3} - \dfrac{1}{2x + 3}\right)}{\left(\dfrac{1}{2x} - \dfrac{1}{2x + 3}\right)}$

94. $\dfrac{\left(\dfrac{1}{x} - \dfrac{1}{y}\right)}{\left(\dfrac{1}{x} + \dfrac{1}{y}\right)}$

Chapter Test

Take this test as you would take a test in class. After you are done, check your work against the answers given in the back of the book.

1. Evaluate the expression $-3x^2 - 5x$ when $x = -3$.

2. Complete the table at the left given that \$3000 is deposited in an account with an annual interest rate of 8%, compounded monthly. What can you conclude from the table?

Year	Balance
5	
10	
15	
20	
25	

Table for 2

In Exercises 3–8, simplify the expression.

3. $8(-2x^2)^3$

4. $3\sqrt{x} - 7\sqrt{x}$

5. $5^{1/4} \cdot 5^{7/4}$

6. $\sqrt{48} - \sqrt{80}$

7. $\sqrt{12x^3}$

8. $\dfrac{2}{5 - \sqrt{7}}$

In Exercises 9 and 10, write the polynomial in standard form.

9. $(3x + 7)^2$

10. $3x(x + 5) - 2x(4x - 7)$

In Exercises 11–14, completely factor the expression.

11. $5x^2 - 80$

12. $4x^2 + 12x + 9$

13. $x^3 - 6x^2 - 3x + 18$

14. $x^3 + 2x^2 - 4x - 8$

15. Simplify: $\dfrac{x^2 - 16}{3x + 12}$.

16. Multiply and simplify: $\dfrac{3x - 5}{x + 3} \cdot \dfrac{x^2 + 7x + 12}{9x^2 - 25}$.

17. Add and simplify: $\dfrac{x}{x - 3} + \dfrac{3x}{x - 4}$.

18. Subtract and simplify: $\dfrac{3}{x + 5} - \dfrac{4}{x - 2}$.

In Exercises 19 and 20, find the domain of the expression.

19. $\sqrt{x + 3}$

20. $\dfrac{1}{x - 2}$

21. Simplify the complex fraction $\dfrac{\left(\dfrac{2x - 9}{x - 1}\right)}{\left(\dfrac{3}{x - 1} + \dfrac{1 - x}{x + 2}\right)}$.

22. The total amount of expenditures (in billions of dollars) for U.S. colleges and universities from 1995 to 2002 can be represented by the polynomial $8.23x + 195.6$, where x represents the year, with $x = 5$ corresponding to 1995. Evaluate the polynomial when $x = 5$ and $x = 12$. Then describe your results in everyday terms. (Source: U.S. National Center for Education Statistics)

R2 Equations and Inequalities

Chapter Sections

R2.1 Linear Equations

R2.2 Mathematical Modeling

R2.3 Quadratic Equations

R2.4 The Quadratic Formula

R2.5 Other Types of Equations

R2.6 Linear Inequalities

R2.7 Other Types of Inequalities

Make a Decision Are Baseball Salaries Out of the Ballpark?

Baseball began in the mid-1800s and spread throughout the United States during the Civil War. In 1876 the National League, consisting of eight professional teams, was formed. The American League began playing with eight teams in 1901.

Today there are 30 teams playing in the National and American Leagues. Attendance for the 2001 season topped 72.5 million.

Salaries for professional baseball players have steadily increased. The average salaries for players from 1995 to 2003 are listed in the table. (Source: Major League Baseball Players Association)

Make a Decision When do you think the average salary will reach $2,500,000?

Year	Average salary (in thousands of dollars)
1995	1111
1996	1120
1997	1337
1998	1399
1999	1611
2000	1896
2001	2139
2002	2296
2003	2372

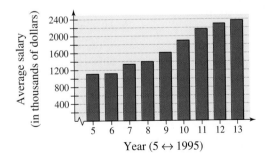

To explore this real-world application further, go to the Make a Decision Project on page R148.

R2.1   Linear Equations

Objectives

- Classify an equation as an identity or a conditional equation.
- Solve a linear equation in one variable.
- Use a linear model to solve an application problem.

Equations and Solutions

An **equation** is a statement that two algebraic expressions are equal. Some examples of equations in x are

$$3x - 5 = 7, \qquad x^2 - x - 6 = 0, \qquad \text{and} \qquad \sqrt{2x} = 4.$$

To **solve** an equation in x means to find all values of x for which the equation is **true.** Such values are called **solutions.** For instance, $x = 4$ is a solution of the equation $3x - 5 = 7$, because $3(4) - 5 = 7$ is a true statement.

An equation that is true for *every* real number in the domain of the variable is called an **identity.** Two examples of identities are

$$x^2 - 9 = (x + 3)(x - 3) \qquad \text{and} \qquad \frac{x}{3x^2} = \frac{1}{3x}, \qquad x \neq 0.$$

The first equation is an identity because it is a true statement for all real values of x. The second is an identity because it is true for all nonzero real values of x.

An equation that is true for just *some* (or even none) of the real numbers in the domain of the variable is called a **conditional equation.** For example, the equation $x^2 - 9 = 0$ is conditional because $x = 3$ and $x = -3$ are the only values in the domain that satisfy the equation.

EXAMPLE 1   CLASSIFYING EQUATIONS

Classify each equation as an identity or a conditional equation.

a. $2(x + 3) = 2x + 6$ **b.** $2(x + 3) = x + 6$ **c.** $2(x + 3) = 2x + 3$

SOLUTION

a. This equation is an identity because it is true for every real value of x.

b. This equation is a conditional equation because there are real values of x (such as $x = 1$) for which the equation is not true.

c. This equation is a conditional equation because there are no real number values of x for which the equation is true.

✓CHECKPOINT   Now try Exercise 1.

Be sure you understand that equations are used in algebra for two distinct purposes: (1) *identities* are usually used to state mathematical properties and (2) *conditional equations* are usually used to model and solve problems that occur in real life.

Linear Equations in One Variable

The most common type of conditional equation is a **linear equation.**

Definition of a Linear Equation

A **linear equation** in one variable x is an equation that can be written in the standard form

$$ax + b = 0$$

where a and b are real numbers with $a \neq 0$.

A linear equation has exactly one solution. To see this, consider the following steps. (Remember that $a \neq 0$.)

$ax + b = 0$	Original equation
$ax = -b$	Subtract b from each side.
$x = -\dfrac{b}{a}$	Divide each side by a.

So, the equation $ax + b = 0$ has exactly one solution, $x = -b/a$.

To solve a linear equation in x, you should isolate x by forming a sequence of **equivalent** (and usually simpler) equations, each having the same solution as the original equation. The operations that yield equivalent equations come from the basic rules of algebra reviewed in Section R1.2.

Forming Equivalent Equations

A given equation can be transformed into an equivalent equation by one or more of the following steps.

	Given Equation	*Equivalent Equation*
1. Remove symbols of grouping, combine like terms, or simplify one or both sides of the equation.	$2x - x = 4$ $3(x - 2) = 5$	$x = 4$ $3x - 6 = 5$
2. Add (or subtract) the same quantity to (from) *each* side of the equation.	$x + 1 = 6$	$x = 5$
3. Multiply (or divide) *each* side of the equation by the same *nonzero* quantity.	$2x = 6$	$x = 3$
4. Interchange sides of the equation.	$2 = x$	$x = 2$

EXAMPLE 2 SOLVING A LINEAR EQUATION

Solve $3x - 6 = 0$.

SOLUTION

$3x - 6 = 0$	Write original equation.
$3x = 6$	Add 6 to each side.
$x = 2$	Divide each side by 3.

✓CHECKPOINT Now try Exercise 19.

After solving an equation, you should **check each solution** in the *original* equation. For instance, in Example 2, you can check that 2 is a solution by substituting 2 for x in the original equation $3x - 6 = 0$, as follows.

CHECK

$3x - 6 = 0$	Write original equation.
$3(2) - 6 \stackrel{?}{=} 0$	Substitute 2 for x.
$6 - 6 = 0$	Solution checks. ✓

STUDY TIP

Students sometimes tell us that a solution looks easy when we work it in class, but they don't see where to begin when trying it alone. Keep in mind that no one—not even great mathematicians—can expect to look at every mathematical problem and immediately know where to begin. Many problems involve some trial and error before a solution is found.

EXAMPLE 3 SOLVING A LINEAR EQUATION

Solve $6(x - 1) + 4 = 3(7x + 1)$.

SOLUTION

$6(x - 1) + 4 = 3(7x + 1)$	Write original equation.
$6x - 6 + 4 = 21x + 3$	Distributive Property
$6x - 2 = 21x + 3$	Simplify.
$-15x = 5$	Add 2 and subtract $21x$.
$x = -\frac{1}{3}$	Divide each side by -15.

The solution is $x = -\frac{1}{3}$. You can check this as follows.

CHECK

$6(x - 1) + 4 = 3(7x + 1)$	Write original equation.
$6\left(-\frac{1}{3} - 1\right) + 4 \stackrel{?}{=} 3\left[7\left(-\frac{1}{3}\right) + 1\right]$	Substitute $-\frac{1}{3}$ for x.
$6\left(-\frac{4}{3}\right) + 4 \stackrel{?}{=} 3\left(-\frac{7}{3} + 1\right)$	Add fractions.
$-8 + 4 \stackrel{?}{=} -7 + 3$	Simplify.
$-4 = -4$	Solution checks. ✓

✓CHECKPOINT Now try Exercise 23.

Equations Involving Fractional Expressions

Use the *table* feature of your graphing utility to check the solution in Example 3. In the equation editor, enter the expression to the left of the equal sign in y_1 and enter the expression to the right of the equal sign in y_2 as follows.

$$y_1 = 6(x - 1) + 4$$

$$y_2 = 3(7x + 1)$$

Set the *table* feature to ASK mode. When you enter the solution $-\frac{1}{3}$ for x, both y_1 and y_2 are -4, as shown.

X	Y₁	Y₂
-.3333	-4	-4

X=-.333333333333

Similarly, a graphing utility can help you determine if a solution is extraneous. For instance, enter the equation from Example 5 into the graphing utility's equation editor. Then, use the *table* feature in ASK mode to enter -2 for x. You will see that the graphing utility displays ERROR in the y_2 column. So, the solution $x = -2$ is extraneous.

To solve an equation involving fractional expressions, you can multiply every term in the equation by the least common denominator (LCD) of the terms.

EXAMPLE 4 AN EQUATION INVOLVING FRACTIONAL EXPRESSIONS

$\dfrac{x}{3} + \dfrac{3x}{4} = 2$	Original equation
$(12)\dfrac{x}{3} + (12)\dfrac{3x}{4} = (12)2$	Multiply each term by least common denominator.
$4x + 9x = 24$	Simplify.
$13x = 24$	Combine like terms.
$x = \dfrac{24}{13}$	Divide each side by 13.

The solution is $x = \frac{24}{13}$. Check this in the original equation.

✓*CHECKPOINT*   Now try Exercise 27.

When multiplying or dividing an equation by a *variable expression*, it is possible to introduce an **extraneous** solution—one that does not satisfy the original equation. In such cases a check is especially important.

EXAMPLE 5 AN EQUATION WITH AN EXTRANEOUS SOLUTION

Solve $\dfrac{1}{x - 2} = \dfrac{3}{x + 2} - \dfrac{6x}{x^2 - 4}$.

SOLUTION

The least common denominator is $x^2 - 4 = (x + 2)(x - 2)$. Multiplying each term by this LCD and simplifying produces the following.

$\dfrac{1}{x - 2} = \dfrac{3}{x + 2} - \dfrac{6x}{x^2 - 4}$	Write original equation.
$\dfrac{1}{x - 2}(x + 2)(x - 2) = \dfrac{3}{x + 2}(x + 2)(x - 2) - \dfrac{6x}{x^2 - 4}(x + 2)(x - 2)$	
$x + 2 = 3(x - 2) - 6x, \quad x \neq \pm 2$	Simplify.
$x + 2 = 3x - 6 - 6x$	Distributive Property
$4x = -8$	Combine like terms and simplify.
$x = -2$	Extraneous solution

By checking $x = -2$, you can see that it yields a denominator of zero for the fraction $3/(x + 2)$. So, $x = -2$ is extraneous, and the equation has *no solution.*

✓*CHECKPOINT*   Now try Exercise 47.

An equation with a *single fraction* on each side can be cleared of denominators by **cross-multiplying,** which is equivalent to multiplying each side of the equation by the least common denominator and then simplifying.

EXAMPLE 6   CROSS-MULTIPLYING TO SOLVE AN EQUATION

Solve $\dfrac{3y - 2}{2y + 1} = \dfrac{6y - 9}{4y + 3}$.

SOLUTION

$$\frac{3y - 2}{2y + 1} = \frac{6y - 9}{4y + 3} \qquad \text{Write original equation.}$$

$$(3y - 2)(4y + 3) = (6y - 9)(2y + 1) \qquad \text{Cross-multiply.}$$

$$12y^2 + y - 6 = 12y^2 - 12y - 9 \qquad \text{Multiply.}$$

$$13y = -3 \qquad \text{Isolate } y\text{-term on left.}$$

$$y = -\frac{3}{13} \qquad \text{Divide each side by 13.}$$

The solution is $y = -\frac{3}{13}$. Check this in the original equation.

✓CHECKPOINT   Now try Exercise 37.

EXAMPLE 7   USING A CALCULATOR TO SOLVE AN EQUATION

Solve $\dfrac{1}{9.38} - \dfrac{3}{x} = \dfrac{5}{0.3714}$.

SOLUTION

Roundoff error will be minimized if you solve for x before performing any calculations. The least common denominator is $(9.38)(0.3714)(x)$.

$$\frac{1}{9.38} - \frac{3}{x} = \frac{5}{0.3714}$$

$$(9.38)(0.3714)(x)\left(\frac{1}{9.38} - \frac{3}{x}\right) = (9.38)(0.3714)(x)\left(\frac{5}{0.3714}\right)$$

$$0.3714x - 3(9.38)(0.3714) = 5(9.38)(x), \quad x \neq 0$$

$$[0.3714 - 5(9.38)]x = 3(9.38)(0.3714)$$

$$x = \frac{3(9.38)(0.3714)}{0.3714 - 5(9.38)}$$

$$x \approx -0.225 \qquad \text{Round to three decimal places.}$$

The solution is $x \approx -0.225$. Check this in the original equation.

✓CHECKPOINT   Now try Exercise 59.

STUDY TIP

Because of roundoff error, a check of a decimal solution may not yield exactly the same values for each side of the original equation. The difference, however, should be quite small.

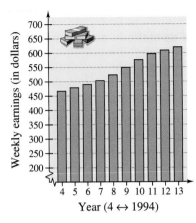

Figure R2.1

Application

EXAMPLE 8   WEEKLY EARNINGS

Make a Decision   The median weekly earnings y (in dollars) for workers in the United States from 1994 to 2003 can be modeled by the linear equation

$$y = 18.6t + 383, \quad 4 \le t \le 13$$

where $t = 4$ represents 1994. See Figure R2.1. According to the model, in what year can you expect the median weekly earnings to reach \$700? Do you think the median weekly earnings will actually reach \$700 in that year?   (Source: U.S. Bureau of Labor Statistics)

SOLUTION

To determine when the median weekly earnings will reach \$700, solve the model for t when $y = 700$.

$y = 18.6t + 383$	Write original model.
$700 = 18.6t + 383$	Substitute 700 for y so you can solve for t.
$317 = 18.6t$	Subtract 383 from each side.
$t = \dfrac{317}{18.6} \approx 17$	Divide each side by 18.6.

Because $t = 4$ corresponds to 1994, it follows that $t = 17$ corresponds to 2007. So, you can expect the median weekly earnings to reach \$700 by 2007. It is possible for the median weekly earnings to reach \$700 in 2007 because the median weekly earnings steadily increased from 1994 to 2003.

✔CHECKPOINT   Now try Exercise 71.

Discussing the Concept | Checking a Solution

The solution $x = 2$ does not check in the following original equation. Why? Is there an error? When a solution does not check, what are the possibilities?

$$\frac{2}{x - 1} = \frac{1}{x - 2} - \frac{1}{(x - 2)(x - 1)} \quad \text{Original equation}$$

$$\frac{2}{x - 1}(x - 2)(x - 1) = \frac{1}{x - 2}(x - 2)(x - 1) - \frac{1}{(x - 2)(x - 1)}(x - 2)(x - 1)$$

$$2(x - 2) = (x - 1) - 1$$

$$2x - 4 = x - 2$$

$$x = 2$$

R2.1 Warm Up

The following warm-up exercises involve skills that were covered in earlier sections. You will use these skills in the exercise set for this section. For additional help, review Section R1.2.

In Exercises 1–10, perform the indicated operations and simplify your answer.

1. $(2x - 4) - (5x + 6)$

2. $(3x - 5) + (2x - 7)$

3. $2(x + 1) - (x + 2)$

4. $-3(2x - 4) + 7(x + 2)$

5. $\dfrac{x}{3} + \dfrac{x}{5}$

6. $x - \dfrac{x}{4}$

7. $\dfrac{1}{x + 1} - \dfrac{1}{x}$

8. $\dfrac{2}{x} + \dfrac{3}{x}$

9. $\dfrac{4}{x} + \dfrac{3}{x - 2}$

10. $\dfrac{1}{x + 1} - \dfrac{1}{x - 1}$

R2.1 Exercises

In Exercises 1–6, determine whether the equation is an identity or a conditional equation.

1. $2(x - 1) = 2x - 2$

2. $3(x + 2) = 3x + 6$

3. $2(x - 1) = 3x + 4$

4. $3(x + 2) = 2x + 4$

5. $2(x + 1) = 2x + 1$

6. $3(x + 4) = 3x + 4$

In Exercises 7–16, determine whether each value of x is a solution of the equation.

Equation	Values
7. $5x - 3 = 3x + 5$	(a) $x = 0$ (b) $x = -5$
	(c) $x = 4$ (d) $x = 10$
8. $7 - 3x = 5x - 17$	(a) $x = -3$ (b) $x = 0$
	(c) $x = 8$ (d) $x = 3$
9. $3x^2 + 2x - 5 = 2x^2 - 2$	(a) $x = -3$ (b) $x = 1$
	(c) $x = 4$ (d) $x = -5$
10. $5x^3 + 2x - 3 = 4x^3 + 2x - 11$	
	(a) $x = 2$ (b) $x = -2$
	(c) $x = 0$ (d) $x = 10$
11. $\dfrac{5}{2x} - \dfrac{4}{x} = 3$	(a) $x = -\frac{1}{2}$ (b) $x = 4$
	(c) $x = 0$ (d) $x = \frac{1}{4}$
12. $3 + \dfrac{1}{x + 2} = 4$	(a) $x = -1$ (b) $x = -2$
	(c) $x = 0$ (d) $x = 5$
13. $(x + 5)(x - 3) = 20$	(a) $x = 3$ (b) $x = -2$
	(c) $x = 0$ (d) $x = -7$

Equation	Values
14. $(3x + 5)(2x - 7) = 0$	(a) $x = -\frac{5}{3}$ (b) $x = -\frac{2}{7}$
	(c) $x = \frac{7}{2}$ (d) $x = \frac{5}{3}$
15. $\sqrt{3x - 2} = 4$	(a) $x = \frac{2}{5}$ (b) $x = -\frac{5}{2}$
	(c) $x = -\frac{1}{3}$ (d) $x = -2$
16. $\sqrt[3]{x - 8} = 3$	(a) $x = 2$ (b) $x = -5$
	(c) $x = 35$ (d) $x = 8$

In Exercises 17–54, solve the equation and check your solution. (Some equations have no solution.)

17. $x + 10 = 15$

18. $7 - x = 18$

19. $7 - 2x = 15$

20. $7x + 2 = 16$

21. $8x - 5 = 3x + 10$

22. $7x + 3 = 3x - 13$

23. $2(x + 5) - 7 = 3(x - 2)$

24. $2(13t - 15) + 3(t - 19) = 0$

25. $6[x - (2x + 3)] = 8 - 5x$

26. $8(x + 2) - 3(2x + 1) = 2(x + 5)$

27. $\dfrac{5x}{4} + \dfrac{1}{2} = x - \dfrac{1}{2}$

28. $\dfrac{x}{5} - \dfrac{x}{2} = 3$

29. $\frac{3}{2}(z + 5) - \frac{1}{4}(z + 24) = 0$

30. $\dfrac{3x}{2} + \dfrac{1}{4}(x - 2) = 10$

31. $0.25x + 0.75(10 - x) = 3$

32. $0.60x + 0.40(100 - x) = 50$

33. $x + 8 = 2(x - 2) - x$

34. $3(x + 3) = 5(1 - x) - 1$

35. $\dfrac{100 - 4u}{3} = \dfrac{5u + 6}{4} + 6$

36. $\dfrac{17 + y}{y} + \dfrac{32 + y}{y} = 100$

37. $\dfrac{5x - 4}{5x + 4} = \dfrac{2}{3}$ **38.** $\dfrac{10x + 3}{5x + 6} = \dfrac{1}{2}$

39. $10 - \dfrac{13}{x} = 4 + \dfrac{5}{x}$ **40.** $\dfrac{15}{x} - 4 = \dfrac{6}{x} + 3$

41. $\dfrac{1}{x - 3} + \dfrac{1}{x + 3} = \dfrac{10}{x^2 - 9}$

42. $\dfrac{1}{x - 2} + \dfrac{3}{x + 3} = \dfrac{4}{x^2 + x - 6}$

43. $\dfrac{x}{x + 4} + \dfrac{4}{x + 4} + 2 = 0$

44. $\dfrac{2}{(x - 4)(x - 2)} = \dfrac{1}{x - 4} + \dfrac{2}{x - 2}$

45. $\dfrac{7}{2x + 1} - \dfrac{8x}{2x - 1} = -4$

46. $\dfrac{4}{u - 1} + \dfrac{6}{3u + 1} = \dfrac{15}{3u + 1}$

47. $\dfrac{3}{x(x - 3)} + \dfrac{4}{x} = \dfrac{1}{x - 3}$

48. $3 = 2 + \dfrac{2}{z + 2}$

49. $(x + 2)^2 + 5 = (x + 3)^2$

50. $(x + 1)^2 + 2(x - 2) = (x + 1)(x - 2)$

51. $(x + 2)^2 - x^2 = 4(x + 1)$

52. $4(x + 1) - 3x = x + 5$

53. $(2x + 1)^2 = 4(x^2 + x + 1)$

54. $(2x - 1)^2 = 4(x^2 - x + 6)$

55. A friend states that the solution to the equation

$$\dfrac{2}{x(x - 2)} + \dfrac{5}{x} = \dfrac{1}{x - 2}$$

is $x = 2$. Describe how you would convince your friend that $x = 2$ is not a solution and that the equation has no solution.

56. Explain why a solution of an equation involving fractional expressions may be extraneous.

57. What method or methods would you suggest for checking a solution of an equation involving fractional expressions?

58. What is meant by "equivalent equations"? Give an example of two equivalent equations.

In Exercises 59–64, use a calculator to solve the equation. (Round your solution to three decimal places.)

59. $0.275x + 0.725(500 - x) = 300$

60. $2.763 - 4.5(2.1x - 5.1432) = 6.32x + 5$

61. $\dfrac{x}{0.6321} + \dfrac{x}{0.0692} = 1000$

62. $(x + 5.62)^2 + 10.83 = (x + 7)^2$

63. $\dfrac{2}{7.398} - \dfrac{4.405}{x} = \dfrac{1}{x}$ **64.** $\dfrac{x}{2.625} + \dfrac{x}{4.875} = 1$

65. What method or methods would you recommend for checking the solutions to Exercises 59–64 using your graphing utility?

66. In Exercises 59–64, your answers are rounded to three decimal places. What effect does rounding have as you check a solution?

In Exercises 67–70, evaluate the expression in two ways. (a) Calculate entirely on your calculator using appropriate parentheses, and then round the answer to two decimal places. (b) Round both the numerator and the denominator to two decimal places before dividing, and then round the final answer to two decimal places. Does the second method introduce an additional roundoff error?

67. $\dfrac{1 + 0.73205}{1 - 0.73205}$ **68.** $\dfrac{1 + 0.86603}{1 - 0.86603}$

69. $\dfrac{333 + \dfrac{1.98}{0.74}}{4 + \dfrac{6.25}{3.15}}$ **70.** $\dfrac{1.73205 - 1.19195}{3 - (1.73205)(1.19195)}$

The icon ▦ indicates an exercise or part of an exercise in which you are instructed to use a graphing utility.

71. **Cable Television Advertising** The amount y (in millions of dollars) of money spent on cable television advertising from 1995 to 2002 can be approximated by the linear equation

$$y = 1531.5t - 1467, \quad 5 \le t \le 12$$

where t represents the year, with $t = 5$ corresponding to 1995. If this linear pattern continues, when will the amount spent on cable television advertising reach $20 billion per year? (Source: McCann-Erickson, Inc.)

72. **Supreme Court** The number C of U.S. Supreme Court cases on docket from 1996 to 2001 can be approximated by the linear equation

$$C = 344.3t + 5400, \quad 6 \le t \le 11$$

where t is the year, with $t = 6$ corresponding to 1996. If this linear pattern continues, in what year will the number of U.S. Supreme Court cases reach 11,000? (Source: Office of the Clerk, Supreme Court of the United States)

Human Height In Exercises 73 and 74, use the following information. The relationship between the length of an adult's femur (thigh bone) and the height of the adult can be approximated by the linear equations

$$y = 0.432x - 10.44 \qquad \text{Female}$$

$$y = 0.449x - 12.15 \qquad \text{Male}$$

where y is the length of the femur in inches and x is the height of the adult in inches (see figure).

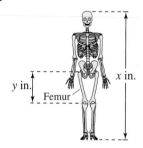

73. An anthropologist discovers a femur belonging to an adult human female. The bone is 16 inches long. Estimate the height of the female.

74. *Make a Decision* From the foot bones of an adult human male, an anthropologist estimates that the male was 69 inches tall. A few feet away from the site where the foot bones were discovered, the anthropologist discovers an adult male femur that is 19 inches long. Is it possible that the leg and foot bones came from the same person? Explain.

Social Security Benefits In Exercises 75 and 76, use the following information. From 1995 to 2002, the average monthly benefit y (in dollars) for retired workers can be approximated by $y = 25.4t + 588$, $5 \le t \le 12$, where t is the year, with $t = 5$ corresponding to 1995. (Source: U.S. Social Security Administration)

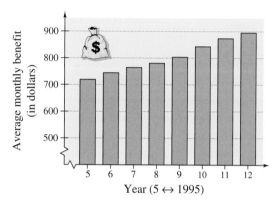

75. In which year was the average monthly benefit approximately $800?

76. *Make a Decision* Do you think the model will continue to be accurate? Explain your reasoning.

Social Security Benefits In Exercises 77 and 78, use the following information. From 1995 to 2002, the total Social Security benefit y (in billions of dollars) can be approximated by $y = 16.9t + 243$, $5 \le t \le 12$, where t is the year, with $t = 5$ corresponding to 1995. (Source: U.S. Social Security Administration)

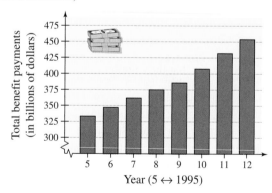

77. In which year was the total benefit paid approximately $400 billion?

78. If the model continues to be accurate, in what year will the total benefit paid reach $500 billion?

R2.2 **Mathematical Modeling**

Objectives

- **Construct a mathematical model from a verbal model.**
- **Model and solve percent and mixture problems.**
- **Use common formulas to solve geometry and simple interest problems.**
- **Develop a general problem-solving strategy.**

Introduction to Problem Solving

In this section, you will study ways of using algebra to solve real-life problems. To do this, you will construct one or more equations that represent the real-life problem. This procedure is called **mathematical modeling.**

A good approach to mathematical modeling is to use two stages. In the first stage, the verbal description of the problem is used to form a *verbal model.* Then, after assigning labels to each of the quantities in the verbal model, you form a *mathematical model* or an *algebraic equation.*

| Verbal description | ⇒ | Verbal model | ⇒ | Algebraic equation |

When you are trying to construct a verbal model, it is helpful to look for a *hidden equality*—a statement that two algebraic expressions are equal. For instance, in the following example the hidden equality equates your annual income to 24 paychecks and one bonus check.

EXAMPLE 1   USING A VERBAL MODEL

You have accepted a job for which your annual salary will be $36,500. This salary includes a year-end bonus of $500. If you are paid twice a month, what will your gross pay be for each paycheck?

SOLUTION

Because there are 12 months in a year and you will be paid twice a month, it follows that you will receive 24 paychecks during the year.

Verbal Model: Income for year = 24 paychecks + Bonus

Labels: Income for year = 36,500   (dollars)
            Amount of each paycheck = x (dollars)
            Bonus = 500   (dollars)

Equation: $36,500 = 24x + 500$

Using the techniques discussed in Section R2.1, you will find that the solution is $x = \$1500$. Check that your answer is reasonable. Is $1500 a reasonable answer if you are paid twice a month and earn $36,500 a year?

✓CHECKPOINT   Now try Exercise 17.

Translating Key Words and Phrases

Key Words and Phrases	Verbal Description	Algebraic Statement
Consecutive		
Next, subsequent	Consecutive integers	$n, n + 1$
Addition		
Sum, plus, greater, increased by, more than, exceeds, total of	The sum of 5 and x Seven more than y	$5 + x$ $y + 7$
Subtraction		
Difference, minus, less than, decreased by, subtracted from, reduced by, the remainder	Four decreased by b Three less than z Five subtracted from w	$4 - b$ $z - 3$ $w - 5$
Multiplication		
Product, multiplied by, twice, times, percent of	Two times x	$2x$
Division		
Quotient, divided by, per	The quotient of x and 8	$\dfrac{x}{8}$

EXAMPLE 2   CONSTRUCTING MATHEMATICAL MODELS

a. A salary of $28,000 is increased by 9%. What is the new salary?

Verbal Model: New salary $=$ 9%(salary) $+$ Original salary

Labels: Original salary $= 28,000$ (dollars)
New salary $= S$ (dollars)
Percent $= 0.09$ (percent in decimal form)

Equation: $S = 0.09(28,000) + 28,000$

b. All computers in a store are reduced by 20%. Find the original price of a laptop selling for $1760.

Verbal Model: Original price $-$ 20%(original price) $=$ Sale price

Labels: Original price $= p$ (dollars)
Sale price $= 1760$ (dollars)
Percent $= 0.2$ (percent in decimal form)

Equation: $p - 0.2p = 1760$

✓CHECKPOINT   Now try Exercise 3.

Using Mathematical Models

Study the next several examples carefully. Your goal should be to develop a *general problem-solving strategy*.

EXAMPLE 3 FINDING THE PERCENT OF A RAISE

You accept a job that pays $8 an hour. You are told that after a two-month probationary period, your hourly wage will be increased to $9 an hour. What percent raise will you receive after the two-month period?

SOLUTION

Verbal Model:	Raise = Percent · Old wage	

Labels:	Old wage = 8	(dollars)
	Raise = 1	(dollar)
	Percent = r	(percent in decimal form)

Equation: $1 = r \cdot 8$

By solving this equation, you will find that you will receive a raise of $\frac{1}{8} = 0.125$, or 12.5%.

✔CHECKPOINT Now try Exercise 39.

EXAMPLE 4 FINDING THE PERCENT OF A SALARY

Your annual salary is $35,000. In addition to your salary, your employer also provides the following benefits. The total of this benefits package is equal to what percent of your annual salary?

Social security (employer's portion):	6.2% of salary	$2170
Worker's compensation:	0.5% of salary	$175
Unemployment compensation:	0.75% of salary	$262.50
Medical insurance:	$2600 per year	$2600
Retirement contribution:	5% of salary	$1750

SOLUTION

Verbal Model:	Benefits package = Percent · Salary	

Labels:	Salary = 35,000	(dollars)
	Benefits package = 6957.50	(dollars)
	Percent = r	(percent in decimal form)

Equation: $6957.50 = r \cdot 35,000$

By solving this equation, you will find that your benefits package is equal to $r = 6957.50/35,000$, or about 19.9% of your salary.

✔CHECKPOINT Now try Exercise 27.

In 2003, 15.2% of the population of the United States had no health insurance. (Source: Centers for Disease Control and Prevention, National Health Interview Survey)

Robert Brenner/PhotoEdit, Inc.

| EXAMPLE 5 | FINDING THE DIMENSIONS OF A ROOM |

A rectangular family room is twice as long as it is wide, and its perimeter is 84 feet. Find the dimensions of the family room.

SOLUTION

For this problem, it helps to sketch a diagram, as shown in Figure R2.2.

Verbal Model: $2 \cdot$ Length $+ 2 \cdot$ Width $=$ Perimeter

Labels: Perimeter = 84   (feet)
            Width = w (feet)
            Length = $l = 2w$ (feet)

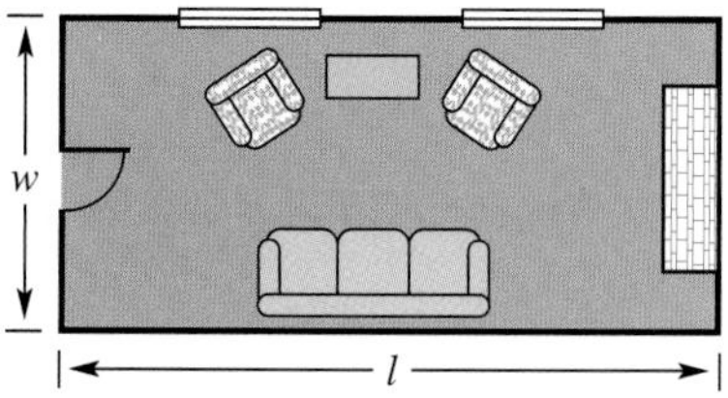

Equation: $2(2w) + 2w = 84$

$4w + 2w = 84$

$6w = 84$

$w = 14$ feet

$l = 2w = 28$ feet

The dimensions of the room are 14 feet by 28 feet.

✓CHECKPOINT   Now try Exercise 29.

Figure R2.2

| EXAMPLE 6 | A DISTANCE PROBLEM |

Make a Decision A plane is flying nonstop from New York to San Francisco, an air distance of about 2600 miles, as shown in Figure R2.3. After 1.5 hours in the air, the plane flies over Chicago (a distance of about 700 miles from New York). How long will it take the plane to fly from New York to San Francisco? At what time will the plane have to leave New York in order to arrive in San Francisco by 8 P.M.? (Assume that the plane flies at a constant speed during the entire flight and that all times are Eastern Standard Time.)

SOLUTION

To solve this problem, use the formula that relates distance, rate, and time. That is, (distance) = (rate)(time). Because it took the plane 1.5 hours to travel a distance of about 700 miles, you can conclude that its rate (or speed) must have been

$$\text{Rate} = \frac{\text{distance}}{\text{time}} = \frac{700 \text{ miles}}{1.5 \text{ hours}} \approx 466.67 \text{ miles per hour.}$$

Because the entire trip is about 2600 miles, the time for the entire trip is

$$\text{Time} = \frac{\text{distance}}{\text{rate}} = \frac{2600 \text{ miles}}{466.67 \text{ miles per hour}} \approx 5.57 \text{ hours.}$$

Because 0.57 hour represents about 34 minutes, you can conclude that the trip will take about 5 hours and 34 minutes. The plane will have to leave New York by 2:26 P.M. in order to arrive in San Francisco by 8 P.M.

✓CHECKPOINT   Now try Exercise 43.

Figure R2.3

Another way to solve the distance problem in Example 6 is to use the concept of **ratio and proportion.** To do this, let x represent the time required to fly from New York to San Francisco, set up the following proportion, and solve for x.

$$\frac{\text{Time to San Francisco}}{\text{Time to Chicago}} = \frac{\text{Distance to San Francisco}}{\text{Distance to Chicago}}$$

$$\frac{x}{1.5} = \frac{2600}{700}$$

$$x = 1.5 \cdot \frac{2600}{700}$$

$$x \approx 5.57$$

Notice how ratio and proportion are used with a result from geometry to solve the problem in the following example.

EXAMPLE 7 AN APPLICATION INVOLVING SIMILAR TRIANGLES

To determine the height of Petronas Tower 1 (in Kuala Lumpur, Malaysia), you measure the shadow cast by the building to be 113 meters long, as shown in Figure R2.4. Then you measure the shadow cast by a one-meter post and find that its shadow is 25 centimeters long. How can you use this information to determine the height of Petronas Tower 1?

SOLUTION

To find the height of the tower, you can use a result from geometry that states that the ratios of corresponding sides of similar triangles are equal.

Verbal Model: $\dfrac{\text{Height of tower}}{\text{Length of tower's shadow}} = \dfrac{\text{Height of post}}{\text{Length of post's shadow}}$

Labels:

Height of tower $= x$	(meters)
Length of tower's shadow $= 113$	(meters)
Height of post $= 100$	(centimeters)
Length of post's shadow $= 25$	(centimeters)

Equation: $\dfrac{x}{113} = \dfrac{100}{25}$

$$x = 113 \cdot \frac{100}{25}$$

$$x = 113 \cdot 4$$

$$x = 452 \text{ meters}$$

The Petronas Tower 1 is 452 meters high.

✓CHECKPOINT Now try Exercise 51.

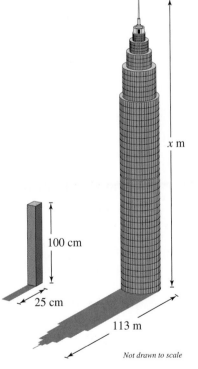

x m

100 cm

25 cm

113 m

Not drawn to scale

Figure R2.4

Mixture Problems

The next example is called a **mixture problem** because it involves two different unknown quantities that are *mixed* in a specific way. Watch for a *hidden product* in the verbal model.

EXAMPLE 8   A SIMPLE INTEREST PROBLEM

A year ago, you invested a total of $10,000 at $4\frac{1}{2}\%$ and $5\frac{1}{2}\%$ simple interest. During 1 year, the two accounts earned $508.75. How much did you invest in each?

SOLUTION

Simple interest problems are based on the formula $I = Prt$, where I is the interest, P is the principal, r is the annual interest rate (in decimal form), and t is the time in years.

Verbal	Interest		Interest		Total
Model:	from $4\frac{1}{2}\%$	$+$	from $5\frac{1}{2}\%$	$=$	interest

You can let x represent the amount invested at $4\frac{1}{2}\%$. Because the total amount invested at $4\frac{1}{2}\%$ and $5\frac{1}{2}\%$ is $10,000, you can let $10,000 - x$ represent the amount invested at $5\frac{1}{2}\%$.

Labels:
Amount invested at $4\frac{1}{2}\% = x$ (dollars)
Amount invested at $5\frac{1}{2}\% = 10,000 - x$ (dollars)
Interest from $4\frac{1}{2}\% = Prt = (x)(0.045)(1)$ (dollars)
Interest from $5\frac{1}{2}\% = Prt = (10,000 - x)(0.055)(1)$ (dollars)
Total interest $= 508.75$ (dollars)

Equation:
$$0.045x + 0.055(10,000 - x) = 508.75$$
$$0.045x + 550 - 0.055x = 508.75$$
$$-0.01x = -41.25$$
$$x = \$4125$$

So, the amount invested at $4\frac{1}{2}\%$ is $4125 and the amount invested at $5\frac{1}{2}\%$ is
$$10,000 - x = 10,000 - 4125$$
$$= \$5875.$$

Check these results in the original statement of the problem, as follows.

CHECK

Interest from $4\frac{1}{2}\%$	Interest from $5\frac{1}{2}\%$	Total interest

$$0.045(4125) + 0.055(10,000 - 4125) \overset{?}{=} 508.75$$
$$185.625 + 323.125 \overset{?}{=} 508.75$$
$$508.75 = 508.75 \qquad \text{Solution checks. ✓}$$

✓CHECKPOINT   Now try Exercise 55.

In Example 8, did you recognize the hidden products in the two terms on the left side of the equation? Both hidden products come from the common formula

Interest = Principal · Rate · Time

$$I = Prt.$$

Common Formulas

Many common types of geometric, scientific, and investment problems use ready-made equations, called **formulas.** Knowing formulas such as those in the following lists will help you translate and solve a wide variety of real-life problems involving perimeter, area, volume, temperature, interest, and distance.

Common Formulas for Area, Perimeter, and Volume

Square	Rectangle	Circle	Triangle
$A = s^2$	$A = lw$	$A = \pi r^2$	$A = \frac{1}{2}bh$
$P = 4s$	$P = 2l + 2w$	$C = 2\pi r$	$P = a + b + c$

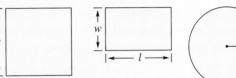

	Rectangular Solid	Circular Cylinder	Sphere
Cube	$V = lwh$	$V = \pi r^2 h$	$V = \frac{4}{3}\pi r^3$
$V = s^3$			

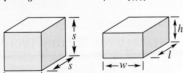 

Miscellaneous Common Formulas

Temperature: F = degrees Fahrenheit, C = degrees Celsius

$$F = \frac{9}{5}C + 32$$

Simple interest: I = interest, P = principal, r = interest rate, t = time

$$I = Prt$$

Distance: d = distance traveled, r = rate, t = time

$$d = rt$$

When working with applied problems, you often need to rewrite common formulas. For instance, the formula $P = 2l + 2w$ for the perimeter of a rectangle can be rewritten or solved for w to produce $w = \frac{1}{2}(P - 2l)$.

←4 cm→

h

Figure R2.5

EXAMPLE 9   USING A FORMULA

A cylindrical can has a volume of 200 cubic centimeters and a radius of 4 centimeters, as shown in Figure R2.5. Find the height of the can.

SOLUTION

The formula for the *volume of a cylinder* is $V = \pi r^2 h$. To find the height of the can, solve for h.

$$h = \frac{V}{\pi r^2}$$

Then, using $V = 200$ and $r = 4$, find the height.

$$h = \frac{200}{\pi(4)^2} \qquad \text{Substitute 200 for } V \text{ and 4 for } r.$$

$$= \frac{200}{16\pi} \qquad \text{Simplify denominator.}$$

$$\approx 3.98 \qquad \text{Use a calculator.}$$

So, the height of the can is about 3.98 centimeters. You can use unit analysis to check that your answer is reasonable.

$$\frac{200 \text{ cm}^3}{16\pi \text{ cm}^2} \approx 3.98 \text{ cm}$$

✓CHECKPOINT   Now try Exercise 61.

Strategy for Solving Word Problems

1. *Search* for the hidden equality—two expressions said to be equal or known to be equal. A sketch may be helpful.

2. *Write* a verbal model that equates these two expressions. Identify any *hidden* products.

3. *Assign* numbers to the known quantities and letters (or algebraic expressions) to the unknown quantities.

4. *Rewrite* the verbal model as an algebraic equation using the assigned labels.

5. *Solve* the resulting algebraic equation.

6. *Check* to see that the answer satisfies the word problem as stated. (Remember that "solving for x" or some other variable may not completely answer the question.)

R2.2 Warm Up

The following warm-up exercises involve skills that were covered in earlier sections. You will use these skills in the exercise set for this section. For additional help, review Section R2.1.

In Exercises 1–10, solve the equation (if possible) and check your answer.

1. $3x - 42 = 0$

2. $64 - 16x = 0$

3. $2 - 3x = 14 + x$

4. $7 + 5x = 7x - 1$

5. $5[1 + 2(x + 3)] = 6 - 3(x - 1)$

6. $2 - 5(x - 1) = 2[x + 10(x - 1)]$

7. $\dfrac{x}{3} + \dfrac{x}{2} = \dfrac{1}{3}$

8. $\dfrac{2}{x} + \dfrac{2}{5} = 1$

9. $1 - \dfrac{2}{z} = \dfrac{z}{z + 3}$

10. $\dfrac{x}{x + 1} - \dfrac{1}{2} = \dfrac{4}{3}$

R2.2 Exercises

Creating a Mathematical Model   In Exercises 1–10, write an algebraic expression for the verbal expression.

1. The sum of two consecutive natural numbers

2. The product of two natural numbers whose sum is 25

3. **Distance Traveled**   The distance traveled in t hours by a car traveling at 50 miles per hour

4. **Travel Time**   The travel time for a plane that is traveling at a rate of r miles per hour for 200 miles

5. **Acid Solution**   The amount of acid in x gallons of a 20% acid solution

6. **Discount**   The sale price of an item that is discounted by 20% of its list price L

7. **Geometry**   The perimeter of a rectangle whose width is x and whose length is twice the width

8. **Geometry**   The area of a triangle whose base is 20 inches and whose height is h inches

9. **Total Cost**   The total cost of producing x units for which the fixed costs are \$1200 and the cost per unit is \$25

10. **Total Revenue**   The total revenue obtained by selling x units at \$3.59 per unit

Using a Mathematical Model   In Exercises 11–16, write a mathematical model for the number problem, and solve the problem.

11. The sum of two consecutive natural numbers is 525. Find the two numbers.

12. Find three consecutive natural numbers whose sum is 804.

13. One positive number is five times another positive number. The difference between the two numbers is 148. Find the numbers.

14. One number is one-fifth of another number. The difference between the two numbers is 76. Find the numbers.

15. Find two consecutive integers whose product is five less than the square of the smaller number.

16. Find two consecutive natural numbers such that the difference of their reciprocals is one-fourth the reciprocal of the smaller number.

17. **Weekly Paycheck**   Your weekly paycheck is 15% *more* than your coworker's. Your two paychecks total $645. Find the amount of each paycheck.

18. **Monthly Profit**   The total profit for a company in February was 20% *higher* than it was in January. The total profit for the 2 months was $157,498. Find the profit for each month.

19. **Weekly Paycheck**   Your weekly paycheck is 15% *less* than your coworker's. Your two paychecks total $645. Find the amount of each paycheck.

20. **Monthly Profit**   The total profit for a company in February was 20% *lower* than it was in January. The total profit for the 2 months was $157,498. Find the profit for each month.

Movie Sequels   In Exercises 21–24, use the following information. The movie industry frequently releases sequels and/or prequels to successful movies. The revenue of each *Star Wars* movie is shown. Compare the revenue of each *Star Wars* movie to its predecessor by finding the percent increase or decrease in the domestic gross. (Source: Infoplease.com)

Movie	Domestic gross (in dollars)
Star Wars (1977)	$460,998,007
The Empire Strikes Back (1980)	$290,271,960
Return of the Jedi (1983)	$309,209,079
Episode I: The Phantom Menace (1999)	$431,088,295
Episode II: Attack of the Clones (2002)	$310,675,583

21. *Star Wars* (1977) to *The Empire Strikes Back* (1980)

22. *The Empire Strikes Back* (1980) to *Return of the Jedi* (1983)

23. *Return of the Jedi* (1983) to *Episode I: The Phantom Menace* (1999)

24. *Episode I: The Phantom Menace* (1999) to *Episode II: Attack of the Clones* (2002)

25. **World Internet Users**   In 2000, there were 385 million Internet users in the world. By 2001, that amount had increased by 29.6%. By 2002, the number of world Internet users had increased 9% over the 2001 total and in 2003, the number of users had increased by an additional 6.8%.   (Source: *International Telecommunications Union Yearbook of Statistics, CIA World Factbook*)

(a) Find the number of users in 2001.

(b) Find the number of users in 2002.

(c) Find the number of users in 2003.

26. **Sporting Goods Sales**   In 1998, sales of sporting goods totaled $69,848,000,000. In 1999, that amount increased by 1.9%. In 2000, the amount increased 4.6% over the 1999 amount and in 2001, the 2000 amount decreased by 1.3%. (Source: National Sporting Goods Association)

(a) Find the sporting goods sales in 1999.

(b) Find the sporting goods sales in 2000.

(c) Find the sporting goods sales in 2001.

27. **Television Owners**   In a survey conducted in 2003, Nielsen Media Research found that 106.7 million U.S. households owned at least one television and 100% of those households had at least one color television. Use the bar graph to determine how many of those households had two televisions, three or more televisions, a VCR, basic cable, and premium cable.   (Source: Nielsen Media Research)

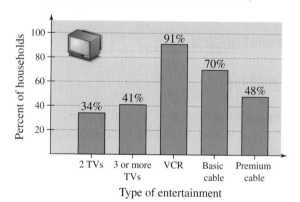

28. U.S. Car Sales In 2002, the number of U.S. car sales was 8,317,954. Use the bar graph to determine how many of each type of car was sold in 2002. (Source: Ward's Communications)

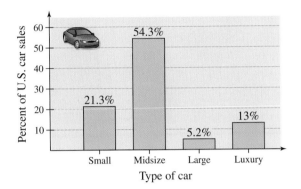

29. Geometry A room is 1.5 times as long as it is wide, and its perimeter is 75 feet (see figure). Find the dimensions of the room.

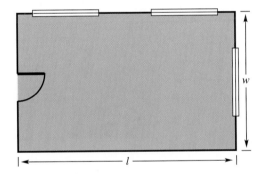

30. Geometry A picture frame has a total perimeter of 3 feet (see figure). The width of the frame is 0.62 times its length. Find the dimensions of the frame.

31. Simple Interest You invest $2500 at 7% simple interest. How many years will it take for the investment to earn $1000 in interest?

32. Simple Interest An investment earns $3200 interest over a seven-year period. What is the rate of interest on a $4800 principal investment?

33. Course Grade To get an A in a course you must have an average of at least 90% on four tests that are worth 100 points each. Your scores on the first three tests were 87, 92, and 84. What must you score on the fourth test to get an A for the course?

34. Course Grade You are taking a course that has four tests. The first three tests are worth 100 points each and the fourth test is worth 200 points. To get an A in the course you must have an average of at least 90% on the four tests. Your scores on the first three tests were 87, 92, and 84. What must you score on the fourth test to get an A for the course?

35. Loan Payments A family has annual loan payments totaling $13,077.75, or 58.6% of its annual income. What is the family's income?

36. Food Budget Your annual budget for food (eating at home and eating at restaurants) is 23.2% of your annual income. During the year, you spent $4832.12 on food. What is your annual income?

37. List Price The price of a swimming pool has been discounted 15%. The sale price is $1200. Find the original list price of the swimming pool.

38. List Price The price of a home theater system has been discounted 10%. The sale price is $499. Find the original price of the system.

39. Discount Rate A satellite radio system for your car has been discounted by $30. The sale price is $119. What percent of the original list price is the discount?

40. Discount Rate The price of a shirt has been discounted by $20. The sale price is $29.95. What percent of the original list price is the discount?

Weekly Salary In Exercises 41 and 42, use the following information to write a mathematical model and solve. Due to economic factors, your employer has reduced your weekly wage by 15%. Before the reduction, your weekly salary was $425.

41. What is your reduced salary?

42. *Make a Decision* What percent raise must you receive to bring your weekly salary back up to $425? Explain why the percent raise is different from the percent reduction.

43. Travel Time You are driving on a freeway to another town that is 150 miles from your home. After 30 minutes, you pass a freeway exit that you know is 25 miles from your home. Assuming that you continue at the same constant speed, how long will it take for the entire trip?

44. Travel Time A plane is flying from Orlando to Denver, a distance of about 1526 miles. After 1 hour and 15 minutes, the plane flies over a town that is 500 miles from Orlando. How long will it take the plane to fly from Orlando to Denver? (Assume that the plane flies at a constant speed the entire flight.)

45. Travel Time Two cars start at the same time at a given point and travel in the same direction at average speeds of 40 miles per hour and 55 miles per hour. How much time must elapse before the cars are 5 miles apart?

46. Catch-Up Time Students are traveling in two cars to a football game 135 miles away. The first car leaves on time and travels at an average speed of 45 miles per hour. The second car starts $\frac{1}{2}$ hour later and travels at an average speed of 55 miles per hour. At these speeds, how long will it take the second car of students to catch up to the first car?

47. Travel Time Two families meet at a park for a picnic. At the end of the day, one family travels east at an average speed of 42 miles per hour and the other family travels west at an average speed of 50 miles per hour. Both families have approximately 160 miles to travel. Find the time it takes each family to get home.

48. Average Speed A truck traveled at an average speed of 55 miles per hour on a 200-mile trip to pick up a load of freight. On the return trip (with the truck fully loaded), the average speed was 40 miles per hour. Find the average speed for the round trip.

49. Radio Waves Radio waves travel at the same speed as light, 3.0×10^8 meters per second. Find the time required for a radio wave to travel from mission control in Houston to NASA astronauts on the surface of the moon 3.84×10^8 meters away.

50. Distance to a Star Find the distance to a star that is 50 light years (distance traveled by light in 1 year) away. (Light travels at 186,000 miles per second.)

51. Height of a Tree You want to measure the height of a tree in your yard. To do this, you measure the tree's shadow and find that it is 25 feet long. You also measure the shadow of a five-foot lamppost and find that it is 2 feet long (see figure). How tall is the tree?

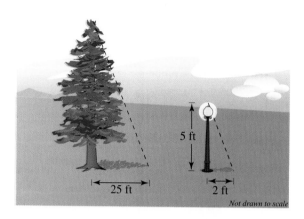

Not drawn to scale

Figure for 51

52. Height of a Building You want to measure the height of a building. To do this, you measure the building's shadow and find that it is 50 feet long. You also measure the shadow of a four-foot stake and find its shadow to be $3\frac{1}{2}$ feet long. How tall is the building?

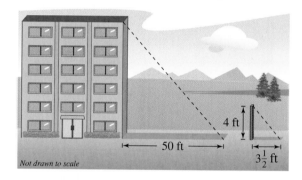

Not drawn to scale

53. Projected Expenses From January through May, a company's expenses totaled $234,980. If the monthly expenses continue at this rate, what will be the total expenses for the year?

54. Projected Revenue From January through August, a company's revenues totaled $345,950. If the monthly revenue continues at this rate, what will be the total revenue for the year?

55. Investment Mix You invest $12,000 in two funds paying $5\frac{1}{2}\%$ and 7% simple interest. The total annual interest is $772.50. How much is invested in each fund?

56. Investment Mix You invest $25,000 in two funds paying 3% and $4\frac{1}{2}\%$ simple interest. The total annual interest is $956.25. How much is invested in each fund?

57. Comparing Investment Returns You invest $12,000 in a fund paying $9\frac{1}{2}\%$ simple interest and $8000 in a fund for which the interest rate is variable. At the end of the year you receive notification that the total interest for both funds is $2054.40. Find the equivalent simple interest rate on the variable rate fund.

58. Comparing Investment Returns You have $10,000 on deposit earning simple interest that is linked to the prime rate. When the prime rate dropped, your rate dropped by $1\frac{1}{2}\%$ for the last quarter of the year. Your total annual interest was $1112.50. What was your interest rate for the first three quarters and for the last quarter?

59. Production Limit A company has fixed costs of $10,000 per month and variable costs of $8.50 per unit manufactured. The company has $85,000 available to cover the monthly costs. How many units can the company manufacture? (*Fixed costs* are those that occur regardless of the level of production. *Variable costs* depend on the level of production.)

60. Production Limit A company has fixed costs of $10,000 per month and variable costs of $9.30 per unit manufactured. The company has $85,000 available to cover the monthly costs. How many units can the company manufacture? (*Fixed costs* are those that occur regardless of the level of production. *Variable costs* depend on the level of production.)

61. Length of a Tank The diameter of a cylindrical propane gas tank is 4 feet (see figure). The total volume of the tank is 603.2 cubic feet. Find the length of the tank.

62. Water Depth A trough is 12 feet long, 3 feet deep, and 3 feet wide (see figure). Find the depth of the water when the trough contains 70 gallons. (1 gallon ≈ 0.13368 cubic foot.)

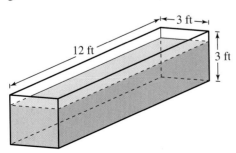

63. Mixture A 55-gallon barrel contains a mixture with a concentration of 40%. How much of this mixture must be withdrawn and replaced by 100% concentrate to bring the mixture up to 75% concentration? (See figure.)

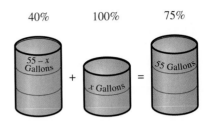

64. Mixture A farmer mixed gasoline and oil to make 2 gallons of mixture for his two-cycle chain saw engine. This mixture was 32 parts gasoline and 1 part two-cycle oil. How much gasoline must be added to bring the mixture to 40 parts gasoline and 1 part oil?

New York City Marathon In Exercises 65 and 66, find the average speed of the record-holding runners in the New York City Marathon. The length of the course is 26 miles, 385 yards. (Note that 1 mile = 5280 feet = 1760 yards.)

65. Men's record time: 2 hours, $7\frac{3}{4}$ minutes

66. Women's record time: 2 hours, $22\frac{1}{2}$ minutes

In Exercises 67–80, solve for the indicated variable.

67. Area of a Triangle
Solve for h in $A = \frac{1}{2}bh$.

68. Perimeter of a Rectangle
Solve for l in $P = 2l + 2w$.

69. Volume of a Rectangular Prism
Solve for l in $V = lwh$.

70. Volume of a Right Circular Cylinder
Solve for h in $V = \pi r^2 h$.

71. Markup
Solve for C in $S = C + RC$.

72. Discount
Solve for L in $S = L - RL$.

73. Investment at Simple Interest
Solve for r in $A = P + Prt$.

74. Investment at Compound Interest

Solve for P in $A = P\left(1 + \dfrac{r}{n}\right)^{nt}$.

75. Area of a Trapezoid

Solve for b in $A = \frac{1}{2}(a + b)h$.

76. Area of a Sector of a Circle

Solve for θ in $A = \dfrac{\pi r^2 \theta}{360}$.

77. Arithmetic Progression

Solve for n in $L = a + (n - 1)d$.

78. Geometric Progression

Solve for r in $S = \dfrac{rL - a}{r - 1}$.

79. Lateral Surface Area of a Cylinder

Solve for h in $A = 2\pi rh$.

80. Surface Area of a Cone

Solve for l in $S = \pi r^2 + \pi rl$.

81. Think About It Sometimes, when developing a mathematical model for a problem, the most accurate equation is not used as the model. Explain why and include factors that influence a mathematical model's selection.

82. *Make a Decision* Applied problems in textbooks usually give precisely the right amount of information that is necessary to solve a given problem. In real life, however, you must often sort through the given information and discard information that is irrelevant to the problem. Such irrelevant information is called a **red herring.** Find any red herrings in the following problem.

From 100 to 200 feet beneath the surface of the ocean, pressure changes at a rate of approximately 4.4 pounds per square inch for every 10-foot change in depth. A diver takes 30 minutes to ascend 25 feet from a depth of 150 feet. What change in pressure does the diver experience?

IN THE NEWS
Chocolate Obsession Leads to Physics Discovery

WASHINGTON (Reuters)- Princeton physicist Paul Chaikin's passion for M&M candies was so well known that his students played a sweet practical joke on him by leaving a 55-gallon drum of the candies in his office. . . .

The barrel full of the oblate little candies made Chaikin think about how well they packed in. A series of studies have shown they pack more tightly than perfect spheres—something that surprises many physicists and Chaikin himself.

"It is a startling and wonderful result," said Sidney Nagel, a physicist at the University of Chicago. "One doesn't normally stop to think about this. If you did, you might have guessed what would happen, but you'd have guessed wrongly."

The issue of how particles pack together has intrigued scientists for centuries and has implications for fields such as the design of high-density ceramic materials for use in aerospace or other industries.

Chaikin and his colleague, chemist Salvatore Torquato. . . said they found that oblate spheroids—such as plain M&Ms—pack surprisingly more densely than regular spheres when poured randomly and shaken.

When poured in, they said, spheres occupy about 64 percent of the space in a container. M&M's manage to pack in at a density of about 68 percent.

"We just stretched a sphere and suddenly things changed dramatically," said Torquato. "To me, it's remarkable that you can take this simple system with common candies and probe one of the deepest problems in condensed matter physics. . . ."

Reprinted with permission per Valeo Clearance Licence 3.5398.3295906-87705.

83. What volume of M&Ms can you fit in a circular cylinder container with a height of 10 inches and a radius of 3 inches?

Quadratic Equations

Objectives

- **Solve a quadratic equation by factoring.**
- **Solve a quadratic equation by extracting square roots.**
- **Construct and use a quadratic model to solve an application problem.**

Solving Quadratic Equations by Factoring

In the first two sections of this chapter, you studied linear equations in one variable. In this and the next section, you will study quadratic equations.

Definition of a Quadratic Equation

A **quadratic equation** in x is an equation that can be written in the general form

$$ax^2 + bx + c = 0$$

where a, b, and c are real numbers with $a \neq 0$. Another name for a quadratic equation in x is a **second-degree polynomial equation in x.**

There are three basic techniques for solving quadratic equations: factoring, extracting square roots, and the *Quadratic Formula*. (The Quadratic Formula is discussed in the next section.) The first technique is based on the following property.

Zero-Factor Property

If $ab = 0$, then $a = 0$ or $b = 0$.

STUDY TIP

The Zero-Factor Property applies *only* to equations written in general form (in which one side of the equation is zero). So, be sure that all terms are collected on one side *before* factoring. For instance, in the equation

$$(x - 5)(x + 2) = 8$$

it is *incorrect* to set each factor equal to 8. Can you solve this equation correctly?

To use this property, rewrite the left side of the general form of a quadratic equation as the product of two linear factors. Then find the solutions of the quadratic equation by setting each linear factor equal to zero.

EXAMPLE 1   SOLVING A QUADRATIC EQUATION BY FACTORING

Solve $x^2 - 3x - 10 = 0$.

SOLUTION

$x^2 - 3x - 10 = 0$	Write original equation.
$(x - 5)(x + 2) = 0$	Factor.
$x - 5 = 0$ ⟹ $x = 5$	Set 1st factor equal to 0.
$x + 2 = 0$ ⟹ $x = -2$	Set 2nd factor equal to 0.

The solutions are $x = 5$ and $x = -2$. Check these in the original equation.

✓CHECKPOINT   Now try Exercise 11.

EXAMPLE 2   SOLVING A QUADRATIC EQUATION BY FACTORING

Solve $6x^2 - 3x = 0$.

SOLUTION

$6x^2 - 3x = 0$	Write original equation.
$3x(2x - 1) = 0$	Factor out common factor.
$3x = 0 \implies x = 0$	Set 1st factor equal to 0.
$2x - 1 = 0 \implies x = \frac{1}{2}$	Set 2nd factor equal to 0.

The solutions are $x = 0$ and $x = \frac{1}{2}$. Check these by substituting in the original equation, as follows.

CHECK

$6x^2 - 3x = 0$	Write original equation.
$6(0)^2 - 3(0) \overset{?}{=} 0$	Substitute 0 for x.
$0 - 0 = 0$	First solution checks. ✓
$6\left(\frac{1}{2}\right)^2 - 3\left(\frac{1}{2}\right) \overset{?}{=} 0$	Substitute $\frac{1}{2}$ for x.
$\frac{6}{4} - \frac{3}{2} = 0$	Second solution checks. ✓

✓**CHECKPOINT**   Now try Exercise 13.

TECHNOLOGY

To check the solution in Example 3 with your graphing utility, you should first write the equation in general form.

$$9x^2 - 6x + 1 = 0$$

Then enter the expression $9x^2 - 6x + 1$ into y_1 of the equation editor. Now you can use the ASK mode of the *table* feature of your graphing utility to check the solution. For instructions on how to use the *table* feature, see Appendix A; for specific keystrokes, go to the text website at *college.hmco.com*.

If the two factors of a quadratic expression are the same, the corresponding solution is a **double** or **repeated** solution.

EXAMPLE 3   A QUADRATIC EQUATION WITH A REPEATED SOLUTION

Solve $9x^2 - 6x = -1$.

SOLUTION

$9x^2 - 6x = -1$	Write original equation.
$9x^2 - 6x + 1 = 0$	Write in general form.
$(3x - 1)^2 = 0$	Factor.
$3x - 1 = 0$	Set repeated factor equal to 0.
$x = \frac{1}{3}$	Solution

The only solution is $x = \frac{1}{3}$. Check this by substituting in the original equation, as follows.

$9x^2 - 6x = -1$	Write original equation.
$9\left(\frac{1}{3}\right)^2 - 6\left(\frac{1}{3}\right) \overset{?}{=} -1$	Substitute $\frac{1}{3}$ for x.
$1 - 2 \overset{?}{=} -1$	Simplify.
$-1 = -1$	Solution checks. ✓

✓**CHECKPOINT**   Now try Exercise 15.

Extracting Square Roots

There is a nice shortcut for solving equations of the form $u^2 = d$, where $d > 0$. By factoring, you can see that this equation has two solutions.

$u^2 = d$	Write original equation.
$u^2 - d = 0$	Write in general form.
$(u + \sqrt{d})(u - \sqrt{d}) = 0$	Factor.
$u + \sqrt{d} = 0 \implies u = -\sqrt{d}$	Set 1st factor equal to 0.
$u - \sqrt{d} = 0 \implies u = \sqrt{d}$	Set 2nd factor equal to 0.

Solving an equation of the form $u^2 = d$ without going through the steps of factoring is called **extracting square roots.**

Extracting Square Roots

The equation $u^2 = d$, where $d > 0$, has exactly two solutions:

$$u = \sqrt{d} \quad \text{and} \quad u = -\sqrt{d}.$$

These solutions can also be written as $u = \pm\sqrt{d}$.

EXAMPLE 4 EXTRACTING SQUARE ROOTS

Solve $4x^2 = 12$.

SOLUTION

$4x^2 = 12$	Write original equation.
$x^2 = 3$	Divide each side by 4.
$x = \pm\sqrt{3}$	Extract square roots.

The solutions are $x = \sqrt{3}$ and $x = -\sqrt{3}$. Check these in the original equation.

✓CHECKPOINT Now try Exercise 23.

EXAMPLE 5 EXTRACTING SQUARE ROOTS

Solve $(x - 3)^2 = 7$.

SOLUTION

$(x - 3)^2 = 7$	Write original equation.
$x - 3 = \pm\sqrt{7}$	Extract square roots.
$x = 3 \pm \sqrt{7}$	Add 3 to each side.

The solutions are $x = 3 \pm \sqrt{7}$. Check these in the original equation.

✓CHECKPOINT Now try Exercise 29.

Applications

Quadratic equations often occur in problems dealing with area. Here is a simple example.

A square room has an area of 144 square feet. Find the dimensions of the room.

To solve this problem, you can let x represent the length of each side of the room. Then, by solving the equation

$$x^2 = 144$$

you can conclude that each side of the room is 12 feet long. Note that although the equation $x^2 = 144$ has two solutions, $x = -12$ and $x = 12$, the negative solution makes no sense (for this problem), so you should choose the positive solution.

EXAMPLE 6   FINDING THE DIMENSIONS OF A ROOM

A sunroom is 3 feet longer than it is wide (see Figure R2.6) and has an area of 154 square feet. Find the dimensions of the room.

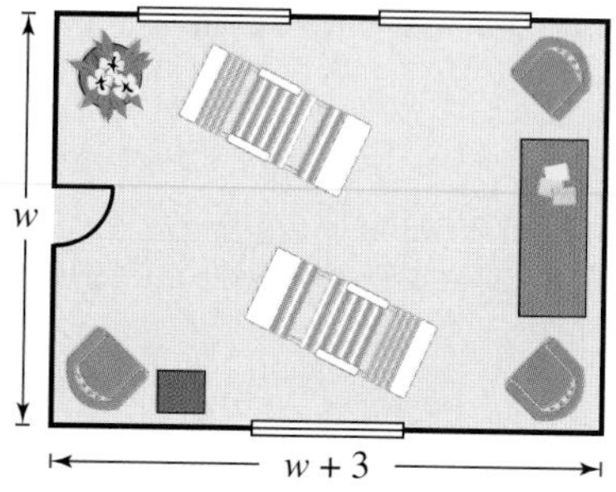

SOLUTION

You can begin by using the same type of problem-solving strategy that was presented in Section R2.2.

Verbal Model: Width of room · Length of room = Area of room

Labels: Area of room = 154   (square feet)
Width of room = w (feet)
Length of room = $w + 3$ (feet)

Equation: $w(w + 3) = 154$

$$w^2 + 3w - 154 = 0$$

$$(w - 11)(w + 14) = 0$$

$$w - 11 = 0 \quad \Longrightarrow \quad w = 11$$

$$w + 14 = 0 \quad \Longrightarrow \quad w = -14$$

Choosing the positive value, you can conclude that the width is 11 feet and the length is $w + 3 = 14$ feet. You can check this in the original statement of the problem as follows.

CHECK

The length of 14 feet is 3 feet more than the width of 11 feet. ✓

The area of the sunroom is $(11)(14) = 154$ square feet. ✓

✓CHECKPOINT   Now try Exercise 61.

Figure R2.6

w

$w + 3$

Another application of quadratic equations involves an object that is falling (or vertically projected into the air). The equation that gives the height of such an object is called a **position equation,** and on Earth's surface it has the form

$$s = -16t^2 + v_0t + s_0.$$

In this equation, s represents the height of the object (in feet), v_0 represents the initial velocity of the object (in feet per second), s_0 represents the initial height of the object (in feet), and t represents the time (in seconds).

EXAMPLE 7   FALLING OBJECT

Make a Decision A construction worker accidentally drops a wrench from a height of 235 feet and yells "Look out below!" (see Figure R2.7). Could a person at ground level hear this warning in time to get out of the way of the falling wrench?

SOLUTION

Because sound travels at about 1100 feet per second, it follows that a person at ground level hears the warning within 1 second of the time the wrench is dropped. To set up a mathematical model for the height of the wrench, use the position equation

$$s = -16t^2 + v_0t + s_0.$$

Because the object is dropped rather than thrown, the initial velocity is $v_0 = 0$ feet per second. Moreover, because the initial height is $s_0 = 235$ feet, you have the following model.

$$s = -16t^2 + (0)t + 235 = -16t^2 + 235$$

After falling for 1 second, the height of the wrench is $-16(1)^2 + 235 = 219$ feet. After falling for 2 seconds, the height of the wrench is $-16(2)^2 + 235 = 171$ feet. To find the number of seconds it takes the wrench to hit the ground, let the height s be zero and solve the equation for t.

$s = -16t^2 + 235$	Write position equation.
$0 = -16t^2 + 235$	Substitute 0 for height.
$16t^2 = 235$	Add $16t^2$ to each side.
$t^2 = \dfrac{235}{16}$	Divide each side by 16.
$t = \dfrac{\sqrt{235}}{4}$	Extract positive square root.
$t \approx 3.83$	Use a calculator.

The wrench will take about 3.83 seconds to hit the ground. If the person hears the warning 1 second after the wrench is dropped, the person still has almost 3 seconds to get out of the way.

✓CHECKPOINT   Now try Exercise 65.

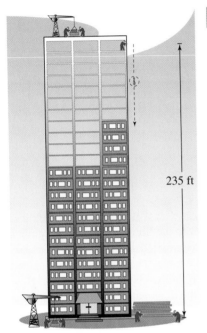

235 ft

Figure R2.7

The position equation used in Example 7 ignores air resistance. This implies that it is appropriate to use the position equation only to model falling objects that have little air resistance and that fall over short distances.

A third type of application that often involves a quadratic equation is one dealing with the hypotenuse of a right triangle. Recall from geometry that the sides of a right triangle are related by a formula called the **Pythagorean Theorem.** This theorem states that if a and b are the lengths of the legs of the triangle and c is the length of the hypotenuse (see Figure R2.8),

$$a^2 + b^2 = c^2. \qquad \text{Pythagorean Theorem}$$

Notice how this formula is used in the next example.

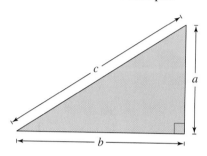

Figure R2.8

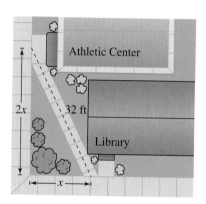

Figure R2.9

EXAMPLE 8   CUTTING ACROSS THE LAWN

An L-shaped sidewalk from the athletic center to the library on a college campus is shown in Figure R2.9. The sidewalk was constructed so that the length of one sidewalk forming the L is twice as long as the other. The length of the diagonal sidewalk that cuts across the grounds between the two buildings is 32 feet. How many feet does a person save by walking on the diagonal sidewalk?

SOLUTION

Using the Pythagorean Theorem, you have

$$a^2 + b^2 = c^2 \qquad \text{Pythagorean Theorem}$$
$$x^2 + (2x)^2 = 32^2 \qquad \text{Substitute for } a, b, \text{ and } c.$$
$$5x^2 = 1024 \qquad \text{Combine like terms.}$$
$$x^2 = 204.8 \qquad \text{Divide each side by 5.}$$
$$x = \pm\sqrt{204.8} \qquad \text{Take the square root of each side.}$$
$$x = \sqrt{204.8}. \qquad \text{Extract positive square root.}$$

The total distance covered by walking on the L-shaped sidewalk is

$$x + 2x = 3x$$
$$= 3\sqrt{204.8}$$
$$\approx 42.9 \text{ feet.}$$

Walking on the diagonal sidewalk saves a person about $42.9 - 32 = 10.9$ feet.

✓**CHECKPOINT**   Now try Exercise 73.

A fourth type of common application of a quadratic equation is one in which a quantity y is changing over time t according to a quadratic model.

EXAMPLE 9 CELLULAR TELEPHONE SUBSCRIBERS

From 1990 to 2002, the number of cellular telephone subscribers in the United States (in millions) can be modeled by

$$S = 0.97t^2 + 7.8, \quad 0 \le t \le 12$$

where $t = 0$ represents 1990. (See Figure R2.10.) If the number of cellular telephone subscribers continues to increase according to this model, in what year will the number of subscribers reach 225,000,000? (Source: Cellular Telecommunications & Internet Association)

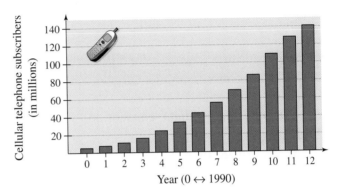

Figure R2.10

SOLUTION

To solve this problem, let the number of subscribers be 225 million and solve the equation for t.

$0.97t^2 + 7.8 = 225$	Substitute 225 for S.
$0.97t^2 = 217.2$	Subtract 7.8 from each side.
$t^2 \approx 223.918$	Divide each side by 0.97.
$t \approx \sqrt{223.918}$	Extract positive square root.

The solution is $t \approx 14.96$. However, because t is greater than 14, you can round up to $t = 15$ years. Because $t = 0$ represents 1990, you can conclude that, according to this model, the number of cellular telephone subscribers in the United States will reach 225,000,000 in the year 2005.

✓CHECKPOINT Now try Exercise 79.

Discussing the Concept | Geometry

A **Pythagorean triple** is a set of three positive integers (a, b, c) that satisfy the Pythagorean Theorem for right triangles, $a^2 + b^2 = c^2$. One such Pythagorean triple is $(3, 4, 5)$, because $3^2 + 4^2 = 5^2$. How many other Pythagorean triples can you find? (Remember that each member of the triple must be an integer.) Make a list of these triples, and discuss the method you used to find them.

R2.3 Warm Up

The following warm-up exercises involve skills that were covered in earlier sections. You will use these skills in the exercise set for this section. For additional help, review Sections R1.4 and R1.6.

In Exercises 1–4, simplify the expression.

1. $\sqrt{\frac{7}{50}}$

2. $\sqrt{32}$

3. $\sqrt{7^2 + 3 \cdot 7^2}$

4. $\sqrt{\frac{1}{4} + \frac{3}{8}}$

In Exercises 5–10, factor the expression.

5. $3x^2 + 7x$

6. $4x^2 - 25$

7. $16 - (x - 11)^2$

8. $x^2 + 7x - 18$

9. $10x^2 + 13x - 3$

10. $6x^2 - 73x + 12$

R2.3 Exercises

In Exercises 1–10, write the quadratic equation in general form.

1. $2x^2 = 3 - 5x$

2. $4x^2 - 2x = 9$

3. $x^2 = 25x$

4. $10x^2 = 90$

5. $(x - 3)^2 = 2$

6. $12 - 3(x + 7)^2 = 0$

7. $x(x + 2) = 3x^2 + 1$

8. $x(x + 5) = 2(x + 5)$

9. $\dfrac{3x^2 - 10}{5} = 12x$

10. $\dfrac{x^2 - 7}{3} = 2x$

In Exercises 11–22, solve the quadratic equation by factoring.

11. $x^2 - 2x - 8 = 0$

12. $x^2 - 10x + 9 = 0$

13. $6x^2 + 3x = 0$

14. $9x^2 - 1 = 0$

15. $x^2 + 10x + 25 = 0$

16. $16x^2 + 56x + 49 = 0$

17. $3 + 5x - 2x^2 = 0$

18. $2x^2 = 19x + 33$

19. $x^2 + 4x = 12$

20. $x^2 + 4x = 21$

21. $-x^2 - 7x = 10$

22. $-x^2 + 8x = 12$

In Exercises 23–36, solve the equation by extracting square roots. List both the exact answer *and* a decimal answer that has been rounded to two decimal places.

23. $x^2 = 16$

24. $x^2 = 144$

25. $x^2 = 7$

26. $x^2 = 27$

27. $3x^2 = 36$

28. $9x^2 = 25$

29. $(x - 12)^2 = 18$

30. $(x + 13)^2 = 21$

31. $(x + 2)^2 = 12$

32. $(x + 5)^2 = 20$

33. $12x^2 = 300$

34. $6x^2 = 250$

35. $3x^2 + 2(x^2 - 4) = 15$

36. $x^2 + 3(x^2 - 5) = 10$

In Exercises 37–58, solve the equation using any convenient method.

37. $x^2 = 64$

38. $7x^2 = 32$

39. $x^2 - 2x + 1 = 0$

40. $x^2 - 6x + 5 = 0$

41. $16x^2 - 9 = 0$

42. $11x^2 + 33x = 0$

43. $4x^2 - 12x + 9 = 0$

44. $x^2 - 14x + 49 = 0$

45. $(x + 3)^2 = 81$

46. $(x - 5)^2 = 8$

47. $4x = 4x^2 - 3$

48. $80 + 6x = 9x^2$

49. $50 + 5x = 3x^2$

50. $144 - 73x + 4x^2 = 0$

51. $12x = x^2 + 27$

52. $26x = 8x^2 + 15$

53. $50x^2 - 60x + 10 = 0$

54. $9x^2 + 12x + 3 = 0$

55. $(x + 3)^2 - 4 = 0$

56. $(x - 2)^2 - 9 = 0$

57. $(x + 1)^2 = x^2$

58. $(x + 1)^2 = 4x^2$

59. Consider the expression $(x + 2)^2$. How would you convince someone in your class that $(x + 2)^2 \neq x^2 + 4$? Give an argument based on the rules of algebra. Give an argument using your graphing utility.

60. Consider the expression $\sqrt{a^2 + b^2}$. How would you convince someone in your class that $\sqrt{a^2 + b^2} \neq a + b$? Give an argument based on the rules of algebra or geometry. Give an argument using your graphing utility.

61. Geometry A one-story building is 14 feet longer than it is wide (see figure). The building has 1632 square feet of floor space. What are the dimensions of the building?

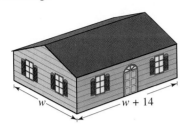

62. Geometry A billboard is 10 feet longer than it is high (see figure). The billboard has 336 square feet of advertising space. What are the dimensions of the billboard?

63. Geometry A triangular sign has a height that is equal to its base. The area of the sign is 4 square feet. Find the base and height of the sign.

64. Geometry The building lot shown in the figure has an area of 8000 square feet. What are the dimensions of the lot?

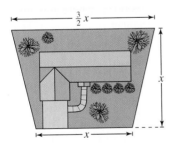

In Exercises 65–68, assume that air resistance is negligible, which implies that the position equation $s = -16t^2 + v_0 t + s_0$ is a reasonable model.

65. Falling Object A rock is dropped from the top of a 200-foot cliff that overlooks the ocean. How long will it take for the rock to hit the water?

66. Royal Gorge Bridge The Royal Gorge Bridge near Canon City, Colorado is the highest suspension bridge in the world. The bridge is 1053 feet above the Arkansas river. A rock is dropped from the bridge. How long does it take the rock to hit the water?

67. Olympic Diver The high-dive platform in the Olympics is 10 meters above the water. A diver wants to perform an armstand dive, which means she will drop to the water from a handstand position. How long will the diver be in the air? (*Hint:* 1 meter ≈ 3.2808 feet)

68. The Owl and the Mouse An owl is circling a field and sees a mouse. The owl folds its wings and begins to dive. If the owl starts its dive from a height of 100 feet, how long does the mouse have to escape?

IN THE NEWS
The Flying ELVI

The "Flying ELVI" is a ten member skydiving team first featured in the fun filled hit movie "Honeymoon in Vegas." They combine a spine tingling aerial skydiving performance of smoke trails, pyrotechnic fireworks, and precision maneuvers with an over-the-top entertaining stage show.

Jumping from altitudes of 5,000–12,500 feet above the earth, they free-fall for up to 9,500 feet at speeds ranging from 120 to 160 miles per hour. Team members are all United States Parachute Association Pro rated skydivers with an average of over 2,000 jumps each. Having thrilled audiences at events across the country for the past nine years, these daredevils of the sky are more popular than ever, and bring a bit of nostalgia to every event as they inspire fond memories of the King himself, Elvis Presley. . . .

The Flying Elvi website, http://www.flyingelvi.com/home.html.

69. If the "Flying ELVI" jump from a height of 12,500 feet at an initial velocity of 0 feet per second and open their parachutes at 3000 feet, how long will the team be in free fall?

70. If the "Flying ELVI" jump from a height of 5000 feet at an initial velocity of 0 feet per second and open their parachutes at 3000 feet, how long will the team be in free fall?

71. Geometry The hypotenuse of an isosceles right triangle is 5 centimeters long. How long are the sides? (An isosceles triangle is one that has two sides of equal length.)

72. Geometry An equilateral triangle has a height of 1 foot. How long are each of its sides? (*Hint:* Use the height of the triangle to partition the triangle into two right triangles of the same size.)

73. Flying Distance A commercial jet flies to three cities whose locations form the vertices of a right triangle (see figure). The air distance from Atlanta to Buffalo is about 703 miles and the air distance from Atlanta to Chicago is about 583 miles. Approximate the air distance from Atlanta to Buffalo *by way of* Chicago.

74. Depth of a Submarine The sonar of a navy cruiser detects a submarine that is 3000 feet from the cruiser. The angle formed by the ocean surface and a line from the cruiser to the submarine is 45° (see figure). How deep is the submarine?

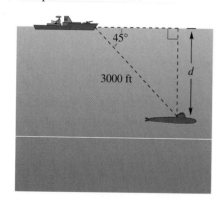

75. Salvage Depth A marine salvage ship detects a wreck 2000 feet off the bow. The angle between the ocean surface and a line from the ship to the wreck is 45°. How deep is the wreck?

76. College Costs The cost C (in dollars) to attend a two-year public college full time from the academic years 1995/1996 to 2000/2001 in the United States can be approximated by the model

$$C = 7.62t^2 + 3985, \quad 6 \le t \le 11$$

where $t = 6$ represents the 1995/1996 academic year (see figure). Assuming the cost to attend a two-year public college full time continues to follow this model, in what year will two-year college costs first exceed $6000? (Source: National Center for Education Statistics)

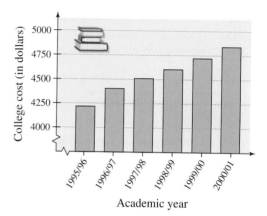

Academic year

77. Total Revenue The demand equation for a product is $p = 20 - 0.0002x$, where p is the price per unit and x is the number of units sold. The total revenue R for selling x units is given by $R = xp = x(20 - 0.0002x)$. How many units must be sold to produce a revenue of $500,000?

78. Total Revenue The demand equation for a product is $p = 30 - 0.0005x$, where p is the price per unit and x is the number of units sold. The total revenue R for selling x units is given by

$$R = xp = x(30 - 0.0005x).$$

How many units must be sold to produce a revenue of $450,000?

79. Production Cost A company determines that the average monthly cost C (in dollars) of raw materials for manufacturing a product line can be modeled by

$$C = 20.75t^2 + 5104, \quad t \ge 0$$

where t is the year, with $t = 0$ corresponding to 2000. If the average monthly cost of raw materials continues to increase according to this model, in what year will the average monthly cost reach $10,000?

80. Monthly Cost A company determines that the average monthly cost C (in dollars) for staffing temporary positions can be modeled by

$$C = 115.35t^2 + 12,072, \quad t \geq 0$$

where t represents the year, with $t = 0$ corresponding to 2000. If the average monthly cost for staffing temporary positions continues to increase according to this model, in what year will the average monthly cost reach $20,000?

81. *Make a Decision*: U.S. Population The resident population P (in thousands) of the United States from 1800 to 1890 can be approximated by the model

$$P = 694.59t^2 + 6179, \quad 0 \leq t \leq 9$$

where t represents the year, with $t = 0$ corresponding to 1800, $t = 1$ corresponding to 1810, and so on (see figure). If this model had continued to be valid up through the present time, in what year would the resident population of the United States have reached 250,000,000? Judging from the figure, would you say that this model is a good representation of the resident population through 1890? How about through 2002, when the United States resident population was approximately 288,369,000 people? (Source: U.S. Census Bureau)

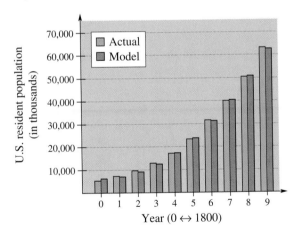

Year (0 ↔ 1800)

82. U.S. Population The resident population P (in thousands) of the United States from 1900 to 2000 can be approximated by the model $P = 1951.00t^2 + 97,551, \quad 0 \leq t \leq 10$, where t represents the year, with $t = 0$ corresponding to 1900, $t = 1$ corresponding to 1910, and so on (see figure). If this model continues to be valid, in what year will the resident population of the United States reach 330,000,000? (Source: U.S. Census Bureau)

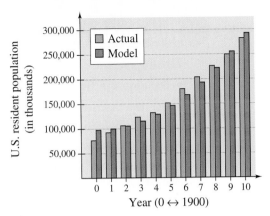

Figure for 82

83. *Make a Decision* The U.S. Census Bureau predicts that the population in 2050 will be 419,854,000. Does the model in Exercise 82 appear to be a valid model for the year 2050?

84. *Make a Decision*: Fast Food The food and beverage sales S (in billions of dollars) for fast-food restaurants from 1998 to 2002 can be approximated by

$$S = 0.220t^2 + 84.7, \quad 8 \leq t \leq 12$$

where t represents the year, with $t = 8$ corresponding to 1998. Assuming this model continues to be valid, estimate fast-food sales for the year 2003. The fast-food industry expected sales to be $120,874,000,000 in 2003. Does your estimate agree with the fast-food industry's expectation? (Source: National Restaurant Association)

85. *Make a Decision*: Head Start The enrollment E (in thousands) in the Head Start Program from 1995 to 2001 can be approximated by the model

$$E = 1.57t^2 + 708, \quad 5 \leq t \leq 11$$

where t represents the year, with $t = 5$ corresponding to 1995. If this model continues to be valid, in what year will Head Start enrollment reach 920,000 students? Could you use this model to estimate the Head Start enrollment for the year 1980? Explain. (Source: U.S. Administration for Children and Families)

R2.4   The Quadratic Formula

Objectives

- Develop the Quadratic Formula by completing the square.
- Use the discriminant to determine the number of real solutions of a quadratic equation.
- Solve a quadratic equation using the Quadratic Formula.
- Use the Quadratic Formula to solve an application problem.

Development of the Quadratic Formula

In Section R2.3 you studied two methods for solving quadratic equations. These two methods are efficient for special quadratic equations that are factorable or that can be solved by extracting square roots. There are, however, many quadratic equations that cannot be solved efficiently by either of these two techniques. Fortunately, there is a general formula that can be used to solve *any* quadratic equation. It is called the **Quadratic Formula.** This formula is derived using a process called **completing the square.**

$$ax^2 + bx + c = 0 \qquad \text{General form, } a \neq 0$$

$$ax^2 + bx = -c \qquad \text{Subtract } c \text{ from each side.}$$

$$x^2 + \frac{b}{a}x = -\frac{c}{a} \qquad \text{Divide each side by } a.$$

$$x^2 + \frac{b}{a}x + \left(\frac{b}{2a}\right)^2 = -\frac{c}{a} + \left(\frac{b}{2a}\right)^2 \qquad \text{Complete the square.}$$

$$\left(\text{half of } \frac{b}{a}\right)^2$$

$$\left(x + \frac{b}{2a}\right)^2 = \frac{b^2 - 4ac}{4a^2} \qquad \text{Simplify.}$$

$$x + \frac{b}{2a} = \pm\sqrt{\frac{b^2 - 4ac}{4a^2}} \qquad \text{Extract square roots.}$$

$$x = \frac{-b \pm \sqrt{b^2 - 4ac}}{2a} \qquad \text{Solutions}$$

STUDY TIP

The Quadratic Formula is one of the most important formulas in algebra, and you should memorize it. We have found that it helps to try to memorize a verbal statement of the rule. For instance, you might try to remember the following verbal statement of the Quadratic Formula: "The opposite of *b*, plus or minus the square root of *b* squared minus 4*ac*, all divided by 2*a*."

The Quadratic Formula

The solutions of

$$ax^2 + bx + c = 0, \quad a \neq 0$$

are given by the **Quadratic Formula,**

$$x = \frac{-b \pm \sqrt{b^2 - 4ac}}{2a}.$$

The Discriminant

In the Quadratic Formula, the quantity under the radical sign, $b^2 - 4ac$, is called the **discriminant** of the quadratic expression $ax^2 + bx + c$.

$b^2 - 4ac$ Discriminant

It can be used to determine the nature of the solutions of a quadratic equation.

Solutions of a Quadratic Equation

The solutions of a quadratic equation

$$ax^2 + bx + c = 0, \quad a \neq 0$$

can be classified by the discriminant, $b^2 - 4ac$, as follows.

1. If $b^2 - 4ac > 0$, the equation has *two* distinct real solutions.

2. If $b^2 - 4ac = 0$, the equation has *one* repeated real solution.

3. If $b^2 - 4ac < 0$, the equation has *no* real solutions.

If the discriminant of a quadratic equation is negative, as in case 3 above, then its square root is imaginary (not a real number) and the Quadratic Formula yields two complex solutions. You will study complex solutions in Section 2.5.

EXAMPLE 1   USING THE DISCRIMINANT

Use the discriminant to determine the number of real solutions of each of the following quadratic equations.

a. $4x^2 - 20x + 25 = 0$

b. $13x^2 + 7x + 1 = 0$

c. $5x^2 = 8x$

SOLUTION

a. Using $a = 4$, $b = -20$, and $c = 25$, the discriminant is

$$b^2 - 4ac = (-20)^2 - 4(4)(25) = 400 - 400 = 0.$$

So, there is *one* repeated real solution.

b. Using $a = 13$, $b = 7$, and $c = 1$, the discriminant is

$$b^2 - 4ac = (7)^2 - 4(13)(1) = 49 - 52 = -3 < 0.$$

So, there are *no* real solutions.

c. In general form, this equation is $5x^2 - 8x = 0$, with $a = 5$, $b = -8$, and $c = 0$, which implies that the discriminant is

$$b^2 - 4ac = (-8)^2 - 4(5)(0) = 64 > 0.$$

So, there are *two* distinct real solutions.

✓*CHECKPOINT*   Now try Exercise 1.

Using the Quadratic Formula

When using the Quadratic Formula, remember that *before* the formula can be applied, you must first write the quadratic equation in general form.

EXAMPLE 2 TWO DISTINCT SOLUTIONS

Solve $x^2 + 3x = 9$.

SOLUTION

$$x^2 + 3x = 9 \qquad \text{Write original equation.}$$

$$x^2 + 3x - 9 = 0 \qquad \text{Write in general form.}$$

$$x = \frac{-3 \pm \sqrt{(3)^2 - 4(1)(-9)}}{2(1)} \qquad \text{Quadratic Formula}$$

$$x = \frac{-3 \pm \sqrt{45}}{2} \qquad \text{Simplify.}$$

$$x = \frac{-3 \pm 3\sqrt{5}}{2} \qquad \text{Simplify.}$$

The solutions are

$$x = \frac{-3 + 3\sqrt{5}}{2} \quad \text{and} \quad x = \frac{-3 - 3\sqrt{5}}{2}.$$

Check these in the original equation.

✓CHECKPOINT   Now try Exercise 9.

TECHNOLOGY

You can check the solutions to Example 2 using a calculator.

```
((-3+3√(5))/2)²+
3((-3+3√(5))/2)
                9
((-3-3√(5))/2)²+
3((-3-3√(5))/2)
                9
```

EXAMPLE 3 ONE REPEATED SOLUTION

Solve $8x^2 - 24x + 18 = 0$.

SOLUTION

Begin by dividing each side by the common factor 2.

$$8x^2 - 24x + 18 = 0 \qquad \text{Write original equation.}$$

$$4x^2 - 12x + 9 = 0 \qquad \text{Divide each side by 2.}$$

$$x = \frac{-(-12) \pm \sqrt{(-12)^2 - 4(4)(9)}}{2(4)} \qquad \text{Quadratic Formula}$$

$$x = \frac{12 \pm \sqrt{0}}{8} \qquad \text{Simplify.}$$

$$x = \frac{3}{2} \qquad \text{Repeated solution}$$

The only solution is $x = \frac{3}{2}$. Check this in the original equation.

✓CHECKPOINT   Now try Exercise 25.

The discriminant in Example 3 is a perfect square (zero in this case), and you could have factored the quadratic as

$$4x^2 - 12x + 9 = 0$$

$$(2x - 3)^2 = 0$$

and concluded that the solution is $x = \frac{3}{2}$. Because factoring is easier than applying the Quadratic Formula, try factoring first when solving a quadratic equation. If, however, factors cannot easily be found, then use the Quadratic Formula. For instance, try solving the quadratic equation

$$x^2 - x - 12 = 0$$

in two ways—by factoring and by the Quadratic Formula—to see that you get the same solutions either way.

When using a calculator with the Quadratic Formula, you should get in the habit of using the memory key to store intermediate steps. This will save steps and minimize roundoff error.

TECHNOLOGY

Try to calculate the value of x in Example 4 by using additional parentheses instead of storing the intermediate result, 196.434777, in your calculator.

| EXAMPLE 4 | USING A CALCULATOR WITH THE QUADRATIC FORMULA |

Solve $16.3x^2 - 197.6x + 7.042 = 0$.

SOLUTION

In this case, $a = 16.3$, $b = -197.6$, $c = 7.042$, and you have

$$x = \frac{-(-197.6) \pm \sqrt{(-197.6)^2 - 4(16.3)(7.042)}}{2(16.3)}.$$

To evaluate these solutions, begin by calculating the square root of the discriminant, as follows.

Scientific Calculator Steps

197.6 $\boxed{+/-}$ $\boxed{x^2}$ $\boxed{-}$ 4 $\boxed{\times}$ 16.3 $\boxed{\times}$ 7.042 $\boxed{=}$ $\boxed{\surd}$

Graphing Calculator Steps

$\boxed{\surd}$ $\boxed{(}$ $\boxed{(}$ $\boxed{(-)}$ 197.6 $\boxed{)}$ $\boxed{x^2}$ $\boxed{-}$ 4 $\boxed{\times}$ 16.3 $\boxed{\times}$ 7.042 $\boxed{)}$ $\boxed{ENTER}$

In either case, the result is 196.434777. Storing this result and using the recall key, you can find the following two solutions.

$$x \approx \frac{197.6 + 196.434777}{2(16.3)} \approx 12.087 \qquad \text{Add stored value.}$$

$$x \approx \frac{197.6 - 196.434777}{2(16.3)} \approx 0.036 \qquad \text{Subtract stored value.}$$

Instead of storing the intermediate result, 196.434777, in your calculator, you could use additional parentheses to evaluate the solutions using just one step.

✓**CHECKPOINT**   Now try Exercise 31.

Applications

In Section R2.3, you studied four basic types of applications involving quadratic equations: area, falling bodies, the Pythagorean Theorem, and quadratic models. The solution to each of these types of problems can involve the Quadratic Formula. For instance, Example 5 shows how the Quadratic Formula can be used to analyze a quadratic model for alternative fuel vehicles.

EXAMPLE 5   **ALTERNATIVE FUEL VEHICLES**

Make a Decision   From 1993 to 2002, the number of alternative fuel vehicles A in use in the United States can be modeled by the quadratic equation

$$A = 2083.66t^2 - 8300.0t + 322,860, \quad 3 \le t \le 12$$

where time $t = 3$ represents 1993. The number of vehicles is shown graphically in Figure R2.11. Assuming that this model continues to be valid, in what year will the number of alternative fuel vehicles in use reach 1,000,000? (Source: Energy Information Administration)

SOLUTION

To find the year in which the number of alternative fuel vehicles will reach 1,000,000, solve the equation

$$1,000,000 = 2083.66t^2 - 8300.0t + 322,860.$$

To begin, write the equation in general form.

$$2083.66t^2 - 8300.0t - 677,140 = 0$$

Then apply the Quadratic Formula with $a = 2083.66$, $b = -8300.0$, and $c = -677,140$.

$$t = \frac{-(-8300.0) \pm \sqrt{(-8300.0)^2 - 4(2083.66)(-677,140)}}{2(2083.66)}$$

Of the two possible solutions, only the positive one makes sense in this context. Choosing the positive solution, you find that

$$t = \frac{-(-8300) + \sqrt{(-8300.0)^2 - 4(2083.66)(-677,140)}}{2(2083.66)}$$

$$= \frac{8300 + \sqrt{5,712,608,130}}{4167.32}$$

$$\approx 20.$$

Because $t = 3$ corresponds to 1993, it follows that $t \approx 20$ corresponds to 2010. So, from the model you can conclude that the number of alternative fuel vehicles in use in the United States will reach 1,000,000 in 2010. A table that shows the number of alternative fuel vehicles for several years can help confirm this result.

✓CHECKPOINT   Now try Exercise 65.

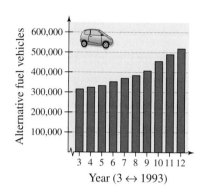

Figure R2.11

TECHNOLOGY

A program can be written to solve equations using the Quadratic Formula. A program for several models of graphing utilities can be found on the website for this text at *college.hmco.com.* Use a program to solve Example 5.

STUDY TIP

Note in the position equation

$$s = -16t^2 + v_0t + s_0$$

that the initial velocity v_0 is positive when an object is rising and negative when an object is falling.

Figure R2.12

In Section R2.3, you learned that the position equation for a falling object is of the form

$$s = -16t^2 + v_0t + s_0$$

where s is the height (in feet) of the object, v_0 is the initial velocity (in feet per second), t is the time (in seconds), and s_0 is the initial height (in feet). This equation is valid only for free-falling objects near Earth's surface. Because of differences in gravitational force, position equations are different on other planets or moons. The next example looks at a position equation for a falling object on our moon.

EXAMPLE 6 THROWING AN OBJECT ON THE MOON

An astronaut standing on the surface of the moon throws a rock vertically into space, as shown in Figure R2.12. The height of the rock (in feet) is given by

$$s = -2.7t^2 + 27t + 6.$$

How much time (in seconds) will elapse before the rock strikes the lunar surface?

SOLUTION

Because s gives the height of the rock at any time t, you can find the time that the rock hits the surface of the moon by setting s equal to zero and solving for t.

$$-2.7t^2 + 27t + 6 = 0 \qquad \text{Substitute 0 for } s.$$

$$t = \frac{-27 \pm \sqrt{(27)^2 - 4(-2.7)(6)}}{2(-2.7)} \qquad \text{Quadratic Formula}$$

$$\approx 10.2 \text{ seconds} \qquad \text{Choose positive solution.}$$

So, about 10.2 seconds will elapse before the rock hits the lunar surface.

✓CHECKPOINT   Now try Exercise 59.

TECHNOLOGY

Use the *last entry* feature of your graphing calculator to find the time in the air on Earth for the rock in Example 6. Simply replace -2.7 with -16 in the expression for t. For specific keystrokes on using the *last entry* feature, go to the text website at *college.hmco.com*.

Discussing the Concept | Methods for Solving a Quadratic Equation

List all of the methods that can be used to solve a quadratic equation. Try not to refer to your book or notes. Identify the types of quadratic equations that are more easily solved by each method. Give an example of each type of equation and use the most efficient method to solve it.

R2.4 (Warm Up

The following warm-up exercises involve skills that were covered in earlier sections. You will use these skills in the exercise set for this section. For additional help, review Sections R1.4 and R2.3.

In Exercises 1–4, simplify the expression.

1. $\sqrt{9 - 4(3)(-12)}$
2. $\sqrt{36 - 4(2)(3)}$
3. $\sqrt{12^2 - 4(3)(4)}$
4. $\sqrt{15^2 + 4(9)(12)}$

In Exercises 5–10, solve the quadratic equation by factoring.

5. $x^2 - x - 2 = 0$
6. $2x^2 + 3x - 9 = 0$
7. $x^2 - 4x = 5$
8. $2x^2 + 13x = 7$
9. $x^2 = 5x - 6$
10. $x(x - 3) = 4$

R2.4 (Exercises

In Exercises 1–8, use the discriminant to determine the number of real solutions of the quadratic equation.

1. $4x^2 - 4x + 1 = 0$
2. $2x^2 - x - 1 = 0$
3. $3x^2 + 4x + 1 = 0$
4. $x^2 + 2x + 4 = 0$
5. $2x^2 - 5x = -5$
6. $3 - 6x = -3x^2$
7. $\frac{1}{5}x^2 + \frac{6}{5}x - 8 = 0$
8. $\frac{1}{3}x^2 - 5x + 25 = 0$

In Exercises 9–30, use the Quadratic Formula to solve the equation.

9. $2x^2 + x - 1 = 0$
10. $2x^2 - x - 1 = 0$
11. $16x^2 + 8x - 3 = 0$
12. $25x^2 - 20x + 3 = 0$
13. $2 + 2x - x^2 = 0$
14. $x^2 - 10x + 22 = 0$
15. $x^2 + 14x + 44 = 0$
16. $6x = 4 - x^2$
17. $x^2 + 8x - 4 = 0$
18. $4x^2 - 4x - 4 = 0$
19. $12x - 9x^2 = -3$
20. $16x^2 + 22 = 40x$
21. $36x^2 + 24x = 7$
22. $3x + x^2 - 1 = 0$
23. $4x^2 + 4x = 7$
24. $16x^2 - 40x + 5 = 0$
25. $28x - 49x^2 = 4$
26. $9x^2 + 24x + 16 = 0$
27. $8t = 5 + 2t^2$
28. $25h^2 + 80h + 61 = 0$
29. $(y - 5)^2 = 2y$
30. $(x + 6)^2 = -2x$

In Exercises 31–36, use a calculator to solve the equation. (Round your answer to three decimal places.)

31. $5.1x^2 - 1.7x - 3.2 = 0$
32. $10.4x^2 + 8.6x + 1.2 = 0$
33. $7.06x^2 - 4.85x + 0.50 = 0$
34. $-0.005x^2 + 0.101x - 0.193 = 0$
35. $422x^2 - 506x - 347 = 0$
36. $2x^2 - 2.50x - 0.42 = 0$

In Exercises 37–46, solve the equation using any convenient method.

37. $3x + 4 = 2x - 7$

38. $x^2 - 2x + 5 = x^2 - 5$

39. $4x^2 - 15 = 25$

40. $4x^2 + 2x + 4 = 2x - 8$

41. $x^2 + 3x + 1 = 0$

42. $x^2 + 3x - 4 = 0$

43. $(x - 1)^2 = 9$

44. $2x^2 - 4x - 6 = 0$

45. $100x^2 - 400 = 0$

46. $2x^2 + 4x - 9 = 2(x - 1)^2$

Writing Real-Life Problems In Exercises 47–50, solve the number problem *and* write a real-life problem that could be represented by this verbal model. For instance, an applied problem that could be represented by Exercise 47 is as follows.

The sum of the length and width of a one-story house is 100 feet. The house has 2500 square feet of floor space. What are the length and width of the house?

47. Find two numbers whose sum is 100 and whose product is 2500.

48. One number is 1 more than another number. The product of the two numbers is 72. Find the numbers.

49. One number is 1 more than another number. The sum of their squares is 113. Find the numbers.

50. One number is 2 more than another number. The product of the two numbers is 440. Find the numbers.

Cost Equation In Exercises 51–54, use the cost equation to find the number of units x that a manufacturer can produce for the cost C. (Round your answer to the nearest positive integer.)

51. $C = 0.125x^2 + 20x + 5000$ $C = \$14,000$

52. $C = 0.5x^2 + 15x + 5000$ $C = \$11,500$

53. $C = 800 + 0.04x + 0.002x^2$ $C = \$1680$

54. $C = 800 - 10x + \dfrac{x^2}{4}$ $C = \$896$

55. Seating Capacity A rectangular classroom seats 72 students. If the seats were rearranged with three more seats in each row, the classroom would have two fewer rows. Find the original number of seats in each row.

56. Geometry A rancher has 200 feet of fencing to enclose two adjacent rectangular corrals (see figure). Find the dimensions such that the total enclosed area will be 1400 square feet.

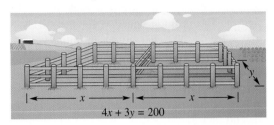

$4x + 3y = 200$

57. Geometry An open box is to be made from a square piece of material by cutting two-inch squares from the corners and turning up the sides (see figure). The volume of the finished box is to be 200 cubic inches. Find the size of the original piece of material.

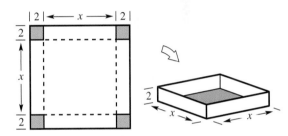

58. Geometry An open box (see figure) is to be constructed from 108 square inches of material. Find the dimensions of the square base.

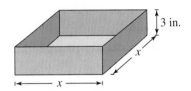

3 in.

59. On the Moon An astronaut on the moon throws a rock straight upward into space. The height of the rock is given by

$$s = -2.7t^2 + 40t + 5$$

where the initial velocity is 40 feet per second and the initial height is 5 feet. How long will it take the rock to hit the surface of the moon? If the rock had been thrown with the same initial velocity and height on Earth, how long would it have remained in the air? (See Example 6.)

60. Hot Air Balloon   Two people are riding in a hot air balloon that is 200 feet above the ground. One person drops a sack of sand and the balloon starts to rise. One second later, the balloon is 20 feet higher and the other person drops a sack of sand (see figure). The position equation for the first sack is $s = -16t^2 + 200$, and (using an initial velocity of -20 feet per second) the position equation for the second sack is $s = -16t^2 - 20t + 220$. Which sack of sand will hit the ground first? (*Hint:* Remember that the first sack was dropped one second before the second sack.)

Falling Objects   In Exercises 61 and 62, use the following information. The position equation for falling objects on Earth is of the form

$$s = -16t^2 + v_0t + s_0$$

where s is the height of the object (in feet), v_0 is the initial velocity (in feet per second), t is the time (in seconds), and s_0 is the initial height (in feet).

61. *Make a Decision*   If a rock were thrown on the surface of Earth with an initial velocity of 27 feet per second from an initial height of 6 feet, would it take a longer or shorter period of time to reach the ground than it would on the surface of the moon? (See Example 6.)

62. Use the Quadratic Formula to find the actual time that the rock would remain in the air.

63. Flying Distance   A small commuter airline flies to three cities whose locations form the vertices of a right triangle (see figure). The total flight distance (from City A to City B to City C and back to City A) is 1400 miles. It is 600 miles between the two cities that are farthest apart. Find the other two distances between cities.

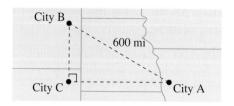

Figure for 63

64. Distance from a Dock   A windlass is used to tow a boat to a dock. The rope is attached to the boat at a point 15 feet below the level of the windlass (see figure). Find the distance from the boat to the dock when there is 75 feet of rope out.

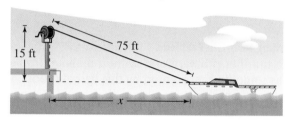

65. *Make a Decision*: Consumer Spending   The total number of dollars C spent per person each year on media use (television, radio, newspaper, movies, Internet, and so on) from 1996 to 2003 can be approximated by the model

$$C = 0.3164t^2 + 33.601t + 281.05, \quad 6 \le t \le 13$$

where t represents the year, with $t = 6$ corresponding to 1996. The figure shows the actual spending and the spending represented by the model. (Source: Veronis Suhler Stevenson, New York)

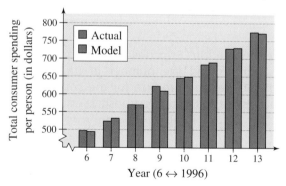

(a) During what year was consumer spending per person approximately $620?

(b) According to the model, in what year will consumer spending per person reach $850?

(c) Consumer spending per person is expected to be $885 per year by 2006. Does the model agree? Explain your reasoning.

66. *Make a Decision*: **Prescription Drugs** The total amount D (in billions of dollars) spent on prescription drugs in the United States from 1990 to 2001 can be approximated by the model

$$D = 0.91t^2 - 1.5t + 44, \quad 0 \le t \le 11$$

where t represents the year, with $t = 0$ corresponding to 1990. The figure shows the actual costs and the costs represented by the model. If this model continues to be valid, in what year will the total amount spent on prescription drugs reach $378,000,000,000? (Source: U.S. Centers for Medicare and Medicaid Services)

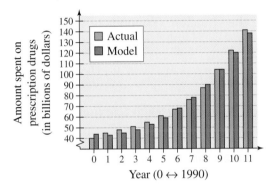

Year (0 ↔ 1990)

67. *Make a Decision*: **Prescription Drugs** The total amount A (in billions of dollars) projected by the industry to be spent on prescription drugs in the United States from 2002 to 2012 can be approximated by the model

$$C = 0.89t^2 + 15.9t + 126, \quad 2 \le t \le 12$$

where t represents the year, with $t = 2$ corresponding to 2002. According to this model, in what year will the total amount spent on prescription drugs reach $374,000,000,000? Does the model from Exercise 66 match the industry's projection for that year? (Source: U.S. Centers for Medicare and Medicaid Services)

68. *Make a Decision*: **Projected Science Spending** The projected amount S (in billions of dollars) of federal spending on science, space, and technology from 2002 to 2007 can be approximated by the model

$$S = 0.011t^2 + 0.47t + 20.8, \quad 2 \le t \le 7$$

where t represents the year, with $t = 2$ corresponding to 2002. According to this model, in what year is the science, space, and technology spending projected to reach $24,000,000,000? (Source: U.S. Office of Management and Budget)

69. Flying Speed Two planes leave simultaneously from the same airport, one flying due east and the other due south (see figure). The eastbound plane is flying 50 miles per hour faster than the southbound plane. After 3 hours the planes are 2440 miles apart. Find the speed of each plane.

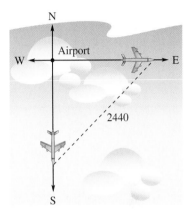

70. Flying Speed Two planes leave simultaneously from the same airport, one flying due east and the other due south. The eastbound plane is flying 100 miles per hour faster than the southbound plane. After 2 hours the planes are 1500 miles apart. Find the speed of each plane.

71. Total Revenue The demand equation for a product is

$$p = 60 - 0.0004x$$

where p is the price per unit and x is the number of units sold. The total revenue R for selling x units is given by $R = xp$. How many units must be sold to produce a revenue of $220,000?

72. Total Revenue The demand equation for a product is

$$p = 50 - 0.0005x$$

where p is the price per unit and x is the number of units sold. The total revenue R for selling x units is given by $R = xp$. How many units must be sold to produce a revenue of $250,000?

73. When the Quadratic Formula is used to solve certain problems, such as the problem in Example 5 on page R108, why is only one solution used?

Mid-Chapter Quiz

Take this quiz as you would take a quiz in class. After you are done, check your work against the answers given in the back of the book.

In Exercises 1–4, solve the equation and check your solution.

1. $3(x - 2) - 4(2x + 5) = 4$

2. $\dfrac{3x + 3}{5x - 2} = \dfrac{3}{4}$

3. $\dfrac{2}{x(x - 1)} + \dfrac{1}{x} = \dfrac{1}{x - 4}$

4. $(x + 3)^2 - x^2 = 6(x + 2)$

In Exercises 5 and 6, solve the equation. (Round your solution to three decimal places.)

5. $\dfrac{x}{2.004} - \dfrac{x}{5.128} = 100$

6. $0.378x + 0.757(500 - x) = 215$

7. Describe how you can check your answers to Exercises 1–4 using your graphing utility.

In Exercises 8 and 9, write an algebraic equation for the verbal description. Find the solution if possible and check.

8. A company has fixed costs of $20,000 per month and variable costs of $7.50 per unit manufactured. The company has $80,000 to cover monthly costs. How many units can the company manufacture?

9. The demand equation for a product is $p = 65 - 0.0002x$, where p is the price per unit and x is the number of units sold. The total revenue R for selling x units is given by $R = xp$. How many units must be sold to produce revenue of $225,000?

In Exercises 10–15, solve the equation by the indicated method.

10. *Factoring:* $3x^2 + 13x = 10$

11. *Extracting roots:* $3x^2 = 15$

12. *Extracting roots:* $(x + 3)^2 = 17$

13. *Quadratic Formula:* $2x + x^2 = 5$

14. *Quadratic Formula:* $3x^2 + 7x - 2 = 0$

15. *Quadratic Formula:* $3x^2 - 4.50x - 0.32 = 0$

In Exercises 16 and 17, use the discriminant to determine the number of real solutions of the quadratic equation.

16. $2x^2 - 4x + 9 = 0$

17. $4x^2 - 12x + 9 = 0$

18. Describe how you would convince a fellow student that $(x + 3)^2 = x^2 + 6x + 9$.

19. A rock is dropped from a height of 250 feet. With no initial velocity, how long will it take the rock to hit the ground?

20. An open box 5 inches deep and 180 cubic inches in volume is to be constructed. Find the dimensions of the square base.

R2.5   Other Types of Equations

Objectives

- **Solve a polynomial equation by factoring.**
- **Rewrite and solve an equation involving radicals or rational exponents.**
- **Rewrite and solve an equation involving fractions or absolute values.**
- **Construct and use a nonquadratic model to solve an application problem.**

Polynomial Equations

In this section you will extend the techniques for solving equations to nonlinear and nonquadratic equations. At this point in the text, you have only three basic methods for solving nonlinear equations—*factoring, extracting roots,* and the *Quadratic Formula.* So the main goal of this section is to learn to *rewrite* nonlinear equations in a form to which you can apply one of these methods.

EXAMPLE 1   SOLVING A POLYNOMIAL EQUATION BY FACTORING

Solve $3x^4 = 48x^2$.

SOLUTION

The basic approach is first to write the polynomial equation in general form with zero on one side, then to factor the other side, and finally to set each factor equal to zero and solve.

$$3x^4 = 48x^2 \qquad \text{Write original equation.}$$
$$3x^4 - 48x^2 = 0 \qquad \text{Write in general form.}$$
$$3x^2(x^2 - 16) = 0 \qquad \text{Factor out common factor.}$$
$$3x^2(x + 4)(x - 4) = 0 \qquad \text{Difference of two squares}$$
$$3x^2 = 0 \quad \Longrightarrow \quad x = 0 \qquad \text{Set 1st factor equal to 0.}$$
$$x + 4 = 0 \quad \Longrightarrow \quad x = -4 \qquad \text{Set 2nd factor equal to 0.}$$
$$x - 4 = 0 \quad \Longrightarrow \quad x = 4 \qquad \text{Set 3rd factor equal to 0.}$$

> **STUDY TIP**
>
> A common mistake that is made in solving an equation such as that in Example 1 is dividing each side of the equation by the variable factor x^2. This loses the solution $x = 0$. When using factoring to solve an equation, be sure to set each variable factor equal to zero. Don't divide each side of an equation by a variable factor in an attempt to simplify the equation.

You can check these solutions by substituting in the original equation, as follows.

CHECK

$$3x^4 = 48x^2 \qquad \text{Write original equation.}$$
$$3(0)^4 = 48(0)^2 \qquad \text{0 checks. ✓}$$
$$3(-4)^4 = 48(-4)^2 \qquad -4 \text{ checks. ✓}$$
$$3(4)^4 = 48(4)^2 \qquad 4 \text{ checks. ✓}$$

After checking, you can conclude that the solutions are $x = 0$, $x = -4$, and $x = 4$.

✓CHECKPOINT   Now try Exercise 3.

| EXAMPLE 2 | SOLVING A POLYNOMIAL EQUATION BY FACTORING |

Solve $x^3 - 3x^2 - 3x + 9 = 0$.

SOLUTION

$$x^3 - 3x^2 - 3x + 9 = 0 \qquad \text{Write original equation.}$$
$$x^2(x - 3) - 3(x - 3) = 0 \qquad \text{Group terms.}$$
$$(x - 3)(x^2 - 3) = 0 \qquad \text{Factor by grouping.}$$
$$x - 3 = 0 \implies x = 3 \qquad \text{Set 1st factor equal to 0.}$$
$$x^2 - 3 = 0 \implies x = \pm\sqrt{3} \qquad \text{Set 2nd factor equal to 0.}$$

The solutions are $x = 3$, $x = \sqrt{3}$, and $x = -\sqrt{3}$. Check these in the original equation. Notice that this polynomial has a degree of 3 and has three solutions.

✓*CHECKPOINT*   Now try Exercise 9.

DISCOVERY

What do you observe about the degrees of the polynomials in Examples 1, 2, and 3 and the possible numbers of solutions of the equations? Does your observation apply to the quadratic equations in Sections R2.3 and R2.4?

Occasionally, mathematical models involve equations that are of **quadratic type.** In general, an equation is of quadratic type if it can be written in the form

$$au^2 + bu + c = 0$$

where $a \neq 0$ and u is an algebraic expression.

| EXAMPLE 3 | SOLVING AN EQUATION OF QUADRATIC TYPE |

Solve $x^4 - 3x^2 + 2 = 0$.

SOLUTION

This equation is of quadratic type with $u = x^2$.

$$(x^2)^2 - 3(x^2) + 2 = 0$$

To solve this equation, you can factor the left side of the equation as the product of two second-degree polynomials.

$$x^4 - 3x^2 + 2 = 0 \qquad \text{Write original equation.}$$
$$(x^2 - 1)(x^2 - 2) = 0 \qquad \text{Partially factor.}$$
$$(x + 1)(x - 1)(x^2 - 2) = 0 \qquad \text{Completely factor.}$$
$$x + 1 = 0 \implies x = -1 \qquad \text{Set 1st factor equal to 0.}$$
$$x - 1 = 0 \implies x = 1 \qquad \text{Set 2nd factor equal to 0.}$$
$$x^2 - 2 = 0 \implies x = \pm\sqrt{2} \qquad \text{Set 3rd factor equal to 0.}$$

The solutions are $x = -1$, $x = 1$, $x = \sqrt{2}$, and $x = -\sqrt{2}$. Check these in the original equation. Notice that this polynomial has a degree of 4 and has four solutions.

✓*CHECKPOINT*   Now try Exercise 13.

Solving Equations Involving Radicals

The steps involved in solving the remaining equations in this section will often introduce *extraneous solutions,* as discussed in Section R2.1. Operations such as squaring each side of an equation, raising each side of an equation to a rational power, or multiplying each side of an equation by a variable quantity all create this potential danger. So, when you use any of these operations, a check is crucial.

EXAMPLE 4   AN EQUATION INVOLVING A RADICAL

Solve $\sqrt{2x + 7} - x = 2$.

SOLUTION

$\sqrt{2x + 7} - x = 2$	Write original equation.
$\sqrt{2x + 7} = x + 2$	Isolate the square root.
$2x + 7 = x^2 + 4x + 4$	Square each side.
$0 = x^2 + 2x - 3$	Write in general form.
$0 = (x + 3)(x - 1)$	Factor.
$x + 3 = 0 \implies x = -3$	Set 1st factor equal to 0.
$x - 1 = 0 \implies x = 1$	Set 2nd factor equal to 0.

By checking these values, you can determine that the only solution is $x = 1$.

✓CHECKPOINT   Now try Exercise 21.

EXAMPLE 5   AN EQUATION INVOLVING A RATIONAL EXPONENT

Solve $4x^{3/2} - 8 = 0$.

SOLUTION

$4x^{3/2} - 8 = 0$	Write original equation.
$4x^{3/2} = 8$	Add 8 to each side.
$x^{3/2} = 2$	Isolate $x^{3/2}$.
$x = 2^{2/3}$	Raise each side to the $\frac{2}{3}$ power.
$x \approx 1.587$	Round to three decimal places.

CHECK

$4x^{3/2} - 8 = 0$	Write original equation.
$4(2^{2/3})^{3/2} \overset{?}{=} 8$	Substitute $2^{2/3}$ for x.
$4(2) \overset{?}{=} 8$	Power of a Power Property
$8 = 8$	Solution checks. ✓

✓CHECKPOINT   Now try Exercise 35.

STUDY TIP

The basic technique used in Example 5 is to isolate the factor with the rational exponent and raise each side to the *reciprocal power.* In Example 4, this is equivalent to isolating the square root and *squaring* each side.

Equations Involving Fractions or Absolute Values

In Section R2.1, you learned how to solve equations involving fractions. Recall that the first step is to multiply each side of the equation by the least common denominator (LCD).

EXAMPLE 6   AN EQUATION INVOLVING FRACTIONS

Solve $\dfrac{2}{x} = \dfrac{3}{x-2} - 1$.

SOLUTION

For this equation, the LCD of the three terms is $x(x - 2)$, so we begin by multiplying each side of the equation by this expression.

$$\frac{2}{x} = \frac{3}{x-2} - 1 \qquad\qquad \text{Write original equation.}$$

$$x(x-2)\frac{2}{x} = x(x-2)\frac{3}{x-2} - x(x-2)(1) \qquad \text{Multiply each side by LCD.}$$

$$2(x-2) = 3x - x(x-2), \quad x \neq 0, 2 \qquad \text{Simplify.}$$

$$2x - 4 = -x^2 + 5x \qquad\qquad \text{Distributive Property}$$

$$x^2 - 3x - 4 = 0 \qquad\qquad \text{Write in general form.}$$

$$(x-4)(x+1) = 0 \implies \qquad \text{Factor.}$$

$$x - 4 = 0 \implies \quad x = 4 \qquad \text{Set 1st factor equal to 0.}$$

$$x + 1 = 0 \qquad\quad x = -1 \qquad \text{Set 2nd factor equal to 0.}$$

Notice that both $x = 4$ and $x = -1$ are possible solutions because the only values excluded from the domain of the fractions are $x = 0$ and $x = 2$. The values are omitted from the domain because they result in division by zero.

CHECK

$$\frac{2}{x} = \frac{3}{x-2} - 1 \qquad\qquad \text{Write original equation.}$$

$$\frac{2}{4} \stackrel{?}{=} \frac{3}{4-2} - 1 \qquad\qquad \text{Substitute 4 for } x.$$

$$\frac{1}{2} = \frac{3}{2} - 1 \qquad\qquad \text{4 checks. } \checkmark$$

$$\frac{2}{-1} \stackrel{?}{=} \frac{3}{-1-2} - 1 \qquad\qquad \text{Substitute } -1 \text{ for } x.$$

$$-2 = -1 - 1 \qquad\qquad -1 \text{ checks. } \checkmark$$

The solutions are $x = 4$ and $x = -1$.

✓CHECKPOINT   Now try Exercise 41.

To solve an equation involving an absolute value, remember that the expression inside the absolute value signs can be positive or negative. This results in *two* separate equations, each of which must be solved. For instance, the equation

$$|x - 2| = 3$$

results in the two equations

$$x - 2 = 3 \quad \text{and} \quad -(x - 2) = 3$$

which implies that the original equation has two solutions: $x = 5$ and $x = -1$. When setting up the negative expression, it is important to remember to place parentheses around the entire expression from inside the absolute value bars. After you set up the two equations, solve each one independently.

EXAMPLE 7 AN EQUATION INVOLVING ABSOLUTE VALUE

Solve $|x^2 - 3x| = -4x + 6$.

SOLUTION

Because the variable expression inside the absolute value signs can be positive or negative, you must solve the following two equations.

First Equation

$x^2 - 3x = -4x + 6$	Use positive expression.
$x^2 + x - 6 = 0$	Write in general form.
$(x + 3)(x - 2) = 0$	Factor.
$x + 3 = 0 \implies x = -3$	Set 1st factor equal to 0.
$x - 2 = 0 \implies x = 2$	Set 2nd factor equal to 0.

Second Equation

$-(x^2 - 3x) = -4x + 6$	Use negative expression.
$x^2 - 7x + 6 = 0$	Write in general form.
$(x - 1)(x - 6) = 0$	Factor.
$x - 1 = 0 \implies x = 1$	Set 1st factor equal to 0.
$x - 6 = 0 \implies x = 6$	Set 2nd factor equal to 0.

CHECK

$|(-3)^2 - 3(-3)| = -4(-3) + 6$	-3 checks. ✓
$|(2)^2 - 3(2)| \neq -4(2) + 6$	2 does not check.
$|(1)^2 - 3(1)| = -4(1) + 6$	1 checks. ✓
$|(6)^2 - 3(6)| \neq -4(6) + 6$	6 does not check.

The solutions are $x = -3$ and $x = 1$.

✓**CHECKPOINT**   Now try Exercise 55.

Applications

It would be impossible to categorize the many different types of applications that involve nonlinear and nonquadratic models. However, from the few examples and exercises given below, we hope you will gain some appreciation for the variety of applications that can occur.

EXAMPLE 8   REDUCED RATES

Make a Decision   A ski club charters a bus for a ski trip at a cost of $700. In an attempt to lower the bus fare per skier, the club invites nonmembers to go along. When five nonmembers join the trip, the fare per skier decreases by $7. How many club members are going on the trip?

SOLUTION

Begin the solution by creating a verbal model and assigning labels, as follows.

Verbal Model: $\boxed{\text{Cost per skier}} \cdot \boxed{\text{Number of skiers}} = \boxed{\text{Cost of trip}}$

Labels:

Cost of trip $= 700$	(dollars)
Number of ski club members $= x$	(people)
Number of skiers $= x + 5$	(people)
Original cost per member $= \dfrac{700}{x}$	(dollars per person)
Cost per skier $= \dfrac{700}{x} - 7$	(dollars per person)

Equation:

$$\left(\frac{700}{x} - 7\right)(x + 5) = 700 \qquad \text{Original equation}$$

$$\left(\frac{700 - 7x}{x}\right)(x + 5) = 700 \qquad \text{Rewrite first factor.}$$

$$(700 - 7x)(x + 5) = 700x, \ x \neq 0 \quad \text{Multiply each side by } x.$$

$$700x + 3500 - 7x^2 - 35x = 700x \qquad \text{Multiply factors.}$$

$$-7x^2 - 35x + 3500 = 0 \qquad \text{Subtract } 700x \text{ from each side.}$$

$$x^2 + 5x - 500 = 0 \qquad \text{Divide each side by } -7.$$

$$(x + 25)(x - 20) = 0 \qquad \text{Factor left side of equation.}$$

$$x + 25 = 0 \implies x = -25 \qquad \text{Set 1st factor equal to 0.}$$

$$x - 20 = 0 \implies x = 20 \qquad \text{Set 2nd factor equal to 0.}$$

At this point, you must decide which value of x makes more sense in the context of the problem. Choosing the positive value of x, you can conclude that 20 ski club members are going on the trip. Check this in the original statement of the problem.

✓CHECKPOINT   Now try Exercise 63.

Interest earned on a savings account is calculated by one of three basic methods: simple interest, interest compounded n times per year, and interest compounded continuously. The next example uses the formula for interest that is compounded n times per year,

$$A = P\left(1 + \frac{r}{n}\right)^{nt}.$$

In this formula, A is the balance in the account, P is the principal (or original deposit), r is the annual percentage rate (in decimal form), n is the number of compoundings per year, and t is the time in years. Later, in Chapter 3, you will study the derivation of this formula for compound interest.

EXAMPLE 9   COMPOUND INTEREST

When you were born, your grandparents deposited $5000 in a long-term investment on which the interest was compounded quarterly. On your 25th birthday the balance in the account is $25,062.59. What is the annual interest rate for this investment?

SOLUTION

Formula: $A = P\left(1 + \dfrac{r}{n}\right)^{nt}$

Labels: Amount $= A = 25{,}062.59$ (dollars)
Principal $= P = 5000$ (dollars)
Time $= t = 25$ (years)
Compoundings per year $= n = 4$ (compoundings)
Annual interest rate $= r$ (percent in decimal form)

Equation: $25{,}062.59 = 5000\left(1 + \dfrac{r}{4}\right)^{4(25)}$ Substitute.

$\dfrac{25{,}062.59}{5000} = \left(1 + \dfrac{r}{4}\right)^{100}$ Divide each side by 5000.

$5.0125 \approx \left(1 + \dfrac{r}{4}\right)^{100}$ Use a calculator.

$(5.0125)^{1/100} \approx 1 + \dfrac{r}{4}$ Raise each side to reciprocal power.

$1.01625 \approx 1 + \dfrac{r}{4}$ Use a calculator.

$0.01625 \approx \dfrac{r}{4}$ Subtract 1 from each side.

$0.065 \approx r$ Multiply each side by 4.

The annual interest rate is about $0.065 = 6.5\%$. Check this in the original statement of the problem.

✔CHECKPOINT   Now try Exercise 65.

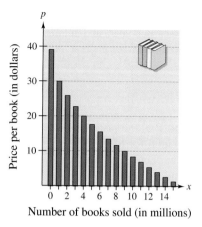

Figure R2.13

EXAMPLE 10   MARKET RESEARCH

The marketing department for a publisher is asked to determine the price of a book. The department determines that the demand for the book depends on the price of the book according to the formula

$$p = 40 - \sqrt{0.0001x + 1}, \quad 0 \le x \le 15{,}990{,}000$$

where p is the price per book in dollars and x is the number of books sold at the given price. For instance, in Figure R2.13, note that if the price were $39, then (according to the model) no one would be willing to buy the book. On the other hand, if the price were $17.60, 5 million copies could be sold. If the publisher set the price at $12.95, how many copies would be sold?

SOLUTION

$$p = 40 - \sqrt{0.0001x + 1}$$ Write given model.

$$12.95 = 40 - \sqrt{0.0001x + 1}$$ Set price at $12.95.

$$\sqrt{0.0001x + 1} = 27.05$$ Isolate radical.

$$0.0001x + 1 = 731.7025$$ Square each side.

$$0.0001x = 730.7025$$ Subtract 1 from each side.

$$x = 7{,}307{,}025$$ Divide each side by 0.0001.

So, by setting the book's price at $12.95, the publisher can expect to sell about 7.3 million copies.

✓**CHECKPOINT**   Now try Exercise 73.

Discussing the Concept | Compound Interest

Suppose that on the day you were born a family member deposited money in an investment that earned interest compounded monthly at the annual rate of 7%. How much money needed to be deposited for the account to be worth $100,000 on your next birthday? How much money needed to be deposited for the account to be worth $1,000,000 on your next birthday?

R2.5 (Warm Up)

The following warm-up exercises involve skills that were covered in earlier sections. You will use these skills in the exercise set for this section. For additional help, review Sections R2.3 and R2.4.

In Exercises 1–10, find the real solutions of the equation.

1. $x^2 - 22x + 121 = 0$

2. $x(x - 20) + 3(x - 20) = 0$

3. $(x + 20)^2 = 625$

4. $5x^2 + x = 0$

5. $3x^2 + 4x - 4 = 0$

6. $12x^2 + 8x - 55 = 0$

7. $x^2 + 4x - 5 = 0$

8. $4x^2 + 4x - 15 = 0$

9. $x^2 - 3x + 1 = 0$

10. $x^2 - 4x + 2 = 0$

R2.5 (Exercises)

In Exercises 1–56, find the real solutions of the equation. Check your solutions.

1. $x^3 - 2x^2 - 3x = 0$ **2.** $20x^3 - 125x = 0$

3. $4x^4 - 18x^2 = 0$

4. $2x^4 - 15x^3 + 18x^2 = 0$

5. $x^4 - 81 = 0$ **6.** $x^6 - 64 = 0$

7. $5x^3 + 30x^2 + 45x = 0$

8. $9x^4 - 24x^3 + 16x^2 = 0$

9. $x^3 - 7x^2 - 4x + 28 = 0$

10. $x^3 + 2x^2 + 3x + 6 = 0$

11. $x^4 - x^3 + x - 1 = 0$

12. $x^4 + 2x^3 - 8x - 16 = 0$

13. $x^4 - 12x^2 + 11 = 0$ **14.** $x^4 - 29x^2 + 100 = 0$

15. $x^4 + 5x^2 - 36 = 0$ **16.** $x^4 - 4x^2 + 3 = 0$

17. $4x^4 - 65x^2 + 16 = 0$ **18.** $36t^4 + 29t^2 - 7 = 0$

19. $x^6 + 7x^3 - 8 = 0$ **20.** $x^6 + 3x^3 + 2 = 0$

21. $\sqrt{2x} - 10 = 0$ **22.** $4\sqrt{x} - 3 = 0$

23. $\sqrt{x - 10} - 4 = 0$ **24.** $\sqrt{5 - x} - 3 = 0$

25. $\sqrt[3]{2x + 5} + 3 = 0$ **26.** $\sqrt[3]{3x + 1} - 5 = 0$

27. $2x + 9\sqrt{x} - 5 = 0$ **28.** $6x - 7\sqrt{x} - 3 = 0$

29. $x = \sqrt{11x - 30}$ **30.** $2x - \sqrt{15 - 4x} = 0$

31. $-\sqrt{26 - 11x} + 4 = x$

32. $x + \sqrt{31 - 9x} = 5$

33. $\sqrt{x + 1} - 3x = 1$ **34.** $\sqrt{2x + 1} + x = 7$

35. $(x - 5)^{2/3} = 16$ **36.** $(x + 3)^{4/3} = 16$

37. $(x + 3)^{3/2} = 8$ **38.** $(x^2 + 2)^{2/3} = 9$

39. $(x^2 - 5)^{2/3} = 16$ **40.** $(x^2 - x - 22)^{4/3} = 16$

41. $\dfrac{1}{x} - \dfrac{1}{x + 1} = 3$ **42.** $\dfrac{x}{x^2 - 4} + \dfrac{1}{x + 2} = 3$

43. $\dfrac{20 - x}{x} = x$ **44.** $\dfrac{4}{x} - \dfrac{5}{3} = \dfrac{x}{6}$

45. $\dfrac{1}{x} = \dfrac{4}{x - 1} + 1$ **46.** $x + \dfrac{9}{x + 1} = 5$

47. $\dfrac{4}{x + 1} - \dfrac{3}{x + 2} = 1$ **48.** $\dfrac{x + 1}{3} - \dfrac{x + 1}{x + 2} = 0$

49. $|x + 1| = 2$ **50.** $|x - 2| = 3$

51. $|2x - 1| = 5$ **52.** $|3x + 2| = 7$

53. $|x| = x^2 + x - 3$ **54.** $|x^2 + 6x| = 3x + 18$

55. $|x - 10| = x^2 - 10x$ **56.** $|x + 1| = x^2 - 5$

57. Error Analysis   Find the error(s) in the solution.

$$\sqrt{3x} = \sqrt{7x + 4}$$

$$3x^2 = 7x + 4$$

$$x = \frac{-7 \pm \sqrt{7^2 - 4(3)(4)}}{2(3)}$$

$$x = -1 \text{ and } x = -\frac{4}{3}$$

58. Error Analysis   Find the error(s) in the solution.

$$\sqrt{6 - 2x} - 3 = 0$$

$$6 - 2x + 9 = 0$$

$$-2x = -15$$

$$x = \frac{15}{2}$$

In Exercises 59–62, use a calculator to find the real solutions of the equation. (Round your answers to three decimal places.)

59. $3.2x^4 - 1.5x^2 - 2.1 = 0$

60. $7.08x^6 + 4.15x^3 - 9.6 = 0$

61. $1.8x - 6\sqrt{x} - 5.6 = 0$

62. $4x + 8\sqrt{x} + 3.6 = 0$

63. *Make a Decision*: **Sharing the Cost** A college charters a bus for $1700 to take a group of students to the Fiesta Bowl. When six more students join the trip, the cost per student drops by $7.50. How many students were in the original group?

64. *Make a Decision*: **Sharing the Cost** Three students plan to rent an apartment and share equally in the rent. By adding a fourth person, each person could save $100 a month. How much is the monthly rent?

65. Compound Interest A deposit of $2500 reaches a balance of $3544.06 after 5 years. The interest on the account is compounded monthly. What is the annual interest rate for this investment?

66. Compound Interest A sales representative describes a "guaranteed investment fund" that is offered to new investors. You are told that if you deposit $10,000 in the fund you will be guaranteed a return of at least $25,000 after 20 years. (a) If after 20 years you received the minimum guarantee, what annual interest rate did you receive? (b) If after 20 years you received $35,000, what annual interest rate did you receive? (Assume that the interest in the fund is compounded quarterly.)

67. Borrowing Money You borrow $100 from a friend and agree to pay the money back, plus $10 in interest, after 6 months. Assuming that the interest is compounded monthly, what annual interest rate are you paying?

68. Cash Advance You take out a cash advance of $500 on a credit card. After 2 months, you owe $515.75. The interest is compounded monthly. What is the annual interest rate for this cash advance?

69. Airline Passengers An airline offers daily flights between Chicago and Denver. The total monthly cost C (in millions of dollars) of these flights is modeled by $C = \sqrt{0.2x + 1}$, where x is the number of passengers flying that month in thousands (see figure). The total cost of the flights for a month is 2.5 million dollars. How many passengers flew that month?

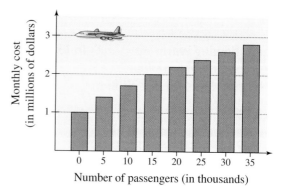

Figure for 69

70. New Truck Sales The number S of new truck sales (in millions) in the United States from 1992 to 2001 can be approximated by the model

$$S = \sqrt{0.15t^2 + 4.71t + 15.53}, \quad 2 \le t \le 11$$

where t represents the year, with $t = 2$ corresponding to 1992. (Source: U.S. Bureau of Economic Analysis)

(a) Determine the number of new truck sales in 1993, 1995, 2000, and 2001.

(b) According to this model, in what year will new truck sales first exceed 20,000,000?

71. *Make a Decision*: **Life Expectancy** The life expectancy table (for ages 48–65) used by the U.S. National Center for Health Statistics is modeled by

$$y = \sqrt{0.93x^2 - 144.86x + 5827.81}$$

where x represents a person's current age and y represents the average number of additional years the person is expected to live. If a person's life expectancy is estimated to be 20 years, how old is the person? Explain why this model is not used for people over the age of 65.

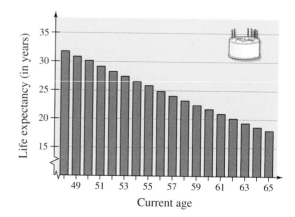

72. Fuel Consumption The amount of fuel F (in millions of gallons) consumed in the United States by motor vehicles from 1995 to 2001 can be modeled by

$$F = 144{,}673 + 2346t - \frac{65{,}758}{t}, \quad 5 \le t \le 11$$

where $t = 5$ represents 1995 (see figure). Predict the year in which the total amount of fuel consumed will reach 176,000,000,000 gallons. (Source: U.S. Department of Transportation)

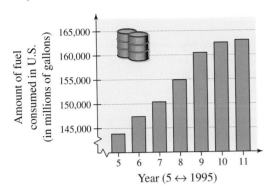

Year ($5 \leftrightarrow 1995$)

73. *Make a Decision*: **Market Research** The demand equation for a product is modeled by

$$p = 30 - \sqrt{0.0001x + 1}$$

where x is the number of units demanded per day and p is the price per unit. Find the demand when the price is set at \$13.95. Explain why this model is only valid for $0 \le x \le 8{,}990{,}000$.

74. Power Line A power station is on one side of a river that is $\frac{1}{2}$ mile wide. A factory is 6 miles downstream on the other side of the river. It costs \$18 per foot to run power lines over land and \$24 per foot to run them under water. The project's cost is \$616,877.27. Find the length x as labeled in the figure.

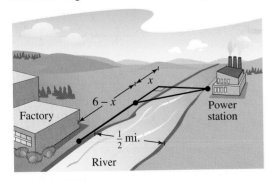

75. Sailboat Stays Two stays for the mast on a sailboat are attached to the boat at two points, as shown in the figure. One point is 10 feet from the mast and the other point is 15 feet from the mast. The total length of the two stays is 35 feet. How high on the mast are the stays attached?

76. Batting Average A softball player has been at bat 120 times and has 20 hits. The player's batting average is $20/120 \approx .167$. How many consecutive hits must the player get to obtain a batting average of .200? [*Hint:* (Batting average) = (total hits) ÷ (total times at bat)]

77. Work Rate With only the cold water valve open, it takes 8 minutes to fill the tub of a washing machine. With both the hot and cold water valves open, it takes 5 minutes. The time it takes for the tub to fill with only the hot water valve open can be modeled by the equation

$$\frac{1}{8} + \frac{1}{t} = \frac{1}{5}$$

where t is the time (in minutes) for the tub to fill. How long does it take for the tub of the washing machine to fill with only the hot water valve open?

78. Community Service You and a friend volunteer to paint a small house as a community service project. Working alone, you can paint the house in 10 hours. Your friend can paint the house in 12 hours working alone. How long will it take both of you, working together, to paint the house?

79. Community Service You and a friend volunteer to paint a large house as a community service project. Working alone, you can paint the house in 25 hours. Your friend can paint the house in 20 hours working alone. How long will it take both of you, working together, to paint the house?

R2.6   Linear Inequalities

Objectives

- Write bounded and unbounded intervals using inequalities or interval notation.
- Solve and graph a linear inequality.
- Construct and use a linear inequality to solve an application problem.

Introduction

Simple inequalities are used to *order* real numbers. The inequality symbols $<, \leq, >,$ and $\geq$ are used to compare two numbers and to denote subsets of real numbers. For instance, the simple inequality

$$x \geq 3$$

denotes all real numbers x that are greater than or equal to 3.

In this section you will expand your work with inequalities to include more involved statements such as

$$5x - 7 > 3x + 9 \quad \text{and} \quad -3 \leq 6x - 1 < 3.$$

As with an equation, you **solve an inequality** in the variable x by finding all values of x for which the inequality is true. Such values are **solutions** and are said to **satisfy** the inequality. The set of all real numbers that are solutions of an inequality is the **solution set** of the inequality.

The set of all points on the real number line that represent the solution set of an inequality is the **graph** of the inequality. Graphs of many types of inequalities consist of intervals on the real number line. The four different types of **bounded** intervals are summarized below.

Bounded Intervals on the Real Number Line

Let a and b be real numbers such that $a < b$. The following intervals on the real number line are **bounded.** The numbers a and b are the **endpoints** of each interval.

Notation	Interval Type	Inequality	Graph
$[a, b]$	Closed	$a \leq x \leq b$	
(a, b)	Open	$a < x < b$	
$[a, b)$		$a \leq x < b$	
$(a, b]$		$a < x \leq b$	

Note that a closed interval contains both of its endpoints and an open interval does not contain either of its endpoints. Often, the solution of an inequality is an interval on the real line that is **unbounded.** For instance, the interval consisting of all positive numbers is unbounded. The symbols ∞, **positive infinity,** and $-\infty$, **negative infinity,** do not represent real numbers. They are simply convenient symbols used to describe the unboundedness of an interval such as $(1, \infty)$.

Unbounded Intervals on the Real Number Line

Let a and b be real numbers. The following intervals on the real number line are **unbounded.**

Notation	Interval Type	Inequality	Graph
$[a, \infty)$		$x \geq a$	
(a, ∞)	Open	$x > a$	
$(-\infty, b]$		$x \leq b$	
$(-\infty, b)$	Open	$x < b$	
$(-\infty, \infty)$	Entire real line	$-\infty < x < \infty$	

EXAMPLE 1   INTERVALS AND INEQUALITIES

Write an inequality to represent each of the following intervals and state whether the interval is bounded or unbounded.

a. $(-3, 5]$ **b.** $(-3, \infty)$

c. $[0, 2]$ **d.** $(-\infty, 0)$

SOLUTION

a. $(-3, 5]$ corresponds to $-3 < x \leq 5$.   Bounded

b. $(-3, \infty)$ corresponds to $x > -3$.   Unbounded

c. $[0, 2]$ corresponds to $0 \leq x \leq 2$.   Bounded

d. $(-\infty, 0)$ corresponds to $x < 0$.   Unbounded

✓CHECKPOINT   Now try Exercise 1.

Properties of Inequalities

The procedures for solving linear inequalities in one variable are much like those for solving linear equations. To isolate the variable, you can make use of the **properties of inequalities.** These properties are similar to the properties of equality, but there are two important exceptions. When each side of an inequality is multiplied or divided by a negative number, the direction of the inequality symbol must be reversed. Here is an example.

$$-2 < 5 \qquad \text{Original inequality}$$

$$(-3)(-2) > (-3)(5) \qquad \begin{array}{l}\text{Multiply each side by } -3 \text{ and} \\ \text{reverse the inequality symbol.}\end{array}$$

$$6 > -15 \qquad \text{Simplify.}$$

Two inequalities that have the same solution set are **equivalent.** For instance, the inequalities

$$x + 2 < 5 \quad \text{and} \quad x < 3$$

are equivalent. To obtain the second inequality from the first, you can subtract 2 from each side of the inequality. The following list describes operations that can be used to create equivalent inequalities.

Properties of Inequalities

Let a, b, c, and d be real numbers.

1. *Transitive Property*

$$a < b \text{ and } b < c \quad \Longrightarrow \quad a < c$$

2. *Addition of Inequalities*

$$a < b \text{ and } c < d \quad \Longrightarrow \quad a + c < b + d$$

3. *Addition of a Constant*

$$a < b \quad \Longrightarrow \quad a + c < b + c$$

4. *Multiplication by a Constant*

$$\text{For } c > 0, a < b \quad \Longrightarrow \quad ac < bc.$$

$$\text{For } c < 0, a < b \quad \Longrightarrow \quad ac > bc. \qquad \begin{array}{l}\text{Reverse direction} \\ \text{of inequality.}\end{array}$$

Each of the properties above is true if the symbol $<$ is replaced by $\leq$ and the symbol $>$ is replaced by $\geq$. For instance, another form of the multiplication property would be as follows.

$$\text{For } c > 0, a \leq b \quad \Longrightarrow \quad ac \leq bc.$$

$$\text{For } c < 0, a \leq b \quad \Longrightarrow \quad ac \geq bc.$$

Solving a Linear Inequality

The simplest type of inequality to solve is a **linear inequality** in a single variable. For instance, $2x + 3 > 4$ is a linear inequality in x.

As you read through the following examples, pay special attention to the steps in which the inequality symbol is reversed. Remember that when you multiply or divide by a negative number, you must reverse the inequality symbol.

EXAMPLE 2   SOLVING A LINEAR INEQUALITY

Solve $5x - 7 > 3x + 9$.

SOLUTION

$5x - 7 > 3x + 9$	Write original inequality.
$2x - 7 > 9$	Subtract $3x$ from each side.
$2x > 16$	Add 7 to each side.
$x > 8$	Divide each side by 2.

The solution set is all real numbers that are greater than 8, which is denoted by $(8, \infty)$. The graph is shown in Figure R2.14.

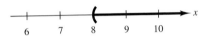

Figure R2.14   Solution Interval: $(8, \infty)$

✓CHECKPOINT   Now try Exercise 27.

Checking the solution set of an inequality is not as simple as checking the solutions of an equation. You can, however, get an indication of the validity of a solution set by substituting a few convenient values of x to see whether the original inequality is satisfied.

EXAMPLE 3   SOLVING A LINEAR INEQUALITY

Solve $1 - \dfrac{3x}{2} \geq x - 4$.

SOLUTION

$1 - \dfrac{3x}{2} \geq x - 4$	Write original inequality.
$2 - 3x \geq 2x - 8$	Multiply each side by 2.
$2 - 5x \geq -8$	Subtract $2x$ from each side.
$-5x \geq -10$	Subtract 2 from each side.
$x \leq 2$	Divide each side by -5 and reverse inequality.

The solution set is all real numbers that are less than or equal to 2, which is denoted by $(-\infty, 2]$. The graph is shown in Figure R2.15.

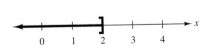

Figure R2.15   Solution Interval: $(-\infty, 2]$

✓CHECKPOINT   Now try Exercise 29.

Sometimes it is convenient to write two inequalities as a **double inequality.** For instance, you can write the two inequalities $-4 \leq 5x - 2$ and $5x - 2 < 7$ more simply as

$$-4 \leq 5x - 2 < 7$$

This enables you to solve the two inequalities together, as demonstrated in Example 4.

EXAMPLE 4   SOLVING A DOUBLE INEQUALITY

Solve $-3 \leq 6x - 1 < 3$.

SOLUTION

To solve a double inequality, you can isolate x as the middle term.

$-3 \leq 6x - 1 < 3$	Write original inequality.
$-3 + 1 \leq 6x - 1 + 1 < 3 + 1$	Add 1 to each part.
$-2 \leq 6x < 4$	Simplify.
$\dfrac{-2}{6} \leq \dfrac{6x}{6} < \dfrac{4}{6}$	Divide each part by 6.
$-\dfrac{1}{3} \leq x < \dfrac{2}{3}$	Simplify.

The solution set is all real numbers that are greater than or equal to $-\frac{1}{3}$ and less than $\frac{2}{3}$. The interval notation for this solution set is

$$\left[-\tfrac{1}{3}, \tfrac{2}{3}\right).$$ Solution set

The graph of this solution set is shown in Figure R2.16.

Figure R2.16   Solution Interval: $\left[-\tfrac{1}{3}, \tfrac{2}{3}\right)$.

✓CHECKPOINT   Now try Exercise 35.

The double inequality in Example 4 could have been solved in two parts as follows.

$$-3 \leq 6x - 1 \quad \text{and} \quad 6x - 1 < 3$$
$$-2 \leq 6x \qquad\qquad\quad 6x < 4$$
$$-\frac{1}{3} \leq x \qquad\qquad\quad x < \frac{2}{3}$$

The solution set consists of all real numbers that satisfy *both* inequalities. In other words, the solution set is the set of all values of x for which $-\frac{1}{3} \leq x < \frac{2}{3}$.

When combining two inequalities to form a double inequality, be sure that the inequalities satisfy the Transitive Property. For instance, it is *incorrect* to combine the inequalities $3 < x$ and $x \leq -1$ as $3 < x \leq -1$. This "inequality" is obviously wrong because 3 is not less than -1.

Inequalities Involving Absolute Value

STUDY TIP

When an absolute value inequality is solved, the solution may consist of a bounded interval on the real number line (Example 5) or two unbounded intervals on the real number line (Example 6).

Solving an Absolute Value Inequality

Let x be a variable or an algebraic expression and let a be a real number such that $a \geq 0$.

1. The solutions of $|x| < a$ are all values of x that lie between $-a$ and a.

   $|x| < a$ if and only if $-a < x < a$.

2. The solutions of $|x| > a$ are all values of x that are less than $-a$ or greater than a.

   $|x| > a$ if and only if $x < -a$ or $x > a$.

These rules are also valid if $<$ is replaced by $\leq$ and $>$ is replaced by $\geq$.

EXAMPLE 5 SOLVING AN ABSOLUTE VALUE INEQUALITY

Solve $|x - 5| < 2$.

SOLUTION

$$|x - 5| < 2 \qquad \text{Write original inequality.}$$
$$-2 < x - 5 < 2 \qquad \text{Equivalent inequality}$$
$$-2 + 5 < x - 5 + 5 < 2 + 5 \qquad \text{Add 5 to each part.}$$
$$3 < x < 7 \qquad \text{Simplify.}$$

The solution set consists of all real numbers that are greater than 3 and less than 7, which is denoted by $(3, 7)$. The graph is shown in Figure R2.17.

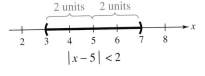

2 units
2 units
x
2
3
4
5
6
7
8
|x − 5| < 2

Figure R2.17

✓CHECKPOINT Now try Exercise 47.

EXAMPLE 6 SOLVING AN ABSOLUTE VALUE INEQUALITY

Solve $|x + 3| \geq 7$.

SOLUTION

$$|x + 3| \geq 7 \qquad \text{Write original inequality.}$$
$$x + 3 \leq -7 \quad \text{or} \quad x + 3 \geq 7 \qquad \text{Equivalent inequalities}$$
$$x + 3 - 3 \leq -7 - 3 \qquad x + 3 - 3 \geq 7 - 3 \qquad \text{Subtract 3 from each side.}$$
$$x \leq -10 \qquad x \geq 4 \qquad \text{Simplify.}$$

The solution set is all real numbers that are less than or equal to -10 *or* greater than or equal to 4, which is denoted by $(-\infty, -10] \cup [4, \infty)$ (see Figure R2.18). The symbol $\cup$ (union) means *or.*

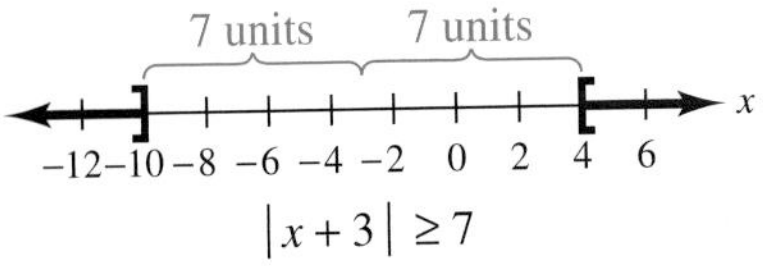

7 units
7 units
x
−12
−10
−8
−6
−4
−2
0
2
4
6
|x + 3| ≥ 7

Figure R2.18

✓CHECKPOINT Now try Exercise 51.

Applications

EXAMPLE 7   COMPARATIVE SHOPPING

Make a Decision   A compact car can be rented from Company A for $200 per week with no extra charge for mileage. A similar car can be rented from Company B for $110 per week, plus $.25 for each mile driven. How many miles must you drive in a week to make the rental fee for Company B more than that for Company A?

SOLUTION

Verbal Model:

Weekly cost for Company B	>	Weekly cost for Company A

Labels: Miles driven in one week = m (miles)
Weekly cost for Company A = 200   (dollars)
Weekly cost for Company B = $110 + 0.25m$ (dollars)

Inequality: $110 + 0.25m > 200$

$$0.25m > 90$$

$$m > 360$$

If you drive more than 360 miles in a week, the rental fee for Company B is more than the rental fee for Company A.

✓CHECKPOINT   Now try Exercise 69.

EXAMPLE 8   EXERCISE PROGRAM

A man begins an exercise and diet program that is designed to reduce his weight by at least 2 pounds per week. At the beginning of the diet the man weighs 225 pounds. Find the maximum number of weeks before the man's weight will reach his goal of 192 pounds.

SOLUTION

Verbal Model:

Desired weight	≤	Current weight	−	2 pounds per week	·	Number of weeks

Labels: Desired weight = 192   (pounds)
Current weight = 225   (pounds)
Number of weeks = x (weeks)

Inequality: $192 \leq 225 - 2x$

$$-33 \leq -2x$$

$$16.5 \geq x$$

Losing at least 2 pounds per week, it will take at most $16\frac{1}{2}$ weeks for the man to reach his goal.

✓CHECKPOINT   Now try Exercise 73.

Americans are increasingly concerned about their diet and exercise. In 2002, 55% of American adults participated in an exercise program. (Source: U.S. National Endowment for the Arts.)

Comstock Images/Alamy Images

EXAMPLE 9   ACCURACY OF A MEASUREMENT

Make a Decision   You go to a candy store to buy chocolates that cost $9.89 per pound. The scale used in the store has a state seal of approval that indicates the scale is accurate to within half an ounce. According to the scale, your purchase weighs one-half pound and costs $4.95. How much might you have been undercharged or overcharged due to an error in the scale?

SOLUTION

To solve this problem, let x represent the *true* weight of the candy. Because the state seal indicates that the scale is accurate to within half an ounce (or $\frac{1}{32}$ of a pound), you can conclude that the absolute value of the difference between the exact weight (x) and the scale weight $\left(\frac{1}{2}\right)$ is less than or equal to $\frac{1}{32}$ of a pound. That is,

$$\left| x - \frac{1}{2} \right| \le \frac{1}{32}.$$

You can solve this inequality as follows.

$$-\frac{1}{32} \le x - \frac{1}{2} \le \frac{1}{32}$$

$$\frac{15}{32} \le x \le \frac{17}{32}$$

$$0.46875 \le x \le 0.53125$$

In other words, your "one-half" pound of candy could have weighed as little as 0.46875 pound (which would have cost $0.46875 \cdot \$9.89 = \$4.64$) or as much as 0.53125 pound (which would have cost $0.53125 \cdot \$9.89 = \$5.25$). So, you could have been undercharged by as much as $0.30 or overcharged by as much as $0.31.

✓CHECKPOINT   Now try Exercise 85.

Discussing the Concept │ Absolute Value Inequalities

Describe the solutions of

$$|3x + 4| > -2.$$

Give an analytical argument. Describe the solutions of

$$|3x + 4| < -2.$$

Give an analytical argument. Explain the difference between the inequalities.

R2.6 **Warm Up**

The following warm-up exercises involve skills that were covered in earlier sections. You will use these skills in the exercise set for this section. For additional help, review Section R1.1.

In Exercises 1–4, determine which of the two numbers is larger.

1. $-\frac{1}{2}, -7$

2. $-\frac{1}{3}, -\frac{1}{6}$

3. $-\pi, -3$

4. $-6, -\frac{13}{2}$

In Exercises 5–8, use inequality notation to denote the statement.

5. x is nonnegative.

6. z is strictly between -3 and 10.

7. P is no more than 2.

8. W is at least 200.

In Exercises 9 and 10, evaluate the expression for the values of x.

9. $|x - 10|, x = 12, x = 3$

10. $|2x - 3|, x = \frac{3}{2}, x = 1$

R2.6 **Exercises**

In Exercises 1–6, write an inequality that represents the interval, and state whether the interval is bounded or unbounded.

1. $[-1, 5]$

2. $(2, 10]$

3. $(11, \infty)$

4. $[-5, \infty)$

5. $(-\infty, -2)$

6. $(-\infty, 7]$

In Exercises 7–14, match the inequality with its graph. [The graphs are labeled (a), (b), (c), (d), (e), (f), (g), and (h).]

(a)

(b)

(c)

(d)

(e)

(f)

(g)

(h)

7. $x < 4$

8. $x \geq 6$

9. $-2 < x \leq 5$

10. $0 \leq x \leq \frac{7}{2}$

11. $|x| < 4$

12. $|x| > 3$

13. $|x - 5| > 2$

14. $|x + 6| < 3$

In Exercises 15–20, determine whether each value of x is a solution of the inequality.

15. $5x - 12 > 0$

(a) $x = 3$ (b) $x = -3$ (c) $x = \frac{5}{2}$ (d) $x = \frac{3}{2}$

16. $x + 1 < \dfrac{2x}{3}$

(a) $x = 0$ (b) $x = 4$ (c) $x = -4$ (d) $x = -3$

17. $0 < \dfrac{x - 2}{4} < 2$

(a) $x = 4$ (b) $x = 10$ (c) $x = 0$ (d) $x = \frac{7}{2}$

18. $-1 < \dfrac{3 - x}{2} \leq 1$

(a) $x = 0$ (b) $x = -5$ (c) $x = 1$ (d) $x = 5$

19. $|x - 10| \geq 3$

(a) $x = 13$ (b) $x = -1$ (c) $x = 14$ (d) $x = 9$

20. $|2x - 3| < 15$

(a) $x = -6$ (b) $x = 0$ (c) $x = 12$ (d) $x = 7$

In Exercises 21–60, solve the inequality and graph the solution on the real number line.

21. $\dfrac{3}{2}x \geq 9$

22. $\dfrac{2}{5}x > 7$

23. $-10x < 40$

24. $-6x > 15$

25. $\frac{3}{5}x - 7 < 8$

26. $\frac{5}{4}x + 1 \leq 11$

27. $2x + 7 < 3 + 4x$

28. $6x - 4 \leq 2 + 8x$

29. $2x - 1 \geq 5x$

30. $3x + 1 \geq 2 + x$

31. $3(x + 2) + 7 < 2x - 5$

32. $2(x + 7) - 4 \geq 5(x - 3)$

33. $-3(x - 1) + 7 < 2x + 8$

34. $5 - 3x > -5(x + 4) + 6$

35. $3 \leq 2x - 1 < 7$

36. $3 > 1 - \frac{x}{2} > -3$

37. $1 < 2x + 3 < 9$

38. $-8 \leq 1 - 3(x - 2) < 13$

39. $-4 < \frac{2x - 3}{3} < 4$

40. $0 \leq \frac{x + 3}{2} < 5$

41. $\frac{3}{4} > x + 1 > \frac{1}{4}$

42. $-1 < -\frac{x}{3} < 1$

43. $|x| < 6$

44. $|x| > 8$

45. $\left|\frac{x}{2}\right| > 3$

46. $|5x| > 10$

47. $|x + 3| < 5$

48. $\left|\frac{2x + 1}{2}\right| < 6$

49. $|x - 20| \leq 4$

50. $|x - 7| < 6$

51. $|2x - 5| > 6$

52. $2|5 - 3x| + 7 < 21$

53. $\left|\frac{x - 3}{2}\right| \geq 5$

54. $\left|1 - \frac{2x}{3}\right| < 1$

55. $|9 - 2x| - 2 < -1$

56. $|x + 14| + 3 > 17$

57. $2|x + 10| \geq 9$

58. $3|4 - 5x| \leq 9$

59. $|x - 5| < 0$

60. $|x - 5| \geq 0$

In Exercises 61–68, use absolute value notation to define the solution set.

61.
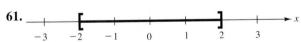

62.

63.

64.

65. All real numbers at most 10 units from 12

66. All real numbers at least 5 units from 8

67. All real numbers whose distances from -3 are more than 5

68. All real numbers whose distances from -6 are no more than 7

69. *Make a Decision*: **Comparative Shopping** You can rent a midsize car from Company A for $250 per week with no extra charge for mileage. A similar car can be rented from Company B for $150 per week, plus 25 cents for each mile driven. How many miles must you drive in a week to make the rental fee for Company B greater than that for Company A?

70. *Make a Decision*: **Comparative Shopping** Your department sends its copying to the photocopy center of your company. The photocopy center bills your department $0.10 per page. You are considering buying a departmental copier for $3000. With your own copier the cost per page would be $0.03. The expected life of the copier is 4 years. How many copies must you make in the four-year period to justify purchasing the copier?

71. **Simple Interest** For $1000 to grow to more than $1250 in 2 years, what must the simple interest rate be?

72. **Simple Interest** For $1000 to grow to more than $1500 in 2 years, what must the simple interest rate be?

73. **Weight Loss Program** A person enrolls in a diet program that guarantees a loss of at least $1\frac{1}{2}$ pounds per week. The person's weight at the beginning of the program is 164 pounds. Find the maximum number of weeks before the person attains a weight of 128 pounds.

74. **Salary Increase** You accept a new job with a starting salary of $24,500. You are told that you will receive an annual raise of at least $1250. What is the maximum number of years you must work before your annual salary will be $30,000?

75. **Break-Even Analysis** The revenue R for selling x units of a product is $R = 115.95x$. The cost C of producing x units is $C = 95x + 750$. In order to obtain a profit, the revenue must be greater than the cost.

(a) Complete the table.

x	10	20	30	40	50
R					
C					

(b) For what values of x will this product return a profit?

76. Break-Even Analysis   The revenue R for selling x units of a product is $R = 24.55x$. The cost C of producing x units is

$$C = 15.4x + 150,000.$$

In order to obtain a profit, the revenue must be greater than the cost. For what values of x will this product return a profit?

77. Annual Operating Cost   A utility company has a fleet of vans. The annual operating cost C per van is

$$C = 0.32m + 2300$$

where m is the number of miles traveled by a van in a year. What number of miles will yield an annual operating cost that is less than $10,000?

78. Daily Sales   A doughnut shop sells a dozen doughnuts for $2.95. Beyond the fixed costs (rent, utilities, and insurance) of $150 per day, it costs $1.45 for enough materials (flour, sugar, and so on) and labor to produce a dozen doughnuts. The daily profit from doughnut sales varies between $50 and $200. Between what levels (in dozens) do the daily sales vary?

79. IQ Scores   The admissions office of a college wants to determine whether there is a relationship between IQ scores x and grade-point averages y after the first year of school. An equation that models the data obtained by the admissions office is

$$y = 0.067x - 5.638.$$

Estimate the values of x that predict a grade-point average of a least 3.0.

80. *Make a Decision*: Weightlifting   You want to determine whether there is a relationship between an athlete's weight x (in pounds) and the athlete's bench-press weight y (in pounds). An equation that models the data you obtained is

$$y = 1.3x - 36.$$

(a) Estimate the values of x that predict a maximum bench-press weight of at least 200 pounds.

(b) Do you think an athlete's weight is a good indicator of the athlete's bench-press weight? What other factors might influence an individual's bench-press weight?

81. Cable Television Subscribers   The number of cable television subscribers S (in thousands) in the United States between 1990 and 2001 can be modeled by

$$S = 1704t + 51,325, \quad 0 \le t \le 11$$

where t represents the year, with $t = 0$ corresponding to 1990 (see figure). According to this model, in what year will the number of subscribers exceed 77 million?   (Source: Kagan World Media)

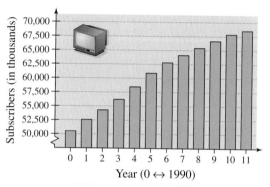

Year ($0 \leftrightarrow 1990$)

82. Public College Enrollment   The projected public college enrollment E (in thousands) in the United States from 2005 to 2012 can be modeled by

$$E = 152.5t + 11,655, \quad 5 \le t \le 12$$

where t represents the year, with $t = 5$ corresponding to 2005 (see figure). According to this model, when will public college enrollment exceed 15,000,000?   (Source: U.S. National Center for Education Statistics)

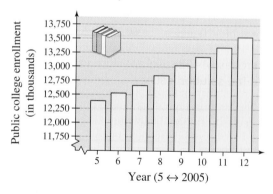

Year ($5 \leftrightarrow 2005$)

83. Geometry   The side of a square is measured as 10.4 inches with a possible error of $\frac{1}{16}$ inch. Using these measurements, determine the interval containing the possible areas of the square.

84. Geometry The side of a square is measured as 24.2 centimeters with a possible error of 0.25 centimeter. Using these measurements, determine the interval containing the possible areas of the square.

85. *Make a Decision*: Accuracy of Measurement You buy six T-bone steaks that cost $14.99 per pound. The weight listed on the package is 5.72 pounds. If the scale that weighed the package is accurate to within $\frac{1}{2}$ ounce, how much money might you have been undercharged or overcharged?

86. *Make a Decision*: Accuracy of Measurement You stop at a self-service gas station to buy 15 gallons of 87-octane gasoline at $1.69 a gallon. If the pump scale is accurate to within one-tenth of a gallon, how much money might you have been undercharged or overcharged?

87. Human Height The heights h of two-thirds of a population satisfy the inequality

$$|h - 68.5| \le 2.7$$

where h is measured in inches. Determine the interval on the real number line in which these heights lie.

88. Time Study A time study was conducted to determine the length of time required to perform a particular task in a manufacturing process. The times required by approximately two-thirds of the workers in the study satisfied the inequality

$$\left| \frac{t - 15.6}{1.9} \right| < 1$$

where t is time in minutes. Determine the interval on the real number line in which these times lie.

89. Humidity Control The specifications for an electronic device state that it is to be operated in a room with relative humidity h defined by

$$|h - 50| \le 30.$$

What are the minimum and maximum relative humidities for the operation of this device?

90. Body Temperature Physicians consider an adult's body temperature x (in degrees Fahrenheit) to be normal if it satisfies the inequality

$$|x - 98.6| \le 1.$$

Determine the range of temperatures that are considered to be normal.

Math *MATTERS*

Man and Mouse

A man who falls from a height of 2000 feet (without a parachute) will strike the ground with lethal force. But a mouse can fall from the same height and simply get up and walk away. Why?

The answer is that the speed at which a falling object hits the ground depends partly on the air resistance of the object, which in turn depends on the object's weight and surface area. If the ratio of an object's surface area to its weight is large, then its air resistance will be large. On the other hand, if the ratio of an object's surface area to its weight is small, then its air resistance will be small. This is why a parachute works—a person with a parachute has a

much larger surface area (for approximately the same weight) than a person without a parachute. So how does this relate to the falling man and mouse? The falling mouse has a much greater air resistance than the falling man because the ratio of the mouse's surface area to its weight is greater than the ratio of the man's surface area to his weight. To convince yourself that the mouse's ratio is greater than the man's, try the following experiment. Find the ratio of surface area to weight for the cubes described below.

Notice that as the cube becomes larger, the ratio of its surface area to its weight becomes smaller. (The answers are given in the back of the book.)

Length of Side	Surface Area	Volume	Density	Weight
1 ft	6 ft^2	1 ft^3	1 lb/ft^3	1 lb
2 ft	24 ft^2	8 ft^3	1 lb/ft^3	8 lb
3 ft	54 ft^2	27 ft^3	1 lb/ft^3	27 lb
4 ft	96 ft^2	64 ft^3	1 lb/ft^3	64 lb

R2.7   Other Types of Inequalities

Objectives

- Use critical numbers to determine test intervals for a polynomial inequality.
- Solve and graph a polynomial inequality.
- Solve and graph a rational inequality.
- Determine the domain of an expression involving a square root.
- Construct and use a polynomial inequality to solve an application problem.

Polynomial Inequalities

To solve a polynomial inequality such as $x^2 - 2x - 3 < 0$, you can use the fact that a polynomial can change signs only at its **zeros** (the x-values that make the polynomial equal to zero). Between two consecutive zeros, a polynomial must be entirely positive or entirely negative. This means that when the real zeros of a polynomial are put in order, they divide the real number line into intervals in which the polynomial has no sign changes. These zeros are the **critical numbers** of the inequality, and the resulting intervals are the **test intervals** for the inequality. For example, the polynomial above factors as

$$x^2 - 2x - 3 = (x + 1)(x - 3)$$

and has two zeros, $x = -1$ and $x = 3$. These zeros divide the real number line into three test intervals:

$$(-\infty, -1), \quad (-1, 3), \quad \text{and} \quad (3, \infty). \qquad \text{(See Figure R2.19.)}$$

So, to solve the inequality $x^2 - 2x - 3 < 0$, you need only test one value from each of these test intervals.

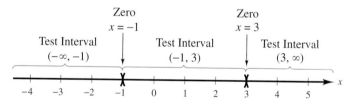

Figure R2.19   Three test intervals for $x^2 - 2x - 3 < 0$

Finding Test Intervals for a Polynomial

To determine the intervals on which the values of a polynomial are entirely negative or entirely positive, use the following steps.

1. Find all real zeros of the polynomial, and arrange the zeros in increasing order. These zeros are the **critical numbers** of the polynomial.

2. Use the critical numbers to determine the **test intervals.**

3. Choose one representative x-value in each test interval and evaluate the polynomial at that value.

| EXAMPLE 1 | SOLVING A POLYNOMIAL INEQUALITY |

Solve $x^2 - x - 6 < 0$.

SOLUTION

By factoring the quadratic as

$$x^2 - x - 6 = (x + 2)(x - 3)$$

you can see that the critical numbers are $x = -2$ and $x = 3$. The critical numbers act as boundaries between the real numbers that satisfy the inequality and the real numbers that do not satisfy the inequality. So, the polynomial's test intervals are

$$(-\infty, -2), \quad (-2, 3), \quad \text{and} \quad (3, \infty). \qquad \text{Test intervals}$$

In each test interval, choose a representative x-value and evaluate the polynomial.

Interval	x-Value	Polynomial Value	Conclusion
$(-\infty, -2)$	$x = -3$	$(-3)^2 - (-3) - 6 = 6$	Positive
$(-2, 3)$	$x = 0$	$(0)^2 - (0) - 6 = -6$	Negative
$(3, \infty)$	$x = 4$	$(4)^2 - (4) - 6 = 6$	Positive

TECHNOLOGY

You can use the *table* feature of your graphing utility to check the sign of the polynomial in each interval.

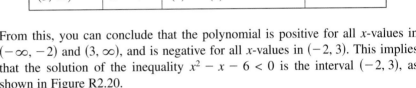

From this, you can conclude that the polynomial is positive for all x-values in $(-\infty, -2)$ and $(3, \infty)$, and is negative for all x-values in $(-2, 3)$. This implies that the solution of the inequality $x^2 - x - 6 < 0$ is the interval $(-2, 3)$, as shown in Figure R2.20.

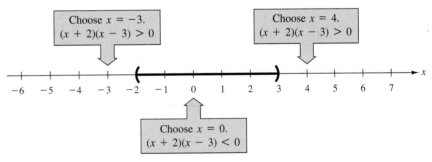

Choose $x = -3$.
$(x + 2)(x - 3) > 0$

Choose $x = 4$.
$(x + 2)(x - 3) > 0$

Choose $x = 0$.
$(x + 2)(x - 3) < 0$

Figure R2.20

✓CHECKPOINT   Now try Exercise 13.

As with linear inequalities, you can check the reasonableness of a solution of a polynomial inequality by substituting x-values into the original inequality. For instance, to check the solution found in Example 1, try substituting several x-values from the interval $(-2, 3)$ into the inequality

$$x^2 - x - 6 < 0.$$

Regardless of which x-values you choose, the inequality should be satisfied.

In Example 1, the polynomial inequality was given in general form. Whenever this is not the case, begin the solution process by writing the inequality in general form—with the polynomial on one side and zero on the other.

EXAMPLE 2   SOLVING A POLYNOMIAL INEQUALITY

Solve $x^3 - 3x^2 > 10x$.

SOLUTION

$$x^3 - 3x^2 > 10x \qquad \text{Write original inequality.}$$
$$x^3 - 3x^2 - 10x > 0 \qquad \text{Write in general form.}$$
$$x(x - 5)(x + 2) > 0 \qquad \text{Factor.}$$

You can see that the critical numbers are $x = -2$, $x = 0$, and $x = 5$, and the test intervals are $(-\infty, -2)$, $(-2, 0)$, $(0, 5)$, and $(5, \infty)$. In each test interval, choose a representative x-value and evaluate the polynomial.

Interval	x-Value	Polynomial Value	Conclusion
$(-\infty, -2)$	$x = -3$	$(-3)^3 - 3(-3)^2 - 10(-3) = -24$	Negative
$(-2, 0)$	$x = -1$	$(-1)^3 - 3(-1)^2 - 10(-1) = 6$	Positive
$(0, 5)$	$x = 2$	$2^3 - 3(2)^2 - 10(2) = -24$	Negative
$(5, \infty)$	$x = 6$	$6^3 - 3(6)^2 - 10(6) = 48$	Positive

From this, you can conclude that the inequality is satisfied on the open intervals $(-2, 0)$ and $(5, \infty)$. So, the solution set consists of all real numbers in the intervals $(-2, 0)$ and $(5, \infty)$, as shown in Figure R2.21.

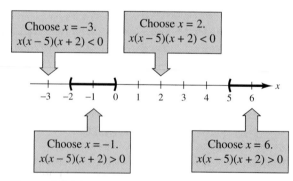

Figure R2.21

✓CHECKPOINT   Now try Exercise 17.

When solving a polynomial inequality, be sure you have accounted for the particular type of inequality symbol given in the inequality. For instance, in Example 2, note that the solution consisted of two *open* intervals because the original inequality contained a "greater than" symbol. If the original inequality had been $x^3 - 3x^2 \geq 10x$, the solution would have consisted of the *closed* interval $[-2, 0]$ and the interval $[5, \infty)$.

Each of the polynomial inequalities in Examples 1 and 2 has a solution set that consists of a single interval or the union of two intervals. When solving the exercises for this section, you should watch for some unusual solution sets, as illustrated in Example 3.

EXAMPLE 3   UNUSUAL SOLUTION SETS

What is unusual about the solution set for each inequality?

a. $x^2 + 2x + 4 > 0$

The solution set of this inequality consists of the entire set of real numbers, $(-\infty, \infty)$. In other words, the value of the quadratic $x^2 + 2x + 4$ is positive for every real value of x.

b. $x^2 + 2x + 1 \le 0$

The solution set of this inequality consists of the single real number $\{-1\}$, because the quadratic $x^2 + 2x + 1$ has one critical number, $x = -1$, and it is the only value that satisfies the inequality.

c. $x^2 + 3x + 5 < 0$

The solution set of this inequality is empty. In other words, the quadratic $x^2 + 3x + 5$ is *not* less than zero for any value of x.

d. $x^2 - 4x + 4 > 0$

The solution set of the following inequality consists of all real numbers *except* the number 2. In interval notation, this solution can be written as $(-\infty, 2) \cup (2, \infty)$.

TECHNOLOGY

Graphs of Inequalities and Graphing Utilities   Most graphing utilities can graph an inequality. Consult your user's guide for specific instructions. Once you know how to graph an inequality, you may check solutions by graphing. (Make sure you use an appropriate viewing window.) For example, the solution to

$$x^2 - 5x < 0$$

is the interval $(0, 5)$. When graphed, the solution occurs as an interval above the horizontal axis on the graphing utility, as shown in Figure R2.22. The graph does not indicate whether 0 and/or 5 are part of the solution. You must determine whether the endpoints are part of the solution based on the inequality sign and the type of inequality.

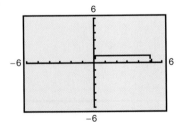

Figure R2.22

Rational Inequalities

The concepts of critical numbers and test intervals can be extended to inequalities involving rational expressions. Use the fact that the value of a rational expression can change sign only at its *zeros* (the *x*-values for which its numerator is zero) and its *undefined values* (the *x*-values for which its denominator is zero). These two types of numbers make up the **critical numbers** of a rational inequality.

EXAMPLE 4   SOLVING A RATIONAL INEQUALITY

Solve $\dfrac{2x - 7}{x - 5} \le 3$.

SOLUTION

$$\dfrac{2x - 7}{x - 5} \le 3 \qquad \text{Write original inequality.}$$

$$\dfrac{2x - 7}{x - 5} - 3 \le 0 \qquad \text{Write in general form.}$$

$$\dfrac{2x - 7 - 3x + 15}{x - 5} \le 0 \qquad \text{Add fractions.}$$

$$\dfrac{-x + 8}{x - 5} \le 0 \qquad \text{Simplify.}$$

Critical numbers: $x = 5, x = 8$

Test intervals: $(-\infty, 5), (5, 8), (8, \infty)$

Test: Is $\dfrac{-x + 8}{x - 5} \le 0$?

After testing these intervals, as shown in Figure R2.23, you can see that the inequality is satisfied on the open intervals $(-\infty, 5)$ and $(8, \infty)$. Moreover, because $(-x + 8)/(x - 5) = 0$ when $x = 8$, you can conclude that the solution set consists of all real numbers in the intervals $(-\infty, 5) \cup [8, \infty)$.

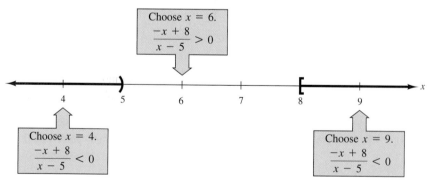

Figure R2.23

✓CHECKPOINT   Now try Exercise 23.

TECHNOLOGY

When using a graphing utility to check an inequality, always set your viewing window so that it includes all of the critical numbers.

Applications

One common application of inequalities comes from business and involves profit, revenue, and cost. The formula that relates these three quantities is

$$\boxed{\text{Profit}} = \boxed{\text{Revenue}} - \boxed{\text{Cost}}$$

$$P = R - C.$$

EXAMPLE 5   INCREASING THE PROFIT FOR A PRODUCT

Make a Decision The marketing department of a calculator manufacturer has determined that the demand for a new model of calculator is given by

$$p = 100 - 10x, \quad 0 \le x \le 10 \qquad \text{Demand equation}$$

where p is the price per calculator in dollars and x represents the number of calculators sold, in millions. (If this model is accurate, no one would be willing to pay \$100 for the calculator. At the other extreme, the company couldn't *give* away more than 10 million calculators.) The revenue, in millions of dollars, for selling x million calculators is given by

$$R = xp = x(100 - 10x). \qquad \text{Revenue equation}$$

See Figure R2.24. The total cost of producing x million calculators is \$10 per calculator plus a one-time development cost of \$2,500,000. So, the total cost, in millions of dollars, is

$$C = 10x + 2.5. \qquad \text{Cost equation}$$

What price should the company charge per calculator to obtain a profit of at least \$190,000,000?

SOLUTION

Verbal Model: $\boxed{\text{Profit}} = \boxed{\text{Revenue}} - \boxed{\text{Cost}}$

Equation: $P = R - C$

$$P = 100x - 10x^2 - (10x + 2.5)$$

$$P = -10x^2 + 90x - 2.5$$

To answer the question, you must solve the inequality

$$-10x^2 + 90x - 2.5 \ge 190.$$

Using the techniques described in this section, you can find the solution to be $3.5 \le x \le 5.5$, as shown in Figure R2.25. The prices that correspond to these x-values are given by

$$100 - 10(3.5) \ge p \ge 100 - 10(5.5)$$

$$45 \le p \le 65$$

The best decision for the company is to charge at least \$45 per calculator and at most \$65 per calculator.

✓CHECKPOINT   Now try Exercise 53.

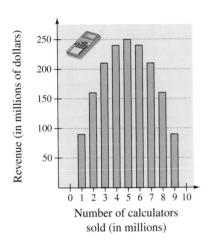

Figure R2.24

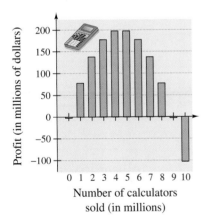

Figure R2.25

Another common application of inequalities is finding the domain of an expression that involves a square root, as shown in Example 6.

EXAMPLE 6   FINDING THE DOMAIN OF AN EXPRESSION

Find the domain of the expression $\sqrt{64 - 4x^2}$.

SOLUTION

Remember that the domain of an expression is the set of all x-values for which the expression is defined. Because $\sqrt{64 - 4x^2}$ is defined (has real values) only if $64 - 4x^2$ is nonnegative, the domain is given by $64 - 4x^2 \geq 0$.

$$64 - 4x^2 \geq 0 \qquad \text{Write in general form.}$$
$$16 - x^2 \geq 0 \qquad \text{Divide each side by 4.}$$
$$(4 - x)(4 + x) \geq 0 \qquad \text{Factor.}$$

So, the inequality has two critical numbers: $x = -4$ and $x = 4$. You can use these two numbers to test the inequality as follows.

Critical numbers: $x = -4, x = 4$

Test intervals: $(-\infty, -4), (-4, 4), (4, \infty)$

Test: Is $(4 - x)(4 + x) \geq 0$?

A test shows that $64 - 4x^2$ is greater than or equal to 0 in the *closed interval* $[-4, 4]$. So, the domain of the expression $\sqrt{64 - 4x^2}$ is the interval $[-4, 4]$, as shown in Figure R2.26.

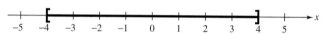

Figure R2.26

✓CHECKPOINT   Now try Exercise 31.

Discussing the Concept | Profit Analysis

Consider the relationship

$$P = R - C$$

described on page R143. Discuss why it might be beneficial to solve $P < 0$ if you owned a business. Use the situation described in Example 5 to illustrate your reasoning.

R2.7 Warm Up

The following warm-up exercises involve skills that were covered in earlier sections. You will use these skills in the exercise set for this section. For additional help, review Section R2.6.

In Exercises 1–10, solve the inequality.

1. $-\dfrac{y}{3} > 2$

2. $-6z < 27$

3. $-3 \le 2x + 3 < 5$

4. $-3x + 5 \ge 20$

5. $10 > 4 - 3(x + 1)$

6. $3 < 1 + 2(x - 4) < 7$

7. $2|x| \le 7$

8. $|x - 3| > 1$

9. $|x + 4| > 2$

10. $|2 - x| \le 4$

R2.7 Exercises

In Exercises 1–30, solve the inequality and graph the solution on the real number line.

1. $x^2 \le 9$

2. $x^2 < 5$

3. $x^2 > 4$

4. $(x - 3)^2 \ge 1$

5. $(x + 2)^2 < 25$

6. $(x + 6)^2 \le 8$

7. $x^2 + 4x + 4 \ge 9$

8. $x^2 - 6x + 9 < 16$

9. $x^2 + x < 6$

10. $x^2 + 2x > 3$

11. $3(x - 1)(x + 1) > 0$

12. $6(x + 2)(x - 1) < 0$

13. $x^2 + 2x - 3 < 0$

14. $x^2 - 4x - 1 > 0$

15. $4x^3 - 6x^2 < 0$

16. $4x^3 - 12x^2 > 0$

17. $x^3 - 4x \ge 0$

18. $2x^3 - x^4 \le 0$

19. $x^3 - 2x^2 - x + 2 \ge 0$

20. $x^3 + 5x^2 - 4x - 20 \le 0$

21. $\dfrac{1}{x} > x$

22. $\dfrac{1}{x} < 4$

23. $\dfrac{x + 6}{x + 1} < 2$

24. $\dfrac{x + 12}{x + 2} \ge 3$

25. $\dfrac{3x - 5}{x - 5} > 4$

26. $\dfrac{5 + 7x}{1 + 2x} < 4$

27. $\dfrac{4}{x + 5} > \dfrac{1}{2x + 3}$

28. $\dfrac{5}{x - 6} > \dfrac{3}{x + 2}$

29. $\dfrac{1}{x - 3} \le \dfrac{9}{4x + 3}$

30. $\dfrac{1}{x} \ge \dfrac{1}{x + 3}$

In Exercises 31–40, find the domain of the expression.

31. $\sqrt[4]{4 - x^2}$

32. $\sqrt{x^2 - 4}$

33. $\sqrt{81 - 4x^2}$

34. $\sqrt{144 - 9x^2}$

35. $\sqrt{x^2 - 7x + 12}$

36. $\sqrt{12 - x - x^2}$

37. $\sqrt{x^2 + 4}$

38. $\sqrt[4]{7 + x^2}$

39. $\sqrt{x^2 - 3x + 3}$

40. $\sqrt[4]{-x^2 + 2x - 2}$

In Exercises 41 and 42, consider the domains of the expressions $\sqrt[3]{x^2 - 7x + 12}$ and $\sqrt{x^2 - 7x + 12}$.

41. Explain why the domain of $\sqrt[3]{x^2 - 7x + 12}$ consists of all real numbers.

42. Explain why the domain of $\sqrt{x^2 - 7x + 12}$ is different from the domain of $\sqrt[3]{x^2 - 7x + 12}$.

In Exercises 43–48, use a calculator to solve the inequality. (Round each number in your answer to two decimal places.)

43. $0.4x^2 + 5.26 < 10.2$

44. $-1.3x^2 + 3.78 > 2.12$

45. $-0.5x^2 + 12.5x + 1.6 > 0$

46. $1.2x^2 + 4.8x + 3.1 < 5.3$

47. $\dfrac{1}{2.3x - 5.2} > 3.4$

48. $\dfrac{2}{3.1x - 3.7} > 5.8$

49. Height of a Projectile A projectile is fired straight upward from ground level with an initial velocity of 160 feet per second. During what time period will its height exceed 384 feet?

50. Height of a Projectile A projectile is fired straight upward from ground level with an initial velocity of 128 feet per second. During what time period will its height be less than 128 feet?

51. Geometry   A rectangular playing field with a perimeter of 100 meters is to have an area of at least 500 square meters (see figure). Within what bounds must the length lie?

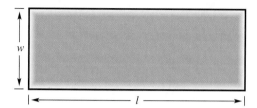

52. Geometry   A rectangular room with a perimeter of 50 feet is to have an area of at least 120 square feet. Within what bounds must the length lie?

53. *Make a Decision*: Company Profits   The revenue R and cost C for a product are given by

$$R = x(50 - 0.0002x)$$

and

$$C = 12x + 150,000$$

where R and C are measured in dollars and x represents the number of units sold (see figure).

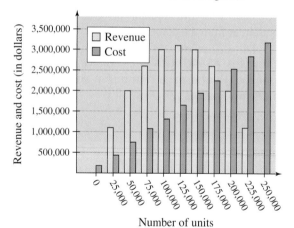

Number of units

(a) How many units must be sold to obtain a profit of at least $1,650,000?

(b) The demand equation for the product is

$$p = 50 - 0.0002x$$

where p is the price per unit. What price per unit will produce a profit of at least $1,650,000?

(c) After first achieving a profit, at how many units sold does the company first see its revenue dip below cost?

54. *Make a Decision*: Company Profits   The revenue R and cost C for a product are given by

$$R = x(75 - 0.0005x)$$

and

$$C = 30x + 250,000$$

where R and C are measured in dollars and x represents the number of units sold (see figure).

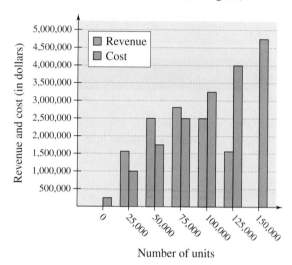

Number of units

(a) How many units must be sold to obtain a profit of at least $750,000?

(b) The demand equation for the product is

$$p = 75 - 0.0005x$$

where p is the price per unit. What price per unit will produce a profit of at least $750,000?

(c) After first achieving a profit, at how many units sold does the company first see its revenue dip below cost?

55. Compound Interest P dollars, invested at interest rate r compounded annually, increases to an amount

$$A = P(1 + r)^2$$

in 2 years. For an investment of $1000 to increase to an amount greater than $1200 in 2 years, the interest rate must be greater than what percent?

56. Compound Interest P dollars, invested at interest rate r compounded annually, increases to an amount

$$A = P(1 + r)^3$$

in 3 years. For an investment of $500 to increase to an amount greater than $600 in 3 years, the interest rate must be greater than what percent?

57. World Population The world population P (in millions) from 1990 to 2002 can be modeled by

$$P = -0.47t^2 + 84.7t + 5285, \quad 0 \le t \le 12$$

where t represents the year, with $t = 0$ corresponding to 1990 (see figure). According to this model, in what year will the world population exceed 6,500,000,000? (Source: U.S. Census Bureau)

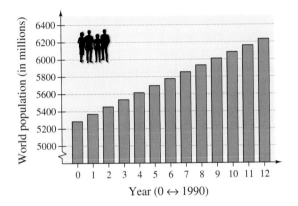

Year (0 ↔ 1990)

58. College Graduates The number B (in thousands) of bachelor's degrees awarded annually in the United States from 1980 to 2000 can be modeled by

$$B = 0.27t^2 + 10.6t + 925, \quad 0 \le t \le 20$$

where t represents the year, with $t = 0$ corresponding to 1980 (see figure). According to this model, in what year will the number of bachelor's degrees awarded annually exceed 1,400,000? (Source: National Center for Education Statistics)

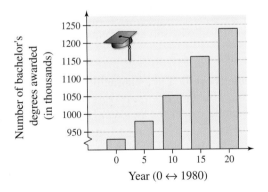

Year (0 ↔ 1980)

59. Higher Education The cost P of higher education at public institutions in the United States for the academic years 1995/1996 to 2001/2002 can be modeled by

$$P = 8.49t^2 + 132.2t + 5181, \quad 6 \le t \le 12$$

where $t = 6$ represents the 1995/1996 academic year (see figure). According to this model, in what academic year will the cost of higher education at public institutions first exceed $10,000? (Source: National Center for Education Statistics)

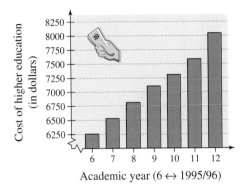

Academic year (6 ↔ 1995/96)

60. *Make a Decision*: **Car Dealerships** The number D of franchised new car dealerships in the United States from 1995 to 2001 can be modeled by

$$D = -32.14t^2 + 360t + 21,771, \quad 5 \le t \le 11$$

where t represents the year, with $t = 5$ corresponding to 1995 (see figure). According to this model, in what year will the number of franchised new car dealerships drop below 20,000? Do you think this model is valid for predicting the number of franchised new car dealerships in the coming decades? Explain your reasoning. (Source: National Automobile Dealers Association)

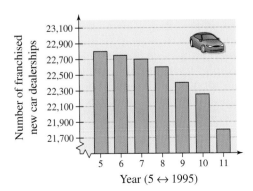

Year (5 ↔ 1995)

Make a Decision **Project: Salaries in Professional Baseball**

Year	Average salary, S
1995	1111
1996	1120
1997	1337
1998	1399
1999	1611
2000	1896
2001	2139
2002	2296
2003	2372

The record-breaking attendance (72.6 million people) for major league baseball was not the only record that was broken in the 2001 season. Barry Bonds (San Francisco) hit 73 home runs, drew 177 walks, and had a slugging percentage of .863 to set records in each category. Rickey Henderson (San Diego) broke the career record for runs with 2248 and extended his career records for walks and stolen bases to 2141 and 1395, respectively. Roger Clemens (New York Yankees) became the first pitcher in baseball history to have a 20–1 start. Albert Pujols (St. Louis) set a new rookie RBI (runs batted in) record with 130. As players reach and exceed previous records, they expect to be compensated accordingly. The average salary for a major league baseball player nearly doubled from 1995 to 2001.

The table at the left shows the average salaries S (in thousands of dollars) for professional baseball players from 1995 to 2003. The average salary S can be modeled by the equation

$$S = 5.84t^2 + 72.8t + 531, \quad 5 \le t \le 13$$

where t represents the year, with $t = 5$ corresponding to 1995. Use this information to investigate the following questions.

1. **Compare the Data**   Make a table that compares the actual average salaries for 1995 to 2003 with the average salaries found using the model.

2. *Make a Decision*   Use the model to predict the average salary in 2004 and in 2005. Use your school's library, the Internet, or some other reference source to find the actual average salaries. How well did the model predict the average salaries?

3. *Make a Decision*   According to this model, when will the average salary reach $4,500,000?

   (a) Answer the question numerically by creating a table of values.

   (b) Answer the question algebraically by solving the equation

   $$4500 = 5.84t^2 + 72.8t + 531.$$

4. **Find a Model**   Enter the table above in a graphing utility. Enter 5 for 1995, enter 6 for 1996, and so on. Use the *quadratic regression* feature of the graping utility to find a quadratic model for the data. Do you get the same model as the one given above?

5. **Research**   Use a reference source to find data that can be closely modeled by a quadratic model. Compare the model with the actual data numerically and graphically.

Chapter Summary

After studying this chapter, you should have acquired the following skills.
These skills are keyed to the Review Exercises that begin on page R150.
Answers to odd-numbered Review Exercises are given in the back of the book.

R2.1 • Classify an equation as an identity or a conditional equation. *Review Exercises 1, 2*

• Determine whether a given value is a solution. *Review Exercises 3, 4*

• Solve a linear equation in one variable. *Review Exercises 5–12*

R2.2 • Use mathematical models to solve word problems. *Review Exercises 13, 15, 23, 24*

• Model and solve percent and mixture problems. *Review Exercises 14, 16, 21, 22, 25, 26*

• Use common formulas to solve geometry and simple interest problems. *Review Exercises 17–20*

R2.3 • Solve a quadratic equation by factoring. *Review Exercises 27–30*

• Solve a quadratic equation by extracting square roots. *Review Exercises 31–34*

• Analyze a quadratic equation. *Review Exercises 35, 36*

• Construct and use a quadratic model to solve an application problem. *Review Exercises 37–40*

R2.4 • Use the discriminant to determine the number of real solutions of a quadratic equation. *Review Exercises 41, 42*

• Solve a quadratic equation using the Quadratic Formula. *Review Exercises 43–50*

• Use the Quadratic Formula to solve an application problem. *Review Exercises 51, 52*

R2.5 • Solve a polynomial equation by factoring. *Review Exercises 53–56*

• Rewrite and solve an equation involving radicals or rational exponents. *Review Exercises 57–62*

• Rewrite and solve an equation involving fractions or absolute values. *Review Exercises 63–66*

• Construct and use a nonquadratic model to solve an application problem. *Review Exercises 67–70*

R2.6 • Solve and graph a linear inequality. *Review Exercises 71–76*

• Construct and use a linear inequality to solve an application problem. *Review Exercises 77, 78*

R2.7 • Solve and graph a polynomial inequality. *Review Exercises 79–81, 85, 86*

• Solve and graph a rational inequality. *Review Exercises 82–84, 87, 88*

• Determine the domain of an expression involving a radical. *Review Exercises 89–94*

• Construct and use a polynomial inequality to solve an application problem. *Review Exercises 95–102*

Review Exercises

In Exercises 1 and 2, determine whether the equation is an identity or a conditional equation.

1. $5(x - 3) = 2x + 9$ **2.** $3(x + 2) = 3x + 6$

In Exercises 3 and 4, determine whether each value of x is a solution of the equation.

3. $3x^2 + 7x + 5 = x^2 + 9$

(a) $x = 0$ (b) $x = \frac{1}{2}$ (c) $x = -4$ (d) $x = -1$

4. $6 + \dfrac{3}{x - 4} = 5$

(a) $x = 5$ (b) $x = 0$ (c) $x = -2$ (d) $x = 7$

In Exercises 5–10, solve the equation (if possible) and check your solution.

5. $4(x + 3) - 3 = 2(4 - 3x) - 4$

6. $\dfrac{3x - 2}{5x - 1} = \dfrac{3}{4}$

7. $(x + 3) + 2(x - 4) = 5(x + 3)$

8. $\dfrac{3}{x - 4} + \dfrac{8}{2x + 5} = \dfrac{11}{2x^2 - 3x - 20}$

9. $\dfrac{x}{x + 3} - \dfrac{4}{x + 3} + 2 = 0$

10. $7 - \dfrac{3}{x} = 8 + \dfrac{5}{x}$

In Exercises 11 and 12, use a calculator to solve the equation. (Round your solution to three decimal places.)

11. $0.375x - 0.75(300 - x) = 200$

12. $\dfrac{x}{0.0645} + \dfrac{x}{0.098} = 2$

13. Three consecutive even integers have a sum of 42. Find the smallest of these integers.

14. Annual Salary   Your annual salary is $24,500. You receive a 6% raise. What is your new annual salary?

15. Fitness   When using a pull-up weight machine, the amount you set is subtracted from your weight and you pull the remaining amount. Write a model that describes the weight x that must be set if a person weighing 130 pounds wishes to pull 100 pounds. Solve for x.

16. e-Retail   In 2003, online holiday shoppers spent $11,248 million on items from the categories shown in the bar graph. What was the online holiday shopping total for each category? (Source: Goldman Sachs, Harris Interactive, and Nielsen//Net Ratings eSpending Report)

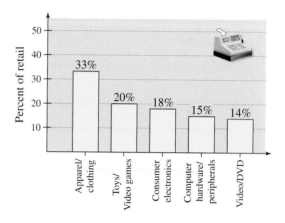

17. Geometry   A volleyball court is twice as long as it is wide, and its perimeter is 177 feet. Find the dimensions of the volleyball court.

18. Geometry   A room is 1.25 times as long as it is wide, and its perimeter is 90 feet. Find the dimensions of the room.

19. Simple Interest   You deposit $400 in a savings account earning 3% simple interest annually. How much interest will you have earned after 1 year?

20. Simple Interest   You deposit $750 in a money market account. One year later the account balance is $783.75. What was the interest rate for 1 year?

21. List Price   The price of an outdoor barbeque grill has been discounted 20%. The sale price is $129. Find the original price of the grill.

22. Discount Rate   The price of a 57-inch HDTV widescreen has been reduced $200. The sale price is $1800. What percent of the original list price is the discount?

23. Travel Time   Two cars start at the same time at a given point and travel in the same direction at average speeds of 45 miles per hour and 50 miles per hour. How much time must elapse before the two cars are 10 miles apart?

24. Projected Revenue From January through June, a company's revenues have totaled $435,112. If the monthly revenues continue at this rate, what will be the total revenue for the year?

25. Mixture A car radiator contains 10 quarts of a 30% antifreeze solution. The car's owner wishes to create a 10-quart solution that is 50% antifreeze. How many quarts will have to be replaced with pure antifreeze?

26. Mixture A three-gallon acid solution contains 5% boric acid. How many gallons of 20% boric acid solution should be added to make a final solution that is 12% boric acid?

In Exercises 27–30, solve the quadratic equation by factoring. Check your solutions.

27. $6x^2 = 5x + 4$

28. $-x^2 = 15x + 36$

29. $x^2 - 11x + 24 = 0$

30. $15 + x - 2x^2 = 0$

In Exercises 31–34, solve the equation by extracting square roots. List both the exact answer and a decimal answer that has been rounded to two decimal places.

31. $x^2 = 11$

32. $16x^2 = 25$

33. $(x + 4)^2 = 18$

34. $(x - 1)^2 = 5$

35. Describe at least two ways you can use a graphing utility to check a solution of a quadratic equation.

36. Error Analysis A student solves Exercise 31 by extracting square roots and states that the two solutions are $x = \sqrt{11}$ and $x \approx 3.32$. What are some of the possible errors the student may have made? Give an analytical argument to persuade the student that there are two *different* solutions to Exercise 31.

37. Geometry A billboard is 12 feet longer than it is high. The billboard has 405 square feet of advertising space. What are the dimensions of the billboard? Use a diagram to help answer the question.

38. Grand Canyon The Grand Canyon is 6000 feet deep at its deepest part. A rock is dropped over the deepest part of the canyon. How long does the rock take to hit the water in the Colorado River below?

39. Total Revenue The demand equation for a product is

$$p = 50 - 0.0001x$$

where p is the price per unit and x is the number of units sold. The total revenue R for selling x units is given by

$$R = xp = x(50 - 0.0001x).$$

How many units must be sold to produce a revenue of $6,000,000?

40. Depth of an Underwater Cable A ship's sonar locates a cable 2000 feet from the ship (see figure). The angle between the surface of the water and a line from the ship to the cable is 45°. How deep is the cable?

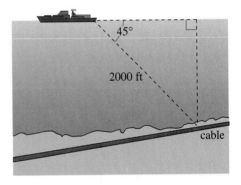

In Exercises 41 and 42, use the discriminant to determine the number of real solutions of the quadratic equation.

41. $x^2 + 11x + 24 = 0$ **42.** $x^2 + 5x + 12 = 0$

In Exercises 43–48, use the Quadratic Formula to solve the equation. Check your solutions.

43. $x^2 - 12x + 30 = 0$

44. $5x^2 + 16x - 12 = 0$

45. $(y + 7)^2 = -5y$

46. $6x = 7 - 2x^2$

47. $x^2 + 6x - 3 = 0$

48. $10x^2 - 11x = 2$

In Exercises 49 and 50, use a calculator to solve the equation. (Round your answers to three decimal places.)

49. $3.6x^2 - 5.7x - 1.9 = 0$

50. $34x^2 - 296x + 47 = 0$

51. On the Moon   An astronaut standing on the edge of a cliff on the moon drops a rock over the cliff. The height of the rock is given by

$$s = -2.7t^2 + 100.$$

The rock's initial velocity is 0 feet per second and the initial height is 100 feet. Determine how long it will take the rock to hit the lunar surface. If the rock were dropped off a similar cliff on Earth, how long would it remain in the air?

52. Geometry   An open box is to be made from a square piece of material by cutting three-inch squares from the corners and turning up the sides (see figure). The volume of the finished box is to be 363 cubic inches. Find the size of the original piece of material.

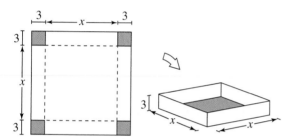

In Exercises 53–66, find the real solutions of the equation. Check your solutions.

53. $3x^3 - 9x^2 - 12x = 0$

54. $x^4 - 5x^2 + 4 = 0$

55. $x^4 + 3x^3 - 5x - 15 = 0$

56. $x^6 + 26x^3 - 27 = 0$

57. $2\sqrt{x} - 5 = 0$ **58.** $\sqrt{3x - 2} + x = 4$

59. $2\sqrt{x - 3} - 4 = 3x$ **60.** $\sqrt[3]{3x + 5} = 5$

61. $(x^2 - 5)^{2/3} = 9$ **62.** $(x^2 - 5x - 6)^{4/3} = 16$

63. $|5x + 4| = 11$ **64.** $|x^2 + 4x| - 2x = 8$

65. $\dfrac{5}{x + 1} + \dfrac{3}{x + 3} = 1$ **66.** $x + \dfrac{3}{x + 2} = 2$

67. Sharing the Cost   Three students are planning to share the expense of renting a condominium at a resort for 2 weeks. By adding a fourth person to the group, each person could save $50 in rental fees. How much is the rent for the two-week period?

68. Sharing the Cost   A college charters a bus for $1800 to take a group to a museum. When four more students join the trip, the cost per student drops by $5. How many students were in the original group?

69. Cash Advance   You take out a cash advance of $400 on a credit card. After 3 months, the amount you owe is $421.56. What is the annual percentage rate for this cash advance? (Assume that the interest is compounded monthly and that you made no payments after the first and second months.)

70. Market Research   The demand equation for a product is given by

$$p = 42 - \sqrt{0.001x + 2}$$

where x is the number of units demanded per day and p is the price per unit. Find the demand when the price is set at $29.95.

In Exercises 71–76, solve the inequality and graph the solution on the real number line.

71. $3(x - 1) < 2x + 8$

72. $-5 \le 2 - 4(x + 2) \le 6$

73. $-3 < \dfrac{2x + 1}{4} < 3$

74. $-1 \le -5 - 3x < 4$

75. $|x + 10| + 3 < 5$

76. $|2x - 3| - 4 > 2$

77. *Make a Decision*: Break-Even Analysis   The revenue R for selling x units of a product is

$$R = 125.95x.$$

The cost C of producing x units is

$$C = 92x + 1200.$$

In order to obtain a profit, the revenue must be greater than the cost. What is the minimum number of units the company should produce in order to return a profit?

78. *Make a Decision*: Accuracy of Measurement   You buy an 18-inch gold chain that costs $8.95 per inch. If the chain is measured accurately to within $\frac{1}{16}$ of an inch, how much money might you have been undercharged or overcharged?

In Exercises 79–84, solve the inequality and graph the solution on the real number line.

79. $5(x + 1)(x - 3) < 0$ **80.** $(x + 4)^2 \le 4$

81. $x^3 - 9x < 0$ **82.** $\dfrac{x + 5}{x + 8} \ge 2$

83. $\dfrac{2 + 3x}{4 - x} < 2$

84. $\dfrac{1}{x + 1} \geq \dfrac{1}{x + 5}$

In Exercises 85–88, use a calculator to solve the inequality. (Round each number in your answer to two decimal places.)

85. $-1.2x^2 + 4.76 > 1.32$

86. $3.5x^2 + 4.9x - 6.1 < 2.4$

87. $\dfrac{1}{3.7x - 6.1} > 2.9$

88. $\dfrac{3}{5.4x - 2.7} < 8.9$

In Exercises 89–94, find the domain of the expression.

89. $\sqrt{x - 10}$

90. $\sqrt[4]{2x + 5}$

91. $\sqrt[3]{2x - 1}$

92. $\sqrt[5]{x^2 - 4}$

93. $\sqrt{x^2 - 15x + 54}$

94. $\sqrt{81 - 4x^2}$

95. Height of a Projectile   A projectile is fired straight upward from ground level with an initial velocity of 134 feet per second. During what time period will its height exceed 276 feet?

96. Height of a Flare   A flare is fired straight upward from ground level with an initial velocity of 100 feet per second. During what time period will its height exceed 150 feet?

97. Compound Interest P dollars, invested at interest rate r compounded annually, increases to an amount

$$A = P(1 + r)^5$$

in 5 years. An investment of $1000 is to increase to an amount greater than $1400 in 5 years. The interest rate must be greater than what percent?

98. Compound Interest P dollars, invested at an interest rate r compounded semiannually, increases to an amount

$$A = P(1 + r/2)^{2 \cdot 8}$$

in 8 years. An investment of $2000 is to increase to an amount greater than $4200 in 8 years. The interest rate must be greater than what percent?

99. Company Profits   The revenue R and cost C for a product are given by

$$R = x(80 - 0.0005x) \quad \text{and} \quad C = 20x + 300,000$$

where R and C are measured in dollars and x represents the number of units sold. How many units must be sold to obtain a profit of at least $900,000?

100. Price of a Product   In Exercise 99, the revenue equation is

$$R = x(80 - 0.0005x)$$

which implies that the demand equation is

$$p = 80 - 0.0005x$$

where p is the price per unit. What price per unit should the company set to obtain a profit of at least $1,200,000?

101. Registered Vehicles   The number R (in thousands) of registered vehicles in the United States from 1990 to 2001 can be approximated by the model

$$R = 3774.4t + 188,175, \quad 0 \leq t \leq 11$$

where t represents the year, with $t = 0$ corresponding to 1990. (Source: U.S. Department of Transportation)

(a) Complete the table. Round each value of R to the nearest whole number.

t	0	5	10	11
R				

(b) According to this model, in what year will the number of registered vehicles be at least 260,000,000?

102. Personal Income   The average personal income I (in dollars) in the United States from 1990 to 2001 can be approximated by the model

$$I = 42.01t^2 + 549.0t + 19,574, \quad 0 \leq t \leq 11$$

where t represents the year, with $t = 0$ corresponding to 1990. (Source: U.S. Bureau of Economic Analysis)

(a) Complete the table. Round each value of I to the nearest whole number.

t	1	3	5	7	9	11
I						

(b) According to this model, in what year will the average income be at least $47,358?

Chapter Test

Take this test as you would take a test in class. After you are done, check your work against the answers given in the back of the book.

1. Solve the equation $3(x + 2) - 8 = 4(2 - 5x) + 7$.

2. Find the domain of (a) $\sqrt[3]{2x + 3}$ and (b) $\sqrt{9 - x^2}$.

3. In May, the total profit for a company was 20% less than it was in April. The total profit for the 2 months was $315,655.20. Find the profit for each month.

In Exercises 4–13, solve the equation. Check your solution(s).

4. *Factoring:* $6x^2 + 7x = 5$ 5. *Factoring:* $12 + 5x - 2x^2 = 0$

6. *Extracting roots:* $x^2 - 5 = 10$

7. *Quadratic Formula:* $(x + 5)^2 = -3x$

8. *Quadratic Formula:* $3x^2 - 11x = 2$

9. *Quadratic Formula:* $5.4x^2 - 3.2x - 2.5 = 0$

10. $|3x + 2| = 8$ 11. $\sqrt{x - 3} + x = 5$

12. $x^4 - 10x^2 + 9 = 0$ 13. $(x^2 - 9)^{2/3} = 9$

14. The demand equation for a product is $p = 40 - 0.0001x$, where p is the price per unit and x is the number of units sold. The total revenue R for selling x units is given by $R = xp$. How many units must be sold to produce a revenue of $2,000,000? Explain your reasoning.

In Exercises 15–18, solve the inequality and graph the solution on the real number line.

15. $\dfrac{3x + 1}{5} < 2$ 16. $|4 - 5x| \geq 24$

17. $\dfrac{x + 3}{x + 7} > 2$ 18. $3x^3 - 12x \leq 0$

19. The revenue R and cost C for a product are given by

$$R = x(100 - 0.0005x) \quad \text{and} \quad C = 30x + 200,000$$

where R and C are measured in dollars and x represents the number of units sold. How many units must be sold to obtain a profit of at least $500,000?

20. The percent P of college freshmen from 1980 to 2000 who had an average high school grade of A–, A, or A+ can be approximated by the model

$$P = 0.044t^2 - 0.08t + 26.9, \quad 0 \leq t \leq 20$$

where t represents the year, with $t = 0$ corresponding to 1980. According to this model, in what year will the percent of college freshmen with an average high school grade of A–, A, or A+ exceed 50%? (Source: The Higher Education Research Institute)

Cumulative Test: Chapters R1–R2

Take this test as you would take a test in class. After you are done, check your work against the answers given in the back of the book.

In Exercises 1–3, simplify the expression.

1. $4(-2x^2)^3$

2. $\sqrt{18x^5}$

3. $\dfrac{2}{3 - \sqrt{5}}$

4. Factor completely: $x^3 - 6x^2 - 3x + 18$.

5. Simplify: $\dfrac{x^2 - 16}{5x - 20}$.

6. Simplify: $\dfrac{\frac{1}{x} - \frac{1}{y}}{\frac{1}{y} + \frac{1}{x}}$.

7. The number N (in millions) of mail order prescriptions filled in the United States from 1995 to 2002 can be approximated by the model

$$N = 12.5t + 22, \quad 5 \le t \le 12$$

where t represents the year, with $t = 5$ corresponding to 1995.   (Source: National Association of Chain Drug Stores)

(a) Estimate the number of mail order prescriptions in 2002.

(b) If this model continues to be valid, in what year will the number of mail order prescriptions exceed 220,000,000?

In Exercises 8–13, solve the equation.

8. *Factoring:* $2x^2 - 11x = -5$

9. *Quadratic Formula:* $5.2x^2 + 1.5x - 3.9 = 0$

10. $|3x + 1| = 9$

11. $\sqrt{2x - 1} + x = 4$

12. $x^4 - 17x^2 = -16$

13. $(x^2 - 14)^{3/2} = 8$

In Exercises 14–16, solve the inequality and graph the solution on the real number line.

14. $-2 < \dfrac{1 - 3x}{5} < 2$

15. $2x^3 - 16x \ge 0$

16. $|5 - 3x| \le 21$

17. The revenue R and cost C for a product are given by

$$R = x(100 - 0.0003x) \quad \text{and} \quad C = 30x + 100{,}000$$

where R and C are measured in dollars and x represents the number of units sold. How many units must be sold to obtain a profit of at least $600,000?

18. The average basic monthly rate R for cable TV in the United States from 1990 to 2002 can be approximated by the model

$$R = 0.0353t^2 + 1.066t + 16.74, \quad 0 \le t \le 12$$

where t represents the year, with $t = 0$ corresponding to 1990. According to this model, in what year will the average basic monthly cable rate exceed $45?   (Source: Kagan World Media)

Functions and Graphs

Chapter 1

Chapter Sections

1.1 Graphs of Equations

1.2 Lines in the Plane

1.3 Linear Modeling and Direct Variation

1.4 Functions

1.5 Graphs of Functions

1.6 Transformations of Functions

1.7 The Algebra of Functions

1.8 Inverse Functions

Make a Decision ➤ How Many People Live in Your House?

Demography is the statistical study of human populations with reference to size, density, distribution, and vital statistics. The United States Census Bureau tracks the demographics of the nation and uses these findings to forecast future trends. For example, the Census Bureau predicts that the U.S. population will have increased to about 420 million by the year 2050.

The table shows the average sizes of U.S. households for selected years from 1950 to 2000. These data are also shown in the graph below the table. (Source: U.S. Census Bureau)

Make a Decision Although the overall population of the United States is increasing, from the graph you can see that the average household size is decreasing. Can you think of any possible explanations for this trend?

Year	People per household
1950	3.37
1955	3.33
1960	3.33
1965	3.29
1970	3.14
1975	2.94
1980	2.76
1985	2.69
1990	2.63
1995	2.65
2000	2.62

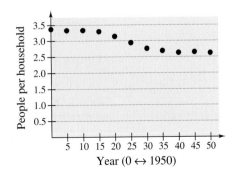

To explore this real-world application further, go to the Make a Decision Project on page 94.

Objectives

- Plot points in the Cartesian plane.
- Find the distance between two points in the plane.
- Use the Midpoint Formula to find the midpoint of a line segment joining two points.
- Determine whether a point is a solution of an equation.
- Sketch the graph of an equation using a table of values.
- Find the *x*- and *y*-intercepts of the graph of an equation.
- Determine the symmetry of a graph.
- Write the equation of a circle in standard form.

The Cartesian Plane

Just as you can represent real numbers by points on a real number line, you can represent ordered pairs of real numbers by points in a plane. This plane is called the **rectangular coordinate system,** or the **Cartesian plane,** named after the French mathematician René Descartes (1596–1650).

The Cartesian plane is formed by using two real number lines intersecting at right angles, as shown in Figure 1.1. The horizontal real number line is usually called the **x-axis,** and the vertical real number line is usually called the **y-axis.** The point of intersection of these two axes is the **origin,** and the two axes divide the plane into four parts called **quadrants.**

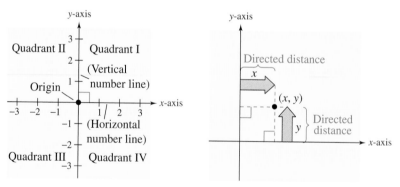

Figure 1.1   **Figure 1.2**

Each point in the plane corresponds to an **ordered pair** (x, y) of real numbers x and y, called the **coordinates** of the point. The **x-coordinate** represents the directed distance from the *y*-axis to the point, and the **y-coordinate** represents the directed distance from the *x*-axis to the point, as shown in Figure 1.2.

$$\underset{\substack{\text{Directed distance} \\ \text{from } y\text{-axis}}}{} \; (x, y) \; \underset{\substack{\text{Directed distance} \\ \text{from } x\text{-axis}}}{}$$

The notation (x, y) denotes both a point in the plane and an open interval on the real number line. The context will tell you which meaning is intended.

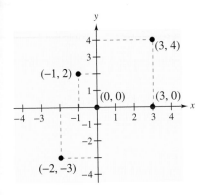

Figure 1.3

EXAMPLE 1 PLOTTING POINTS IN THE CARTESIAN PLANE

Plot the points $(-1, 2)$, $(3, 4)$, $(0, 0)$, $(3, 0)$, and $(-2, -3)$.

SOLUTION

To plot the point $(-1, 2)$, imagine a vertical line through -1 on the x-axis and a horizontal line through 2 on the y-axis. The intersection of these two lines is the point $(-1, 2)$. The other four points can be plotted in a similar way, as shown in Figure 1.3.

✓CHECKPOINT   Now try Exercise 5(a).

The Distance and Midpoint Formulas

Recall from the Pythagorean Theorem that, for a right triangle with hypotenuse of length c and legs of lengths a and b, you have $a^2 + b^2 = c^2$, as shown in Figure 1.4. (The converse is also true. That is, if $a^2 + b^2 = c^2$, then the triangle is a right triangle.)

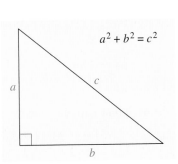

Figure 1.4

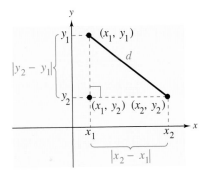

Figure 1.5

Suppose you want to determine the distance d between two points (x_1, y_1) and (x_2, y_2) that do not lie on the same horizontal or vertical line. With these two points, a right triangle can be formed, as shown in Figure 1.5. The length of the vertical side of the triangle is $|y_2 - y_1|$, and the length of the horizontal side is $|x_2 - x_1|$. By the Pythagorean Theorem, you can write

$$d^2 = |x_2 - x_1|^2 + |y_2 - y_1|^2$$
$$d = \sqrt{|x_2 - x_1|^2 + |y_2 - y_1|^2} \qquad \text{Choose positive square root.}$$
$$d = \sqrt{(x_2 - x_1)^2 + (y_2 - y_1)^2}.$$

The result is the **Distance Formula**.

The Distance Formula

The distance d between the points (x_1, y_1) and (x_2, y_2) in the coordinate plane is

$$d = \sqrt{(x_2 - x_1)^2 + (y_2 - y_1)^2}.$$

The following formula shows how to find the *midpoint* of the line segment that joins two points.

The Midpoint Formula

The **midpoint** of the line segment joining the points (x_1, y_1) and (x_2, y_2) in the coordinate plane is

$$\left(\frac{x_1 + x_2}{2}, \frac{y_1 + y_2}{2}\right).$$

EXAMPLE 2   USING THE DISTANCE AND MIDPOINT FORMULAS

Find (a) the distance between and (b) the midpoint of the line segment joining the points $(-2, 1)$ and $(3, 4)$.

SOLUTION

a. Let $(x_1, y_1) = (-2, 1)$ and $(x_2, y_2) = (3, 4)$, and apply the Distance Formula.

$$d = \sqrt{[3 - (-2)]^2 + (4 - 1)^2} \qquad \text{Distance Formula}$$

$$= \sqrt{5^2 + 3^2} \qquad \text{Simplify.}$$

$$= \sqrt{34} \approx 5.83 \qquad \text{Simplify.}$$

See Figure 1.6.

b. By the Midpoint Formula, you have

$$\text{Midpoint} = \left(\frac{x_1 + x_2}{2}, \frac{y_1 + y_2}{2}\right) \qquad \text{Midpoint Formula}$$

$$= \left(\frac{-2 + 3}{2}, \frac{1 + 4}{2}\right) \qquad \text{Substitute for } x_1, x_2, y_1, \text{ and } y_2.$$

$$= \left(\frac{1}{2}, \frac{5}{2}\right). \qquad \text{Simplify.}$$

See Figure 1.7.

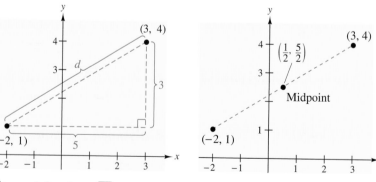

Figure 1.6 $d = \sqrt{34} \approx 5.83$ **Figure 1.7**

✓CHECKPOINT   Now try Exercises 5(b) and 5(c).

The Graph of an Equation

Frequently, a relationship between two quantities is written in the form of an equation. In the remainder of this section, you will study a procedure for sketching the graph of an equation. For an equation in the variables x and y, a point (a, b) is a **solution** if the substitution $x = a$ and $y = b$ satisfies the equation.

EXAMPLE 3   SOLUTION OF AN EQUATION

Determine whether $(-1, 0)$ is a solution of the equation $y = 2x^2 - 4x - 6$.

SOLUTION

$$y = 2x^2 - 4x - 6 \qquad \text{Write original equation.}$$
$$0 \overset{?}{=} 2(-1)^2 - 4(-1) - 6 \qquad \text{Substitute } -1 \text{ for } x \text{ and } 0 \text{ for } y.$$
$$0 = 0 \qquad \text{Simplify.}$$

Both sides of the equation are equivalent, so the point $(-1, 0)$ is a solution.

✓CHECKPOINT   Now try Exercise 21.

Most equations have *infinitely* many solutions. The **graph of an equation** is the set of all points that are solutions of the equation.

EXAMPLE 4   SKETCHING THE GRAPH OF AN EQUATION

Sketch the graph of $3x + y = 5$.

SOLUTION

First rewrite the equation as $y = 5 - 3x$ with y isolated on the left. Next, construct a table of values by choosing several values of x and calculating the corresponding values of y.

x	-1	0	1	2	3
$y = 5 - 3x$	8	5	2	-1	-4

From the table, it follows that $(-1, 8)$, $(0, 5)$, $(1, 2)$, $(2, -1)$, and $(3, -4)$ are solution points of the equation. After plotting these points and connecting them, you can see that they appear to lie on a line, as shown in Figure 1.8.

✓CHECKPOINT   Now try Exercise 25.

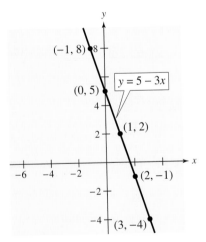

Figure 1.8

The Point-Plotting Method of Graphing

1. If possible, isolate one of the variables.

2. Construct a table of values showing several solution points.

3. Plot these points on a rectangular coordinate system.

4. Connect the points with a smooth curve or line.

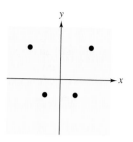

Figure 1.9

Step 4 of the point-plotting method can be difficult. For instance, how would you connect the four points in Figure 1.9? Without further information about the equation, any one of the three graphs in Figure 1.10 would be reasonable. These graphs show that with too few solution points, you can misrepresent the graph of an equation. Throughout this course, you will study many ways to improve your graphing techniques. For now, you should plot enough points to reveal the essential behavior of the graph. It is important to use negative values, zero, and positive values for x when constructing a table.

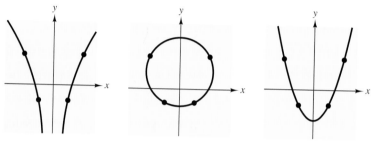

Figure 1.10

EXAMPLE 5   SKETCHING THE GRAPH OF AN EQUATION

Sketch the graph of $y = x^2 - 2$.

SOLUTION

First, construct a table of values by choosing several convenient values of x and calculating the corresponding values of y.

x	-3	-2	-1	0	1	2	3
$y = x^2 - 2$	7	2	-1	-2	-1	2	7

Next, plot the corresponding solution points. Finally, connect the points with a smooth curve, as shown in Figure 1.11.

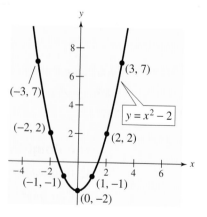

Figure 1.11

✔CHECKPOINT   Now try Exercise 26.

Intercepts of a Graph

When you are sketching a graph, two types of points that are especially useful are those for which either the *y*-coordinate or the *x*-coordinate is zero.

Definition of Intercepts

1. The **x-intercepts** of a graph are the points at which the graph intersects the *x*-axis. To find the *x*-intercepts, let *y* be zero and solve for *x*.

2. The **y-intercepts** of a graph are the points at which the graph intersects the *y*-axis. To find the *y*-intercepts, let *x* be zero and solve for *y*.

Some texts denote the *x*-intercept as the *x*-coordinate of the point $(a, 0)$ rather than the point itself. Unless it is necessary to make a distinction, we will use the term *intercept* to mean either the point or the coordinate.

Graphs may have no intercepts, one intercept, or several intercepts. For instance, consider the three graphs in Figure 1.12.

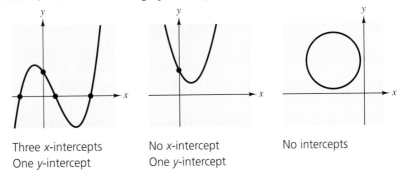

Three *x*-intercepts   No *x*-intercept   No intercepts
One *y*-intercept   One *y*-intercept

Figure 1.12

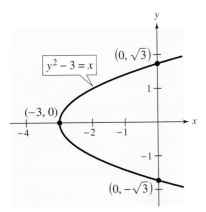

Figure 1.13

EXAMPLE 6   FINDING X- AND Y-INTERCEPTS

Find the *x*- and *y*-intercepts of the graph of

$$y^2 - 3 = x.$$

SOLUTION

To find the *x*-intercept, let $y = 0$. This produces $-3 = x$, which implies that the graph has one *x*-intercept, which occurs at

$$(-3, 0). \qquad \text{\small x-intercept}$$

To find the *y*-intercept, let $x = 0$. This produces $y^2 - 3 = 0$, which has two solutions: $y = \pm\sqrt{3}$. So, the graph has two *y*-intercepts, which occur at

$$\left(0, \sqrt{3}\right) \quad \text{and} \quad \left(0, -\sqrt{3}\right). \qquad \text{\small y-intercepts}$$

See Figure 1.13.

✓CHECKPOINT   Now try Exercise 29.

Symmetry

Symmetry with respect to the *x*-axis means that if the Cartesian plane were folded along the *x*-axis, the portion of the graph above the *x*-axis would coincide with the portion below the *x*-axis. Symmetry with respect to the *y*-axis or the origin can be described in a similar manner. Symmetry with respect to the origin means that if the Cartesian plane were rotated 180° about the origin, the portion of the graph to the right of the origin would coincide with the portion to the left of the origin. (See Figure 1.14.)

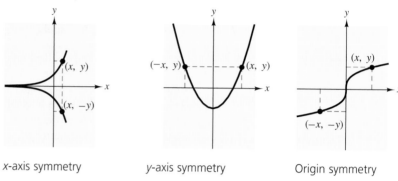

x-axis symmetry   *y*-axis symmetry   Origin symmetry

Figure 1.14

Knowing the symmetry of a graph *before* attempting to sketch it is helpful, because then you need only half as many solution points to sketch the graph. The three basic types of symmetry are described as follows.

Definition of Symmetry

1. A graph is **symmetric with respect to the *x*-axis** if, whenever (x, y) is on the graph, $(x, -y)$ is also on the graph.

2. A graph is **symmetric with respect to the *y*-axis** if, whenever (x, y) is on the graph, $(-x, y)$ is also on the graph.

3. A graph is **symmetric with respect to the origin** if, whenever (x, y) is on the graph, $(-x, -y)$ is also on the graph.

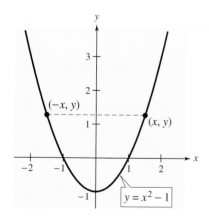

Figure 1.15 *y*-Axis Symmetry

You can apply this definition of symmetry to the graph of the equation $y = x^2 - 1$. Replacing *x* with $-x$ produces the following.

$y = x^2 - 1$	Write original equation.
$y = (-x)^2 - 1$	Replace *x* with $-x$.
$y = x^2 - 1$	Replacement yields equivalent equation.

Because the substitution did not change the equation, it follows that if (x, y) is a solution of the equation, then $(-x, y)$ must also be a solution. So, the graph of $y = x^2 - 1$ is symmetric with respect to the *y*-axis, as shown in Figure 1.15.

Tests for Symmetry

1. The graph of an equation is symmetric with respect to the *x-axis* if replacing *y* with $-y$ yields an equivalent equation.

2. The graph of an equation is symmetric with respect to the *y-axis* if replacing *x* with $-x$ yields an equivalent equation.

3. The graph of an equation is symmetric with respect to the *origin* if replacing *x* with $-x$ *and y* with $-y$ yields an equivalent equation.

EXAMPLE 7   USING SYMMETRY AS A SKETCHING AID

Describe the symmetry of the graph of $x - y^2 = 1$.

SOLUTION

Of the three tests for symmetry, the only one that is satisfied by this equation is the test for *x*-axis symmetry.

$$x - y^2 = 1 \qquad \text{Write original equation.}$$
$$x - (-y)^2 = 1 \qquad \text{Replace } y \text{ with } -y.$$
$$x - y^2 = 1 \qquad \text{Replacement yields equivalent equation.}$$

So, the graph is symmetric with respect to the *x*-axis. To sketch the graph, plot the points above the *x*-axis and use symmetry to complete the graph, as shown in Figure 1.16.

✓CHECKPOINT   Now try Exercise 43.

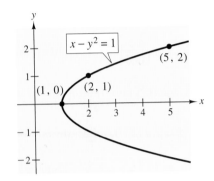

Figure 1.16

The Equation of a Circle

So far in this section, you have studied the point-plotting method and two additional concepts (intercepts and symmetry) that can be used to streamline the graphing procedure. Another graphing aid is *equation recognition,* which is the ability to recognize the general shape of a graph simply by looking at its equation.

Figure 1.17 shows a circle of *radius r* with *center* at the point (h, k). The point (x, y) is on this circle if and only if its distance from the center (h, k) is *r*. This means that a **circle** in the plane consists of all points (x, y) that are a given positive distance *r* from a fixed point (h, k). Using the Distance Formula, you can conclude that the point (x, y) lies on the circle if and only if

$$\sqrt{(x - h)^2 + (y - k)^2} = r.$$

By squaring each side of this equation, you obtain the **standard form of the equation of a circle.** For example, a circle with its center at the origin, $(h, k) = (0, 0)$, and radius $r = 4$ is given by

$$\sqrt{(x - 0)^2 + (y - 0)^2} = 4 \qquad \text{Substitute for } h, k, \text{ and } r.$$
$$\sqrt{x^2 + y^2} = 4 \qquad \text{Simplify.}$$
$$x^2 + y^2 = 16. \qquad \text{Square each side.}$$

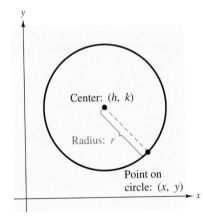

Figure 1.17

STUDY TIP

To find the correct h and k given the standard form of the equation of the circle in Example 8, it may be helpful to rewrite the quantities $(x + 1)^2$ and $(y - 2)^2$ using subtraction.

$$(x + 1)^2 = [x - (-1)]^2,$$

$$h = -1$$

$$(y - 2)^2 = [y - (2)]^2,$$

$$k = 2$$

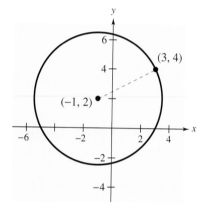

Figure 1.18

Standard Form of the Equation of a Circle

The **standard form of the equation of a circle** is

$$(x - h)^2 + (y - k)^2 = r^2.$$

The point (h, k) is called the **center** of the circle, and the positive number r is called the **radius** of the circle. The standard form of the equation of a circle whose center is the *origin* is $x^2 + y^2 = r^2$.

EXAMPLE 8   FINDING THE EQUATION OF A CIRCLE

The point $(3, 4)$ lies on a circle whose center is at $(-1, 2)$, as shown in Figure 1.18. Write the standard form of the equation of this circle.

SOLUTION

The radius of the circle is the distance between $(-1, 2)$ and $(3, 4)$.

$$r = \sqrt{(x - h)^2 + (y - k)^2} \qquad \text{Distance Formula}$$

$$r = \sqrt{[3 - (-1)]^2 + (4 - 2)^2} \qquad \text{Substitute for } x, y, h, \text{ and } k.$$

$$= \sqrt{4^2 + 2^2} \qquad \text{Simplify.}$$

$$= \sqrt{16 + 4} \qquad \text{Simplify.}$$

$$= \sqrt{20} \qquad \text{Radius}$$

Using $(h, k) = (-1, 2)$ and $r = \sqrt{20}$, the equation of the circle is

$$(x - h)^2 + (y - k)^2 = r^2 \qquad \text{Equation of circle}$$

$$[x - (-1)]^2 + (y - 2)^2 = \left(\sqrt{20}\right)^2 \qquad \text{Substitute for } h, k, \text{ and } r.$$

$$(x + 1)^2 + (y - 2)^2 = 20. \qquad \text{Standard form}$$

✔CHECKPOINT   Now try Exercise 81.

If you remove the parentheses in the standard equation in Example 8, you obtain the following.

$$(x + 1)^2 + (y - 2)^2 = 20 \qquad \text{Standard form}$$

$$x^2 + 2x + 1 + y^2 - 4y + 4 = 20 \qquad \text{Expand terms.}$$

$$x^2 + y^2 + 2x - 4y - 15 = 0 \qquad \text{General form}$$

The last equation is the **general form of the equation of a circle.**

$$Ax^2 + Ay^2 + Dx + Ey + F = 0, \qquad A \neq 0$$

The general form of the equation of a circle is less useful than the standard form. For instance, it is not immediately apparent from the general equation shown above that the center is $(-1, 2)$ and the radius is $\sqrt{20}$. To graph the equation of a circle, it is best to write the equation in standard form. You can do this by **completing the square,** as demonstrated in Example 9.

EXAMPLE 9   COMPLETING THE SQUARE TO SKETCH A CIRCLE

Sketch the circle given by $4x^2 + 4y^2 + 20x - 16y + 37 = 0$.

SOLUTION

Begin by writing the original equation in standard form by completing the square for both the x-terms *and* the y-terms.

$$4x^2 + 4y^2 + 20x - 16y + 37 = 0 \qquad \text{Write original equation.}$$

$$x^2 + y^2 + 5x - 4y + \frac{37}{4} = 0 \qquad \text{Divide by 4.}$$

$$\left(x^2 + 5x + \quad\right) + \left(y^2 - 4y + \quad\right) = -\frac{37}{4} \qquad \text{Group terms.}$$

$$\left[x^2 + 5x + \left(\frac{5}{2}\right)^2\right] + (y^2 - 4y + 2^2) = -\frac{37}{4} + \frac{25}{4} + 4 \quad \text{Complete the square.}$$

$$\left(x + \frac{5}{2}\right)^2 + (y - 2)^2 = 1 \qquad \text{Standard form}$$

STUDY TIP

Recall that to complete the square, you add the square of half the coefficient of the linear term to each side.

So, the center of the circle is $\left(-\frac{5}{2}, 2\right)$ and the radius of the circle is 1. Using this information, you can sketch the circle, as shown in Figure 1.19.

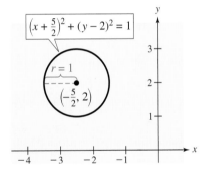

Figure 1.19

✓CHECKPOINT   Now try Exercise 85.

Discussing the Concept │ Interpreting an Intercept

Create a graph depicting each real-life situation, and find the indicated intercept. Explain the practical significance of the intercept.

a. (y-intercept) The trade-in value y of a used car x years after its purchase is $y = 12{,}000 - 3000x$.

b. (x-intercept) The number of families y demanding child care services each week is related to the price per hour x of child care by $y = 7850 - 20x^2$.

Find an example of a real-life situation for which it makes sense for there to be no x-intercept.

1.1 Warm Up

The following warm-up exercises involve skills that were covered in earlier sections. You will use these skills in the exercise set for this section. For additional help, review Sections R1.2, R1.4, and R2.5.

In Exercises 1–6, simplify the expression.

1. $\sqrt{(2-6)^2 + [1-(-2)]^2}$

2. $\sqrt{(1-4)^2 + (-2-1)^2}$

3. $\dfrac{4+(-2)}{2}$

4. $\dfrac{-1+(-3)}{2}$

5. $\sqrt{18} + \sqrt{45}$

6. $\sqrt{12} + \sqrt{44}$

In Exercises 7–10, solve the equation.

7. $\sqrt{(4-x)^2 + (5-2)^2} = \sqrt{58}$

8. $\sqrt{(8-6)^2 + (y-5)^2} = 2\sqrt{5}$

9. $x^3 - 9x = 0$

10. $x^4 - 8x^2 + 16 = 0$

1.1 Exercises

In Exercises 1–12, (a) plot the points, (b) find the distance between the points, and (c) find the midpoint of the line segment joining the points.

1. $(2, -5), (-6, 1)$

2. $(1, 12), (6, 0)$

3. $(3, -11), (-12, -3)$

4. $(-7, 3), (2, -9)$

5. $(-1, 2), (5, 4)$

6. $(2, 10), (10, 2)$

7. $\left(\frac{1}{2}, 1\right), \left(-\frac{5}{2}, \frac{4}{3}\right)$

8. $\left(-\frac{1}{3}, -\frac{1}{3}\right), \left(-\frac{1}{6}, -\frac{1}{2}\right)$

9. $(1.7, 8.5), (-3.2, 5.3)$

10. $(21.7, 10.2), (7.9, -2.3)$

11. $(-36, -18), (48, -72)$

12. $(1.451, 3.051), (5.906, 11.360)$

In Exercises 13–16, find the length of the hypotenuse in two ways: (a) use the Pythagorean Theorem and (b) use the Distance Formula.

13.
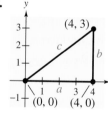

14.
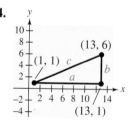

15.
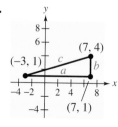

16.
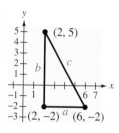

In Exercises 17 and 18, find x such that the distance between the points is 15.

17. $(3, -4), (x, 5)$

18. $(x, 8), (-9, -4)$

In Exercises 19 and 20, find y such that the distance between the points is 20.

19. $(-15, y), (-3, -7)$

20. $(6, -1), (-10, y)$

In Exercises 21–24, determine whether each point is a solution of the equation.

	Equation	Points	
21.	$y = 2x^2 - 3$	(a) $(-1, -1)$	(b) $(2, 5)$
22.	$4y - 3x + 7 = 0$	(a) $(1, -2)$	(b) $(4, 3)$
23.	$y = \sqrt{x - 8}$	(a) $(9, 1)$	(b) $(17, 3)$
24.	$y = \dfrac{2x}{6 - x}$	(a) $(-3, 0)$	(b) $(-4, -8)$

In Exercises 25 and 26, complete the table. Use the resulting solution points to sketch the graph of the equation.

25. $y = \frac{3}{4}x - 1$

x	-2	0	1	$\frac{4}{3}$	2
y					

26. $y = 5 - x^2$

x	-2	-1	0	1	2
y					

In Exercises 27–34, find the *x*- and *y*-intercepts of the graph of the equation.

27. $y = 2x - 1$

28. $y = (x - 4)(x + 2)$

29. $y = x^2 + x - 2$

30. $y = 4 - x^2$

31. $y = x\sqrt{x + 2}$

32. $y = x\sqrt{x + 3}$

33. $2y - xy + 3x = 4$

34. $x^2y - x^2 + 4y = 0$

35. Use your knowledge of the Cartesian plane and intercepts to explain why you let *y* equal zero when you are finding the *x*-intercepts of the graph of an equation, and why you let *x* equal zero when you are finding the *y*-intercepts of the graph of an equation.

36. Is it possible for a graph to have no *x*-intercepts? no *y*-intercepts? no *x*-intercepts and no *y*-intercepts? Give examples to support your answers.

In Exercises 37–46, check for symmetry with respect to both axes and the origin.

37. $x^4 - 2y = 0$

38. $y = x^4 - x^2 + 3$

39. $x - y^2 = 0$

40. $x^2 + y^2 = 9$

41. $y^2 = x + 3$

42. $y = \sqrt{4 - x^2}$

43. $y = \sqrt{9 - x^2}$

44. $x^3y = 1$

45. $xy = 4$

46. $y = \dfrac{x}{x^2 + 1}$

In Exercises 47–50, use symmetry to complete the graph of the equation.

47. *y*-axis symmetry

$y = -x^2 + 4$

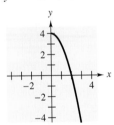

48. *x*-axis symmetry

$y^2 = -x + 4$

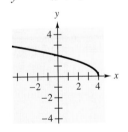

49. Origin symmetry

$y = -x^3 + x$

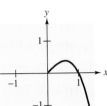

50. *y*-axis symmetry

$y = |x| - 2$

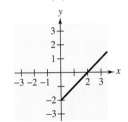

In Exercises 51–56, match the equation with its graph. [The graphs are labeled (a), (b), (c), (d), (e), and (f).]

(a)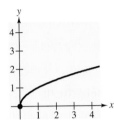

(b)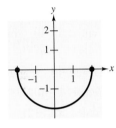

(c)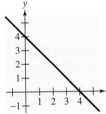

(d)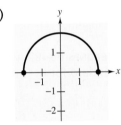

(e)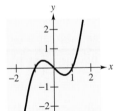

(f)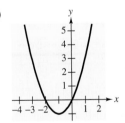

51. $y = 4 - x$

52. $y = \sqrt{4 - x^2}$

53. $y = x^2 + 2x$

54. $y = \sqrt{x}$

55. $y = x^3 - x$

56. $y = -\sqrt{4 - x^2}$

In Exercises 57–76, sketch the graph of the equation. Identify any intercepts and test for symmetry.

57. $y = 5 - 3x$

58. $y = 2x - 3$

59. $y = 1 - x^2$

60. $y = x^2 - 1$

61. $y = x^2 - 4x + 3$

62. $y = -x^2 - 4x$

63. $y = x^3 + 2$

64. $y = x^3 - 1$

65. $y = \dfrac{8}{x^2 + 4}$

66. $y = \dfrac{4}{x^2 + 1}$

67. $y = \sqrt{x + 1}$

68. $y = \sqrt{1 - x}$

69. $y = \sqrt[3]{x}$

70. $y = \sqrt[3]{x + 1}$

71. $y = |x - 4|$

72. $y = |x| - 3$

73. $x = y^2 - 1$

74. $x = y^2 - 4$

75. $x^2 + y^2 = 4$

76. $x^2 + y^2 = 16$

In Exercises 77–84, find the standard form of the equation of the specified circle.

77. Center: $(0, 0)$; radius: 3

78. Center: $(0, 0)$; radius: 5

79. Center: $(-4, 1)$; radius: $\sqrt{2}$

80. Center: $\left(0, \frac{1}{3}\right)$; radius: $\frac{1}{3}$

81. Center: $(-1, 2)$; point on circle: $(0, 0)$

82. Center: $(3, -2)$; point on circle: $(-1, 1)$

83. Endpoints of a diameter: $(-3, 4)$, $(5, -2)$

84. Endpoints of a diameter: $(-4, -1)$, $(4, 1)$

In Exercises 85–92, write the equation of the circle in standard form. Then sketch the circle.

85. $x^2 + y^2 - 6x + 4y - 3 = 0$

86. $x^2 + y^2 - 2x + 6y - 15 = 0$

87. $x^2 + y^2 - 2x + 6y + 10 = 0$

88. $5x^2 + 5y^2 + 10x + 1 = 0$

89. $2x^2 + 2y^2 - 2x - 2y - 3 = 0$

90. $4x^2 + 4y^2 - 4x + 2y - 1 = 0$

91. $16x^2 + 16y^2 + 16x + 40y - 7 = 0$

92. $x^2 + y^2 - 4x + 2y + 3 = 0$

In Exercises 93 and 94, an equation of a circle is written in standard form. Indicate the coordinates of the center of the circle and determine the radius of the circle. Rewrite the equation of the circle in general form.

93. $(x - 2)^2 + (y + 3)^2 = 16$

94. $\left(x + \frac{1}{2}\right)^2 + (y + 1)^2 = 5$

Gold Prices   In Exercises 95 and 96, use the figure, which shows the average price of gold from 1975 to 2002.   (Source: U.S. Bureau of Mines; U.S. Geological Survey)

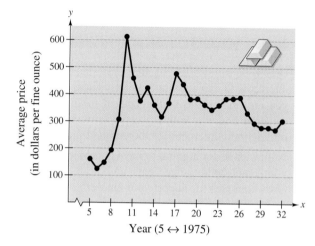

95. What is the highest price of gold shown in the graph? When did this occur?

96. What is the lowest price of gold shown in the graph? When did this occur?

Fuel Efficiency   In Exercises 97 and 98, use the figure, which shows the average fuel efficiency for automobiles in the United States from 1980 to 2001.   (Source: U.S. Federal Highway Administration)

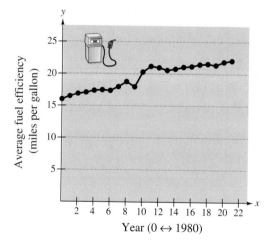

97. Estimate the percent increase in fuel efficiency from 1980 to 1990.

98. Estimate the percent increase in fuel efficiency from 1980 to 2001.

99. Life Expectancy The table shows the life expectancy of a child (at birth) for selected years from 1970 to 2000. (Source: U.S. National Center for Health Statistics)

Year	1970	1975	1980
Life expectancy, *y*	70.8	72.6	73.7

Year	1985	1990	1995	2000
Life expectancy, *y*	74.7	75.4	75.8	77.0

A mathematical model for the life expectancy during this period is

$$y = \frac{71.072 + 1.545t}{1 + 0.018t}, \quad 0 \le t \le 30$$

where *y* represents the life expectancy in years and $t = 0$ represents 1970.

(a) Create a table to compare the actual data to the values given by the model.

(b) Sketch a scatter plot of the data. Then sketch a graph of the model on the same set of coordinate axes.

(c) Use the model to predict the life expectancy of a child born in the year 2005.

100. *Make a Decision*: **Earnings per Share** The earnings per share *y* (in dollars) for Microsoft Corporation from 1990 to 2002 can be modeled by

$$y = 0.0074t^2 - 0.004t + 0.03, \quad 0 \le t \le 12$$

where *t* represents the year, with $t = 0$ corresponding to 1990. Sketch the graph of this equation. In 2002, Valueline projected that the earnings per share for Microsoft Corporation in 2003 and 2004 would be $0.97 and $0.88, respectively. Compare Valueline's projections for 2003 and 2004 with the earnings per share predicted by the model. Do Valueline's estimates agree with the model? (Source: Microsoft Corporation and Valueline)

101. *Make a Decision*: **Earnings per Share** The earnings per share *y* (in dollars) for Home Depot stores from 1990 to 2002 can be modeled by

$$y = 0.0107t^2 - 0.010t + 0.13, \quad 0 \le t \le 12$$

where *t* represents the year, with $t = 0$ corresponding to 1990. Sketch the graph of this equation. In 2002, Valueline projected that the earnings per share for Home Depot in 2003 and 2004 would be $1.80 and $2.05, respectively. Compare Valueline's projections for 2003 and 2004 with the earnings per share predicted by the model. Do Valueline's estimates agree with the model? (Source: Home Depot, Inc. and Valueline)

IN THE NEWS

New Study Finds Tortillas Are The Second Most Popular Bread Type in America

The $5.2 Billion Tortilla Industry Trails White Bread Sales by Two Percent; Far Surpasses Wheat Bread Product Sales

DALLAS, TX—The Tortilla Industry Association (TIA) has announced the findings of its biennial market research study, the State of the Tortilla Industry Survey: 2002, conducted by Aspex Research. Findings show that the tortilla's popularity has reached record heights, making this staple literally the best thing since sliced bread. Having cornered 32 percent of the sales for the U.S. Bread Industry, tortillas trail white bread sales by only two percent—making them the second most popular bread type in America with sales that far surpass those of whole wheat bread, bagels and rolls. Over the past two years, annual growth for the tortilla industry has expanded nine percent, with 2002, U.S. tortilla industry sales reaching $5.2 billion.

The Tortilla Industry Association, "New study finds tortillas are the second most popular bread type in America," www.tortilla-info.com/prrevenue03.htm.

A mathematical model that approximates the sales *y* (in billions of dollars) of the tortilla industry from 1996 to 2002 is

$$y = 0.42t + 0.2, \quad 6 \le t \le 12$$

where *t* represents the year, with $t = 6$ corresponding to 1996. (Source: Tortilla Industry Association, RIA, Mintel report)

102. Sketch a graph of the model.

103. Use the model to predict tortilla sales in 2006.

1.2   Lines in the Plane

Objectives

- **Find the slope of a line passing through two points.**
- **Use the point-slope form to find the equation of a line.**
- **Use the slope-intercept form to sketch a line.**
- **Use slope to determine if lines are parallel or perpendicular, and write the equation of a line parallel or perpendicular to a given line.**

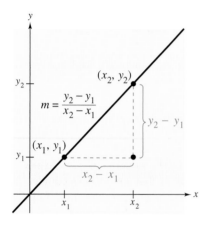

Figure 1.20

The Slope of a Line

The **slope** of a nonvertical line is a measure of the steepness of the line. Specifically, the slope represents the number of units the line rises or falls vertically for each unit of horizontal change from left to right. For instance, consider the two points (x_1, y_1) and (x_2, y_2) on the line shown in Figure 1.20. As you move from left to right along this line, a change of $y_2 - y_1$ units in the vertical direction corresponds to a change of $x_2 - x_1$ units in the horizontal direction. That is,

$$y_2 - y_1 = \text{the change in } y \quad \text{and} \quad x_2 - x_1 = \text{the change in } x.$$

The slope of the line is defined as the quotient of these two changes.

Definition of the Slope of a Line

The **slope** m of the nonvertical line passing through the points (x_1, y_1) and (x_2, y_2) is

$$m = \frac{y_2 - y_1}{x_2 - x_1} = \frac{\text{change in } y}{\text{change in } x}$$

where $x_1 \neq x_2$.

The change in x is sometimes called the *run* and the change in y is sometimes called the *rise*.

When this formula is used for slope, the *order of subtraction* is important. Given two points on a line, you are free to label either one of them as (x_1, y_1) and the other as (x_2, y_2). However, once this is done, you must form the numerator and denominator using the same order of subtraction.

$$m = \frac{y_2 - y_1}{x_2 - x_1} \qquad m = \frac{y_1 - y_2}{x_1 - x_2} \qquad m = \frac{y_2 - y_1}{x_1 - x_2}$$

Correct   Correct   Incorrect

In real-life problems, such as finding the steepness of a ramp or the increase in the value of a product, the slope of a line can be interpreted as either a *ratio* or a *rate*. If the x-axis and the y-axis have the same units of measure, then the slope has no units and is a *ratio*. If the x-axis and the y-axis have different units of measure, then the slope is a *rate* or *rate of change*. You will learn more about rates of change in Section 1.3.

EXAMPLE 1   FINDING THE SLOPE OF A LINE THROUGH TWO POINTS

Find the slope of the line passing through each pair of points.

a. $(-2, 0)$ and $(3, 1)$ **b.** $(-1, 2)$ and $(2, 2)$

c. $(0, 4)$ and $(1, -1)$ **d.** $(3, 4)$ and $(3, 1)$

SOLUTION

a. Letting $(x_1, y_1) = (-2, 0)$ and $(x_2, y_2) = (3, 1)$, you obtain a slope of

$$m = \frac{y_2 - y_1}{x_2 - x_1}$$ ◁ Difference in y-values
◁ Difference in x-values

$$= \frac{1 - 0}{3 - (-2)}$$

$$= \frac{1}{5}.$$

b. The slope of the line passing through $(-1, 2)$ and $(2, 2)$ is

$$m = \frac{2 - 2}{2 - (-1)} = \frac{0}{3} = 0.$$

c. The slope of the line passing through $(0, 4)$ and $(1, -1)$ is

$$m = \frac{-1 - 4}{1 - 0} = \frac{-5}{1} = -5.$$

d. The slope of the line passing through $(3, 4)$ and $(3, 1)$ is undefined. Applying the formula for slope, you have

$$m = \frac{1 - 4}{3 - 3} = \frac{-3}{0}.$$ Division by zero is undefined.

Because division by zero is not defined, the slope of a vertical line is not defined.

The graphs of the four lines are shown in Figure 1.21.

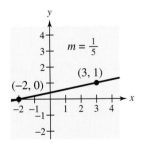

(a) If m is positive, the line rises from left to right.

(b) If m is zero, the line is horizontal.

(c) If m is negative, the line falls from left to right.

(d) If the line is vertical, the slope is undefined.

Figure 1.21

✓**CHECKPOINT**   Now try Exercise 11.

From the slopes of the lines shown in Example 1, you can make the following generalizations about the slope of a line.

Slope of a Line

1. A line with positive slope ($m > 0$) *rises* from left to right.
2. A line with negative slope ($m < 0$) *falls* from left to right.
3. A line with zero slope ($m = 0$) is *horizontal*.
4. A line with undefined slope is *vertical*.

The Point-Slope Form

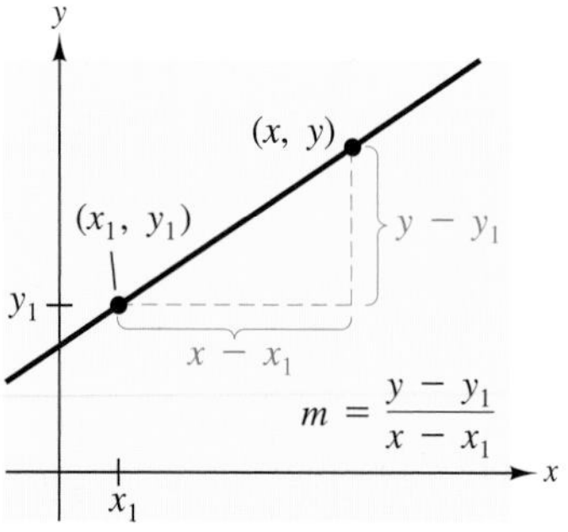

If you know the slope of a line and you also know the coordinates of one point on the line, you can find an equation for the line. For instance, in Figure 1.22, let (x_1, y_1) be a given point on the line whose slope is m. If (x, y) is any *other* point on the line, it follows that

$$\frac{y - y_1}{x - x_1} = m.$$

This equation in the variables x and y can be rewritten to produce the following **point-slope form** of the equation of a line.

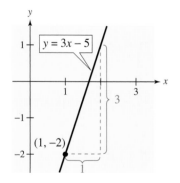

Figure 1.22 Any two points on a line can be used to determine the slope of the line.

Point-Slope Form of the Equation of a Line

The **point-slope form** of the equation of the line that passes through the point (x_1, y_1) and has a slope of m is

$$y - y_1 = m(x - x_1).$$

EXAMPLE 2   THE POINT-SLOPE FORM OF THE EQUATION OF A LINE

Find an equation of the line that passes through $(1, -2)$ and has a slope of 3.

SOLUTION

Use the point-slope form with $(x_1, y_1) = (1, -2)$ and $m = 3$.

$$y - y_1 = m(x - x_1) \qquad \text{Point-slope form}$$
$$y - (-2) = 3(x - 1) \qquad \text{Substitute } y_1 = -2, x_1 = 1, \text{ and } m = 3.$$
$$y + 2 = 3x - 3 \qquad \text{Simplify.}$$
$$y = 3x - 5 \qquad \text{Equation of line}$$

The graph of this line is shown in Figure 1.23.

✓CHECKPOINT   Now try Exercise 25.

Figure 1.23

The point-slope form can be used to find the equation of a line passing through two points (x_1, y_1) and (x_2, y_2). First, use the formula for the slope of a line passing through two points. Then, use the point-slope form to obtain

$$y - y_1 = \frac{y_2 - y_1}{x_2 - x_1}(x - x_1).$$

This is sometimes called the **two-point form** of the equation of a line.

EXAMPLE 3 A LINEAR MODEL FOR SALES PREDICTION

During the first two quarters of the year, a jewelry company had sales of $3.4 million and $3.7 million, respectively.

a. Write a linear equation giving the sales y in terms of the quarter x.

b. *Make a Decision* Use the equation to predict the sales during the fourth quarter. Can you assume that sales will follow this linear pattern?

SOLUTION

a. Let $(1, 3.4)$ and $(2, 3.7)$ be two points on the line representing the total sales. The slope of the line passing through these two points is

$$m = \frac{3.7 - 3.4}{2 - 1} = 0.3.$$

By the point-slope form, the equation of the line is as follows.

$$y - 3.4 = 0.3(x - 1) \qquad \text{\small Substitute for } y_1, m, \text{ and } x_1 \text{ in point-slope form.}$$

$$y = 0.3x + 3.1 \qquad \text{\small Write in slope-intercept form.}$$

b. Using the equation from part (a), the fourth-quarter sales $(x = 4)$ should be $y = 0.3(4) + 3.1 = \$4.3$ million. See Figure 1.24. Without more data, you cannot assume that the sales pattern will be linear. Many factors, such as seasonal demand and past sales history, help to determine the sales pattern.

✓CHECKPOINT   Now try Exercise 75.

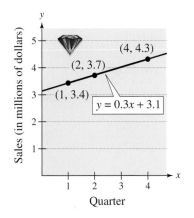

Figure 1.24

The estimation method illustrated in Example 3 is called **linear extrapolation.** Note in Figure 1.25(a) that for linear extrapolation, the estimated point lies to the *right* of the given points. When the estimated point lies *between* two given points, the procedure is called **linear interpolation,** as shown in Figure 1.25(b).

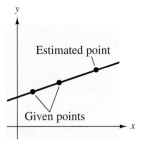

(a) Linear extrapolation

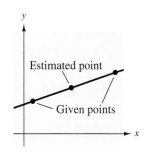

(b) Linear interpolation

Figure 1.25

Sketching Graphs of Lines

You have seen that to *find the equation of a line* it is convenient to use the point-slope form. This formula, however, is not particularly useful for *sketching the graph of a line.* The form that is better suited to graphing linear equations is the **slope-intercept form** of the equation of a line. To derive the slope-intercept form, write the following.

$$y - y_1 = m(x - x_1) \qquad \text{Point-slope form}$$

$$y = mx - mx_1 + y_1 \qquad \text{Solve for } y.$$

$$= mx + (y_1 - mx_1) \qquad \text{Commutative Property of Addition}$$

$$y = mx + b \qquad \text{Slope-intercept form}$$

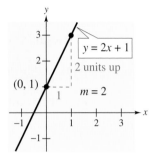

(a) When m is positive, the line rises from left to right.

Slope-Intercept Form of the Equation of a Line

The graph of the equation

$$y = mx + b$$

is a line whose slope is m and whose y-intercept is $(0, b)$.

EXAMPLE 4   SKETCHING THE GRAPHS OF LINEAR EQUATIONS

Sketch the graph of each linear equation.

a. $y = 2x + 1$

b. $y = 2$

c. $x + y = 2$

SOLUTION

a. Because $b = 1$, the y-intercept is $(0, 1)$. Moreover, because the slope is $m = 2$, this line *rises* two units for each unit it moves to the right, as shown in Figure 1.26(a).

b. By writing the equation $y = 2$ in the form

$$y = (0)x + 2$$

you can see that the y-intercept is $(0, 2)$ and the slope is zero. A zero slope implies that the line is horizontal, as shown in Figure 1.26(b).

c. By writing the equation $x + y = 2$ in slope-intercept form

$$y = -x + 2$$

you can see that the y-intercept is $(0, 2)$. Moreover, because the slope is $m = -1$, this line *falls* one unit for each unit it moves to the right, as shown in Figure 1.26(c).

✓CHECKPOINT   Now try Exercise 37.

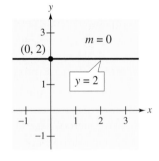

(b) When m is zero, the line is horizontal.

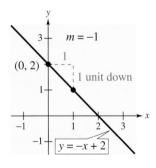

(c) When m is negative, the line falls from left to right.

Figure 1.26

From the slope-intercept form of the equation of a line, you can see that a horizontal line ($m = 0$) has an equation of the form

$$y = (0)x + b \quad \text{or} \quad y = b. \qquad \text{Horizontal line}$$

This is consistent with the fact that each point on a horizontal line through $(0, b)$ has a y-coordinate of b, as shown in Figure 1.27.

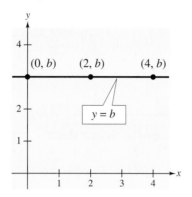

Figure 1.27 Horizontal Line   **Figure 1.28** Vertical Line

Similarly, each point on a vertical line through $(a, 0)$ has an x-coordinate of a, as shown in Figure 1.28. So, a vertical line has an equation of the form

$$x = a. \qquad \text{Vertical line}$$

This equation cannot be written in slope-intercept form because the slope of a vertical line is undefined. However, *every* line has an equation that can be written in the **general form**

$$Ax + By + C = 0 \qquad \text{General form}$$

where A and B are not *both* zero. If $A = 0$ (and $B \neq 0$), the general equation can be reduced to the form $y = b$, which represents a horizontal line. If $B = 0$ (and $A \neq 0$), the general equation can be reduced to the form $x = a$, which represents a vertical line.

Summary of Equations of Lines

1. General form: $Ax + By + C = 0$

2. Vertical line: $x = a$

3. Horizontal line: $y = b$

4. Slope-intercept form: $y = mx + b$

5. Point-slope form: $y - y_1 = m(x - x_1)$

Parallel and Perpendicular Lines

The slope of a line is a convenient tool for determining whether two lines are parallel or perpendicular.

DISCOVERY

Use a graphing utility to graph each equation in the same viewing window.

$$y_1 = \frac{3}{2}x - 1$$

$$y_2 = \frac{3}{2}x$$

$$y_3 = \frac{3}{2}x + 2$$

What is true about the graphs? What do you notice about the slopes of the equations?

Parallel Lines

Two distinct nonvertical lines are **parallel** if and only if their slopes are equal.

EXAMPLE 5   EQUATIONS OF PARALLEL LINES

Find an equation of the line that passes through the point $(2, -1)$ and is parallel to the line $2x - 3y = 5$, as shown in Figure 1.29.

SOLUTION

Writing the original equation in slope-intercept form produces the following.

$$2x - 3y = 5 \qquad \text{Write original equation.}$$

$$-3y = -2x + 5 \qquad \text{Subtract } 2x \text{ from each side.}$$

$$y = \frac{2}{3}x - \frac{5}{3} \qquad \text{Write in slope-intercept form.}$$

So, the given line has a slope of $m = \frac{2}{3}$. Because any line parallel to the given line must also have a slope of $\frac{2}{3}$, the required line through $(2, -1)$ has the following equation.

$$y - (-1) = \frac{2}{3}(x - 2) \qquad \text{Write in point-slope form.}$$

$$y + 1 = \frac{2}{3}x - \frac{4}{3} \qquad \text{Simplify.}$$

$$y = \frac{2}{3}x - \frac{4}{3} - 1 \qquad \text{Solve for } y.$$

$$y = \frac{2}{3}x - \frac{7}{3} \qquad \text{Write in slope-intercept form.}$$

Notice the similarity between the slope-intercept form of the original equation and the slope-intercept form of the parallel equation.

✓CHECKPOINT   Now try Exercise 63(a).

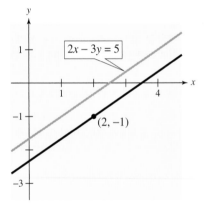

Figure 1.29

You have seen that two nonvertical lines are parallel if and only if they have the same slope. Two nonvertical lines are *perpendicular* if and only if their slopes are negative reciprocals of each other. For instance, the lines $y = 2x$ and $y = -\frac{1}{2}x$ are perpendicular because one has a slope of $2 = \frac{2}{1}$ and the other has a slope of $-\frac{1}{2}$.

DISCOVERY

Use a graphing utility to graph each equation in the same viewing window.

$$y_1 = \frac{2}{3}x + \frac{5}{2}$$

$$y_2 = -\frac{3}{2}x + 2$$

When you examine the graphs with a square setting, what do you observe? What do you notice about the slopes of the two lines?

Perpendicular Lines

Two nonvertical lines are **perpendicular** if and only if their slopes are negative reciprocals of each other. That is,

$$m_1 = -\frac{1}{m_2}.$$

EXAMPLE 6 EQUATIONS OF PERPENDICULAR LINES

Find an equation of the line that passes through the point $(2, -1)$ and is perpendicular to the line $2x - 3y = 5$, as shown in Figure 1.30.

SOLUTION

By writing the equation of the original line in the slope-intercept form $y = \frac{2}{3}x - \frac{5}{3}$, you can see that the line has a slope of $\frac{2}{3}$. So, any line that is perpendicular to this line must have a slope of $-\frac{3}{2}$ (because $-\frac{3}{2}$ is the negative reciprocal of $\frac{2}{3}$). The required line through the point $(2, -1)$ has the following equation.

$$y - (-1) = -\frac{3}{2}(x - 2) \qquad \text{Write in point-slope form.}$$

$$y + 1 = -\frac{3}{2}x + 3 \qquad \text{Simplify.}$$

$$y = -\frac{3}{2}x + 3 - 1 \qquad \text{Solve for } y.$$

$$y = -\frac{3}{2}x + 2 \qquad \text{Write in slope-intercept form.}$$

✓CHECKPOINT Now try Exercise 63(b).

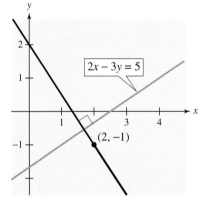

Figure 1.30

Discussing the Concept | Interpreting Slope

You think the relationship between daily high temperature y (in degrees Fahrenheit) and time x (in days) is linear. You observe the high temperature on day 0 to be 60°F and the high temperature 15 days later to be 68°F.

a. Find the slope of the line passing through these two points.

b. Interpret the meaning of the slope in this situation.

c. Use linear extrapolation to predict the high temperatures on days 30, 60, 105, and 180. Are all of these predictions reasonable? Why or why not?

1.2 Warm Up

The following warm-up exercises involve skills that were covered in earlier sections. You will use these skills in the exercise set for this section. For additional help, review Sections R1.2 and R2.1.

In Exercises 1–4, simplify the expression.

1. $\dfrac{4 - (-5)}{-3 - (-1)}$

2. $\dfrac{-5 - 8}{0 - (-3)}$

3. Find $-1/m$ for $m = 4/5$.

4. Find $-1/m$ for $m = -2$.

In Exercises 5–10, solve for y in terms of x.

5. $2x - 3y = 5$

6. $4x + 2y = 0$

7. $y - (-4) = 3[x - (-1)]$

8. $y - 7 = \frac{2}{3}(x - 3)$

9. $y - (-1) = \dfrac{3 - (-1)}{2 - 4}(x - 4)$

10. $y - 5 = \dfrac{3 - 5}{0 - 2}(x - 2)$

1.2 Exercises

In Exercises 1–6, estimate the slope of the line.

1.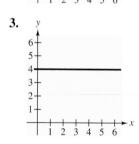

2.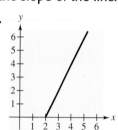

3.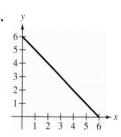

4.

5.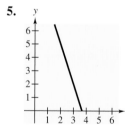

6.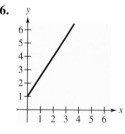

In Exercises 7 and 8, determine if a line with the following description has a positive slope, a negative slope, or an undefined slope.

7. Line rises from left to right

8. Vertical line

In Exercises 9 and 10, sketch the lines through the point with the indicated slopes on the same set of coordinate axes.

	Point	Slopes	
9.	$(5, -1)$	(a) -2	(b) 1
		(c) 0	(d) $\frac{3}{2}$
10.	$(-4, -3)$	(a) -1	(b) 2
		(c) $\frac{1}{4}$	(d) undefined

In Exercises 11–16, plot the points and find the slope of the line passing through the pair of points.

11. $(6, 9), (-4, -1)$

12. $(2, 4), (4, -4)$

13. $(-6, -1), (-6, 4)$

14. $(0, -10), (-4, 0)$

15. $\left(-\frac{1}{3}, 1\right), \left(-\frac{2}{3}, \frac{5}{6}\right)$

16. $\left(\frac{7}{8}, \frac{3}{4}\right), \left(\frac{5}{4}, -\frac{1}{4}\right)$

In Exercises 17–24, use the point on the line and the slope of the line to find three additional points through which the line passes. (There are many correct answers.)

	Point	Slope
17.	$(5, -2)$	$m = 0$
18.	$(-3, 4)$	$m = 0$
19.	$(2, -5)$	m is undefined.
20.	$(-1, 3)$	m is undefined.
21.	$(5, -6)$	$m = 1$
22.	$(10, -6)$	$m = -1$
23.	$(-6, -1)$	$m = \frac{1}{2}$
24.	$(7, -5)$	$m = -\frac{2}{3}$

In Exercises 25–36, find an equation of the line that passes through the point and has the indicated slope. Sketch the line.

Point	Slope
25. $(7, 0)$	$m = 1$
26. $(-2, 0)$	$m = -4$
27. $(-3, 6)$	$m = -2$
28. $(-8, 3)$	$m = -\frac{1}{2}$
29. $(4, 0)$	$m = -\frac{1}{3}$
30. $(-2, -5)$	$m = \frac{3}{4}$
31. $(6, -1)$	m is undefined.
32. $(3, -2)$	m is undefined.
33. $(-2, -7)$	$m = 0$
34. $(-10, 4)$	$m = 0$
35. $\left(4, \frac{5}{2}\right)$	$m = \frac{4}{3}$
36. $\left(-\frac{1}{2}, \frac{3}{2}\right)$	$m = -3$

In Exercises 37–42, find the slope and y-intercept (if possible) of the line specified by the equation. Then sketch the line.

37. $4x - y - 6 = 0$ **38.** $2x + 3y - 9 = 0$

39. $8 - 3x = 0$ **40.** $3y + 5 = 0$

41. $7x + 6y - 30 = 0$ **42.** $x - y - 10 = 0$

In Exercises 43–50, find an equation of the line passing through the points.

43. $(7, -4), (-7, 3)$ **44.** $(4, 3), (-4, -4)$

45. $(-9, 11), (-9, 14)$ **46.** $(-1, 4), (6, 4)$

47. $\left(2, \frac{1}{2}\right), \left(\frac{1}{2}, \frac{5}{4}\right)$ **48.** $\left(1, 1\right), \left(6, -\frac{2}{3}\right)$

49. $(1, 0.6), (-2, -0.6)$ **50.** $(-8, 0.6), (2, -2.4)$

51. A fellow student does not understand why the slope of a vertical line is undefined. Describe how you would help this student understand the concept of undefined slope.

52. Another student overhears your conversation in Exercise 51 and states, "I do not understand why a horizontal line has zero slope and how that is different from undefined or no slope." Describe how you would explain the concepts of zero slope and undefined slope and how they are different from each other.

In Exercises 53–56, use the *intercept form* to find the equation of the line with the given intercepts. The intercept form of the equation of a line with given intercepts $(a, 0)$ and $(0, b)$ is

$$\frac{x}{a} + \frac{y}{b} = 1, \quad a \neq 0, \quad b \neq 0.$$

53. x-intercept: $(1, 0)$ **54.** x-intercept: $(-3, 0)$
   y-intercept: $(0, -4)$ y-intercept: $(0, 4)$

55. x-intercept: $\left(-\frac{1}{6}, 0\right)$ **56.** x-intercept: $\left(-\frac{2}{3}, 0\right)$
   y-intercept: $\left(0, -\frac{2}{3}\right)$ y-intercept: $\left(0, \frac{1}{2}\right)$

In Exercises 57–62, determine if the lines L_1 and L_2 passing through the pairs of points are parallel, perpendicular, or neither.

57. L_1: $(0, -1), (5, 9)$; L_2: $(0, 3), (4, 1)$

58. L_1: $(3, 6), (-6, 0)$; L_2: $\left(0, -1\right), \left(5, \frac{7}{3}\right)$

59. L_1: $(-2, -1), (1, 5)$; L_2: $(1, 3), (5, -5)$

60. L_1: $(4, 8), (-4, 2)$; L_2: $\left(3, -5\right), \left(-1, \frac{1}{3}\right)$

61. L_1: $(-1, 7), (-6, 4)$; L_2: $(0, 1), (5, 4)$

62. L_1: $(-1, 3), (2, -5)$; L_2: $(3, 0), (2, -7)$

In Exercises 63–68, write an equation of the line through the point (a) parallel to the given line and (b) perpendicular to the given line.

Point	Line
63. $(6, 2)$	$y - 2x = -1$
64. $(-5, 4)$	$x + y = 8$
65. $\left(\frac{1}{4}, -\frac{2}{3}\right)$	$2x - 3y = 5$
66. $\left(\frac{7}{8}, \frac{3}{4}\right)$	$5x + 3y = 0$
67. $(-1, 0)$	$y = -3$
68. $(2, 5)$	$x = 4$

69. Temperature   Find an equation of the line that gives the relationship between the temperature in degrees Celsius C and the temperature in degrees Fahrenheit F. Remember that water freezes at $0°$ Celsius ($32°$ Fahrenheit) and boils at $100°$ Celsius ($212°$ Fahrenheit).

70. Temperature   Use the result of Exercise 69 to complete the table. Is there a temperature for which the Fahrenheit reading is the same as the Celsius reading? If so, what is it?

C		$-10°$	$10°$			$177°$
F	$0°$			$68°$	$90°$	

71. Simple Interest   A person deposits P dollars in an account that pays simple interest. After 2 months, the balance in the account is $759 and after 3 months, the balance in the account is $763.50. Find an equation that gives the relationship between the balance A and the time t in months.

72. Simple Interest   Use the result of Exercise 71 to complete the table.

A			$759.00	$763.50			
t	0	1			4	5	6

73. *Make a Decision*: Wheelchair Ramp   The maximum recommended slope of a wheelchair ramp is $\frac{1}{12}$. A business is installing a wheelchair ramp that rises 22 inches over a horizontal length of 24 feet. Is the ramp steeper than recommended?   (Source: Americans with Disabilities Act Handbook)

74. *Make a Decision*: Revenue   A line representing daily revenues y in terms of time x in days has a slope of $m = 100$. Interpret the change in daily revenues for a one-day increase in time.

75. *Make a Decision*: Fourth-Quarter Sales   During the first and second quarters of the year, a business had sales of $145,000 and $152,000, respectively. From these data, can you assume that the sales follow a linear growth pattern? If the pattern is linear, what will the sales be during the fourth quarter?

76. College Enrollment   A small college had 3254 students in 2003 and 3752 students in 2005. The enrollment follows a linear growth pattern. How many students will the college have in 2006?

77. Annual Salary   Your salary was $29,500 in 2003 and $32,800 in 2005. Your salary follows a linear growth pattern. What salary will you be making in 2006?

78. Athletic Footwear   In 2000, the total amount spent on athletic footwear in the United States was $13,026 million. In 2002, $14,107 million was spent in the United States on athletic footwear. Assume the total amount spent on athletic footwear in the United States continues to increase in a linear pattern. What will the amount spent on athletic footwear total in 2005?   (Source: National Sporting Goods Association)

79. *Make a Decision*: American Express Revenue   In 1998, American Express had a total revenue of $19.132 billion. By 2001, the total revenue had reached $22.582 billion. If the total revenue had continued to increase in a linear pattern, how much would the revenue have been in 2002? The actual revenue in 2002 was $23.807 billion. Do you think the increase in revenue was approximately linear?   (Source: American Express Company)

80. *Make a Decision*: Shopping Centers   In 1995, the number of shopping centers in the United States was 41,235. By 2000, the number had increased to 45,115. If the number of shopping centers had continued to increase in a linear pattern, what would the number have been in 2002? The actual number of shopping centers in 2002 was 46,438. Do you think the increase in the number of shopping centers was approximately linear?   (Source: National Research Bureau)

81. *Make a Decision*: Lottery   In 1995, the sales from Lotto games in the United States totaled $10.6 billion. By 2000, the sales from Lotto games dropped to $9.2 billion. If the sales from Lotto games had continued to follow a linear pattern, what would have been the total sales amount in 2002? The actual sales from Lotto games totaled $9.6 billion in 2002. Do you think the pattern of sales from Lotto games was approximately linear?   (Source: TLF Publications, Inc.)

82. Apartment Rent   The rent for a two-bedroom apartment was $800 in 2002 and $1200 in 2004. The rent for a two-bedroom apartment appears to follow a linear growth pattern. What is the rate of change per year for the rent? What will the rent be in 2005?

83. Scuba Diving   Pressure increases at the rate of 1 atmosphere (the weight of air pressure on a body at sea level) for each additional 33 feet of depth under water. At sea level the pressure is 1 atmosphere. At 66 feet below the surface, the total pressure is 3 atmospheres. Write a linear equation to describe the pressure p in terms of the depth d below the surface of the sea. What is the rate of change of pressure with respect to depth?   (Source: PADI Open Water Diver Manual)

Math*MATTERS*

The Pythagorean Theorem is one of the most famous theorems in mathematics. Its name comes from Pythagoras of Samos (*c.* 580–500 B.C.), a Greek mathematician who founded a school of mathematics and philosophy at Crotona, in what is now southern Italy. The Pythagorean Theorem states a relationship among the three sides of a right triangle. Specifically, the theorem states that the sum of the squares of the two legs of a right triangle is equal to the square of the hypotenuse. This result is usually written as

$$a^2 + b^2 = c^2.$$ Pythagorean Theorem

One way to prove the Pythagorean Theorem is to use a geometric argument, as shown in the series of figures below. In the figure at the top left, the areas of the three squares represent the squares of the sides of a right triangle. To prove that $a^2 + b^2 = c^2$, you need to show that the sum of the areas of the two smaller squares is equal to the area of the larger square. One way to do this is to make four copies of the given right triangle, and use the four triangles to form a square whose sides have lengths of $a + b$. By doing this in two different ways, you can conclude that $a^2 + b^2 = c^2$. Do you see why this is true?

According to the *Guinness Book of World Records 1995*, the Pythagorean Theorem is the "most proven theorem" in mathematics. This publication claims that "*The Pythagorean Proposition* published in 1940 contained 370 different proofs of Pythagoras' theorem, including one by President James Garfield."

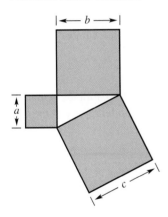

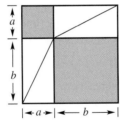

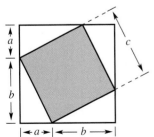

1.3 **Linear Modeling and Direct Variation**

Objectives

- Use a mathematical model to approximate a set of data points.
- Construct a linear model to relate quantities that vary directly.
- Construct and use a linear model with slope as the rate of change.
- Use a scatter plot to find a linear model that fits a set of data.

Introduction

The primary objective of applied mathematics is to find equations or **mathematical models** that describe real-world phenomena. In developing a mathematical model to represent actual data, you should strive for two (often conflicting) goals—accuracy and simplicity. That is, you want the model to be simple enough to be workable, yet accurate enough to produce meaningful results.

You have already studied some techniques for fitting models to data. For instance, in Section 1.2, you learned how to find the equation of a line that passes through two points. In this section, you will study other techniques for fitting models to data: *direct variation, rates of change,* and *linear regression.*

EXAMPLE 1 A MATHEMATICAL MODEL

The average number of hours per year y that Americans 12 years of age and older spent on the Internet from 1996 to 2001 is shown in the table. A linear model that approximates these data is

$$y = 24.7t - 140, \quad 6 \le t \le 11$$

where t represents the year, with $t = 6$ corresponding to 1996. Plot the actual data *and* the model on the same graph. How closely does the model represent the data? (Source: Veronis Suhler Stevenson)

Year	Hours, y
6	10
7	34
8	54
9	82
10	106
11	134

SOLUTION

The actual data are plotted in Figure 1.31, along with the graph of the linear model. From the figure, it appears that the model is a "good fit" for the actual data. You can see how well the model fits the data by comparing the actual values of y with the values of y given by the model (these values are labeled y^* in the table below).

t	6	7	8	9	10	11
y	10	34	54	82	106	134
y^*	8.2	32.9	57.6	82.3	107	131.7

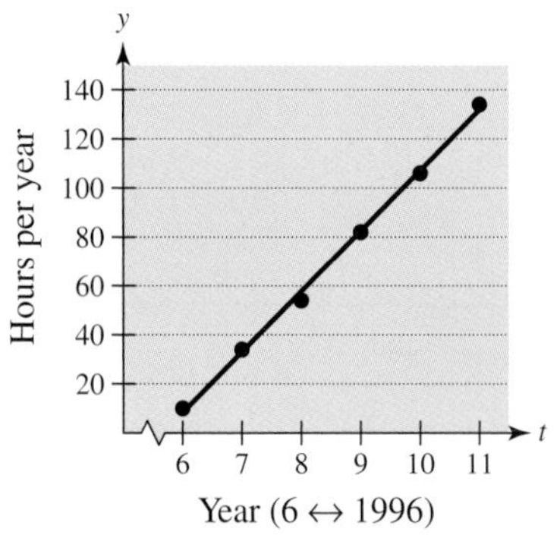

Figure 1.31

 CHECKPOINT   Now try Exercise 1.

Direct Variation

There are two basic types of linear models. The more general model has a *y*-intercept that is nonzero: $y = mx + b, b \neq 0$. The simpler model, $y = mx$, has a *y*-intercept that is zero. In the simpler model, *y* is said to **vary directly** as *x*, or to be **directly proportional** to *x*.

Direct Variation

The following statements are equivalent.

1. *y* **varies directly** as *x*.

2. *y* is **directly proportional** to *x*.

3. $y = mx$ for some nonzero constant *m*.

m is the **constant of variation** or the **constant of proportionality.**

EXAMPLE 2 STATE INCOME TAX

In Pennsylvania, state income tax is directly proportional to *gross income*. You work in Pennsylvania and your state income tax deduction is $76.75 for a gross monthly income of $2500.00. Find a mathematical model that gives the Pennsylvania state income tax in terms of gross income.

SOLUTION

Verbal Model: $\boxed{\text{State income tax}} = \boxed{m} \cdot \boxed{\text{Gross income}}$

Labels: State income tax = *y*   (dollars)
Gross income = *x*   (dollars)
Income tax rate = *m*   (percent in decimal form)

Equation: $y = mx$

To find *m*, substitute the given information into the equation $y = mx$, and then solve for *m*.

$$y = mx \qquad \text{Direct variation model}$$
$$76.75 = m(2500) \qquad \text{Substitute } y = 76.75 \text{ and } x = 2500.$$
$$0.0307 = m \qquad \text{Income tax rate}$$

So, the equation (or model) for state income tax in Pennsylvania is $y = 0.0307x$. In other words, Pennsylvania has a state income tax rate of 3.07% of gross income. The graph of this equation is shown in Figure 1.32.

✓CHECKPOINT   Now try Exercise 11.

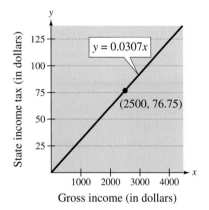

Figure 1.32

Most measurements in the English system and metric system are directly proportional. The next example shows how to use a direct proportion to convert between miles per hour and kilometers per hour.

EXAMPLE 3   THE ENGLISH AND METRIC SYSTEMS

You are traveling at a rate of 64 miles per hour. You switch your speedometer reading to metric units and notice that the speed is 103 kilometers per hour. Use this information to find a mathematical model that relates miles per hour to kilometers per hour.

SOLUTION

If you let y represent the speed in miles per hour and x represent the speed in kilometers per hour, you know that y and x are related by the equation

$$y = mx.$$

You are given that $y = 64$ when $x = 103$. By substituting these values into the equation $y = mx$, you can find the value of m.

$y = mx$	Direct variation model
$64 = m(103)$	Substitute $y = 64$ and $x = 103$.
$\dfrac{64}{103} = m$	Divide each side by 103.
$0.62136 \approx m$	Use a calculator.

So, the conversion factor from kilometers per hour to miles per hour is approximately 0.62136, and the model is

$$y = 0.62136x.$$

The graph of this equation is shown in Figure 1.33.

✔CHECKPOINT   Now try Exercise 15.

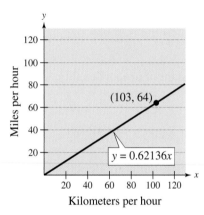

$y = 0.62136x$

Figure 1.33

Once you have found a model that converts kilometers per hour to miles per hour, you can use the model to convert any speed from the metric system to the English system, as shown in the table.

Kilometers per hour	Miles per hour
20	12.4
40	24.9
60	37.3
80	49.7
100	62.1
120	74.6

The conversion equation

$$y = 0.62136x$$

can be approximated by the simpler equation $y = \frac{5}{8}x$ because $\frac{5}{8} = 0.625$. For instance, to convert 40 kilometers per hour to miles per hour, divide by 8 and multiply by 5 to obtain 25 miles per hour.

Rates of Change

A second common type of linear model is one that involves a known rate of change. In the linear equation

$$y = mx + b$$

you know that m represents the slope of the line. In real-life problems, the slope can often be interpreted as the **rate of change** of y with respect to x. Rates of change should always be listed in appropriate units of measure.

| EXAMPLE 4 | HEIGHT OF A MOUNTAIN CLIMBER | |

A mountain climber is climbing up a 500-foot cliff. By 1 P.M., the mountain climber has climbed 115 feet up the cliff. By 4 P.M., the climber has reached a height of 280 feet, as shown in Figure 1.34.

a. Find the average rate of change of the climber and use this rate of change to find the equation that relates the height of the climber to the time.

b. *Make a Decision*   Will the climber reach the top of the cliff by 7:30 P.M.?

SOLUTION

a. Let y represent the height of the climber and let t represent the time. Then the two points that represent the climber's two positions are

$$(t_1, y_1) = (1, 115) \quad \text{and} \quad (t_2, y_2) = (4, 280).$$

So, the average rate of change of the climber is

$$\text{Average rate of change} = \frac{y_2 - y_1}{t_2 - t_1}$$

$$= \frac{280 - 115}{4 - 1}$$

$$= 55 \text{ feet per hour.}$$

An equation that relates the height of the climber to the time is

$$y - y_1 = m(t - t_1) \qquad \text{Point-slope form}$$

$$y - 115 = 55(t - 1) \qquad \text{Substitute } y_1 = 115, t_1 = 1, \text{ and } m = 55.$$

$$y = 55t + 60. \qquad \text{Linear model}$$

If you had chosen to use the point (t_2, y_2) to determine the equation, you would have obtained a different equation initially: $y - 280 = 55(t - 4)$. However, simplifying this equation yields the same linear model $y = 55t + 60$.

b. To find the time when the climber reaches the top of the cliff, let $y = 500$ and solve for t to obtain

$$t = 8.$$

Continuing at the same rate, the climber will *not* reach the top of the cliff by 7:30 P.M. because $t = 8$ corresponds to 8 P.M.

✓CHECKPOINT   Now try Exercise 25.

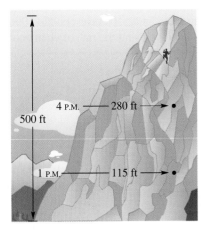

Figure 1.34

EXAMPLE 5   POPULATION OF ORLANDO, FLORIDA

Between 1990 and 2002, the population of the city of Orlando, Florida increased at an average rate of approximately 2583 people per year. In 1990, the population was 163,000. Find a mathematical model that gives the population of Orlando in terms of the year, and use the model to estimate the population in 2007. (Source: U.S. Census Bureau)

SOLUTION

Let y represent the population of Orlando, and let t represent the calendar year, with $t = 0$ corresponding to 1990. It is convenient to let $t = 0$ correspond to 1990 because you were given the population in 1990. Now, using the rate of change of 2583 people per year, you have

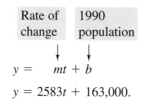

Rate of   1990
change   population
$$\downarrow \qquad \downarrow$$
$$y = \quad mt + b$$
$$y = 2583t + 163,000.$$

Using this model, you can predict the 2007 population to be

$$2007 \text{ population} = 2583(17) + 163,000$$
$$= 206,911.$$

The graph is shown in Figure 1.35.

✓CHECKPOINT   Now try Exercise 37.

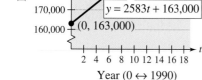

Figure 1.35

EXAMPLE 6   STRAIGHT-LINE DEPRECIATION

A race team buys a $4750 welder that has a useful life of 10 years. The salvage value at the end of the 10 years is $400. Write a linear equation that describes the book value of the welder each year.

SOLUTION

Let V represent the value of the welder (in dollars) at the end of the year t. You can represent the initial value of the welder by the ordered pair $(0, 4750)$ and the salvage value by the ordered pair $(10, 400)$. The slope of the line is

$$m = \frac{400 - 4750}{10 - 0}$$
$$= -435$$

which represents the annual depreciation in *dollars per year.* Using the slope-intercept form, you can write the equation of the line as follows.

$$V = -435t + 4750 \qquad \text{Slope-intercept form}$$

The graph of the equation is shown in Figure 1.36.

✓CHECKPOINT   Now try Exercise 27.

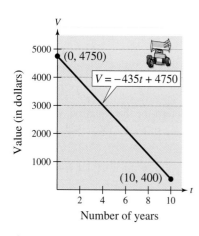

Figure 1.36

Scatter Plots and Regression Analysis

TECHNOLOGY

When you use the *regression* feature of your graphing utility, you may obtain an "*r*-value," which gives a measure of how well the model fits the data (see figure).

```
LinReg
 y=ax+b
 a=199.5824176
 b=6781.274725
 r²=.9506631171
 r=.9750195471
```

The closer the value of $|r|$ is to 1, the better the fit. For the data in Example 7, $r \approx 0.975$, which implies that the model is a good fit. For instructions on how to use the *regression* feature, see Appendix A; for specific keystrokes, go to the text website at *college.hmco.com*.

Another type of linear modeling is a graphical approach that is commonly used in statistics. To find a mathematical model that approximates a set of actual data points, plot the points on a rectangular coordinate system. This collection of points is called a **scatter plot.** Your graphing utility may have a built-in statistical program to calculate the equation of the best fitting line for linear data. The statistical method of fitting a line to a collection of points is called **linear regression.** A discussion of linear regression is beyond the scope of this text, but the program in most graphing utilities is easy to use and allows you to analyze linear data that may not be convenient to graph by hand.

EXAMPLE 7   AIRCRAFT DEPARTURES

The table shows the annual number y of aircraft departures (in thousands) from 1990 to 2002.   (Source: Air Transport Association of America)

Year	x	Departures, y	Year	x	Departures, y
1990	0	6924	1997	7	8127
1991	1	6783	1998	8	8292
1992	2	7051	1999	9	8627
1993	3	7245	2000	10	9035
1994	4	7531	2001	11	8788
1995	5	8062	2002	12	9029
1996	6	8230			

a. Use the *regression* feature of a graphing utility to find a linear model for the data. Let $x = 0$ represent 1990.

b. Use a graphing utility to graph the linear model along with a scatter plot of the data.

c. Use the linear model to predict the number of aircraft departures in 2005.

SOLUTION

a. Enter the data into a graphing utility. Then, using the *regression* feature of the graphing utility, you should obtain a linear model for the data that is similar to the following.

$$y = 199.6x + 6781, \quad 0 \le x \le 12$$

b. The graph of the equation along with the scatter plot is shown in Figure 1.37.

c. Substituting $x = 15$ into the equation $y = 199.6x + 6781$, you get $y = 9775$. So, according to the model, there will be approximately 9775 thousand aircraft departures in 2005.

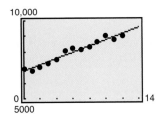

Figure 1.37

✓CHECKPOINT   Now try Exercise 45.

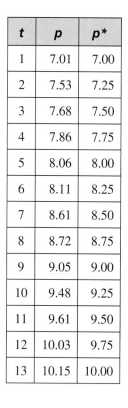

Year	Prize money, p
1991	7.01
1992	7.53
1993	7.68
1994	7.86
1995	8.06
1996	8.11
1997	8.61
1998	8.72
1999	9.05
2000	9.48
2001	9.61
2002	10.03
2003	10.15

EXAMPLE 8   PRIZE MONEY AT THE INDIANAPOLIS 500

The total prize money p (in millions of dollars) awarded at the Indianapolis 500 from 1991 to 2003 is shown in the table. Construct a scatter plot that represents the data and find a linear model that approximates the data.   (Source: INDY Racing League)

Figure 1.38

SOLUTION

Let t represent the year, with $t = 1$ corresponding to 1991. The scatter plot for the points is shown in Figure 1.38. From the scatter plot, draw a line that approximates the data. Then, to find the equation of the line, approximate two points *on the line*: $(5, 8)$ and $(9, 9)$. The slope of this line is

$$m \approx \frac{p_2 - p_1}{t_2 - t_1} = \frac{9 - 8}{9 - 5} = 0.25.$$

Using the point-slope form, you can determine that the equation of the line is

$$p - 8 = 0.25(t - 5) \qquad \text{Point-slope form}$$

$$p = 0.25t + 6.75. \qquad \text{Slope-intercept form}$$

To check this model, compare the actual p-values with the p-values given by the model (these values are labeled as p^* in the table at the left).

✓CHECKPOINT   Now try Exercise 47.

t	p	p^*
1	7.01	7.00
2	7.53	7.25
3	7.68	7.50
4	7.86	7.75
5	8.06	8.00
6	8.11	8.25
7	8.61	8.50
8	8.72	8.75
9	9.05	9.00
10	9.48	9.25
11	9.61	9.50
12	10.03	9.75
13	10.15	10.00

Discussing the Concept ｜ Gathering and Analyzing Data

Measure the height h and forearm f of each person in the class with a tape measure. Gather the data in the form (h, f) and plot the points on a set of coordinate axes. Do the points appear to follow a linear model? Find an equation for a line that approximately represents the points.

1.3 (Warm Up

The following warm-up exercises involve skills that were covered in earlier sections. You will use these skills in the exercise set for this section. For additional help, review Sections 1.1 and 1.2.

In Exercises 1–4, sketch the line.

1. $y = 2x$

2. $y = \frac{1}{2}x$

3. $y = 2x + 1$

4. $y = \frac{1}{2}x + 1$

In Exercises 5 and 6, find an equation of the line that has the given slope and y-intercept.

5. Slope: 1; y-intercept: $(0, 2)$

6. Slope: $\frac{3}{2}$; y-intercept: $(0, 3)$

In Exercises 7–10, find an equation of the line that passes through the two points.

7. $(1, 3)$ and $(6, 8)$

8. $(0, 4)$ and $(7, 10)$

9. $(1, 5.2)$ and $(5, 4.7)$

10. $(2, 6.5)$ and $(8, 3.6)$

1.3 (Exercises

1. Employment The total number of people employed (in millions) in the United States from 1991 to 2002 is given by the following ordered pairs.

(1991, 118)	(1995, 125)	(1999, 133)
(1992, 118)	(1996, 127)	(2000, 137)
(1993, 120)	(1997, 130)	(2001, 137)
(1994, 123)	(1998, 131)	(2002, 136)

A linear model that approximates these data is $y = 115 + 2.0t$, $1 \le t \le 12$, where y represents the number of people employed (in millions) and t represents the year, with $t = 1$ corresponding to 1991. Plot the actual data *and* the model on the same graph. How closely does the model represent the data?   (Source: U. S. Bureau of Labor Statistics)

2. Olympic Swimming The winning times (in minutes) in the women's 400-meter freestyle swimming event in the Olympics from 1948 to 2000 are given by the following ordered pairs.

(1948, 5.30)	(1968, 4.53)	(1988, 4.06)
(1952, 5.20)	(1972, 4.32)	(1992, 4.12)
(1956, 4.91)	(1976, 4.16)	(1996, 4.12)
(1960, 4.84)	(1980, 4.15)	(2000, 4.10)
(1964, 4.72)	(1984, 4.12)	

A linear model that approximates these data is $y = 5.30 - 0.024t$, $8 \le t \le 60$, where y represents the winning time (in minutes) and t represents the year, with $t = 8$ corresponding to 1948. Plot the actual data *and* the model on the same graph. How closely does the model represent the data?   (Source: *World Almanac and Book of Facts 2004*)

3. *Make a Decision* In Exercise 2, the winning times (in minutes) for the years 1948 to 1972 are relatively high compared with the winning times from 1976 to 2000. Consider the winning times for the years 1976 to 2000 only and use a graphing utility to create a scatter plot. Then use the *regression* feature of the graphing utility to find a best-fitting line for the data. Graph the line with the scatter plot. Use the equation of the line to predict the winning time in this event at the 2004 Summer Olympics. Is your prediction reasonable?

4. *Make a Decision* Use a graphing utility and the winning times for the years 1948 to 1972 from Exercise 2 to create a scatter plot. Then use the *regression* feature of the graphing utility to find a best-fitting line for the data. Graph the line with the scatter plot. Use the equation of the line to predict the winning time for this event at the 2000 Summer Olympics. Is your prediction reasonable? How does your prediction compare with actual winning time in 2000?

The icon ⊞ indicates an exercise or part of an exercise in which you are instructed to use a graphing utility.

Direct Variation   In Exercises 5–10, y is proportional to x. Use the x- and y-values to find a linear model that relates y and x.

5. $x = 8$, $y = 3$ **6.** $x = 5$, $y = 9$

7. $x = 15$, $y = 300$ **8.** $x = 12$, $y = 204$

9. $x = 7$, $y = 3.2$ **10.** $x = 11$, $y = 1.5$

11. Simple Interest   The simple interest that a person receives from an investment is directly proportional to the amount of the investment. By investing $2500 in a bond issue, you obtained an interest payment of $187.50 at the end of 1 year. Find a mathematical model that gives the interest I at the end of 1 year in terms of the amount invested P.

12. Simple Interest   The simple interest that a person receives from an investment is directly proportional to the amount of the investment. By investing $5000 in a municipal bond, you obtained interest of $337.50 at the end of 1 year. Find a mathematical model that gives the interest I at the end of 1 year in terms of the amount invested P.

13. Property Tax   Your property tax is based on the assessed value of your property. (The assessed value is often lower than the actual value of the property.) A house that has an assessed value of $150,000 has a property tax of $5520.

(a) Find a mathematical model that gives the amount of property tax y in terms of the assessed value x of the property.

(b) Use the model to find the property tax on a house that has an assessed value of $185,000.

14. State Sales Tax   An item that sells for $145.99 has a sales tax of $10.22.

(a) Find a mathematical model that gives the amount of sales tax y in terms of the retail price x.

(b) Use the model to find the sales tax on a purchase that has a retail price of $540.50.

15. Centimeters and Inches   On a yardstick, you notice that 13 inches is the same length as 33 centimeters.

(a) Use this information to find a mathematical model that relates centimeters to inches.

(b) Use the model to complete the table.

Inches	5	10	20	25	30
Centimeters					

16. Liters and Gallons   You are buying gasoline and notice that 14 gallons of gasoline is the same as 53 liters.

(a) Use this information to find a linear model that relates gallons to liters.

(b) Use the model to complete the table.

Gallons	5	10	20	25	30
Liters					

IN THE NEWS
Thin Car Travels Far

Volkswagen's canoe-skinny mini could do New York to D.C. on 1 gallon of gas. By John Matras

. . . Forget power, space, and speed: Volkswagen AG's latest idea-on-wheels does not address the requirements of the average American family driver. What it can do is travel more than 100 kilometers on a single liter of fuel. Translation: 235 miles per gallon.

The car's designers combined highly tuned aerodynamics, exotic materials, and a 0.3-liter diesel engine to achieve 0.99 liters per 100 kilometers. The project, the brainchild of engineer Thomas Gänsicke, is an engineering exercise and therefore has rather whimsical features. Most noticeable are the car's canoe-like proportions: It's 4 feet wide and 11 feet long. Occupants sit tandem, the passenger straddling the driver's seat, both wedged under a 4-foot-long gullwing canopy. . . .

It can, he promises, "swim with the usual traffic." Who better to emphasize that point than Ferdinand Piëch, chairman of VW? For the most recent board meeting in April, Piëch drove the 1-liter car from Wolfsburg to Hamburg, 110 miles, averaging 264 miles per gallon on the way. . . .

"Thin Car Travels Far," by John Matras, from Popular Science, *August 2002*

17. Use the fact that 100 kilometers per 0.99 liter is the same as 235 miles per gallon to find a mathematical model that relates kilometers per liter to miles per gallon.

18. Use the model to convert 264 miles per gallon to kilometers per liter.

In Exercises 19–24, you are given the 2005 value of a product *and* the rate at which the value is expected to change during the next 5 years. Use this information to write a linear equation that gives the dollar value of the product in terms of the year. (Let $t = 5$ represent 2005.)

2005 Value	Rate
19. $2540	$125 increase per year
20. $156	$4.50 increase per year
21. $20,400	$2000 decrease per year
22. $45,000	$2800 decrease per year
23. $154,000	$12,500 increase per year
24. $245,000	$5600 increase per year

25. Height of a Parachutist   After opening the parachute, the descent of a parachutist follows a linear model. At 2:08 P.M. the height of the parachutist is 7000 feet. At 2:10 P.M. the height is 4600 feet.

(a) Write a linear equation that gives the height of the parachutist in terms of the time t. (Let $t = 0$ represent 2:08 P.M. and let t be measured in seconds.)

(b) Use the equation in part (a) to find the time when the parachutist will reach the ground.

26. Distance Traveled by a Car   You are driving at a constant speed on a freeway. At 4:30 P.M. you drive by a sign that gives the distance to Montgomery, Alabama as 84 miles. At 4:59 P.M. you drive by another sign that gives the distance to Montgomery as 56 miles.

(a) Write a linear equation that gives your distance from Montgomery in terms of time t. (Let $t = 0$ represent 4:30 P.M. and let t be measured in minutes.)

(b) Use the equation in part (a) to find the time when you will reach Montgomery.

27. Straight-Line Depreciation   A business purchases a piece of equipment for $875. After 5 years the equipment will have no value. Write a linear equation giving the value V of the equipment during the 5 years.

28. Straight-Line Depreciation   A business purchases a piece of equipment for $25,000. The equipment will be replaced in 10 years, at which time its salvage value is expected to be $2000. Write a linear equation giving the value V of the equipment during the 10 years.

29. Sale Price and List Price   A store is offering a 15% discount on all items. Write a linear equation giving the sale price S for an item with a list price L.

30. Sale Price and List Price   A store is offering a 25% discount on all shirts. Write a linear equation giving the sale price S for a shirt with a list price L.

31. Hourly Wages   A manufacturer pays its assembly line workers $11.50 per hour. In addition, workers receive a piecework rate of $0.75 per unit produced. Write a linear equation for the hourly wages W in terms of the number of units x produced per hour.

32. Sales Commission   A salesperson receives a monthly salary of $2500 plus a commission of 7% of sales. Write a linear equation for the salesperson's monthly wage W in terms of the person's monthly sales S.

In Exercises 33–36, match the description with one of the graphs. Also find the slope of the graph and describe how it is interpreted in the real-life situation. [The graphs are labeled (a), (b), (c), and (d).]

(a) (b)

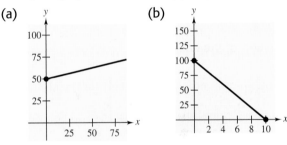

(c) (d)

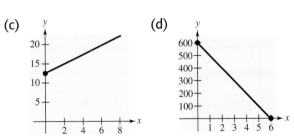

33. A person is paying $10 per week to a friend to repay a $100 loan.

34. An employee is paid $12.50 per hour plus $1.50 for each unit produced per hour.

35. A sales representative receives $50 per day for food, plus $0.25 for each mile traveled.

36. A digital camera that was purchased for $600 depreciates $100 per year.

37. High School Enrollment   A high school had an enrollment of 1200 students in 1995. During the next 10 years, the enrollment increased by approximately 50 students per year.

(a) Write a linear equation giving the enrollment N in terms of the year t. (Let $t = 5$ correspond to the year 1995.)

(b) If this constant rate of growth continues, predict the enrollment in the year 2010.

38. Cost of Renting   The monthly rent in an office building is related to the office size. The rent for a 600-square-foot office is $750. The rent for a 900-square-foot office is $1150.

(a) Write a linear equation giving the monthly rent r in terms of the square footage x.

(b) Use the equation in part (a) to find the monthly rent for an office that has 1300 square feet of floor space.

In Exercises 39–44, can the data be approximated by a linear model? If so, sketch the line that you think best approximates the data. Then find an equation of the line.

39.
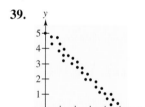

40.
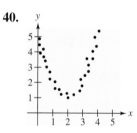

41.
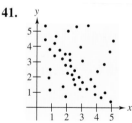

42.
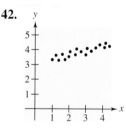

43.
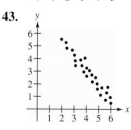

44.
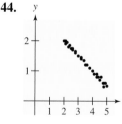

45. *Make a Decision*: Operating Revenue   The total operating revenue R (in millions of dollars) for the United States airline industry for the years 1992 to 2000 is shown in the table.   (Source: Air Transport Association of America)

Year	1992	1993	1994	1995	1996
Revenue, R	78,357	84,559	88,313	95,117	101,937

Year	1997	1998	1999	2000
Revenue, R	109,917	113,810	119,455	130,839

(a) Use a graphing utility to create a scatter plot for these data. Let t represent the year, with $t = 2$ corresponding to 1992. Do the data appear linear?

(b) Use the *regression* feature of a graphing utility to find a linear model for predicting total operating revenue.

(c) State the slope of the graph of the linear model from part (b) and interpret its meaning in the context of the problem.

(d) Use your linear model from part (b) to predict total operating revenues for 2001 and 2003. Are your predictions reasonable?

46. Think About It   A linear mathematical model for predicting prize winnings in a race is based on data from only 3 years. Discuss the potential accuracy or inaccuracy of such a model.

47. *Make a Decision*: Discus Throw   The lengths (in feet) of the winning men's discus throws in the Olympics from 1904 to 2000 are given by the following ordered pairs.   (Source: *World Almanac and Book of Facts 2004*)

(1904, 128.8), (1908, 134.1), (1912, 148.3)
(1920, 146.6), (1924, 151.3), (1928, 155.3)
(1932, 162.3), (1936, 165.6), (1948, 173.2)
(1952, 180.5), (1956, 184.9), (1960, 194.2)
(1964, 200.1), (1968, 212.5), (1972, 211.3)
(1976, 221.4), (1980, 218.7), (1984, 218.5)
(1988, 225.8), (1992, 213.7), (1996, 227.7)
(2000, 227.3)

(a) Create a scatter plot of the data. (Let y represent the length of the winning discus throw in feet and let t represent the year, with $t = 4$ corresponding to 1904.)

(b) Use two points on the scatter plot to find a linear model that approximates the data.

(c) State the slope of the graph of the model and interpret its meaning in the context of the problem.

(d) Use the equation in part (b) to predict the length of the winning men's discus throw in 2004.

(e) Use your school's library, the Internet, or some other reference source to find the length of the winning men's discus throw in 2004. How accurate was your prediction in part (d)?

(f) Do you think this is an example in which a linear model can be used to approximate the data for future years? What do you think the length of the winning discus throw will be in the year 2100?

48. *Make a Decision*: **Total Sales** The total sales (in millions of dollars) for Microsoft Corporation from 1995 to 2002 are given by the following ordered pairs. (Source: Microsoft Corp.)

(1995, 5937), (1996, 8671), (1997, 11,358),

(1998, 14,484), (1999, 19,747), (2000, 22,956),

(2001, 25,296), (2002, 28,365)

(a) Use a graphing utility to create a scatter plot of the data. (Let y represent the total sales in millions of dollars and let t represent the year, with $t = 5$ corresponding to 1995.)

(b) Use two points on the scatter plot to find an equation of the line that approximates the data.

(c) Use the *regression* feature of a graphing utility to predict the sales of Microsoft in 2003. Compare the result with the prediction given by the equation in part (b).

(d) The actual sales for Microsoft Corporation in 2003 were $32,187 million. How accurate were your predictions?

49. Real Estate Rentals A real estate office handles an apartment complex with 50 units. When the rent per unit is $520 per month, all 50 units are occupied. However, when the rent is $575 per month, the average number of occupied units drops to 47. Assume that the relationship between the monthly rent p and the demand x is linear.

(a) Write the equation of the line giving the demand x in terms of the rent p.

(b) Use the equation in part (a) to predict the number of units occupied when the rent is raised to $630.

50. *Make a Decision*: **Health Care** The total health care expenditures E (in billions of dollars) in the United States from 1990 to 2001 are shown in the table. (Source: U.S. Centers for Medicare and Medicaid Services)

Year	1990	1991	1992	1993	1994	1995
Expenditures, E	696	762	827	888	937	990

Year	1996	1997	1998	1999	2000	2001
Expenditures, E	1039	1093	1150	1220	1310	1425

(a) Use a graphing utility to create a scatter plot for these data. Let t represent the year, with $t = 0$ corresponding to 1990. Do the data appear linear?

(b) Use the *regression* feature of a graphing utility to find a linear model for predicting the total annual expenditures for health care in the United States.

(c) Use your linear model from part (b) to predict the total health care expenditures for 2002 and 2003. Do your predictions seem reasonable?

51. *Make a Decision*: **Sunday Newspapers** The circulation C (in millions) of Sunday newspapers from 1995 to 2002 is shown in the table below. (Source: Editor & Publisher Co.)

Year	1995	1996	1997	1998	1999
Circulation, C	61.5	60.8	60.5	60.1	59.9

Year	2000	2001	2002
Circulation, C	59.4	59.1	58.8

(a) Use a graphing utility to create a scatter plot for the data. Let t represent the year, with $t = 5$ corresponding to 1995. Do the data appear linear?

(b) Use the *regression* feature of a graphing utility to find a linear model for predicting the annual circulation of Sunday newspapers in the United States.

(c) Use the linear model from part (b) to predict circulation in 2003 and 2005.

(d) What trend do you see in the data and in your predictions? What might account for this trend? As manager of the newspaper, what strategies would you use to address this trend?

1.4   Functions

Objectives

- Determine if an equation or a set of ordered pairs represents a function.
- Use function notation and evaluate a function.
- Find the domain of a function.
- Write a function that relates quantities in an application problem.

Introduction to Functions

Many everyday phenomena involve two quantities that are related to each other by some rule of correspondence. Here are some examples.

1. The simple interest I earned on $1000 for 1 year is related to the annual interest rate r by the formula $I = 1000r$.

2. The distance d traveled on a bicycle in 2 hours is related to the speed s of the bicycle by the formula $d = 2s$.

3. The area A of a circle is related to its radius r by the formula $A = \pi r^2$.

Not all correspondences between two quantities have simple mathematical formulas. For instance, people commonly match up athletes with jersey numbers and hours of the day with temperature. In each of these cases, however, there is some rule of correspondence that matches each item from one set with exactly one item from a different set. Such a rule of correspondence is called a **function.**

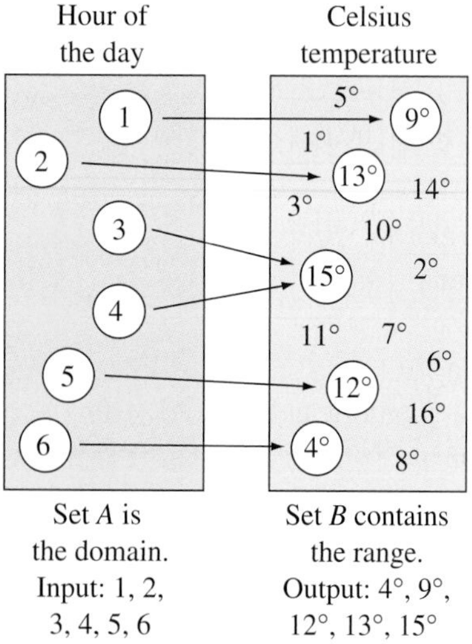

Hour of the day   Celsius temperature

Set A is the domain.
Input: 1, 2, 3, 4, 5, 6

Set B contains the range.
Output: 4°, 9°, 12°, 13°, 15°

Figure 1.39 Function from Set A to Set B

Definition of a Function

A **function** f from a set A to a set B is a rule of correspondence that assigns to each element x in the set A exactly one element y in the set B. The set A is the **domain** (or set of inputs) of the function f, and the set B contains the **range** (or set of outputs).

To get a better idea of this definition, look at the function illustrated in Figure 1.39. This function can be represented by the following set of ordered pairs.

$$\{(1, 9°), (2, 13°), (3, 15°), (4, 15°), (5, 12°), (6, 4°)\}$$

In each ordered pair, the first coordinate is the input and the second coordinate is the output. In this example, note the following characteristics of a function.

1. Each element of A must be matched with an element of B.

2. Some elements of B may not be matched with any element of A.

3. Two or more elements of A may be matched with the same element of B.

4. An element of A (the domain) cannot be matched with two different elements of B.

In the following two examples, you are asked to decide whether different correspondences are functions. To do this, you must decide whether each element of the domain A is matched with exactly one element of the range B. If any element of A is matched with two or more elements of B, the correspondence is not a function. For example, people are not a function of their birthday month because many people are born in a single month.

EXAMPLE 1 TESTING FOR FUNCTIONS

Let $A = \{a, b, c\}$ and $B = \{1, 2, 3, 4, 5\}$. Which of the following sets of ordered pairs or figures represent functions from set A to set B?

a. $\{(a, 2), (b, 3), (c, 4)\}$

b. $\{(a, 4), (b, 5)\}$

c.

d.

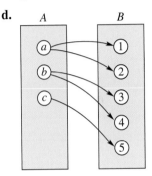

SOLUTION

a. This collection of ordered pairs *does* represent a function from A to B. Each element of A is matched with exactly one element of B.

b. This collection of ordered pairs *does not* represent a function from A to B. Not every element of A is matched with an element of B.

c. This figure *does* represent a function from A to B. It does not matter that each element of A is matched with the same element of B.

d. This figure *does not* represent a function from A to B. The element a of A is matched with *two* elements, 1 and 2, of B. This is also true of the element b.

✓CHECKPOINT Now try Exercise 1.

Representing functions by sets of ordered pairs is a common practice in *discrete mathematics*. In algebra, however, it is more common to represent functions by equations or formulas involving two variables. For instance, the equation

$$y = x^2 \qquad \text{\small y is a function of x.}$$

represents the variable y as a function of the variable x. In this equation, x is the **independent variable** and y is the **dependent variable.** The domain of the function is the set of all values taken on by the independent variable x, and the range of the function is the set of all values taken on by the dependent variable y.

EXAMPLE 2　TESTING FOR FUNCTIONS REPRESENTED BY EQUATIONS

Which of the equations represent(s) y as a function of x?

a. $x^2 + y = 1$

b. $-x + y^2 = 1$

SOLUTION

To determine whether y is a function of x, try to solve for y in terms of x.

a. Solving for y yields

$$x^2 + y = 1 \qquad \text{Write original equation.}$$
$$y = 1 - x^2. \qquad \text{Solve for } y.$$

To each value of x there corresponds exactly one value of y. So, y *is* a function of x.

b. Solving for y yields

$$-x + y^2 = 1 \qquad \text{Write original equation.}$$
$$y^2 = 1 + x \qquad \text{Add } x \text{ to each side.}$$
$$y = \pm\sqrt{1 + x}. \qquad \text{Solve for } y.$$

The $\pm$ indicates that to a given value of x there correspond two values of y. So, y *is not* a function of x.

✔CHECKPOINT　Now try Exercise 17.

Function Notation

When an equation is used to represent a function, it is convenient to name the function so that it can be referenced easily. For example, you know that the equation $y = 1 - x^2$ describes y as a function of x. Suppose you give this function the name "f." Then you can use the following **function notation.**

Input	Output	Equation
x	$f(x)$	$f(x) = 1 - x^2$

The symbol $f(x)$ is read as the **value of f at x** or simply **f of x.** The symbol $f(x)$ corresponds to the y-value for a given x. So, you can write $y = f(x)$. Keep in mind that f is the *name* of the function, whereas $f(x)$ is the *value* of the function at x. For instance, the function given by

$$f(x) = 3 - 2x$$

has *function values* denoted by $f(-1)$, $f(0)$, $f(2)$, and so on. To find these values, substitute the specified input values into the given equation.

For $x = -1$, $\quad f(-1) = 3 - 2(-1) = 3 + 2 = 5$.

For $x = 0$, $\quad\quad f(0) = 3 - 2(0) = 3 - 0 = 3$.

For $x = 2$, $\quad\quad f(2) = 3 - 2(2) = 3 - 4 = -1$.

Although f is often used as a convenient function name and x is often used as the independent variable, you can use other letters. For instance,

$$f(x) = x^2 - 4x + 7, \quad f(t) = t^2 - 4t + 7, \quad \text{and} \quad g(s) = s^2 - 4s + 7$$

all define the same function. In fact, the role of the independent variable in a function is simply that of a "placeholder." Consequently, the function above could be described by the form

$$f(\quad) = (\quad)^2 - 4(\quad) + 7.$$

EXAMPLE 3   EVALUATING A FUNCTION

Let $g(x) = -x^2 + 4x + 1$ and find the following.

a. $g(2)$ **b.** $g(t)$ **c.** $g(x + 2)$

SOLUTION

a. Replacing x with 2 in $g(x) = -x^2 + 4x + 1$ yields the following.

$$g(2) = -(2)^2 + 4(2) + 1 = -4 + 8 + 1 = 5$$

b. Replacing x with t yields the following.

$$g(t) = -(t)^2 + 4(t) + 1 = -t^2 + 4t + 1$$

c. Replacing x with $x + 2$ yields the following.

$$g(x + 2) = -(x + 2)^2 + 4(x + 2) + 1$$
$$= -(x^2 + 4x + 4) + 4x + 8 + 1$$
$$= -x^2 - 4x - 4 + 4x + 8 + 1$$
$$= -x^2 + 5$$

✔**CHECKPOINT**   Now try Exercise 31.

STUDY TIP

In Example 3(c), note that $g(x + 2)$ is not equal to $g(x) + g(2)$. In general, $g(u + v) \neq g(u) + g(v)$.

A function defined by two or more equations over a specified domain is called a **piecewise-defined function.**

EXAMPLE 4   A PIECEWISE-DEFINED FUNCTION

Evaluate the function when $x = -1, 0$, and 1.

$$f(x) = \begin{cases} x^2 + 1, & x < 0 \\ x - 1, & x \geq 0 \end{cases}$$

SOLUTION

Because $x = -1$ is less than 0, use $f(x) = x^2 + 1$ to obtain

$$f(-1) = (-1)^2 + 1 = 2.$$

For $x = 0$, use $f(x) = x - 1$ to obtain $f(0) = (0) - 1 = -1$. For $x = 1$, use $f(x) = x - 1$ to obtain $f(1) = (1) - 1 = 0$. The graph of the function is shown in Figure 1.40.

✔**CHECKPOINT**   Now try Exercise 41.

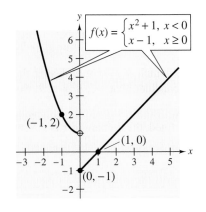

Figure 1.40

DISCOVERY

Use a graphing utility to graph $y = \sqrt{4 - x^2}$. What is the domain of this function? Then graph $y = \sqrt{x^2 - 4}$. What is the domain of this function? Do the domains of these two functions overlap? If so, for what values?

Finding the Domain of a Function

The domain of a function can be described explicitly or it can be *implied* by the expression used to define the function. The **implied domain** is the set of all real numbers for which the expression is defined. For instance, the function given by

$$f(x) = \frac{1}{x^2 - 4} \qquad \text{Domain excludes } x\text{-values that result in division by zero.}$$

has an implied domain that consists of all real x other than $x = \pm 2$. These two values are excluded from the domain because division by zero is undefined. Another common type of implied domain is that used to avoid even roots of negative numbers. For example, the function given by

$$f(x) = \sqrt{x} \qquad \text{Domain excludes } x\text{-values that result in even roots of negative numbers.}$$

is defined only for $x \geq 0$. So, its implied domain is the interval $[0, \infty)$. In general, the domain of a function *excludes* values that would cause division by zero *or* result in the even root of a negative number.

EXAMPLE 5   FINDING THE DOMAIN OF A FUNCTION

Find the domain of each function.

a. f: $\{(-3, 0), (-1, 4), (0, 2), (2, 2), (4, -1)\}$ **b.** $g(x) = \dfrac{1}{x + 5}$

c. Volume of a sphere: $V = \frac{4}{3}\pi r^3$ **d.** $h(x) = \sqrt{4 - x^2}$

e. $r(x) = \sqrt[3]{x + 3}$

SOLUTION

a. The domain of f consists of all first coordinates in the set of ordered pairs.

$$\text{Domain} = \{-3, -1, 0, 2, 4\}$$

b. Excluding x-values that yield zero in the denominator, the domain of g is the set of all real numbers x such that $x \neq -5$.

c. Because this function represents the volume of a sphere, the values of the radius r must be positive. So, the domain is the set of all real numbers r such that $r > 0$.

d. This function is defined only for x-values for which $4 - x^2 \geq 0$. Using the methods described in Section R2.7, you can conclude that $-2 \leq x \leq 2$. So, the domain of h is the interval $[-2, 2]$.

e. Because the cube root of any real number is defined, the domain of r is the set of all real numbers, or $(-\infty, \infty)$.

✓CHECKPOINT   Now try Exercise 51.

In Example 5(c), note that the domain of a function may be implied by the physical context. For instance, from the equation $V = \frac{4}{3}\pi r^3$, you would have no reason to restrict r to positive values, but the physical context implies that a sphere cannot have a negative or zero radius.

Applications

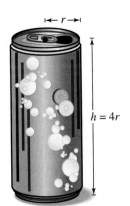

Figure 1.41

EXAMPLE 6 THE DIMENSIONS OF A CONTAINER

You work in the marketing department of a soft-drink company and are experimenting with a new soft-drink can that is slightly narrower and taller than a standard can. For your experimental can, the height is 4 times the radius, as shown in Figure 1.41.

a. Write the volume of the can as a function of the radius r.

b. Write the volume of the can as a function of the height h.

SOLUTION

a. $V = \pi r^2 h = \pi r^2(4r) = 4\pi r^3$ $\qquad$ *V* is a function of *r*.

b. $V = \pi\left(\dfrac{h}{4}\right)^2 h = \dfrac{\pi h^3}{16}$ $\qquad$ *V* is a function of *h*.

✓CHECKPOINT $\qquad$ Now try Exercise 67.

EXAMPLE 7 THE PATH OF A BASEBALL

Make a Decision $\quad$ A baseball is hit 3 feet above ground at a velocity of 100 feet per second and at an angle of 45°. The path of the baseball is given by the function

$$y = -0.0032x^2 + x + 3$$

where y and x are measured in feet. Will the baseball clear a 10-foot fence located 300 feet from home plate?

SOLUTION

When $x = 300$, the height of the baseball is given by

$$y = -0.0032(300)^2 + 300 + 3 = 15 \text{ feet.}$$

The ball will clear the fence, as shown in Figure 1.42.

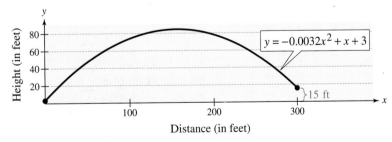

Figure 1.42

Notice that in Figure 1.42, the baseball is not at the point $(0, 0)$ before it is hit. This is because the original problem states that the baseball was hit 3 feet above the ground.

✓CHECKPOINT $\qquad$ Now try Exercise 71.

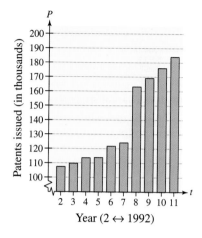

Figure 1.43

Patents issued (in thousands)

EXAMPLE 8   PATENTS

The number P (in thousands) of patents issued increased in a linear pattern from 1992 to 1997. Then, in 1998, the number of patents issued took a jump and, until 2001, increased in a *different* linear pattern (see Figure 1.43). These two patterns can be approximated by the function

$$P = \begin{cases} 3.42t + 99.7, & 2 \le t \le 7 \\ 6.96t + 106.9, & 8 \le t \le 11 \end{cases}$$

where t represents the year, with $t = 2$ corresponding to 1992. Use this function to approximate the total number of patents issued between 1992 and 2001. (Source: U.S. Patent and Trademark Office)

SOLUTION

For 1992 to 1997, use the equation $P = 3.42t + 99.7$ to approximate the number of patents issued, as shown in the table.

Year	t	P	
1992	2	106.5	For 1992 to 1997, use $P = 3.42t + 99.7$.
1993	3	110.0	
1994	4	113.4	
1995	5	116.8	
1996	6	120.2	
1997	7	123.6	
1998	8	162.6	For 1998 to 2001, use $P = 6.96t + 106.9$.
1999	9	169.5	
2000	10	176.5	
2001	11	183.5	

For 1998 to 2001, use the equation $P = 6.96t + 106.9$ to approximate the number of patents issued, as shown in the table above. To approximate the total number of patents issued from 1992 to 2001, add the amounts for each of the years, as follows.

$$106.5 + 110.0 + 113.4 + 116.8 + 120.2 + 123.6 + 162.6 + 169.5$$
$$+ 176.5 + 183.5 = 1382.6$$

Because the number of patents issued is measured in thousands, you can conclude that the total number of patents issued between 1992 and 2001 was approximately 1,382,600.

✓CHECKPOINT   Now try Exercise 73.

Summary of Function Terminology

Function: A **function** is a relationship between two variables such that to each value of the independent variable there corresponds exactly one value of the dependent variable.

Function Notation: $y = f(x)$

f is the **name** of the function.

y is the **dependent variable.**

x is the **independent variable.**

$f(x)$ is the **value of the function at x.**

Domain: The **domain** of a function is the set of all values (inputs) of the independent variable for which the function is defined. If x is in the domain of f, then f is said to be **defined** at x. If x is not in the domain of f, then f is said to be **undefined** at x.

Range: The **range** of a function is the set of all values (outputs) assumed by the dependent variable (that is, the set of all function values).

Implied Domain: If f is defined by an algebraic expression and the domain is not specified, the **implied domain** consists of all real numbers for which the expression is defined.

Discussing the Concept | Everyday Functions

In groups of two or three, identify common real-life functions. Consider everyday activities, events, and expenses, such as long distance telephone calls and car insurance. Here are a couple of examples.

a. The statement, "Your happiness is a function of the grade you receive in this course" is *not* a correct mathematical use of the word *function*. The word *happiness* is ambiguous.

b. The statement, "Your federal income tax is a function of your adjusted gross income" is a correct mathematical use of the word *function*. Once you have determined your adjusted gross income, your income tax can be determined.

Describe your function in words. Avoid using ambiguous words. Give an example of a piecewise-defined function.

1.4   Warm Up

The following warm-up exercises involve skills that were covered in earlier sections. You will use these skills in the exercise set for this section. For additional help, review Sections R1.2, R2.1, R2.5, and R2.7.

In Exercises 1–4, simplify the expression.

1. $2(-3)^3 + 4(-3) - 7$

2. $4(-1)^2 - 5(-1) + 4$

3. $(x + 1)^2 + 3(x + 1) - 4 - (x^2 + 3x - 4)$

4. $(x - 2)^2 - 4(x - 2) - (x^2 - 4)$

In Exercises 5 and 6, solve for y in terms of x.

5. $2x + 5y - 7 = 0$

6. $y^2 = x^2$

In Exercises 7–10, solve the inequality.

7. $x^2 - 4 \geq 0$

8. $9 - x^2 \geq 0$

9. $x^2 + 2x + 1 \geq 0$

10. $x^2 - 3x + 2 \geq 0$

1.4   Exercises

In Exercises 1–4, decide whether the set of figures represents a function from A to B.

$A = \{a, b, c\}$ and $B = \{1, 2, 3, 4\}$

Give reasons for your answers.

1. 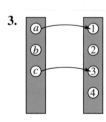   2.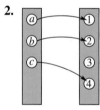

3. 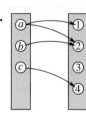   4.

In Exercises 5–8, decide whether the set of ordered pairs represents a function from A to B.

$A = \{0, 1, 2, 3\}$ and $B = \{-2, -1, 0, 1, 2\}$

Give reasons for your answers.

5. $\{(0, 1), (1, -2), (2, 0), (3, 2)\}$

6. $\{(0, -1), (2, 2), (1, -2), (3, 0), (1, 1)\}$

7. $\{(0, 0), (1, 0), (2, 0), (3, 0)\}$

8. $\{(0, 2), (3, 0), (1, 1)\}$

In Exercises 9–12, decide whether the set of ordered pairs represents a function from A to B.

$A = \{a, b, c\}$ and $B = \{0, 1, 2, 3\}$

Give reasons for your answers.

9. $\{(a, 1), (c, 2), (c, 3), (b, 3)\}$

10. $\{(a, 1), (b, 2), (c, 3)\}$

11. $\{(1, a), (0, a), (2, c), (3, b)\}$

12. $\{(c, 0), (b, 0), (a, 3)\}$

In Exercises 13–16, the domain of f is the set $A = \{-2, -1, 0, 1, 2\}$. Write the function as a set of ordered pairs.

13. $f(x) = x^2$

14. $f(x) = \dfrac{2x}{x^2 + 1}$

15. $f(x) = \sqrt{x + 2}$

16. $f(x) = |x + 1|$

In Exercises 17–26, decide whether the equation represents y as a function of x.

17. $x^2 + y^2 = 4$

18. $x = y^2$

19. $x^2 + y = 4$

20. $x + y^2 = 4$

21. $2x + 3y = 4$

22. $x^2 + y^2 - 2x - 4y + 1 = 0$

23. $y^2 = x^2 - 1$

24. $y = \sqrt{x + 5}$

25. $x^2 y - x^2 + 4y = 0$

26. $xy - y - x - 2 = 0$

In Exercises 27–30, evaluate the function.

27. $f(x) = 6 - 4x$

(a) $f(3) = 6 - 4(\quad)$

(b) $f(-7) = 6 - 4(\quad)$

(c) $f(t) = 6 - 4(\quad)$

(d) $f(c + 1) = 6 - 4(\quad)$

28. $f(s) = \dfrac{1}{s + 1}$

(a) $f(4) = \dfrac{1}{(\quad) + 1}$

(b) $f(0) = \dfrac{1}{(\quad) + 1}$

(c) $f(4x) = \dfrac{1}{(\quad) + 1}$

(d) $f(x + 1) = \dfrac{1}{(\quad) + 1}$

29. $g(x) = \dfrac{1}{x^2 - 2x}$

(a) $g(1) = \dfrac{1}{(\quad)^2 - 2(\quad)}$

(b) $g(-3) = \dfrac{1}{(\quad)^2 - 2(\quad)}$

(c) $g(t) = \dfrac{1}{(\quad)^2 - 2(\quad)}$

(d) $g(t + 1) = \dfrac{1}{(\quad)^2 - 2(\quad)}$

30. $f(t) = \sqrt{25 - t^2}$

(a) $f(3) = \sqrt{25 - (\quad)^2}$

(b) $f(5) = \sqrt{25 - (\quad)^2}$

(c) $f(x + 5) = \sqrt{25 - (\quad)^2}$

(d) $f(2x) = \sqrt{25 - (\quad)^2}$

In Exercises 31–42, evaluate the function at each specified value of the independent variable and simplify.

31. $f(x) = 2x - 3$

(a) $f(1)$ (b) $f(-3)$

(c) $f(x - 1)$ (d) $f\left(\frac{1}{4}\right)$

32. $g(y) = 7 - 3y$

(a) $g(0)$ (b) $g\left(\frac{7}{3}\right)$

(c) $g(s)$ (d) $g(s + 2)$

33. $h(t) = t^2 - 2t$

(a) $h(2)$ (b) $h(-1)$

(c) $h(x + 2)$ (d) $h(1.5)$

34. $V(r) = \frac{4}{3}\pi r^3$

(a) $V(3)$ (b) $V(0)$

(c) $V\left(\frac{3}{2}\right)$ (d) $V(2r)$

35. $f(y) = 3 - \sqrt{y}$

(a) $f(4)$ (b) $f(100)$

(c) $f(4x^2)$ (d) $f(0.25)$

36. $f(x) = \sqrt{x + 8} + 2$

(a) $f(-8)$ (b) $f(1)$

(c) $f(x - 8)$ (d) $f(x + 8)$

37. $q(x) = \dfrac{1}{x^2 - 9}$

(a) $q(4)$ (b) $q(0)$

(c) $q(3)$ (d) $q(y + 3)$

38. $q(t) = \dfrac{2t^2 + 3}{t^2}$

(a) $q(2)$ (b) $q(0)$

(c) $q(x)$ (d) $q(-x)$

39. $f(x) = \dfrac{|x|}{x}$

(a) $f(2)$ (b) $f(-2)$

(c) $f(x^2)$ (d) $f(x - 1)$

40. $f(x) = |x| + 4$

(a) $f(2)$ (b) $f(-2)$

(c) $f(x^2)$ (d) $f(x + 2)$

41. $f(x) = \begin{cases} 2x + 1, & x < 0 \\ 2x + 2, & x \geq 0 \end{cases}$

(a) $f(-1)$ (b) $f(0)$

(c) $f(1)$ (d) $f(2)$

42. $f(x) = \begin{cases} x^2 + 2, & x \leq 1 \\ 2x^2 + 2, & x > 1 \end{cases}$

(a) $f(-2)$ (b) $f(0)$

(c) $f(1)$ (d) $f(2)$

In Exercises 43–50, find all real values of x such that $f(x) = 0$.

43. $f(x) = 15 - 3x$ **44.** $f(x) = \dfrac{3x - 4}{5}$

45. $f(x) = x^2 - 9$ **46.** $f(x) = 2x^2 - 11x + 5$

47. $f(x) = x^3 - x$

48. $f(x) = x^3 - 3x^2 - 4x + 12$

49. $f(x) = \dfrac{3}{x - 1} + \dfrac{4}{x - 2}$

50. $f(x) = 2 + \dfrac{3}{x}$

In Exercises 51–64, find the domain of the function.

51. $g(x) = 1 - 2x^2$

52. $f(x) = 5x^2 + 2x - 1$

53. $h(t) = \dfrac{4}{t}$

54. $s(y) = \dfrac{3y}{y + 5}$

55. $g(y) = \sqrt[3]{y - 10}$

56. $f(t) = \sqrt[3]{t + 4}$

57. $f(x) = \sqrt[4]{1 - x^2}$

58. $g(x) = \sqrt{x + 1}$

59. $g(x) = \dfrac{1}{x} - \dfrac{3}{x + 2}$

60. $h(x) = \dfrac{10}{x^2 - 2x}$

61. $f(x) = \dfrac{\sqrt{x + 1}}{x - 2}$

62. $f(s) = \dfrac{\sqrt{s - 1}}{s - 4}$

63. $f(x) = \dfrac{x - 4}{\sqrt{x}}$

64. $f(x) = \dfrac{x - 5}{\sqrt{x^2 - 9}}$

65. Consider $f(x) = \sqrt{x - 2}$ and $g(x) = \sqrt[3]{x - 2}$. Why are the domains of f and g different?

66. A student says that the domain of

$$f(x) = \dfrac{\sqrt{x + 1}}{x - 3}$$

is all real numbers except $x = 3$. Is the student correct? Explain.

67. Volume of a Box An open box is to be made from a square piece of material 12 inches on a side by cutting equal squares from the corners and turning up the sides (see figure).

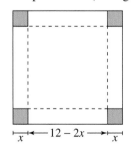
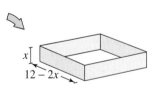

(a) Write the volume V of the box as a function of its height x.

(b) What is the domain of the function?

(c) Determine the volume of a box with a height of 5 inches.

68. Volume of a Package A rectangular package with a square cross section to be sent by a postal service can have a maximum combined length and girth (perimeter of a cross section) of 108 inches (see figure).

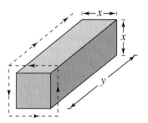

(a) Write the volume V of a package with maximum length and girth as a function of x.

(b) What is the domain of the function?

(c) Determine the volume of a package with a height of 15 inches.

69. Height of a Balloon A balloon carrying a transmitter ascends vertically from a point 2000 feet from the receiving station (see figure). Let d be the distance between the balloon and the receiving station. Write the height of the balloon as a function of d. What is the domain of this function?

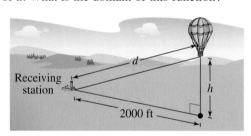

70. Cost, Revenue, and Profit A company produces a product for which the variable cost is \$12.30 per unit and the fixed costs are \$98,000. The product sells for \$17.98. Let x be the number of units produced and sold.

(a) The total cost for a business is the sum of the variable cost and the fixed costs. Write the total cost C as a function of the number of units produced.

(b) Write the revenue R as a function of the number of units sold.

(c) Write the profit P as a function of the number of units sold. (*Note:* $P = R - C$.)

71. *Make a Decision*: **Path of a Ball** The height y (in feet) of a baseball thrown by a child is given by

$$y = -\frac{1}{10}x^2 + 3x + 6$$

where x is the horizontal distance (in feet) from where the ball was thrown. Will the ball fly over the head of another child 30 feet away trying to catch the ball? (Assume that the child who is trying to catch the ball holds a baseball glove at a height of 5 feet.)

72. National Defense The national defense budget outlay V (in billions of dollars) for veterans in the United States from 1990 to 2000 can be approximated by the model

$$V = \begin{cases} -0.234t^2 + 3.01t + 28.9, & 0 \le t \le 5 \\ 2.41t + 22.4, & 6 \le t \le 10 \end{cases}$$

where t represents the year, with $t = 0$ corresponding to 1990. Use the model to find total veteran outlays in 1994 and 2000. (Source: U. S. Office of Management and Budget)

73. Mobile Homes The number N (in thousands) of mobile homes manufactured for residential use in the United States from 1991 to 2001 can be approximated by the model

$$N = \begin{cases} 33.93t + 144.0, & 1 \le t \le 6 \\ -19.59t^2 + 314.5t - 900, & 7 \le t \le 11 \end{cases}$$

where t represents the year, with $t = 1$ corresponding to 1991 (see figure). Use the model to find the total number of mobile homes manufactured between 1991 and 2001. (Source: U.S. Census Bureau)

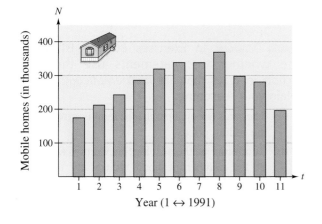

Year (1 ↔ 1991)

74. *Make a Decision*: **College Textbooks** The total consumer expenditures E (in millions of dollars) for college textbooks in the United States from 1995 to 2002 are shown in the table. (Source: Book Industry Study Group, Inc.)

Year	Expenditures, E
1995	2708
1996	2920
1997	3110
1998	3365
1999	3773
2000	3905
2001	4187
2002	4706

(a) Use a graphing utility to create a scatter plot of the data. Let t represent the year, with $t = 5$ corresponding to 1995.

(b) Use the *regression* feature of a graphing utility to find a linear model and a quadratic model for the data.

(c) Use each model to approximate the expenditures for the years 1995 to 2002. Compare the values generated by each model with the actual values shown in the table. Which model is a better fit? Justify your answer.

75. *Make a Decision*: **Cost of Housing** The average annual expenditures E (in dollars) per household for housing in the United States from 1994 to 2002 are shown in the table. (Source: U.S. Bureau of Labor Statistics)

Year	Expenditures, E
1994	10,106
1995	10,458
1996	10,747
1997	11,272
1998	11,713
1999	12,057
2000	12,319
2001	13,011
2002	13,283

(a) Use a graphing utility to create a scatter plot of the data. Let t represent the year, with $t = 4$ corresponding to 1994.

(b) Use the *regression* feature of a graphing utility to find a linear model and a quadratic model for the data.

(c) Use a graphing utility to graph the linear model with the scatter plot. Then graph the quadratic model with the scatter plot. Which model appears to fit the data better? Is one model easier to use than the other? Which model would you choose to use? Explain your reasoning.

76. *Make a Decision*: **Average Cost**   The inventor of a new game believes that the variable cost for producing the game is $0.95 per unit and the fixed costs are $6000. The inventor sells each game for $1.69. Let x be the number of games sold.

(a) Write the total cost C as a function of the number of games sold.

(b) Write the average cost per unit $\overline{C} = C/x$ as a function of x.

(c) Complete the table.

x	100	1000	10,000	100,000
$\overline{C}$				

(d) Write a paragraph analyzing the data in the table. What do you observe about the average cost per unit as x gets larger?

77. *Make a Decision*: **Average Cost**   A manufacturer determines that the variable cost for a new product is $1.15 per unit and the fixed costs are $35,000. The product is to be sold for $2.19. Let x be the number of units sold.

(a) Write the total cost C as a function of the number of units sold.

(b) Write the average cost per unit $\overline{C} = C/x$ as a function of x.

(c) Complete the table.

x	100	1000	10,000	100,000
$\overline{C}$				

(d) Write a paragraph analyzing the data in the table. What do you observe about the average cost per unit as x gets larger?

78. *Make a Decision*: **Charter Bus Fares**   For groups of 80 or more people, a charter bus company determines the rate per person (in dollars) according to the formula

$$\text{Rate} = 8 - 0.05(n - 80) \quad n \geq 80$$

where n is the number of people in the group.

(a) Write the total revenue R for the bus company as a function of n.

(b) Complete the table.

n	90	100	110	120	130	140	150
R							

(c) Write a paragraph analyzing the data in the table.

79. *Make a Decision*: **Ripples in a Pond**   A stone is thrown into the middle of a calm pond, causing ripples to form in concentric circles. The radius r of the outermost ripple increases at the rate of 0.8 foot per second.

(a) Write a function for the radius r of the circle formed by the outermost ripple in terms of time t.

(b) Write a function for the area A enclosed by the outermost ripple. Complete the table.

Time, t	1	2	3	4	5
Radius, r					
Area, A					

(c) Compare the ratios $A(2)/A(1)$ and $A(4)/A(2)$. What do you observe? Based on your observation, predict the area when $t = 8$. Verify by checking $t = 8$ in the area function.

Make a Decision   In Exercises 80 and 81, determine whether the statements use the word *function* in ways that are *mathematically* correct. Explain your reasoning.

80. (a) The sales tax on a purchased item is a function of the selling price.

(b) Your score on the next algebra exam is a function of the number of hours you study the night before the exam.

81. (a) The amount in your savings account is a function of your salary.

(b) The speed at which a free-falling baseball strikes the ground is a function of the height from which it was dropped.

Mid-Chapter Quiz

Take this quiz as you would take a quiz in class. After you are done, check your work against the answers given in the back of the book.

In Exercises 1–3, (a) plot the points, (b) find the distance between the points, and (c) find the midpoint of the line segment joining the points.

1. $(-3, 2)$, $(4, -5)$ **2.** $(1.3, -4.5)$, $(-3.7, 0.7)$ **3.** $(4, -2)$, $\left(-1, -\frac{5}{2}\right)$

4. A business had sales of \$1,330,000 in 2001 and \$1,890,000 in 2005. Predict the sales in 2007. Explain your reasoning.

In Exercises 5–8, find an equation of the line that passes through the given point and has the indicated slope. Sketch the line.

	Point	Slope		Point	Slope
5.	$(3, 5)$	$m = \frac{2}{3}$	**6.**	$(-2, 4)$	$m = 0$
7.	$(2, -3)$	m is undefined.	**8.**	$(-2, -5)$	$m = -2$

In Exercises 9–11, sketch the graph of the equation. Identify any intercepts and symmetry.

9. $y = 9 - x^2$ **10.** $y = x\sqrt{x + 4}$ **11.** $y = |x - 3|$

In Exercises 12 and 13, find the standard form of the equation of the circle.

12. Center: $(2, -3)$;   radius: 4   **13.** Center: $\left(0, -\frac{1}{2}\right)$;   radius: 2

14. Write the equation $x^2 + y^2 - 2x + 4y - 4 = 0$ in standard form. Then sketch the circle.

In Exercises 15 and 16, evaluate the function as indicated and simplify.

15. $f(x) = 3(x + 2) - 4$
  (a) $f(0)$ (b) $f(-3)$

16. $g(t) = 2t^3 - t^2$
  (a) $g(1)$ (b) $g(-2)$

In Exercises 17 and 18, find the domain of the function.

17. $h(x) = \sqrt{x + 2}$ **18.** $f(x) = 3x^2 + 5x + 1$

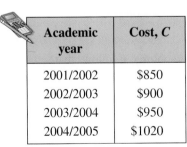

Academic year	Cost, C
2001/2002	\$850
2002/2003	\$900
2003/2004	\$950
2004/2005	\$1020

Table for 19 and 20

In Exercises 19 and 20, the Scholarship Workshop Online's estimates for the cost C of books and supplies for college students are shown in the table.

19. Let $t = 1$ represent the 2001/2002 academic year. Use a graphing utility to create a scatter plot of the data and use the *regression* feature to find a linear model for the data.

20. Use the model you found in Exercise 19 to predict the cost of books and supplies in the 2006/2007 academic year.

21. Write the area A of a circle as a function of its circumference C.

1.5   Graphs of Functions

Objectives
- **Find the domain and range using the graph of a function.**
- **Identify the graph of a function using the Vertical Line Test.**
- **Describe the increasing and decreasing behavior of a function.**
- **Find the relative minima and relative maxima of the graph of a function.**
- **Classify a function as even or odd.**
- **Identify six common graphs and use them to sketch the graph of a function.**

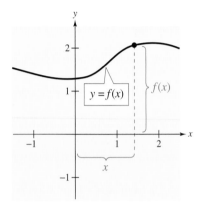

Figure 1.44

The Graph of a Function

In Section 1.4, you studied functions from an algebraic point of view. In this section, you will study functions from a graphical perspective.

The **graph of a function** f is the collection of ordered pairs $(x, f(x))$ such that x is in the domain of f. As you study this section, remember that

$$x = \text{the directed distance from the } y\text{-axis}$$

$$f(x) = \text{the directed distance from the } x\text{-axis}$$

as shown in Figure 1.44. If the graph of a function has an x-intercept at $(a, 0)$, then a is a **zero** of the function. In other words, the zeros of a function are the values of x for which $f(x) = 0$. For instance, the function given by $f(x) = x^2 - 4$ has two zeros: -2 and 2.

The **range** of a function (the set of values assumed by the dependent variable) is often more easily determined graphically than algebraically. This technique is illustrated in Example 1.

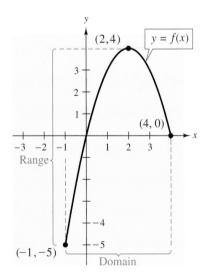

Figure 1.45

| EXAMPLE 1 | FINDING THE DOMAIN AND RANGE OF A FUNCTION |

Use the graph of the function f, shown in Figure 1.45, to find (a) the domain of f, (b) the function values $f(-1)$ and $f(2)$, and (c) the range of f.

SOLUTION

a. Because the graph does not extend beyond $x = -1$ (on the left) and $x = 4$ (on the right), the domain of f is all x in the interval $[-1, 4]$.

b. Because $(-1, -5)$ is a point on the graph of f, it follows that

$$f(-1) = -5.$$

Similarly, because $(2, 4)$ is a point on the graph of f, it follows that

$$f(2) = 4.$$

c. Because the graph does not extend below $f(-1) = -5$ or above $f(2) = 4$, the range of f is the interval $[-5, 4]$.

✓ **CHECKPOINT**   Now try Exercise 1.

By the definition of a function, at most one y-value corresponds to a given x-value. This means that the graph of a function cannot have two or more different points with the same x-coordinate, and no two points on the graph of a function can be vertically above or below each other. It follows, then, that a vertical line can intersect the graph of a function at most once. This observation provides a convenient visual test called the **Vertical Line Test** for functions.

Vertical Line Test for Functions

A set of points in a coordinate plane is the graph of y as a function of x if and only if no vertical line intersects the graph at more than one point.

EXAMPLE 2   VERTICAL LINE TEST FOR FUNCTIONS

Use the Vertical Line Test to decide whether the graphs in Figure 1.46 represent y as a function of x.

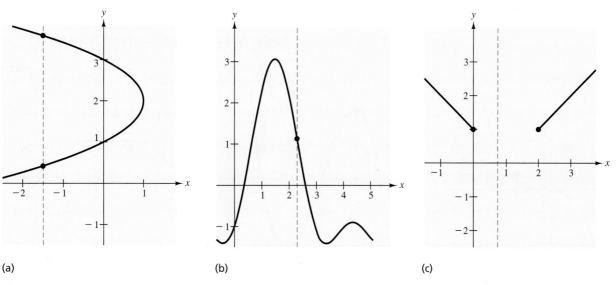

(a)   (b)   (c)

Figure 1.46

SOLUTION

a. This *is not* a graph of y as a function of x because you can find a vertical line that intersects the graph twice. That is, for a particular input x, there is more than one output y.

b. This *is* a graph of y as a function of x because every vertical line intersects the graph at most once. That is, for a particular input x, there is at most one output y.

c. This *is* a graph of y as a function of x. (Note that if a vertical line does not intersect the graph, it simply means that the function is undefined for that particular value of x.) That is, for a particular input x, there is at most one output y.

✓**CHECKPOINT**   Now try Exercise 9.

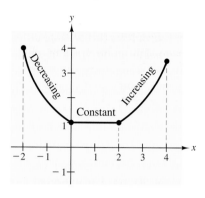

Figure 1.47

Increasing and Decreasing Functions

The more you know about the graph of a function, the more you know about the function itself. Consider the graph shown in Figure 1.47. As you move from *left to right,* this graph decreases, then is constant, and then increases.

Increasing, Decreasing, and Constant Functions

A function f is **increasing** on an interval if, for any x_1 and x_2 in the interval, $x_1 < x_2$ implies $f(x_1) < f(x_2)$.

A function f is **decreasing** on an interval if, for any x_1 and x_2 in the interval, $x_1 < x_2$ implies $f(x_1) > f(x_2)$.

A function f is **constant** on an interval if, for any x_1 and x_2 in the interval, $f(x_1) = f(x_2)$.

EXAMPLE 3   INCREASING AND DECREASING FUNCTIONS

Describe the increasing or decreasing behavior of each function shown in Figure 1.48.

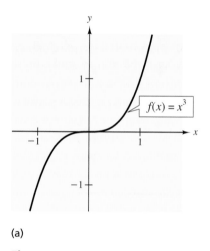

(a)

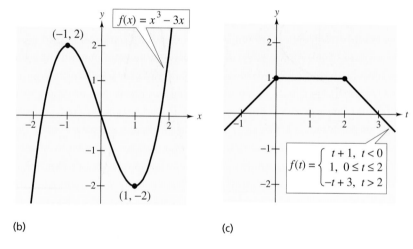

(b) (c)

Figure 1.48

SOLUTION

a. This function is increasing over the entire real line.

b. This function is increasing on the interval $(-\infty, -1)$, decreasing on the interval $(-1, 1)$, and increasing on the interval $(1, \infty)$.

c. This function is increasing on the interval $(-\infty, 0)$, constant on the interval $(0, 2)$, and decreasing on the interval $(2, \infty)$.

✓CHECKPOINT   Now try Exercise 15.

The points at which a function changes its increasing, decreasing, or constant behavior are helpful in determining the **relative minimum** or **relative maximum** values of the function.

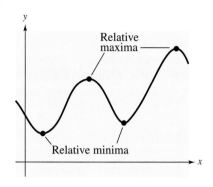

Figure 1.49

Definition of Relative Minimum and Relative Maximum

A function value $f(a)$ is called a **relative minimum** of f if there exists an interval (x_1, x_2) that contains a such that

$$x_1 < x < x_2 \quad \text{implies} \quad f(a) \leq f(x).$$

A function value $f(a)$ is called a **relative maximum** of f if there exists an interval (x_1, x_2) that contains a such that

$$x_1 < x < x_2 \quad \text{implies} \quad f(a) \geq f(x).$$

Figure 1.49 shows several examples of relative minima and relative maxima. In Section 2.1, you will study a technique for finding the *exact point* at which a second-degree polynomial function has a relative minimum or relative maximum. For the time being, however, you can use a graphing utility to find reasonable approximations of these points.

EXAMPLE 4 APPROXIMATING A RELATIVE MINIMUM

Use a graphing utility to approximate the relative minimum of the function given by $f(x) = 3x^2 - 4x - 2$.

SOLUTION

The graph of f is shown in Figure 1.50. By using the *zoom* and *trace* features of a graphing utility, you can estimate that the function has a relative minimum at the point

$$(0.67, -3.33). \qquad \text{Relative minimum}$$

Later, in Section 2.1, you will be able to determine that the exact point at which the relative minimum occurs is $\left(\frac{2}{3}, -\frac{10}{3}\right)$.

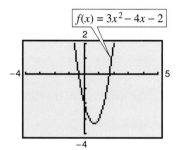

Figure 1.50

✓CHECKPOINT   Now try Exercise 21.

You can also use the *table* feature of a graphing utility to approximate numerically the relative minimum of the function in Example 4. Using a table that begins at 0.6 and increments the value of x by 0.01, you can approximate the minimum of $f(x) = 3x^2 - 4x - 2$ to be -3.33, which occurs at $(0.67, -3.33)$. A third way to find the relative minimum is to use the *minimum* feature of a graphing utility.

TECHNOLOGY

For instructions on how to use the *table* feature and the *minimum* feature, see Appendix A; for specific keystrokes, go to the text website at *college.hmco.com*.

TECHNOLOGY

If you use a graphing utility to estimate the x- and y-values of a relative minimum or relative maximum, the *zoom* feature will often produce graphs that are nearly flat. To overcome this problem, you can manually change the vertical setting of the viewing window. The graph will stretch vertically if the values of Ymin and Ymax are closer together.

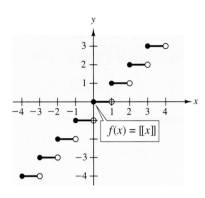

Figure 1.51 Greatest Integer Function

Step Functions

The **greatest integer function** is denoted by $[\![x]\!]$ and is defined as

$$f(x) = [\![x]\!] = \text{the greatest integer less than or equal to } x.$$

The graph of this function is shown in Figure 1.51. Note that the graph of the greatest integer function jumps vertically one unit at each integer and is constant (a horizontal line segment) between each pair of consecutive integers. Because of the jumps in its graph, the greatest integer function is an example of a type of function called a **step function.** Some values of the greatest integer function are as follows.

$$[\![-1]\!] = -1 \qquad [\![-0.5]\!] = -1$$
$$[\![0]\!] = 0 \qquad [\![0.5]\!] = 0$$
$$[\![1]\!] = 1 \qquad [\![1.5]\!] = 1$$

The range of the greatest integer function is the set of all integers.

If you use a graphing utility to graph a step function, you should set the utility to *dot* mode rather than *connected* mode.

EXAMPLE 5   THE PRICE OF A TELEPHONE CALL

The cost of a long-distance telephone call is $0.10 for up to, but not including, the first minute and $0.05 for each additional minute (or portion of a minute). The greatest integer function

$$C = 0.10 + 0.05[\![t]\!], \quad t > 0$$

can be used to model the cost of this call, where C is the total cost of the call (in dollars) and t is the length of the call (in minutes).

a. Sketch the graph of this function.

b. *Make a Decision* How long can you talk without spending more than $1?

SOLUTION

a. For calls up to, but not including, 1 minute, the cost is $0.10. For calls between 1 and 2 minutes, the cost is $0.15, and so on.

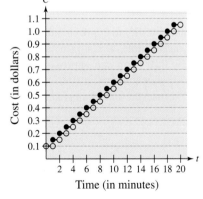

Figure 1.52

Length of call, t	Cost of call, C
$0 < t < 1$	$0.10
$1 \le t < 2$	$0.15
$2 \le t < 3$	$0.20
$\vdots$	$\vdots$
$19 \le t < 20$	$1.05

Using these and other values, you can sketch the graph shown in Figure 1.52.

b. From the graph, you can see that your phone call must be less than 19 minutes to avoid spending more than $1.

✓**CHECKPOINT**   Now try Exercise 69.

Even and Odd Functions

In Section 1.1, you studied different types of symmetry of a graph. In the terminology of functions, a function is said to be **even** if its graph is symmetric with respect to the *y*-axis and **odd** if its graph is symmetric with respect to the origin. The symmetry tests in Section 1.1 yield the following tests for even and odd functions. Even though symmetry with respect to the *x*-axis is introduced in Section 1.1, it will not be discussed here because a graph that is symmetric about the *x*-axis is not a function.

Tests for Even and Odd Functions

A function given by $y = f(x)$ is **even** if, for each x in the domain of f,
$$f(-x) = f(x).$$
A function given by $y = f(x)$ is **odd** if, for each x in the domain of f,
$$f(-x) = -f(x).$$

EXAMPLE 6   EVEN AND ODD FUNCTIONS

Decide whether each function is even, odd, or neither.

a. $g(x) = x^3 - x$ **b.** $h(x) = x^2 + 1$

SOLUTION

a. The function given by $g(x) = x^3 - x$ is odd because
$$g(-x) = (-x)^3 - (-x) = -x^3 + x = -(x^3 - x) = -g(x).$$

b. The function given by $h(x) = x^2 + 1$ is even because
$$h(-x) = (-x)^2 + 1 = x^2 + 1 = h(x).$$

The graphs of the two functions are shown in Figure 1.53.

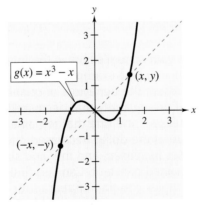

(a) Odd function (symmetric about origin)

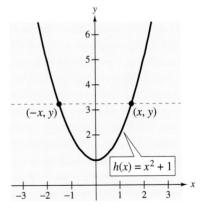

(b) Even function (symmetric about y-axis)

Figure 1.53

CHECKPOINT   Now try Exercise 27.

Common Graphs

Figure 1.54 shows the graphs of six common functions. You need to be familiar with these graphs. They can be used as an aid to sketching other graphs. For instance, the graph of the absolute value function given by

$$f(x) = |x - 2|$$

is $\vee$-shaped.

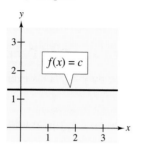

(a) Constant function

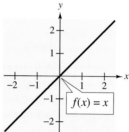

(b) Identity function

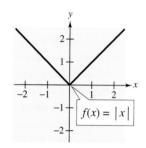

(c) Absolute value function

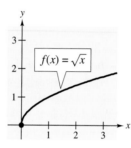

(d) Square root function

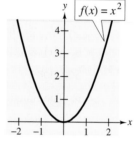

(e) Squaring function

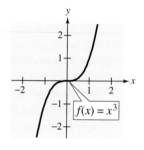

(f) Cubing function

Figure 1.54

Discussing the Concept | Increasing and Decreasing Functions

Use your school's library, the Internet, or some other reference source to find examples of three different functions that represent the behaviors of quantities between 1992 and 2002. Find one quantity that decreased during the decade, one that increased, and one that remained constant. For instance, the value of the dollar decreased, the population of the United States increased, and the land size of the United States remained constant. Can you find three other examples? If the examples you find appear to represent *linear growth or decline*, use the methods described in Section 1.3 to find a linear function $f(x) = ax + b$ that approximates the data.

1.5 Warm Up

The following warm-up exercises involve skills that were covered in earlier sections. You will use these skills in the exercise set for this section. For additional help, review Sections R2.4, R2.5, and 1.4.

1. Find $f(2)$ for $f(x) = -x^3 + 5x$.   **2.** Find $f(6)$ for $f(x) = x^2 - 6x$.

3. Find $f(-x)$ for $f(x) = 3/x$.   **4.** Find $f(-x)$ for $f(x) = x^2 + 3$.

In Exercises 5 and 6, solve the equation.

5. $x^3 - 16x = 0$ **6.** $2x^2 - 3x + 1 = 0$

In Exercises 7–10, find the domain of the function.

7. $g(x) = 4(x - 4)^{-1}$ **8.** $f(x) = 2x/(x^2 - 9x + 20)$

9. $h(t) = \sqrt[4]{5 - 3t}$ **10.** $f(t) = t^3 + 3t - 5$

1.5 Exercises

In Exercises 1–8, find the domain and range of the function. Then evaluate f at the given x-value.

1. $f(x) = \sqrt{x - 1}$,
$x = 1$

2. $f(x) = \sqrt{x^2 - 4}$,
$x = -2$

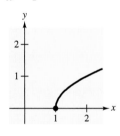

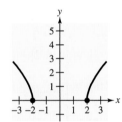

3. $f(x) = 4 - x^2$, $x = 0$ **4.** $f(x) = |x - 2|$, $x = 2$

5. $f(x) = x^3 - 1$, $x = 0$ **6.** $f(x) = \dfrac{|x|}{x}$, $x = 5$

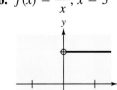

7. $f(x) = \sqrt{25 - x^2}$,
$x = 0$

8. $f(x) = \sqrt{x^2 - 9}$,
$x = 3$

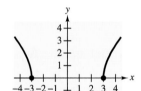

In Exercises 9–12, use the Vertical Line Test to decide whether y is a function of x.

9. $y = x^2$ **10.** $x - y^2 = 0$

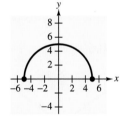

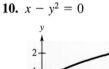

11. $x^2 + y^2 = 9$ **12.** $x^2 = xy - 1$

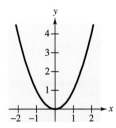

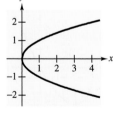

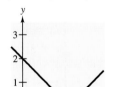

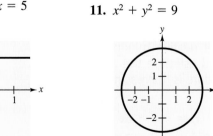

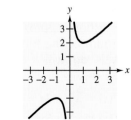

In Exercises 13–20, describe the increasing and decreasing behavior of the function. Find the point or points where the behavior of the function changes.

13. $f(x) = 2x$

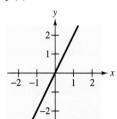

14. $f(x) = x^2 - 2x$

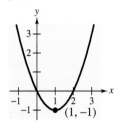

15. $f(x) = x^3 - 3x^2$

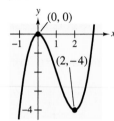

16. $f(x) = \sqrt{x^2 - 4}$

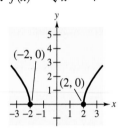

17. $f(x) = 3x^4 - 6x^2$

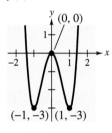

18. $f(x) = x^{2/3}$

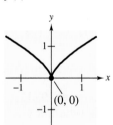

19. $y = x\sqrt{x + 3}$

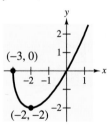

20. $y = |x + 1| + |x - 1|$

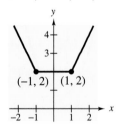

In Exercises 21–26, use a graphing utility to graph the function, approximate the relative minimum/maximum of the function, and estimate the open intervals on which the function is increasing or decreasing.

21. $f(x) = x^2 - 4x + 1$
22. $f(x) = -x^2 + 6x + 3$
23. $f(x) = x^3 - 3x^2$
24. $f(x) = -x^3 + 3x + 1$

25. $f(x) = \frac{1}{4}(-4x^4 - 5x^3 + 10x^2 + 8x + 6)$
26. $f(x) = \frac{1}{4}(x^4 + x^3 - 10x^2 + 2x - 15)$

In Exercises 27–32, decide whether the function is even, odd, or neither.

27. $f(x) = x^6 - 2x^2 + 3$ **28.** $h(x) = x^3 - 5$
29. $g(x) = x^3 - 5x$ **30.** $f(x) = x\sqrt{1 - x^2}$
31. $f(t) = t^2 + 2t - 3$ **32.** $g(s) = 4s^{2/3}$

In Exercises 33–46, sketch the graph of the function and determine whether the function is even, odd, or neither.

33. $f(x) = 3$

34. $g(x) = x$

35. $f(x) = 5 - 3x$

36. $h(x) = x^2 - 4$

37. $g(s) = \dfrac{s^3}{4}$

38. $f(t) = -t^4$

39. $f(x) = \sqrt{1 - x}$

40. $f(x) = x^{3/2}$

41. $g(t) = \sqrt[3]{t - 1}$

42. $f(x) = |x + 2|$

43. $f(x) = \begin{cases} 2x + 1, & x \le -1 \\ x^2 - 2, & x > -1 \end{cases}$

44. $f(x) = \begin{cases} 3x + 2, & x \le 0 \\ x^2 - 4, & x > 0 \end{cases}$

45. $f(x) = \begin{cases} x + 3, & x \le 0 \\ 3, & 0 < x \le 2 \\ 2x - 1, & x > 2 \end{cases}$

46. $f(x) = \begin{cases} x + 3, & x \le 1 \\ 4, & 1 < x < 3 \\ 3x - 5, & x \ge 3 \end{cases}$

In Exercises 47–60, sketch the graph of the function.

47. $f(x) = 4 - x$ **48.** $f(x) = 4x + 2$
49. $f(x) = x^2 - 9$ **50.** $f(x) = x^2 - 4x$
51. $f(x) = 1 - x^4$ **52.** $f(x) = \sqrt{x + 2}$
53. $f(x) = x^2 + 1$ **54.** $f(x) = -1(1 + |x|)$
55. $f(x) = -5$ **56.** $f(x) = \frac{1}{2}(2 + |x|)$
57. $f(x) = -[\![x]\!]$ **58.** $f(x) = 2[\![x]\!]$
59. $f(x) = [\![x - 1]\!]$ **60.** $f(x) = [\![x + 1]\!]$

61. *Make a Decision*: **Population of Connecticut** The population P (in thousands) of Connecticut for the years 1991 to 2002 can be approximated by the model

$$P = -0.185t^3 + 4.75t^2 - 16.9t + 3294,$$
$$1 \le t \le 12$$

where t represents the year, with $t = 1$ corresponding to 1991. Use a graphing utility to find the maximum and minimum population of Connecticut between 1991 and 2002. During what years was the population increasing? decreasing? Is it realistic to assume that the population will continue to follow this model? (Source: U.S. Census Bureau)

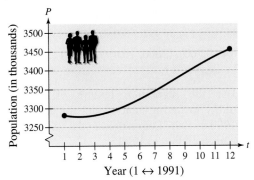

Year (1 ↔ 1991)

62. *Make a Decision*: **Population of North Dakota** The population P (in thousands) of North Dakota for the years 1994 to 2002 can be approximated by the model

$$P = 0.127t^3 - 3.73t^2 + 32.4t + 562,$$
$$4 \le t \le 12$$

where t represents the year, with $t = 4$ corresponding to 1994. Use a graphing utility to find the maximum and minimum population of North Dakota between 1994 and 2002. During what years was the population increasing? decreasing? Is it realistic to assume that the population will continue to follow this model? (Source: U.S. Census Bureau)

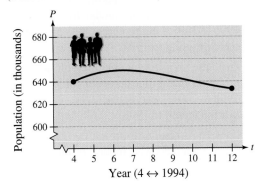

Year (4 ↔ 1994)

63. *Make a Decision*: **Company Revenue** You work for a company whose total revenue R (in hundreds of thousands of dollars) for the years 1990 to 2005 can be approximated by the function

$$R = -0.266t^2 + 5.70t + 17, \quad 0 \le t \le 15$$

where t represents the year, with $t = 0$ corresponding to 1990. Your boss asks you to estimate the years in which the revenue was increasing and the years in which the revenue was decreasing. Graph the revenue function with a graphing utility and use the *trace* feature to solve the problem. Find the maximum revenue for the years 1990 to 2005. Find the minimum revenue for the years 1990 to 2005.

64. **Radio Stations** From 1994 to 2003, the total number R of radio stations in the United States that operated with a country format can be approximated by the function

$$R = 0.993t^3 - 24.27t^2 + 116.9t + 2506,$$
$$4 \le t \le 13$$

where t represents the year, with $t = 4$ corresponding to 1994 (see figure). (Source: M Street Corporation)

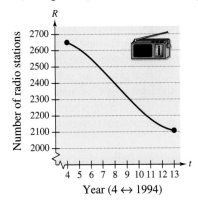

Year (4 ↔ 1994)

(a) Estimate the maximum number of country stations from 1994 to 2005.

(b) Estimate the minimum number of country stations from 1994 to 2005.

(c) Verify your estimates from parts (a) and (b) with a graphing utility.

65. **Reasoning** When finding a maximum or minimum value in Exercises 61–64, why should you also check the endpoints of the function?

66. *Make a Decision*: **Maximum Profit** The marketing department of a company estimates that the demand for a product is given by $p = 100 - 0.0001x$, where p is the price per unit and x is the number of units. The cost C of producing x units is given by $C = 350,000 + 30x$, and the profit P for producing and selling x units is given by

$$P = R - C = xp - C.$$

Sketch the graph of the profit function and estimate the number of units that would produce a maximum profit. Verify your estimate using a graphing utility.

67. *Make a Decision*: **Maximum Profit** The marketing department of a company estimates that the demand for a product is given by $p = 120 - 0.0001x$, where p is the price per unit and x is the number of units. The cost C of producing x units is given by $C = 450,000 + 50x$, and the profit P for producing and selling x units is given by

$$P = R - C = xp - C.$$

Sketch the graph of the profit function and estimate the number of units that would produce a maximum profit. Verify your estimate using a graphing utility.

68. Cost of Overnight Delivery The cost of sending an overnight package from New York to Atlanta is $9.80 for up to, but not including, the first pound and $2.50 for each additional pound (or portion of a pound). A model for the total cost C of sending the package is $C = 9.8 + 2.5[\![x]\!]$, $x > 0$, where x is the weight of the package (in pounds). Sketch the graph of this function.

69. Cost of Overnight Delivery The cost of sending an overnight package from Los Angeles to Miami is $10.75 for up to, but not including, the first pound and $3.95 for each additional pound (or portion of a pound). A model for the total cost C of sending the package is $C = 10.75 + 3.95[\![x]\!]$, $x > 0$, where x is the weight of the package (in pounds). Sketch the graph of this function.

70. Research and Development The total amount C (in billions of dollars) spent on research and development in the United States from 1960 to 2002 can be approximated by the model

$$C = \begin{cases} 0.089t^2 + 0.16t + 15.7, & 0 \le t \le 19 \\ 9.36t - 127.7, & 20 \le t \le 42 \end{cases}$$

where t represents the year, with $t = 0$ corresponding to 1960. Sketch the graph of this function. (Source: U.S. National Science Foundation)

71. Grade Level Salaries The 2004 salary S (in dollars) for federal employees at the Step 1 level can be approximated by the model

$$S = \begin{cases} 2696.9x + 11,287, & x = 1, 2, \ldots, 10 \\ 10,677.8x - 75,220, & x = 11, \ldots, 15 \end{cases}$$

where x represents the "GS" grade. Sketch a *bar graph* that represents this function. (Source: U.S. Office of Personnel Management)

72. Cable TV Systems The numbers of cable television systems in the United States from 1990 to 2002 are given by the following ordered pairs. (Source: *Television and Cable Factbook*)

(1990, 9575), (1991, 10,704), (1992, 11,035),

(1993, 11,108), (1994, 11,214), (1995, 11,218),

(1996, 11,119), (1997, 10,950), (1998, 10,845),

(1999, 10,700), (2000, 10,400), (2001, 9924),

(2002, 9947)

(a) Use the *regression* feature of a graphing utility to find a cubic model for the data from 1990 to 1996. Let t represent the year, with $t = 0$ corresponding to 1990.

(b) Use the *regression* feature of a graphing utility to find a quadratic model for the data from 1997 to 2002. Let t represent the year, with $t = 7$ corresponding to 1997.

(c) Use your results from parts (a) and (b) to construct a piecewise model for all of the data.

73. Average Miles per Gallon The average number of miles per gallon for passenger cars in the United States from 1993 to 2001 is given by the following ordered pairs. (Source: U.S. Federal Highway Administration)

(1993, 20.6), (1994, 20.8), (1995, 21.1),

(1996, 21.2), (1997, 21.5), (1998, 21.6),

(1999, 21.4), (2000, 21.9), (2001, 22.1)

(a) Use the *regression* feature of a graphing utility to find a cubic model for the data from 1993 to 1997. Let t represent the year, with $t = 3$ corresponding to 1993.

(b) Use the *regression* feature of a graphing utility to find a linear model for the data from 1998 to 2001. Let t represent the year, with $t = 8$ corresponding to 1998.

(c) Use your results from parts (a) and (b) to construct a piecewise model for all of the data.

1.6   Transformations of Functions

Objectives
- Use vertical and horizontal shifts to sketch graphs of functions.
- Use reflections to sketch graphs of functions.
- Use nonrigid translations to sketch graphs of functions.

Vertical and Horizontal Shifts

Many functions have graphs that are simple transformations of the common graphs that are summarized on page 60. For example, you can obtain the graph of $h(x) = x^2 + 2$ by shifting the graph of $f(x) = x^2$ *upward* two units, as shown in Figure 1.55. In function notation, h and f are related as follows.

$$h(x) = x^2 + 2 = f(x) + 2 \qquad \text{Upward shift of two units}$$

Similarly, you can obtain the graph of $g(x) = (x - 2)^2$ by shifting the graph of $f(x) = x^2$ to the *right* two units, as shown in Figure 1.56. In this case, the functions g and f have the following relationship.

$$g(x) = (x - 2)^2 = f(x - 2) \qquad \text{Right shift of two units}$$

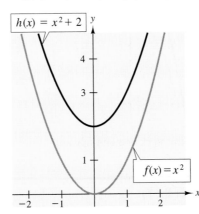

Figure 1.55 Vertical Shift Upward   **Figure 1.56** Horizontal Shift to the Right

Vertical and Horizontal Shifts

Let c be a positive real number. **Vertical** and **horizontal shifts** in the graph of $y = f(x)$ are represented as follows.

1. Vertical shift c units **upward**: $h(x) = f(x) + c$

2. Vertical shift c units **downward**: $h(x) = f(x) - c$

3. Horizontal shift c units to the **right:** $h(x) = f(x - c)$

4. Horizontal shift c units to the **left:** $h(x) = f(x + c)$

STUDY TIP

In items 3 and 4, be sure you see that $h(x) = f(x - c)$ corresponds to a *right* shift and $h(x) = f(x + c)$ corresponds to a *left* shift.

Some graphs can be obtained from a combination of vertical and horizontal shifts. This is demonstrated in Example 1(b). Vertical and horizontal shifts generate a *family of functions,* each with the same shape but at different locations in the plane.

EXAMPLE 1   SHIFTS IN THE GRAPH OF A FUNCTION

Use the graph of $f(x) = x^3$ to sketch the graph of each function.

a. $g(x) = x^3 + 1$ **b.** $h(x) = (x + 2)^3 + 1$

SOLUTION

a. Relative to the graph of $f(x) = x^3$, the graph of $g(x) = x^3 + 1$ is an upward shift of one unit, as shown in Figure 1.57(a).

b. Relative to the graph of $f(x) = x^3$, the graph of $h(x) = (x + 2)^3 + 1$ involves a left shift of two units *and* an upward shift of one unit, as shown in Figure 1.57(b).

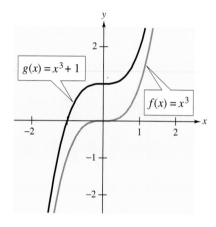

(a) Vertical shift: one unit upward

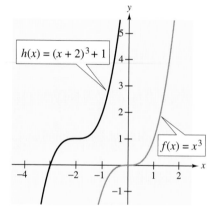

(b) Horizontal shift: two units left;
Vertical shift: one unit upward

Figure 1.57

Note that the functions f, g, and h belong to the family of cubic functions.

✔CHECKPOINT   Now try Exercise 5.

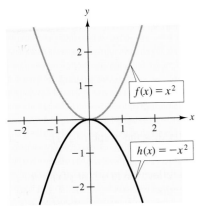

Figure 1.58 Reflection

Reflections

The second common type of transformation is a **reflection.** For instance, if you consider the x-axis to be a mirror, the graph of $h(x) = -x^2$ is the mirror image (or reflection) of the graph of $f(x) = x^2$, as shown in Figure 1.58.

Reflections in the Coordinate Axes

Reflections in the coordinate axes of the graph of $y = f(x)$ are represented as follows.

1. **Reflection in the x-axis:** $g(x) = -f(x)$
2. **Reflection in the y-axis:** $h(x) = f(-x)$

EXAMPLE 2 REFLECTIONS OF THE GRAPH OF A FUNCTION

Compare the graph of each function with the graph of $f(x) = \sqrt{x}$.

a. $g(x) = -\sqrt{x}$ **b.** $h(x) = \sqrt{-x}$

SOLUTION

a. The graph of g is a reflection of the graph of f in the x-axis because

$$g(x) = -\sqrt{x} = -f(x). \qquad \text{See Figure 1.59(a).}$$

b. The graph of h is a reflection of the graph of f in the y-axis because

$$h(x) = \sqrt{-x} = f(-x). \qquad \text{See Figure 1.59(b).}$$

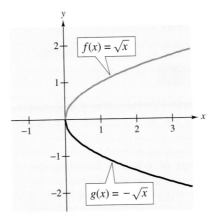

(a) Reflection in x-axis

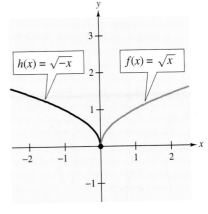

(b) Reflection in y-axis

Figure 1.59

✓CHECKPOINT   Now try Exercise 25.

When sketching the graph of a function involving square roots, remember that the domain must be restricted to exclude negative numbers inside the radical. For instance, here are the domains of the functions in Example 2.

Domain of $g(x) = -\sqrt{x}$: $x \geq 0$ Domain of $h(x) = \sqrt{-x}$: $x \leq 0$

TECHNOLOGY

You will find programs for several models of graphing utilities that will give you practice working with reflections, horizontal shifts, and vertical shifts at the website for this text at *college.hmco.com.* These programs will graph the function

$$y = R(x + H)^2 + V$$

where $R = \pm 1$, H is an integer between -6 and 6, and V is an integer between -3 and 3. Each time you run the program, different values of R, H, and V are possible. From the graph, you should be able to determine the values of R, H, and V.

EXAMPLE 3   REFLECTIONS AND SHIFTS

Use the graph of $f(x) = x^2$ to sketch the graph of each function.

a. $g(x) = -(x - 3)^2$

b. $h(x) = -x^2 + 2$

SOLUTION

a. To sketch the graph of $g(x) = -(x - 3)^2$, first shift the graph of $f(x) = x^2$ to the right three units. Then reflect the result in the x-axis.

b. To sketch the graph of $h(x) = -x^2 + 2$, first reflect the graph of $f(x) = x^2$ in the x-axis. Then shift the result upward two units.

The graphs of both functions are shown in Figure 1.60.

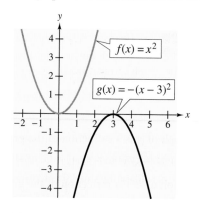

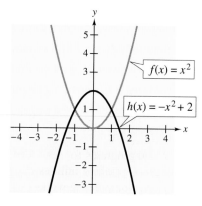

(a) Shift and then reflect in x-axis   (b) Reflect in x-axis and then shift

Figure 1.60

✓CHECKPOINT   Now try Exercise 29.

EXAMPLE 4   FINDING EQUATIONS FROM GRAPHS

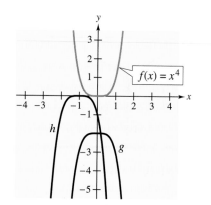

Figure 1.61

The graphs labeled g and h in Figure 1.61 are transformations of the graph of $f(x) = x^4$. Find an equation for each function.

SOLUTION

The graph of g is a reflection in the x-axis *followed* by a downward shift of two units of the graph of $f(x) = x^4$. So, the equation for g is $g(x) = -x^4 - 2$. The graph of h is a horizontal shift of one unit to the left *followed* by a reflection in the x-axis of the graph of $f(x) = x^4$. So, the equation for h is $h(x) = -(x + 1)^4$.

✓CHECKPOINT   Now try Exercise 43.

Can you think of another way to find an equation for g in Example 4? If you were to shift the graph of f upward two units and then reflect the graph in the x-axis, you would obtain the equation $g(x) = -(x^4 + 2)$. The Distributive Property yields $g(x) = -x^4 - 2$, which is the same equation obtained in Example 4.

Nonrigid Transformations

Horizontal shifts, vertical shifts, and reflections are **rigid** transformations because the basic shape of the graph is unchanged. These transformations change only the *position* of the graph in the xy-plane. A **nonrigid** transformation is one that causes a *distortion*—a change in the shape of the original graph. For instance, a nonrigid transformation of the graph of $y = f(x)$ is represented by $g(x) = cf(x)$, where the transformation is a **vertical stretch** if $c > 1$ and a **vertical shrink** if $0 < c < 1$.

DISCOVERY

Use a graphing utility to graph $f(x) = 2x^2$. Compare this graph with the graph of $h(x) = x^2$. Describe the effect of multiplying x^2 by a number greater than 1. Then graph $g(x) = \frac{1}{2}x^2$. Compare this with the graph of $h(x) = x^2$. Describe the effect of multiplying x^2 by a number greater than 0 but less than 1. Can you think of an easy way to remember this generalization? Use the *table* feature of a graphing utility to compare the values for $f(x)$, $g(x)$, and $h(x)$. What do you notice? How does this relate to the vertical stretch or vertical shrink of the graph of a function?

| EXAMPLE 5 | NONRIGID TRANSFORMATIONS |

Compare the graph of each function with the graph of $f(x) = |x|$.

a. $h(x) = 3|x|$

b. $g(x) = \dfrac{1}{3}|x|$

SOLUTION

a. Relative to the graph of $f(x) = |x|$, the graph of

$$h(x) = 3|x| = 3f(x)$$

is a vertical stretch (each y-value is multiplied by 3) of the graph of f.

b. Similarly, the graph of

$$g(x) = \frac{1}{3}|x| = \frac{1}{3}f(x)$$

is a vertical shrink $\left(\text{each } y\text{-value is multiplied by } \tfrac{1}{3}\right)$ of the graph of f.

The graphs of both functions are shown in Figure 1.62.

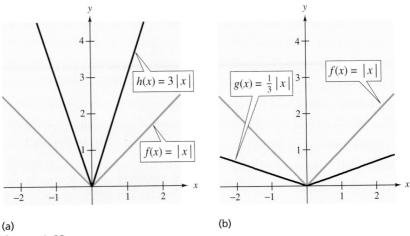

(a)   (b)

Figure 1.62

✓CHECKPOINT   Now try Exercise 33.

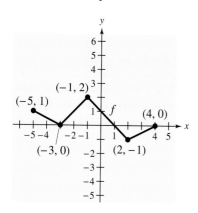

Figure 1.63

EXAMPLE 6   RIGID AND NONRIGID TRANSFORMATIONS

Use the graph of f shown in Figure 1.63 to sketch each graph.

a. $g(x) = f(x - 2) + 1$ **b.** $h(x) = \frac{1}{2}f(x)$

SOLUTION

a. The graph of g is a horizontal shift to the right two units and a vertical shift upward one unit of the graph of f. The graph of g is shown in Figure 1.64(a).

b. The graph of h is a vertical shrink of the graph of f. The graph of h is shown in Figure 1.64(b).

For $x = -5$, $h(-5) = \frac{1}{2}f(-5) = \frac{1}{2}(1) = \frac{1}{2}$.

For $x = -3$, $h(-3) = \frac{1}{2}f(-3) = \frac{1}{2}(0) = 0$.

For $x = -1$, $h(-1) = \frac{1}{2}f(-1) = \frac{1}{2}(2) = 1$.

For $x = 2$, $h(2) = \frac{1}{2}f(2) = \frac{1}{2}(-1) = -\frac{1}{2}$.

For $x = 4$, $h(4) = \frac{1}{2}f(4) = \frac{1}{2}(0) = 0$.

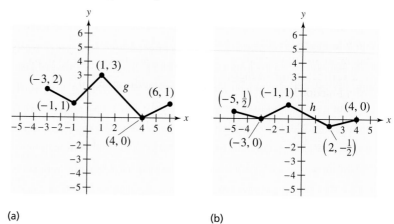

(a)   (b)

Figure 1.64

✓CHECKPOINT   Now try Exercise 45.

Discussing the Concept | Constructing Transformations

Use a graphing utility to graph $f(x) = 3 - x^3$. Describe how to alter this function to produce each transformation described below. Graph each transformation in the same viewing window with f. Confirm that the transformation moves f as described.

a. Graph of f shifted upward four units

b. Graph of f shifted to the left two units

c. Graph of f shifted downward two units and to the right one unit

d. Graph of f reflected in the x-axis

1.6 Warm Up

In Exercises 1 and 2, evaluate the function at the indicated value.

1. Find $f(3)$ for $f(x) = x^2 - 4x + 15$.
2. Find $f(-x)$ for $f(x) = 2x/(x - 3)$.

In Exercises 3 and 4, solve the equation.

3. $-x^3 + 10x = 0$
4. $3x^2 + 2x - 8 = 0$

In Exercises 5–10, sketch the graph of the function.

5. $f(x) = -2$
6. $f(x) = -x$
7. $f(x) = x + 5$
8. $f(x) = 2 - x$
9. $f(x) = 3x - 4$
10. $f(x) = 9x + 10$

1.6 Exercises

In Exercises 1–8, describe the sequence of transformations from $f(x) = x^2$ to g. Then sketch the graph of g by hand. Verify with a graphing utility.

1. $g(x) = x^2 + 3$
2. $g(x) = x^2 - 2$
3. $g(x) = (x + 4)^2$
4. $g(x) = (x - 3)^2$
5. $g(x) = (x - 2)^2 + 2$
6. $g(x) = (x + 1)^2 - 3$
7. $g(x) = -x^2 + 1$
8. $g(x) = -(x - 2)^2$

In Exercises 9–16, describe the sequence of transformations from $f(x) = |x|$ to g. Then sketch the graph of g by hand. Verify with a graphing utility.

9. $g(x) = |x| + 2$
10. $g(x) = |x| - 3$
11. $g(x) = |x - 1|$
12. $g(x) = |x + 4|$
13. $g(x) = -|x| + 3$
14. $g(x) = |x + 2| - 3$
15. $g(x) = 4 - |x - 2|$
16. $g(x) = |x - 2| + 2$

In Exercises 17–24, describe the sequence of transformations from $f(x) = \sqrt{x}$ to g. Then sketch the graph of g by hand. Verify with a graphing utility.

17. $g(x) = \sqrt{x - 2}$
18. $g(x) = \sqrt{x + 3}$
19. $g(x) = \sqrt{x - 3} + 1$
20. $g(x) = \sqrt{x + 5} - 2$
21. $g(x) = 2 - \sqrt{x - 4}$
22. $g(x) = \sqrt{2x}$
23. $g(x) = \sqrt{-x} + 1$
24. $g(x) = \sqrt{2x} - 5$

In Exercises 25–34, describe the sequence of transformations from $f(x) = \sqrt[3]{x}$ to y. Then sketch the graph of y by hand. Verify with a graphing utility.

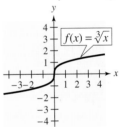

25. $y = \sqrt[3]{-x}$
26. $y = -\sqrt[3]{x}$
27. $y = \sqrt[3]{x} - 1$
28. $y = \sqrt[3]{x} + 1$
29. $y = 2 - \sqrt[3]{x + 1}$
30. $y = -\sqrt[3]{x - 1} - 4$
31. $y = \sqrt[3]{x + 1} - 1$
32. $y = \frac{1}{2}\sqrt[3]{x}$
33. $y = \frac{1}{2}\sqrt[3]{x} - 3$
34. $y = 2\sqrt[3]{x - 2} + 1$

In Exercises 35–40, identify the transformation shown in the graph and the associated common function. Write the equation for the graphed function.

35.
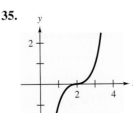

36.
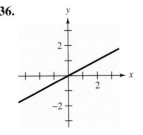

37.

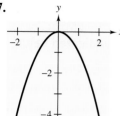

38.

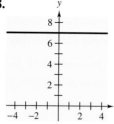

39.

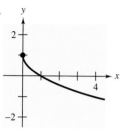

40.

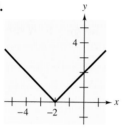

 41. Use a graphing utility to graph f for $c = -2, 0,$ and 2 in the same viewing window.

(a) $f(x) = \frac{1}{2}x + c$

(b) $f(x) = \frac{1}{2}(x - c)$

(c) $f(x) = \frac{1}{2}(cx)$

In each case, compare the graph with the graph of $y = \frac{1}{2}x$.

42. Use a graphing utility to graph f for $c = -2, 0,$ and 2 in the same viewing window.

(a) $f(x) = x^3 + c$

(b) $f(x) = (x - c)^3$

(c) $f(x) = (x - 2)^3 + c$

In each case, compare the graph with the graph of $y = x^3$.

43. Use the graph of $f(x) = x^2$ to write equations for the functions whose graphs are shown.

(a)

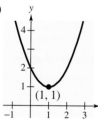

(b)

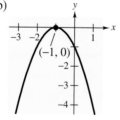

44. Use the graph of $f(x) = x^3$ to write equations for the functions whose graphs are shown.

(a)

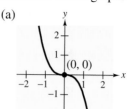

(b)

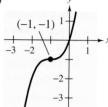

45. Use the graph of f (see figure) to sketch each graph.

(a) $y = f(x) + 2$ (b) $y = -f(x)$

(c) $y = f(x - 2)$ (d) $y = f(x + 3)$

(e) $y = 2f(x)$ (f) $y = f(-x)$

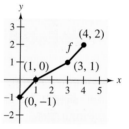

Figure for 45   Figure for 46

46. Use the graph of f (see figure) to sketch each graph.

(a) $y = f(x) - 1$ (b) $y = f(x + 1)$

(c) $y = f(x - 1)$ (d) $y = -f(x - 2)$

(e) $y = f(-x)$ (f) $y = \frac{1}{2}f(x)$

47. Use the graph of f (see figure) to sketch each graph.

(a) $y = f(-x)$ (b) $y = f(x) + 4$

(c) $y = 2f(x)$ (d) $y = -f(x - 4)$

(e) $y = f(x) - 3$ (f) $y = -f(x) - 1$

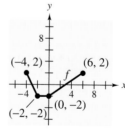

Figure for 47   Figure for 48

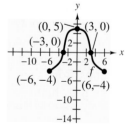

48. Use the graph of f (see figure) to sketch each graph.

(a) $y = f(x - 5)$ (b) $y = -f(x) + 3$

(c) $y = \frac{1}{3}f(x)$ (d) $y = -f(x + 1)$

(e) $y = f(-x)$ (f) $y = f(x) - 5$

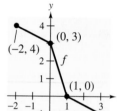

In Exercises 49–52, consider the graph of $f(x) = |x|$. Use your knowledge of rigid and nonrigid transformations to write an equation for each of the following descriptions. Verify with a graphing utility.

49. The graph of f is shifted two units to the right and one unit upward.

50. The graph of f is reflected about the x-axis, shifted one unit to the left, and shifted two units upward.

51. The graph of f is vertically stretched by a factor of 3 and reflected about the x-axis.

52. The graph of f is vertically shrunk by a factor of $\frac{1}{2}$ and shifted two units to the right.

In Exercises 53–56, consider the graph of $g(x) = \sqrt{x}$. Use your knowledge of rigid and nonrigid transformations to write an equation for each of the following descriptions. Verify with a graphing utility.

53. The graph of g is shifted three units to the left and two units upward.

54. The graph of g is reflected about the x-axis, shifted two units to the right, and shifted one unit downward.

55. The graph of g is vertically stretched by a factor of 4 and reflected about the x-axis.

56. The graph of g is vertically shrunk by a factor of $\frac{1}{2}$ and shifted three units to the left.

In Exercises 57 and 58, use the graph of $f(x) = x^3 - 3x^2$ to write an equation for the function g shown in the graph.

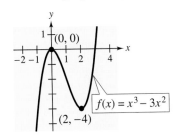

$f(x) = x^3 - 3x^2$
$(0, 0)$
$(2, -4)$

57.

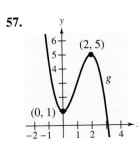

$(2, 5)$
g
$(0, 1)$

58.

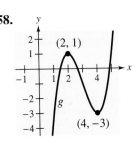

$(2, 1)$
g
$(4, -3)$

59. The point $(3, 9)$ on the graph of $f(x) = x^2$ has been shifted to the point $(4, 7)$ after a rigid transformation. Identify the shift and write the new function g in terms of f.

60. The point $(8, 2)$ on the graph of $f(x) = \sqrt[3]{x}$ has been shifted to the point $(5, 0)$ after a rigid transformation. Identify the shift and write the new function h in terms of f.

61. Profit The weekly profit P (in hundreds of dollars) for a product is given by the model

$$P(x) = 80 + 20x - 0.5x^2, \qquad 0 \le x \le 20$$

where x is the amount (in hundreds of dollars) spent on advertising.

(a) Use a graphing utility to graph the profit function.

(b) The business estimates that taxes and operating costs will increase by an average of $2500 per week during the next year. Rewrite the profit equation to reflect this expected decrease in profits. Identify the type of transformation applied to the graph of the equation.

(c) Rewrite the profit equation so that x measures advertising expenditures in dollars. [Find $P(x/100)$.] Identify the type of transformation applied to the graph of the profit function.

62. Automobile Aerodynamics The number of horsepower H required to overcome wind drag on an automobile is approximated by

$$H(x) = 0.002x^2 + 0.005x - 0.029, \quad 10 \le x \le 100$$

where x is the speed of the car (in miles per hour).

(a) Use a graphing utility to graph the function.

(b) Rewrite the horsepower function so that x represents the speed in kilometers per hour. [Find $H(x/1.6)$.] Identify the type of transformation applied to the graph of the horsepower function.

63. Make a Decision: Exploration Use a graphing utility to graph each function. Describe any similarities and differences you observe among the graphs.

(a) $y = x$ (b) $y = x^2$

(c) $y = x^3$ (d) $y = x^4$

(e) $y = x^5$ (f) $y = x^6$

64. Make a Decision Use the results of Exercise 63 to make a conjecture about the shapes of the graphs of $y = x^7$ and $y = x^8$. Use a graphing utility to verify your conjecture.

1.7 The Algebra of Functions ◀

Objectives

- **Find the sum, difference, product, and quotient of two functions.**
- **Form the composition of two functions and determine the domain.**
- **Identify a function as the composition of two functions.**
- **Use combinations and compositions of functions to solve an application problem.**

Arithmetic Combinations of Functions

Just as two real numbers can be combined by the operations of addition, subtraction, multiplication, and division to form other real numbers, two *functions* can be combined to create new functions. For example, the functions given by $f(x) = 2x - 3$ and $g(x) = x^2 - 1$ can be combined as follows.

$$f(x) + g(x) = (2x - 3) + (x^2 - 1) = x^2 + 2x - 4 \qquad \text{Sum}$$

$$f(x) - g(x) = (2x - 3) - (x^2 - 1) = -x^2 + 2x - 2 \qquad \text{Difference}$$

$$f(x)g(x) = (2x - 3)(x^2 - 1) = 2x^3 - 3x^2 - 2x + 3 \qquad \text{Product}$$

$$\frac{f(x)}{g(x)} = \frac{2x - 3}{x^2 - 1}, \quad x \neq \pm 1, \quad g(x) \neq 0 \qquad \text{Quotient}$$

The domain of an arithmetic combination of the functions f and g consists of all real numbers that are common to the domains of f and g.

Sum, Difference, Product, and Quotient of Functions

Let f and g be two functions with overlapping domains. Then, for all x common to both domains, the **sum, difference, product,** and **quotient** of f and g are defined as follows.

1. *Sum:* $\qquad\qquad (f + g)(x) = f(x) + g(x)$

2. *Difference:* $\qquad (f - g)(x) = f(x) - g(x)$

3. *Product:* $\qquad\quad (fg)(x) = f(x) \cdot g(x)$

4. *Quotient:* $\qquad\quad \left(\dfrac{f}{g}\right)(x) = \dfrac{f(x)}{g(x)}, \quad g(x) \neq 0$

EXAMPLE 1 FINDING THE SUM OF TWO FUNCTIONS

Given $f(x) = 2x + 1$ and $g(x) = x^2 + 2x - 1$, find $(f + g)(x)$.

SOLUTION

$$(f + g)(x) = f(x) + g(x) = (2x + 1) + (x^2 + 2x - 1) = x^2 + 4x$$

✓CHECKPOINT   Now try Exercise 5(a).

| EXAMPLE 2 | FINDING THE DIFFERENCE OF TWO FUNCTIONS |

Given $f(x) = 2x + 1$ and $g(x) = x^2 + 2x - 1$, find $(f - g)(x)$. Then evaluate the difference when $x = 2$.

SOLUTION

The difference of the functions f and g is given by

$$(f - g)(x) = f(x) - g(x) \qquad \text{Definition of difference of two functions}$$
$$= (2x + 1) - (x^2 + 2x - 1) \qquad \text{Substitute for } f(x) \text{ and } g(x).$$
$$= -x^2 + 2. \qquad \text{Simplify.}$$

When $x = 2$, the value of this difference is

$$(f - g)(2) = -(2)^2 + 2 = -2.$$

✓**CHECKPOINT**   Now try Exercise 21.

STUDY TIP

Note that in Example 2, $(f - g)(2)$ can also be evaluated as follows.

$$(f - g)(2) = f(2) - g(2)$$
$$= [2(2) + 1]$$
$$- [2^2 + 2(2) - 1]$$
$$= 5 - 7$$
$$= -2$$

In Examples 1 and 2, both f and g have domains that consist of all real numbers. So, the domains of $(f + g)$ and $(f - g)$ are also the set of all real numbers. Remember that any restrictions on the domains of f and g must be considered when forming the sum, difference, product, or quotient of f and g.

| EXAMPLE 3 | THE QUOTIENT OF TWO FUNCTIONS |

Find the domains of $\left(\dfrac{f}{g}\right)(x)$ and $\left(\dfrac{g}{f}\right)(x)$ for the functions

$$f(x) = \sqrt{x} \qquad \text{and} \qquad g(x) = \sqrt{4 - x^2}.$$

SOLUTION

The quotient of f and g is given by

$$\left(\frac{f}{g}\right)(x) = \frac{f(x)}{g(x)} = \frac{\sqrt{x}}{\sqrt{4 - x^2}}$$

and the quotient of g and f is given by

$$\left(\frac{g}{f}\right)(x) = \frac{g(x)}{f(x)} = \frac{\sqrt{4 - x^2}}{\sqrt{x}}.$$

The domain of f is $[0, \infty)$ and the domain of g is $[-2, 2]$. The intersection of these two domains is $[0, 2]$, which implies that the domains of f/g and g/f are as follows.

Domain of $\dfrac{f}{g}$: $[0, 2)$

Domain of $\dfrac{g}{f}$: $(0, 2]$

Can you see why these two domains differ slightly?

✓**CHECKPOINT**   Now try Exercise 5(d).

Composition of Functions

Another way to combine two functions is to form the **composition** of one with the other. For instance, if $f(x) = x^2$ and $g(x) = x + 1$, the composition of f with g is given by

$$f(g(x)) = f(x + 1) = (x + 1)^2.$$

This composition is denoted as $f \circ g$ and read as "f composed with g."

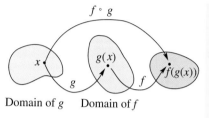

$f \circ g$

Domain of g Domain of f

Figure 1.65

Definition of the Composition of Two Functions

The **composition** of the functions f and g is given by

$$(f \circ g)(x) = f(g(x)).$$

The domain of $f \circ g$ is the set of all x in the domain of g such that $g(x)$ is in the domain of f. (See Figure 1.65.)

From the definition above, it follows that the domain of $f \circ g$ is always a subset of the domain of g, and the range of $f \circ g$ is always a subset of the range of f.

EXAMPLE 4   COMPOSITION OF FUNCTIONS

Given $f(x) = x + 2$ and $g(x) = 4 - x^2$, find the following.

a. $(f \circ g)(x)$

b. $(g \circ f)(x)$

SOLUTION

a. The composition of f with g is as follows.

$$
\begin{aligned}
(f \circ g)(x) &= f(g(x)) && \text{Definition of } f \circ g \\
&= f(4 - x^2) && \text{Definition of } g(x) \\
&= (4 - x^2) + 2 && \text{Definition of } f(x) \\
&= -x^2 + 6 && \text{Simplify.}
\end{aligned}
$$

b. The composition of g with f is as follows.

$$
\begin{aligned}
(g \circ f)(x) &= g(f(x)) && \text{Definition of } g \circ f \\
&= g(x + 2) && \text{Definition of } f(x) \\
&= 4 - (x + 2)^2 && \text{Definition of } g(x) \\
&= 4 - (x^2 + 4x + 4) && \text{Expand.} \\
&= -x^2 - 4x && \text{Simplify.}
\end{aligned}
$$

Note that, in this case, $(f \circ g)(x) \neq (g \circ f)(x)$.

✔CHECKPOINT   Now try Exercise 29.

> **EXAMPLE 5**   FINDING THE DOMAIN OF A COMPOSITE FUNCTION

Find the composition $(f \circ g)(x)$ for the functions given by

$$f(x) = x^2 - 9 \qquad \text{and} \qquad g(x) = \sqrt{9 - x^2}.$$

Then find the domain of $(f \circ g)$.

SOLUTION

The composition of the functions is as follows.

$$
\begin{aligned}
(f \circ g)(x) &= f(g(x)) \\
&= f\left(\sqrt{9 - x^2}\right) \\
&= \left(\sqrt{9 - x^2}\right)^2 - 9 \\
&= 9 - x^2 - 9 \\
&= -x^2
\end{aligned}
$$

From this result, it might appear that the domain of the composition is the set of all real numbers. However, because the domain of f is the set of all real numbers and the domain of g is $[-3, 3]$, the domain of $f \circ g$ is $[-3, 3]$.

✓ CHECKPOINT   Now try Exercise 37.

TECHNOLOGY

In Example 5, the domain of the composite function is $[-3, 3]$. To convince yourself of this, use a graphing utility to graph

$$y = \left(\sqrt{9 - x^2}\right)^2 - 9$$

as shown in the figure below. Notice that the graphing utility does not extend the graph to the left of $x = -3$ or to the right of $x = 3$.

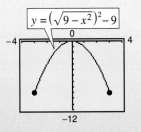

In Examples 4 and 5, you formed the composition of two functions. To "decompose" a composite function, look for an "inner" function and an "outer" function. For instance, the function h given by

$$h(x) = (3x - 5)^3$$

is the composition of f with g, where $f(x) = x^3$ and $g(x) = 3x - 5$. That is,

$$h(x) = (3x - 5)^3 = [g(x)]^3 = f(g(x)).$$

In the function h, $g(x) = 3x - 5$ is the *inner* function and $f(x) = x^3$ is the *outer* function.

> **EXAMPLE 6**   IDENTIFYING A COMPOSITE FUNCTION

Write the function given by $h(x) = \dfrac{1}{(x - 2)^2}$ as a composition of two functions.

SOLUTION

One way to write h as a composition of two functions is to take the inner function to be $g(x) = x - 2$ and the outer function to be

$$f(x) = \frac{1}{x^2} = x^{-2}.$$

Then you can write

$$h(x) = \frac{1}{(x - 2)^2} = (x - 2)^{-2} = f(x - 2) = f(g(x)).$$

✓ CHECKPOINT   Now try Exercise 45.

Applications

| EXAMPLE 7 | POLITICAL MAKEUP OF THE U.S. SENATE | |

Consider three functions R, D, and I that represent the numbers of Republicans, Democrats, and Independents, respectively, in the U.S. Senate from 1965 to 2003. Sketch the graphs of R, D, and I and the sum of R, D, and I in the same coordinate plane. The numbers of Republicans and Democrats in the Senate are shown below.

The Capitol building in Washington, D.C., is where each state's Congressional representatives convene. In recent years, no party has had a strong majority, which can make it difficult to pass legislation.

Year	R	D		Year	R	D
1965	32	68		1985	53	47
1967	36	64		1987	45	55
1969	43	57		1989	45	55
1971	44	54		1991	44	56
1973	42	56		1993	43	57
1975	37	60		1995	52	48
1977	38	61		1997	55	45
1979	41	58		1999	55	45
1981	53	46		2001	50	50
1983	54	46		2003	51	48

SOLUTION

The graphs of R, D, and I are shown in Figure 1.66. Note that the sum of R, D, and I is the constant function $R + D + I = 100$. This follows from the fact that the number of senators in the United States is 100 (two from each state).

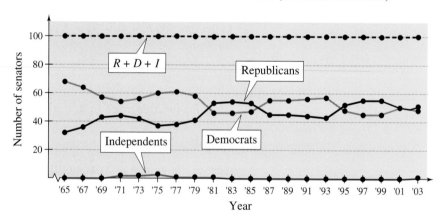

Figure 1.66 Number of U.S. Senators (by political party)

✓CHECKPOINT   Now try Exercise 53.

EXAMPLE 8   BACTERIA COUNT

Make a Decision   The number of bacteria in a refrigerated food is given by

$$N(T) = 20T^2 - 80T + 500, \qquad 2 \le T \le 14$$

where T is the temperature of the food in degrees Celsius. When the food is removed from refrigeration, the temperature of the food is given by

$$T(t) = 4t + 2, \qquad 0 \le t \le 3$$

where t is the time in hours. Find (a) the composition $N(T(t))$, (b) the number of bacteria in the food when $t = 2$ hours, and (c) how long the food can remain unrefrigerated before the bacteria count reaches 2000.

SOLUTION

a. $N(T(t)) = 20(4t + 2)^2 - 80(4t + 2) + 500$

$$= 20(16t^2 + 16t + 4) - 320t - 160 + 500$$

$$= 320t^2 + 320t + 80 - 320t - 160 + 500$$

$$= 320t^2 + 420$$

b. When $t = 2$, the number of bacteria is

$$N(T(2)) = 320(2)^2 + 420 = 1280 + 420 = 1700.$$

c. The bacteria count will reach $N = 2000$ when $320t^2 + 420 = 2000$. By solving this equation, you can determine that the bacteria count will reach 2000 when

$$t \approx 2.2 \text{ hours.}$$

So, the food can remain unrefrigerated for about 2 hours and 12 minutes.

✓**CHECKPOINT**   Now try Exercise 61.

Discussing the Concept ┃ Composition of Functions

The suggested retail price of a new car is p dollars. The dealer has advertised a factory rebate of $1500 and a 7% discount.

a. Write a function R, in terms of p, giving the cost of the car after receiving the rebate.

b. Write a function D, in terms of p, giving the cost of the car after receiving the discount.

c. Form the composite functions $(R \circ D)(p)$ and $(D \circ R)(p)$. Explain what each composite function represents.

d. Find $(R \circ D)(30,000)$ and $(D \circ R)(30,000)$. Which function yields the lower price for the car? Explain.

e. Research the rebates and discounts offered by several car dealers in your area. How do they compare with one another?

1.7 Warm Up

The following warm-up exercises involve skills that were covered in earlier sections. You will use these skills in the exercise set for this section. For additional help, review Section R1.7.

In Exercises 1–10, perform the indicated operations and simplify the result.

1. $\dfrac{1}{x} + \dfrac{1}{1 - x}$

2. $\dfrac{2}{x + 3} - \dfrac{2}{x - 3}$

3. $\dfrac{3}{x - 2} - \dfrac{2}{x(x - 2)}$

4. $\dfrac{x}{x - 5} + \dfrac{1}{3}$

5. $(x - 1)\left(\dfrac{1}{\sqrt{x^2 - 1}}\right)$

6. $\left(\dfrac{x}{x^2 - 4}\right)\left(\dfrac{x^2 - x - 2}{x^2}\right)$

7. $(x^2 - 4) \div \left(\dfrac{x + 2}{5}\right)$

8. $\left(\dfrac{x}{x^2 + 3x - 10}\right) \div \left(\dfrac{x^2 + 3x}{x^2 + 6x + 5}\right)$

9. $\dfrac{(1/x) + 5}{3 - (1/x)}$

10. $\dfrac{(x/4) - (4/x)}{x - 4}$

1.7 Exercises

In Exercises 1–4, use the graphs of f and g to graph $h(x) = (f + g)(x)$.

1.
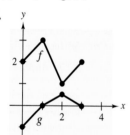

2.
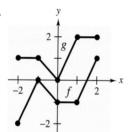

3.
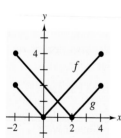

4.
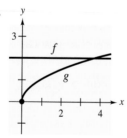

In Exercises 5–12, find (a) $(f + g)(x)$, (b) $(f - g)(x)$, (c) $(fg)(x)$, and (d) $(f/g)(x)$. What is the domain of f/g?

5. $f(x) = x + 1$, $g(x) = x - 1$

6. $f(x) = 2x - 3$, $g(x) = 1 - x$

7. $f(x) = x^2$, $g(x) = 1 - x$

8. $f(x) = 2x - 5$, $g(x) = 5$

9. $f(x) = x^2 + 5$, $g(x) = \sqrt{1 - x}$

10. $f(x) = \sqrt{x^2 - 4}$, $g(x) = \dfrac{x^2}{x^2 + 1}$

11. $f(x) = \dfrac{1}{x}$, $g(x) = \dfrac{1}{x^2}$

12. $f(x) = \dfrac{x}{x + 1}$, $g(x) = x^3$

In Exercises 13–24, evaluate the function for $f(x) = x^2 + 1$ and $g(x) = x - 4$.

13. $(f + g)(3)$

14. $(f - g)(-2)$

15. $(f - g)(2t)$

16. $(f + g)(t - 1)$

17. $(fg)(-2)$

18. $(fg)(-6)$

19. $\left(\dfrac{f}{g}\right)(5)$

20. $\left(\dfrac{f}{g}\right)(0)$

21. $(f - g)(0)$

22. $(f + g)(1)$

23. $\left(\dfrac{f}{g}\right)(-1) - g(3)$

24. $(2f)(5) + (3g)(-4)$

In Exercises 25–28, find (a) $f \circ g$, (b) $g \circ f$, and (c) $f \circ f$.

25. $f(x) = x^2$, $g(x) = x - 1$

26. $f(x) = 4x$, $g(x) = 2x + 1$

27. $f(x) = 3x + 5$, $g(x) = 5 - x$

28. $f(x) = x^3$, $\quad g(x) = \dfrac{1}{x}$

In Exercises 29–36, find (a) $f \circ g$ and (b) $g \circ f$.

29. $f(x) = \sqrt{x + 4}$, $\quad g(x) = x^2$

30. $f(x) = \sqrt[3]{x - 1}$, $\quad g(x) = x^3 + 1$

31. $f(x) = \frac{1}{3}x - 3$, $\quad g(x) = 3x + 1$

32. $f(x) = x^4$, $\quad g(x) = x^4$

33. $f(x) = \sqrt{x}$, $\quad g(x) = \sqrt{x}$

34. $f(x) = 2x - 3$, $\quad g(x) = 2x - 3$

35. $f(x) = |x|$, $\quad g(x) = x + 6$

36. $f(x) = x^{2/3}$, $\quad g(x) = x^6$

In Exercises 37–40, determine the domain of (a) f, (b) g, and (c) $f \circ g$.

37. $f(x) = \sqrt{x}$, $\quad g(x) = x^2 + 1$

38. $f(x) = \dfrac{1}{x}$, $\quad g(x) = x + 3$

39. $f(x) = \dfrac{3}{x^2 - 1}$, $\quad g(x) = x + 1$

40. $f(x) = 2x + 3$, $\quad g(x) = \dfrac{x}{2}$

In Exercises 41–44, use the graphs of f and g to evaluate the functions.

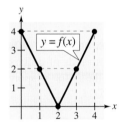
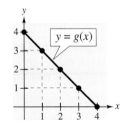

41. (a) $(f + g)(3)$ (b) $\left(\dfrac{f}{g}\right)(2)$

42. (a) $(f - g)(1)$ (b) $(fg)(4)$

43. (a) $(f \circ g)(2)$ (b) $(g \circ f)(2)$

44. (a) $(f \circ g)(0)$ (b) $(g \circ f)(3)$

In Exercises 45–52, find two functions f and g such that $(f \circ g)(x) = h(x)$. (There are many correct answers.)

45. $h(x) = (2x + 1)^2$

46. $h(x) = (1 - x)^3$

47. $h(x) = \sqrt[3]{x^2 - 4}$

48. $h(x) = \sqrt{9 - x}$

49. $h(x) = \dfrac{1}{x + 2}$

50. $h(x) = \dfrac{4}{(5x + 2)^2}$

51. $h(x) = (x + 4)^2 + 2(x + 4)$

52. $h(x) = (x + 3)^{3/2}$

53. Stopping Distance While traveling in a car at x miles per hour, you are required to stop quickly to avoid an accident. The distance the car travels (in feet) during your reaction time is given by $R(x) = \frac{3}{4}x$. The distance the car travels while you are braking (in feet) is given by

$$B(x) = \dfrac{1}{15}x^2.$$

Find the function that represents the total stopping distance T. (*Hint: $T = R + B$.*) Graph the functions R, B, and T on the same set of coordinate axes for $0 \le x \le 60$.

54. Cost The weekly cost C of producing x units in a manufacturing process is given by the function

$$C(x) = 60x + 750.$$

The number of units x produced in t hours is given by $x(t) = 50t$. Find and interpret $(C \circ x)(t)$.

55. Cost The weekly cost C of producing x units in a manufacturing process is given by the function

$$C(x) = 70x + 375.$$

The number of units x produced in t hours is given by $x(t) = 40t$. Find and interpret $(C \circ x)(t)$.

56. *Make a Decision*: Comparing Profits A company has two manufacturing plants, one in New Jersey and the other in California. From 2000 to 2005, the profits for the manufacturing plant in New Jersey have been decreasing according to the function

$$P_1 = 17.92 - 0.45t, \quad t = 0, 1, 2, 3, 4, 5$$

where P_1 represents the profits (in millions of dollars) and t represents the year, with $t = 0$ corresponding to 2000. On the other hand, the profits for the manufacturing plant in California have been increasing according to the function

$$P_2 = 14.51 + 0.56t, \quad t = 0, 1, 2, 3, 4, 5.$$

Write a function that represents the overall company profits during the six-year period. Use the *stacked bar graph* in the figure on the next page, which represents the total profits for the company during this six-year period, to determine whether the overall company profits have been increasing or decreasing.

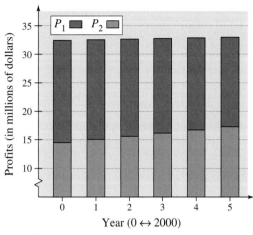

Figure for 56

57. *Make a Decision*: **Comparing Sales** You own two fast-food restaurants. During the years 2000 to 2005, the sales for the first restaurant have been decreasing according to the function

$$R_1 = 490 - 0.75t^2, \qquad t = 0, 1, 2, 3, 4, 5$$

where R_1 represents the sales (in thousands of dollars) and t represents the year, with $t = 0$ corresponding to 2000. During the same six-year period, the sales for the second restaurant have been increasing according to the function

$$R_2 = 252.8 + 0.67t, \qquad t = 0, 1, 2, 3, 4, 5.$$

Write a function that represents the total sales for the two restaurants. Use the *stacked bar graph* in the figure, which represents the total sales during this six-year period, to determine whether the total sales have been increasing or decreasing.

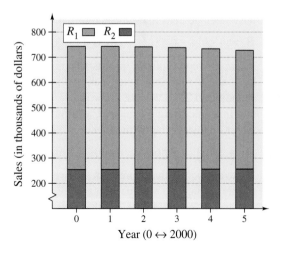

58. Female Labor Force The table shows the marital status of women in the civilian labor force from 1994 to 2002. The numbers (in millions) of working women whose status is single, married, or other (widowed, divorced, or separated) are represented by the variables $y_1, y_2,$ and y_3, respectively. (Source: U.S. Bureau of Labor Statistics)

Year	y_1	y_2	y_3
1994	15.3	32.9	12.0
1995	15.5	33.4	12.1
1996	15.8	33.6	12.4
1997	16.5	33.8	12.7
1998	17.1	33.9	12.8
1999	17.6	34.4	12.9
2000	17.8	35.1	13.3
2001	18.0	35.2	13.6
2002	18.2	35.5	13.7

(a) Create a stacked bar graph for the data.

(b) Use the *regression* feature of a graphing utility to find linear models for $y_1, y_2,$ and y_3. Let t represent the year, with $t = 4$ corresponding to 1994.

(c) Use a graphing utility to graph the models for $y_1, y_2, y_3,$ and $y_4 = y_1 + y_2 + y_3$ in the same viewing window. Use y_4 to predict the total number of women in the work force in 2005 and 2007.

59. *Make a Decision*: **Cost, Revenue, and Profit** The table shows the revenue y_1 (in thousands of dollars) and cost y_2 (in thousands of dollars) for a sports memorabilia store from 1994 to 2004.

Year	y_1	y_2
1994	25.1	20.9
1995	29.8	22.3
1996	34.4	24.5
1997	38.3	25.9
1998	42.0	27.8
1999	49.8	30.0
2000	54.1	31.7
2001	58.7	33.5
2002	63.2	35.8
2003	68.9	38.3
2004	74.0	40.5

(a) Use the *regression* feature of a graphing utility to find linear models for y_1 and y_2. Let t represent the year, with $t = 4$ corresponding to 1994.

(b) Use a graphing utility to graph the models for y_1, y_2, and $y_3 = y_1 - y_2$ in the same viewing window. What does y_3 represent in the context of the problem? Determine the value of y_3 in 2007.

(c) Create a stacked bar graph for y_2 and y_3. What do the heights of the bars represent?

60. Bacteria Count The number of bacteria in a refrigerated food product is given by

$$N(T) = 10T^2 - 20T + 600, \qquad 1 \le T \le 20$$

where T is the temperature of the food. When the food is removed from the refrigerator, the temperature of the food is given by

$$T(t) = 3t + 1$$

where t is the time in hours. Find (a) the composite function $N(T(t))$ and (b) the time when the bacteria count reaches 1500.

61. Bacteria Count The number of bacteria in a refrigerated food product is given by

$$N(T) = 25T^2 - 50T + 300, \qquad 2 \le T \le 20$$

where T is the temperature of the food. When the food is removed from the refrigerator, the temperature of the food is given by

$$T(t) = 2t + 1$$

where t is the time in hours. Find (a) the composite function $N(T(t))$ and (b) the time when the bacteria count reaches 750.

62. Troubled Waters A pebble is dropped into a calm pond, causing ripples in the form of concentric circles (see figure). The radius (in feet) of the outermost ripple is given by $r(t) = 0.6t$, where t is time in seconds after the pebble strikes the water. The area of the outermost circle is given by the function $A(r) = \pi r^2$. Find and interpret $(A \circ r)(t)$.

Price-Earnings Ratio In Exercises 63 and 64, the average annual price-earnings ratio for a corporation's stock is defined as the average price of the stock divided by the earnings per share. The average price of a corporation's stock is given as the function P and the earnings per share is given as the function E. Find the price-earnings ratio, P/E, for the years 1992 to 2002.

63. *McDonald's Corporation*

Year	P	E
1992	$11.05	$0.65
1993	$13.07	$0.73
1994	$14.36	$0.84
1995	$18.81	$0.99
1996	$23.87	$1.11
1997	$24.15	$1.15
1998	$31.12	$1.26
1999	$42.26	$1.39
2000	$33.43	$1.46
2001	$28.42	$1.36
2002	$24.16	$1.32

(Source: McDonald's Corporation)

64. *Walt Disney Company*

Year	P	E
1992	$11.42	$0.51
1993	$13.66	$0.54
1994	$14.35	$0.68
1995	$17.14	$0.84
1996	$20.13	$0.74
1997	$25.21	$0.92
1998	$33.84	$0.90
1999	$30.36	$0.66
2000	$35.55	$0.90
2001	$29.79	$0.98
2002	$20.46	$0.55

(Source: The Walt Disney Company)

65. Find the domains of $(f/g)(x)$ and $(g/f)(x)$ for the functions

$$f(x) = \sqrt{x} \quad \text{and} \quad g(x) = \sqrt{9 - x^2}.$$

Why do the two domains differ?

1.8 Inverse Functions

Objectives
- **Determine if a function has an inverse function.**
- **Find the inverse function of a function.**
- **Graph a function and its inverse function.**

Inverse Functions

Recall from Section 1.4 that a function can be represented by a set of ordered pairs. For instance, the function $f(x) = x + 4$ from the set $A = \{1, 2, 3, 4\}$ to the set $B = \{5, 6, 7, 8\}$ can be written as follows.

$$f(x) = x + 4: \{(1, 5), (2, 6), (3, 7), (4, 8)\}$$

By interchanging the first and second coordinates of each of these ordered pairs, you can form the **inverse function** of f, which is denoted by f^{-1}. It is a function from the set B to the set A, and it can be written as follows.

$$f^{-1}(x) = x - 4: \{(5, 1), (6, 2), (7, 3), (8, 4)\}$$

Note that the domain of f is equal to the range of f^{-1} and vice versa, as shown in Figure 1.67. Also note that the functions f and f^{-1} have the effect of "undoing" each other. In other words, when you form the composition of f with f^{-1} or the composition of f^{-1} with f, you obtain the identity function, as follows.

$$f(f^{-1}(x)) = f(x - 4) = (x - 4) + 4 = x$$
$$f^{-1}(f(x)) = f^{-1}(x + 4) = (x + 4) - 4 = x$$

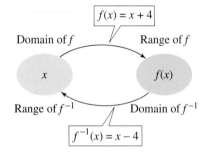

$f(x) = x + 4$

Domain of f Range of f

x $f(x)$

Range of f^{-1} Domain of f^{-1}

$f^{-1}(x) = x - 4$

Figure 1.67

EXAMPLE 1   FINDING INVERSE FUNCTIONS INFORMALLY

Find the inverse function of $f(x) = 4x$. Then verify that both $f(f^{-1}(x))$ and $f^{-1}(f(x))$ are equal to the identity function.

SOLUTION

The given function *multiplies* each input by 4. To "undo" this function, you need to *divide* each input by 4. So, the inverse function of $f(x) = 4x$ is

$$f^{-1}(x) = \frac{x}{4}.$$

You can verify that both $f(f^{-1}(x))$ and $f^{-1}(f(x))$ are equal to the identity function as follows.

$$f(f^{-1}(x)) = f\left(\frac{x}{4}\right) = 4\left(\frac{x}{4}\right) = x$$

$$f^{-1}(f(x)) = f^{-1}(4x) = \frac{4x}{4} = x$$

✓CHECKPOINT   Now try Exercise 1.

EXAMPLE 2	FINDING INVERSE FUNCTIONS INFORMALLY

Find the inverse function of $f(x) = x - 6$. Then verify that both $f(f^{-1}(x))$ and $f^{-1}(f(x))$ are equal to the identity function.

SOLUTION

The given function *subtracts* 6 from each input. To "undo" this function, you need to *add* 6 to each input. So, the inverse function of $f(x) = x - 6$ is

$$f^{-1}(x) = x + 6.$$

You can verify that both $f(f^{-1}(x))$ and $f^{-1}(f(x))$ are equal to the identity function as follows.

$$f(f^{-1}(x)) = f(x + 6)$$ Substitute $x + 6$ for $f^{-1}(x)$.

$$= (x + 6) - 6$$ Substitute $x + 6$ into $f(x)$.

$$= x$$ Identity function

$$f^{-1}(f(x)) = f^{-1}(x - 6)$$ Substitute $x - 6$ for $f(x)$.

$$= (x - 6) + 6$$ Substitute $x - 6$ into $f^{-1}(x)$.

$$= x$$ Identity function

✓CHECKPOINT   Now try Exercise 3.

The formal definition of inverse function is as follows.

Definition of Inverse Function

Let f and g be two functions such that

$$f(g(x)) = x \qquad \text{for every } x \text{ in the domain of } g$$

and

$$g(f(x)) = x \qquad \text{for every } x \text{ in the domain of } f.$$

Under these conditions, the function g is the **inverse function** of the function f. The function g is denoted by f^{-1} (read "f-inverse"). So,

$$f(f^{-1}(x)) = x \qquad \text{and} \qquad f^{-1}(f(x)) = x.$$

The domain of f must be equal to the range of f^{-1}, and the range of f must be equal to the domain of f^{-1}.

Don't be confused by the use of -1 to denote the inverse function f^{-1}. In this text, whenever f^{-1} is written it *always* refers to the inverse function of the function f and *not* to the reciprocal of $f(x)$. That is,

$$f^{-1}(x) \neq \frac{1}{f(x)}.$$

If the function g is the inverse function of the function f, it must also be true that the function f is the inverse function of the function g. For this reason, you can say that the functions f and g are *inverse functions of each other.*

EXAMPLE 3   **VERIFYING INVERSE FUNCTIONS**

Show that the following functions are inverse functions.

$$f(x) = 2x^3 - 1 \quad \text{and} \quad g(x) = \sqrt[3]{\frac{x + 1}{2}}$$

SOLUTION

$$f(g(x)) = f\left(\sqrt[3]{\frac{x + 1}{2}}\right) = 2\left(\sqrt[3]{\frac{x + 1}{2}}\right)^3 - 1$$

$$= 2\left(\frac{x + 1}{2}\right) - 1$$

$$= x + 1 - 1$$

$$= x$$

$$g(f(x)) = g(2x^3 - 1) = \sqrt[3]{\frac{(2x^3 - 1) + 1}{2}}$$

$$= \sqrt[3]{\frac{2x^3}{2}}$$

$$= \sqrt[3]{x^3}$$

$$= x$$

DISCOVERY

Graph the equations from Example 3 and the equation $y = x$ on a graphing utility using a square viewing window.

$$y_1 = 2x^3 - 1$$

$$y_2 = \sqrt[3]{\frac{x + 1}{2}}$$

$$y_3 = x$$

What do you observe about the graphs of y_1 and y_2?

✓**CHECKPOINT**   Now try Exercise 5(a).

EXAMPLE 4   **VERIFYING INVERSE FUNCTIONS**

Make a Decision Which of the functions given by

$$g(x) = \frac{x - 2}{5} \quad \text{and} \quad h(x) = \frac{5}{x} + 2$$

is the inverse function of $f(x) = \frac{5}{x - 2}$?

SOLUTION

By forming the composition of f with g, you can see that

$$f(g(x)) = f\left(\frac{x - 2}{5}\right) = \frac{5}{[(x - 2)/5] - 2} = \frac{25}{x - 12} \neq x.$$

Because this composition is not equal to the identity function x, it follows that g is *not* the inverse function of f. By forming the composition of f with h, you have

$$f(h(x)) = f\left(\frac{5}{x} + 2\right) = \frac{5}{[(5/x) + 2] - 2} = \frac{5}{5/x} = x.$$

So, it appears that h is the inverse function of f. You can confirm this result by showing that the composition of h with f is also equal to the identity function. (Try doing this.)

✓**CHECKPOINT**   Now try Exercise 7(a).

Finding Inverse Functions

For simple functions (such as the ones in Examples 1 and 2), you can find inverse functions by inspection. For more complicated functions, however, it is best to use the following guidelines. The key step in these guidelines is switching the roles of x and y. This step corresponds to the fact that inverse functions have ordered pairs with the coordinates reversed.

STUDY TIP

Note in Step 3 of the guidelines for finding the inverse function of a function that it is possible for a function to have no inverse function. For instance, the function given by $f(x) = x^2$ has no inverse function.

Finding Inverse Functions

1. In the equation for $f(x)$, replace $f(x)$ by y.

2. Interchange the roles of x and y.

3. Solve the new equation for y. If the new equation does not represent y as a function of x, the function f does not have an inverse function. If the new equation does represent y as a function of x, continue to Step 4.

4. Replace y by $f^{-1}(x)$ in the new equation.

5. Verify that f and f^{-1} are inverse functions of each other by showing that the domain of f is equal to the range of f^{-1}, the range of f is equal to the domain of f^{-1}, and $f(f^{-1}(x)) = x = f^{-1}(f(x))$.

EXAMPLE 5   FINDING INVERSE FUNCTIONS

Find the inverse function of $f(x) = \dfrac{5 - 3x}{2}$.

SOLUTION

$$f(x) = \frac{5 - 3x}{2} \qquad \text{Write original function.}$$

$$y = \frac{5 - 3x}{2} \qquad \text{Replace } f(x) \text{ by } y.$$

$$x = \frac{5 - 3y}{2} \qquad \text{Interchange } x \text{ and } y.$$

$$2x = 5 - 3y \qquad \text{Multiply each side by 2 to eliminate the fraction.}$$

$$3y = 5 - 2x \qquad \text{Isolate the } y\text{-term.}$$

$$y = \frac{5 - 2x}{3} \qquad \text{Solve for } y \text{ by dividing each side by 3.}$$

$$f^{-1}(x) = \frac{5 - 2x}{3} \qquad \text{Replace } y \text{ by } f^{-1}(x).$$

Note that both f and f^{-1} have domains and ranges that consist of the entire set of real numbers. Check that $f(f^{-1}(x)) = x$ and $f^{-1}(f(x)) = x$.

✓**CHECKPOINT**   Now try Exercise 25.

The Graph of an Inverse Function

The graphs of a function *f* and its inverse function f^{-1} are related to each other in the following way. If the point (a, b) lies on the graph of *f*, then the point (b, a) must lie on the graph of f^{-1}, and vice versa. This means that the graph of f^{-1} is a *reflection* of the graph of *f* in the line $y = x$, as shown in Figure 1.68.

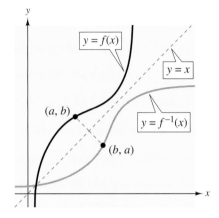

The graph of f^{-1} is a reflection of the graph of *f* in the line $y = x$.

Figure 1.68

EXAMPLE 6   THE GRAPHS OF *F* AND F^{-1}

Sketch the graphs of the inverse functions given by

$$f(x) = 2x - 3 \quad \text{and} \quad f^{-1}(x) = \tfrac{1}{2}(x + 3)$$

on the same rectangular coordinate system and show that the graphs are reflections of each other in the line $y = x$.

SOLUTION

The graphs of *f* and f^{-1} are shown in Figure 1.69. Visually, it appears that the graphs are reflections of each other in the line $y = x$. You can further verify this reflective property by testing a few points on each graph. Note in the following list that if the point (a, b) is on the graph of *f*, then the point (b, a) is on the graph of f^{-1}.

Graph of $f(x) = 2x - 3$	Graph of $f^{-1}(x) = \dfrac{1}{2}(x + 3)$
$(-1, -5)$	$(-5, -1)$
$(0, -3)$	$(-3, 0)$
$(1, -1)$	$(-1, 1)$
$(2, 1)$	$(1, 2)$
$(3, 3)$	$(3, 3)$

✓CHECKPOINT   Now try Exercise 35.

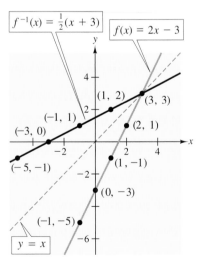

Figure 1.69

The Study Tip on page 87 mentioned that the function given by

$$f(x) = x^2$$

has no inverse function. What this really means is that *assuming the domain of f is the entire real line*, the function given by $f(x) = x^2$ has no inverse function. If, however, the domain of f is restricted to the nonnegative real numbers, then f does have an inverse function, as demonstrated in Example 7.

EXAMPLE 7   THE GRAPHS OF F AND F^{-1}

Sketch the graphs of the inverse functions given by

$$f(x) = x^2, \quad x \geq 0, \quad \text{and} \quad f^{-1}(x) = \sqrt{x}$$

on the same rectangular coordinate system and show that the graphs are reflections of each other in the line $y = x$.

SOLUTION

The graphs of f and f^{-1} are shown in Figure 1.70. Visually, it appears that the graphs are reflections of each other in the line $y = x$. You can further verify this reflective property by testing a few points on each graph. Note in the following list that if the point (a, b) is on the graph of f, then the point (b, a) is on the graph of f^{-1}.

Graph of $f(x) = x^2, \quad x \geq 0$	*Graph of* $f^{-1}(x) = \sqrt{x}$
$(0, 0)$	$(0, 0)$
$(1, 1)$	$(1, 1)$
$(2, 4)$	$(4, 2)$
$(3, 9)$	$(9, 3)$

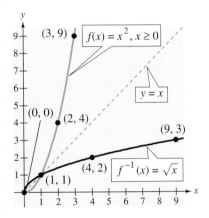

Figure 1.70

To verify algebraically that the functions are inverse functions of each other, show that $f(f^{-1}(x)) = x$ and $f^{-1}(f(x)) = x$ as follows.

$$f(f^{-1}(x)) = f(\sqrt{x}) = (\sqrt{x})^2 = x, \quad \text{if } x \geq 0$$

$$f^{-1}(f(x)) = f^{-1}(x^2) = \sqrt{x^2} = x, \quad \text{if } x \geq 0$$

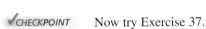✓CHECKPOINT   Now try Exercise 37.

The guidelines for finding the inverse function of a function include an *algebraic* test for determining whether a function has an inverse function. The reflective property of the graphs of inverse functions gives you a nice *geometric* test for determining whether a function has an inverse function. This test is called the **Horizontal Line Test** for inverse functions.

Horizontal Line Test for Inverse Functions

A function f has an inverse function if and only if no *horizontal* line intersects the graph of f at more than one point.

EXAMPLE 8   APPLYING THE HORIZONTAL LINE TEST

Use the graph of f to determine whether the function has an inverse function.

a. $f(x) = x^3 - 1$ **b.** $f(x) = x^2 - 1$

SOLUTION

a. The graph of the function given by $f(x) = x^3 - 1$ is shown in Figure 1.71(a). Because no horizontal line intersects the graph of f at more than one point, you can conclude that f *does* have an inverse function.

b. The graph of the function given by $f(x) = x^2 - 1$ is shown in Figure 1.71(b). Because it is possible to find a horizontal line that intersects the graph of f at more than one point, you can conclude that f *does not* have an inverse function.

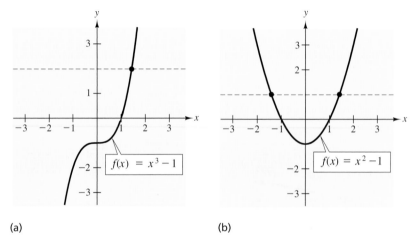

(a)   (b)

Figure 1.71

✓CHECKPOINT   Now try Exercise 49.

Discussing the Concept | Visualizing Inverse Functions

Sketch the graphs of $f(x) = x^2$ and $g(x) = x$ on graph paper. Fold the graph paper on the line $g(x) = x$ and sketch the reflection of $f(x) = x^2$ about $g(x) = x$. What do you notice about the resulting graph? Does it pass the Vertical Line Test for functions?

Does $f(x)$ pass the Vertical Line Test for functions? That is, is $f(x)$ a function? Does $f(x)$ pass the Horizontal Line Test for inverse functions? That is, does $f(x)$ have an inverse?

Write a short paragraph discussing what information can be derived from the Vertical Line Test for functions and the Horizontal Line Test for inverse functions.

1.8 Warm Up

The following warm-up exercises involve skills that were covered in earlier sections. You will use these skills in the exercise set for this section. For additional help, review Sections R1.2, R1.4, R2.1, R2.5, and 1.4.

In Exercises 1–4, find the domain of the function.

1. $f(x) = \sqrt[3]{x + 1}$

2. $f(x) = \sqrt{x + 1}$

3. $g(x) = \dfrac{2}{x^2 - 2x}$

4. $h(x) = \dfrac{x}{3x + 5}$

In Exercises 5–8, simplify the expression.

5. $2\left(\dfrac{x + 5}{2}\right) - 5$

6. $7 - 10\left(\dfrac{7 - x}{10}\right)$

7. $\sqrt[3]{2\left(\dfrac{x^3}{2} - 2\right) + 4}$

8. $\sqrt[5]{(x + 2)^5} - 2$

In Exercises 9 and 10, solve for x in terms of y.

9. $y = \dfrac{2x - 6}{3}$

10. $y = \sqrt[3]{2x - 4}$

1.8 Exercises

In Exercises 1–4, find the inverse function informally. Verify that $f(f^{-1}(x)) = x$ and $f^{-1}(f(x)) = x$.

1. $f(x) = 2x$

2. $f(x) = -\dfrac{x}{4}$

3. $f(x) = x - 5$

4. $f(x) = x + 7$

In Exercises 5–12, show that f and g are inverse functions by (a) using the definition of inverse functions and (b) using a graphing utility.

5. $f(x) = 5x + 1, \qquad g(x) = \dfrac{x - 1}{5}$

6. $f(x) = 3 - 4x, \qquad g(x) = \dfrac{3 - x}{4}$

7. $f(x) = x^3, \qquad g(x) = \sqrt[3]{x}$

8. $f(x) = \dfrac{1}{x}, \qquad g(x) = \dfrac{1}{x}$

9. $f(x) = \sqrt{x - 4}, \qquad g(x) = x^2 + 4, \quad x \geq 0$

10. $f(x) = 9 - x^2, \quad x \geq 0$
$\qquad g(x) = \sqrt{9 - x}, \quad x \leq 9$

11. $f(x) = 1 - x^3, \qquad g(x) = \sqrt[3]{1 - x}$

12. $f(x) = \dfrac{1}{1 + x}, \quad x \geq 0$

$\qquad g(x) = \dfrac{1 - x}{x}, \quad 0 < x \leq 1$

In Exercises 13–16, use the graph of f to complete the table and to sketch the graph of f^{-1}.

13.

x	0	1	2	3	4
$f^{-1}(x)$					

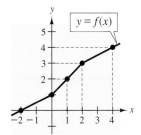

14.

x	0	2	4	6
$f^{-1}(x)$				

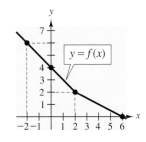

15.

x	−2	0	2	3
$f^{-1}(x)$				

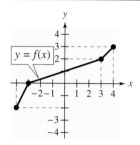

16.

x	1	2	3	4
$f^{-1}(x)$				

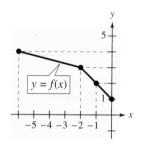

In Exercises 17–32, determine whether the function has an inverse function. If it does, find its inverse function.

17. $f(x) = x^4$

18. $f(x) = \dfrac{1}{x^2}$

19. $g(x) = \dfrac{x}{8}$

20. $f(x) = 3x + 5$

21. $p(x) = -4$

22. $f(x) = \dfrac{3x + 4}{5}$

23. $f(x) = (x + 3)^2,\ x \geq -3$

24. $q(x) = (x - 5)^2$

25. $h(x) = \dfrac{1}{x}$

26. $f(x) = |x - 2|,\ x \leq 2$

27. $f(x) = \sqrt{2x + 3}$

28. $f(x) = \sqrt{x - 2}$

29. $g(x) = x^2 - x^4$

30. $f(x) = \dfrac{x^2}{x^2 + 1}$

31. $f(x) = 25 - x^2,\ x \leq 0$

32. $f(x) = 36 + x^2,\ x \leq 0$

Error Analysis   In Exercises 33 and 34, a student has handed in the answer to a problem on a quiz. Find the error(s) in each solution and discuss how to explain each error to the student.

33. Find the inverse function f^{-1} of $f(x) = \sqrt{2x - 5}$.

$$f(x) = \sqrt{2x - 5},\ \text{so}$$
$$f^{-1}(x) = \frac{1}{\sqrt{2x - 5}}$$

34. Find the inverse function f^{-1} of $f(x) = \frac{3}{5}x + \frac{1}{3}$.

$$f(x) = \frac{3}{5}x + \frac{1}{3},\ \text{so}$$
$$f^{-1}(x) = \frac{5}{3}x - 3$$

In Exercises 35–44, find the inverse function f^{-1} of the function f. Then, using a graphing utility, graph both f and f^{-1} in the same viewing window.

35. $f(x) = 2x - 3$

36. $f(x) = 3x$

37. $f(x) = x^5$

38. $f(x) = x^3 + 1$

39. $f(x) = \sqrt{x}$

40. $f(x) = x^2,\ x \geq 0$

41. $f(x) = \sqrt{4 - x^2},\ 0 \leq x \leq 2$

42. $f(x) = \dfrac{4}{x}$

43. $f(x) = \sqrt[3]{x - 1}$

44. $f(x) = x^{3/5}$

In Exercises 45–48, does the function have an inverse function? Explain your reasoning.

45.

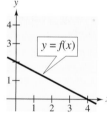

46.

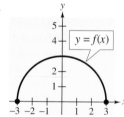

47.

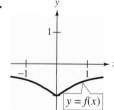

48.

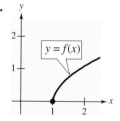

In Exercises 49–54, graph the function and use the Horizontal Line Test to determine whether the function has an inverse function.

49. $g(x) = \dfrac{4 - x}{6}$

50. $f(x) = 10$

51. $h(x) = |x + 4|$ **52.** $g(x) = (x + 5)^3$

53. $f(x) = -\sqrt{16 - x^2}$ **54.** $f(x) = (x + 2)^2$

In Exercises 55–58, use the functions given by

$$f(x) = \frac{1}{8}x - 3 \quad \text{and} \quad g(x) = x^3$$

to find the value.

55. $(f^{-1} \circ g^{-1})(1)$ **56.** $(g^{-1} \circ f^{-1})(-3)$

57. $(f^{-1} \circ f^{-1})(6)$ **58.** $(g^{-1} \circ g^{-1})(-4)$

In Exercises 59–62, use the functions given by

$$f(x) = x + 4 \quad \text{and} \quad g(x) = 2x - 5$$

to find the composition of functions.

59. $g^{-1} \circ f^{-1}$ **60.** $f^{-1} \circ g^{-1}$

61. $(f \circ g)^{-1}$ **62.** $(g \circ f)^{-1}$

63. Cost The cost C of making x pizzas is given by $C(x) = 6.50x + 1200$. Find the inverse function $C^{-1}(x)$ and explain what it computes. Describe the domains of $C(x)$ and $C^{-1}(x)$.

64. Profit A company's profit P for producing x units is given by $P(x) = 32x - 4256$. Find the inverse function $P^{-1}(x)$ and explain what it means. Describe the domains of $P(x)$ and $P^{-1}(x)$.

65. *Make a Decision*: Movie Theaters The average price of admission y (in dollars) to a movie theater from 1995 to 2001 is shown in the table. (Source: Motion Picture Association of America, Inc.)

Year	1995	1996	1997	1998	1999	2000	2001
Admission price, y	4.35	4.42	4.59	4.69	5.08	5.39	5.66

(a) Use a graphing utility to create a scatter plot of the data. Let t represent the year, with $t = 5$ corresponding to 1995.

(b) Use the *regression* feature of a graphing utility to find a linear model for the data.

(c) Algebraically find the inverse function of the model in part (b). Explain what this inverse function represents in a real-life context.

(d) Use the inverse function you found in part (c) to estimate the year in which the average admission price to a movie theater will reach $7.00.

66. *Make a Decision*: Credit Cards The total amount of revolving credit C (in billions of dollars) in the United States from 1994 to 2002 is shown in the table. (Source: Board of Governors of the Federal Reserve System)

Year	1994	1995	1996	1997	1998
Amount, C	365.6	443.1	498.9	531.0	562.5

Year	1999	2000	2001	2002
Amount, C	598.0	668.0	701.4	712.9

(a) Use a graphing utility to create a scatter plot of the data. Let t represent the year, with $t = 4$ corresponding to 1994.

(b) Use the *regression* feature of a graphing utility to find a linear model for the data.

(c) Algebraically find the inverse function of the model in part (b). Explain what this inverse function represents in a real-life context.

(d) Use the inverse function you found in part (c) to estimate the year in which total revolving credit may reach $950,000,000,000.

67. Reasoning You are helping a friend to find the inverse function of a one-to-one function. He states that interchanging the roles of x and y is "cheating." Explain how you would use the graphs of $f(x) = x^2$, $x \geq 0$ and $f^{-1}(x) = \sqrt{x}$ to justify that particular step in the process of finding an inverse function.

68. Diesel Mechanics The function given by

$$y = 0.03x^2 + 245.5, \qquad 0 < x < 100$$

approximates the exhaust temperature y for a diesel engine in degrees Fahrenheit, where x is the percent load for the diesel engine. Solve the equation for x in terms of y and use the result to find the percent load for a diesel engine when the exhaust temperature is 410°F.

69. Earnings-Dividend Ratio From 1992 to 2002, the earnings per share for Wal-Mart Stores were approximately related to the dividends per share by the function given by

$$f(x) = \sqrt{0.054x - 0.023}, \qquad 0.44 \leq x \leq 1.81$$

where f represents the dividends per share (in dollars) and x represents the earnings per share (in dollars). In 2000, Wal-Mart paid dividends of $0.23 per share. Find the inverse function of f and use the inverse function to approximate the earnings per share in 2000. (Source: Wal-Mart Stores, Inc.)

Make a Decision ▶ **Project: Demographics**

Year	People per household
1950	3.37
1955	3.29
1960	3.33
1965	3.29
1970	3.14
1975	2.94
1980	2.76
1985	2.69
1990	2.63
1995	2.65
2000	2.62

The United States Census Bureau tracks statistics on the social, political, and economic organization of the nation. These data are published in *The Statistical Abstract of the United States,* and are also available on the Internet. The Census Bureau gathers the information by taking surveys and accessing other government and private statistical publications. Governments, businesses, research groups, companies, organizations, and individuals use the information to construct mathematical models that enable them to make predictions about future trends.

The table at the left shows the average number of people in U.S. households for selected years from 1950 to 2000. It is not always easy to see from a table what type of pattern, if any, is developing in the data. Mathematical modeling is a way to find an equation that represents a set of real-life data. In this chapter, you studied one of the simplest mathematical models to use—the line.

1. *Make a Decision* Look at the table at the left. What possible pattern do you see in the data?

2. *Make a Decision* Create a scatter plot for the data. How does the scatter plot compare with your answer to Question 1? Do the points appear to resemble a linear pattern?

3. *Make a Decision* Use the *regression* feature of a graphing utility to find a linear model for the data.

   (a) Write the linear model.

   (b) What is the slope of the line? What does the slope of the line tell you about the average size of U.S. households?

   (c) What is the *x*-intercept of the line? Does an *x*-intercept make sense in the context of the data? Explain.

4. **Comparing Actual Data with the Model**   Enter the linear model from Question 3(a) into the equation editor of your graphing utility. Use the *table* feature to evaluate the values that your model gives for average household size for different years. How do the model values compare with the actual values from the table?

5. *Make a Decision* Collect a set of data by yourself or with a group of students from your class. For example, you could ask students in your class how much time they spent studying for an exam and their scores on the exam. Another source of data could be the Census Bureau's publications or website. Analyze the data you have collected by creating a scatter plot and using the statistical and *table* features of your graphing utility. Can you find a linear model that fits the data reasonably well? Do the intercepts of your model make sense? What conclusions or predictions can you make about the subject of your investigation?

Chapter Summary

After studying this chapter, you should have acquired the following skills.
These skills are keyed to the Review Exercises that begin on page 96. Answers
to odd-numbered Review Exercises are given in the back of the book.

1.1 • Plot points in the Cartesian plane, find the distance between two
points, and find the midpoint of a line segment joining two points.
Review Exercises 1–6

• Determine whether a point is a solution of an equation.
Review Exercises 7, 8

• Sketch the graph of an equation using a table of values.
Review Exercises 9, 10

• Find the x- and y-intercepts of and determine the symmetry of
the graph of an equation.
Review Exercises 11–16

• Write the equation of a circle in standard form.
Review Exercises 17–20

1.2 • Find the slope of a line passing through two points.
Review Exercises 21–24

• Use the point-slope form to find the equation of a line.
Review Exercises 25–28

• Use the slope-intercept form to sketch a line.
Review Exercises 29, 30

• Use slope to determine if lines are parallel or perpendicular and
write the equation of a line parallel or perpendicular to a given line.
Review Exercises 31–36

1.3 • Construct a linear model that relates quantities that vary directly
and that uses slope as the rate of change.
Review Exercises 37–45

• Use a scatter plot to find a linear model that fits a set of data.
Review Exercise 46

1.4 • Determine if an equation or a set of ordered pairs represents a function.
Review Exercises 47–52

• Use function notation, evaluate a function, and find the domain
of a function.
Review Exercises 53–61

• Write a function that relates quantities in an application problem.
Review Exercises 62–64

1.5 • Find the domain and range using the graph of a function.
Review Exercises 65–68

• Identify the graph of a function using the Vertical Line Test.
Review Exercises 69–74

• Describe the increasing and decreasing behavior of a function.
Review Exercises 65–68, 84

• Find the relative maxima and relative minima of the graph
of a function.
Review Exercises 65–68

• Classify a function as even or odd.
Review Exercises 65–68

• Identify six common graphs and use them to sketch the graph
of a function.
Review Exercises 75–83

1.6 • Use vertical and horizontal shifts, reflections, and nonrigid
transformations to sketch graphs of functions.
Review Exercises 85–92

1.7 • Find the sum, difference, product, and quotient of two functions.
Review Exercises 93–98

• Form the composition of two functions and determine the domain.
Review Exercises 99–102

• Identify a function as the composition of two functions.
Review Exercises 103–106

• Use combinations and compositions of functions to solve an
application problem.
Review Exercises 107–110

1.8 • Determine if a function has an inverse function and find and graph
inverse functions.
Review Exercises 111–121

Review Exercises

In Exercises 1–4, (a) plot the points, (b) find the distance between the points, and (c) find the midpoint of the line segment joining the points.

1. $(3, 2), (-3, -5)$

2. $(-9, 3), (5, 7)$

3. $(3.45, 6.55), (-1.06, -3.87)$

4. $(-6.7, -3.9), (5.1, 8.2)$

In Exercises 5 and 6, find x such that the distance between the points is 25.

5. $(10, 10), (x, -5)$ **6.** $(x, -5), (-15, 10)$

In Exercises 7 and 8, determine whether the point is a solution of the equation.

7. $y = x^2 - 7x - 10$ (a) $(1, 4)$ (b) $(2, 0)$

8. $y = \sqrt{25 - x^2}$ (a) $(3, -2)$ (b) $(0, 5)$

In Exercises 9 and 10, complete the table. Use the resulting solution points to sketch the graph of the equation.

9. $y = -\frac{1}{2}x + 2$

x	−2	0	2	3	4
y					

10. $y = x^2 - 3x$

x	−1	0	1	2	3
y					

In Exercises 11–16, sketch the graph of the equation. Identify any intercepts and test for symmetry.

11. $y = x^2 + 3$ **12.** $y^2 = x$

13. $y = 3x - 4$ **14.** $y = \sqrt{9 - x}$

15. $y = x^3 + 1$ **16.** $y = |x - 3|$

In Exercises 17 and 18, find the standard form of the equation of the specified circle.

17. Center: $(1, 3)$; radius: 5

18. Endpoints of a diameter: $(2, 5), (4, 3)$

In Exercises 19 and 20, write the equation of the circle in standard form and sketch its graph.

19. $x^2 + y^2 - 6x + 4y - 3 = 0$

20. $4x^2 + 4y^2 + 8x - 16y - 44 = 0$

In Exercises 21–24, plot the points and find the slope of the line passing through the points.

21. $(3, 7), (2, -1)$ **22.** $(3, -2), (-1, -2)$

23. $(3, 4), (3, -2)$ **24.** $(-1, 5), (2, -3)$

In Exercises 25–28, find an equation of the line that passes through the point and has the indicated slope. Sketch the line.

	Point	Slope
25.	$(0, -5)$	$m = \frac{3}{2}$
26.	$(-2, 6)$	$m = 0$
27.	$(3, 0)$	$m = -\frac{2}{3}$
28.	$(5, 4)$	m is undefined.

In Exercises 29 and 30, find the slope and y-intercept (if possible) of the line specified by the equation. Then sketch the line.

29. $5x - 4y + 11 = 0$

30. $3y - 2 = 0$

In Exercises 31–34, determine whether the lines L_1 and L_2 passing through the pairs of points are parallel, perpendicular, or neither.

31. L_1: $(0, 3), (-2, 1)$; L_2: $(-8, -3), (4, 9)$

32. L_1: $(-3, -1), (2, 5)$; L_2: $(2, 1), (8, 6)$

33. L_1: $(3, 6), (-1, -5)$; L_2: $(-2, 3), (4, 7)$

34. L_1: $(-1, 2), (-1, 4)$; L_2: $(7, 3), (4, 7)$

In Exercises 35 and 36, write an equation of the line through the point (a) parallel to the given line and (b) perpendicular to the given line.

	Point	Line
35.	$(3, -2)$	$5x - 4y = 8$
36.	$(-8, 3)$	$2x + 3y = 5$

Direct Variation   In Exercises 37–40, y is proportional to x. Use the x- and y-values to find a linear model that relates y and x.

37. $x = 3$, $y = 7$

38. $x = 5$, $y = 3.8$

39. $x = 10$, $y = 3480$

40. $x = 14$, $y = 1.95$

41. Property Tax   The property tax in a city is based on the assessed value of the property. A house that has an assessed value of $40,000 has a property tax of $1620. Find a mathematical model that gives the amount of property tax y in terms of the assessed value of the property x. Use the model to find the property tax on a house that has an assessed value of $65,000.

42. Feet and Meters   You are driving on the highway and you notice a billboard that indicates it is 1000 feet or 305 meters to the next restaurant of a national fast-food chain. Use this information to find a linear model that relates feet to meters. Use the model to complete the table.

Feet	20	50	100	120
Meters				

43. Fourth-Quarter Sales   During the second and third quarters of the year, a business had sales of $160,000 and $185,000, respectively. Assume the growth of the sales follows a linear pattern. What will sales be during the fourth quarter?

44. Dollar Value   The dollar value of a product in 2002 is $85 and the item is expected to increase in value at a rate of $3.75 per year. Write a linear equation that gives the dollar value of the product in terms of the year. Use this model to estimate the dollar value of the product in 2007. (Let $t = 2$ represent 2002.)

45. Straight-Line Depreciation   A small business purchases a piece of equipment for $30,000. After 10 years, the equipment will have to be replaced. Its salvage value at that time is expected to be $2500. Write a linear equation giving the value V of the equipment during the 10 years it will be used.

46. *Make a Decision*: **Occupied Housing Units**   The number H of occupied housing units (in millions) in the United States from 1997 to 2002 is shown in the table.   (Source: U.S. Census Bureau)

Year	Housing units, H
1997	102.2
1998	103.5
1999	104.9
2000	105.7
2001	107.0
2002	108.5

(a) Use a graphing utility to create a scatter plot for the data. Let t represent the year, with $t = 0$ corresponding to 1990. Do the data appear linear?

(b) Use the *regression* feature of a graphing utility to find a linear model for the number of occupied housing units in a given year.

(c) Use the linear model from part (b) to predict the number of occupied housing units in 2003 and 2005. Are your predictions reasonable?

In Exercises 47–50, decide whether the equation represents y as a function of x.

47. $3x - 4y = 12$

48. $y^2 = x^2 - 9$

49. $y = \sqrt{x + 3}$

50. $x^2 + y^2 - 6x + 8y = 0$

In Exercises 51 and 52, decide whether the set of ordered pairs represents a function from A to B.
$A = \{1, 2, 3\}$ $B = \{-3, -4, -7\}$
Give reasons for your answer.

51. $\{(1, -3), (2, -7), (3, -3)\}$

52. $\{(1, -4), (2, -3), (3, -9)\}$

In Exercises 53 and 54, evaluate the function at each specified value of the independent variable and simplify.

53. $f(x) = \sqrt{x + 9} - 3$

   (a) $f(7)$ (b) $f(0)$ (c) $f(-5)$ (d) $f(x + 2)$

54. $f(x) = \begin{cases} x^2 + 1, & x \le 2 \\ x + 3, & x > 2 \end{cases}$

   (a) $f(0)$ (b) $f(2)$ (c) $f(3)$ (d) $f(-5)$

In Exercises 55–60, find the domain of the function.

55. $f(x) = 3x^2 + 8x + 4$

56. $g(t) = \dfrac{5}{t^2 - 9}$

57. $h(x) = \sqrt{x + 9}$

58. $f(t) = \sqrt[3]{t - 5}$

59. $g(t) = \dfrac{\sqrt{t - 3}}{t - 5}$

60. $h(x) = \sqrt[4]{9 - x^2}$

61. Reasoning A student has difficulty understanding why the domains of

$$h(x) = \dfrac{x^2 - 4}{x} \quad \text{and} \quad k(x) = \dfrac{x}{x^2 - 4}$$

are different. How would you explain their respective domains algebraically? How could you use a graphing utility to explain their domains?

62. Volume of a Box An open box is to be made from a square piece of material 20 inches on a side by cutting equal squares from the corners and turning up the sides (see figure).

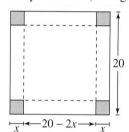

(a) Write the volume V of the box as a function of its height x.

(b) What is the domain of this function?

(c) Use a graphing utility to graph the function.

63. Balance in an Account A person deposits $5000 in an account that pays 6.25% interest compounded quarterly.

(a) Write the balance of the account in terms of the time t that the principal is left in the account.

(b) What is the domain of this function?

64. Vertical Motion The velocity v (in feet per second) of a ball thrown vertically upward from ground level is given by $v(t) = -32t + 48$, where t is the time (in seconds).

(a) Find the velocity when $t = 1$.

(b) Find the time when the ball reaches its maximum height. [*Hint:* Find the time when $v(t) = 0$.]

(c) Find the velocity when $t = 2$.

In Exercises 65–68, (a) determine the domain and range of the function, (b) determine the intervals over which the function is increasing, decreasing, or constant, (c) determine if the function is even, odd, or neither, and (d) approximate any relative minimum or relative maximum values of the function.

65. $f(x) = x^2 + 1$

66. $f(x) = \sqrt{x^2 - 9}$

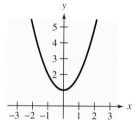

67. $f(x) = x^3 - 4x^2$

68. $f(x) = |x - 2|$

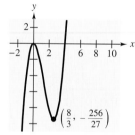

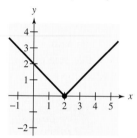

In Exercises 69–74, use the Vertical Line Test to decide whether y is a function of x.

69. $y = \frac{1}{2}x^2$

70. $y = \frac{1}{4}x^3$

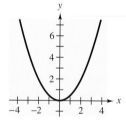

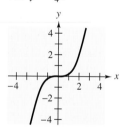

71. $x - y^2 = 1$

72. $x^2 + y^2 = 25$

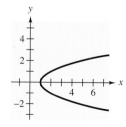

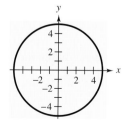

73. $x^2 = 2xy - 1$ **74.** $x = |y + 2|$

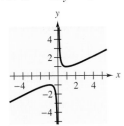   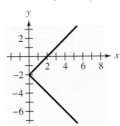

In Exercises 75–82, sketch the graph of the function.

75. $f(x) = |x + 3|$ **76.** $g(x) = \sqrt{x^2 - 16}$

77. $h(x) = 2[\![x]\!] + 1$ **78.** $f(x) = 3$

79. $g(x) = \begin{cases} x + 2, & x < 0 \\ 2, & x = 0 \\ x^2 + 2, & x > 0 \end{cases}$

80. $g(x) = \begin{cases} 3x + 1, & x < -1 \\ x^2 - 3, & x \geq -1 \end{cases}$

81. $h(x) = x^2 - 3x$ **82.** $f(x) = \sqrt{9 - x^2}$

83. Cost of Overnight Delivery The cost of sending an overnight package from Los Angeles to Dallas is $10.25 for up to, but not including, the first pound and $2.75 for each additional pound (or portion of a pound). A model for the total cost C of sending the package is

$$C = 10.25 + 2.75[\![x]\!], \quad x > 0$$

where x is the weight of the package (in pounds). Sketch the graph of this function.

84. Revenue A company determines that the total revenue R (in hundreds of thousands of dollars) for the years 1990 to 2002 can be approximated by the function

$$R = -0.035t^3 + 0.60t^2 - 1.2t + 4.75,$$
$$0 \leq t \leq 12$$

where t represents the year, with $t = 0$ corresponding to 1990. Graph the revenue function using a graphing utility and use the *trace* feature to estimate the years during which the revenue was increasing and the years during which the revenue was decreasing.

In Exercises 85 and 86, describe the sequence of transformations from $f(x) = x^2$ to g. Then sketch the graph of g.

85. $g(x) = -(x - 1)^2 - 2$

86. $g(x) = -x^2 + 3$

In Exercises 87 and 88, describe the sequence of transformations from $f(x) = \sqrt{x}$ to g. Then sketch the graph of g.

87. $g(x) = -\sqrt{x - 2}$

88. $g(x) = \sqrt{x} + 2$

In Exercises 89 and 90, describe the sequence of transformations from $f(x) = \sqrt[3]{x}$ to g. Then sketch the graph of g.

89. $g(x) = \sqrt[3]{x + 2}$

90. $g(x) = 2\sqrt[3]{x}$

In Exercises 91 and 92, identify the transformation shown in the graph and the associated common function. Write the equation for the graphed function.

91.   **92.**

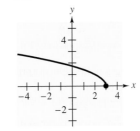   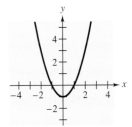

In Exercises 93 and 94, find $(f + g)(x)$, $(f - g)(x)$, $(fg)(x)$, and $(f/g)(x)$. What is the domain of f/g?

93. $f(x) = \sqrt{x^2 - 4}, \quad g(x) = 3$

94. $f(x) = 3x + 2, \quad g(x) = x^2 - 4x$

In Exercises 95–98, evaluate the function for $f(x) = x^2 - 1$ and $g(x) = x + 2$.

95. $(f + g)(2)$ **96.** $(f - g)(-1)$

97. $(fg)(3)$ **98.** $\left(\dfrac{f}{g}\right)(0)$

In Exercises 99–102, find and determine the domain of (a) $f \circ g$ and (b) $g \circ f$.

99. $f(x) = x^2, \quad g(x) = x + 3$

100. $f(x) = 2x - 5, \quad g(x) = x^2 + 2$

101. $f(x) = \dfrac{1}{x}, \quad g(x) = 3x + x^2$

102. $f(x) = \dfrac{1}{x}, \quad g(x) = x^3$

In Exercises 103–106, find two functions f and g such that $(f \circ g)(x) = h(x)$. (There are many correct answers.)

103. $h(x) = (6x - 5)^2$ **104.** $h(x) = \sqrt[3]{x + 2}$

105. $h(x) = \dfrac{1}{(x - 1)^2}$

106. $h(x) = (x - 3)^3 + 2(x - 3)$

107. *Make a Decision*: **Comparing Sales**   You own two dry cleaning establishments. From 1997 to 2001, the sales for one of the establishments have been decreasing according to the function

$$R_1 = 399.7 - 0.6t - 0.3t^2, \quad t = 7, 8, 9, 10, 11$$

where R_1 represents the sales (in thousands of dollars) and t represents the year, with $t = 7$ corresponding to 1997. During the same five-year period, the sales for the second establishment have been increasing according to the function

$$R_2 = 200.82 + 0.82t, \quad t = 7, 8, 9, 10, 11.$$

Write a function that represents the total sales for the two establishments. Make a stacked bar graph to represent the total sales during this five-year period. Are total sales increasing or decreasing?

108. Area   A square concrete foundation is prepared as a base for a large cylindrical gasoline tank (see figure).

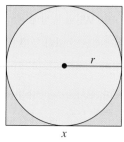

(a) Write the radius r of the tank as a function of the length x of the sides of the square.

(b) Write the area A of the circular base of the tank as a function of the radius r.

(c) Find and interpret $(A \circ r)(x)$.

109. *Make a Decision*   You are a sales representative for an automobile manufacturer. You are paid an annual salary plus a bonus of 3% of your sales over $500,000. Consider the two functions given by $f(x) = x - 500,000$ and $g(x) = 0.03x$. If x is greater than $500,000$, does $f(g(x))$ or $g(f(x))$ represent your bonus? Explain.

110. Cost   The weekly cost C of producing x units in a manufacturing process is $C(x) = 50x + 900$. The number of units produced in t hours is $x(t) = 100t$. Find and interpret $(C \circ x)(t)$.

In Exercises 111 and 112, show that f and g are inverse functions of each other.

111. $f(x) = 3x + 5, \qquad g(x) = \dfrac{x - 5}{3}$

112. $f(x) = \sqrt[3]{x - 3}, \qquad g(x) = x^3 + 3$

In Exercises 113–116, determine whether the function has an inverse function. If it does, find the inverse function and graph f and f^{-1} in the same coordinate plane.

113. $f(x) = 3x^2$ **114.** $f(x) = \sqrt[3]{x + 1}$

115. $f(x) = \dfrac{1}{x}$ **116.** $f(x) = \dfrac{x^2}{x^2 - 9}$

In Exercises 117–120, (a) find f^{-1}, (b) sketch the graphs of f and f^{-1} on the same coordinate system, and (c) verify that $f^{-1}(f(x)) = x$ and $f(f^{-1}(x)) = x$.

117. $f(x) = \frac{1}{2}x - 3$ **118.** $f(x) = \sqrt{x + 1}$

119. $f(x) = x^2, \quad x \ge 0$ **120.** $f(x) = \sqrt[3]{x - 1}$

121. Projected Energy Consumption   The projected annual energy consumption in quadrillion Btu (British thermal units) in the United States for selected years is shown in the table.   (Source: U.S. Energy Information Administration)

Year	Consumption
2005	103.16
2010	113.26
2015	121.91
2020	130.12

(a) Use a graphing utility to create a scatter plot for the data. Let t represent the year, with $t = 5$ corresponding to 2005.

(b) Use the *regression* feature of a graphing utility to find a linear model for the data.

(c) If the data can be modeled by a one-to-one function, find the inverse function of the model and use it to predict in what year energy consumption will reach 114 quadrillion Btu.

Chapter Test

Take this test as you would take a test in class. After you are done, check your work against the answers given in the back of the book.

In Exercises 1 and 2, find the distance between the points and the midpoint of the line segment connecting the points.

1. $(-3, 2), (5, -2)$ **2.** $(3.25, 7.05), (-2.37, 1.62)$

3. Find the intercepts of the graph of $y = (x + 5)(x - 3)$.

4. Describe the symmetry of the graph of $y = \dfrac{x}{x^2 - 4}$.

5. Find an equation of the line through $(4, -3)$ with a slope of $\frac{3}{4}$.

6. Write the equation of the circle in standard form and sketch its graph.
$$x^2 + y^2 - 4x - 2y - 4 = 0$$

In Exercises 7 and 8, decide whether the statement is true or false. Explain.

7. The equation $2x - 3y = 5$ identifies y as a function of x.

8. If $A = \{3, 4, 5\}$ and $B = \{-1, -2, -3\}$, the set $\{(3, -9), (4, -2), (5, -3)\}$ represents a function from A to B.

In Exercises 9 and 10, (a) find the domain and range of the function, (b) determine the intervals over which the function is increasing, decreasing, or constant, (c) determine whether the function is even or odd, and (d) approximate any relative minimum or relative maximum values of the function.

9. $f(x) = 2 - x^2$ (See figure.)   **10.** $g(x) = \sqrt{x^2 - 4}$ (See figure.)

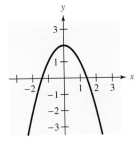

Figure for 9

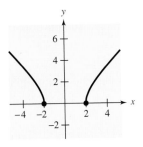

Figure for 10

In Exercises 11 and 12, sketch the graph of the function.

11. $g(x) = \begin{cases} x + 1, & x < 0 \\ 1, & x = 0 \\ x^2 + 1, & x > 0 \end{cases}$ **12.** $h(x) = (x - 3)^2 + 4$

In Exercises 13–16, use $f(x) = x^2 + 2$ and $g(x) = 2x - 1$ to find the function.

13. $(f - g)(x)$ **14.** $(fg)(x)$ **15.** $(f \circ g)(x)$ **16.** $g^{-1}(x)$

17. A business purchases a piece of equipment for \$25,000. After 5 years, the equipment will be worth only \$5000. Write a linear equation that gives the value V of the equipment during the 5 years.

18. The total sales S (in billions of dollars) for electronics and appliance stores in the United States for the years 1995 to 2002 are listed in the table. Use a graphing utility to create a scatter plot of the data and find a linear model for the data. Let t represent the year, with $t = 5$ corresponding to 1995. (Source: U.S. Census Bureau)

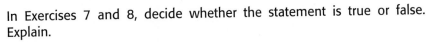

Year	Sales, S
1995	64.9
1996	68.5
1997	70.2
1998	76.0
1999	81.9
2000	86.4
2001	85.2
2002	90.1

Table for 18

2 Polynomial and Rational Functions

© Owaki-Kulla/CORBIS

Chapter **2**

Chapter Sections

2.1 Quadratic Functions and Models

2.2 Polynomial Functions of Higher Degree

2.3 Polynomial Division

2.4 Real Zeros of Polynomial Functions

2.5 Complex Numbers

2.6 The Fundamental Theorem of Algebra

2.7 Rational Functions

Make a Decision ➤ **Has the Sun Set on Solar Power?**

Solar power and wind power are two types of renewable energy sources that can be cost-effective and do not pollute the environment.

From 1990 to 2002, the solar energy consumption *s* (in trillion Btu) in the United States rose and then declined, as shown in the top graph. Wind energy consumption was consistent from 1990 to 1997. Then, wind energy consumption *w* (in trillion Btu) surged from 1998 to 2002, as shown in the bottom graph. (Source: Energy Information Administration)

Make a Decision What type of model does each data set appear to fit?

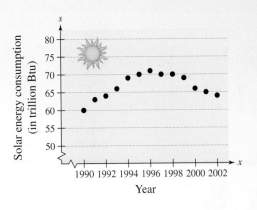

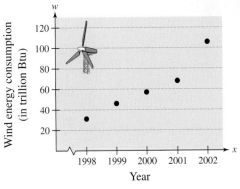

To explore this real-world application further, go to the Make a Decision Project on page 181.

2.1   Quadratic Functions and Models

Objectives

- Sketch the graph of a quadratic function and identify its vertex and intercepts.
- Find a quadratic function given its vertex and a point on its graph.
- Construct and use a quadratic model to solve an application problem.

The Graph of a Quadratic Function

In this and the next section, you will study the graphs of polynomial functions.

Definition of a Polynomial Function

Let n be a nonnegative integer and let $a_n, a_{n-1}, \ldots, a_2, a_1, a_0$ be real numbers with $a_n \neq 0$. The function given by

$$f(x) = a_n x^n + a_{n-1} x^{n-1} + \cdots + a_2 x^2 + a_1 x + a_0$$

is called a **polynomial function of x with degree n.**

Polynomial functions are classified by degree. Recall that the degree of a polynomial is the highest degree of its terms. For instance, the polynomial function

$$f(x) = a, \quad a \neq 0 \qquad \text{Constant function}$$

has degree 0 and is called a **constant function.** In Chapter 1, you learned that the graph of this type of function is a horizontal line. The polynomial function

$$f(x) = ax + b, \quad a \neq 0 \qquad \text{Linear function}$$

has degree 1 and is called a **linear function.** In Chapter 1, you learned that the graph of the linear function $f(x) = ax + b$ is a line whose slope is a and whose y-intercept is $(0, b)$. In this section, you will study second-degree polynomial functions, which are called **quadratic functions.**

For instance, each of the following functions is a quadratic function.

$$f(x) = x^2 + 6x + 2$$

$$g(x) = 2(x + 1)^2 - 3$$

$$h(x) = (x - 2)(x + 1)$$

Definition of a Quadratic Function

Let a, b, and c be real numbers with $a \neq 0$. The function of x given by

$$f(x) = ax^2 + bx + c \qquad \text{Quadratic function}$$

is called a **quadratic function.**

The graph of a quadratic function is called a **parabola.** It is "$\cup$"-shaped and can open upward or downward.

All parabolas are symmetric with respect to a line called the **axis of symmetry,** or simply the **axis** of the parabola. The point at which the axis intersects the parabola is the **vertex** of the parabola, as shown in Figure 2.1. If the leading coefficient is positive, the graph of $f(x) = ax^2 + bx + c$ is a parabola that opens upward, and if the leading coefficient is negative, the graph of $f(x) = ax^2 + bx + c$ is a parabola that opens downward.

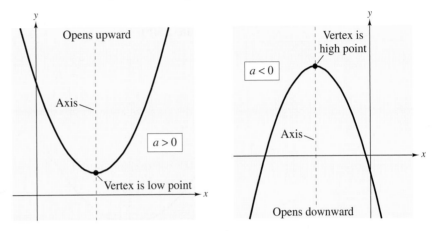

Figure 2.1

The simplest type of quadratic function is

$$f(x) = ax^2.$$

Its graph is a parabola whose vertex is $(0, 0)$. If $a > 0$, the vertex is the *minimum* point on the graph, and if $a < 0$, the vertex is the *maximum* point on the graph, as shown in Figure 2.2.

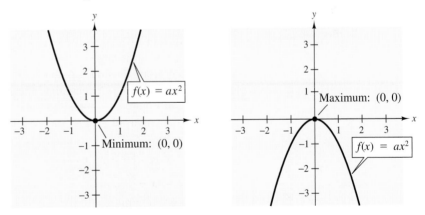

$a > 0$: Parabola opens upward $\qquad$ $a < 0$: Parabola opens downward

Figure 2.2

When sketching the graph of $f(x) = ax^2$, it is helpful to use the graph of $y = x^2$ as a reference, as discussed in Section 1.6. There you saw that when $a > 1$, the graph of $y = af(x)$ is a vertical stretch of the graph of $y = f(x)$. When $0 < a < 1$, the graph of $y = af(x)$ is a vertical shrink of the graph of $y = f(x)$. This is demonstrated again in Example 1.

EXAMPLE 1   SKETCHING THE GRAPH OF A QUADRATIC FUNCTION

a. Compared with the graph of $y = x^2$, each output of $f(x) = \frac{1}{3}x^2$ vertically "shrinks" the graph by a factor of $\frac{1}{3}$, creating the broader parabola shown in Figure 2.3(a).

b. Compared with the graph of $y = x^2$, each output of $g(x) = 2x^2$ vertically "stretches" the graph by a factor of 2, creating the narrower parabola shown in Figure 2.3(b).

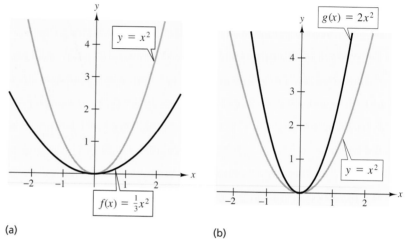

(a)   (b)

Figure 2.3

✓CHECKPOINT   Now try Exercise 15.

In Example 1, note that the coefficient a determines how widely the parabola given by $f(x) = ax^2$ opens. If $|a|$ is small, the parabola opens more widely than if $|a|$ is large.

Recall from Section 1.6 that the graphs of $y = f(x \pm c)$, $y = f(x) \pm c$, $y = -f(x)$, and $y = f(-x)$ are rigid transformations of the graph of $y = f(x)$. For instance, in Figure 2.4, notice how the graph of $y = x^2$ can be transformed to produce the graphs of $f(x) = -x^2 + 1$ and $g(x) = (x + 2)^2 - 3$.

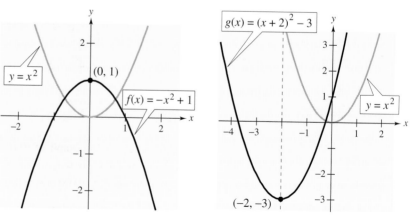

Figure 2.4

The Standard Form of a Quadratic Function

The **standard form** of a quadratic function is

$$f(x) = a(x - h)^2 + k.$$

This form is especially convenient for sketching a parabola because it identifies the vertex of the parabola.

Standard Form of a Quadratic Function

The quadratic function

$$f(x) = a(x - h)^2 + k, \qquad a \neq 0$$

is said to be in **standard form.** The graph of f is a parabola whose axis is the vertical line $x = h$ and whose vertex is the point (h, k). If $a > 0$, the parabola opens upward, and if $a < 0$, the parabola opens downward.

To write a quadratic function in standard form, you can use the process of *completing the square,* as illustrated in Example 2.

EXAMPLE 2   WRITING A QUADRATIC FUNCTION IN STANDARD FORM

Sketch the graph of $f(x) = 2x^2 + 8x + 7$ and identify the vertex.

SOLUTION

Begin by writing the quadratic function in standard form. Notice that the first step in completing the square is to factor out any coefficient of x^2 that is not 1.

$f(x) = 2x^2 + 8x + 7$	Write original function.
$= 2(x^2 + 4x) + 7$	Factor 2 out of x terms.
$= 2(x^2 + 4x + 4 - 4) + 7$	Add and subtract 4 within parentheses to complete the square.

$$(4/2)^2$$

After adding and subtracting 4 within the parentheses, you must now regroup the terms to form a perfect square trinomial. The -4 can be removed from inside the parentheses. But, because of the 2 outside of the parentheses, you must multiply -4 by 2 as shown below.

$f(x) = 2(x^2 + 4x + 4) - 2(4) + 7$	Regroup terms.
$= 2(x^2 + 4x + 4) - 8 + 7$	Simplify.
$= 2(x + 2)^2 - 1$	Standard form

From this form, you can see that the graph of f is a parabola that opens upward with vertex $(-2, -1)$. This corresponds to a left shift of two units and a downward shift of one unit relative to the graph of $y = 2x^2$, as shown in Figure 2.5.

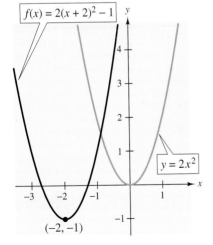

$f(x) = 2(x + 2)^2 - 1$

$y = 2x^2$

$(-2, -1)$

Figure 2.5

✓CHECKPOINT   Now try Exercise 21.

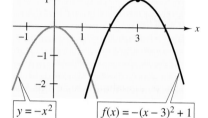

Figure 2.6

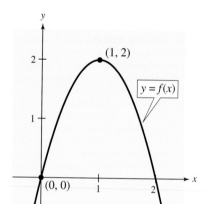

Figure 2.7

| EXAMPLE 3 | WRITING A QUADRATIC FUNCTION IN STANDARD FORM |

Sketch the graph of $f(x) = -x^2 + 6x - 8$ and identify the vertex.

SOLUTION

As in Example 2, begin by writing the quadratic function in standard form.

$$f(x) = -x^2 + 6x - 8 \qquad \text{Write original function.}$$
$$= -(x^2 - 6x) - 8 \qquad \text{Factor } -1 \text{ out of } x \text{ terms.}$$
$$= -(x^2 - 6x + 9 - 9) - 8 \qquad \text{Add and subtract 9 within parentheses to complete the square.}$$

$$\underset{(-6/2)^2}{\underline{\qquad\uparrow}}$$

$$= -(x^2 - 6x + 9) - (-9) - 8 \qquad \text{Regroup terms.}$$
$$= -(x - 3)^2 + 1 \qquad \text{Standard form}$$

So, the graph of f is a parabola that opens downward with vertex at $(3, 1)$, as shown in Figure 2.6.

✔CHECKPOINT   Now try Exercise 27.

| EXAMPLE 4 | FINDING THE EQUATION OF A PARABOLA |

Find an equation for the parabola whose vertex is $(1, 2)$ and that passes through the point $(0, 0)$, as shown in Figure 2.7.

SOLUTION

Because the parabola has a vertex at $(h, k) = (1, 2)$, the equation must have the form

$$f(x) = a(x - 1)^2 + 2. \qquad \text{Standard form}$$

Because the parabola passes through the point $(0, 0)$, it follows that when $x = 0$, $f(x)$ must equal 0. Substitute 0 for x and 0 for $f(x)$ to obtain the equation

$$0 = a(0 - 1)^2 + 2.$$

This equation can be solved easily for a, and you can see that

$$a = -2.$$

You can now write an equation for the parabola.

$$f(x) = -2(x - 1)^2 + 2 \qquad \text{Substitute for } a, h, \text{ and } k \text{ in standard form.}$$
$$= -2x^2 + 4x \qquad \text{Simplify.}$$

✔CHECKPOINT   Now try Exercise 33.

To find the x-intercepts of the graph of $f(x) = ax^2 + bx + c$, you must solve the equation

$$ax^2 + bx + c = 0.$$

If $ax^2 + bx + c$ does not factor, you can use the Quadratic Formula to find the x-intercepts. Remember, however, that a parabola may have no x-intercepts.

Applications

Many applications involve finding the maximum or minimum value of a quadratic function. By writing the quadratic function $f(x) = ax^2 + bx + c$ in standard form, you can determine that the vertex occurs at $x = -b/2a$.

EXAMPLE 5   THE MAXIMUM HEIGHT OF A BASEBALL

A baseball is hit 3 feet above ground at a velocity of 100 feet per second and at an angle of 45° with respect to the ground. The path of the baseball is given by the function

$$f(x) = -0.0032x^2 + x + 3$$

where $f(x)$ is the height of the baseball (in feet) and x is the distance from home plate (in feet). What is the maximum height reached by the baseball?

SOLUTION

For this quadratic function, you have

$$f(x) = ax^2 + bx + c$$
$$= -0.0032x^2 + x + 3.$$

So, $a = -0.0032$ and $b = 1$. Because the function has a maximum when $x = -b/2a$, you can conclude that the baseball reaches its maximum height when it is

$$x = -\frac{b}{2a} = -\frac{1}{2(-0.0032)} = 156.25 \text{ feet}$$

from home plate. At this distance, the maximum height is

$$f(156.25) = -0.0032(156.25)^2 + 156.25 + 3 = 81.125 \text{ feet}.$$

The path of the baseball is shown in Figure 2.8.

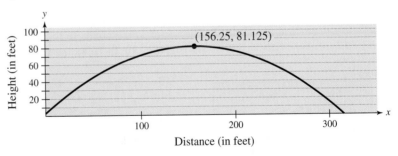

Figure 2.8

✓CHECKPOINT   Now try Exercise 43.

In Section 1.3 you plotted data points in the coordinate plane and estimated the best-fitting line. Fitting a quadratic model by this same process would be difficult. Most graphing utilities have a built-in statistical program that easily calculates the best-fitting quadratic model for a set of data points. Refer to the user's guide of your graphing utility for the required steps.

EXAMPLE 6   FITTING A QUADRATIC FUNCTION TO DATA

Make a Decision The total value y of manufacturers' shipments of cassettes (not including singles), in millions of dollars, from 1986 to 2001 is shown in the table. Use a graphing utility to plot the data and find the quadratic model that best fits the data. Find the vertex of the graph of the quadratic model and interpret its meaning in the context of the problem. Let $x = 6$ represent the year 1986. (Source: Recording Industry Association of America)

Year	x	Value, y	Year	x	Value, y
1986	6	2499.5	1994	14	2976.4
1987	7	2959.7	1995	15	2303.6
1988	8	3385.1	1996	16	1905.3
1989	9	3345.8	1997	17	1522.7
1990	10	3472.4	1998	18	1419.9
1991	11	3019.6	1999	19	1061.6
1992	12	3116.3	2000	20	626.0
1993	13	2915.8	2001	21	363.4

SOLUTION

Begin by entering the data into your graphing utility and displaying the scatter plot. From the scatter plot in Figure 2.9(a) you can see that the points have a parabolic trend. Use the *quadratic regression* feature to find the quadratic function that best fits the data. The quadratic equation that best fits the data is

$$y = -23.597x^2 + 450.56x + 1025.4, \qquad 6 \le x \le 21.$$

Graph the data and the equation in the same viewing window, as shown in Figure 2.9(b). By using the *maximum* feature of your graphing utility, you can see that the vertex of the graph is approximately $(10, 3176)$, as shown in Figure 2.9(c). The vertex corresponds to the year in which the value of cassettes shipped was the greatest. So, in 1990, the value of cassettes shipped reached a maximum.

TECHNOLOGY

For instructions on how to use the *regression* feature, see Appendix A; for specific keystrokes, go to the text website at *college.hmco.com*.

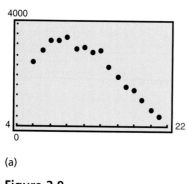

(a)

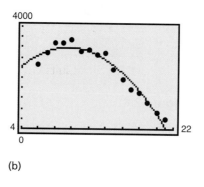

(b)

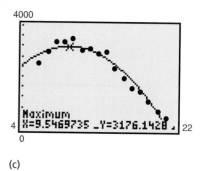

(c)

Figure 2.9

✓CHECKPOINT   Now try Exercise 55.

EXAMPLE 7 CHARITABLE CONTRIBUTIONS

The percent of income that a particular city gives to charities is related to the amount of income. For families in this city with annual incomes between $5000 and $100,000, the percent P can be modeled by

$$P(x) = 0.0014x^2 - 0.1529x + 5.855, \quad 5 \le x \le 100$$

where x is the annual income (in thousands of dollars). According to this model, what income level corresponds to the minimum percent of income given to charitable contributions?

SOLUTION

One way to answer the question is to sketch the graph of the quadratic function, as shown in Figure 2.10. From this graph, it appears that the minimum percent corresponds to an income level of about $55,000. Another way to answer the question is to use the fact that the minimum point of the parabola occurs when $x = -b/2a$.

$$x = -\frac{b}{2a}$$

$$= -\frac{-0.1529}{2(0.0014)}$$

$$\approx 54.6$$

From this x-value, you can conclude that the minimum percent corresponds to an income level of about $54,600.

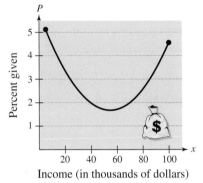

Percent given / Income (in thousands of dollars)

Figure 2.10

✓CHECKPOINT Now try Exercise 49.

Discussing the Concept | Approximating Intercepts and Extreme Points

A graphing utility with *zoom* and *trace* features can be a useful tool in approximating the intercepts and vertices of quadratic functions to acceptable degrees of accuracy. Use a graphing utility to locate and approximate the intercepts and vertices of the following quadratic functions to two decimal places. Confirm the coordinates of the vertices algebraically. How do the estimates and the algebraic results compare?

a. $y = x^2 - 3x + 4$ **b.** $y = x^2 - 17x + 64$

c. $y = 7.52x^2 + 4.16x$ **d.** $y = -\frac{1}{2}x^2 + 5x - \frac{13}{2}$

e. $y = \frac{x^2}{19} + \frac{13x}{3} + \frac{4}{21}$ **f.** $y = 14x^2 - 5x + 2$

2.1 Warm Up

The following warm-up exercises involve skills that were covered in earlier sections. You will use these skills in the exercise set for this section. For additional help, review Sections R2.3 and R2.4.

In Exercises 1–4, solve the equation by factoring.

1. $2x^2 + 11x - 6 = 0$ **2.** $5x^2 - 12x - 9 = 0$

3. $3 + x - 2x^2 = 0$ **4.** $x^2 + 20x + 100 = 0$

In Exercises 5–10, use the Quadratic Formula to solve the equation.

5. $x^2 - 6x + 4 = 0$ **6.** $x^2 + 4x + 1 = 0$

7. $2x^2 - 16x + 25 = 0$ **8.** $3x^2 + 30x + 74 = 0$

9. $x^2 + 3x + 1 = 0$

10. $x^2 + 3x - 3 = 0$

2.1 Exercises

In Exercises 1–8, match the quadratic function with its graph. [The graphs are labeled (a), (b), (c), (d), (e), (f), (g), and (h).]

(a)
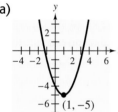

(b)
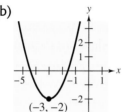

(g)
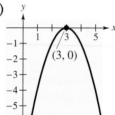

(h)
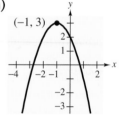

(c)
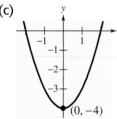

(d)
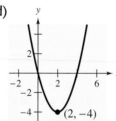

1. $f(x) = -(x - 3)^2$ **2.** $f(x) = (x + 5)^2$

3. $f(x) = x^2 - 4$ **4.** $f(x) = 5 - x^2$

5. $f(x) = (x + 3)^2 - 2$ **6.** $f(x) = (x - 1)^2 - 5$

7. $f(x) = -(x + 1)^2 + 3$ **8.** $f(x) = (x - 2)^2 - 4$

(e)
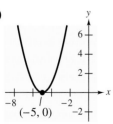

(f)
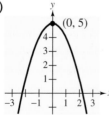

In Exercises 9–14, find an equation of the parabola.

9.
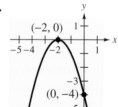

10.
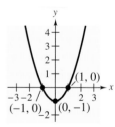

11.
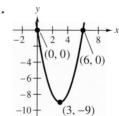

12.
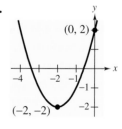

13.

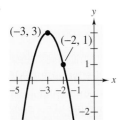

14.

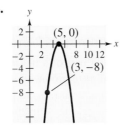

In Exercises 15–32, sketch the graph of the quadratic function. Identify the vertex and intercepts.

15. $f(x) = 3x^2$

16. $f(x) = x^2 - 4$

17. $f(x) = 16 - x^2$

18. $h(x) = 25 - x^2$

19. $f(x) = (x + 5)^2 - 6$

20. $f(x) = (x - 6)^2 + 3$

21. $g(x) = x^2 + 2x + 1$

22. $h(x) = x^2 - 4x + 2$

23. $f(x) = -(x^2 + 2x - 3)$

24. $f(x) = -(x^2 + 6x - 3)$

25. $f(x) = x^2 - x + \frac{5}{4}$

26. $f(x) = x^2 + 3x + \frac{1}{4}$

27. $f(x) = -x^2 + 2x + 5$

28. $f(x) = -x^2 - 4x + 1$

29. $h(x) = 4x^2 - 4x + 21$

30. $f(x) = 2x^2 - x + 1$

31. $f(x) = \frac{1}{4}(x^2 - 16x + 32)$

32. $g(x) = \frac{1}{2}(x^2 + 4x - 2)$

In Exercises 33–36, find the quadratic function that has the indicated vertex and whose graph passes through the given point.

33. Vertex: $(2, -1)$; point: $(4, -3)$

34. Vertex: $(-3, 5)$; point: $(-6, -1)$

35. Vertex: $(5, 12)$; point: $(7, 15)$

36. Vertex: $(-2, -2)$; point: $(-1, 0)$

In Exercises 37–42, find two quadratic functions whose graphs have the given x-intercepts. Find one function that has a graph that opens upward and another that has a graph that opens downward. (There are many correct answers.)

37. $(2, 0), (-1, 0)$

38. $(-4, 0), (0, 0)$

39. $(0, 0), (10, 0)$

40. $(4, 0), (8, 0)$

41. $(-3, 0), \left(-\frac{1}{2}, 0\right)$

42. $\left(-\frac{5}{2}, 0\right), (2, 0)$

43. *Make a Decision*: **Optimal Area** The perimeter of a rectangle is 100 feet. Let x represent the width of the rectangle and write a quadratic function for the area of the rectangle in terms of its width. Find the vertex of the graph of the quadratic function and interpret its meaning in the context of the problem.

44. *Make a Decision*: **Optimal Area** The perimeter of a rectangle is 400 feet. Let x represent the width of the rectangle and write a quadratic function for the area of the rectangle in terms of its width. Find the vertex of the graph of the quadratic function and interpret its meaning in the context of the problem.

45. Optimal Area A rancher has 200 feet of fencing with which to enclose two adjacent rectangular corrals (see figure). What measurements will produce a maximum enclosed area?

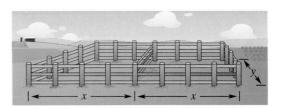

46. Optimal Area An indoor physical-fitness room consists of a rectangular region with a semicircle on each end (see figure). The perimeter of the room is to be a 200-meter running track. What measurements will produce a maximum area of the rectangle?

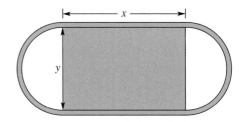

Optimal Revenue In Exercises 47 and 48, find the number of units that produces a maximum revenue. The revenue R is measured in dollars and x is the number of units produced.

47. $R = 900x - 0.01x^2$

48. $R = 100x - 0.0002x^2$

49. Optimal Cost A manufacturer of lighting fixtures has daily production costs C (in dollars) of

$$C(x) = 800 - 10x + 0.25x^2$$

where x is the number of units produced. How many fixtures should be produced each day to yield a minimum cost?

50. Optimal Profit The profit P (in dollars) for a manufacturer of sound systems is given by

$$P(x) = -0.0002x^2 + 140x - 250,000$$

where x is the number of units produced. What production level will yield a maximum profit?

51. Maximum Height of a Diver The path of a diver is given by

$$y = -\frac{4}{9}x^2 + \frac{24}{9}x + 10$$

where y is the height (in feet) and x is the horizontal distance from the end of the diving board (in feet) (see figure). Use a graphing utility and the *trace* or *maximum* feature to find the maximum height of the diver.

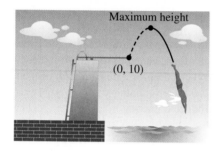

Maximum height

(0, 10)

52. Maximum Height The winning women's shot put in the 2000 Summer Olympics was recorded by Yanina Korolchik of Belarus. The path of her winning toss is approximately given by

$$y = -0.01707x^2 + 1.0775x + 5$$

where y is the height of the shot (in feet) and x is the horizontal distance (in feet). Use a graphing utility and the *trace* or *maximum* feature to find the length of the winning toss and the maximum height of the shot.

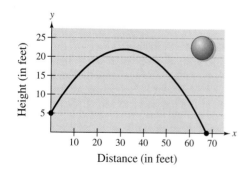

Distance (in feet)

IN THE NEWS

Commerce Bancorp, Inc.

Being part of a mature industry, banks face quite a challenge producing the kind of growth that NAIC-style investors seek. In addition, bank results can be cyclical; the companies are vulnerable to economic downturns and shifts in interest rates.

Commerce Bancorp, Inc., appears to be meeting this challenge, however. The regional banking company is aggressively expanding its network of retail branches, and it boasts strong, steady growth well above industry averages.

Earnings per share have shown average annual increases of 31 percent over the past five years, for example. . . .

Lamiman, Kevin, "Stock to Study: Featured Company: Commerce Bancorp, Inc." Better Investing, March 2004: 42.

53. The table shows the earnings per share y of Commerce Bancorp, Inc. stock from 1994 to 2002. (Source: Commerce Bancorp, Inc.)

Year	Earnings per share, y
1994	0.71
1995	0.67
1996	0.68
1997	0.82
1998	0.90
1999	1.09
2000	1.25
2001	1.51
2002	2.04

Use the *regression* feature of a graphing utility to find a quadratic model for the data. Let t represent the year, with $t = 4$ corresponding to 1994.

54. According to the model you found in Exercise 53, when were earnings per share at a relative minimum? Does this result agree with the actual data?

55. *Make a Decision*: **Regression Problem** Let x be the number of units (in tens of thousands) that a computer company produces and let $p(x)$ be the profit (in hundreds of thousands of dollars). The table shows the profit for different levels of production.

Units, x	2	4	6	8	10
Profit, $p(x)$	270.5	307.8	320.1	329.2	325.0

Units, x	12	14	16	18	20
Profit, $p(x)$	311.2	287.8	254.8	212.2	160.0

(a) Use a graphing utility to create a scatter plot of the data.

(b) Use the *regression* feature of a graphing utility to find a quadratic model for $p(x)$.

(c) Use a graphing utility to graph your model for $p(x)$ with the scatter plot of the data.

(d) Find the vertex of the graph of the model in part (c). Interpret its meaning in the context of the problem.

(e) With these data and this model, the profit begins to decrease. Discuss how it is possible for production to increase and profit to decrease.

56. *Make a Decision*: **Regression Problem** Let x be the angle (in degrees) at which a baseball is hit with no spin at an initial speed of 40 meters per second and let $d(x)$ be the distance (in meters) the ball travels. The table shows the distances for the different angles at which the ball is hit. (Source: *The Physics of Sports*)

Angle, x	10	15	30	36	42
Distance, $d(x)$	58.3	79.7	126.9	136.6	140.6

Angle, x	44	45	48	54	60
Distance, $d(x)$	140.9	140.9	139.3	132.5	120.5

(a) Use a graphing utility to create a scatter plot of the data.

(b) Use the *regression* feature of a graphing utility to find a quadratic model for $d(x)$.

(c) Use a graphing utility to graph your model for $d(x)$ with the scatter plot of the data.

(d) Find the vertex of the graph of the model in part (c). Interpret its meaning in the context of the problem.

Math *MATTERS*

Circles and Pi

Pi is a special number that represents a relationship that is found in *all* circles. The relationship is this: The ratio of the circumference of *any* circle to the diameter of the circle always produces the same number—the number pi, denoted by the Greek symbol π.

Calculation of the decimal representation of the number π has consumed the time of many mathematicians (and computers). Because the number π is irrational, its decimal representation does not repeat or terminate. Here is a decimal representation of π that is accurate to 27 decimal places.

$$\pi \approx 3.141592653589793238462643383$$

The number π is used in most formulas that deal with circles. For instance, the area of a circle is given by

$$A = \pi r^2$$

where r is the radius of the circle. Use this formula to determine which of the colored regions inside the circle has the greater area. Is the area of the red circle larger than that of the blue ring, or do both regions have the same area? (The answer is given in the back of the book.)

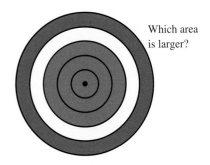

Which area is larger?

2.2   Polynomial Functions of Higher Degree

Objectives

- Sketch a transformation of a monomial function.
- Determine end behavior of graphs of polynomial functions.
- Find the real zeros of a polynomial function.
- Sketch the graph of a polynomial function.
- Use a polynomial model to solve an application problem.

Graphs of Polynomial Functions

In this section, you will study basic characteristics of the graphs of polynomial functions. The first characteristic is that the graph of a polynomial function is **continuous**. Essentially, this means that the graph of a polynomial function has no breaks, as shown in Figure 2.11(a). Functions with graphs that are not continuous are not polynomial functions, as shown in Figure 2.11(b).

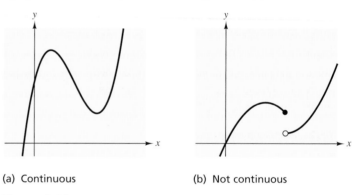

(a) Continuous   (b) Not continuous

Figure 2.11

The second characteristic is that the graph of a polynomial function has only smooth, rounded turns, as shown in Figure 2.12(a). A polynomial function cannot have a sharp turn, as shown in Figure 2.12(b).

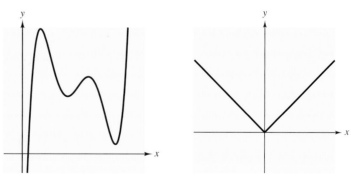

(a) Polynomial functions have   (b) Polynomial functions do not
    smooth, rounded graphs.   have sharp turns.

Figure 2.12

The polynomial functions that have the simplest graphs are monomials of the form $f(x) = x^n$, where n is an integer greater than zero. From Figure 2.13, you can see that when n is *even*, the graph is similar to the graph of $f(x) = x^2$, and when n is *odd*, the graph is similar to the graph of $f(x) = x^3$. Moreover, the greater the value of n, the flatter the graph near the origin.

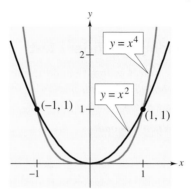

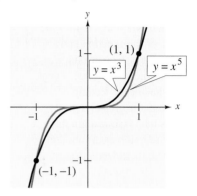

(a) If n is even, the graph of $y = x^n$ touches the axis at the x-intercept.

(b) If n is odd, the graph of $y = x^n$ crosses the axis at the x-intercept.

Figure 2.13

EXAMPLE 1   SKETCHING TRANSFORMATIONS OF MONOMIAL FUNCTIONS

Sketch each graph.

a. $f(x) = -x^5$ **b.** $h(x) = (x + 1)^4$

SOLUTION

a. Because the degree of $f(x) = -x^5$ is odd, its graph is similar to the graph of $y = x^3$. In Figure 2.14(a), note that the negative coefficient has the effect of reflecting the graph about the x-axis.

b. The graph of $h(x) = (x + 1)^4$ is a left shift, by one unit, of the graph of $y = x^4$, as shown in Figure 2.14(b).

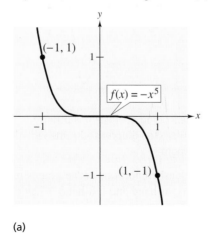

(a)

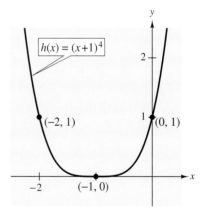

(b)

Figure 2.14

✓*CHECKPOINT*   Now try Exercise 11.

The Leading Coefficient Test

In Example 1, note that both graphs eventually rise or fall without bound as x moves to the right. Whether the graph of a polynomial eventually rises or falls can be determined by the function's degree (even or odd) and by its leading coefficient, as indicated in the **Leading Coefficient Test.**

Leading Coefficient Test

As x moves without bound to the left or to the right, the graph of the polynomial function $f(x) = a_n x^n + \cdots + a_1 x + a_0$ eventually rises or falls in the following manner.

1. When n is *odd:*

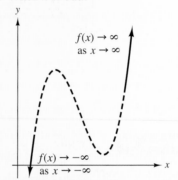

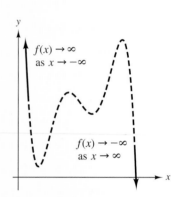

If the leading coefficient is positive ($a_n > 0$), the graph falls to the left and rises to the right.

If the leading coefficient is negative ($a_n < 0$), the graph rises to the left and falls to the right.

2. When n is *even:*

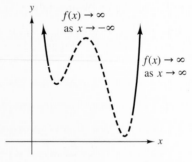

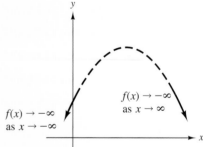

If the leading coefficient is positive ($a_n > 0$), the graph rises to the left and right.

If the leading coefficient is negative ($a_n < 0$), the graph falls to the left and right.

The dashed portions of the graphs indicate that the test determines *only* the right-hand and left-hand behavior of the graph.

DISCOVERY

For each function, identify the degree of the function and whether it is even or odd. Identify the leading coefficient, and whether the leading coefficient is greater than 0 or less than 0. Use a graphing utility to graph each function. Describe the relationship between the function's degree and the sign of its leading coefficient and the left- and right-hand behavior of the graph of the function.

(a) $y = x^3 - 2x^2 - x + 1$

(b) $y = 2x^5 + 2x^2 - 5x + 1$

(c) $y = -2x^5 - x^2 + 5x + 3$

(d) $y = -x^3 + 5x - 2$

(e) $y = 2x^2 + 3x - 4$

(f) $y = x^4 - 3x^2 + 2x - 1$

(g) $y = x^2 + 3x + 2$

(h) $y = -x^6 - x^2 - 5x + 4$

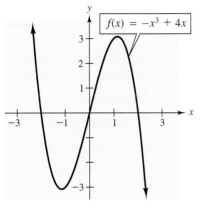

$f(x) = -x^3 + 4x$

Figure 2.15

EXAMPLE 2 APPLYING THE LEADING COEFFICIENT TEST

Describe the right-hand and left-hand behavior of the graph of

$$f(x) = -x^3 + 4x.$$

SOLUTION

Because the degree is odd and the leading coefficient is negative, the graph rises to the left and falls to the right, as shown in Figure 2.15.

✔**CHECKPOINT** Now try Exercise 17.

In Example 2, note that the Leading Coefficient Test tells you only whether the graph *eventually* rises or falls to the right or left. Other characteristics of the graph, such as intercepts, relative minima, and relative maxima, must be determined by means of other tests. For example, later you will use the number of real zeros of a polynomial function to determine how many times the graph of the function crosses the x-axis.

EXAMPLE 3 APPLYING THE LEADING COEFFICIENT TEST

Describe the right-hand and left-hand behavior of the graphs of the following functions.

a. $f(x) = x^4 - 5x^2 + 4$ **b.** $f(x) = x^5 - x$

SOLUTION

a. Because the degree is even and the leading coefficient is positive, the graph rises to the left and right, as shown in Figure 2.16(a).

b. Because the degree is odd and the leading coefficient is positive, the graph falls to the left and rises to the right, as shown in Figure 2.16(b).

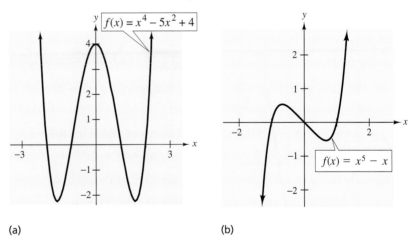

(a)   (b)

Figure 2.16

✔**CHECKPOINT** Now try Exercise 19.

Real Zeros of Polynomial Functions

It can be shown that for a polynomial function f of degree n, the following statements are true. Remember that the **zeros** of a function are the x-values for which the function is zero.

1. The graph of f has, at most, $n - 1$ *turning points*. Turning points are points at which the graph changes from increasing to decreasing or vice versa. For instance, the graph of $f(x) = x^4 - 1$ has at most $4 - 1 = 3$ turning points.

2. The function f has, at most, n real zeros. For instance, the function given by $f(x) = x^4 - 1$ has at most $n = 4$ real zeros. (You will study this result in detail in Section 2.6 on the Fundamental Theorem of Algebra.)

Finding the zeros of polynomial functions is one of the most important problems in algebra. There is a strong interplay between graphical and algebraic approaches to this problem. Sometimes you can use information about the graph of a function to help find its zeros, and in other cases you can use information about the zeros of a function to help sketch its graph.

Real Zeros of Polynomial Functions

If f is a polynomial function and a is a real number, then the following statements are equivalent.

1. $x = a$ is a *zero* of the function f.

2. $x = a$ is a *solution* of the polynomial equation $f(x) = 0$.

3. $(x - a)$ is a *factor* of the polynomial $f(x)$.

4. $(a, 0)$ is an *x-intercept* of the graph of f.

In the equivalent statements above, notice that finding zeros of polynomial functions is closely related to factoring and finding x-intercepts.

EXAMPLE 4   FINDING ZEROS OF A POLYNOMIAL FUNCTION

Find all real zeros of $f(x) = x^3 - x^2 - 2x$.

SOLUTION

By factoring, you obtain the following.

$$f(x) = x^3 - x^2 - 2x \qquad \text{Write original function.}$$
$$= x(x^2 - x - 2) \qquad \text{Remove common monomial factor.}$$
$$= x(x - 2)(x + 1) \qquad \text{Factor completely.}$$

So, the real zeros are $x = 0$, $x = 2$, and $x = -1$, and the corresponding x-intercepts are $(0, 0)$, $(2, 0)$, and $(-1, 0)$, as shown in Figure 2.17. In the figure, note that the graph has two turning points. This is consistent with the fact that a third-degree polynomial can have *at most* $3 - 1 = 2$ turning points.

✓**CHECKPOINT**   Now try Exercise 27.

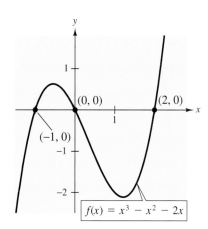

Figure 2.17

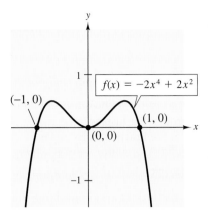

Figure 2.18

EXAMPLE 5   FINDING ZEROS OF A POLYNOMIAL FUNCTION

Find all real zeros of $f(x) = -2x^4 + 2x^2$.

SOLUTION

In this case, the polynomial factors as follows.

$$f(x) = -2x^2(x^2 - 1) = -2x^2(x - 1)(x + 1)$$

So, the real zeros are $x = 0$, $x = 1$, and $x = -1$, and the corresponding x-intercepts are $(0, 0)$, $(1, 0)$, and $(-1, 0)$, as shown in Figure 2.18. Note in the figure that the graph has three turning points, which is consistent with the fact that a fourth-degree polynomial can have *at most* three turning points.

✓**CHECKPOINT**   Now try Exercise 35.

In Example 5, the real zero arising from $-2x^2 = 0$ is called a **repeated zero.** In general, a factor $(x - a)^k$ yields a repeated zero $x = a$ of **multiplicity** k. If k is odd, the graph *crosses* the x-axis at $x = a$. If k is even, the graph *touches* (but does not cross) the x-axis at $x = a$. This is illustrated in Figure 2.18.

EXAMPLE 6   SKETCHING THE GRAPH OF A POLYNOMIAL FUNCTION

Sketch the graph of $f(x) = 3x^4 - 4x^3$.

SOLUTION

Because the leading coefficient is positive and the degree is even, you know that the graph eventually rises to the left and to the right, as shown in Figure 2.19(a). By factoring $f(x) = 3x^4 - 4x^3$ as $f(x) = x^3(3x - 4)$, you can see that the zeros of f are $x = 0$ and $x = \frac{4}{3}$ (both of odd multiplicity). So, the x-intercepts occur at $(0, 0)$ and $\left(\frac{4}{3}, 0\right)$. To sketch the graph by hand, find a few additional points, as shown in the table. Then plot the points and draw a continuous curve through the points to complete the graph, as shown in Figure 2.19(b). If you are unsure of the shape of a portion of a graph, plot some additional points.

x	f(x)
-1	7
0.5	-0.3125
1	-1
1.5	1.6875

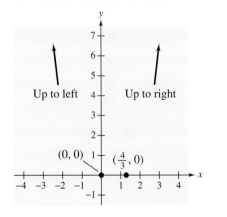

(a)

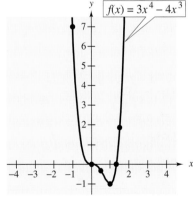

(b)

Figure 2.19

✓**CHECKPOINT**   Now try Exercise 49.

TECHNOLOGY

Example 6 uses an algebraic approach to describe the graph of the function. A graphing utility is a valuable complement to this approach. Remember that when using a graphing utility, it is important that you find a viewing window that shows all important parts of the graph. For instance, the graph below shows the important parts of the graph of the function in Example 6.

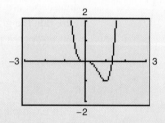

Application

EXAMPLE 7   CHARITABLE CONTRIBUTIONS REVISITED

Example 7 in Section 2.1 discussed the model

$$P(x) = 0.0014x^2 - 0.1529x + 5.855, \qquad 5 \le x \le 100$$

where P is the percent of annual income given and x is the annual income (in thousands of dollars). Note that this model gives the charitable contributions as a *percent* of annual income. To find the average *amount* that a family gives to charity, you can multiply the given model by the income $1000x$ (and divide by 100 to change from percent to decimal form) to obtain

$$A(x) = 0.014x^3 - 1.529x^2 + 58.55x, \qquad 5 \le x \le 100$$

where A represents the amount of charitable contributions (in dollars). Sketch the graph of this function and use the graph to estimate the annual salary for a family that gives $1000 a year to charities.

SOLUTION

Because the leading coefficient is positive and the degree is odd, you know that the graph eventually falls to the left and rises to the right. To sketch the graph by hand, find a few points, as shown in the table. Then plot the points and complete the graph, as shown in Figure 2.20.

x	5	25	45	65	85	100
$A(x)$	256.28	726.88	814.28	1190.48	2527.48	4565.00

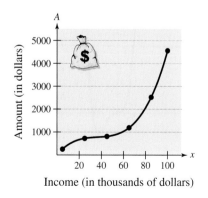

Income (in thousands of dollars)

Figure 2.20

From the graph you see that an annual contribution of $1000 corresponds to an annual income of about $59,000.

✔CHECKPOINT   Now try Exercise 61.

2.2 Warm Up

The following warm-up exercises involve skills that were covered in earlier sections. You will use these skills in the exercise set for this section. For additional help, review Sections R1.6, R2.3, R2.4, and R2.5.

In Exercises 1–6, factor the expression completely.

1. $12x^2 + 7x - 10$

2. $25x^3 - 60x^2 + 36x$

3. $12z^4 + 17z^3 + 5z^2$

4. $y^3 + 125$

5. $x^3 + 3x^2 - 4x - 12$

6. $x^3 + 2x^2 + 3x + 6$

In Exercises 7–10, find all real solutions of the equation.

7. $5x^2 + 8 = 0$

8. $x^2 - 6x + 4 = 0$

9. $4x^2 + 4x - 11 = 0$

10. $x^4 - 18x^2 + 81 = 0$

2.2 Exercises

In Exercises 1–8, match the polynomial function with its graph. [The graphs are labeled (a), (b), (c), (d), (e), (f), (g), and (h).]

(a)

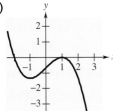

(b)

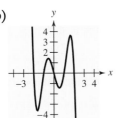

(c)

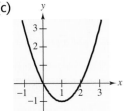

(d)

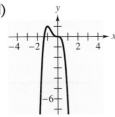

(e)

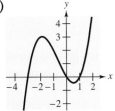

(f)

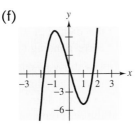

(g)

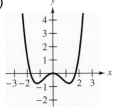

(h)

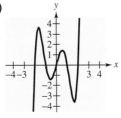

1. $f(x) = \frac{1}{2}(x^3 + 2x^2 - 3x)$

2. $f(x) = x^2 - 2x$

3. $f(x) = \frac{1}{3}x^4 - x^2$

4. $f(x) = -3x^4 - 4x^3$

5. $f(x) = 3x^3 - 9x + 1$

6. $f(x) = x^5 - 5x^3 + 4x$

7. $f(x) = -\frac{1}{3}x^3 + x - \frac{2}{3}$

8. $f(x) = -x^5 + 5x^3 - 4x$

In Exercises 9–12, use the graph of $y = x^3$ to sketch the graph of the function.

9. $f(x) = x^3 + 5$

10. $f(x) = (x - 4)^3$

11. $f(x) = (x - 2)^3 - 3$

12. $f(x) = -(x + 1)^3 + 5$

In Exercises 13–16, use the graph of $y = x^4$ to sketch the graph of the function.

13. $f(x) = (x + 3)^4$

14. $f(x) = x^4 - 4$

15. $f(x) = 3 - x^4$

16. $f(x) = \frac{1}{2}(x - 1)^4$

In Exercises 17–26, describe the right-hand and left-hand behavior of the graph of the polynomial function.

17. $f(x) = -x^3 + 1$

18. $f(x) = \frac{1}{3}x^3 + 5x$

19. $g(x) = 6 - 4x^2 + x - 3x^5$

20. $f(x) = 2x^5 - 5x + 7.5$

21. $f(x) = 4x^8 - 2$

22. $h(x) = 1 - x^6$

23. $f(x) = 2 + 5x - x^2 - x^3 + 2x^4$

24. $f(x) = \dfrac{3x^4 - 2x + 5}{4}$

25. $h(t) = -\frac{2}{3}(t^2 - 5t + 3)$

26. $f(s) = -\frac{7}{8}(s^3 + 5s^2 - 7s + 1)$

Algebraic and Graphical Approaches In Exercises 27–42, find all real zeros of the function algebraically. Then use a graphing utility to confirm your results.

27. $f(x) = x^2 - 16$

28. $f(x) = 64 - x^2$

29. $h(t) = t^2 + 8t + 16$

30. $f(x) = x^2 - 12x + 36$

31. $f(x) = \frac{1}{3}x^2 + \frac{1}{3}x - \frac{2}{3}$

32. $f(x) = \frac{1}{2}x^2 + \frac{5}{2}x - \frac{3}{2}$

33. $f(x) = 2x^2 + 4x + 6$

34. $g(x) = -5(x^2 + 2x - 4)$

35. $f(t) = t^3 - 4t^2 + 4t$

36. $f(x) = x^4 - x^3 - 20x^2$

37. $g(t) = \frac{1}{2}t^4 - \frac{1}{2}$

38. $f(x) = x^5 + x^3 - 6x$

39. $f(x) = 2x^4 - 2x^2 - 40$

40. $g(t) = t^5 - 6t^3 + 9t$

41. $f(x) = x^3 - 3x^2 + 2x - 6$

42. $f(x) = x^3 - 4x^2 - 25x + 100$

Analyzing a Graph In Exercises 43–54, analyze the graph of the function algebraically and use the results to sketch the graph *by hand.* Then use a graphing utility to confirm your sketch.

43. $f(x) = -\frac{3}{2}$

44. $h(x) = \frac{1}{3}x - 3$

45. $f(t) = \frac{1}{2}(t^2 - 4t - 1)$

46. $g(x) = -x^2 + 10x - 16$

47. $f(x) = 4x^2 - x^3$

48. $f(x) = 1 - x^3$

49. $f(x) = x^3 - 9x$

50. $f(x) = \frac{1}{4}x^4 - 2x^2$

51. $g(t) = -\frac{1}{4}(t - 2)^2(t + 2)^2$

52. $f(x) = x(x - 2)^2(x + 1)$

53. $f(x) = 1 - x^6$

54. $g(x) = 1 - (x + 1)^6$

55. Modeling Polynomials Sketch the graph of a polynomial function that is of fourth degree, has a zero of multiplicity 2, and has a negative leading coefficient. Sketch another graph under the same conditions but with a positive leading coefficient.

56. Modeling Polynomials Sketch the graph of a polynomial function that is of fifth degree, has a zero of multiplicity 2, and has a negative leading coefficient. Sketch another graph under the same conditions but with a positive leading coefficient.

57. Modeling Polynomials Determine the equation of the fourth-degree polynomial function $f(x)$ whose graph is shown.

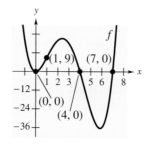

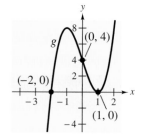

Figure for 57   Figure for 58

58. Modeling Polynomials Determine the equation of the third-degree polynomial function $g(x)$ whose graph is shown.

59. *Make a Decision*: **Ice Cream** The number of gallons G (in millions) of regular ice cream produced in the United States from 1994 to 2002 is shown in the table. (Source: U.S. Department of Agriculture)

Year	Gallons, G	Year	Gallons, G
1994	876	1999	972
1995	862	2000	980
1996	879	2001	970
1997	914	2002	989
1998	935		

(a) Use a graphing utility to create a scatter plot of the data. Let t represent the year, with $t = 4$ corresponding to 1994.

(b) Use the *regression* feature of a graphing utility to find a quadratic model and a quartic model (a fourth-degree polynomial model) for G. Then use the graphing utility to graph each model with the data.

(c) Make a table to compare the actual data values to the values given by each model.

(d) Based on your table from part (c), decide which model represents the data more accurately. Explain your reasoning.

60. *Make a Decision*: Alternative Fuel Consumption

The gasoline-equivalent gallons E (in thousands) consumed by vehicles that run on 85% ethanol in the United States from 1994 to 2002 are shown in the table. (Source: Energy Information Administration)

Year	Gallons consumed, E
1994	80
1995	190
1996	694
1997	1280
1998	1727
1999	3916
2000	12,071
2001	14,623
2002	17,783

(a) Use a graphing utility to create a scatter plot of the data. Let t represent the year, with $t = 4$ corresponding to 1994.

(b) Use the *regression* feature of a graphing utility to find a quadratic model and a quartic model for E. Then use the graphing utility to graph each model with the data.

(c) Make a table to compare the actual data values to the values given by each model.

(d) Based on your table from part (c), decide which model represents the data more accurately. Explain your reasoning.

61. Advertising Expenses The total revenue R (in millions of dollars) for a soft-drink company is related to its advertising expenses by the function

$$R = \frac{1}{50,000}(-x^3 + 600x^2), \qquad 0 \le x \le 400$$

where x is the amount spent on advertising (in tens of thousands of dollars). Use the graph of this function to estimate the point on the graph at which the function is increasing most rapidly. This point is called the *point of diminishing returns* because any expenses above this amount will yield less return per dollar invested in advertising.

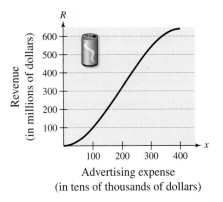

Figure for 61

62. Advertising Expenses The total revenue R (in millions of dollars) for a hotel corporation is related to its advertising expense by the function

$$R = -0.148x^3 + 4.889x^2 - 17.778x + 125.185,$$

$$0 \le x \le 20$$

where x is the amount spent on advertising (in millions of dollars). Use the graph of this function to estimate the point on the graph at which the function is increasing most rapidly. This point is called the *point of diminishing returns* because any expenses above this amount will yield less return per dollar invested in advertising.

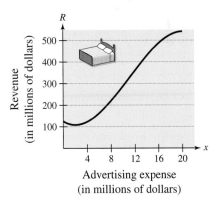

63. *Make a Decision* Use a graphing utility to graph the functions given by $f(x) = x^2$, $g(x) = x^4$, and $h(x) = x^6$. Do the three functions have a common shape? Are their graphs identical? Why or why not?

64. *Make a Decision* Use a graphing utility to graph the functions given by $f(x) = x^3$, $g(x) = x^5$, and $h(x) = x^7$. Do the three functions have a common shape? Are their graphs identical? Why or why not?

2.3   Polynomial Division

Objectives

- **Divide one polynomial by a second polynomial using long division.**
- **Simplify a rational expression using long division.**
- **Use synthetic division to divide two polynomials.**
- **Use the Remainder Theorem and synthetic division to evaluate a polynomial.**
- **Use the Factor Theorem to factor a polynomial function.**
- **Use polynomial division to solve an application problem.**

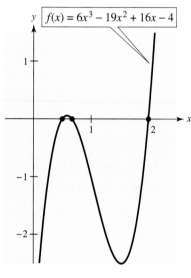

y $f(x) = 6x^3 - 19x^2 + 16x - 4$

Figure 2.21

Long Division of Polynomials

In this section, you will study two procedures for *dividing* polynomials. These procedures are especially valuable in factoring and finding the zeros of polynomial functions. To begin, consider the graph shown in Figure 2.21. Notice that a zero of f occurs at $x = 2$. Because $x = 2$ is a zero of the polynomial function f, you know that $(x - 2)$ is a factor of $f(x)$. This means that there exists a second-degree polynomial $q(x)$ such that

$$f(x) = (x - 2) \cdot q(x).$$

To find $q(x)$, you can use **long division**, as illustrated in Example 1.

EXAMPLE 1   LONG DIVISION OF POLYNOMIALS

Divide the polynomial $6x^3 - 19x^2 + 16x - 4$ by $x - 2$, and use the result to factor the polynomial completely.

SOLUTION

$$
\begin{array}{r}
6x^2 - 7x + 2 \\
x - 2 \overline{)\, 6x^3 - 19x^2 + 16x - 4} \\
\underline{6x^3 - 12x^2} \qquad\qquad\qquad\qquad \\
-7x^2 + 16x \qquad\qquad \\
\underline{-7x^2 + 14x} \qquad\qquad \\
2x - 4 \\
\underline{2x - 4} \\
0
\end{array}
$$

Multiply: $6x^2(x - 2)$.

Subtract and bring down $16x$.

Multiply: $-7x(x - 2)$.

Subtract and bring down -4.

Multiply: $2(x - 2)$.

Subtract.

From this division, you can conclude that

$$6x^3 - 19x^2 + 16x - 4 = (x - 2)(6x^2 - 7x + 2)$$

and by factoring the quadratic $6x^2 - 7x + 2$, you have

$$6x^3 - 19x^2 + 16x - 4 = (x - 2)(2x - 1)(3x - 2).$$

✓CHECKPOINT   Now try Exercise 1.

Note that the factorization shown in Example 1 agrees with the graph shown in Figure 2.21 in that the three x-intercepts occur at $x = 2$, $x = \frac{1}{2}$, and $x = \frac{2}{3}$.

In Example 1, $x - 2$ is a factor of the polynomial $6x^3 - 19x^2 + 16x - 4$, and the long division process produces a remainder of zero. Often, long division will produce a nonzero remainder. For instance, if you divide $x^2 + 3x + 5$ by $x + 1$, you obtain the following.

$$
\begin{array}{r}
x + 2 \quad \longleftarrow \text{Quotient} \\
x + 1 \overline{)\, x^2 + 3x + 5} \quad \longleftarrow \text{Dividend} \\
\underline{x^2 + x} \\
2x + 5 \\
\underline{2x + 2} \\
3 \quad \longleftarrow \text{Remainder}
\end{array}
$$

Divisor $\Longrightarrow$

In fractional form, you can write this result as follows.

$$
\underbrace{\frac{\overbrace{x^2 + 3x + 5}^{\text{Dividend}}}{\underbrace{x + 1}_{\text{Divisor}}}} = \overbrace{x + 2}^{\text{Quotient}} + \frac{\overbrace{3}^{\text{Remainder}}}{\underbrace{x + 1}_{\text{Divisor}}}
$$

This example illustrates the following well-known theorem called the **Division Algorithm.**

The Division Algorithm

If $f(x)$ and $d(x)$ are polynomials such that $d(x) \neq 0$, and the degree of $d(x)$ is less than or equal to the degree of $f(x)$, there exist unique polynomials $q(x)$ and $r(x)$ such that

$$
f(x) = d(x)q(x) + r(x)
$$

Dividend | Quotient
Divisor | Remainder

where $r(x) = 0$ or the degree of $r(x)$ is less than the degree of $d(x)$. If the remainder $r(x)$ is zero, $d(x)$ **divides evenly** into $f(x)$.

The Division Algorithm can also be written as

$$
\frac{f(x)}{d(x)} = q(x) + \frac{r(x)}{d(x)}.
$$

In the Division Algorithm, the rational expression $f(x)/d(x)$ is **improper** because the degree of $f(x)$ is greater than or equal to the degree of $d(x)$. On the other hand, the rational expression $r(x)/d(x)$ is **proper** because the degree of $r(x)$ is less than the degree of $d(x)$.

Before you apply the Division Algorithm, follow these steps.

1. Write the dividend and divisor in descending powers of the variable.

2. Insert placeholders with zero coefficients for each missing power of the variable.

EXAMPLE 2   LONG DIVISION OF POLYNOMIALS

Divide $x^3 - 1$ by $x - 1$.

SOLUTION

Because there is no x^2-term or x-term in the dividend, you need to line up the subtraction by using zero coefficients (or leaving spaces) for the missing terms.

$$
\begin{array}{r}
x^2 + x + 1 \\
x - 1 \overline{)\ x^3 + 0x^2 + 0x - 1} \\
\underline{x^3 - x^2} \\
x^2 + 0x \\
\underline{x^2 - x} \\
x - 1 \\
\underline{x - 1} \\
0
\end{array}
$$

Insert $0x^2$ and $0x$.

Multiply x^2 by $(x - 1)$.

Subtract and bring down $0x$.

Multiply x by $(x - 1)$.

Subtract and bring down -1.

Multiply 1 by $(x - 1)$.

Subtract.

So, $x - 1$ divides evenly into $x^3 - 1$ and you can write

$$\frac{x^3 - 1}{x - 1} = x^2 + x + 1.$$

✓CHECKPOINT   Now try Exercise 13.

You can check the result of a division problem by multiplying. For instance, in Example 2, try checking that $(x - 1)(x^2 + x + 1) = x^3 - 1$.

EXAMPLE 3   LONG DIVISION OF POLYNOMIALS

Divide $2x^4 + 4x^3 - 5x^2 + 3x - 2$ by $x^2 + 2x - 3$.

SOLUTION

$$
\begin{array}{r}
2x^2 + 1 \\
x^2 + 2x - 3 \overline{)\ 2x^4 + 4x^3 - 5x^2 + 3x - 2} \\
\underline{2x^4 + 4x^3 - 6x^2} \\
x^2 + 3x - 2 \\
\underline{x^2 + 2x - 3} \\
x + 1
\end{array}
$$

Multiply $2x^2$ by $(x^2 + 2x - 3)$.

Subtract and bring down $3x - 2$.

Multiply 1 by $(x^2 + 2x - 3)$.

Subtract.

Note that the first subtraction eliminated two terms from the dividend. When this happens, the quotient skips a term. So, you can write

$$\frac{2x^4 + 4x^3 - 5x^2 + 3x - 2}{x^2 + 2x - 3} = 2x^2 + 1 + \frac{x + 1}{x^2 + 2x - 3}.$$

✓CHECKPOINT   Now try Exercise 15.

Synthetic Division

There is a nice shortcut for long division by polynomials of the form $x - k$. The shortcut is called **synthetic division.** We summarize the pattern for synthetic division of a cubic polynomial as follows. (The pattern for higher-degree polynomials is similar.)

Synthetic Division (for a Cubic Polynomial)

To divide $ax^3 + bx^2 + cx + d$ by $x - k$, use the following pattern.

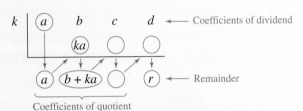

Vertical pattern: Add terms in columns.
Diagonal pattern: Multiply results by k.

STUDY TIP

Synthetic division works only for divisors of the form $x - k$. You cannot use synthetic division to divide a polynomial by a quadratic, a cubic, or any other higher-degree polynomial.

EXAMPLE 4   USING SYNTHETIC DIVISION

Use synthetic division to divide $x^4 - 10x^2 - 2x + 4$ by $x + 3$.

SOLUTION

You should set up the array as follows. Note that a zero is included for each missing term in the dividend.

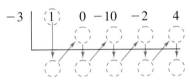

Then, use the synthetic division pattern by adding terms in columns and multiplying the results by -3.

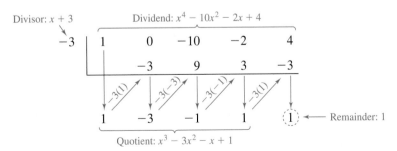

So, you have $\dfrac{x^4 - 10x^2 - 2x + 4}{x + 3} = x^3 - 3x^2 - x + 1 + \dfrac{1}{x + 3}$.

✓CHECKPOINT   Now try Exercise 25.

Remainder and Factor Theorems

The remainder obtained in the synthetic division process has an important interpretation, given in the Remainder Theorem.

The Remainder Theorem

If a polynomial $f(x)$ is divided by $x - k$, the remainder is

$$r = f(k).$$

The Remainder Theorem tells you that synthetic division can be used to evaluate a polynomial function. That is, to evaluate a polynomial function $f(x)$ at $x = k$, divide $f(x)$ by $x - k$. The remainder will be $f(k)$, as illustrated in Example 5.

EXAMPLE 5   USING THE REMAINDER THEOREM

Use the Remainder Theorem to evaluate the following function when $x = -2$.

$$f(x) = 3x^3 + 8x^2 + 5x - 7$$

SOLUTION

Using synthetic division, you obtain the following.

$$
\begin{array}{r|rrrr}
-2 & 3 & 8 & 5 & -7 \\
   &   & -6 & -4 & -2 \\
\hline
   & 3 & 2 & 1 & -9
\end{array}
$$

Because the remainder is $r = -9$, you can conclude that

$$f(-2) = -9.$$

This means that $(-2, -9)$ is a point on the graph of f. Try checking this by substituting $x = -2$ in the original function.

✓CHECKPOINT   Now try Exercise 45.

Another important theorem is the Factor Theorem, which is stated below.

Factor Theorem

A polynomial $f(x)$ has a factor $(x - k)$ if and only if $f(k) = 0$.

You can think of the Factor Theorem as stating that if $(x - k)$ is a factor of $f(x)$, then $f(k) = 0$. Conversely, if $f(k) = 0$, then $(x - k)$ is a factor of $f(x)$.

TECHNOLOGY

Remember, you can also evaluate a function with your graphing utility by entering the function in the equation editor and using the *table* feature in ASK mode. For instructions on how to use the *table* feature, see Appendix A; for specific keystrokes, go to the text website at *college.hmco.com*.

EXAMPLE 6   FACTORING A POLYNOMIAL

Show that $(x - 2)$ and $(x + 3)$ are factors of the polynomial function given by

$$f(x) = 2x^4 + 7x^3 - 4x^2 - 27x - 18.$$

Then find the remaining factors of $f(x)$.

SOLUTION

Using synthetic division with the factor $(x - 2)$, you obtain the following.

$$
\begin{array}{r|rrrrr}
2 & 2 & 7 & -4 & -27 & -18 \\
  &   & 4 & 22 & 36 & 18 \\
\hline
  & 2 & 11 & 18 & 9 & 0
\end{array}
$$

⟹ 0 remainder, so $f(2) = 0$ and $(x - 2)$ is a factor.

Take the result of this division and perform synthetic division again using the factor $(x + 3)$.

$$
\begin{array}{r|rrrr}
-3 & 2 & 11 & 18 & 9 \\
   &   & -6 & -15 & -9 \\
\hline
   & 2 & 5 & 3 & 0
\end{array}
$$

⟹ 0 remainder, so $f(-3) = 0$ and $(x + 3)$ is a factor.

Quadratic: $2x^2 + 5x + 3$

Because the resulting quadratic expression factors as

$$2x^2 + 5x + 3 = (2x + 3)(x + 1)$$

the complete factorization of $f(x)$ is

$$f(x) = (x - 2)(x + 3)(2x + 3)(x + 1).$$

Note that this factorization implies that f has four real zeros:

$$2, -3, -\tfrac{3}{2}, \text{ and } -1.$$

This is confirmed by the graph of f, which is shown in Figure 2.22.

✓CHECKPOINT   Now try Exercise 51.

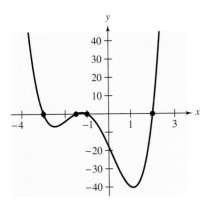

Figure 2.22

Results of Synthetic Division

The remainder r obtained in the synthetic division of $f(x)$ by $x - k$ provides the following information.

1. The remainder r gives the value of f at $x = k$. That is, $r = f(k)$.

2. If $r = 0$, $(x - k)$ is a factor of $f(x)$.

3. If $r = 0$, $(k, 0)$ is an x-intercept of the graph of f.

Throughout this text, we have emphasized the importance of developing several problem-solving strategies. In the exercises for this section, try using more than one strategy to solve several of the exercises. For instance, if you find that $x - k$ divides evenly into $f(x)$, try sketching the graph of f. You should find that $(k, 0)$ is an x-intercept of the graph.

Application

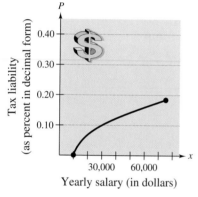

Yearly salary (in dollars)

Figure 2.23

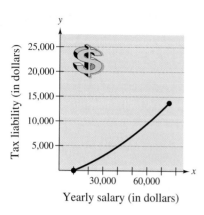

Yearly salary (in dollars)

Figure 2.24

EXAMPLE 7 TAX LIABILITY

The 2001 federal income tax liability for an employee who was single and claimed no dependents is given by the function

$$y = 0.0000014x^2 + 0.092x - 1048, \quad 10,000 \le x \le 75,000$$

where y represents the tax liability (in dollars) and x represents the employee's yearly salary (in dollars) (see Figure 2.23). (Source: U.S. Department of the Treasury)

a. Find a function that gives the tax liability as a *percent* of the yearly salary.

b. *Make a Decision* Graph the function from part (a). What conclusions can you make from the graph?

SOLUTION

a. Because the yearly salary is given by x and the tax liability is given by y, the percent (in decimal form) of yearly salary that the person owes in federal income tax is

$$P = \frac{y}{x}$$

$$= \frac{0.0000014x^2 + 0.092x - 1048}{x}$$

$$= 0.0000014x + 0.092 - \frac{1048}{x}.$$

b. The graph of the function P is shown in Figure 2.24. From the graph you can see that as a person's yearly salary increases, the percent that he or she must pay in federal income tax also increases.

✔CHECKPOINT   Now try Exercise 74.

Discussing the Concept | Understanding the Remainder Theorem

The Remainder Theorem states: "If a polynomial $f(x)$ is divided by $x - k$, the remainder is $r = f(k)$." Given

$$f(x) = x^3 + 6x^2 + 5x - 12$$

with the real zeros -4, -3, and 1, explain algebraically and graphically what it means if $r = 0$. Explain algebraically and graphically what it means if $r \ne 0$. What is the relationship between the Remainder Theorem and the zeros and factors of $f(x)$?

2.3 Warm Up

The following warm-up exercises involve skills that were covered in earlier sections. You will use these skills in the exercise set for this section. For additional help, review Sections R1.5 and R1.6.

In Exercises 1–4, write the expression in standard polynomial form.

1. $(x - 1)(x^2 + 2) + 5$
2. $(x^2 - 3)(2x + 4) + 8$
3. $(x^2 + 1)(x^2 - 2x + 3) - 10$
4. $(x + 6)(2x^3 - 3x) - 5$

In Exercises 5–10, factor the polynomial.

5. $x^2 - 4x + 3$
6. $8x^2 - 24x - 80$
7. $3x^2 + 2x - 5$
8. $9x^2 - 24x + 16$
9. $4x^3 - 10x^2 + 6x$
10. $6x^3 + 7x^2 + 2x$

2.3 Exercises

In Exercises 1–18, use long division to divide.

	Dividend	Divisor
1.	$3x^2 - 7x + 4$	$x - 1$
2.	$5x^2 - 17x - 12$	$x - 4$
3.	$2x^2 + 10x + 12$	$x + 3$
4.	$2x^2 + x - 11$	$x + 5$
5.	$2x^3 + 6x^2 - x - 3$	$2x^2 - 1$
6.	$3x^3 - 12x^2 - 2x + 8$	$3x^2 - 2$
7.	$x^4 + 5x^3 + 6x^2 - x - 2$	$x + 2$
8.	$x^4 + 2x^3 - 3x^2 - 8x - 4$	$x^2 - 4$
9.	$7x + 3$	$x + 4$
10.	$8x - 5$	$2x + 3$
11.	$6x^3 + 10x^2 + x + 8$	$2x^2 + 1$
12.	$2x^3 - 8x^2 + 3x - 9$	$x - 4$
13.	$x^3 - 27$	$x^2 - 1$
14.	$x^3 - 9$	$x^2 + 1$
15.	$x^3 - 4x^2 + 5x - 2$	$x + 2$
16.	$x^3 - x^2 + 2x - 8$	$x - 2$
17.	$2x^5 - 8x^3 + 4x - 1$	$x^2 - 2x + 1$
18.	$x^5 + 7$	$x^3 - 1$

In Exercises 19–36, use synthetic division to divide.

	Dividend	Divisor
19.	$2x^3 + 5x^2 - 7x + 20$	$x + 4$
20.	$3x^3 - 23x^2 - 12x + 32$	$x - 8$
21.	$4x^3 - 9x + 8x^2 - 18$	$x + 2$
22.	$9x^3 - 16x - 18x^2 + 32$	$x - 2$
23.	$-x^3 + 75x - 250$	$x + 10$
24.	$3x^3 - 16x^2 - 72$	$x - 6$
25.	$5x^3 - 6x^2 + 8$	$x - 4$
26.	$6x^4 - 15x^3 - 11x$	$x + 2$
27.	$10x^4 - 50x^3 - 800$	$x - 6$
28.	$x^5 - 13x^4 - 120x + 80$	$x + 3$
29.	$2x^5 - 30x^3 - 37x + 13$	$x - 4$
30.	$5x^3$	$x + 3$
31.	$-3x^4$	$x - 2$
32.	$-3x^4$	$x + 2$
33.	$5 - 3x + 2x^2 - x^3$	$x + 1$
34.	$180x - x^4$	$x - 6$
35.	$4x^3 + 16x^2 - 23x - 15$	$x + \frac{1}{2}$
36.	$3x^3 - 4x^2 + 5$	$x - \frac{3}{2}$

In Exercises 37–44, write the function in the form

$$f(x) = (x - k)q(x) + r$$

for the given value of k, and demonstrate that $f(k) = r$.

37. $f(x) = x^3 - x^2 - 14x + 11$, $k = 4$
38. $f(x) = x^3 - 5x^2 - 11x + 8$, $k = -2$
39. $f(x) = 2x^3 - 9x^2 + 10x - 9$, $k = \frac{1}{2}$
40. $f(x) = \frac{1}{3}(15x^4 + 10x^3 - 6x^2 + 17x + 14)$,
    $k = -\frac{2}{3}$

41. $f(x) = x^3 + 3x^2 - 2x - 14,$ $k = \sqrt{2}$

42. $f(x) = x^3 + 2x^2 - 5x - 4,$ $k = -\sqrt{5}$

43. $f(x) = -3x^3 + 8x^2 + 10x - 8,$ $k = 2 + \sqrt{2}$

44. $f(x) = 4x^3 - 6x^2 - 12x - 4,$ $k = 1 - \sqrt{3}$

In Exercises 45–50, use synthetic division to find each function value.

45. $f(x) = 2x^5 - 3x^2 - 4x - 1$

  (a) $f(-2)$ (b) $f(-4)$

  (c) $f(1)$ (d) $f(3)$

46. $g(x) = x^6 - 4x^4 + 3x^2 + 2$

  (a) $g(2)$ (b) $g(-4)$

  (c) $g(7)$ (d) $g(-1)$

47. $f(x) = 1.2x^3 - 0.5x^2 - 2.1x - 2.4$

  (a) $f(2)$ (b) $f(-6)$

  (c) $f\left(\frac{2}{3}\right)$ (d) $f(1)$

48. $f(x) = 0.4x^4 - 1.6x^3 + 0.7x^2 - 2$

  (a) $f(1)$ (b) $f(-2)$

  (c) $f(5)$ (d) $f(-10)$

49. $f(x) = x^3 - 2x^2 - 11x + 52$

  (a) $f(5)$ (b) $f(-5)$

  (c) $f(1.2)$ (d) $f(2)$

50. $g(x) = x^3 - x^2 + 25x - 25$

  (a) $g(5)$ (b) $g\left(\frac{1}{5}\right)$

  (c) $g(-1.5)$ (d) $g(-1)$

In Exercises 51–56, (a) verify the given factors of the function f, (b) find the remaining factors of f, (c) use your results to write the complete factorization of f, (d) list all real zeros of f, and (e) confirm your results by using a graphing utility to graph the function.

Function	*Factors*
51. $f(x) = x^3 - 12x - 16$	$(x + 2), (x - 4)$
52. $f(x) = x^3 - 28x - 48$	$(x + 4), (x - 6)$
53. $f(x) = 2x^3 + 3x^2 - 17x + 12$	$(2x - 3), (x - 1)$
54. $f(x) = 4x^3 + x^2 - 16x - 4$	$(4x + 1), (x + 2)$
55. $f(x) = x^3 + 2x^2 - 3x - 6$	$(x - \sqrt{3}), (x + 2)$
56. $f(x) = x^3 + 2x^2 - 2x - 4$	$(x - \sqrt{2}), (x + 2)$

57. You divide a polynomial by another polynomial. The remainder is zero. What conclusions can you make?

58. Suppose that the remainder obtained in a polynomial division by $x - k$ is zero. How is the divisor related to the graph of the dividend?

In Exercises 59–62, match the function with its graph and use the result to find all real solutions of $f(x) = 0$. [The graphs are labeled (a), (b), (c), and (d).]

(a)
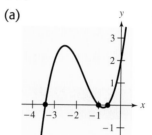

(b)
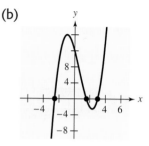

(c)
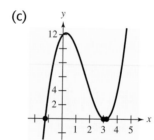

(d)
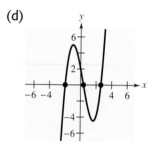

59. $f(x) = x^3 - 2x^2 - 7x + 12$

60. $f(x) = x^3 - x^2 - 5x + 2$

61. $f(x) = x^3 + 5x^2 + 6x + 2$

62. $f(x) = x^3 - 5x^2 + 2x + 12$

63. *Make a Decision*: **Modeling Polynomials** A third-degree polynomial function $f(x)$ has real zeros -1, 2, and $\frac{10}{3}$. Find two different polynomial functions, one with a positive leading coefficient and one with a negative leading coefficient, that could be $f(x)$. How many different polynomial functions are possible for $f(x)$?

64. *Make a Decision*: **Modeling Polynomials** A fourth-degree polynomial function $g(x)$ has real zeros -2, 0, 1, and 5. Find two different polynomial functions, one with a positive leading coefficient and one with a negative leading coefficient, that could be $g(x)$. How many different polynomial functions are possible for $g(x)$?

In Exercises 65–70, simplify the rational expression.

65. $\dfrac{4x^3 - x^2 - 13x + 10}{4x - 5}$

66. $\dfrac{2x^3 + 7x^2 - 10x + 3}{2x - 1}$

67. $\dfrac{3x^3 - 7x^2 - 13x - 28}{x - 4}$

68. $\dfrac{2x^3 + 3x^2 - 3x - 2}{x - 1}$

69. $\dfrac{x^4 + 6x^3 + 11x^2 + 6x}{x^2 + 3x + 2}$

70. $\dfrac{x^4 + 9x^3 - 5x^2 - 36x + 4}{x^2 - 4}$

71. Dimensions of a Room   A rectangular room has a volume of

$$x^3 + 11x^2 + 34x + 24$$

cubic feet. The height of the room is $x + 1$ feet (see figure). Find the number of square feet of floor space in the room.

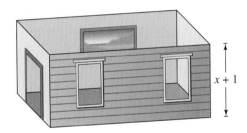

72. Dimensions of a House   A rectangular house has a volume of

$$x^3 + 55x^2 + 650x + 2000$$

cubic feet (the space in the attic is not counted). The height of the house is $x + 5$ feet (see figure). Find the number of square feet of floor space *on the first floor* of the house.

73. Profit   A company making compact discs estimates that the profit P (in dollars) from selling a particular disc is given by

$$P = -153.6x^3 + 5760x^2 - 100,000,$$
$$0 \le x \le 35$$

where x is the advertising expense (in tens of thousands of dollars). For this disc, the advertising expense was \$300,000 ($x = 30$) and the profit was \$936,800.

(a) From the graph shown in the figure, it appears that the company could have obtained the same profit by spending less on advertising. Use the graph to estimate another amount the company could have spent on advertising that would have produced the same profit.

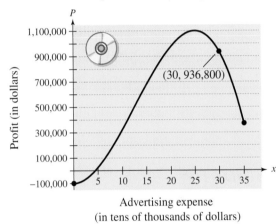

Advertising expense
(in tens of thousands of dollars)

(b) Use synthetic division with $x = 30$ to confirm algebraically the result of part (a).

74. *Make a Decision*: **Profit**   A company that produces calculators estimates that the profit P (in dollars) from selling a particular model of calculator is

$$P = -125x^3 + 6450x^2 - 150,000, \quad 0 \le x \le 50$$

where x is the advertising expense (in tens of thousands of dollars). For this model of calculator, the advertising expense was \$400,000 ($x = 40$) and the profit was \$2,170,000.

(a) Use a graphing utility to graph the profit function.

(b) Could the company have obtained the same profit by spending less on advertising? Explain your reasoning.

Objectives

- **Find all possible rational zeros of a function using the Rational Zero Test.**
- **Find all real zeros of a function.**
- **Approximate the real zeros of a polynomial function using the Intermediate Value Theorem.**
- **Approximate the real zeros of a polynomial function using a graphing utility.**
- **Apply techniques for approximating real zeros to solve an application problem.**

The Rational Zero Test

The **Rational Zero Test** relates the possible rational zeros of a polynomial (having integer coefficients) to the leading coefficient and to the constant term of the polynomial.

The Rational Zero Test

If the polynomial $f(x) = a_n x^n + a_{n-1} x^{n-1} + \cdots + a_2 x^2 + a_1 x + a_0$ has *integer* coefficients, then every rational zero of f has the form

$$\text{Possible rational zeros} = \frac{\text{factors of constant term}}{\text{factors of leading coefficient}} = \frac{p}{q}$$

where p and q have no common factors other than 1,

$p = $ a factor of the constant term a_0, and

$q = $ a factor of the leading coefficient a_n.

STUDY TIP

When the leading coefficient is 1, the possible rational zeros are simply the factors of the constant term.

Having formed this list of *possible rational zeros,* use a trial-and-error method to determine which, if any, are actual zeros of the polynomial.

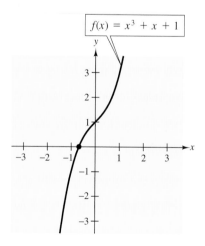

Figure 2.25

| EXAMPLE 1 | RATIONAL ZERO TEST WITH LEADING COEFFICIENT OF 1

Find the rational zeros of $f(x) = x^3 + x + 1$. (See Figure 2.25.)

SOLUTION

Because the leading coefficient is 1, the possible rational zeros are the factors of the constant term, 1 and -1. By testing these possible zeros, you can see that neither checks.

$$f(1) = (1)^3 + 1 + 1 = 3 \qquad f(-1) = (-1)^3 + (-1) + 1 = -1$$

So, you can conclude that the given polynomial has *no* rational zeros. Note from the graph of f in Figure 2.25 that f does have one real zero (between -1 and 0). By the Rational Zero Test, you know that this real zero is *not* a rational number.

✓CHECKPOINT   Now try Exercise 1.

| EXAMPLE 2 | RATIONAL ZERO TEST WITH LEADING COEFFICIENT OF 1 |

Find the rational zeros of $f(x) = x^4 - x^3 + x^2 - 3x - 6$.

SOLUTION

Because the leading coefficient is 1, the possible rational zeros are the factors of the constant term.

Possible rational zeros: $\pm 1, \pm 2, \pm 3, \pm 6$

A test of these possible zeros shows that $x = -1$ and $x = 2$ are the only two that check. Check the others to be sure.

✓**CHECKPOINT**   Now try Exercise 3.

If the leading coefficient of a polynomial is not 1, the list of possible rational zeros can increase dramatically. In such cases the search can be shortened in several ways: (1) a programmable calculator can be used to speed up the calculations; (2) a graph, created either by hand or with a graphing utility, can give a good estimate of the locations of the zeros; and (3) synthetic division can be used to test the possible rational zeros.

TECHNOLOGY

There are several ways to use your graphing utility to locate the zeros of a polynomial after listing the possible rational zeros. You can use the *table* feature by setting the increments of x to the smallest difference between possible rational zeros, or use the *table* feature in ASK mode. In either case the value in the function column will be 0 when x is a zero of the function. Another way to locate zeros is to graph the function. Be sure that your viewing window contains all the possible rational zeros.

To see how to use synthetic division to test the possible rational zeros, let's take another look at the function

$$f(x) = x^4 - x^3 + x^2 - 3x - 6$$

given in Example 2. To test that $x = -1$ and $x = 2$ are zeros of f, you can apply synthetic division, as follows.

$$
\begin{array}{r|rrrrr}
-1 & 1 & -1 & 1 & -3 & -6 \\
   &   & -1 & 2 & -3 & 6 \\
\hline
   & 1 & -2 & 3 & -6 & 0
\end{array}
\qquad\Longrightarrow\qquad
\begin{array}{r|rrrr}
2 & 1 & -2 & 3 & -6 \\
  &   & 2 & 0 & 6 \\
\hline
  & 1 & 0 & 3 & 0
\end{array}
$$

So, you have

$$f(x) = (x + 1)(x - 2)(x^2 + 3).$$

Because the factor $(x^2 + 3)$ produces no real zeros, you can conclude that $x = -1$ and $x = 2$ are the only *real* zeros of f. This is verified in the graph of f shown in Figure 2.26.

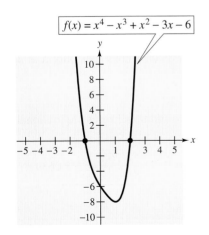

$f(x) = x^4 - x^3 + x^2 - 3x - 6$

Figure 2.26

Finding the first zero is often the hardest part. After that, the search is simplified by using the lower-degree polynomial obtained in synthetic division. Once the lower-degree polynomial is quadratic, either factoring or the Quadratic Formula can be used to find the remaining zeros.

EXAMPLE 3   USING THE RATIONAL ZERO TEST

Find the rational zeros of $f(x) = 2x^3 + 3x^2 - 8x + 3$.

SOLUTION

The leading coefficient is 2 and the constant term is 3.

$$\text{Possible rational zeros: } \frac{\text{Factors of 3}}{\text{Factors of 2}} = \frac{\pm 1, \pm 3}{\pm 1, \pm 2} = \pm 1, \pm 3, \pm \frac{1}{2}, \pm \frac{3}{2}$$

By synthetic division, you can determine that $x = 1$ is a zero.

$$\begin{array}{r|rrrr} 1 & 2 & 3 & -8 & 3 \\ & & 2 & 5 & -3 \\ \hline & 2 & 5 & -3 & 0 \end{array}$$

So, $f(x)$ factors as

$$f(x) = (x - 1)(2x^2 + 5x - 3)$$
$$= (x - 1)(2x - 1)(x + 3)$$

and you can conclude that the zeros of f are $x = 1$, $x = \frac{1}{2}$, and $x = -3$.

✓CHECKPOINT   Now try Exercise 5.

EXAMPLE 4   USING THE RATIONAL ZERO TEST

Find all the real zeros of $f(x) = 10x^3 - 15x^2 - 16x + 12$.

SOLUTION

The leading coefficient is 10 and the constant term is 12.

$$\text{Possible rational zeros: } \frac{\text{Factors of 12}}{\text{Factors of 10}} = \frac{\pm 1, \pm 2, \pm 3, \pm 4, \pm 6, \pm 12}{\pm 1, \pm 2, \pm 5, \pm 10}$$

With so many possibilities (32, in fact), it is worth your time to stop and sketch a graph. From Figure 2.27, it looks like three reasonable choices would be $x = -\frac{6}{5}$, $x = \frac{1}{2}$, and $x = 2$. Testing these by synthetic division shows that only $x = 2$ checks. So, you have

$$f(x) = (x - 2)(10x^2 + 5x - 6).$$

Using the Quadratic Formula, you find that the two additional zeros are irrational numbers.

$$x = \frac{-5 + \sqrt{265}}{20} \approx 0.5639 \quad \text{and} \quad x = \frac{-5 - \sqrt{265}}{20} \approx -1.0639$$

✓CHECKPOINT   Now try Exercise 21.

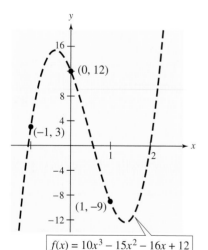

$$f(x) = 10x^3 - 15x^2 - 16x + 12$$

Figure 2.27

The Intermediate Value Theorem

The next theorem, called the **Intermediate Value Theorem,** tells you of the existence of real zeros of polynomial functions. The theorem implies that if $(a, f(a))$ and $(b, f(b))$ are two points on the graph of a polynomial such that $f(a) \neq f(b)$, then for any number d between $f(a)$ and $f(b)$ there must be a number c between a and b such that $f(c) = d$. (See Figure 2.28.)

Intermediate Value Theorem

Let a and b be real numbers such that $a < b$. If f is a polynomial function such that $f(a) \neq f(b)$, then, in the interval $[a, b]$, f takes on every value between $f(a)$ and $f(b)$.

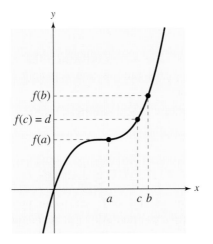

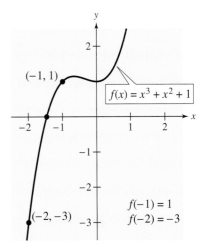

$$f(x) = x^3 + x^2 + 1$$

$$f(-1) = 1$$
$$f(-2) = -3$$

Figure 2.28 If d lies between $f(a)$ and $f(b)$, there exists c between a and b such that $f(c) = d$.

Figure 2.29 f has a zero between -2 and -1.

The Intermediate Value Theorem helps you locate the real zeros of a polynomial function in the following way. If you can find a value $x = a$ where a polynomial function is positive, and another value $x = b$ where it is negative, you can conclude that the function has at least one real zero between these two values. For example, the function

$$f(x) = x^3 + x^2 + 1$$

is negative when $x = -2$ and positive when $x = -1$. So, it follows from the Intermediate Value Theorem that f must have a real zero somewhere between -2 and -1, as shown in Figure 2.29. By continuing this line of reasoning, you can approximate any real zeros of a polynomial function to any desired level of accuracy. This concept is further demonstrated in Example 5.

APPROXIMATING A ZERO OF A POLYNOMIAL FUNCTION

Use the Intermediate Value Theorem to approximate the real zero of

$$f(x) = x^3 - x^2 + 1.$$

SOLUTION

Begin by computing a few function values, as follows.

x	-2	-1	0	1
$f(x)$	-11	-1	1	1

Because $f(-1)$ is negative and $f(0)$ is positive, you can apply the Intermediate Value Theorem to conclude that the function has a zero between -1 and 0. To pinpoint this zero more closely, divide the interval $[-1, 0]$ into tenths and evaluate the function at each point. When you do this, you will find that

$$f(-0.8) = -0.152 \quad \text{and} \quad f(-0.7) = 0.167.$$

So, f must have a zero between -0.8 and -0.7, as shown in Figure 2.30. By continuing this process you can approximate this zero to any desired level of accuracy.

✓CHECKPOINT   Now try Exercise 23.

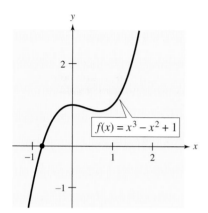

Figure 2.30 f has a zero between -0.8 and -0.7.

Approximating Zeros of Polynomial Functions

There are several different techniques for approximating the zeros of a polynomial function. All such techniques are better suited to computers or graphing utilities than they are to "hand calculations." In this section, you will study two techniques that can be used with a graphing utility. The first is called the **zoom-and-trace** technique.

Zoom-and-Trace Technique

To approximate a real zero of a function with a graphing utility, use the following steps.

1. Graph the function so that the real zero you want to approximate appears as an x-intercept on the screen.

2. Move the cursor near the x-intercept and use the *zoom* feature to zoom in to get a better look at the intercept.

3. Use the *trace* feature to find the x-values that occur just before and just after the x-intercept. If the difference in these values is sufficiently small, use their average as the approximation. If not, continue zooming in until the approximation reaches the desired level of accuracy.

STUDY TIP

To help you visually determine when you have zoomed in enough times to reach the desired level of accuracy, set the X-scale of the viewing window to the accuracy you need and zoom in repeatedly. For instance, to approximate the zero to the nearest hundredth, set the X-scale to 0.01.

The amount that a graphing utility zooms in is determined by the *zoom factor*. The zoom factor is a positive number greater than or equal to 1 that gives the ratio of the larger screen to the smaller screen. For instance, if you zoom in with a zoom factor of 2, you will obtain a screen in which the *x*- and *y*-values are half their original values. This text uses a zoom factor of 4.

EXAMPLE 6   APPROXIMATING THE ZEROS OF A POLYNOMIAL FUNCTION

Approximate the real zeros of the polynomial function

$$f(x) = x^3 + 4x + 2.$$

to the nearest thousandth.

SOLUTION

To begin, use a graphing utility to graph the function, as shown in Figure 2.31(a). Set the X-scale to 0.001 and zoom in several times until the tick marks on the *x*-axis become visible. The final screen should be similar to the one shown in Figure 2.31(b).

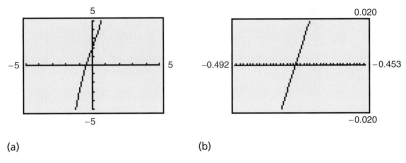

(a)   (b)

Figure 2.31

At this point, you can use the *trace* feature to determine that the *x*-values just to the left and right of the *x*-intercept are

$$x \approx -0.4735 \quad \text{and} \quad x \approx -0.4733.$$

So, to the nearest thousandth, you can approximate the zero of the function to be

$$x \approx -0.473.$$

To check this, try substituting -0.473 into the function. You should obtain a result that is approximately zero.

✓CHECKPOINT   Now try Exercise 35.

___

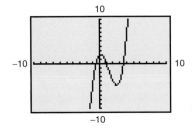

Figure 2.32

In Example 6, the cubic polynomial function has only one real zero. Remember that functions can have two or more real zeros. In such cases, you can use the zoom-and-trace technique for each zero separately. For instance, the function

$$f(x) = x^3 - 4x^2 + x + 2$$

has three real zeros, as shown in Figure 2.32. Using a zoom-and-trace approach for each real zero, you can approximate the real zeros to be

$$-0.562, \quad 1.000, \quad \text{and} \quad 3.562.$$

The second technique that can be used with some graphing utilities is to use the graphing utility's *zero* or *root* feature. The name of this feature differs with different calculators. Consult your user's guide to determine if this feature is available.

EXAMPLE 7   APPROXIMATING THE ZEROS OF A POLYNOMIAL FUNCTION

Approximate the real zeros of $f(x) = x^3 - 2x^2 - x + 1$.

SOLUTION

To begin, use a graphing utility to graph the function, as shown in the first screen of Figure 2.33. Notice that the graph has three *x*-intercepts. To approximate the leftmost intercept, find an appropriate viewing window and use the *zero* feature, as shown below. The calculator should display an approximation of $x \approx -0.8019377$, which is accurate to seven decimal places.

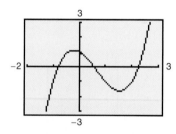

Find an appropriate viewing window, then use the *zero* feature.

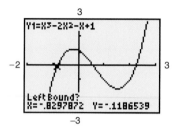

Cursor to the left of the intercept and press "Enter."

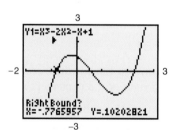

Cursor to the right of the intercept and press "Enter."

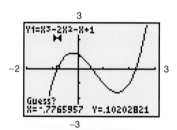

Cursor near the intercept and press "Enter."

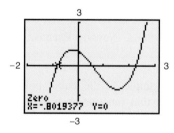

Figure 2.33

TECHNOLOGY

For instructions on how to use the *zoom*, *trace*, *zero*, and *root* features, see Appendix A; for specific keystrokes, go to the text website at *college.hmco.com*.

By repeating this process, you can determine that the other two zeros are $x \approx 0.555$ and $x \approx 2.247$.

✓CHECKPOINT   Now try Exercise 37.

You may be wondering why we spend so much time in algebra trying to find the zeros of a function. The reason is that if you have a technique that will enable you to solve the equation $f(x) = 0$, you can use the same technique to solve the more general equation

$$f(x) = c$$

where c is any real number. This procedure is demonstrated in Example 8.

EXAMPLE 8   Solving the Equation $f(x) = c$

Find a value of x such that $f(x) = 30$ for the function given by

$$f(x) = x^3 - 4x + 4.$$

SOLUTION

The graph of $f(x) = x^3 - 4x + 4$ is shown in Figure 2.34. Note from the graph that $f(x) = 30$ when x is about 3.5. To use the zoom-and-trace technique to approximate this x-value more closely, consider the equation

$$x^3 - 4x + 4 = 30$$

$$x^3 - 4x - 26 = 0.$$

So, the *solutions* of the equation $f(x) = 30$ are precisely the same x-values as the *zeros* of $g(x) = x^3 - 4x - 26$. Using the graph of g, as shown in Figure 2.35, you can approximate the zero of g to be

$$x \approx 3.41.$$

You can check this value by substituting $x = 3.41$ into the original function.

$$f(3.41) = (3.41)^3 - 4(3.41) + 4$$

$$\approx 30.01 \; \checkmark$$

Remember that with decimal approximations, a check usually will not produce an exact value.

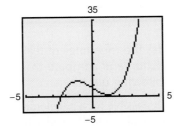

Figure 2.34   **Figure 2.35**

✔CHECKPOINT   Now try Exercise 47.

Application

EXAMPLE 9   PROFIT AND ADVERTISING EXPENSES

Make a Decision   A company that produces sports clothes estimates that the profit from selling a particular line of sportswear is given by

$$P = -0.014x^3 + 0.752x^2 - 40, \qquad 0 \le x \le 50$$

where P is the profit (in tens of thousands of dollars) and x is the advertising expense (in tens of thousands of dollars). According to this model, how much money should the company spend on advertising to obtain a profit of $2,750,000?

SOLUTION

From Figure 2.36, it appears that there are two different values of x between 0 and 50 that will produce a profit of $2,750,000. However, because of the context of the problem, it is clear that the better answer is the smaller of the two numbers. So, to solve the equation

$$-0.014x^3 + 0.752x^2 - 40 = 275$$

$$-0.014x^3 + 0.752x^2 - 315 = 0$$

find the zeros of the function $g(x) = -0.014x^3 + 0.752x^2 - 315$. Using the zoom-and-trace technique, you can find that the leftmost zero is $x \approx 32.8$. You can check this solution by substituting $x = 32.8$ into the original function.

$$P = -0.014(32.8)^3 + 0.752(32.8)^2 - 40 \approx 275$$

The company should spend about $328,000 on advertising for the line of sportswear.

✓CHECKPOINT   Now try Exercise 57.

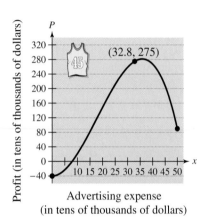

Advertising expense
(in tens of thousands of dollars)

Figure 2.36

Discussing the Concept | Comparing Real Zeros and Rational Zeros

Discuss the meanings of real zeros and rational zeros of a polynomial function and compare these two types of zeros. Then answer the following questions.

a. Is it possible for a polynomial function to have no rational zeros but to have real zeros? If so, give an example.

b. If a polynomial function has three real zeros, and only one of them is a rational number, must the other two zeros be irrational numbers?

c. Consider a cubic polynomial function $f(x) = ax^3 + bx^2 + cx + d$, where $a \ne 0$. Is it possible that f has no real zeros? Is it possible that f has no rational zeros?

d. Is it possible that a second-degree polynomial function with integer coefficients has one rational zero and one irrational zero? If so, give an example.

2.4 (Warm Up)

The following warm-up exercises involve skills that were covered in earlier sections. You will use these skills in the exercise set for this section. For additional help, review Sections R2.5 and 2.3.

In Exercises 1 and 2, find a polynomial function with integer coefficients having the given zeros.

1. $-1, \frac{2}{3}, 3$

2. $-2, 0, \frac{3}{4}, 2$

In Exercises 3 and 4, use synthetic division to divide.

3. $\dfrac{x^5 - 9x^3 + 5x + 18}{x + 3}$

4. $\dfrac{3x^4 + 17x^3 + 10x^2 - 9x - 8}{x + \frac{2}{3}}$

In Exercises 5–8, use the given zero to find all the real zeros of f.

5. $f(x) = 2x^3 + 11x^2 + 2x - 4, \ x = \frac{1}{2}$

6. $f(x) = 6x^3 - 47x^2 - 124x - 60, \ x = 10$

7. $f(x) = 4x^3 - 13x^2 - 4x + 6, \ x = -\frac{3}{4}$

8. $f(x) = 10x^3 + 51x^2 + 48x - 28, \ x = \frac{2}{5}$

In Exercises 9 and 10, find all real solutions of the equation.

9. $x^4 - 3x^2 + 2 = 0$

10. $x^4 - 7x^2 + 12 = 0$

2.4 (Exercises)

In Exercises 1 and 2, use the Rational Zero Test to list all possible rational zeros of f. Then use a graphing utility to graph the function. Use the graph to help determine which of the possible rational zeros are actual zeros of the function.

1. $f(x) = x^3 + x^2 - 4x - 4$

2. $f(x) = 2x^4 - x^2 - 6$

In Exercises 3–6, find the rational zeros of the polynomial function.

3. $f(x) = x^3 - \frac{3}{2}x^2 - \frac{23}{2}x + 6$

4. $f(x) = x^3 + 3x^2 - x - 3$

5. $f(x) = 4x^4 - 17x^2 + 4$

6. $f(x) = -2x^4 + 13x^3 - 21x^2 + 2x + 8$

In Exercises 7–14, find all real zeros of the function.

7. $f(x) = x^3 - 6x^2 + 11x - 6$

8. $g(x) = x^3 - 4x^2 - x + 4$

9. $h(t) = t^3 + 12t^2 + 21t + 10$

10. $f(x) = x^3 - 4x^2 + 5x - 2$

11. $C(x) = 2x^3 + 3x^2 - 1$

12. $f(x) = 3x^3 - 19x^2 + 33x - 9$

13. $f(x) = x^4 - 11x^2 + 18$

14. $P(t) = t^4 - 19t^2 + 48$

In Exercises 15–20, find all real solutions of the polynomial equation.

15. $z^4 - z^3 - 2z - 4 = 0$

16. $x^4 - 13x^2 - 12x = 0$

17. $2y^4 + 7y^3 - 26y^2 + 23y - 6 = 0$

18. $2x^4 - 11x^3 - 6x^2 + 64x + 32 = 0$

19. $x^5 - x^4 - 3x^3 + 5x^2 - 2x = 0$

20. $x^5 - 7x^4 + 10x^3 + 14x^2 - 24x = 0$

In Exercises 21 and 22, (a) list the possible rational zeros of f, (b) sketch the graph of f so that some of the possible zeros in part (a) can be discarded, and (c) determine all real zeros of f.

21. $f(x) = 32x^3 - 52x^2 + 17x + 3$

22. $f(x) = 4x^3 + 7x^2 - 11x - 18$

In Exercises 23–26, use the Intermediate Value Theorem to approximate the zero of f in the interval $[a, b]$. Give your approximation to the nearest tenth. (If you have a graphing utility, use it to help you approximate the zero.)

23. $f(x) = x^3 + x - 1$, $[0, 1]$

24. $f(x) = x^5 + x + 1$, $[-1, 0]$

25. $f(x) = x^4 - 10x^2 - 11$, $[3, 4]$

26. $f(x) = -x^3 + 3x^2 + 9x - 2$, $[4, 5]$

In Exercises 27–32, match the function with its graph. Then approximate the real zeros of the function to three decimal places. [The graphs are labeled (a), (b), (c), (d), (e), and (f).]

(a)
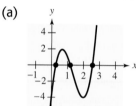

(b)
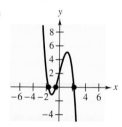

(c)
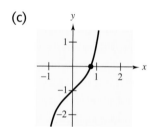

(d)
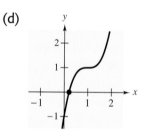

(e)
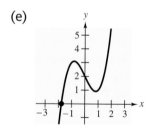

(f)
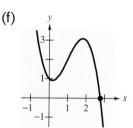

27. $f(x) = x^3 - 2x + 2$

28. $f(x) = x^5 + x - 1$

29. $f(x) = 2x^3 - 6x^2 + 6x - 1$

30. $f(x) = 5x^3 - 20x^2 + 20x - 4$

31. $f(x) = -x^3 + 3x^2 - x + 1$

32. $f(x) = -x^3 + 4x + 2$

In Exercises 33–36, use the *zoom* and *trace* features of a graphing utility to approximate the real zeros of f. Give your approximations to the nearest thousandth.

33. $f(x) = x^4 - x - 3$ **34.** $f(x) = 4x^3 + 14x - 8$

35. $f(x) = x^3 - 3.9x^2 + 4.79x - 1.881$

36. $f(x) = -x^3 + 2x^2 + 4x + 5$

In Exercises 37–40, use the *zero* and *root* features of a graphing utility to approximate the real zeros of f. Give your approximations to the nearest thousandth.

37. $f(x) = x^4 + x - 3$

38. $f(x) = -x^4 + 2x^3 + 4$

39. $f(x) = 7x^4 - 42x^3 + 43x^2 + 216x - 324$

40. $f(x) = 3x^4 - 12x^3 + 27x^2 + 4x - 4$

In Exercises 41–44, match the cubic function with the numbers of rational and irrational zeros.

(a) Rational zeros: 0; Irrational zeros: 1

(b) Rational zeros: 3; Irrational zeros: 0

(c) Rational zeros: 1; Irrational zeros: 2

(d) Rational zeros: 1; Irrational zeros: 0

41. $f(x) = x^3 - 1$ **42.** $f(x) = x^3 - 2$

43. $f(x) = x^3 - x$ **44.** $f(x) = x^3 - 2x$

45. *Make a Decision*: **Dimensions of a Box**   An open box is to be made from a rectangular piece of material, 12 inches by 10 inches, by cutting equal squares from the corners and turning up the sides (see figure).

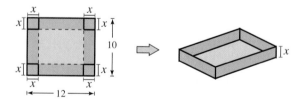

(a) Write the volume V of the box as a function of x. Determine the domain of the function.

(b) Sketch the graph of the function and approximate the dimensions of the box that yield a maximum volume.

(c) Find values of x such that $V = 96$. Which of these values is a physical impossibility in the construction of the box? Explain.

46. *Make a Decision*: **Dimensions of a Box** An open box is to be made from a rectangular piece of material, 15 inches by 9 inches, by cutting equal squares from the corners and turning up the sides (see figure).

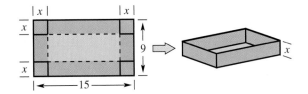

(a) Write the volume V of the box as a function of x. Determine the domain of the function.

(b) Sketch the graph of the function and approximate the dimensions of the box that yield a maximum volume.

(c) Find values of x such that $V = 56$. Which of these values is a physical impossibility in the construction of the box? Explain.

47. Dimensions of a Package A rectangular package with a square cross section to be sent by a postal service has a combined length and girth (perimeter of a cross section) of 108 inches (see figure). Find the dimensions of the package, given that the volume is 11,664 cubic inches.

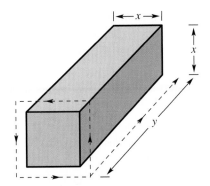

Figure for 47 and 48

48. Dimensions of a Package A rectangular package to be sent by a postal service has a combined length and girth (perimeter of a cross section) of 120 inches (see figure). Find the dimensions of the package, given that the volume is 16,000 cubic inches.

49. Medicine The concentration C of a chemical in the bloodstream t hours after injection into muscle tissue is given by

$$C = \frac{3t^2 + t}{t^3 + 50}, \qquad t \geq 0.$$

The concentration is greatest when $3t^4 + 2t^3 - 300t - 50 = 0$. Approximate this time to the nearest tenth of an hour.

50. Transportation Cost The transportation cost C (in thousands of dollars) of the components used in manufacturing prefabricated homes is given by

$$C = 100\left(\frac{200}{x^2} + \frac{x}{x + 30}\right), \qquad x \geq 1$$

where x is the order size (in hundreds). The cost is a minimum when

$$3x^3 - 40x^2 - 2400x - 36,000 = 0.$$

Approximate the optimal order size to the nearest hundred units.

51. *Make a Decision*: **Recreation Spending** The total amount R (in billions of dollars) spent on recreation in the United States from 1990 to 2001 is shown in the table. (Source: U.S. Department of Commerce, U.S. Bureau of Economic Analysis)

Year	Amount, R	Year	Amount, R
1990	284.9	1996	429.6
1991	292.0	1997	456.6
1992	310.8	1998	489.1
1993	340.1	1999	526.5
1994	368.7	2000	564.7
1995	401.6	2001	593.9

(a) Use a graphing utility to create a scatter plot of the data. Let t represent the year, with $t = 0$ corresponding to 1990.

(b) Use the *regression* feature of a graphing utility to find a linear model, a quadratic model, a cubic model, and a quartic model for the data.

(c) Use a graphing utility to graph each model separately with the data in the same viewing window.

(d) Choose the model that "best fits" the data. Explain why you chose that model. Use your model to estimate the year in which recreation spending will reach $700 billion.

52. *Make a Decision*: **Nitrogen Dioxide Emissions**
The amount N (in thousands of tons) of nitrogen dioxide emissions in the United States from 1990 to 2001 is shown in the table. (Source: U.S. Environmental Protection Agency)

Year	Amount, N	Year	Amount, N
1990	25,530	1996	24,790
1991	25,179	1997	24,712
1992	25,260	1998	24,349
1993	25,357	1999	23,671
1994	25,349	2000	23,199
1995	24,956	2001	22,349

(a) Use a graphing utility to create a scatter plot of the data. Let t represent the year, with $t = 0$ corresponding to 1990.

(b) Use the *regression* feature of a graphing utility to find a quadratic model, a cubic model, and a quartic model for the data.

(c) Use a graphing utility to graph each model separately with the data in the same viewing window.

(d) Choose the model that "best fits" the data. Explain why you chose that model. Use your model to estimate the year in which the nitrogen dioxide emissions will fall to 20,000,000 tons.

53. *Make a Decision*: **Cost of Dental Care** The amount that $100 worth of dental care at 1982–1984 prices would cost at a different time is given by a CPI (Consumer Price Index). The CPIs for dental care in the United States from 1990 to 2002 are shown in the table. (Source: U.S. Bureau of Labor Statistics)

Year	CPI	Year	CPI
1990	155.8	1997	226.6
1991	167.4	1998	236.2
1992	178.7	1999	247.2
1993	188.1	2000	258.5
1994	197.1	2001	269.0
1995	206.8	2002	281.0
1996	216.5		

(a) Use a graphing utility to create a scatter plot of the data. Let t represent the year, with $t = 0$ corresponding to 1990.

(b) Use the *regression* feature of a graphing utility to find a linear model, a quadratic model, a cubic model, and a quartic model for the data.

(c) Use a graphing utility to graph each model separately with the data in the same viewing window. How well does each model fit the data?

(d) Use each model to estimate the CPI for dental care in 2005 and 2007. Are the estimates from each model reasonable for years after 2002?

54. *Make a Decision*: **Domestic Travelers** The total number N (in millions) of leisure trips taken by families in the United States from 1994 to 2002 is shown in the table. (Source: Travel Industry Association of America)

Year	Trips, T	Year	Trips, T
1994	96.6	1999	106.5
1995	97.7	2000	109.2
1996	100.1	2001	113.8
1997	103.4	2002	118.6
1998	106.3		

(a) Use a graphing utility to create a scatter plot of the data. Let t represent the year, with $t = 4$ corresponding to 1994.

(b) Use the *regression* feature of a graphing utility to find a linear model, a quadratic model, a cubic model, and a quartic model for the data.

(c) Use a graphing utility to graph each model separately with the data in the same viewing window. How well does each model fit the data?

(d) Use each model to estimate the number of leisure trips families will take in 2005 and 2007. Are the estimates from each model reasonable?

(e) Discuss the appropriateness of each model for estimating the total number of leisure trips taken by families in the United States after 2002.

55. Advertising Cost A company that produces portable CD players estimates that the profit P (in dollars) for selling a new model is given by

$$P = -76x^3 + 4830x^2 - 320,000, \quad 0 \le x \le 60$$

where x is the advertising expense (in tens of thousands of dollars). Using this model, find the smaller of two advertising amounts that will yield a profit of $2,500,000.

56. Advertising Cost A company that manufactures bicycles estimates that the profit P (in dollars) for selling a new model is given by

$$P = -45x^3 + 2500x^2 - 275{,}000, \quad 0 \le x \le 50$$

where x is the advertising expense (in tens of thousands of dollars). Using this model, find the smaller of two advertising amounts that will yield a profit of $800,000.

57. *Make a Decision*: Demand Function A company that produces cell phones estimates that the demand D for a new model of phone is given by

$$D = -x^3 + 54x^2 - 140x - 3000, \quad 10 \le x \le 50$$

where x is the price of the phone in dollars.

(a) Use a graphing utility to graph D. Use the *trace* feature to determine for what values of x the demand will be 14,400 phones.

(b) You may also determine the price x of the phones that will yield a demand of 14,400 by setting D equal to 14,400 and solving for x with a graphing utility. Discuss this alternative solution method. Of the solutions that lie within the given interval, what price would you recommend the company charge for the phones?

58. *Make a Decision*: Demand Function A company that produces handheld organizers estimates that the demand D for a new model of organizer is given by

$$D = -0.005x^3 + 2.65x^2 - 70x - 2500,$$

$$50 \le x \le 500$$

where x is the price of the organizer in dollars.

(a) Use a graphing utility to graph D. Use the *trace* feature to determine for what values of x the demand will be 80,000 organizers.

(b) You may also determine the price x of the organizers that will yield a demand of 80,000 by setting D equal to 80,000 and solving for x with a graphing utility. Discuss this alternative solution method. Of the solutions that lie within the given interval, what price would you recommend the company charge for the organizers?

59. Use the information in the table.

Interval	Value of $f(x)$
$(-\infty, -2)$	Positive
$(-2, 1)$	Negative
$(1, 4)$	Negative
$(4, \infty)$	Positive

(a) What are the three real zeros of the polynomial function f?

(b) What can be said about the behavior of the graph of f at $x = 1$?

(c) What is the least possible degree of f? Explain. Can the degree of f ever be odd? Explain.

(d) Is the leading coefficient of f positive or negative? Explain.

(e) Write an equation for f. (There are many correct answers.)

(f) Sketch a graph of the equation you wrote in part (e).

60. Graphical Reasoning The graph of one of the following functions is shown below. Identify the function shown in the graph. Explain why each of the others is not the correct function. Use a graphing utility to verify your result.

(a) $f(x) = x^2(x + 2)(x - 3.5)$

(b) $g(x) = (x + 2)(x - 3.5)$

(c) $h(x) = (x + 2)(x - 3.5)(x^2 + 1)$

(d) $k(x) = (x + 1)(x + 2)(x - 3.5)$

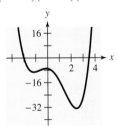

Mid-Chapter Quiz

Take this quiz as you would take a quiz in class. After you are done, check your work against the answers given in the back of the book.

In Exercises 1 and 2, sketch the graph of the quadratic function. Identify the vertex and the intercepts.

1. $f(x) = (x + 1)^2 - 2$ **2.** $f(x) = 25 - x^2$

In Exercises 3 and 4, describe the right-hand and left-hand behavior of the graph of the polynomial function. Verify with a graphing utility.

3. $f(x) = 5x^3 - 7x^2 + 2$ **4.** $f(x) = -x^4 + 5x^2 - 4$

5. Use synthetic division to evaluate $f(x) = 3x^3 - 5x^2 + 9$ at $x = 2$.

In Exercises 6 and 7, write the function in the form $f(x) = (x - k)q(x) + r$ for the given value of k, and demonstrate that $f(k) = r$.

6. $f(x) = x^4 - 5x^2 + 4,$ $k = 1$
7. $f(x) = x^3 + 5x^2 - 2x - 24,$ $k = -3$

8. Simplify $\dfrac{2x^4 + 9x^3 - 32x^2 - 99x + 180}{x^2 + 2x - 15}$.

In Exercises 9–12, find the real zeros of the function.

9. $f(x) = -2x^3 - 7x^2 + 10x + 35$ **10.** $f(x) = 4x^4 - 37x^2 + 9$
11. $f(x) = 3x^4 + 4x^3 - 3x - 4$ **12.** $f(x) = 2x^3 - 3x^2 + 2x - 3$

13. The profit P (in dollars) for a clothing company is

$$P = -95x^3 + 5650x^2 - 250{,}000, \quad 0 \le x \le 55$$

where x is the advertising expense (in tens of thousands of dollars). What is the profit for an advertising expense of $450,000? Use a graphing utility to approximate another advertising expense that would yield the same profit.

14. The worldwide land area A (in millions of hectares) of genetically modified soybean crops from 1996 to 2003 is shown in table at the left.   (Source: Clive James)

(a) Use a graphing utility to create a scatter plot of the data. Let t represent the year, with $t = 6$ corresponding to 1996.

(b) Use the *regression* feature of a graphing utility to find a linear model, a quadratic model, a cubic model, and a quartic model for the data.

(c) Use a graphing utility to graph each model separately with the data in the same viewing window. How well does each model fit the data?

(d) Use each model to predict the area in 2005 and 2007. Discuss the reasonableness of each model's estimate.

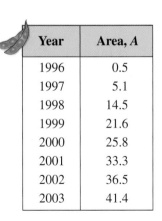

Year	Area, A
1996	0.5
1997	5.1
1998	14.5
1999	21.6
2000	25.8
2001	33.3
2002	36.5
2003	41.4

Table for 14

2.5   Complex Numbers

Objectives

- Perform operations with complex numbers and write the results in standard form.
- Find the complex conjugate of a complex number.
- Solve a polynomial equation.
- Plot a complex number in the complex plane.
- Determine whether a complex number is in the Mandelbrot Set.

The Imaginary Unit *i*

Some quadratic equations have no real solutions. For instance, the quadratic equations

$$x^2 + 1 = 0 \quad \text{and} \quad x^2 = -5 \qquad \text{Equations with no real solution}$$

have no real solution because there is no real number x that can be squared to produce a negative number. To overcome this deficiency, mathematicians created an expanded system of numbers using the **imaginary unit *i*,** defined as

$$i = \sqrt{-1} \qquad \text{Imaginary unit}$$

where $i^2 = -1$. By adding real numbers to real multiples of this imaginary unit, we obtain the set of **complex numbers.** Each complex number can be written in the **standard form $a + bi$.**

Definition of a Complex Number

If a and b are real numbers, the number $a + bi$ is a **complex number,** and it is said to be written in **standard form.** If $b = 0$, the number $a + bi = a$ is a real number. If $b \neq 0$, the number $a + bi$ is called an **imaginary number.** A number of the form bi, where $b \neq 0$, is called a **pure imaginary number.**

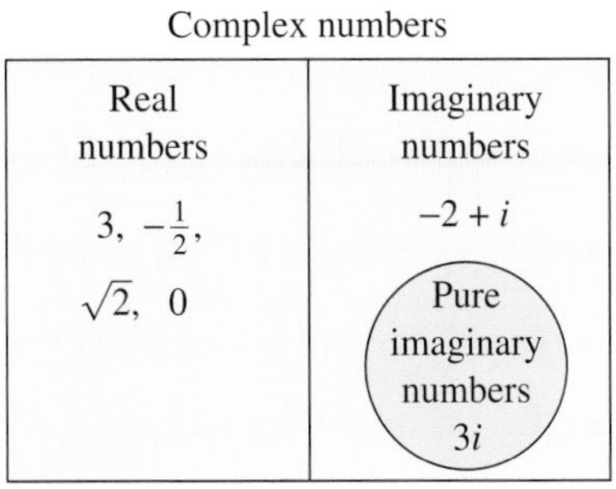

Complex numbers

Real numbers	Imaginary numbers
$3, -\frac{1}{2},$	$-2 + i$
$\sqrt{2}, \ 0$	Pure imaginary numbers $3i$

Figure 2.37

The set of real numbers is a subset of the set of complex numbers, as shown in Figure 2.37. This is true because every real number a can be written as a complex number using $b = 0$. That is, for every real number a, we can write $a = a + 0i$.

Equality of Complex Numbers

Two complex numbers $a + bi$ and $c + di$ written in standard form are **equal** to each other,

$$a + bi = c + di \qquad \text{Equality of two complex numbers}$$

if and only if $a = c$ and $b = d$.

Operations with Complex Numbers

To add (or subtract) two complex numbers, you add (or subtract) the real and imaginary parts of the numbers separately.

Addition and Subtraction of Complex Numbers

If $a + bi$ and $c + di$ are two complex numbers written in standard form, their sum and difference are defined as follows.

$$\text{Sum: } (a + bi) + (c + di) = (a + c) + (b + d)i$$
$$\text{Difference: } (a + bi) - (c + di) = (a - c) + (b - d)i$$

The **additive identity** in the complex number system is zero (the same as in the real number system). Furthermore, the **additive inverse** of the complex number $a + bi$ is

$$-(a + bi) = -a - bi. \qquad\qquad \text{Additive inverse}$$

So, you have

$$(a + bi) + (-a - bi) = 0 + 0i = 0.$$

EXAMPLE 1   ADDING AND SUBTRACTING COMPLEX NUMBERS

Perform the indicated operation(s) and write each result in standard form.

a. $(3 - i) + (2 + 3i)$ **b.** $2i + (-4 - 2i)$ **c.** $3 - (-2 + 3i) + (-5 + i)$

SOLUTION

a. $(3 - i) + (2 + 3i) = 3 - i + 2 + 3i$ Remove parentheses.

$\qquad\qquad\qquad\quad = 3 + 2 - i + 3i$ Group real and imaginary terms.

$\qquad\qquad\qquad\quad = (3 + 2) + (-1 + 3)i$

$\qquad\qquad\qquad\quad = 5 + 2i$ Standard form

b. $2i + (-4 - 2i) = 2i - 4 - 2i$ Remove parentheses.

$\qquad\qquad\qquad\quad = -4 + 2i - 2i$ Group real and imaginary terms.

$\qquad\qquad\qquad\quad = -4$ Standard form

c. $3 - (-2 + 3i) + (-5 + i) = 3 + 2 - 3i - 5 + i$

$\qquad\qquad\qquad\qquad\qquad = 3 + 2 - 5 - 3i + i$

$\qquad\qquad\qquad\qquad\qquad = 0 - 2i$

$\qquad\qquad\qquad\qquad\qquad = -2i$

✓CHECKPOINT   Now try Exercise 19.

Note in Example 1(b) that the sum of two imaginary numbers can be a real number.

Many of the properties of real numbers are valid for complex numbers as well. Here are some examples.

Associative Property of Addition and Multiplication

Commutative Property of Addition and Multiplication

Distributive Property of Multiplication Over Addition

Notice how these properties are used when two complex numbers are multiplied.

$$(a + bi)(c + di) = a(c + di) + bi(c + di) \qquad \text{Distributive}$$
$$= ac + (ad)i + (bc)i + (bd)i^2 \qquad \text{Distributive}$$
$$= ac + (ad)i + (bc)i + (bd)(-1) \qquad \text{Definition of } i$$
$$= ac - bd + (ad)i + (bc)i \qquad \text{Commutative}$$
$$= (ac - bd) + (ad + bc)i \qquad \text{Associative}$$

Rather than trying to memorize this multiplication rule, we suggest that you simply remember how the distributive property is used to multiply two complex numbers. The procedure is similar to multiplying two polynomials and combining like terms (as in the FOIL Method).

EXAMPLE 2   **MULTIPLYING COMPLEX NUMBERS**

Find each product.

a. $(i)(-3i)$ **b.** $(2 - i)(4 + 3i)$ **c.** $(3 + 2i)(3 - 2i)$ **d.** $(3 + 2i)^2$

SOLUTION

a. $(i)(-3i) = -3i^2$ Multiply.

$\qquad\qquad = -3(-1)$ $i^2 = -1$

$\qquad\qquad = 3$ Simplify.

b. $(2 - i)(4 + 3i) = 8 + 6i - 4i - 3i^2$ Product of binomials

$\qquad\qquad\qquad = 8 + 6i - 4i - 3(-1)$ $i^2 = -1$

$\qquad\qquad\qquad = 8 + 3 + 6i - 4i$ Collect terms.

$\qquad\qquad\qquad = 11 + 2i$ Standard form

c. $(3 + 2i)(3 - 2i) = 9 - 6i + 6i - 4i^2$ Product of binomials

$\qquad\qquad\qquad = 9 - 4(-1)$ $i^2 = -1$

$\qquad\qquad\qquad = 9 + 4$ Simplify.

$\qquad\qquad\qquad = 13$ Simplify.

d. $(3 + 2i)^2 = 9 + 6i + 6i + 4i^2$ Product of binomials

$\qquad\qquad\quad = 9 + 4(-1) + 12i$ $i^2 = -1$

$\qquad\qquad\quad = 9 - 4 + 12i$ Simplify.

$\qquad\qquad\quad = 5 + 12i$ Simplify.

✓*CHECKPOINT*   Now try Exercise 33.

Complex Conjugates

Notice in Example 2(c) that the product of two imaginary numbers can be a real number. This occurs with pairs of complex numbers of the form $a + bi$ and $a - bi$, called **complex conjugates.** In general, the product of two complex conjugates can be written as follows.

$$(a + bi)(a - bi) = a^2 - abi + abi - b^2i^2$$
$$= a^2 - b^2(-1) = a^2 + b^2$$

Complex conjugates can be used to write the quotient of $a + bi$ and $c + di$ in standard form, where c and d are not both zero. To do this, multiply the numerator and denominator by the conjugate of the denominator to obtain

$$\frac{a + bi}{c + di} = \frac{a + bi}{c + di}\left(\frac{c - di}{c - di}\right) = \frac{(ac + bd) + (bc - ad)i}{c^2 + d^2}.$$

TECHNOLOGY

Some graphing utilities can perform operations with complex numbers. For specific keystrokes, go to the text website at *college.hmco.com.*

> **EXAMPLE 3**   WRITING QUOTIENTS OF COMPLEX NUMBERS IN STANDARD FORM

Write each quotient in standard form.

a. $\dfrac{1}{1 + i}$ **b.** $\dfrac{2 + 3i}{4 - 2i}$

SOLUTION

a.

$$\frac{1}{1 + i} = \frac{1}{1 + i}\left(\frac{1 - i}{1 - i}\right) \qquad \text{Multiply numerator and denominator by complex conjugate of denominator.}$$

$$= \frac{1 - i}{1^2 - i^2} \qquad \text{Expand.}$$

$$= \frac{1 - i}{1 - (-1)} \qquad i^2 = -1$$

$$= \frac{1 - i}{2} \qquad \text{Simplify.}$$

$$= \frac{1}{2} - \frac{1}{2}i \qquad \text{Write in standard form.}$$

b.

$$\frac{2 + 3i}{4 - 2i} = \frac{2 + 3i}{4 - 2i}\left(\frac{4 + 2i}{4 + 2i}\right) \qquad \text{Multiply numerator and denominator by complex conjugate of denominator.}$$

$$= \frac{8 + 4i + 12i + 6i^2}{16 - 4i^2} \qquad \text{Expand.}$$

$$= \frac{8 - 6 + 16i}{16 + 4} \qquad i^2 = -1$$

$$= \frac{2 + 16i}{20} \qquad \text{Simplify.}$$

$$= \frac{1}{10} + \frac{4}{5}i \qquad \text{Write in standard form.}$$

✓CHECKPOINT   Now try Exercise 45.

Complex Solutions

When using the Quadratic Formula to solve a quadratic equation, you often obtain a result such as $\sqrt{-3}$, which you know is not a real number. By factoring out $i = \sqrt{-1}$, you can write this number in standard form.

$$\sqrt{-3} = \sqrt{3(-1)} = \sqrt{3}\sqrt{-1} = \sqrt{3}\,i$$

The number $\sqrt{3}\,i$ is called the *principal square root* of -3.

Principal Square Root of a Negative Number

If a is a positive number, the **principal square root** of the negative number $-a$ is defined as

$$\sqrt{-a} = \sqrt{a}\,i.$$

STUDY TIP

The definition of principal square root uses the rule

$$\sqrt{ab} = \sqrt{a}\sqrt{b}$$

for $a > 0$ and $b < 0$. This rule is not valid if *both* a and b are negative. For example,

$$\sqrt{(-5)(-5)} = \sqrt{25} = 5.$$

whereas

$$\sqrt{-5}\sqrt{-5} = 5i^2 = -5$$

To avoid problems with multiplying square roots of negative numbers, be sure to convert to standard form *before* multiplying.

| **EXAMPLE 4** | Writing Complex Numbers in Standard Form |

a. $\sqrt{-3}\sqrt{-12} = \sqrt{3}\,i\sqrt{12}\,i = \sqrt{36}\,i^2 = 6(-1) = -6$

b. $\sqrt{-48} - \sqrt{-27} = \sqrt{48}\,i - \sqrt{27}\,i = 4\sqrt{3}\,i - 3\sqrt{3}\,i = \sqrt{3}\,i$

c. $\left(-1 + \sqrt{-3}\right)^2 = \left(-1 + \sqrt{3}\,i\right)^2$

$$= (-1)^2 - 2\sqrt{3}\,i + \left(\sqrt{3}\right)^2 (i^2)$$

$$= 1 - 2\sqrt{3}\,i + 3(-1)$$

$$= -2 - 2\sqrt{3}\,i$$

✓**CHECKPOINT**   Now try Exercise 7.

| **EXAMPLE 5** | Complex Solutions of a Quadratic Equation |

Solve the equation $3x^2 - 2x + 5 = 0$.

SOLUTION

$$x = \frac{-(-2) \pm \sqrt{(-2)^2 - 4(3)(5)}}{2(3)} \qquad \text{Quadratic Formula}$$

$$= \frac{2 \pm \sqrt{-56}}{6} \qquad \text{Simplify.}$$

$$= \frac{2 \pm 2\sqrt{14}\,i}{6} \qquad \text{Write in } i\text{-form.}$$

$$= \frac{1}{3} \pm \frac{\sqrt{14}}{3}\,i \qquad \text{Standard form}$$

✓**CHECKPOINT**   Now try Exercise 59.

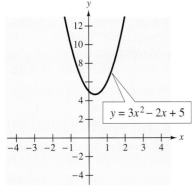

Figure 2.38

The graph of $f(x) = 3x^2 - 2x + 5$, shown in Figure 2.38, does not touch or cross the x-axis. This confirms that the equation in Example 5 has no real solutions.

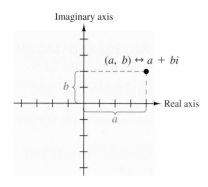

Figure 2.39

Applications

Most applications involving complex numbers are either theoretical (see the next section) or very technical, and so are not appropriate for inclusion in this text. However, to give you some idea of how complex numbers can be used in applications, we give a general description of their use in **fractal geometry.**

To begin, consider a coordinate system called the **complex plane.** Just as every real number corresponds to a point on the real line, every complex number corresponds to a point in the complex plane, as shown in Figure 2.39. In this figure, note that the vertical axis is the **imaginary axis** and the horizontal axis is the **real axis.** The point that corresponds to the complex number $a + bi$ is (a, b).

| Complex number $a + bi$ | ⟹ | Ordered pair (a, b) |

From Figure 2.39, you can see that i is called the imaginary unit because it is located one unit from the origin on the imaginary axis of the complex plane.

EXAMPLE 6   PLOTTING COMPLEX NUMBERS IN THE COMPLEX PLANE

Plot each complex number in the complex plane.

a. $2 + 3i$ **b.** $-1 + 2i$ **c.** 4

SOLUTION

a. To plot the complex number $2 + 3i$, move (from the origin) two units to the right on the real axis and then three units up, as shown in Figure 2.40. In other words, plotting the complex number $2 + 3i$ in the complex plane is comparable to plotting the point $(2, 3)$ in the Cartesian plane.

b. The complex number $-1 + 2i$ corresponds to the point $(-1, 2)$, as shown in Figure 2.40.

c. The complex number 4 corresponds to the point $(4, 0)$, as shown in Figure 2.40.

✓CHECKPOINT   Now try Exercise 71.

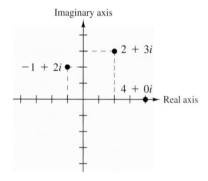

Figure 2.40

In the hands of a person who understands "fractal geometry," the complex plane can become an easel on which stunning pictures, called **fractals,** can be drawn. The most famous such picture is called the **Mandelbrot Set,** named after the Polish-born mathematician Benoit Mandelbrot. To draw the Mandelbrot Set, consider the following sequence of numbers.

$$c, \quad c^2 + c, \quad (c^2 + c)^2 + c, \quad [(c^2 + c)^2 + c]^2 + c, \quad \ldots$$

The behavior of this sequence depends on the value of the complex number c. For some values of c this sequence is **bounded,** which means that the absolute value of each number $\left(|a + bi| = \sqrt{a^2 + b^2}\right)$ in the sequence is less than some fixed number N. For other values it is **unbounded,** which means that the absolute values of the terms of the sequence become infinitely large. If the sequence is bounded, the complex number c is in the Mandelbrot Set, and if the sequence is unbounded, the complex number c is not in the Mandelbrot Set.

EXAMPLE 7 MEMBERS OF THE MANDELBROT SET

Make a Decision Decide which of the following complex numbers are members of the Mandelbrot Set.

a. -2 **b.** i **c.** $1 + i$

SOLUTION

a. For $c = -2$, the corresponding Mandelbrot sequence is

$$-2, \quad 2, \quad 2, \quad 2, \quad 2, \quad 2, \ldots$$

Because the sequence is bounded, the complex number -2 is in the Mandelbrot Set.

b. For $c = i$, the corresponding Mandelbrot sequence is

$$i, \quad -1 + i, \quad -i, \quad -1 + i, \quad -i, \quad -1 + i, \ldots$$

Because the sequence is bounded, the complex number i is in the Mandelbrot Set.

c. For $c = 1 + i$, the corresponding Mandelbrot sequence is

$$1 + i, \quad 1 + 3i, \quad -7 + 7i, \quad 1 - 97i, \quad -9407 - 193i,$$
$$88454401 + 3631103i, \ldots$$

Because the sequence is unbounded, the complex number $1 + i$ is *not* in the Mandelbrot Set.

✓CHECKPOINT   Now try Exercise 77.

With this definition, a picture of the Mandelbrot Set would have only two colors: one color for points that are in the set (the sequence is bounded) and one color for points that are outside the set (the sequence is unbounded). Figure 2.41 shows a black and yellow picture of the Mandelbrot Set. The points that are black are in the Mandelbrot Set and the points that are yellow are not.

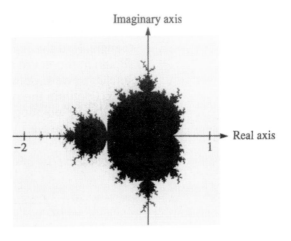

Figure 2.41 Mandelbrot Set

To add more interest to the picture, computer scientists discovered that the points that are not in the Mandelbrot Set can be assigned a variety of colors, depending on "how quickly" their sequences diverge. Figure 2.42 shows three different appendages of the Mandelbrot Set using a spectrum of colors. (The colored portions of the picture represent points that are *not* in the Mandelbrot Set.)

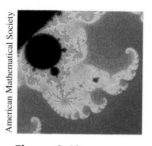 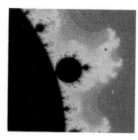 

American Mathematical Society

Figure 2.42

Figures 2.43, 2.44, and 2.45 show other types of fractal sets. From these pictures, you can see why fractals have fascinated people since their discovery (around 1980).

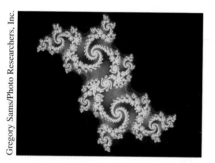  

Gregory Sams/Photo Researchers, Inc.

Fred Espenak/Photo Researchers, Inc.

Gregory Sams/Photo Researchers, Inc.

Figure 2.43   **Figure 2.44**   **Figure 2.45**

Discussing the Concept ┃ Building Your Own Fractal

On a large sheet of paper, use a ruler to draw a large equilateral triangle. Find the midpoint of each side. Connect the midpoints as shown in the figure on the left. Repeat the process for each of the smaller upward-pointing triangles, as shown in the figure on the right. Continue the process, always connecting the midpoints of the upward-pointing triangles. This fractal triangle is called the Sierpinski triangle.

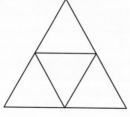 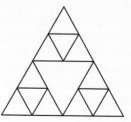

2.5 Warm Up

The following warm-up exercises involve skills that were covered in earlier sections. You will use these skills in the exercise set for this section. For additional help, review Sections R1.4 and R2.4.

In Exercises 1–8, simplify the expression.

1. $\sqrt{12}$

2. $\sqrt{500}$

3. $\sqrt{20} - \sqrt{5}$

4. $\sqrt{27} - \sqrt{243}$

5. $\sqrt{24}\sqrt{6}$

6. $2\sqrt{18}\sqrt{32}$

7. $\dfrac{1}{\sqrt{3}}$

8. $\dfrac{2}{\sqrt{2}}$

In Exercises 9 and 10, solve the quadratic equation.

9. $x^2 + x - 1 = 0$

10. $x^2 + 2x - 1 = 0$

2.5 Exercises

1. Write out the first 16 positive integer powers of i $(i, i^2, i^3, \ldots, i^{16})$, and write each as i, $-i$, 1, or -1. What pattern do you observe?

2. Use the pattern you found in Exercise 1 to help you write each power of i as i, $-i$, 1, or -1.
   (a) i^{28} (b) i^{37} (c) i^{127} (d) i^{82}

In Exercises 3–6, find the real numbers a and b such that the equation is true.

3. $a + bi = 7 + 12i$

4. $a + bi = -2 - 5i$

5. $(a + 3) + (b - 1)i = 7 - 4i$

6. $(a + 6) + 2bi = 6 - 5i$

In Exercises 7–18, write the complex number in standard form and find its complex conjugate.

7. $9 + \sqrt{-16}$

8. $2 + \sqrt{-25}$

9. $-3 - \sqrt{-12}$

10. $1 + \sqrt{-8}$

11. -21

12. 45

13. $-6i + i^2$

14. $4i^2 - 2i^3$

15. $-5i^5$

16. $(-i)^3$

17. $\left(\sqrt{-6}\right)^2 + 3$

18. $\left(\sqrt{-4}\right)^2 - 5$

In Exercises 19–44, perform the indicated operation and write the result in standard form.

19. $(-4 + 3i) + (6 - 2i)$

20. $(13 - 2i) + (-5 + 6i)$

21. $(12 + 5i) - (7 - i)$

22. $(3 + 2i) - (6 + 13i)$

23. $\left(-2 + \sqrt{-8}\right) + \left(5 - \sqrt{-50}\right)$

24. $\left(8 + \sqrt{-18}\right) - \left(4 + 3\sqrt{2}\,i\right)$

25. $-\left(\frac{3}{2} + \frac{5}{2}i\right) + \left(\frac{5}{3} + \frac{11}{3}i\right)$

26. $(1.6 + 3.2i) + (-5.8 + 4.3i)$

27. $(2 + 3i)^2 + (2 - 3i)^2$

28. $(1 - 2i)^2 - (1 + 2i)^2$

29. $\sqrt{-3} \cdot \sqrt{-8}$

30. $\sqrt{-5} \cdot \sqrt{-10}$

31. $\left(\sqrt{-10}\right)^2$

32. $\left(\sqrt{-75}\right)^3$

33. $(1 + i)(3 - 2i)$

34. $(9 - 4i)(1 - i)$

35. $(4 + 5i)(4 - 5i)$

36. $(6 + 7i)(6 - 7i)$

37. $7i(6 - 3i)$

38. $-8i(9 + 4i)$

39. $(4 + 5i)^2$

40. $(2 - 3i)^3$

41. $\left(\sqrt{5} - \sqrt{3}\,i\right)\left(\sqrt{5} + \sqrt{3}\,i\right)$

42. $\left(\sqrt{14} + \sqrt{10}\,i\right)\left(\sqrt{14} - \sqrt{10}\,i\right)$

43. $\left(2 - \sqrt{-8}\right)\left(8 + \sqrt{-6}\right)$

44. $\left(3 + \sqrt{-5}\right)\left(7 - \sqrt{-10}\right)$

In Exercises 45–56, write the quotient in standard form.

45. $\dfrac{2 + i}{2 - i}$

46. $\dfrac{3}{1 - i}$

47. $\dfrac{5}{4 - 2i}$

48. $\dfrac{8 - 7i}{1 - 2i}$

49. $\dfrac{6 - 7i}{i}$

50. $\dfrac{8 + 20i}{2i}$

51. $\dfrac{1}{(2i)^3}$

52. $\dfrac{1}{i^3}$

53. $\dfrac{5}{(1 + i)^3}$

54. $\dfrac{1}{(4 - 5i)^2}$

55. $\dfrac{(21 - 7i)(4 + 3i)}{2 - 5i}$

56. $\dfrac{(2 - 3i)(5i)}{2 + 3i}$

Error Analysis   In Exercises 57 and 58, a student has handed in the following problem. Find the error(s) and discuss how to explain the error(s) to the student.

57. Write $\dfrac{5}{3 - 2i}$ in standard form.

$$\dfrac{5}{3 - 2i} \cdot \dfrac{3 + 2i}{3 + 2i} \quad \dfrac{15 + 10i}{9 - 4} = 3 + 2i$$

58. Multiply $\left(\sqrt{-4} + 3\right)\left(i - \sqrt{-3}\right)$.

$$\left(\sqrt{-4} + 3\right)\left(i - \sqrt{-3}\right)$$
$$= i\sqrt{-4} - \sqrt{-4}\sqrt{-3} + 3i - 3\sqrt{-3}$$
$$= -2i - \sqrt{12} + 3i - 3i\sqrt{3}$$
$$= \left(1 - 3\sqrt{3}\right)i - 2\sqrt{3}$$

In Exercises 59–66, solve the equation.

59. $x^2 - 2x + 2 = 0$

60. $x^2 + 6x + 10 = 0$

61. $4x^2 + 16x + 17 = 0$

62. $9x^2 - 6x + 37 = 0$

63. $4x^2 + 16x + 15 = 0$

64. $9x^2 - 6x + 35 = 0$

65. $16t^2 - 4t + 3 = 0$

66. $5s^2 + 6s + 3 = 0$

In Exercises 67–70, solve the quadratic equation and then use a graphing utility to graph the related quadratic function in the standard viewing window. Discuss how the graph of the quadratic function relates to the solutions of the quadratic equation.

	Equation	*Function*
67.	$x^2 + x + 2 = 0$	$y = x^2 + x + 2$
68.	$-x^2 + 3x - 5 = 0$	$y = -x^2 + 3x - 5$
69.	$x^2 + 3x - 5 = 0$	$y = x^2 + 3x - 5$
70.	$-x^2 - 3x + 4 = 0$	$y = -x^2 - 3x + 4$

In Exercises 71–76, plot the complex number.

71. $-2 + i$

72. i

73. 3

74. $-2 - 3i$

75. $1 - 2i$

76. $-2i$

Make a Decision   In Exercises 77–82, decide whether the number is in the Mandelbrot Set. Explain your reasoning.

77. $c = 0$

78. $c = 2$

79. $c = 1$

80. $c = -1$

81. $c = \frac{1}{2}i$

82. $c = -i$

Math *MATTERS*

Acceptance of Imaginary Numbers

Imaginary numbers were given the name "imaginary" because many mathematicians had a difficult time accepting such numbers. The following excerpt, written by the famous mathematician Carl Gauss (1777–1855), shows that, even in the early 1800s, some people were hesitant to accept the use of imaginary numbers.

"Our general arithmetic, which far surpasses the extent of the geometry of the ancients, is entirely the creation of modern times. Starting with the notion of whole numbers, it has gradually enlarged its domain. To whole numbers have been added fractions; to rational numbers have been added irrational; to positive numbers have been added negative; and to real numbers have been added imaginary numbers.

This advance, however, has always been made with timid steps. The early algebraists called negative solutions of equations false solutions (and this is indeed the case when the problem to which they relate has been stated in such a way that negative solutions have no meaning). The reality of negative numbers is sufficiently justified since there are many cases where they have meaningful interpretations. This has long been admitted, but the imaginary numbers (formerly and occasionally now called impossible numbers) are still rather tolerated than fully accepted."

2.6   The Fundamental Theorem of Algebra

Objectives

- Use the Fundamental Theorem of Algebra and the Linear Factorization Theorem to write a polynomial as the product of linear factors.
- Find a polynomial with real coefficients whose zeros are given.
- Factor a polynomial over the rational, real, and complex numbers.
- Find all real and complex zeros of a polynomial function.

The Fundamental Theorem of Algebra

You have been using the fact that an nth-degree polynomial can have at most n real zeros. In the complex number system, this statement can be improved. That is, in the complex number system, every nth-degree polynomial function has *precisely* n zeros. This important result is derived from the **Fundamental Theorem of Algebra,** first proved by the famous German mathematician Carl Friedrich Gauss (1777–1855).

The Fundamental Theorem of Algebra

If $f(x)$ is a polynomial of degree n, where $n > 0$, then f has at least one zero in the complex number system.

Using the Fundamental Theorem of Algebra and the equivalence of zeros and factors, you obtain the following theorem.

Linear Factorization Theorem

If $f(x)$ is a polynomial of degree n

$$f(x) = a_n x^n + a_{n-1} x^{n-1} + \cdots + a_1 x + a_0$$

where $n > 0$, then f has precisely n linear factors

$$f(x) = a_n(x - c_1)(x - c_2) \cdots (x - c_n)$$

where $c_1, c_2, \ldots, c_n$ are complex numbers and a_n is the leading coefficient of $f(x)$.

Note that neither the Fundamental Theorem of Algebra nor the Linear Factorization Theorem tells you *how* to find the zeros or factors of a polynomial. Such theorems are called **existence theorems.** To find the zeros of a polynomial function, you still must rely on the techniques developed in the earlier parts of the text.

Remember that the n zeros of a polynomial function can be real or complex, and they may be repeated. Example 1 illustrates several cases.

| EXAMPLE 1 | ZEROS OF POLYNOMIAL FUNCTIONS |

Determine the number of zeros of each polynomial function.

a. $f(x) = x - 2$ **b.** $f(x) = x^2 - 6x + 9$

c. $f(x) = x^3 + 4x$ **d.** $f(x) = x^4 - 1$

SOLUTION

a. The first-degree polynomial $f(x) = x - 2$ has exactly *one* zero: $x = 2$.

b. Counting multiplicity, the second-degree polynomial function

$$f(x) = x^2 - 6x + 9 = (x - 3)(x - 3)$$

has exactly *two* zeros: $x = 3$ and $x = 3$.

c. The third-degree polynomial function

$$f(x) = x^3 + 4x = x(x - 2i)(x + 2i)$$

has exactly *three* zeros: $x = 0$, $x = 2i$, and $x = -2i$.

d. The fourth-degree polynomial function

$$f(x) = x^4 - 1 = (x - 1)(x + 1)(x - i)(x + i)$$

has exactly *four* zeros: $x = 1$, $x = -1$, $x = i$, and $x = -i$.

✓CHECKPOINT   Now try Exercise 5.

TECHNOLOGY

Remember that when you use a graphing utility to locate the zeros of a function, the only zeros that appear as x-intercepts are the *real zeros*. Compare the graphs below with the four polynomial functions in Example 1. Which zeros appear on the graphs?

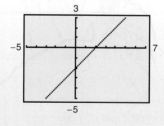

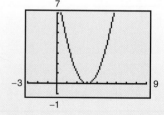

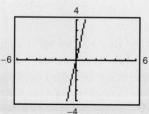

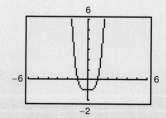

Example 2 shows how you can use the methods described in Sections 2.3 and 2.4 (the Rational Zero Test, synthetic division, and factoring) to find all the zeros of a polynomial function, including the complex zeros.

EXAMPLE 2 FINDING THE ZEROS OF A POLYNOMIAL FUNCTION

Write the polynomial function given by

$$f(x) = x^5 + x^3 + 2x^2 - 12x + 8$$

as the product of linear factors and list all of its zeros.

SOLUTION

From the Rational Zero Test, the possible rational zeros are $\pm 1, \pm 2, \pm 4,$ and ± 8. Synthetic division produces the following.

$$
\begin{array}{r|rrrrrr}
1 & 1 & 0 & 1 & 2 & -12 & 8 \\
  &   & 1 & 1 & 2 & 4 & -8 \\
\hline
  & 1 & 1 & 2 & 4 & -8 & 0
\end{array}
$$ ⟹ 1 is a zero.

$$
\begin{array}{r|rrrrr}
1 & 1 & 1 & 2 & 4 & -8 \\
  &   & 1 & 2 & 4 & 8 \\
\hline
  & 1 & 2 & 4 & 8 & 0
\end{array}
$$ ⟹ 1 is a repeated zero.

$$
\begin{array}{r|rrrr}
-2 & 1 & 2 & 4 & 8 \\
   &   & -2 & 0 & -8 \\
\hline
   & 1 & 0 & 4 & 0
\end{array}
$$ ⟹ −2 is a zero.

So, you have

$$f(x) = x^5 + x^3 + 2x^2 - 12x + 8$$
$$= (x - 1)(x - 1)(x + 2)(x^2 + 4).$$

By factoring $x^2 + 4$ as the difference of two squares over the imaginary numbers

$$x^2 - (-4) = \left(x - \sqrt{-4}\right)\left(x + \sqrt{-4}\right)$$
$$= (x - 2i)(x + 2i)$$

you obtain

$$f(x) = (x - 1)(x - 1)(x + 2)(x - 2i)(x + 2i)$$

which gives the following five zeros of f.

$$1, \quad 1, \quad -2, \quad 2i, \quad \text{and} \quad -2i$$

Note from the graph of f shown in Figure 2.46 that the *real* zeros are the only ones that appear as x-intercepts.

✓CHECKPOINT   Now try Exercise 25.

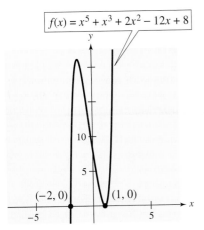

$f(x) = x^5 + x^3 + 2x^2 - 12x + 8$

$(-2, 0)$ $(1, 0)$

Figure 2.46

Conjugate Pairs

In Example 2, note that the two imaginary zeros are **conjugates.** That is, they are of the form $a + bi$ and $a - bi$.

Complex Zeros Occur in Conjugate Pairs

Let $f(x)$ be a polynomial function that has *real coefficients*. If $a + bi$, where $b \neq 0$, is a zero of the function, then the conjugate $a - bi$ is also a zero of the function.

Be sure you see that this result is true only if the polynomial function has *real coefficients*. For instance, the result applies to the function $f(x) = x^2 + 1$, but not to the function $g(x) = x - i$.

You have been using the Rational Zero Test, synthetic division, and factoring to find the zeros of polynomials. The Linear Factorization Theorem enables you to reverse this process and find a polynomial function when its zeros are given.

EXAMPLE 3 FINDING A POLYNOMIAL WITH GIVEN ZEROS

Find a *fourth-degree* polynomial function with real coefficients that has -1, -1, and $3i$ as zeros.

SOLUTION

Because $3i$ is a zero *and* the polynomial is stated to have real coefficients, you know that the conjugate $-3i$ must also be a zero. So, -1, -1, $3i$, and $-3i$ are the four zeros and from the Linear Factorization Theorem, $f(x)$ can be written as a product of linear factors, as shown.

$$f(x) = a(x + 1)(x + 1)(x - 3i)(x + 3i)$$

For simplicity, let $a = 1$. Then multiply the factors with real coefficients to get $(x^2 + 2x + 1)$ and multiply the complex conjugates to get $(x^2 + 9)$. So, you obtain the following fourth-degree polynomial function.

$$f(x) = (x^2 + 2x + 1)(x^2 + 9)$$
$$= x^4 + 2x^3 + 10x^2 + 18x + 9$$

✓CHECKPOINT Now try Exercise 35.

Factoring a Polynomial

The Linear Factorization Theorem shows that you can write any nth-degree polynomial as the product of n linear factors.

$$f(x) = a(x - c_1)(x - c_2)(x - c_3) \cdots (x - c_n)$$

However, this result includes the possibility that some of the values of c_i are complex. The following result implies that even if you do not want to get involved with "imaginary factors," you can still write $f(x)$ as the product of linear and/or quadratic factors.

Factors of a Polynomial

Every polynomial of degree $n > 0$ with real coefficients can be written as the product of linear and quadratic factors with real coefficients, where the quadratic factors have no real zeros.

A quadratic factor with no real zeros is said to be **irreducible over the reals.** Be sure you see that this is not the same as being *irreducible over the rationals.* For example, the quadratic

$$x^2 + 1 = (x - i)(x + i)$$

is irreducible over the reals (and therefore over the rationals). On the other hand, the quadratic

$$x^2 - 2 = \left(x - \sqrt{2}\right)\left(x + \sqrt{2}\right)$$

is irreducible over the rationals, but it is *reducible* over the reals.

EXAMPLE 4   FACTORING A POLYNOMIAL

Use the polynomial $f(x) = x^4 - x^2 - 20$ to complete the following.

a. Write the polynomial as the product of factors that are irreducible over the *rationals.*

b. Write the polynomial as the product of linear factors and quadratic factors that are irreducible over the *reals.*

c. Write the polynomial in completely factored form.

d. *Make a Decision* How many of the zeros are rational, irrational, or imaginary?

SOLUTION

a. Begin by factoring the polynomial into the product of two quadratic polynomials.

$$x^4 - x^2 - 20 = (x^2 - 5)(x^2 + 4)$$

Both of these factors are irreducible over the rationals.

b. By factoring over the reals, you have

$$x^4 - x^2 - 20 = \left(x + \sqrt{5}\right)\left(x - \sqrt{5}\right)(x^2 + 4)$$

where the quadratic factor is irreducible over the reals.

c. In completely factored form, you have

$$x^4 - x^2 - 20 = \left(x + \sqrt{5}\right)\left(x - \sqrt{5}\right)(x - 2i)(x + 2i).$$

d. Using the results from parts (a)–(c), you can conclude that there are no rational zeros, two irrational zeros $\left(\pm\sqrt{5}\right)$, and two imaginary zeros $(\pm 2i)$.

✓CHECKPOINT   Now try Exercise 39.

EXAMPLE 5   FINDING THE ZEROS OF A POLYNOMIAL FUNCTION

Find all the zeros of $f(x) = x^4 - 3x^3 + 6x^2 + 2x - 60$, given that $1 + 3i$ is a zero of f.

SOLUTION

Because imaginary zeros occur in conjugate pairs, you know that $1 - 3i$ is also a zero of f. This means that both

$$[x - (1 + 3i)] \quad \text{and} \quad [x - (1 - 3i)]$$

are factors of $f(x)$. Multiplying these two factors produces

$$[x - (1 + 3i)][x - (1 - 3i)] = [(x - 1) - 3i][(x - 1) + 3i]$$
$$= (x - 1)^2 - 9i^2$$
$$= x^2 - 2x + 10.$$

Using long division, you can divide $x^2 - 2x + 10$ into $f(x)$ to obtain the following.

$$
\begin{array}{r}
x^2 - x - 6 \\
x^2 - 2x + 10 \overline{)\ x^4 - 3x^3 + 6x^2 + 2x - 60} \\
\underline{x^4 - 2x^3 + 10x^2} \\
-x^3 - 4x^2 + 2x \\
\underline{-x^3 + 2x^2 - 10x} \\
-6x^2 + 12x - 60 \\
\underline{-6x^2 + 12x - 60} \\
0
\end{array}
$$

So, you have

$$f(x) = (x^2 - 2x + 10)(x^2 - x - 6)$$
$$= (x^2 - 2x + 10)(x - 3)(x + 2)$$

and you can conclude that the zeros of f are $1 + 3i$, $1 - 3i$, 3, and -2.

✔CHECKPOINT   Now try Exercise 43.

Discussing the Concept | Factoring a Polynomial

Compile a list of all the various techniques that can be used to find the zeros of a polynomial with real coefficients. Include the techniques that make use of a graphing utility. Give an example illustrating each technique and discuss when the use of each technique is appropriate.

2.6 ⟨ Warm Up

The following warm-up exercises involve skills that were covered in earlier sections. You will use these skills in the exercise set for this section. For additional help, review Section 2.5.

In Exercises 1–4, write the complex number in standard form and find its complex conjugate.

1. $4 - \sqrt{-29}$

2. $-5 - \sqrt{-144}$

3. $-1 + \sqrt{-32}$

4. $6 + \sqrt{-1/4}$

In Exercises 5–10, perform the indicated operation and write the result in standard form.

5. $(-3 + 6i) - (10 - 3i)$

6. $(12 - 4i) + 20i$

7. $(4 - 2i)(3 + 7i)$

8. $(2 - 5i)(2 + 5i)$

9. $\dfrac{1 + i}{1 - i}$

10. $(3 + 2i)^3$

2.6 ⟨ Exercises

In Exercises 1–28, find all the zeros of the function and write the polynomial as a product of linear factors.

1. $f(x) = x^2 + 16$

2. $f(x) = x^2 + 36$

3. $h(x) = x^2 - 5x + 5$

4. $g(x) = x^2 + 10x + 23$

5. $f(x) = x^4 - 256$

6. $f(t) = t^4 - 16$

7. $g(x) = x^3 + 5x$

8. $g(x) = x^3 + 7x$

9. $h(x) = x^3 - 11x^2 - 15x + 325$

10. $h(x) = x^3 - 3x^2 + 4x - 2$

11. $g(x) = x^3 - 6x^2 + 13x - 10$

12. $f(x) = x^3 - 2x^2 - 11x + 52$

13. $f(t) = t^3 - 3t^2 - 15t + 125$

14. $f(x) = x^3 + 8x^2 + 20x + 13$

15. $f(x) = x^3 + 24x^2 + 214x + 740$

16. $h(x) = x^3 - x + 6$

17. $h(x) = x^3 + 9x^2 + 27x + 35$

18. $f(s) = 2s^3 - 5s^2 + 12s - 5$

19. $f(x) = 16x^3 - 20x^2 - 4x + 15$

20. $f(x) = 9x^3 - 15x^2 + 11x - 5$

21. $f(x) = 5x^3 - 9x^2 + 28x + 6$

22. $g(x) = 3x^3 - 4x^2 + 8x + 8$

23. $g(x) = x^4 - 4x^3 + 8x^2 - 16x + 16$

24. $h(x) = x^4 + 6x^3 + 10x^2 + 6x + 9$

25. $f(x) = x^4 + 10x^2 + 9$

26. $f(x) = x^4 + 29x^2 + 100$

27. $f(t) = t^5 + 5t^4 - 7t^3 - 43t^2 - 8t - 48$

28. $g(x) = x^5 - 8x^4 + 28x^3 - 56x^2 + 64x - 32$

In Exercises 29–38, find a polynomial with real coefficients that has the given zeros. (There are many correct answers.)

29. $-3, 6i, -6i$

30. $3i, -3i$

31. $1, 2 + i, 2 - i$

32. $6, -5 + 2i, -5 - 2i$

33. $-4, 3i, -3i, 2i, -2i$

34. $2, 2, 2, 4i, -4i$

35. $-5, -5, 1 + \sqrt{3}i$

36. $0, 0, 4, 1 + i$

37. $\frac{2}{3}, -1, 3 + \sqrt{2}i$

38. $\frac{3}{4}, -2, -\frac{1}{2} + i$

In Exercises 39–42, write the polynomial (a) as the product of factors that are irreducible over the *rationals,* (b) as the product of linear and quadratic factors that are irreducible over the *reals,* and (c) in completely factored form.

39. $f(x) = x^4 - 7x^2 - 8$

40. $f(x) = x^4 - 4x^3 - 2x^2 + 32x - 48$
(*Hint:* One factor is $x^2 - 8$.)

41. $f(x) = x^4 - 4x^3 + 5x^2 - 2x - 6$
(*Hint:* One factor is $x^2 - 2x - 2$.)

42. $f(x) = x^4 - 3x^3 - x^2 - 12x - 20$
(*Hint:* One factor is $x^2 + 4$.)

In Exercises 43–52, use the given zero of f to find all the zeros of f.

43. $f(x) = 3x^3 - 5x^2 + 48x - 80, \quad 4i$

44. $f(x) = 2x^3 + 3x^2 + 8x + 12$, $2i$

45. $f(x) = x^4 - 2x^3 + 37x^2 - 72x + 36$, $6i$

46. $f(x) = x^3 - 7x^2 - x + 87$, $5 + 2i$

47. $f(x) = 4x^3 + 23x^2 + 34x - 10$, $-3 + i$

48. $f(x) = 3x^3 - 4x^2 + 8x + 8$, $1 - \sqrt{3}i$

49. $f(x) = x^4 + 3x^3 - 5x^2 - 21x + 22$, $-3 + \sqrt{2}i$

50. $f(x) = x^3 + 4x^2 + 14x + 20$, $-1 - 3i$

51. $f(x) = 8x^3 - 14x^2 + 18x - 9$, $\frac{1}{2}(1 - \sqrt{5}i)$

52. $f(x) = 25x^3 - 55x^2 - 54x - 18$, $\frac{1}{5}(-2 + \sqrt{2}i)$

53. *Make a Decision* Solve $x^4 - 5x^2 + 4 = 0$.
Describe the solutions you found. Use a graphing utility to graph $y = x^4 - 5x^2 + 4$. What is the connection between the solutions you found and the intercepts of the graph?

54. *Make a Decision* Solve $x^4 + 5x^2 + 4 = 0$.
Describe the solutions you found. Use a graphing utility to graph $y = x^4 + 5x^2 + 4$. What is the connection between the solutions you found and the intercepts of the graph?

55. Graphical Analysis Find a fourth-degree polynomial that has four real zeros. Find a fourth-degree polynomial that has two real zeros. Find a fourth-degree polynomial that has no real zeros. Graph the polynomials and describe the similarities and differences among them.

56. Graphical Analysis Find a sixth-degree polynomial that has six real zeros. Find a sixth-degree polynomial that has four real zeros. Find a sixth-degree polynomial that has two real zeros. Find a sixth-degree polynomial that has no real zeros. Graph the polynomials and describe the similarities and differences among them.

57. Profit The demand and cost equations for a television are given by

$$p = 140 - 0.0001x \quad \text{and} \quad C = 80x + 150{,}000$$

where p is the unit price (in dollars), C is the total cost (in dollars), and x is the number of units. The total profit P (in dollars) obtained by producing and selling x units is given by

$$P = R - C = xp - C.$$

Determine a price p that would yield a profit of $9 million. Use a graphing utility to explain why this is not possible.

58. Revenue The demand equation for a television is given by $p = 140 - 0.0001x$, where p is the unit price (in dollars) and x is the number of units sold. The total revenue R obtained by producing and selling x units is given by

$$R = xp.$$

Determine a price p that would yield a revenue of $50 million. Use a graphing utility to explain why this is not possible.

59. *Make a Decision* The imaginary number $2i$ is a zero of $f(x) = x^3 - 2ix^2 - 4x + 8i$, but the complex conjugate of $2i$ is not a zero of $f(x)$. Is this a contradiction of the conjugate pairs statement on page 164? Explain.

60. *Make a Decision* The imaginary number $1 - 2i$ is a zero of $f(x) = x^3 - (1 - 2i)x^2 - 9x + 9(1 - 2i)$, but $1 + 2i$ is not a zero of $f(x)$. Is this a contradiction of the conjugate pairs statement on page 164? Explain.

61. Reasoning Let f be a fourth-degree polynomial function with real coefficients. Three of the zeros of f are 3, $1 + i$, and $1 - i$. Explain how you know that the fourth zero must be a real number.

62. Reasoning Let f be a fourth-degree polynomial function with real coefficients. Three of the zeros of f are -1, 2, and $3 + 2i$. What is the fourth zero? Explain.

63. Reasoning Let f be a third-degree polynomial function with real coefficients. Explain how you know that f must have at least one zero that is a real number.

64. Reasoning Let f be a fifth-degree polynomial function with real coefficients. Explain how you know that f must have at least one zero that is a real number.

65. Think About It A student claims that a third-degree polynomial function with real coefficients can have three complex zeros. Describe how you could use a graphing utility and the Leading Coefficient Test (Section 2.2) to convince the student otherwise.

66. *Make a Decision* A student claims that the polynomial $f(x) = x^4 - 7x^2 + 12$ may be factored over the rational numbers as

$$f(x) = \left(x - \sqrt{3}\right)\left(x + \sqrt{3}\right)(x - 2)(x + 2).$$

Do you agree with this claim? Explain your answer.

2.7 Rational Functions

Objectives

- **Find the domain of a rational function.**
- **Find the vertical and horizontal asymptotes of the graph of a rational function.**
- **Sketch the graph of a rational function.**
- **Sketch the graph of a rational function that has a slant asymptote.**
- **Use a rational function model to solve an application problem.**

Introduction

A **rational function** is one that can be written in the form

$$f(x) = \frac{p(x)}{q(x)}$$

where $p(x)$ and $q(x)$ are polynomials and $q(x)$ is not the zero polynomial. In this section we assume that $p(x)$ and $q(x)$ have no common factors. Unlike polynomial functions, whose domains consist of all real numbers, rational functions often have restricted domains. In general, the *domain* of a rational function of x includes all real numbers except x-values that make the denominator zero.

EXAMPLE 1 FINDING THE DOMAIN OF A RATIONAL FUNCTION

Find the domain of $f(x) = \dfrac{1}{x}$ and discuss its behavior near any excluded x-values.

SOLUTION

The domain of f is all real numbers except $x = 0$. To determine the behavior of f near this x-value, evaluate $f(x)$ to the left and right of $x = 0$.

x approaches 0 from the left

x	−1	−0.5	−0.1	−0.01	−0.001	→ 0
f(x)	−1	−2	−10	−100	−1000	→ −∞

x approaches 0 from the right

x	0	←	0.001	0.01	0.1	0.5	1
f(x)	∞	←	1000	100	10	2	1

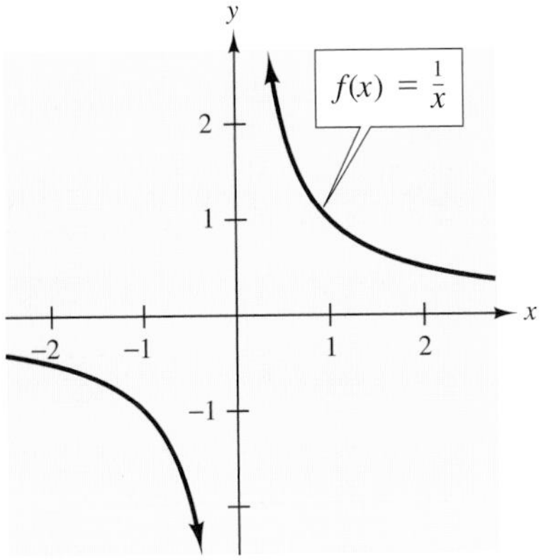

Note that as x approaches 0 *from the left*, $f(x)$ decreases without bound, whereas as x approaches 0 *from the right*, $f(x)$ increases without bound. The graph of f is shown in Figure 2.47.

✓CHECKPOINT   Now try Exercise 1(a).

Figure 2.47

The graph shows $f(x) = \dfrac{1}{x}$.

Horizontal and Vertical Asymptotes

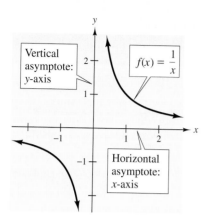

Figure 2.48

In Example 1, the behavior of $f(x) = 1/x$ near $x = 0$ is described by saying that the line $x = 0$ is a **vertical asymptote** of the graph of f, as shown in Figure 2.48. In this figure, note that the graph of f also has a **horizontal asymptote**—the line $y = 0$. This means that the values of $f(x) = 1/x$ approach zero as x increases or decreases without bound.

Definition of Vertical and Horizontal Asymptotes

1. The line $x = a$ is a **vertical asymptote** of the graph of f if

$$f(x) \to \infty \qquad \text{or} \qquad f(x) \to -\infty$$

as $x \to a$, either from the right or from the left.

2. The line $y = b$ is a **horizontal asymptote** of the graph of f if

$$f(x) \to b$$

as $x \to \infty$ or $x \to -\infty$.

The graph of a rational function can never intersect its vertical asymptote. It may or may not intersect its horizontal asymptote. In either case, the distance between the horizontal asymptote and the points on the graph must approach zero (as $x \to \infty$ or $x \to -\infty$). Figure 2.49 shows the horizontal and vertical asymptotes of the graphs of three rational functions.

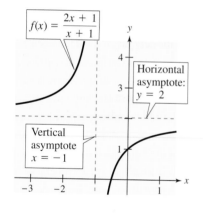

(a)

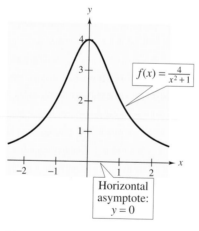

(b)

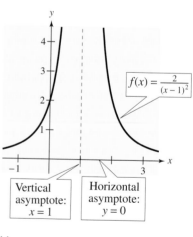

(c)

Figure 2.49

The graphs of

$$f(x) = \frac{1}{x} \quad \text{and} \quad f(x) = \frac{2x + 1}{x + 1} \qquad \text{See Figure 2.48 and Figure 2.49(a).}$$

are called **hyperbolas.**

Horizontal Asymptote of a Rational Function

Let f be the rational function given by

$$f(x) = \frac{a_n x^n + a_{n-1} x^{n-1} + \cdots + a_1 x + a_0}{b_m x^m + b_{m-1} x^{m-1} + \cdots + b_1 x + b_0}, \quad a_n \neq 0, b_m \neq 0.$$

1. If $n < m$, the x-axis is a horizontal asymptote.

2. If $n = m$, the line $y = a_n/b_m$ is a horizontal asymptote.

3. If $n > m$, there is no horizontal asymptote.

EXAMPLE 2   HORIZONTAL ASYMPTOTES OF RATIONAL FUNCTIONS

Tell whether each rational function has a horizontal asymptote. If it does, find the horizontal asymptote.

a. $f(x) = \dfrac{2x}{3x^2 + 1}$ **b.** $g(x) = \dfrac{2x^2}{3x^2 + 1}$ **c.** $h(x) = \dfrac{2x^3}{3x^2 + 1}$

SOLUTION

a. Because the degree of the numerator is *less than* the degree of the denominator, the graph of f has the x-axis as a horizontal asymptote, as shown in Figure 2.50(a).

b. Because the degree of the numerator is *equal to* the degree of the denominator, the graph of g has the line $y = \frac{2}{3}$ as a horizontal asymptote, as shown in Figure 2.50(b).

c. Because the degree of the numerator is *greater than* the degree of the denominator, the graph of h has no horizontal asymptote. See Figure 2.50(c).

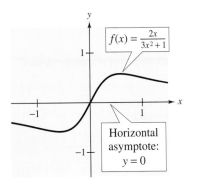

(a)

Figure 2.50

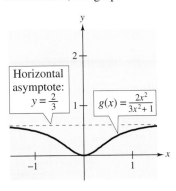

(b)

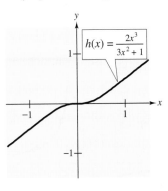

(c)

✓CHECKPOINT   Now try Exercise 1(b).

Sketching the Graph of a Rational Function

DISCOVERY

Consider the rational function

$$f(x) = \frac{x^2 - 4}{x - 2}.$$

Is $x = 2$ in the domain of $f(x)$? Graph $f(x)$ on a graphing utility. Is there a vertical asymptote at $x = 2$? Describe the graph of $f(x)$. Factor the numerator and reduce the rational function. Describe the resulting function. Under what conditions will a rational function have no vertical asymptote?

Guidelines for Graphing Rational Functions

Let $f(x) = p(x)/q(x)$, where $p(x)$ and $q(x)$ are polynomials with no common factors.

1. Find and plot the y-intercept (if any) by evaluating $f(0)$.

2. Find the zeros of the numerator (if any) by solving the equation $p(x) = 0$. Then plot the corresponding x-intercepts.

3. Find the zeros of the denominator (if any) by solving the equation $q(x) = 0$. Then sketch the corresponding vertical asymptotes.

4. Find and sketch the horizontal asymptote (if any) by using the rule for finding the horizontal asymptote of a rational function.

5. Test for symmetry.

6. Plot at least one point both *between and beyond* each x-intercept and vertical asymptote.

7. Use smooth curves to complete the graph between and beyond the vertical asymptotes.

Testing for symmetry can be useful, especially for simple rational functions. For example, the graph of $f(x) = 1/x$ is symmetric with respect to the origin, and the graph of $g(x) = 1/x^2$ is symmetric with respect to the y-axis.

EXAMPLE 3   SKETCHING THE GRAPH OF A RATIONAL FUNCTION

Sketch the graph of $g(x) = \dfrac{3}{x - 2}$.

SOLUTION

Begin by noting that the numerator and denominator have no common factors.

y-Intercept:	$\left(0, -\frac{3}{2}\right)$, because $g(0) = -\frac{3}{2}$
x-Intercept:	None, numerator has no zeros.
Vertical asymptote:	$x = 2$, zero of denominator
Horizontal asymptote:	$y = 0$, degree of $p(x) <$ degree of $q(x)$
Additional points:	

x	-4	1	2	3	5
$g(x)$	-0.5	-3	Undefined	3	1

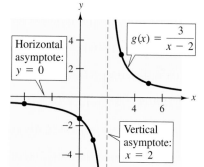

Figure 2.51

By plotting the intercepts, asymptotes, and a few additional points, you can obtain the graph shown in Figure 2.51. In the figure, note that the graph of g is a vertical stretch and a right shift of the graph of $y = 1/x$.

✓CHECKPOINT   Now try Exercise 31.

Note that in the examples in this section, the vertical asymptotes are included in the table of additional points. This is done to emphasize numerically the behavior of the graph of the funciton.

EXAMPLE 4 SKETCHING THE GRAPH OF A RATIONAL FUNCTION

Sketch the graph of $f(x) = \dfrac{x}{x^2 - x - 2}$.

SOLUTION

Factor the denominator to determine more easily the zeros of the denominator.

$$f(x) = \frac{x}{x^2 - x - 2} = \frac{x}{(x + 1)(x - 2)}$$

y-Intercept: $(0, 0)$, because $f(0) = 0$

x-Intercept: $(0, 0)$

Vertical asymptotes: $x = -1, x = 2$, zeros of denominator

Horizontal asymptote: $y = 0$, degree of $p(x) <$ degree of $q(x)$

Additional points:

x	-3	-1	-0.5	1	2	3
f(x)	-0.3	Undefined	0.4	-0.5	Undefined	0.75

The graph is shown in Figure 2.52. Confirm the graph with your graphing utility.

✓CHECKPOINT   Now try Exercise 53.

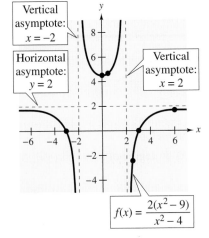

Vertical asymptote: $x = 2$

Vertical asymptote: $x = -1$

Horizontal asymptote: $y = 0$

$f(x) = \dfrac{x}{x^2 - x - 2}$

Figure 2.52

EXAMPLE 5 SKETCHING THE GRAPH OF A RATIONAL FUNCTION

Sketch the graph of $f(x) = \dfrac{2(x^2 - 9)}{x^2 - 4}$.

SOLUTION

By factoring the numerator and denominator, you have

$$f(x) = \frac{2(x^2 - 9)}{x^2 - 4} = \frac{2(x - 3)(x + 3)}{(x - 2)(x + 2)}.$$

y-Intercept: $\left(0, \frac{9}{2}\right)$, because $f(0) = \frac{9}{2}$

x-Intercepts: $(-3, 0)$ and $(3, 0)$

Vertical asymptotes: $x = -2, x = 2$, zeros of denominator

Horizontal asymptote: $y = 2$, degree of $p(x) =$ degree of $q(x)$

Symmetry: With respect to y-axis, because $f(-x) = f(x)$

Additional points:

x	-2	0.5	2	2.5	6
f(x)	Undefined	4.67	Undefined	-2.44	1.6875

The graph is shown in Figure 2.53.

✓CHECKPOINT   Now try Exercise 47.

Vertical asymptote: $x = -2$

Horizontal asymptote: $y = 2$

Vertical asymptote: $x = 2$

$f(x) = \dfrac{2(x^2 - 9)}{x^2 - 4}$

Figure 2.53

Slant Asymptotes

Consider a rational function whose denominator is of degree 1 or greater. If the degree of the numerator is exactly *one more* than the degree of the denominator, the graph of the function has a **slant** (or **oblique**) **asymptote.** For example, the graph of

$$f(x) = \frac{x^2 - x}{x + 1}$$

has a slant asymptote, as shown in Figure 2.54. To find the equation of a slant asymptote, use long division. For instance, by dividing $x + 1$ into $x^2 - x$, you have

$$f(x) = \frac{x^2 - x}{x + 1} = \underbrace{x - 2}_{} + \frac{2}{x + 1}.$$

Slant asymptote
$(y = x - 2)$

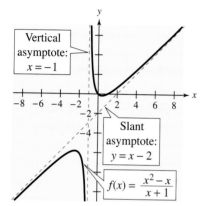

Vertical asymptote: $x = -1$

Slant asymptote: $y = x - 2$

$f(x) = \dfrac{x^2 - x}{x + 1}$

Figure 2.54

As x increases or decreases without bound, the remainder term $2/(x + 1)$ approaches 0, so the graph of f approaches the line $y = x - 2$, as shown in Figure 2.54.

EXAMPLE 6   A RATIONAL FUNCTION WITH A SLANT ASYMPTOTE

Sketch the graph of $f(x) = \dfrac{x^2 - x - 2}{x - 1}$.

SOLUTION

First write $f(x)$ in two different ways. Factoring the numerator enables you to recognize the x-intercepts.

$$f(x) = \frac{x^2 - x - 2}{x - 1} = \frac{(x - 2)(x + 1)}{x - 1}$$

Then long division enables you to recognize that the line $y = x$ is a slant asymptote of the graph.

$$f(x) = \frac{x^2 - x - 2}{x - 1} = x - \frac{2}{x - 1}$$

y-Intercept: $(0, 2)$, because $f(0) = 2$

x-Intercepts: $(-1, 0)$ and $(2, 0)$

Vertical asymptote: $x = 1$, zero of denominator

Horizontal asymptote: None; degree of $p(x) >$ degree of $q(x)$

Slant asymptote: $y = x$

Additional points:

x	-2	0.5	1	1.5	3
f(x)	$-1.\overline{3}$	4.5	Undefined	-2.5	2

Slant asymptote: $y = x$

Vertical asymptote: $x = 1$

$f(x) = \dfrac{x^2 - x - 2}{x - 1}$

Figure 2.55

The graph is shown in Figure 2.55.

✓**CHECKPOINT**   Now try Exercise 11.

Applications

There are many examples of asymptotic behavior in business and biology. For instance, the following two examples describe the asymptotic behavior related to the cost of removing smokestack emissions and the average cost of producing a product.

| EXAMPLE 7 | COST-BENEFIT MODEL |

Make a Decision A utility company burns coal to generate electricity. The cost of removing a certain *percent* of the pollutants from the stack emission is typically not a linear function. That is, if it costs C dollars to remove 25% of the pollutants, it would cost more than $2C$ dollars to remove 50% of the pollutants. As the percent of removed pollutants approaches 100%, the cost tends to become prohibitive. The cost C of removing p percent of the smokestack pollutants is given by

$$C = \frac{80{,}000p}{100 - p}.$$

Suppose that you are a member of a state legislature that is considering a law that will require utility companies to remove 90% of the pollutants from their smokestack emissions. The current law requires 85% removal.

a. How much additional expense is the new law asking the utility company to incur?

b. According to the model, would it be possible to remove 100% of the pollutants?

SOLUTION

a. The graph of this function is shown in Figure 2.56. Note that the graph has a vertical asymptote at $p = 100$. Because the current law requires 85% removal, the current cost to the utility company is

$$C = \frac{80{,}000(85)}{100 - 85} \qquad \text{Substitute 85 for } p.$$

$$\approx \$453{,}333. \qquad \text{Use a calculator.}$$

If the new law increases the percent removal to 90%, the cost to the utility company will be

$$C = \frac{80{,}000(90)}{100 - 90} \qquad \text{Substitute 90 for } p.$$

$$= \$720{,}000. \qquad \text{Use a calculator.}$$

The new law would require the utility company to spend about

$$\$720{,}000 - \$453{,}333 = \$266{,}667$$

more.

b. From Figure 2.56, you can see that the graph has a vertical asymptote at $p = 100$. Because the graph of a rational function can never intersect its vertical asymptote, you can conclude that it is not possible for the company to remove 100% of the pollutants from the stack emissions.

✔CHECKPOINT   Now try Exercise 61.

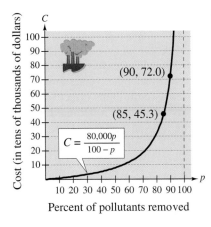

Figure 2.56

EXAMPLE 8   AVERAGE COST OF PRODUCING A PRODUCT

A plastics company has a cost of $C = 0.5x + 5000$, where C is measured in dollars and x is the number of units produced. The *average cost per unit* is given by

$$\overline{C} = \frac{C}{x} = \frac{0.5x + 5000}{x}.$$

Find the average cost per unit when $x = 1000$, $10,000$, and $100,000$. What is the horizontal asymptote for this function, and what does it represent?

SOLUTION

When $x = 1000$, the average cost per unit is

$$\overline{C} = \frac{0.5(1000) + 5000}{1000} = \$5.50.$$

When $x = 10,000$, the average cost per unit is

$$\overline{C} = \frac{0.5(10,000) + 5000}{10,000} = \$1.00.$$

When $x = 100,000$, the average cost per unit is

$$\overline{C} = \frac{0.5(100,000) + 5000}{100,000} = \$0.55.$$

As shown in Figure 2.57, the horizontal asymptote is $\overline{C} = 0.50$. So, as the number of units produced increases without bound, the average production cost per unit approaches $0.50. This points out one of the problems of a small business—it is difficult to have competitively low prices with low levels of production.

✓*CHECKPOINT*   Now try Exercise 65.

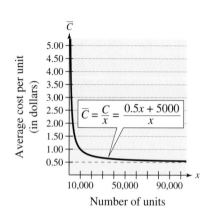

Figure 2.57 As $x \to \infty$, the average cost per unit approaches $0.50.

EXAMPLE 9   PER CAPITA LAND AREA

A model for the population P (in millions) of the United States from 1960 to 2002 is $P = 2.4604t + 179.853$, where t represents the calendar year, with $t = 0$ corresponding to 1960. A model for the land area A (in millions of acres) of the United States from 1960 to 2002 is $A = 2263.961$. Construct a rational function for per capita land area L. Sketch a graph of the rational function. If the model continues to be accurate for the years beyond 2002, what will be the per capita land area in 2010?   (Source: U.S. Census Bureau)

SOLUTION

The rational function for the per capita land area L is

$$L = \frac{A}{P} = \frac{2263.961}{2.4604t + 179.853}.$$

The graph of the function is shown in Figure 2.58. To find the per capita land area in 2010, substitute $t = 50$ into L.

$$L = \frac{2263.961}{2.4604t + 179.853} = \frac{2263.961}{2.4604(50) + 179.853} = \frac{2263.961}{302.873} \approx 7.47$$

The per capita land area will be approximately 7.5 acres per person in 2010.

✓*CHECKPOINT*   Now try Exercise 71.

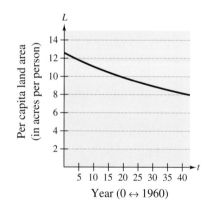

Figure 2.58

2.7 (Warm Up

The following warm-up exercises involve skills that were covered in earlier sections. You will use these skills in the exercise set for this section. For additional help, review Sections R1.6 and 1.2.

In Exercises 1–6, factor the polynomial.

1. $x^2 - 4x$

2. $2x^3 - 6x$

3. $x^2 - 3x - 10$

4. $x^2 - 7x + 10$

5. $x^3 + 4x^2 + 3x$

6. $x^3 - 4x^2 - 2x + 8$

In Exercises 7–10, sketch the graph of the equation.

7. $y = 2$

8. $x = -1$

9. $y = x - 2$

10. $y = -x + 1$

2.7 (Exercises

In Exercises 1–8, find (a) the domain and (b) any horizontal asymptotes of the function.

1. $f(x) = \dfrac{3}{(x + 4)^2}$

2. $f(x) = \dfrac{1}{(x - 2)^2}$

3. $f(x) = \dfrac{x - 7}{5 - x}$

4. $f(x) = \dfrac{1 - 5x}{1 + 2x}$

5. $f(x) = \dfrac{3x^2 + 1}{x^2 + 9}$

6. $f(x) = \dfrac{3x^2 + x - 5}{x^2 + 1}$

7. $f(x) = \dfrac{5x^4}{x^2 + 1}$

8. $f(x) = \dfrac{2x}{x + 1}$

In Exercises 9–12, find any (a) vertical, (b) horizontal, and (c) slant asymptotes of the graph of the function. Then sketch the graph of f.

9. $f(x) = \dfrac{x^2 - 7x + 12}{x - 3}$

10. $f(x) = \dfrac{x + 3}{x^2 - 9}$

11. $f(x) = \dfrac{x^2}{x + 1}$

12. $f(x) = \dfrac{x^3 + x}{x^2 - 1}$

In Exercises 13–18, match the function with its graph. [The graphs are labeled (a), (b), (c), (d), (e), and (f).]

(a)

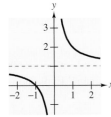

(b)

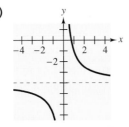

(c)

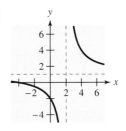

(d)

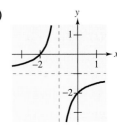

(e)

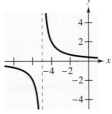

(f)

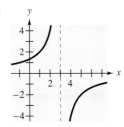

13. $f(x) = -\dfrac{4}{x - 3}$

14. $f(x) = \dfrac{2}{x + 5}$

15. $f(x) = \dfrac{x + 1}{x}$

16. $f(x) = \dfrac{3 - 4x}{x}$

17. $f(x) = \dfrac{x + 4}{x - 2}$

18. $f(x) = -\dfrac{x + 2}{x + 1}$

In Exercises 19–22, compare the graph of $f(x) = 1/x$ with the graph of g.

19. $g(x) = f(x) - 2 = \dfrac{1}{x} - 2$

20. $g(x) = f(x + 1) = \dfrac{1}{x + 1}$

21. $g(x) = -f(x) = -\dfrac{1}{x}$

22. $g(x) = -f(x + 1) = -\dfrac{1}{x + 1}$

In Exercises 23–26, compare the graph of $f(x) = 4/x^2$ with the graph of g.

23. $g(x) = f(x) + 3 = \dfrac{4}{x^2} + 3$

24. $g(x) = f(x - 1) = \dfrac{4}{(x - 1)^2}$

25. $g(x) = -f(x) = -\dfrac{4}{x^2}$ **26.** $g(x) = \dfrac{1}{8}f(x) = \dfrac{1}{2x^2}$

In Exercises 27–30, compare the graph of $f(x) = 8/x^3$ with the graph of g.

27. $g(x) = f(x) + 5 = \dfrac{8}{x^3} + 5$

28. $g(x) = f(x - 3) = \dfrac{8}{(x - 3)^3}$

29. $g(x) = -f(x) = -\dfrac{8}{x^3}$ **30.** $g(x) = \dfrac{1}{4}f(x) = \dfrac{2}{x^3}$

In Exercises 31–54, sketch the graph of the rational function. As sketching aids, check for intercepts, symmetry, vertical asymptotes, and horizontal asymptotes.

31. $f(x) = \dfrac{1}{x - 4}$ **32.** $f(x) = \dfrac{1}{x + 6}$

33. $f(x) = \dfrac{-1}{x + 1}$ **34.** $f(x) = \dfrac{-1}{2 - x}$

35. $f(x) = \dfrac{x + 4}{x - 5}$ **36.** $f(x) = \dfrac{x - 2}{x - 3}$

37. $f(x) = \dfrac{2 + x}{1 - x}$ **38.** $f(x) = \dfrac{3 - x}{2 - x}$

39. $f(t) = \dfrac{3t + 1}{t}$ **40.** $f(t) = \dfrac{1 - 2t}{t}$

41. $C(x) = \dfrac{5 + 2x}{1 + x}$ **42.** $P(x) = \dfrac{1 - 3x}{1 - x}$

43. $g(x) = \dfrac{1}{x + 2} + 2$ **44.** $h(x) = \dfrac{1}{x - 3} + 1$

45. $f(x) = \dfrac{1}{x^2} + 2$ **46.** $f(x) = 2 - \dfrac{3}{x^2}$

47. $h(x) = \dfrac{x^2}{x^2 - 9}$ **48.** $h(t) = \dfrac{4}{t^2 + 1}$

49. $g(s) = \dfrac{s}{s^2 + 1}$ **50.** $g(x) = \dfrac{x}{x^2 - 9}$

51. $f(x) = \dfrac{x}{x^2 - 3x - 4}$

52. $f(x) = \dfrac{-x}{x^2 + x - 6}$

53. $f(x) = \dfrac{3x}{x^2 - x - 2}$

54. $f(x) = \dfrac{2x}{x^2 + x - 2}$

In Exercises 55–58, find a counterexample to show that the statement is incorrect.

55. Every rational function has a vertical asymptote.

56. Every rational function has at least one asymptote.

57. A rational function can have only one vertical asymptote.

58. The graph of a rational function with a slant asymptote cannot cross its slant asymptote.

59. Is it possible for a rational function to have all three types of asymptotes (vertical, horizontal, and slant)? Why or why not?

60. Is it possible for a rational function to have more than one horizontal asymptote? Why or why not?

61. *Make a Decision*: **Seizure of Illegal Drugs** The cost C (in millions of dollars) for the federal government to seize p percent of an illegal drug as it enters the country is

$$C = \dfrac{528p}{100 - p}, \quad 0 \le p < 100.$$

(a) Find the cost of seizing 25%, 50%, and 75% of the drug.

(b) According to this model, would it be possible to seize 100% of the drug? Explain.

62. *Make a Decision*: **Water Pollution** The cost C (in millions of dollars) of removing p percent of the industrial and municipal pollutants discharged into a river is

$$C = \dfrac{255p}{100 - p}, \quad 0 \le p < 100.$$

(a) Find the cost of removing 15%, 50%, and 80% of the pollutants.

(b) According to the model, would it be possible to remove 100% of the pollutants? Explain.

63. Population of Deer The Game Commission introduces 50 deer into newly acquired state game lands. The population N of the herd is given by

$$N = \frac{10(5 + 3t)}{1 + 0.04t}, \quad t \geq 0$$

where t is time (in years).

(a) Find the population when t is 5, 10, and 25.

(b) What is the limiting size of the herd as time progresses?

64. Population of Elk The Game Commission introduces 40 elk into newly acquired state game lands. The population N of the herd is given by

$$N = \frac{10(4 + 2t)}{1 + 0.03t}, \quad t \geq 0$$

where t is time (in years).

(a) Find the population when t is 5, 10, and 25.

(b) What is the limiting size of the herd as time progresses?

65. Average Cost The cost C (in dollars) of producing x tennis balls is

$$C = 150{,}000 + 0.25x.$$

The average cost $\overline{C}$ per tennis ball is

$$\overline{C} = \frac{C}{x} = \frac{150{,}000 + 0.25x}{x}, \quad x > 0.$$

(a) Sketch the graph of the average cost function.

(b) Find the average cost of producing 1000, 10,000 and 100,000 tennis balls. What can you conclude?

66. Average Cost The cost C (in dollars) of producing x basketballs is $C = 250{,}000 + 3x$. The average cost $\overline{C}$ per basketball is

$$\overline{C} = \frac{C}{x} = \frac{250{,}000 + 3x}{x}, \quad x > 0.$$

(a) Sketch the graph of the average cost function.

(b) Find the average cost of producing 1000, 10,000, and 100,000 basketballs. What can you conclude?

67. Human Memory Model Psychologists have developed mathematical models to predict memory performance as a function of the number of trials n of a certain task. Consider the learning curve modeled by

$$P = \frac{0.5 + 0.9(n - 1)}{1 + 0.9(n - 1)}, \quad n > 0$$

where P is the percent of correct responses (in decimal form) after n trials.

(a) Complete the table.

n	1	2	3	4	5	6	7	8	9	10
P										

(b) According to this model, what is the limiting percent of correct responses as n increases?

68. Human Memory Model Consider the learning curve modeled by

$$P = \frac{0.6 + 0.85(n - 1)}{1 + 0.85(n - 1)}, \quad n > 0$$

where P is the percent of correct responses (in decimal form) after n trials.

(a) Complete the table.

n	1	2	3	4	5	6	7	8	9	10
P										

(b) According to this model, what is the limiting percent of correct responses as n increases?

69. Average Recycling Cost The cost C (in dollars) of recycling a waste product is

$$C = 350{,}000 + 5x, \quad x > 0$$

where x is the number of pounds of waste. The average recycling cost $\overline{C}$ per pound is

$$\overline{C} = \frac{C}{x} = \frac{350{,}000 + 5x}{x}, \quad x > 0.$$

(a) Use a graphing utility to graph $\overline{C}$.

(b) Find the average cost of recycling 1000, 10,000, and 100,000 pounds of waste. What can you conclude?

70. *Make a Decision*: **Drug Concentration** The concentration C of a medication in the bloodstream t minutes after sublingual (under the tongue) application is given by

$$C(t) = \frac{3t - 1}{2t^2 + 5}, \quad t > 0.$$

(a) Use a graphing utility to graph the function. Estimate when the concentration is greatest.

(b) Does this function have a horizontal asymptote? If so, discuss the meaning of the asymptote in terms of the concentration of the medication.

IN THE NEWS

Health care spending up to $1.6 trillion in 2002

WASHINGTON—Health care spending in the United States surged to $1.6 trillion in 2002—about $5,440 for every American—and outpaced growth in the rest of the economy for a fourth straight year.

Hospital spending and prescription drug costs fueled the 9.3 percent increase over 2001, the federal Centers for Medicare and Medicaid Services said yesterday.

"This continued acceleration injects pressure into the health care system, and everyone—from businesses, to government, to consumers—is affected," Katharine Levit, a CMS official and the lead author of the report, said.

Early indications, however, are that growth in spending slowed in 2003, according to the report, published in the journal Health Affairs. . . .

Associated Press, "Health care spending up to $1.6 trillion in 2002," Honolulu Star Bulletin, January 8, 2004.

71. Health care spending H (in millions of dollars) in the United States from 1990 to 2002 can be modeled by

$$H = 2384.62t^2 + 36,846.2t + 728,154$$

where t represents the year, with $t = 0$ corresponding to 1990. The population P (in millions) of the United States from 1990 to 2002 can be modeled by

$$P = 3.1981t + 250.452$$

where t represents the year, with $t = 0$ corresponding to 1990. (Source: U.S. Centers for Medicare and Medicaid Services, U.S. Department of Commerce, U.S. Census Bureau)

(a) Create a rational function S for per capita health care spending.

(b) Use a graphing utility to graph the rational function S.

(c) If the model continues to be accurate beyond 2002, what will be the per capita health care spending in 2007?

72. *Make a Decision*: **Full-Length Cassette Sales** The total sales S (in millions of dollars) of full-length cassettes in the United States from 1990 to 2003 can be approximated by the model

$$S = \frac{3957.54 - 293.68t}{1 - 0.101t + 0.011t^2}, \quad 0 \le t \le 13$$

where t represents the year, with $t = 0$ corresponding to 1990. (Source: Recording Industry Association of America)

(a) From the figure, notice the drop in sales of full-length cassettes after about 1992. How would you account for this drop?

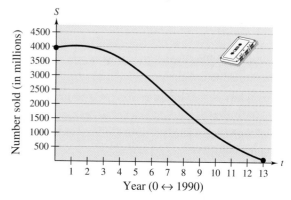

(b) Does the model have a horizontal asymptote? If so, does the function cross its horizontal asymptote?

73. *Make a Decision*: **100-Meter Freestyle** The winning times for the men's 100-meter freestyle swim at the Olympics from 1904 to 2000 can be approximated by the quadratic model

$$y = 1.09 - 0.003t + 0.000004t^2, \quad 4 \le t \le 100$$

where y is the winning time (in minutes) and t represents the year, with $t = 4$ corresponding to 1904. (Source: *The World Almanac and Book of Facts 2003*)

(a) Use a graphing utility to graph the model.

(b) Use the model to predict the winning time for 2008.

(c) Does this model have a horizontal asymptote? Do you think that a model for this type of data should have a horizontal asymptote?

Make a Decision ▸ Project: Renewable Energy

Year	Solar energy consumption, s
1990	60
1991	63
1992	64
1993	66
1994	69
1995	70
1996	71
1997	70
1998	70
1999	69
2000	66
2001	65
2002	64

Year	Wind energy consumption, w
1998	31
1999	46
2000	57
2001	68
2002	106

Solar energy uses photovoltaic cells that convert sunlight directly into electricity. These photovoltaic cells once were used only in space and in solar powered watches and calculators. The energy that wind contains is converted to electricity as air flows past a wind turbine rotor. One-third of the United States has enough wind to generate electricity.

The tables at the left show the solar energy consumption s (in trillion Btu) in the United States from 1990 to 2002 and the wind energy consumption w (in trillion Btu) in the United States from 1998 to 2002. Use this information to investigate the following questions. (Source: Energy Information Administration)

1. **Inspecting Data** Study the tables at the left. What pattern do you see in each set of data?

2. *Make a Decision* Use a graphing utility to create a scatter plot of each data set. How do the scatter plots compare to your answer to Question 1? Do the points appear to resemble a quadratic pattern or a cubic pattern? Do you expect the model to have a negative or a positive leading coefficient?

3. **Fitting a Model to Data** The standard form of a quadratic equation is $y = a(x - h)^2 + k$, where (h, k) is the vertex. Use the vertex (h, k) and any other ordered pair (x, y) to determine a value for a in the quadratic equation for solar energy consumption. Write the quadratic model in the general form $y = ax^2 + bx + c$.

4. **Fitting a Model to Data** Use the *regression* feature of a graphing utility to find another quadratic model for the solar energy consumption data. Write the quadratic model in the general form $y = ax^2 + bx + c$. Then find a cubic model for the wind energy consumption data.

5. **Finding the Vertex of a Model** Algebraically find the vertex of the quadratic model for solar energy consumption found in Question 4. Then use the *trace* feature or the *maximum* feature of a graphing utility to find the vertex. Does the vertex of the model found in Question 4 agree with the actual data given in the table?

6. **Comparing Actual Data with Models** Use the *table* feature of a graphing utility to evaluate the solar energy consumption values that both models give for each year. How do the two models' values compare to the actual values from the table?

7. *Make a Decision* Of the models that you found in Questions 3 and 4, select the one that appears to best fit the actual data. Use a graphing utility to graph the equation with the original data.

8. **Graphing Models** Use a graphing utility to graph the models you found in Question 4 in the same viewing window. Use the *trace* feature of the graphing utility to approximate the year in which wind energy consumption surpassed solar energy consumption.

9. **Research** Use your school's library, the Internet, or some other reference source to research the advantages and disadvantages of using renewable energy.

Chapter Summary

After studying this chapter, you should have acquired the following skills. These skills are keyed to the Review Exercises that begin on page 183. Answers to odd-numbered Review Exercises are given in the back of the book.

2.1
- Sketch the graph of a quadratic function and identify its vertex and intercepts.   *Review Exercises 1–4*
- Find a quadratic function given its vertex and a point on its graph.   *Review Exercises 5, 6*
- Construct and use a quadratic model to solve an application problem.   *Review Exercises 7–12*

2.2
- Determine end behavior of graphs of polynomial functions.   *Review Exercises 13–16*
- Find the real zeros of a polynomial function.   *Review Exercises 17–20*

2.3
- Divide one polynomial by a second polynomial using long division.   *Review Exercises 21, 22*
- Simplify a rational expression using long division.   *Review Exercises 23, 24*
- Use synthetic division to divide two polynomials.   *Review Exercises 25, 26, 31, 32*
- Use the Remainder Theorem and synthetic division to evaluate a polynomial.   *Review Exercises 27, 28*
- Use the Factor Theorem to factor a polynomial function.   *Review Exercises 29, 30*

2.4
- Find all possible rational zeros of a function using the Rational Zero Test.   *Review Exercises 33, 34*
- Find all real zeros of a function.   *Review Exercises 35–40*
- Approximate the real zeros of a polynomial function using the Intermediate Value Theorem.   *Review Exercises 41, 42, 46*
- Approximate the real zeros of a polynomial function using a graphing utility.   *Review Exercises 43, 44*
- Apply techniques for approximating real zeros to solve an application problem.   *Review Exercise 45*

2.5
- Find the complex conjugate of a complex number.   *Review Exercises 47–50*
- Perform operations with complex numbers and write the results in standard form.   *Review Exercises 51–66*
- Solve a polynomial equation.   *Review Exercises 67–70*
- Plot a complex number in the complex plane.   *Review Exercises 71, 72*

2.6
- Use the Fundamental Theorem of Algebra and the Linear Factorization Theorem to write a polynomial as the product of linear factors.   *Review Exercises 73–78*
- Find a polynomial with real coefficients whose zeros are given.   *Review Exercises 79, 80*
- Factor a polynomial over the rational, real, and complex numbers.   *Review Exercises 81, 82*
- Find all real and complex zeros of a polynomial function.   *Review Exercises 83–86*

2.7
- Find the domain of a rational function.   *Review Exercises 87–90*
- Find the vertical and horizontal asymptotes of the graph of a rational function.   *Review Exercises 87–90*
- Sketch the graph of a rational function, including graphs with slant asymptotes.   *Review Exercises 91–96*
- Use a rational function model to solve an application problem.   *Review Exercises 97–101*

Review Exercises

In Exercises 1–4, sketch the graph of the quadratic function. Identify the vertex and intercepts.

1. $f(x) = (x + 3)^2 - 5$ 2. $g(x) = -x^2 + 6x + 8$
3. $h(x) = 3x^2 - 12x + 11$ 4. $f(x) = \frac{1}{3}(x^2 + 5x - 4)$

In Exercises 5 and 6, find the quadratic function that has the indicated vertex and whose graph passes through the given point.

5. Vertex: $(-5, -1)$; point: $(-2, 6)$
6. Vertex: $(2, 5)$; point: $(4, 7)$

7. **Optimal Area** The perimeter of a rectangle is 200 feet. Let x represent the width of the rectangle and write a quadratic function for the area of the rectangle in terms of its width. Of all possible rectangles with perimeters of 200 feet, what are the measurements of the one with the greatest area?

8. **Optimal Revenue** Find the number of units that produces a maximum revenue R (in dollars) for

$R = 800x - 0.01x^2$

where x is the number of units produced.

9. **Optimal Cost** A toy manufacturer has daily production costs C (in dollars) of

$C = 20,000 - 40x + 0.055x^2$

where x is the number of units produced.

(a) Use a graphing utility to graph the cost function.
(b) Graphically estimate the number of units that should be produced to yield a minimum cost.
(c) Explain how to confirm the result of part (b) algebraically.

10. **Optimal Profit** The profit P (in dollars) for an electronics company is given by

$P = -0.0001x^2 + 130x - 300,000$

where x is the number of units produced.

(a) Use a graphing utility to graph the profit function.
(b) Graphically estimate the number of units that should be produced to yield a maximum profit.
(c) Explain how to confirm the result of part (b) algebraically.

11. **Regression Problem** Let x be the angle (in degrees) at which a baseball is hit with a 30-hertz backspin at an initial speed of 40 meters per second and let $d(x)$ be the distance (in meters) the ball travels. The table shows the distances traveled for the different angles at which the ball is hit. (Source: *The Physics of Sports*)

x	10	15	30	36	42	43
$d(x)$	61.2	83.0	130.4	139.4	143.2	143.3

x	44	45	48	54	60
$d(x)$	142.8	142.7	140.7	132.8	119.7

(a) Use a graphing utility to create a scatter plot of the data.
(b) Use the *regression* feature of a graphing utility to find a quadratic model for $d(x)$.
(c) Use a graphing utility to graph your model for $d(x)$ with the scatter plot of the data.
(d) Use the model to approximate the angle that will result in a maximum distance.

12. **Regression Problem** Let $p(t)$ be the percentage of software piracy (unauthorized copying, distributing, or downloading of copyrighted software) worldwide, where t represents the year, with $t = 4$ corresponding to 1994. The table shows the percentage for each year from 1994 to 2001. (Source: Business Software Alliance)

Year, t	4	5	6	7	8	9	10	11
Percentage, $p(t)$	49	46	43	40	38	36	37	40

(a) Use a graphing utility to create a scatter plot of the data.
(b) Use the *regression* feature of a graphing utility to find a quadratic model for $p(t)$.
(c) Use a graphing utility to graph your model for $p(t)$ with the scatter plot of the data.
(d) Use the model to approximate the year in which the percentage of piracy was at a minimum. How does this result compare with the actual data?

In Exercises 13–16, describe the right-hand and left-hand behavior of the graph of the polynomial function.

13. $f(x) = \frac{1}{2}x^3 + 2x$

14. $f(x) = 5 - 3x^2 + 4x^4 - x^6$

15. $f(x) = -x^5 + 3x^2 + 11$

16. $f(x) = \frac{3}{4}(x^4 + 3x^2 + 2)$

In Exercises 17–20, find all real zeros of the function.

17. $f(x) = 16 - x^2$

18. $f(x) = x^4 - 6x^2 + 8$

19. $f(x) = x^3 - 7x^2 + 10x$

20. $f(x) = x^3 - 6x^2 - 3x + 18$

In Exercises 21 and 22, use long division to divide.

Dividend	Divisor
21. $2x^3 - 5x^2 - x$	$2x + 1$
22. $x^4 - 5x^3 + 10x^2 - 12$	$x^2 - 2x + 4$

In Exercises 23 and 24, simplify the rational expression.

23. $\dfrac{x^3 + 9x^2 + 2x - 48}{x - 2}$

24. $\dfrac{x^4 + 5x^3 - 20x - 16}{x^2 - 4}$

In Exercises 25 and 26, use synthetic division to divide.

Dividend	Divisor
25. $x^3 - 6x + 9$	$x + 3$
26. $x^5 - x^4 + x^3 - 13x^2 + x + 6$	$x - 2$

In Exercises 27 and 28, use synthetic division to find each function value.

27. $f(x) = 6 + 2x^2 - 3x^3$ (a) $f(2)$ (b) $f(-1)$

28. $f(x) = 2x^4 + 3x^3 + 6$ (a) $f\left(\frac{1}{2}\right)$ (b) $f(-1)$

In Exercises 29 and 30, (a) verify the given factors of the function f, (b) find the remaining factors of f, (c) use your results to write the complete factorization of f, (d) list all real zeros of f, and (e) confirm your results by using a graphing utility to graph the function.

Function	Factors
29. $f(x) = x^3 - 4x^2 - 11x + 30$	$(x - 5), (x + 3)$
30. $f(x) = 3x^3 + 23x^2 + 37x - 15$	$(3x - 1), (x + 5)$

31. Dimensions of a Room   A rectangular room has a volume of

$$x^3 + 13x^2 + 50x + 56$$

cubic feet. The height of the room is $x + 2$ feet. Find the number of square feet of floor space in the room.

32. Profit   The profit P (in dollars) for selling a novel is given by

$$P = -130x^3 + 6500x^2 - 200,000, \quad 0 \le x \le 50$$

where x is the advertising expense (in tens of thousands of dollars). For this novel, the advertising expense was \$400,000 ($x = 40$), and the profit was \$1,880,000.

(a) Use a graphing utility to graph the function and use the result to find another advertising expense that would have produced the same profit.

(b) Use synthetic division with $x = 40$ to confirm algebraically the result of part (a).

In Exercises 33 and 34, use the Rational Zero Test to list all possible rational zeros of f. Then use a graphing utility to graph the function. Use the graph to help determine which of the possible rational zeros are actual zeros of the function.

33. $f(x) = -4x^3 + 8x^2 - 3x + 15$

34. $f(x) = 3x^4 + 4x^3 - 5x^2 + 10x - 8$

In Exercises 35–40, find all real zeros of the function.

35. $f(x) = x^3 + 2x^2 - 5x - 6$

36. $g(x) = 2x^3 - 15x^2 + 24x + 16$

37. $h(x) = 3x^4 - 27x^2 + 60$

38. $f(x) = x^5 - 4x^3 + 3x$

39. $C(x) = 3x^4 + 3x^3 - 7x^2 - x + 2$

40. $p(x) = x^4 - x^3 - 2x - 4$

In Exercises 41 and 42, use the Intermediate Value Theorem to approximate the zero of f in the interval $[a, b]$. Give your approximation to the nearest tenth.

41. $f(x) = x^3 - 4x + 3, \quad [-3, -2]$

42. $f(x) = x^5 + 5x^2 + x - 1, \quad [0, 1]$

In Exercises 43 and 44, use a graphing utility to approximate the real zeros of f. Give your approximations to the nearest thousandth.

43. $f(x) = 5x^3 - 11x - 3$

44. $f(x) = 2x^4 - 9x^3 - 5x^2 + 10x + 12$

45. Advertising Costs A company that manufactures motorcycles estimates that the profit P (in dollars) from selling the top-of-the-line model is given by

$$P = -35x^3 + 2000x^2 - 27,500, \quad 10 \le x \le 55$$

where x is the advertising expense (in tens of thousands of dollars). Using this model, find the smaller of two advertising amounts that will yield a profit of $538,000.

46. *Make a Decision*: Genetically Modified Crops The worldwide land area A (in millions of hectares) of genetically modified crops from 1996 to 2003 is shown in the table.   (Source: Clive James)

Year	Area, A	Year	Area, A
1996	1.7	2000	44.2
1997	11.0	2001	52.6
1998	27.8	2002	58.7
1999	39.9	2003	67.7

(a) Use a graphing utility to create a scatter plot of the data. Let t represent the year, with $t = 6$ corresponding to 1996.

(b) Use the *regression* feature of a graphing utility to find a quadratic model, a cubic model, and a quartic model for the data.

(c) Use a graphing utility to graph each model separately with the data in the same viewing window. How well does each model fit the data?

(d) Use the graph of the cubic model and the *trace* feature of the graphing utility to predict the year during which the land area of genetically modified crops will reach 95 million hectares.

(e) Use each model to estimate the area in 2005 and 2007. Does each estimate seem reasonable?

In Exercises 47–50, write the complex number in standard form and find its complex conjugate.

47. $\sqrt{-32}$ **48.** 12

49. $-3 + \sqrt{-16}$ **50.** $2 - \sqrt{-18}$

In Exercises 51–62, perform the indicated operation and write the result in standard form.

51. $(7 - 4i) + (-2 + 5i)$

52. $(14 + 6i) - (-1 - 2i)$

53. $\left(1 + \sqrt{-12}\right)\left(5 - \sqrt{-3}\right)$

54. $\left(3 - \sqrt{-4}\right)\left(4 - \sqrt{-49}\right)$

55. $(5 + 8i)(5 - 8i)$ **56.** $\left(\frac{1}{2} + \frac{3}{4}i\right)\left(\frac{1}{2} - \frac{3}{4}i\right)$

57. $-2i(4 - 5i)$ **58.** $-3(-2 + 4i)$

59. $(3 + 4i)^2$ **60.** $(2 - 5i)^2$

61. $(3 + 2i)^2 + (3 - 2i)^2$ **62.** $(1 + i)^2 - (1 - i)^2$

In Exercises 63–66, write the quotient in standard form.

63. $\dfrac{8 - i}{2 + i}$ **64.** $\dfrac{3 - 4i}{1 - 5i}$

65. $\dfrac{4 - 3i}{i}$ **66.** $\dfrac{2}{(1 + i)^2}$

In Exercises 67–70, solve the equation.

67. $2x^2 - x + 3 = 0$ **68.** $3x^2 + 6x + 11 = 0$

69. $4x^2 + 11x + 3 = 0$ **70.** $9x^2 - 2x + 5 = 0$

In Exercises 71 and 72, plot the complex number.

71. $-3 + 2i$ **72.** $-1 - 4i$

In Exercises 73–78, find all the zeros of the function and write the polynomial as a product of linear factors.

73. $f(x) = x^4 - 81$

74. $h(x) = 2x^3 - 5x^2 + 4x - 10$

75. $f(t) = t^3 + 5t^2 + 3t + 15$

76. $h(x) = x^4 + 17x^2 + 16$

77. $g(x) = 4x^3 - 8x^2 + 9x - 18$

78. $f(x) = x^5 - 2x^4 + x^3 - x^2 + 2x - 1$

In Exercises 79 and 80, find a polynomial with real coefficients that has the given zeros. (There are many correct answers.)

79. $3, 4i, -4i$ **80.** $2, -3, 1 - 2i, 1 + 2i$

In Exercises 81 and 82, write the polynomial (a) as the product of factors that are irreducible over the *rationals*, (b) as the product of linear and quadratic factors that are irreducible over the *reals*, and (c) in completely factored form.

81. $f(x) = x^4 + 5x^2 - 24$

82. $f(x) = x^4 - 2x^3 - 2x^2 - 14x - 63$

   (*Hint:* One factor is $x^2 + 7$.)

In Exercises 83–86, use the given zero of f to find all the zeros of f.

83. $f(x) = 4x^3 - x^2 + 64x - 16$, $-4i$

84. $f(x) = 50 - 75x + 2x^2 - 3x^3$, $5i$

85. $f(x) = x^4 + 7x^3 + 24x^2 + 58x + 40$, $-1 + 3i$

86. $f(x) = x^4 + 4x^3 + 8x^2 + 4x + 7$, $-2 - \sqrt{3}i$

In Exercises 87–90, find the domain of the function and identify any horizontal or vertical asymptotes.

87. $f(x) = \dfrac{5}{x - 6}$

88. $f(x) = \dfrac{2x^2 + 5x - 3}{x^2 + 2}$

89. $f(x) = \dfrac{x^2}{x^2 - 4}$

90. $f(x) = \dfrac{2x}{x^2 - 2x - 8}$

In Exercises 91–94, sketch the graph of the rational function. As sketching aids, check for intercepts, symmetry, vertical asymptotes, and horizontal asymptotes.

91. $P(x) = \dfrac{3 - x}{x + 2}$

92. $f(x) = \dfrac{4}{(x - 1)^2}$

93. $g(x) = \dfrac{1}{x^2 - 4} + 2$

94. $h(x) = \dfrac{-3x}{2x^2 + 3x - 5}$

In Exercises 95 and 96, find all possible asymptotes (vertical, horizontal, and/or slant) of the given function. Sketch the graph of f.

95. $f(x) = \dfrac{x^3}{x^2 - 1}$

96. $f(x) = \dfrac{x^2 - 4}{x + 2}$

97. Average Cost   The cost C (in dollars) of producing x lawn chairs is $C = 100,000 + 0.9x$. The average cost $\overline{C}$ per lawn chair is

$$\overline{C} = \frac{C}{x} = \frac{100,000 + 0.9x}{x}, \quad x > 0.$$

(a) Sketch the graph of the average cost function.

(b) Find the average cost of producing 1000, 10,000, and 100,000 lawn chairs. What can you conclude?

98. Average Recycling Cost   The cost C (in dollars) of recycling a waste product is

$$C = 450,000 + 6x, \quad x > 0$$

where x is the number of pounds of waste. The average recycling cost $\overline{C}$ per pound is

$$\overline{C} = \frac{C}{x} = \frac{450,000 + 6x}{x}, \quad x > 0.$$

(a) Sketch the graph of $\overline{C}$.

(b) Find the average cost of recycling 1000, 10,000, and 100,000 pounds of waste. What can you conclude?

99. Population of Fish   The Wildlife Commission introduces 50,000 game fish into a large lake. The population N (in thousands) of the fish is

$$N = \frac{20(4 + 3t)}{1 + 0.05t}, \quad t \geq 0$$

where t is time (in years).

(a) Find the population when $t = 5$, 10, and 25.

(b) What is the limiting number of fish in the lake as time progresses?

100. Human Memory Model   Consider the learning curve modeled by

$$P = \frac{0.5 + 0.9(n - 1)}{1 + 0.9(n - 1)}, \quad n > 0$$

where P is the percent of correct responses (in decimal form) after n trials.

(a) Complete the table.

n	1	2	3	4	5	6	7	8	9	10
P										

(b) According to this model, what is the limiting percent of correct responses as n increases?

101. *Make a Decision*: Smokestack Emission   The cost C (in dollars) of removing p percent of the air pollutants in the stack emission of a utility company that burns coal to generate electricity is

$$C = \frac{95,000p}{100 - p}, \quad 0 \leq p < 100.$$

(a) Find the cost of removing 25% of the pollutants.

(b) Find the cost of removing 60% of the pollutants.

(c) Find the cost of removing 99% of the pollutants.

(d) According to the model, would it be possible to remove 100% of the pollutants? Explain.

Chapter Test

Take this test as you would take a test in class. After you are done, check your work against the answers given in the back of the book.

1. Sketch the graph of the quadratic function given by $f(x) = -(x + 1)^2 + 2$. Identify the vertex and intercepts.

2. Describe the right-hand and left-hand behavior of the graph of f.
   (a) $f(x) = 12x^3 - 5x^2 - 49x + 15$
   (b) $f(x) = 5x^4 - 3x^3 + 2x^2 + 11x + 12$

3. Simplify $\dfrac{x^4 + 4x^3 - 19x^2 - 106x - 120}{x^2 - 3x - 10}$.

4. List all the possible rational zeros of $f(x) = 2x^3 - 3x^2 - 29x + 60$. Use synthetic division to show that $x = \frac{5}{2}$ is a zero of f. Using this result, completely factor the polynomial.

In Exercises 5–8, perform the indicated operation and write the result in standard form.

5. $(12 + 3i) + (4 - 6i)$

6. $(10 - 2i) - (3 + 7i)$

7. $\left(5 + \sqrt{-12}\right)\left(3 - \sqrt{-12}\right)$

8. $(4 + 3i)(2 - 5i)$

9. Write the quotient in standard form: $\dfrac{1 + i}{1 - i}$.

In Exercises 10 and 11, solve the quadratic equation.

10. $x^2 + 5x + 7 = 0$

11. $2x^2 - 5x + 11 = 0$

12. Find a polynomial with real coefficients that has 2, 5, $3i$, and $-3i$ as zeros.

13. Find all the zeros of $f(x) = x^3 + 2x^2 + 5x + 10$, given that $\sqrt{5}i$ is a zero.

14. Sketch the graph of $f(x) = \dfrac{2x}{x + 1}$. Label any intercepts and asymptotes. What is the domain of f?

15. An interior decorating company estimates that the profit P (in dollars) from selling its product is

$$P = -11x^3 + 900x^2 - 50{,}000, \quad x \geq 0$$

where x is the advertising expense (in tens of thousands of dollars). How much should the company spend on advertising to obtain a profit of \$222,000? Explain your reasoning.

16. The average cost $\overline{C}$ (in dollars) of recycling x pounds of a waste product is

$$\overline{C}(x) = \frac{450{,}000 + 5x}{x}, \quad x > 0.$$

Find the average cost of recycling 10,000, 100,000, and 1,000,000 pounds of waste. What can you conclude?

Many people consult an interior decorator for help with determining the best color, fabric, and accessories for their home.

3 Exponential and Logarithmic Functions

Chapter 3

Chapter Sections

3.1 Exponential Functions

3.2 Logarithmic Functions

3.3 Properties of Logarithms

3.4 Solving Exponential and Logarithmic Equations

3.5 Exponential and Logarithmic Models

Make a Decision ► How Quickly Will Your Hot Water Cool?

When a warm object is placed in a cool room, its temperature will gradually change to the temperature of the room. (The same is true for a cool object placed in a warm room.)

According to Newton's Law of Cooling, the rate at which the object's temperature T changes is proportional to the difference between its temperature and the temperature of the room L. This statement can be translated as

$$T = L + Ce^{kt}$$

where k is a constant and $L + C$ is the original temperature of the object.

The table and the scatter plot at the right show the temperature T (in degrees Fahrenheit) for several times t (in minutes) for 300 milliliters of hot water in a ceramic cup.

Make a Decision Will the temperature of the water ever actually reach the temperature of the room?

Time, t	Temperature, T
0	182
2	180
8	160
10	156
21	138
24	134
28	128

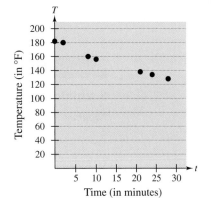

To explore this real-world application further, go to the Make a Decision Project on page 244.

3.1 Exponential Functions

Objectives

- **Evaluate an exponential expression.**
- **Sketch the graph of an exponential function.**
- **Evaluate and sketch the graph of the natural exponential function.**
- **Use the compound interest formula.**
- **Use an exponential model to solve an application problem.**

Exponential Functions

So far, this text has dealt only with **algebraic functions,** which include polynomial functions and rational functions. In this chapter, you will study two types of nonalgebraic functions—*exponential* functions and *logarithmic* functions. These functions are examples of **transcendental functions.**

Definition of Exponential Function

The **exponential function** f **with base** a is denoted by

$$f(x) = a^x$$

where $a > 0$, $a \neq 1$, and x is any real number.

The base $a = 1$ is excluded because it yields $f(x) = 1^x = 1$. This is a constant function, not an exponential function.

You already know how to evaluate a^x for integer and rational values of x. For example, you know that $4^3 = 64$ and $4^{1/2} = 2$. However, to evaluate 4^x for any real number x, you need to interpret forms with *irrational* exponents. For the purposes of this text, it is sufficient to think of

$$a^{\sqrt{2}} \quad \left(\text{where } \sqrt{2} \approx 1.414214\right)$$

as that value having the successively closer approximations

$$a^{1.4}, a^{1.41}, a^{1.414}, a^{1.4142}, a^{1.41421}, a^{1.414214}, \ldots$$

EXAMPLE 1   EVALUATING AN EXPONENTIAL EXPRESSION

Scientific Calculator

Number	Keystrokes	Display
$2^{-\pi}$	2 $\boxed{y^x}$ π $\boxed{+/-}$ $\boxed{=}$	0.113314732

Graphing Calculator

Number	Keystrokes	Display
$2^{-\pi}$	2 $\boxed{\wedge}$ $\boxed{(-)}$ π $\boxed{\text{ENTER}}$	.1133147323

✓*CHECKPOINT*   Now try Exercise 1.

Graphs of Exponential Functions

The graphs of all exponential functions have similar characteristics, as shown in Examples 2, 3, and 4.

EXAMPLE 2   GRAPHS OF $y = a^x$

In the same coordinate plane, sketch the graph of each function.

a. $f(x) = 2^x$ **b.** $g(x) = 4^x$

SOLUTION

The table below lists some values for each function, and Figure 3.1 shows their graphs. Note that both graphs are increasing. Moreover, the graph of $g(x) = 4^x$ is increasing more rapidly than the graph of $f(x) = 2^x$.

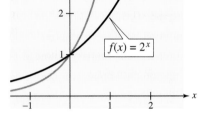

$g(x) = 4^x$

$f(x) = 2^x$

Figure 3.1

x	−2	−1	0	1	2	3
$f(x) = 2^x$	$\frac{1}{4}$	$\frac{1}{2}$	1	2	4	8
$g(x) = 4^x$	$\frac{1}{16}$	$\frac{1}{4}$	1	4	16	64

CHECKPOINT   Now try Exercise 19.

EXAMPLE 3   GRAPHS OF $y = a^{-x}$

In the same coordinate plane, sketch the graph of each function.

a. $F(x) = 2^{-x}$ **b.** $G(x) = 4^{-x}$

SOLUTION

The table below lists some values for each function, and Figure 3.2 shows their graphs. Note that both graphs are decreasing. Moreover, the graph of $G(x) = 4^{-x}$ is decreasing more rapidly than the graph of $F(x) = 2^{-x}$.

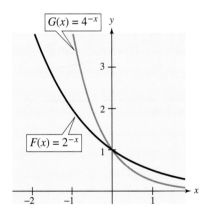

$G(x) = 4^{-x}$

$F(x) = 2^{-x}$

Figure 3.2

x	−3	−2	−1	0	1	2
$F(x) = 2^{-x}$	8	4	2	1	$\frac{1}{2}$	$\frac{1}{4}$
$G(x) = 4^{-x}$	64	16	4	1	$\frac{1}{4}$	$\frac{1}{16}$

CHECKPOINT   Now try Exercise 21.

The tables in Examples 2 and 3 were evaluated by hand. You could, of course, use the *table* feature of a graphing utility to construct tables with even more values.

In Example 3, note that the functions given by $F(x) = 2^{-x}$ and $G(x) = 4^{-x}$ can be rewritten with positive exponents.

$$F(x) = 2^{-x} = \left(\tfrac{1}{2}\right)^x \quad \text{and} \quad G(x) = 4^{-x} = \left(\tfrac{1}{4}\right)^x$$

Comparing the functions in Examples 2 and 3, observe that

$$F(x) = 2^{-x} = f(-x) \quad \text{and} \quad G(x) = 4^{-x} = g(-x).$$

Consequently, the graph of F is a reflection (in the y-axis) of the graph of f. The graphs of G and g have the same relationship.

The graphs in Figures 3.1 and 3.2 are typical of the exponential functions $y = a^x$ and $y = a^{-x}$. They have one y-intercept and one horizontal asymptote (the x-axis), and they are continuous. The basic characteristics of these exponential functions are summarized in Figures 3.3 and 3.4.

Characteristics of Exponential Functions

Graph of $y = a^x$, $a > 1$

- Domain: $(-\infty, \infty)$
- Range: $(0, \infty)$
- Intercept: $(0, 1)$
- Increasing
- x-axis is a horizontal asymptote ($a^x \to 0$ as $x \to -\infty$)
- Continuous

Graph of $y = a^{-x}$, $a > 1$

- Domain: $(-\infty, \infty)$
- Range: $(0, \infty)$
- Intercept: $(0, 1)$
- Decreasing
- x-axis is a horizontal asymptote ($a^{-x} \to 0$ as $x \to \infty$)
- Continuous
- Reflection of graph of $y = a^x$ about y-axis

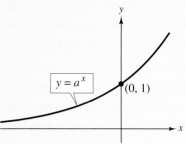

Figure 3.3

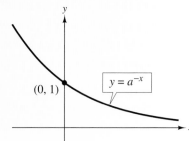

Figure 3.4

DISCOVERY

Use a graphing utility to graph $y = a^x$ for $a = 3, 5,$ and 7 in the same viewing window. (Use a viewing window in which $-2 \le x \le 1$ and $0 \le y \le 2$.) How do the graphs compare with each other? Which graph is on the top in the interval $(-\infty, 0)$? Which is on the bottom? Which graph is on the top in the interval $(0, \infty)$? Which is on the bottom?

Repeat this experiment with the graphs of $y = b^x$ for $b = \frac{1}{3}, \frac{1}{5},$ and $\frac{1}{7}$. (Use a viewing window in which $-1 \le x \le 2$ and $0 \le y \le 2$.) What can you conclude about the shape of the graph of $y = b^x$ and the value of b?

In the following example, notice how the graph of $y = a^x$ is used to sketch the graphs of functions of the form $f(x) = b \pm a^{x+c}$.

EXAMPLE 4   TRANSFORMATIONS OF GRAPHS OF EXPONENTIAL FUNCTIONS

Each of the following graphs is a transformation of the graph of $f(x) = 3^x$, as shown in Figure 3.5.

a. Because $g(x) = 3^{x+1} = f(x + 1)$, the graph of g can be obtained by shifting the graph of f one unit to the *left*.

b. Because $h(x) = 3^x - 2 = f(x) - 2$, the graph of h can be obtained by shifting the graph of f *downward* two units.

c. Because $k(x) = -3^x = -f(x)$, the graph of k can be obtained by *reflecting* the graph of f in the x-axis.

d. Because $j(x) = 3^{-x} = f(-x)$, the graph of j can be obtained by *reflecting* the graph of f in the y-axis.

STUDY TIP

Notice in Example 4(b) that shifting the graph downward two units also shifts the horizontal asymptote of $f(x) = 3^x$ from the x-axis ($y = 0$) to the line $y = -2$.

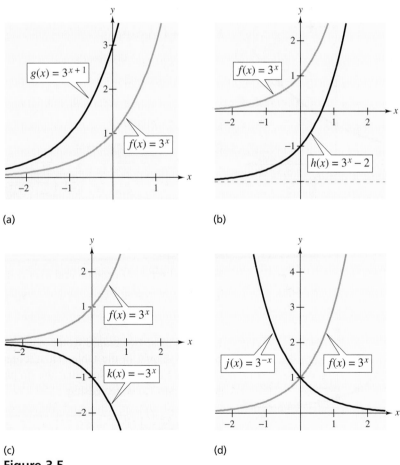

(a)

(b)

(c)

(d)

Figure 3.5

✔CHECKPOINT   Now try Exercise 23.

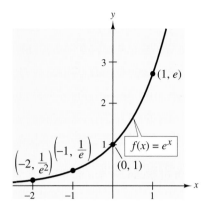

Figure 3.6

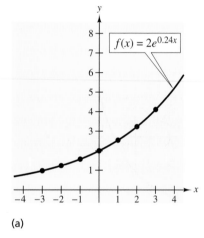

(a)

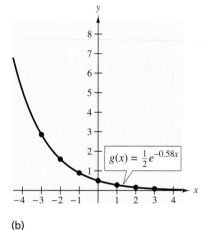

(b)

Figure 3.7

The Natural Base e

In many applications, the most convenient choice for a base is the irrational number

$$e = 2.718281828 \ldots$$

called the **natural base.** The function given by $f(x) = e^x$ is called the **natural exponential function.** Its graph is shown in Figure 3.6. The graph of the natural exponential function has the same basic characteristics as the graph of the exponential function given by $f(x) = a^x$ (see page 192). Be sure you see that for the exponential function given by $f(x) = e^x$, e is the constant $2.718281828 \ldots$, whereas x is the variable.

EXAMPLE 5   EVALUATING THE NATURAL EXPONENTIAL FUNCTION

Use a calculator to evaluate the function given by $f(x) = e^x$ for $x = 2$ and $x = -1$.

SOLUTION

Scientific Calculator

Number	Keystrokes	Display
e^2	2 [2nd] [e^x]	7.389056099
e^{-1}	1 [+/−] [2nd] [e^x]	0.367879441

Graphing Calculator

Number	Keystrokes	Display
e^2	[2nd] [e^x] 2 [)] [ENTER]	7.389056099
e^{-1}	[2nd] [e^x] [(−)] 1 [)] [ENTER]	.3678794412

✓CHECKPOINT   Now try Exercise 7.

EXAMPLE 6   GRAPHING NATURAL EXPONENTIAL FUNCTIONS

Sketch the graph of each natural exponential function.

a. $f(x) = 2e^{0.24x}$ **b.** $g(x) = \frac{1}{2}e^{-0.58x}$

SOLUTION

To sketch these two graphs, you can use a calculator to plot several points on each graph, as shown in the table. Then, connect the points with smooth curves, as shown in Figure 3.7. Note that the graph in part (a) is increasing, whereas the graph in part (b) is decreasing.

x	-3	-2	-1	0	1	2	3
$f(x) = 2e^{0.24x}$	0.974	1.238	1.573	2	2.542	3.232	4.109
$g(x) = \frac{1}{2}e^{-0.58x}$	2.849	1.595	0.893	0.5	0.280	0.157	0.088

✓CHECKPOINT   Now try Exercise 31.

Compound Interest

One of the most familiar examples of exponential growth is that of an investment earning **continuously compounded interest.** The formula for the balance in an account that is compounded n times per year is $A = P(1 + r/n)^{nt}$, where A is the balance in the account, P is the initial deposit, r is the annual interest rate (in decimal form), and t is the number of years. Using exponential functions, you will *develop* this formula and show how it leads to continuous compounding.

Suppose a principal P is invested at an annual interest rate r, compounded once a year. The principal at the end of the first year, P_1, is equal to the initial deposit P plus the interest earned, Pr. So,

$$P_1 = P + Pr.$$

This can be rewritten by factoring out P from each term as follows.

$$P_1 = P + Pr$$
$$= P(1 + r)$$

This pattern of multiplying the previous principal by $1 + r$ is then repeated each successive year, as shown below.

Year	Balance After Each Compounding
0	$P = P$
1	$P_1 = P(1 + r)$
2	$P_2 = P_1(1 + r) = P(1 + r)(1 + r) = P(1 + r)^2$
3	$P_3 = P_2(1 + r) = P(1 + r)^2(1 + r) = P(1 + r)^3$
$\vdots$	
t	$P_t = P(1 + r)^t$

To accommodate more frequent (quarterly, monthly, or daily) compounding of interest, let n be the number of compoundings per year and let t be the number of years. Then the rate per compounding is r/n and the account balance after t years is

$$A = P\left(1 + \frac{r}{n}\right)^{nt}. \qquad \text{Amount (balance) with } n \text{ compoundings per year}$$

If you let the number of compoundings n increase without bound, the process approaches what is called **continuous compounding.** In the formula for n compoundings per year, let $m = n/r$. This produces

$$A = P\left(1 + \frac{r}{n}\right)^{nt} \qquad \text{Amount with } n \text{ compoundings per year}$$

$$= P\left(1 + \frac{1}{m}\right)^{mrt} \qquad \text{Substitute } mr \text{ for } n \text{ and simplify.}$$

$$= P\left[\left(1 + \frac{1}{m}\right)^m\right]^{rt}. \qquad \text{Property of exponents}$$

As m increases without bound, it can be shown that $\left[1 + (1/m)\right]^m$ approaches e. From this, you can conclude that the formula for continuous compounding is $A = Pe^{rt}$.

DISCOVERY

Use a calculator and the formula $A = P(1 + r/n)^{nt}$ to calculate the amount in an account when $P = \$3000$, $r = 6\%$, t is 10 years, and the number of compoundings is (1) by the day, (2) by the hour, (3) by the minute, and (4) by the second. Use these results to present an argument that increasing the number of compoundings does not mean unlimited growth of the amount in the account.

Formulas for Compound Interest

After t years, the balance A in an account with principal P and annual interest rate r (in decimal form) is given by the following formulas.

1. For n compoundings per year: $A = P\left(1 + \dfrac{r}{n}\right)^{nt}$

2. For continuous compounding: $A = Pe^{rt}$

Be sure that the annual interest rate is written in decimal form. For instance, 6% should be written as 0.06 when using compound interest formulas.

EXAMPLE 7   COMPOUND INTEREST

Make a Decision   You invest $12,000 at an annual rate of 3%. Find the balance after 5 years if the interest is compounded (a) quarterly, (b) monthly, and (c) continuously. Which type of compounding earns more money?

SOLUTION

a. For quarterly compounding, you have $n = 4$. So, in 5 years at 3%, the balance is

$$A = P\left(1 + \frac{r}{n}\right)^{nt}$$ Formula for compound interest

$$= 12{,}000\left(1 + \frac{0.03}{4}\right)^{4(5)}$$ Substitute for P, r, n, and t.

$$\approx \$13{,}934.21.$$ Use a calculator.

b. For monthly compounding, you have $n = 12$. So, in 5 years at 3%, the balance is

$$A = P\left(1 + \frac{r}{n}\right)^{nt}$$ Formula for compound interest

$$= 12{,}000\left(1 + \frac{0.03}{12}\right)^{12(5)}$$ Substitute for P, r, n and t.

$$\approx \$13{,}939.40$$ Use a calculator.

c. For continuous compounding, the balance is

$$A = Pe^{rt}$$ Formula for continuous compounding

$$= 12{,}000e^{0.03(5)}$$ Substitute for P, r, and t.

$$\approx \$13{,}942.01$$ Use a calculator.

Note that continuous compounding yields more than quarterly and monthly compounding. This is typical of the two types of compounding. That is, for a given principal, interest rate, and time, continuous compounding will always yield a larger balance than compounding n times a year.

✓**CHECKPOINT**   Now try Exercise 45.

Another Application

EXAMPLE 8 RADIOACTIVE DECAY

Make a Decision In 1986, a nuclear reactor accident occurred in Chernobyl in what was then the Soviet Union. The explosion spread highly toxic radioactive chemicals, such as plutonium, over hundreds of square miles, and the government evacuated the city and the surrounding area. Consider the model

$$P = 10\left(\tfrac{1}{2}\right)^{t/24,100}$$

which represents the amount of plutonium P that remains (from an initial amount of 10 pounds) after t years. Sketch the graph of this function over the interval from $t = 0$ to $t = 100,000$, where $t = 0$ represents 1986. How much of the 10 pounds of plutonium will remain in the year 2007? How much of the 10 pounds will remain after 100,000 years? Why is this city uninhabited?

SOLUTION

The graph of this function is shown in Figure 3.8. Note from this graph that plutonium has a *half-life* of about 24,100 years. That is, after 24,100 years, *half* of the original amount of plutonium will remain. After another 24,100 years, one-quarter of the original amount will remain, and so on. In the year 2007 ($t = 21$), there will still be

$$P = 10\left(\tfrac{1}{2}\right)^{21/24,100} \approx 10\left(\tfrac{1}{2}\right)^{0.0008714} \approx 9.994 \text{ pounds}$$

of the original amount of plutonium remaining. After 100,000 years, there will still be

$$P = 10\left(\tfrac{1}{2}\right)^{100,000/24,100} \approx 10\left(\tfrac{1}{2}\right)^{4.149} \approx 0.564 \text{ pound}$$

of the original amount of plutonium remaining. This city is uninhabited because much of the original amount of radioactive plutonium still remains in the city.

✓CHECKPOINT   Now try Exercise 57.

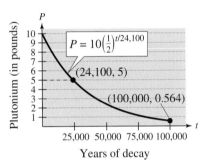

Figure 3.8

Discussing the Concept | Exponential Growth

The pattern 3, 6, 9, 12, 15, . . . is given by $f(n) = 3n$ and is an example of **linear growth**. The pattern 3, 9, 27, 81, 243, . . . is given by $f(n) = 3^n$ and is an example of **exponential growth**. Explain the difference between these two types of growth.
For each of the following patterns, indicate whether the pattern represents linear growth or exponential growth, and find a linear or exponential function that represents the pattern. Give several other examples of linear and exponential growth.

a. $\tfrac{1}{2}, \tfrac{1}{4}, \tfrac{1}{8}, \tfrac{1}{16}, \tfrac{1}{32}, \ldots$ b. 4, 8, 12, 16, 20, . . .

c. $\tfrac{2}{3}, \tfrac{4}{3}, 2, \tfrac{8}{3}, \tfrac{10}{3}, 4, \ldots$ d. 5, 25, 125, 625, . . .

3.1 (Warm Up

The following warm-up exercises involve skills that were covered in earlier sections. You will use these skills in the exercise set for this section. For additional help, review Sections R1.3 and R1.4.

In Exercises 1–10, use the properties of exponents to simplify the expression.

1. $5^{2x}(5^{-x})$

2. $3^{-x}(3^{3x})$

3. $\dfrac{4^{5x}}{4^{2x}}$

4. $\dfrac{10^{2x}}{10^{x}}$

5. $(4^{x})^{2}$

6. $(4^{2x})^{5}$

7. $\left(\dfrac{2^{x}}{3^{x}}\right)^{-1}$

8. $(4^{6x})^{1/2}$

9. $(2^{3x})^{-1/3}$

10. $(16^{x})^{1/4}$

3.1 (Exercises

In Exercises 1–10, use a calculator to evaluate the expression. Round your result to three decimal places.

1. $(2.6)^{1.3}$

2. $(1.07)^{50}$

3. $100(1.03)^{-1.4}$

4. $1500(2^{-5/2})$

5. $6^{-\sqrt{2}}$

6. $1.3^{\sqrt{5}}$

7. e^{4}

8. e^{-5}

9. $e^{2/3}$

10. $e^{-2.7}$

In Exercises 11–18, match the function with its graph. [The graphs are labeled (a), (b), (c), (d), (e), (f), (g), and (h).]

(a)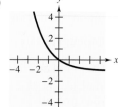

(b)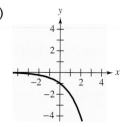

(e)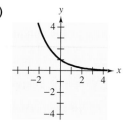

(f)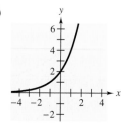

(g)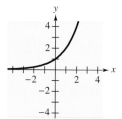

(h)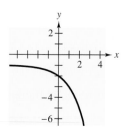

11. $f(x) = 2^{x}$

12. $f(x) = 2^{-x}$

13. $f(x) = -2^{x}$

14. $f(x) = -2^{x} - 1$

15. $f(x) = 2^{x} + 3$

16. $f(x) = 2^{-x} - 1$

17. $f(x) = 2^{x+1}$

18. $f(x) = 2^{x-3}$

(c)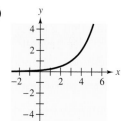

(d)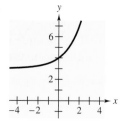

In Exercises 19–36, sketch the graph of the function.

19. $g(x) = 4^{x}$

20. $f(x) = \left(\frac{3}{2}\right)^{x}$

21. $f(x) = 4^{-x}$

22. $h(x) = \left(\frac{3}{2}\right)^{-x}$

23. $h(x) = 4^{x-3}$

24. $g(x) = \left(\frac{3}{2}\right)^{x+2}$

25. $g(x) = 4^{-x} - 2$

26. $f(x) = \left(\frac{3}{2}\right)^{-x} + 2$

27. $y = 2^{-x^{2}}$

28. $y = 3^{-x^{2}}$

29. $y = e^{-0.1x}$

30. $y = e^{0.2x}$

31. $f(x) = 2e^{0.12x}$

32. $f(x) = 3e^{-0.2x}$

33. $f(x) = e^{2x}$

34. $h(x) = e^{x-2}$

35. $g(x) = 1 + e^{-x}$

36. $N(t) = 1000e^{-0.2t}$

Compound Interest In Exercises 37–40, complete the table to find the balance A for P dollars invested at rate r for t years, compounded n times per year.

n	1	2	4	12	365	Continuous
A						

37. $P = \$5000$, $r = 8\%$, $t = 5$ years

38. $P = \$1000$, $r = 10\%$, $t = 10$ years

39. $P = \$2500$, $r = 12\%$, $t = 20$ years

40. $P = \$1000$, $r = 10\%$, $t = 40$ years

Compound Interest In Exercises 41–44, complete the table to find the amount P that must be invested at rate r to obtain a balance of $A = \$100,000$ in t years.

t	1	10	20	30	40	50
P						

41. $r = 9\%$, compounded continuously

42. $r = 12\%$, compounded continuously

43. $r = 10\%$, compounded monthly

44. $r = 7\%$, compounded daily

45. *Make a Decision*: Compound Interest A bank offers two types of interest accounts. The first account receives 5% interest compounded quarterly. The second account receives 3% interest compounded continuously. Which account earns more money? Why?

46. *Make a Decision*: Compound Interest A bank offers two types of interest accounts. The first account receives 6% interest compounded monthly. The second account receives 5% interest compounded continuously. Which account earns more money? Why?

47. *Make a Decision*: Cash Settlement You invest a cash settlement of $2000 for 5 years. You have a choice between an account that pays 5.75% interest compounded monthly with a monthly online access fee of $5 and an account that pays 4.25% interest compounded continuously with free online access. Which account should you choose? Explain your reasoning.

48. *Make a Decision*: Sales Commission You invest a sales commission of $5000 for 8 years. You have a choice between an account that pays 4.75% interest compounded monthly with a monthly online access fee of $3 and an account that pays 4.125% interest compounded continuously with free online access. Which account should you choose? Explain your reasoning.

49. Compound Interest On the day a child was born, a lump sum was deposited in a trust fund paying 6.5% interest compounded continuously. How much money should have been deposited so that the trust will be worth (a) $100,000, (b) $500,000, and (c) $1,000,000 on the child's 25th birthday?

50. Compound Interest The day you were born, money was deposited in a trust fund paying 7.5% interest compounded continuously. How much money should have been deposited so that your trust fund will be worth (a) $100,000, (b) $500,000, and (c) $1,000,000 on your next birthday? (*Note:* Answers will vary depending on your age.)

51. *Make a Decision*: Demand Function The demand function for a limited edition die cast model car is given by

$$p = 4000\left(1 - \frac{6}{6 + e^{-0.003x}}\right).$$

(a) Find the price p for a demand of $x = 1000$ units.

(b) Find the price p for a demand of $x = 1500$ units.

(c) Use a graphing utility to graph the demand function.

(d) Does the graph indicate what a good price for the sports car would be? Explain your reasoning.

52. *Make a Decision*: Demand Function The demand function for an entertainment center is given by

$$p = 5000\left(1 - \frac{4}{4 + e^{-0.002x}}\right).$$

(a) Find the price p for a demand of $x = 100$ units.

(b) Find the price p for a demand of $x = 500$ units.

(c) Use a graphing utility to graph the demand function.

(d) Does the graph indicate what a good price for the entertainment center would be? Explain your reasoning.

53. Bacteria Growth A certain type of bacteria increases according to the model $P(t) = 100e^{0.2197t}$, where t is time (in hours). Find (a) $P(0)$, (b) $P(5)$, and (c) $P(10)$.

IN THE NEWS

Radioactive Wrecks

Sunken nuclear subs pose no immediate threat, but they could be long-term ecological time bombs

Many environmental groups became alarmed this past August after the sinking of the Russian submarine *Kursk* in the Barents Sea. Greenpeace International warned that the pristine Arctic waters could become contaminated by radioactive materials leaking from the submarine's two nuclear reactors. Because the vessel lies in relatively shallow water—only 108 meters below the surface—ocean currents could spread the deadly isotopes to the Barents's rich fishing grounds. . . .

Nuclear engineers who are familiar with submarine reactors agree that the danger of leakage exists, but in all likelihood the contamination will not occur for a long, long time. . . .

Unless the explosion cracked or smashed some of the *Kursk's* fuel rods, the highly radioactive by-products of uranium fission will probably not leak out until well into the millennium. By then, many of the most dangerous isotopes—such as strontium 90 and cesium 137, which have half-lives of about 30 years—will have decayed away. . . .

Thomas Pigford, professor emeritus of nuclear engineering at the University of California at Berkeley, believes the most hazardous contaminant in the long run may be neptunium 237, which has a half-life of 2.1 million years. "It can get into the food chain if fish or shellfish ingest it," Pigford says, "and if it gets into your body, it can have some very bad effects." . . .

From "Radioactive Wrecks," by Mark Alpert. Copyright © 2000 by Scientific American, Inc. All rights reserved. Reprinted with permission.

---

54. Strontium-90 has a half-life of 29.1 years. The amount S of 100 kilograms of strontium-90 present after t years is given by $S = 100e^{-0.0238t}$. How much of the 100 kilograms will remain after 50 years?

55. Neptunium-237 has a half-life of 2.1 million years. The amount N of 200 kilograms of neptunium-237 present after t years is given by $N = 200e^{-0.00000033007t}$. How much of the 200 kilograms will remain after 20,000 years?

56. Population Growth The population P of a town increases according to the model $P(t) = 3500e^{0.0293t}$, where t is time in years, with $t = 5$ corresponding to 1995. Use the model to approximate the population in (a) 1999, (b) 2003, and (c) 2007.

57. Radioactive Decay Five pounds of the element plutonium (^{230}Pu) is released in a nuclear accident. The amount of plutonium P that is present after t months is given by $P = 5e^{-0.1507t}$.

(a) Use a graphing utility to graph this function over the interval from $t = 0$ to $t = 10$.

(b) How much of the 5 pounds of plutonium will remain after 10 months?

(c) Use the graph to estimate the half-life of ^{230}Pu. Explain your reasoning.

58. Radioactive Decay One hundred grams of radium (^{226}Ra) is stored in a container. The amount of radium present after t years is given by $R = 100e^{-0.0004335t}$.

(a) Use a graphing utility to graph this function over the interval from $t = 0$ to $t = 10{,}000$.

(b) How much of the 100 grams of radium will remain after 10,000 years?

(c) Use the graph to estimate the half-life of ^{226}Ra. Explain your reasoning.

59. New York Stock Exchange The total number y (in millions) of shares of stocks traded on the New York Stock Exchange between 1990 and 2002 can be approximated by the function

$$y = 33{,}787.3e^{0.2001t} + 1800, \quad 0 \le t \le 12$$

where t represents the year, with $t = 0$ corresponding to 1990. (Source: New York Stock Exchange, Inc.)

(a) Use the graph to estimate *graphically* the total number of shares traded in 1994, 1997, and 2000.

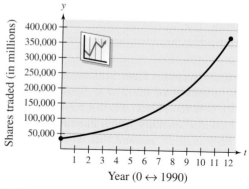

(b) Use the model to confirm *algebraically* the estimates obtained in part (a).

60. Yankee Candle Sales The total sales y (in millions of dollars) for the Yankee Candle Company from 1998 to 2003 can be approximated by the function

$$y = 3293.52e^{0.0162t} - 3555.87, \quad 8 \leq t \leq 13$$

where t represents the year, with $t = 8$ corresponding to 1998 (see figure). (Source: The Yankee Candle Company)

(a) Use the graph to estimate *graphically* the total sales for the Yankee Candle Company in 1998, 2000, and 2003.

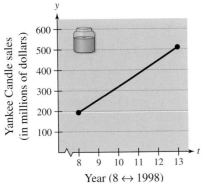

Year (8 ↔ 1998)

(b) Use the model to confirm *algebraically* the estimates obtained in part (a).

61. Age at First Marriage From 1980 to 2002, the median age A of an American woman at her first marriage can be approximated by the model

$$A = 12.39 + \frac{13.81}{1 + e^{-0.0813t - 0.8457}}, \quad 0 \leq t \leq 22$$

where t represents the year, with $t = 0$ corresponding to 1980. (Source: U.S. Census Bureau)

(a) Use the graph to estimate *graphically* the median age of an American woman at her first marriage in 1980, 1990, and 2002.

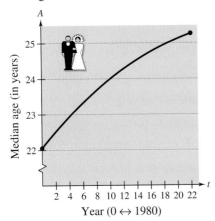

Year (0 ↔ 1980)

(b) Use the model to confirm *algebraically* the estimates obtained in part (a).

62. Age at First Marriage From 1980 to 2002, the median age A of an American man at his first marriage can be approximated by the model

$$A = 23.89 + \frac{3.14}{1 + e^{-0.2059t + 0.9349}}, \quad 0 \leq t \leq 22$$

where t represents the year, with $t = 0$ corresponding to 1980. (Source: U.S. Census Bureau)

(a) Use the graph to estimate *graphically* the median age of an American man at his first marriage in 1980, 1990, and 2002.

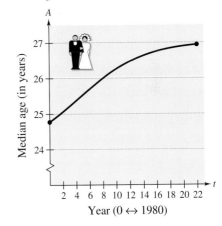

Year (0 ↔ 1980)

(b) Use the model to confirm *algebraically* the results obtained in part (a).

63. *Make a Decision* Compare the results of Exercises 61 and 62. What can you conclude about the differences in men's and women's ages at first marriage?

64. Writing Determine whether $e = \dfrac{271,801}{99,990}$. Justify your answer.

3.2   Logarithmic Functions

Objectives

- **Recognize and evaluate a logarithmic function with base *a*.**
- **Sketch the graph of a logarithmic function.**
- **Recognize and evaluate the natural logarithmic function.**
- **Use a logarithmic model to solve an application problem.**

Logarithmic Functions

In Section 1.8, you studied inverse functions. There, you learned that if a function has the property that no horizontal line intersects the graph of the function more than once, the function must have an inverse function. By looking back at the graphs of the exponential functions introduced in Section 3.1, you will see that every function of the form $f(x) = a^x$ passes the Horizontal Line Test and therefore must have an inverse function. This inverse function is called the **logarithmic function with base *a*.**

Definition of a Logarithmic Function

For $x > 0$ and $0 < a \ne 1$,

$$y = \log_a x \text{ if and only if } x = a^y.$$

The function given by

$$f(x) = \log_a x$$

is called the **logarithmic function with base *a*.**

The equations $y = \log_a x$ and $x = a^y$ are equivalent. The first equation is in logarithmic form and the second is in exponential form.

When evaluating logarithms, remember that *a logarithm is an exponent.* This means that $\log_a x$ is the exponent to which a must be raised to obtain x. For instance, $\log_2 8 = 3$ because 2 must be raised to the third power to obtain 8.

STUDY TIP

By the definition of logarithmic function,

$$\overset{a}{\underset{}{}} \overset{y}{\underset{}{}} \overset{x}{\underset{}{}}$$
$$3^4 = 81$$

can be written as

$$\overset{a}{\underset{}{}} \overset{x}{\underset{}{}} \overset{y}{\underset{}{}}$$
$$\log_3 81 = 4.$$

| EXAMPLE 1 | EVALUATING LOGARITHMIC EXPRESSIONS |

a. $\log_2 32 = 5$ because $2^5 = 32.$

b. $\log_4 2 = \dfrac{1}{2}$ because $4^{1/2} = \sqrt{4} = 2.$

c. $\log_{10} \dfrac{1}{100} = -2$ because $10^{-2} = \dfrac{1}{10^2} = \dfrac{1}{100}.$

d. $\log_3 1 = 0$ because $3^0 = 1.$

✓**CHECKPOINT**   Now try Exercise 21.

The logarithmic function with base 10 is called the **common logarithmic function.** On most calculators, this function is denoted by LOG.

STUDY TIP

Because $\log_a x$ is the inverse function of a^x, it follows that the domain of $\log_a x$ is the range of a^x, $(0, \infty)$. In other words, $\log_a x$ is defined only if x is positive.

EXAMPLE 2   EVALUATING LOGARITHMIC EXPRESSIONS ON A CALCULATOR

Scientific Calculator

Number	Keystrokes	Display
a. $\log_{10} 10$	10 LOG	1
b. $2 \log_{10} 2.5$	2.5 LOG × 2 =	0.795880017
c. $\log_{10}(-2)$	2 +/– LOG	ERROR

Graphing Calculator

Number	Keystrokes	Display
a. $\log_{10} 10$	LOG 10) ENTER	1
b. $2 \log_{10} 2.5$	2 LOG 2.5) ENTER	.7958800173
c. $\log_{10}(-2)$	LOG (−) 2) ENTER	ERROR

Many calculators display an error message (or a complex number) when you try to evaluate $\log_{10}(-2)$. This is because the domain of every logarithmic function is the set of *positive real numbers.* In other words, there is no real number power to which 10 can be raised to obtain -2.

✓**CHECKPOINT**   Now try Exercise 37.

The following properties follow directly from the definition of the logarithmic function with base a.

Properties of Logarithms

1. $\log_a 1 = 0$ because $a^0 = 1$.

2. $\log_a a = 1$ because $a^1 = a$.

3. $\log_a a^x = x$ and $a^{\log_a x} = x$ Inverse Properties

4. If $\log_a x = \log_a y$, then $x = y$.   One-to-One Property

---

EXAMPLE 3   USING PROPERTIES OF LOGARITHMS

a. Solve the equation $\log_2 x = \log_2 3$ for x.

b. Solve the equation $\log_5 x = 1$ for x.

SOLUTION

a. Using the One-to-One Property (Property 4), you can conclude that $x = 3$.

b. Using Property 2, you can conclude that $x = 5$.

✓**CHECKPOINT**   Now try Exercise 27.

Graphs of Logarithmic Functions

To sketch the graph of $y = \log_a x$, you can use the fact that the graphs of inverse functions are reflections of each other in the line $y = x$.

> **EXAMPLE 4**   GRAPHS OF EXPONENTIAL AND LOGARITHMIC FUNCTIONS

In the same coordinate plane, sketch the graph of each function.

a. $f(x) = 2^x$

b. $g(x) = \log_2 x$

SOLUTION

a. For $f(x) = 2^x$, construct a table of values, as follows.

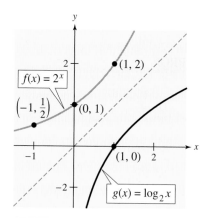

$f(x) = 2^x$

$\left(-1, \frac{1}{2}\right)$ $(0, 1)$ $(1, 2)$ $(1, 0)$ $g(x) = \log_2 x$

Figure 3.9 Inverse Functions

x	-2	-1	0	1	2	3
$f(x) = 2^x$	$\frac{1}{4}$	$\frac{1}{2}$	1	2	4	8

By plotting these points and connecting them with a smooth curve, you obtain the graph shown in Figure 3.9.

b. Because $g(x) = \log_2 x$ is the inverse function of $f(x) = 2^x$, the graph of g is obtained by plotting the points $(f(x), x)$ and connecting them with a smooth curve. The graph of g is a reflection of the graph of f in the line $y = x$, as shown in Figure 3.9.

✓**CHECKPOINT**   Now try Exercise 49.

Before you can confirm the result of Example 4 with a graphing utility, you need to know how to enter $\log_2 x$. You will learn how to do this using the *change-of-base formula* discussed in Section 3.3.

> **EXAMPLE 5**   SKETCHING THE GRAPH OF A LOGARITHMIC FUNCTION

Sketch the graph of the common logarithmic function $f(x) = \log_{10} x$.

SOLUTION

Begin by constructing a table of values. Note that some of the values can be obtained without a calculator by using the Inverse Property of logarithms. Others require a calculator. Next, plot the points and connect them with a smooth curve, as shown in Figure 3.10.

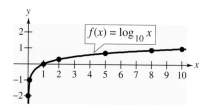

$f(x) = \log_{10} x$

Figure 3.10

x	Without Calculator				With Calculator		
	$\frac{1}{100}$	$\frac{1}{10}$	1	10	2	5	8
$f(x) = \log_{10} x$	-2	-1	0	1	0.301	0.699	0.903

✓**CHECKPOINT**   Now try Exercise 59.

The nature of the graph in Figure 3.10 is typical of functions of the form $f(x) = \log_a x$, $a > 1$. They have one x-intercept and one vertical asymptote. Notice how slowly the graph rises for $x > 1$. The basic characteristics of logarithmic graphs are summarized in Figure 3.11. Note that the vertical asymptote occurs at $x = 0$, where $\log_a x$ is *undefined*.

Characteristics of Logarithmic Functions

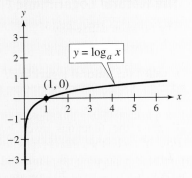

Figure 3.11

Graph of $y = \log_a x$, $a > 1$

- Domain: $(0, \infty)$
- Range: $(-\infty, \infty)$
- Intercept: $(1, 0)$
- Increasing
- One-to-one; therefore has an inverse function
- y-axis is a vertical asymptote $(\log_a x \to -\infty$ as $x \to 0^+)$
- Continuous
- Reflection of graph of $y = a^x$ about the line $y = x$

EXAMPLE 6 SKETCHING THE GRAPHS OF LOGARITHMIC FUNCTIONS

The graph of each of the functions is similar to the graph of $f(x) = \log_{10} x$.

a. Because $g(x) = \log_{10}(x - 1) = f(x - 1)$, the graph of g can be obtained by shifting the graph of f one unit to the *right*. See Figure 3.12(a).

b. Because $h(x) = 2 + \log_{10} x = 2 + f(x)$, the graph of h can be obtained by shifting the graph of f two units *upward*. See Figure 3.12(b).

STUDY TIP

Notice in Example 6(a) that shifting the graph of $f(x)$ one unit to the right also shifts the vertical asymptote from the y-axis ($x = 0$) to the line $x = 1$.

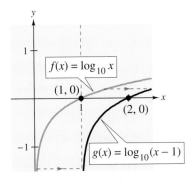

(a) Right shift of one unit

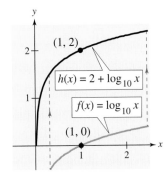

(b) Upward shift of two units

Figure 3.12

✓CHECKPOINT Now try Exercise 61.

The Natural Logarithmic Function

By looking back at the graph of the natural exponential function introduced in Section 3.1, you will see that $f(x) = e^x$ is one-to-one and so has an inverse function. This inverse function is called the **natural logarithmic function** and is denoted by the special symbol ln x, read as "el en of x."

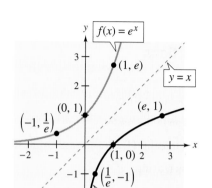

Figure 3.13

The Natural Logarithmic Function

The function defined by

$$f(x) = \log_e x = \ln x, \quad x > 0$$

is called the **natural logarithmic function.**

Because the functions given by $f(x) = e^x$ and $g(x) = \ln x$ are inverse functions of each other, their graphs are reflections of each other in the line $y = x$. This reflective property is illustrated in Figure 3.13. The four properties of logarithms listed on page 203 are also valid for natural logarithms.

Properties of Natural Logarithms

1. $\ln 1 = 0$ because $e^0 = 1$.

2. $\ln e = 1$ because $e^1 = e$.

3. $\ln e^x = x$ and $e^{\ln x} = x$ Inverse Properties

4. If $\ln x = \ln y$, then $x = y$.   One-to-One Property

EXAMPLE 7   USING PROPERTIES OF NATURAL LOGARITHMS

Evaluate each logarithmic expression.

a. $\ln \dfrac{1}{e}$ **b.** $e^{\ln 5}$ **c.** $\dfrac{\ln 1}{3}$ **d.** $2 \ln e$

SOLUTION

a. $\ln \dfrac{1}{e} = \ln e^{-1} = -1$ Inverse Property

b. $e^{\ln 5} = 5$ Inverse Property

c. $\dfrac{\ln 1}{3} = \dfrac{0}{3} = 0$ Property 1

d. $2 \ln e = 2(1) = 2$ Property 2

✓CHECKPOINT   Now try Exercise 33.

On most calculators, the natural logarithm is denoted by ⏹LN, as illustrated in Example 8.

EXAMPLE 8   Evaluating the Natural Logarithmic Function

Use a calculator to evaluate each expression.

a. ln 2   **b.** ln 0.3   **c.** ln e^2 **d.** ln(−1)

SOLUTION

Scientific Calculator

Number	Keystrokes	Display
a. ln 2	2 ⏹LN	0.69314718
b. ln 0.3	.3 ⏹LN	−1.203972804
c. ln e^2	2 ⏹2nd ⏹[e^x] ⏹LN	2
d. ln(−1)	1 ⏹+/− ⏹LN	ERROR

Graphing Calculator

Number	Keystrokes	Display
a. ln 2	⏹LN 2 ⏹) ⏹ENTER	.6931471806
b. ln 0.3	⏹LN .3 ⏹) ⏹ENTER	−1.203972804
c. ln e^2	⏹LN ⏹2nd ⏹[e^x] 2 ⏹) ⏹) ⏹ENTER	2
d. ln(−1)	⏹LN ⏹(−) 1 ⏹) ⏹ENTER	ERROR

✓CHECKPOINT   Now try Exercise 45.

In Example 8, be sure you see that ln(−1) gives an error message on most calculators. This occurs because the domain of ln x is the set of positive real numbers (see Figure 3.13). So, ln(−1) is undefined.

EXAMPLE 9   Finding the Domains of Logarithmic Functions

Find the domain of each function.

a. $f(x) = \ln(x - 2)$

b. $g(x) = \ln(2 - x)$

c. $h(x) = \ln x^2$

SOLUTION

a. Because ln(x − 2) is defined only if $x - 2 > 0$, it follows that the domain of f is (2, ∞).

b. Because ln(2 − x) is defined only if $2 - x > 0$, it follows that the domain of g is (−∞, 2). The graph of g is shown in Figure 3.14.

c. Because ln x^2 is defined only if $x^2 > 0$, it follows that the domain of h is all real numbers except $x = 0$.

✓CHECKPOINT   Now try Exercise 65.

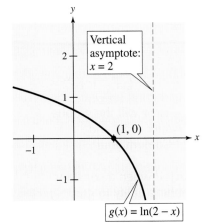

Figure 3.14

Application

EXAMPLE 10   HUMAN MEMORY MODEL

Make a Decision   Students participating in a psychological experiment attended several lectures on a subject. Every month for a year after that, the students were tested to see how much of the material they remembered. The average scores for the group were given by the *human memory model*

$$f(t) = 75 - 6 \ln(t + 1), \quad 0 \le t \le 12$$

where t is the time (in months). Based on the results of the experiment, how many months can a student wait before retaking the exam and still expect to score 60 or better? (Do not count portions of months.)

SOLUTION

To determine how many months a student can wait before retaking the exam and still expect to score 60 or better, use the model to create a table of values showing the scores for several months.

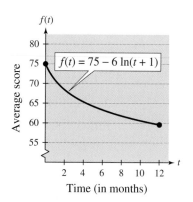

Figure 3.15

Month, t	0	1	2	3	4	5	6
Score, $f(t)$	75	70.84	68.41	66.68	65.34	64.25	63.32

Month, t	7	8	9	10	11	12
Score, $f(t)$	62.52	61.82	61.18	60.61	60.09	59.61

From the table, you can see that a student would need to retake the exam by the 11th month in order to score 60 or better. The graph of f is shown in Figure 3.15.

✓CHECKPOINT   Now try Exercise 71.

Discussing the Concept | Transforming Logarithmic Functions

Use a graphing utility to graph $f(x) = \ln x$. How will the graphs of $h(x) = \ln x + 5$, $j(x) = \ln(x + 5)$, $k(x) = \ln(x - 3)$, and $l(x) = \ln x - 4$ differ from the graph of f? Confirm your predictions with a graphing utility.

How will the basic graph of f be affected when a constant c is introduced: $g(x) = c \ln x$? Use a graphing utility to graph g with several different positive values of c, and summarize the effect of c.

3.2 **Warm Up**

The following warm-up exercises involve skills that were covered in earlier sections. You will use these skills in the exercise set for this section. For additional help, review Sections 1.6 and 3.1.

In Exercises 1–4, solve for x.

1. $2^x = 8$ 2. $4^x = 1$
3. $10^x = 0.1$ 4. $e^x = e$

In Exercises 5 and 6, evaluate the expression. (Round the result to three decimal places.)

5. e^2 6. e^{-1}

In Exercises 7–10, describe how the graph of g is related to the graph of f.

7. $g(x) = f(x + 2)$ 8. $g(x) = -f(x)$
9. $g(x) = -1 + f(x)$ 10. $g(x) = f(-x)$

3.2 **Exercises**

In Exercises 1–10, use the definition of a logarithm to write the equation in logarithmic form. For example, the logarithmic form of $2^3 = 8$ is $\log_2 8 = 3$.

1. $4^4 = 256$ 2. $7^3 = 343$
3. $81^{1/4} = 3$ 4. $9^{3/2} = 27$
5. $6^{-2} = \frac{1}{36}$ 6. $10^{-3} = 0.001$
7. $e^1 = e$ 8. $e^4 = 54.5981\ldots$
9. $e^x = 4$ 10. $e^{-x} = 2$

In Exercises 11–20, use the definition of a logarithm to write the equation in exponential form. For example, the exponential form of $\log_5 125 = 3$ is $5^3 = 125$.

11. $\log_4 16 = 2$ 12. $\log_{10} 1000 = 3$
13. $\log_2 \frac{1}{2} = -1$ 14. $\log_3 \frac{1}{9} = -2$
15. $\ln e = 1$ 16. $\ln \frac{1}{e} = -1$
17. $\log_5 0.2 = -1$ 18. $\log_{10} 0.1 = -1$
19. $\log_{27} 3 = \frac{1}{3}$ 20. $\log_8 2 = \frac{1}{3}$

In Exercises 21–36, evaluate the expression without using a calculator.

21. $\log_3 9$ 22. $\log_5 125$
23. $\log_2\left(\frac{1}{16}\right)$ 24. $\log_6\left(\frac{1}{36}\right)$
25. $\log_8 2$ 26. $\log_{64} 4$

27. $\log_7 7$ 28. $\log_{12} 1$
29. $\log_{10} 0.0001$ 30. $\log_{10} 100$
31. $\ln e$ 32. $\ln e^{10}$
33. $\ln e^{-4}$ 34. $\ln \frac{1}{e^3}$
35. $\log_a a^5$ 36. $\log_a 1$

In Exercises 37–48, use a calculator to evaluate the logarithm. Round your result to three decimal places.

37. $\log_{10} 345$ 38. $\log_{10} 145$
39. $\log_{10}\left(\frac{4}{5}\right)$ 40. $\log_{10} 12.5$
41. $\log_{10} \sqrt{8}$ 42. $\log_{10} \sqrt{3}$
43. $\ln 7$ 44. $2 \ln 9$
45. $\ln 18.42$ 46. $\ln 36.7$
47. $\ln \sqrt{3}$ 48. $\ln \sqrt{5}$

In Exercises 49–52, sketch the graphs of f and g in the same coordinate plane.

49. $f(x) = 3^x$, $g(x) = \log_3 x$
50. $f(x) = 5^x$, $g(x) = \log_5 x$
51. $f(x) = e^x$, $g(x) = \ln x$
52. $f(x) = 10^x$, $g(x) = \log_{10} x$

In Exercises 53–58, match the function with its graph. [The graphs are labeled (a), (b), (c), (d), (e), and (f).]

(a)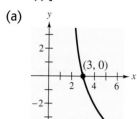

(b)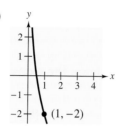

(c)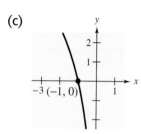

(d)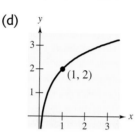

(e)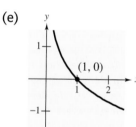

(f)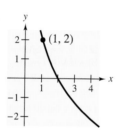

53. $f(x) = \ln x + 2$

54. $f(x) = -\ln x$

55. $f(x) = -3 \ln(x - 2)$

56. $f(x) = 4 \ln(-x)$

57. $f(x) = -3 \ln x + 2$

58. $f(x) = -3 \ln x - 2$

In Exercises 59–68, find the domain, vertical asymptote, and *x*-intercept of the logarithmic function. Then sketch its graph.

59. $f(x) = \log_2 x$

60. $g(x) = \log_8 x$

61. $h(x) = \log_2 (x + 4)$

62. $f(x) = \log_8(x - 2)$

63. $f(x) = -\log_2 x$

64. $h(x) = -\log_8(x + 1)$

65. $g(x) = \ln(-x)$

66. $f(x) = \ln(3 - x)$

67. $h(x) = \ln(x + 1)$

68. $f(x) = 2 + \ln x$

69. Population Growth   The population of a town will double in

$$t = \frac{10 \ln 2}{\ln 67 - \ln 50}$$

years. Find *t*.

70. Work   The work W (in foot-pounds) done in compressing a volume of 9 cubic feet at a pressure of 15 pounds per square inch to a volume of 3 cubic feet is $W = 19{,}440(\ln 9 - \ln 3)$. Find W.

71. *Make a Decision*: Human Memory Model   Students in a mathematics class were given an exam and then retested monthly with an equivalent exam. The average score g for the class can be approximated by the human memory model

$$g(t) = 80 - 17 \log_{10}(t + 1), \quad 0 \le t \le 12$$

where t is the time (in months).

(a) What was the average score on the original exam?

(b) What was the average score after 4 months?

(c) When did the average score drop below 70?

72. *Make a Decision*: Human Memory Model   Students in a seventh-grade class were given an exam. During the next 2 years, the same students were retested several times. The average score g can be approximated by the model

$$g(t) = 90 - 15 \log_{10}(t + 1), \quad 0 \le t \le 24$$

where t is the time (in months).

(a) What was the average score on the original exam?

(b) What was the average score after 6 months?

(c) When did the average score drop below 70?

73. Investment Time   A principal P, invested at $6\frac{1}{2}\%$ interest and compounded continuously, increases to an amount that is K times the principal after t years, where t is given by

$$t = \frac{\ln K}{0.065}.$$

(a) Complete the table.

K	1	2	4	6	8	10	12
t							

(b) Use the table in part (a) to graph the function.

74. Investment Time   A principal P, invested at 7.75% interest and compounded continuously, increases to an amount that is K times the principal after t years, where t is given by

$$t = \frac{\ln K}{0.0775}.$$

Use a graphing utility to graph this function.

Skill Retention Model In Exercises 75 and 76, participants in an industrial psychology study were taught a simple mechanical task and tested monthly on this mechanical task for a period of 1 year. The average scores for the participants are given by the model

$$f(t) = 95 - 12 \log_{10}(t + 1), \quad 0 \le t \le 12$$

where t is the time (in months).

75. Use a graphing utility to graph this function. Use the graph to discuss the domain and range of this function.

76. *Make a Decision* Based on the graph of f, do you think the study's participants practiced the simple mechanical task very often? Cite the behavior of the graph to justify your answer.

77. **Median Age of U.S. Population** The model

$$A = 17.64 - 0.035t + 5.387 \ln t, \quad 10 \le t \le 80$$

approximates the median age A of the United States population from 1980 to 2050. In the model, t represents the year, with $t = 10$ corresponding to 1980 (see figure). (Source: U.S. Census Bureau)

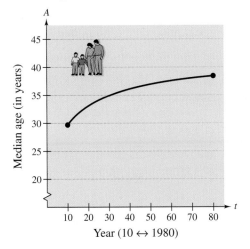

Year (10 ↔ 1980)

(a) Use the model to approximate the median age in the United States in 1980.

(b) Use the model to approximate the median age in the United States in 1990.

(c) Use the model to approximate the change in median age in the United States from 1980 to 2000.

(d) Use the model to project the change in median age in the United States from 1980 to 2050.

78. *Make a Decision*: **Monthly Payment** The model

$$t = 12.542 \ln\left(\frac{x}{x - 1000}\right), \quad x > 1000$$

approximates the length of a home mortgage of $150,000 at 8% interest in terms of the monthly payment. In the model, t is the length of the mortgage (in years) and x is the monthly payment (in dollars) (see figure).

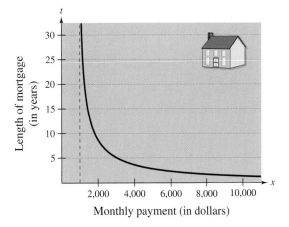

Monthly payment (in dollars)

(a) Use the model to approximate the length of a $150,000 mortgage at 8% interest when the monthly payment is $1100.65 and when the monthly payment is $1254.68.

(b) Approximate the total amount paid over the term of the mortgage with a monthly payment of $1100.65 and with a monthly payment of $1254.68.

(c) Approximate the total interest charge for a monthly payment of $1100.65 and for a monthly payment of $1254.68.

(d) What is the vertical asymptote for the model? Interpret its meaning in the context of the problem.

Properties of Logarithms ←

Objectives

- **Evaluate a logarithm using the change-of-base formula.**
- **Use properties of logarithms to evaluate or rewrite a logarithmic expression.**
- **Use properties of logarithms to expand or condense a logarithmic expression.**
- **Use logarithmic functions to model and solve real-life applications.**

Change of Base

Most calculators have only two types of log keys, one for common logarithms (base 10) and one for natural logarithms (base e). Although common logs and natural logs are the most frequently used, you may occasionally need to evaluate logarithms with other bases. To do this, you can use the following **change-of-base formula.**

Change-of-Base Formula

Let a, b, and x be positive real numbers such that $a \neq 1$ and $b \neq 1$. Then $\log_a x$ can be converted to a different base as follows.

Base b	*Base 10*	*Base e*
$\log_a x = \dfrac{\log_b x}{\log_b a}$	$\log_a x = \dfrac{\log_{10} x}{\log_{10} a}$	$\log_a x = \dfrac{\ln x}{\ln a}$

EXAMPLE 1   CHANGING BASES USING COMMON LOGARITHMS

a. $\log_4 30 = \dfrac{\log_{10} 30}{\log_{10} 4} \approx \dfrac{1.47712}{0.60206} \approx 2.4534$

b. $\log_2 14 = \dfrac{\log_{10} 14}{\log_{10} 2} \approx \dfrac{1.14613}{0.30103} \approx 3.8074$

✓*CHECKPOINT*   Now try Exercise 25.

EXAMPLE 2   CHANGING BASES USING NATURAL LOGARITHMS

a. $\log_4 30 = \dfrac{\ln 30}{\ln 4} \approx \dfrac{3.40120}{1.386294} \approx 2.4534$

b. $\log_2 14 = \dfrac{\ln 14}{\ln 2} \approx \dfrac{2.63906}{0.693147} \approx 3.8074$

✓*CHECKPOINT*   Now try Exercise 27.

Notice in Examples 1 and 2 that the result is the same whether the common logarithm or the natural logarithm is used in the change-of-base formula.

Properties of Logarithms

You know from the preceding section that the logarithmic function with base a is the *inverse function* of the exponential function with base a. So, it makes sense that the properties of exponents should have corresponding properties involving logarithms. For instance, the exponential property $a^0 = 1$ has the corresponding logarithmic property $\log_a 1 = 0$.

Properties of Logarithms

Let a be a positive number such that $a \neq 1$, and let n be a real number. If u and v are positive real numbers, then the following properties are true.

Logarithm with Base a	*Natural Logarithm*	
$\log_a(uv) = \log_a u + \log_a v$	$\ln(uv) = \ln u + \ln v$	Product Rule
$\log_a \dfrac{u}{v} = \log_a u - \log_a v$	$\ln \dfrac{u}{v} = \ln u - \ln v$	Quotient Rule
$\log_a u^n = n \log_a u$	$\ln u^n = n \ln u$	Power Rule

STUDY TIP

There is no general property that can be used to rewrite $\log_a(u \pm v)$. Specifically, $\log_a(x + y)$ is *not* equal to $\log_a x + \log_a y$.

EXAMPLE 3 USING PROPERTIES OF LOGARITHMS

Write each logarithm in terms of $\ln 2$ and $\ln 3$.

a. $\ln 6$ **b.** $\ln \dfrac{2}{27}$

SOLUTION

a. $\ln 6 = \ln(2 \cdot 3)$ Rewrite 6 as $2 \cdot 3$.

$\quad\quad\; = \ln 2 + \ln 3$ Product Rule

b. $\ln \dfrac{2}{27} = \ln 2 - \ln 27$ Quotient Rule

$\quad\quad\quad = \ln 2 - \ln 3^3$ Rewrite 27 as 3^3.

$\quad\quad\quad = \ln 2 - 3 \ln 3$ Power Rule

✓**CHECKPOINT**   Now try Exercise 37.

EXAMPLE 4 USING PROPERTIES OF LOGARITHMS

Use the properties of logarithms to verify that $-\log_{10} \frac{1}{100} = \log_{10} 100$.

SOLUTION

$$-\log_{10} \frac{1}{100} = -\log_{10}(100^{-1}) = -(-1)\log_{10} 100 = \log_{10} 100$$

Try checking this result on your calculator.

✓**CHECKPOINT**   Now try Exercise 57.

Rewriting Logarithmic Expressions

The properties of logarithms are useful for rewriting logarithmic expressions in forms that simplify the operations of algebra. This is true because these properties convert complicated products, quotients, and exponential forms into simpler sums, differences, and products, respectively.

EXAMPLE 5   EXPANDING LOGARITHMIC EXPRESSIONS

Expand each logarithmic expression.

a. $\log_4 5x^3 y$ **b.** $\ln \dfrac{\sqrt{3x-5}}{7}$

SOLUTION

a. $\log_4 5x^3 y = \log_4 5 + \log_4 x^3 + \log_4 y$ Product Rule

$\qquad\qquad = \log_4 5 + 3\log_4 x + \log_4 y$ Power Rule

b. $\ln \dfrac{\sqrt{3x-5}}{7} = \ln \dfrac{(3x-5)^{1/2}}{7}$ Rewrite using rational exponent.

$\qquad\qquad = \ln(3x-5)^{1/2} - \ln 7$ Quotient Rule

$\qquad\qquad = \dfrac{1}{2}\ln(3x-5) - \ln 7$ Power Rule

 CHECKPOINT   Now try Exercise 73.

DISCOVERY

Use a calculator to approximate $\ln \sqrt[3]{e^2}$. Now find the exact value by rewriting $\ln \sqrt[3]{e^2}$ with a rational exponent using the properties of logarithms. How do the two values compare?

In Example 5, the properties of logarithms were used to *expand* logarithmic expressions. In Example 6, this procedure is reversed and the properties of logarithms are used to *condense* logarithmic expressions.

EXAMPLE 6   CONDENSING LOGARITHMIC EXPRESSIONS

Condense each logarithmic expression.

a. $\frac{1}{2}\log_{10} x + 3\log_{10}(x+1)$ **b.** $2\ln(x+2) - \ln x$

c. $\frac{1}{3}[\log_2 x + \log_2(x+1)]$

SOLUTION

a. $\frac{1}{2}\log_{10} x + 3\log_{10}(x+1) = \log_{10} x^{1/2} + \log_{10}(x+1)^3$ Power Rule

$\qquad\qquad = \log_{10}\left[\sqrt{x}\,(x+1)^3\right]$ Product Rule

b. $2\ln(x+2) - \ln x = \ln(x+2)^2 - \ln x$ Power Rule

$\qquad\qquad = \ln \dfrac{(x+2)^2}{x}$ Quotient Rule

c. $\frac{1}{3}[\log_2 x + \log_2(x+1)] = \frac{1}{3}\{\log_2[x(x+1)]\}$ Product Rule

$\qquad\qquad = \log_2[x(x+1)]^{1/3}$ Power Rule

$\qquad\qquad = \log_2 \sqrt[3]{x(x+1)}$ Rewrite with a radical.

STUDY TIP

When applying the properties of logarithms to a logarithmic function, you should be careful to check the domain of the function. For example, the domain of $f(x) = \ln x^2$ is all real $x \neq 0$, whereas the domain of $g(x) = 2\ln x$ is all real $x > 0$.

✓CHECKPOINT   Now try Exercise 95.

Applications

One method of determining how the x- and y-values for a set of nonlinear data are related begins by taking the natural logarithm of each of the x- and y-values. If the points are graphed and fall on a straight line, then you can determine that the x- and y-values are related by the equation

$$\ln y = m \ln x$$

where m is the slope of the straight line.

| EXAMPLE 7 | FINDING A MATHEMATICAL MODEL | |

The table shows the mean distance from the sun x and the period (the time it takes a planet to orbit the sun) y for each of the six planets that are closest to the sun. In the table, the mean distance is given in astronomical units (where Earth's mean distance is defined as 1.0), and the period is given in years. Find an equation that relates y and x.

Planet	Mean distance, x	Period, y
Mercury	0.387	0.241
Venus	0.723	0.615
Earth	1.000	1.000
Mars	1.524	1.881
Jupiter	5.203	11.863
Saturn	9.537	29.447

SOLUTION

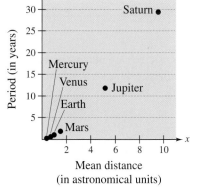

Figure 3.16

The points in the table are plotted in Figure 3.16. From this figure it is not clear how to find an equation that relates y and x. To solve this problem, take the natural logarithm of each of the x- and y-values in the table. This produces the following results.

Planet	Mercury	Venus	Earth	Mars	Jupiter	Saturn
ln x	-0.949	-0.324	0.000	0.421	1.649	2.255
ln y	-1.423	-0.486	0.000	0.632	2.473	3.383

Now, by plotting the points in the second table, you can see that all six of the points appear to lie in a line (see Figure 3.17). Choose any two points to determine the slope of the line. Using the two points $(0.421, 0.632)$ and $(0, 0)$, you can determine that the slope of the line is

$$m = \frac{0.632 - 0}{0.421 - 0} \approx 1.5 = \frac{3}{2}.$$

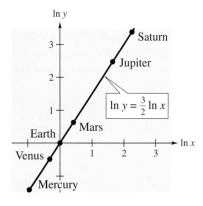

Figure 3.17

By the point-slope form, the equation of the line is $Y = \frac{3}{2}X$, where $Y = \ln y$ and $X = \ln x$. You can therefore conclude that $\ln y = \frac{3}{2} \ln x$.

✓CHECKPOINT   Now try Exercise 105.

Bill Aron/PhotoEdit, Inc.

| EXAMPLE 8 | SOUND INTENSITY | |

Make a Decision   The level of sound L (in decibels) with an intensity of I (in watts per square meter) is given by

$$L = 10 \log_{10} \frac{I}{I_0}$$

where I_0 represents the faintest sound that can be heard by the human ear, and is approximately equal to 10^{-12} watt per square meter. You and your roommate are playing your stereos at the same time and at the same intensity. How much louder is the music when both stereos are playing compared with when just one stereo is playing?

SOLUTION

Let L_1 represent the level of sound when one stereo is playing and let L_2 represent the level of sound when both stereos are playing. Using the formula for level of sound, you can express L_1 as

$$L_1 = 10 \log_{10} \frac{I}{10^{-12}}.$$

For L_2, multiply I by 2 as shown below

$$L_2 = 10 \log_{10} \frac{2I}{10^{-12}}$$

because L_2 represents the level of sound when *two* stereos are playing at the same intensity I. To determine the increase in loudness, subtract L_1 from L_2 as follows.

$$
\begin{aligned}
L_2 - L_1 &= 10 \log_{10} \frac{2I}{10^{-12}} - 10 \log_{10} \frac{I}{10^{-12}} \\
&= 10 \left(\log_{10} \frac{2I}{10^{-12}} - \log_{10} \frac{I}{10^{-12}} \right) \\
&= 10 \left(\log_{10} 2 + \log_{10} \frac{I}{10^{-12}} - \log_{10} \frac{I}{10^{-12}} \right) \\
&= 10 \log_{10} 2 \\
&\approx 3
\end{aligned}
$$

So, the music is about 3 decibels louder. Notice that the variable I drops out of the equation when it is simplified. This means that the loudness increases by 3 decibels when both stereos are played at the same intensity, regardless of the individual intensities of the stereos.

✓**CHECKPOINT**   Now try Exercise 107.

3.3 Warm Up

The following warm-up exercises involve skills that were covered in earlier sections. You will use these skills in the exercise set for this section. For additional help, review Sections R1.3, R1.4, and 3.2.

In Exercises 1–4, evaluate the expression without using a calculator.

1. $\log_7 49$ **2.** $\log_2\left(\frac{1}{32}\right)$ **3.** $\ln\dfrac{1}{e^2}$ **4.** $\log_{10} 0.001$

In Exercises 5–8, simplify the expression.

5. $e^2 e^3$ **6.** $\dfrac{e^2}{e^3}$ **7.** $(e^2)^3$ **8.** $(e^2)^0$

In Exercises 9 and 10, rewrite the expression in exponential form.

9. $\dfrac{1}{x^2}$ **10.** $\sqrt{x}$

3.3 Exercises

In Exercises 1–12, write the logarithm in terms of common logarithms.

1. $\log_5 8$ **2.** $\log_7 12$

3. $\ln 30$ **4.** $\ln 20$

5. $\log_3 n$ **6.** $\log_4 m$

7. $\log_{1/5} x$ **8.** $\log_{1/3} x$

9. $\log_x \frac{3}{10}$ **10.** $\log_x \frac{3}{4}$

11. $\log_{2.6} x$ **12.** $\log_{7.1} x$

In Exercises 13–24, write the logarithm in terms of natural logarithms.

13. $\log_5 8$ **14.** $\log_7 12$

15. $\log_{10} 5$ **16.** $\log_{10} 20$

17. $\log_3 n$ **18.** $\log_2 m$

19. $\log_{1/5} x$ **20.** $\log_{1/3} x$

21. $\log_x \frac{3}{10}$ **22.** $\log_x \frac{3}{4}$

23. $\log_{2.6} x$ **24.** $\log_{7.1} x$

In Exercises 25–36, evaluate the logarithm. Round your result to three decimal places.

25. $\log_2 6$ **26.** $\log_8 3$

27. $\log_{27} 35$ **28.** $\log_{19} 42$

29. $\log_{15} 1250$ **30.** $\log_{20} 125$

31. $\log_5\left(\frac{1}{3}\right)$ **32.** $\log_9 0.4$

33. $\log_{1/4} 10$ **34.** $\log_{1/3} 5$

35. $\log_{1/2} 0.2$ **36.** $\log_{1/3} 0.015$

In Exercises 37–50, approximate the logarithm using the properties of logarithms, given $\log_b 2 \approx 0.3562$, $\log_b 3 \approx 0.5646$, and $\log_b 5 \approx 0.8271$.

37. $\log_b 10$ **38.** $\log_b 15$

39. $\log_b\left(\frac{2}{3}\right)$ **40.** $\log_b\left(\frac{5}{3}\right)$

41. $\log_b 8$ **42.** $\log_b 18$

43. $\log_b \sqrt{2}$ **44.** $\log_b\left(\frac{9}{2}\right)$

45. $\log_b 40$ **46.** $\log_b \sqrt[3]{75}$

47. $\log_b(2b)^{-2}$ **48.** $\log_b(3b^2)$

49. $\log_b \sqrt{5b}$ **50.** $\log_b \sqrt[3]{3b}$

In Exercises 51–56, find the *exact* value of the logarithm.

51. $\log_4 \sqrt[3]{4}$ **52.** $\log_8 \sqrt[4]{8}$

53. $\ln \dfrac{1}{\sqrt{e}}$ **54.** $\ln \sqrt[4]{e^3}$

55. $\log_5\left(\frac{1}{125}\right)$ **56.** $\log_7 \frac{49}{343}$

In Exercises 57–64, use the properties of logarithms to simplify the given logarithmic expression.

57. $\log_9 \frac{1}{18}$ **58.** $\log_5\left(\frac{1}{15}\right)$

59. $\log_7 \sqrt{70}$ **60.** $\log_2(4^2 \cdot 3^4)$

61. $\log_5\left(\frac{1}{250}\right)$ **62.** $\log_{10}\left(\frac{9}{300}\right)$

63. $\ln(5e^6)$ **64.** $\ln \dfrac{6}{e^2}$

In Exercises 65–84, use the properties of logarithms to expand the expression as a sum, difference, and/or multiple of logarithms. (Assume all variables are positive.)

65. $\log_3 4n$

66. $\log_6 6x$

67. $\log_5 \dfrac{x}{25}$

68. $\log_{10} \dfrac{y}{2}$

69. $\log_2 x^4$

70. $\log_2 z^{-3}$

71. $\ln \sqrt{z}$

72. $\ln \sqrt[3]{t}$

73. $\ln xyz$

74. $\ln \dfrac{xy}{z}$

75. $\ln \sqrt{a-1}, \ a > 1$

76. $\ln\left(\dfrac{x^2-1}{x^3}\right)$

77. $\ln\left[\dfrac{(z-1)^2}{z}\right]$

78. $\ln\left(\dfrac{x}{\sqrt{x^2+1}}\right)$

79. $\log_6 \dfrac{x^2}{y^3}$

80. $\log_9 \dfrac{\sqrt{y}}{z^2}$

81. $\ln \sqrt[3]{\dfrac{x}{y}}$

82. $\ln \sqrt{\dfrac{x^2}{y^3}}$

83. $\ln \sqrt[4]{x^3(x^2+3)}$

84. $\ln \sqrt{x^2(x+2)}$

In Exercises 85–100, condense the expression to the logarithm of a single quantity.

85. $\log_3 x + \log_3 5$

86. $\log_5 y + \log_5 x$

87. $\log_4 8 - \log_4 x$

88. $\log_{10} 4 - \log_{10} z$

89. $2 \log_{10}(x+4)$

90. $-4 \log_{10} 2x$

91. $-\ln x - 3 \ln 6$

92. $2 \ln 8 + 5 \ln z$

93. $\frac{1}{3} \ln 5x - \ln(x+1)$

94. $\frac{3}{2} \ln(z-2) + \ln z$

95. $\log_8(x-2) - \log_8(x+2)$

96. $3 \log_7 x + 2 \log_7 y - 4 \log_7 z$

97. $2 \ln 3 - \frac{1}{2} \ln(x^2+1)$

98. $\frac{3}{2} \ln t^6 - \frac{3}{4} \ln t^4$

99. $\ln x - \ln(x+2) - \ln(x-2)$

100. $\ln(x+1) + 2 \ln(x-1) + 3 \ln x$

Curve Fitting   In Exercises 101–104, find a logarithmic equation that relates x and y. Explain the steps used to find the equation.

101.

x	1	2	3	4	5	6
y	1	1.189	1.316	1.414	1.495	1.565

102.

x	1	2	3	4	5	6
y	1	1.587	2.080	2.520	2.924	3.302

103.

x	1	2	3	4	5	6
y	2.5	2.102	1.9	1.768	1.672	1.597

104.

x	1	2	3	4	5	6
y	0.5	2.828	7.794	16	27.951	44.091

105. Nail Length The approximate lengths and diameters (in inches) of common nails are shown in the table. Find a logarithmic equation that relates the diameter y of a common nail to its length x.

Length, x	Diameter, y	Length, x	Diameter, y
1	0.070	4	0.176
2	0.111	5	0.204
3	0.146	6	0.231

106. Galloping Speeds of Animals Four-legged animals run with two different types of motion: trotting and galloping. An animal that is trotting has at least one foot on the ground at all times, whereas an animal that is galloping has all four feet off the ground at some point in its stride. The number of strides per minute at which an animal breaks from a trot to a gallop depends on the weight of the animal. Use the table to find a logarithmic equation that relates an animal's lowest galloping speed y (in strides per minute) to its weight x (in pounds).

Weight, x	Galloping speed, y	Weight, x	Galloping speed, y
25	191.5	75	164.2
35	182.7	500	125.9
50	173.8	1000	114.2

107. Sound Intensity Use the level of sound equation in Example 8 to find the difference in loudness between an average office with an intensity of 1.26×10^{-7} watt per square meter and a broadcast studio with an intensity of 3.16×10^{-10} watt per square meter.

108. *Make a Decision*: **Graphical Analysis** Use a graphing utility to graph

$$f(x) = \ln 5x \quad \text{and} \quad g(x) = \ln 5 + \ln x$$

in the same viewing window. What do you observe about the two graphs? What property of logarithms is being demonstrated graphically?

109. *Make a Decision* You are helping another student learn the properties of logarithms. How would you use a graphing utility to demonstrate to this student the logarithmic property

$$\log_a u^v = v \log_a u$$

(u is a positive number, v is a real number, and a is a positive number such that $a \neq 1$)? What two functions could you use? Briefly describe your explanation of this property using these functions and their graphs.

110. **Reasoning** An algebra student claims that the following is true:

$$\log_a \frac{x}{y} = \frac{\log_a x}{\log_a y} = \log_a x - \log_a y.$$

Discuss how you would use a graphing utility to demonstrate that this claim is not true. Describe how to demonstrate the actual property of logarithms that is hidden in this faulty claim.

111. Complete the proof of the logarithmic property

$$\log_a \frac{u}{v} = \log_a u - \log_a v.$$

Let $\log_a u = x$ and $\log_a v = y.$

$a^x = \boxed{}$ and $a^y = \boxed{}$ Rewrite in exponential form.

$\dfrac{u}{v} = \dfrac{\boxed{}}{\boxed{}} = a^{\boxed{}}$ Divide and substitute for u and v.

$\boxed{} = x - y$ Rewrite in logarithmic form.

$\log_a \dfrac{u}{v} = \boxed{} - \boxed{}$ Substitute for x and y.

Math *MATTERS*

Archimedes and the Gold Crown

Archimedes was a citizen of Syracuse, a city on the island of Sicily. He was born in 287 B.C. and is often regarded (with Isaac Newton and Carl Gauss) as one of the three greatest mathematicians who ever lived. The following story gives some idea of Archimedes's ingenuity.

The king of Syracuse ordered a crown to be made of pure gold. When the crown was delivered, the king suspected that some silver had been mixed with the gold. He asked Archimedes to find a way of checking whether the crown was pure gold (without destroying the crown). One day, while Archimedes was at a public bath, he noticed that the level of the water rose when he stepped into the bath. He suddenly realized how he could solve the problem, and became so excited that he ran home, stark naked, shouting "Eureka!" (which means "I have found it!").

Archimedes's idea was that he could measure the volume of the crown by putting it into a bowl of water. If the crown contained any silver, the volume of the crown would be greater than the volume of an equal weight of pure gold, as shown in the accompanying figure.

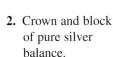

1. Crown and block of pure gold balance.

2. Crown and block of pure silver balance.

3. Gold block makes water level rise.

4. Silver block makes water level rise higher.

5. Crown makes water rise more than level gold block.

Mid-Chapter Quiz

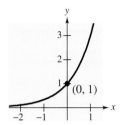

Figure for 1–4

Take this quiz as you would take a quiz in class. After you are done, check your work against the answers given in the back of the book.

In Exercises 1–4, use the graph of $f(x) = 3^x$ to sketch the graph of the function.

1. $g(x) = 3^x - 2$ **2.** $h(x) = 3^{-x}$

3. $k(x) = \log_3 x$ **4.** $j(x) = \log_3 (x - 1)$

5. For $P = \$5000$, $r = 6.5\%$, and $t = 10$ years, find the balance in an account if interest is compounded monthly and continuously.

6. The number of shares of stock y (in billions) sold on the New York Stock Exchange from 1990 to 2002 can be approximated by the model

$$y = 38.44(1.21)^t, \quad 0 \le t \le 12$$

where t represents the year, with $t = 0$ corresponding to 1990. (Source: U.S. Securities and Exchange Commission)

(a) Use the model to approximate the numbers of shares sold in 1997 and 2002.

(b) Use the model to predict the numbers of shares sold in 2006 and 2007.

7. The size of a bacteria population is modeled by $P(t) = 100e^{0.3012t}$, where t is the time in hours. Find (a) $P(0)$, (b) $P(6)$, and (c) $P(12)$.

8. Use the demand function $p = 6000\left(1 - \dfrac{9}{9 + e^{-0.002x}}\right)$ to find the price for a demand of $x = 500$ DVD players.

In Exercises 9–12, evaluate the expression without using a calculator.

9. $\log_{10} 100$ **10.** $\ln e^4$ **11.** $\log_4 \frac{1}{16}$ **12.** $\ln 1$

13. Sketch the graphs of $f(x) = 3^x$ and $g(x) = \log_3 x$ in the same coordinate plane. Identify the domains of f and g. Discuss the special relationship between f and g that is demonstrated by their graphs.

In Exercises 14 and 15, find the *exact* value of the logarithm.

14. $\log_7 \sqrt{343}$ **15.** $\ln \sqrt[5]{e^6}$

In Exercises 16 and 17, expand the logarithmic expression.

16. $\log_{10} \sqrt[3]{\dfrac{xy}{z}}$ **17.** $\ln\left(\dfrac{x^2 + 3}{x^3}\right)$

In Exercises 18 and 19, condense the logarithmic expression.

18. $\ln x + \ln y - \ln 3$ **19.** $-3 \log_{10} 4 - 3 \log_{10} x$

x	y
1	1
2	1.260
3	1.442
4	1.587
6	1.817
8	2.000

Table for 20

20. Find a logarithmic equation that relates x and y (see table at the left).

3.4 Solving Exponential and Logarithmic Equations

Objectives

- Solve an exponential equation.
- Solve a logarithmic equation.
- Use an exponential or a logarithmic model to solve an application problem.

Introduction

So far in this chapter, you have studied the definitions, graphs, and properties of exponential and logarithmic functions. In this section, you will study procedures for *solving equations* involving these exponential and logarithmic functions.

There are two basic strategies for solving exponential or logarithmic equations. The first is based on the One-to-One Properties and the second is based on the Inverse Properties. For $a > 0$ and $a \neq 1$, the following properties are true for all x and y for which $\log_a x$ and $\log_a y$ are defined.

One-to-One Properties	*Inverse Properties*
$a^x = a^y$ if and only if $x = y$.	$a^{\log_a x} = x$
$\log_a x = \log_a y$ if and only if $x = y$.	$\log_a a^x = x$

EXAMPLE 1 SOLVING SIMPLE EQUATIONS

Original Equation	Rewritten Equation	Solution	Property
a. $2^x = 32$	$2^x = 2^5$	$x = 5$	One-to-One
b. $\ln x - \ln 3 = 0$	$\ln x = \ln 3$	$x = 3$	One-to-One
c. $e^x = 7$	$\ln e^x = \ln 7$	$x = \ln 7$	Inverse
d. $\ln x = -3$	$e^{\ln x} = e^{-3}$	$x = e^{-3}$	Inverse
e. $\log_{10} x = -1$	$10^{\log_{10} x} = 10^{-1}$	$x = 10^{-1} = \frac{1}{10}$	Inverse

✓*CHECKPOINT*   Now try Exercise 1.

Strategies for Solving Exponential and Logarithmic Equations

1. Rewrite the original equation in a form that allows the use of the One-to-One Properties of exponential or logarithmic functions.

2. Rewrite an *exponential* equation in logarithmic form and apply the Inverse Property of logarithmic functions.

3. Rewrite a *logarithmic* equation in exponential form and apply the Inverse Property of exponential functions.

Solving Exponential Equations

EXAMPLE 2   SOLVING EXPONENTIAL EQUATIONS

Solve each equation and approximate the result to three decimal places.

a. $4^x = 72$ **b.** $3(2^x) = 42$

SOLUTION

a.

$4^x = 72$	Write original equation.
$\log_4 4^x = \log_4 72$	Take logarithm (base 4) of each side.
$x = \log_4 72$	Inverse Property
$x = \dfrac{\ln 72}{\ln 4} \approx 3.085$	Change-of-base formula

The solution is $x = \log_4 72 \approx 3.085$. Check this in the original equation.

b.

$3(2^x) = 42$	Write original equation.
$2^x = 14$	Divide each side by 3 to isolate the exponential expression.
$\log_2 2^x = \log_2 14$	Take log (base 2) of each side.
$x = \log_2 14$	Inverse Property
$x = \dfrac{\ln 14}{\ln 2} \approx 3.807$	Change-of-base formula

The solution is $x = \log_2 14 \approx 3.807$. Check this in the original equation.

✓CHECKPOINT   Now try Exercise 23.

In Example 2(a), the exact solution is $x = \log_4 72$ and the approximate solution is $x \approx 3.085$. An exact answer is preferred when the solution is an intermediate step in a larger problem. For a final answer, an approximate solution is easier to comprehend.

EXAMPLE 3   SOLVING AN EXPONENTIAL EQUATION

Solve $e^x + 5 = 60$ and approximate the result to three decimal places.

SOLUTION

$e^x + 5 = 60$	Write original equation.
$e^x = 55$	Subtract 5 from each side to isolate the exponential expression.
$\ln e^x = \ln 55$	Take natural log of each side.
$x = \ln 55$	Inverse Property
$x \approx 4.007$	Use a calculator.

The solution is $x = \ln 55 \approx 4.007$. Check this in the original equation.

✓CHECKPOINT   Now try Exercise 27.

TECHNOLOGY

When solving an exponential or logarithmic equation, remember that you can check your solution graphically by "graphing the left and right sides separately" and using the *intersect* feature of your graphing utility to determine the point of intersection. For instance, to check the solution of the equation in Example 2(a), you can graph

$$y_1 = 4^x \quad \text{and} \quad y_2 = 72$$

in the same viewing window, as shown below. Using the *intersect* feature of your graphing utility, you can determine that the graphs intersect when $x \approx 3.085$, which confirms the solution found in Example 2(a).

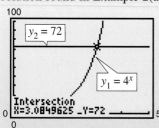

For instructions on how to use the *intersect* feature, see Appendix A; for specific keystrokes, go to the text website at *college.hmco.com*.

EXAMPLE 4   SOLVING AN EXPONENTIAL EQUATION

Solve $2(3^{2t-5}) - 4 = 11$ and approximate the result to three decimal places.

SOLUTION

$2(3^{2t-5}) - 4 = 11$	Write original equation.
$2(3^{2t-5}) = 15$	Add 4 to each side.
$3^{2t-5} = \dfrac{15}{2}$	Divide each side by 2.
$\log_3 3^{2t-5} = \log_3 \dfrac{15}{2}$	Take log (base 3) of each side.
$2t - 5 = \log_3 \dfrac{15}{2}$	Inverse Property
$2t = 5 + \log_3 7.5$	Add 5 to each side.
$t = \dfrac{5}{2} + \dfrac{1}{2}\log_3 7.5$	Divide each side by 2.
$t \approx 3.417$	Use a calculator.

The solution is $t = \frac{5}{2} + \frac{1}{2}\log_3 7.5 \approx 3.417$. Check this in the original equation.

✓*CHECKPOINT*   Now try Exercise 47.

When an equation involves two or more exponential expressions, you can still use a procedure similar to that demonstrated in Examples 2, 3, and 4. However, the algebra is a bit more complicated.

EXAMPLE 5   SOLVING AN EXPONENTIAL EQUATION OF QUADRATIC TYPE

Solve $e^{2x} - 3e^x + 2 = 0$.

SOLUTION

$e^{2x} - 3e^x + 2 = 0$		Write original equation.
$(e^x)^2 - 3e^x + 2 = 0$		Write in quadratic form.
$(e^x - 2)(e^x - 1) = 0$		Factor.
$e^x - 2 = 0$	➡ $x = \ln 2$	Set 1st factor equal to 0.
$e^x - 1 = 0$	➡ $x = 0$	Set 2nd factor equal to 0.

The solutions are $x = \ln 2$ and $x = 0$. Check these in the original equation. Or, perform a graphical check by graphing $y = e^{2x} - 3e^x + 2$ using a graphing utility. The graph should have two x-intercepts: $x = \ln 2$ and $x = 0$, as shown in Figure 3.18.

✓*CHECKPOINT*   Now try Exercise 49.

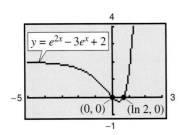

Figure 3.18

Solving Logarithmic Equations

To solve a logarithmic equation such as

$$\ln x = 3 \qquad \text{Logarithmic form}$$

write the equation in exponential form as follows.

$$e^{\ln x} = e^3 \qquad \text{Exponentiate each side.}$$

$$x = e^3 \qquad \text{Exponential form}$$

This procedure is called **exponentiating** each side of an equation.

EXAMPLE 6   SOLVING A LOGARITHMIC EQUATION

a. Solve $\ln x = 2$.

b. Solve $\log_3(5x - 1) = \log_3(x + 7)$.

SOLUTION

a.

$$\ln x = 2 \qquad \text{Write original equation.}$$

$$e^{\ln x} = e^2 \qquad \text{Exponentiate each side.}$$

$$x = e^2 \qquad \text{Inverse Property}$$

The solution is $x = e^2$. Check this in the original equation.

b.

$$\log_3(5x - 1) = \log_3(x + 7) \qquad \text{Write original equation.}$$

$$5x - 1 = x + 7 \qquad \text{One-to-One Property}$$

$$4x = 8 \qquad \text{Add } -x \text{ and 1 to each side.}$$

$$x = 2 \qquad \text{Divide each side by 4.}$$

The solution is $x = 2$. Check this in the original equation.

✓CHECKPOINT   Now try Exercise 61.

EXAMPLE 7   SOLVING A LOGARITHMIC EQUATION

Solve $2 \log_5 3x = 4$.

SOLUTION

$$2 \log_5 3x = 4 \qquad \text{Write original equation.}$$

$$\log_5 3x = 2 \qquad \text{Divide each side by 2.}$$

$$5^{\log_5 3x} = 5^2 \qquad \text{Exponentiate each side (base 5).}$$

$$3x = 25 \qquad \text{Inverse Property}$$

$$x = \frac{25}{3} \qquad \text{Divide each side by 3.}$$

The solution is $x = \frac{25}{3}$. Check this in the original equation.

✓CHECKPOINT   Now try Exercise 69.

| EXAMPLE 8 | SOLVING A LOGARITHMIC EQUATION |

Solve $5 + 2 \ln x = 4$ and approximate the result to three decimal places.

SOLUTION

$5 + 2 \ln x = 4$	Write original equation.
$2 \ln x = -1$	Subtract 5 from each side.
$\ln x = -\dfrac{1}{2}$	Divide each side by 2.
$e^{\ln x} = e^{-1/2}$	Exponentiate each side.
$x = e^{-1/2}$	Inverse Property
$x \approx 0.607$	Use a calculator.

The solution is $x = e^{-1/2} \approx 0.607$. Check this in the original equation.

✓**CHECKPOINT**   Now try Exercise 73.

Because the domain of a logarithmic function generally does not include all real numbers, you should be sure to check for extraneous solutions of logarithmic equations.

| EXAMPLE 9 | CHECKING FOR EXTRANEOUS SOLUTIONS |

Solve $\log_{10} 5x + \log_{10}(x - 1) = 2$.

SOLUTION

$\log_{10} 5x + \log_{10}(x - 1) = 2$	Write original equation.
$\log_{10}[5x(x - 1)] = 2$	Product Property of logarithms
$10^{\log_{10}(5x^2 - 5x)} = 10^2$	Exponentiate each side (base 10).
$5x^2 - 5x = 100$	Inverse Property
$x^2 - x - 20 = 0$	Write in general form.
$(x - 5)(x + 4) = 0$	Factor.
$x - 5 = 0$	Set 1st factor equal to 0.
$x = 5$	Solution
$x + 4 = 0$	Set 2nd factor equal to 0.
$x = -4$	Solution

The solutions appear to be $x = 5$ and $x = -4$. However, when you check these in the original equation, you can see that $x = 5$ is the only solution.

✓**CHECKPOINT**   Now try Exercise 75.

In Example 9, the domain of $\log_{10} 5x$ is $x > 0$ and the domain of $\log_{10}(x - 1)$ is $x > 1$, so the domain of the original equation is $x > 1$. Because the domain is all real numbers greater than 1, the solution $x = -4$ is extraneous.

TECHNOLOGY

You can use a graphing utility to verify that the equation in Example 9 has $x = 5$ as its only solution. Graph

$$y_1 = \log_{10} 5x + \log_{10}(x - 1)$$

and

$$y_2 = 2$$

in the same viewing window. From the graph shown below, it appears that the graphs of the two equations intersect at one point. Use the *intersect* feature or the *zoom* and *trace* features to determine that $x = 5$ is an approximate solution. You can verify this algebraically by substituting $x = 5$ into the original equation.

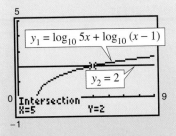

For instructions on how to use the *zoom* and *trace* features, see Appendix A; for specific keystrokes, go to the text website at *college.hmco.com*.

Applications

EXAMPLE 10   DOUBLING AND TRIPLING AN INVESTMENT

Make a Decision   You have deposited $500 in an account that pays 6.75% interest, compounded continuously.

a. How long will it take your money to double?

b. How long will it take your money to triple?

SOLUTION

Using the formula for compound interest with continuous compounding, you can find that the balance in the account is given by

$$A = Pe^{rt}$$

$$= 500e^{0.0675t}.$$

a. To find the time required for the balance to double, let $A = 1000$ and solve the resulting equation for t.

$500e^{0.0675t} = 1000$	Write original equation.
$e^{0.0675t} = 2$	Divide each side by 500.
$\ln e^{0.0675t} = \ln 2$	Take natural log of each side.
$0.0675t = \ln 2$	Inverse Property of logs and exponents
$t = \dfrac{1}{0.0675} \ln 2$	Divide each side by 0.0675.
$t \approx 10.27$	Use a calculator.

The balance in the account will double after approximately 10.27 years.

b. To find the time required for the balance to triple, let $A = 1500$ and solve the resulting equation for t.

$500e^{0.0675t} = 1500$	Write original equation.
$e^{0.0675t} = 3$	Divide each side by 500.
$\ln e^{0.0675t} = \ln 3$	Take natural log of each side.
$0.0675t = \ln 3$	Inverse Property of logs and exponents
$t = \dfrac{1}{0.0675} \ln 3$	Divide each side by 0.0675.
$t \approx 16.28$	Use a calculator.

The balance in the account will triple after approximately 16.28 years.

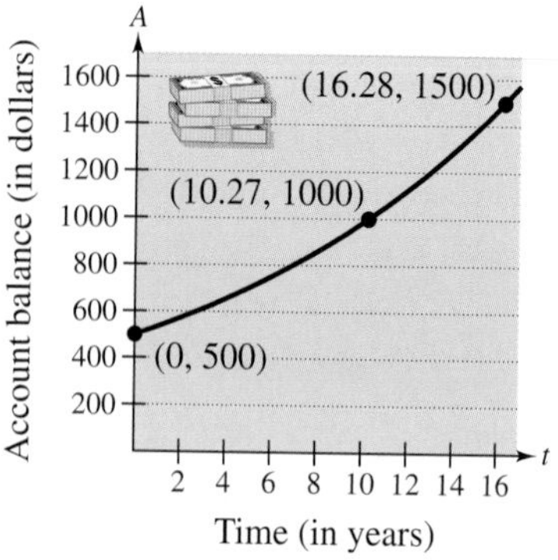

Notice that it took 10.27 years to earn the first $500 and only 6.01 years to earn the second $500. This result is graphically demonstrated in Figure 3.19.

✓CHECKPOINT   Now try Exercise 93.

Figure 3.19

Account balance (in dollars) / Time (in years)

(0, 500)   (10.27, 1000)   (16.28, 1500)

| EXAMPLE 11 | NEW YORK STOCK EXCHANGE |

From 1980 to 2002, the number of shares y (in millions) listed on the New York Stock Exchange can be approximated by

$$y = 31{,}669.99e^{0.11t}$$

where t represents the year, with $t = 0$ corresponding to 1980 (see Figure 3.20). (Source: New York Stock Exchange)

a. According to the model, during which year did the number of shares reach 175,000,000,000?

b. *Make a Decision* Do you think this model can be used to predict the number of shares listed on the New York Stock Exchange after the year 2002?

SOLUTION

a.

$31{,}669.99e^{0.11t} = y$	Write original model.
$31{,}669.99e^{0.11t} = 175{,}000$	Substitute 175,000 for y.
$e^{0.11t} \approx 5.526$	Divide each side by 31,669.99.
$\ln e^{0.11t} \approx \ln 5.526$	Take natural log of each side.
$0.11t \approx 1.709$	Inverse Property
$t \approx 16$	Divide each side by 0.11.

The solution is $t \approx 16$ years. Because $t = 0$ represents 1980, it follows that the number of shares reached 175,000,000,000 in 1996.

b. Recall that the graph of the exponential function given by $f(x) = e^x$ is always increasing, so you know that the given model is always increasing. Also, because the number of shares listed was increasing from 1980 to 2002, you can assume that this increasing trend will continue. So, the model could be used to predict the number of shares listed after the year 2002.

✓**CHECKPOINT**   Now try Exercise 105.

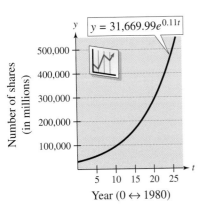

Figure 3.20

Discussing the Concept | Analyzing Relationships Numerically

Use a calculator to fill in the table row-by-row. Discuss the resulting pattern. What can you conclude? Find two equations that summarize the relationships you discovered.

x	$\frac{1}{2}$	1	2	10	25	50
e^x						
$\ln(e^x)$						
$\ln x$						
$e^{\ln x}$						

3.4 Warm Up

The following warm-up exercises involve skills that were covered in earlier sections. You will use these skills in the exercise set for this section. For additional help, review Sections R2.1, R2.4, 2.5, and 3.2.

In Exercises 1–6, solve for x.

1. $x \ln 2 = \ln 3$

2. $(x - 1) \ln 4 = 2$

3. $2xe^2 = e^3$

4. $4xe^{-1} = 8$

5. $x^2 - 4x + 5 = 0$

6. $2x^2 - 3x + 1 = 0$

In Exercises 7–10, simplify the expression.

7. $\log_{10} 10^x$

8. $\log_{10} 10^{2x}$

9. $\ln e^{2x}$

10. $\ln e^{-x^2}$

3.4 Exercises

In Exercises 1–10, solve for x.

1. $5^x = 125$

2. $2^x = 64$

3. $7^x = \frac{1}{49}$

4. $4^x = \frac{1}{256}$

5. $4^{2x-1} = 64$

6. $3^{x-1} = 27$

7. $\log_4 x = 3$

8. $\log_5 5x = 2$

9. $\log_{10} x = -1$

10. $\ln(2x - 1) = 0$

In Exercises 11–22, apply the Inverse Property of logarithmic or exponential functions to simplify the expression.

11. $\ln e^{x^2}$

12. $\ln e^{2x-1}$

13. $\log_{10} 10^{x^2} + 1$

14. $\log_{10} 10^{2x+3}$

15. $\log_5 5^{x^3} - 7$

16. $\log_8 8^{x^5} + 1$

17. $-8 + e^{\ln x^3}$

18. $-1 + \ln e^{2x}$

19. $10^{\log_{10}(x+5)}$

20. $10^{\log_{10}(x^2 + 7x + 10)}$

21. $2^{\log_2 x^2}$

22. $9^{\log_9(3x+7)}$

In Exercises 23–60, solve the exponential equation algebraically. Approximate the result to three decimal places.

23. $4(3^x) = 20$

24. $2(5^x) = 32$

25. $2e^x = 10$

26. $4e^x = 91$

27. $e^x - 9 = 19$

28. $6^x + 10 = 47$

29. $3^{2x} = 80$

30. $6^{5x} = 3000$

31. $5^{-t/2} = 0.20$

32. $4^{-3t} = 0.10$

33. $3^{x-1} = 28$

34. $2^{x-3} = 31$

35. $2^{3-x} = 565$

36. $8^{-2-x} = 431$

37. $8(10^{3x}) = 12$

38. $5(10^{x-6}) = 7$

39. $3(5^{x-1}) = 21$

40. $8(3^{6-x}) = 40$

41. $e^{3x} = 12$

42. $e^{2x} = 50$

43. $500e^{-x} = 300$

44. $1000e^{-4x} = 75$

45. $7 - 2e^x = 6$

46. $-14 + 3e^x = 11$

47. $6(2^{3x-1}) - 7 = 9$

48. $8(4^{6-2x}) + 13 = 41$

49. $e^{2x} - 4e^x - 5 = 0$

50. $e^{2x} - 5e^x + 6 = 0$

51. $e^{2x} - 3e^x - 4 = 0$

52. $e^{2x} - 9e^x - 36 = 0$

53. $\dfrac{500}{100 - e^{x/2}} = 20$

54. $\dfrac{400}{1 + e^{-x}} = 350$

55. $\dfrac{3000}{2 + e^{2x}} = 2$

56. $\dfrac{119}{e^{6x} - 14} = 7$

57. $\left(1 + \dfrac{0.065}{365}\right)^{365t} = 4$

58. $\left(4 - \dfrac{2.471}{40}\right)^{9t} = 21$

59. $\left(1 + \dfrac{0.10}{12}\right)^{12t} = 2$

60. $\left(16 - \dfrac{0.878}{26}\right)^{3t} = 30$

In Exercises 61–88, solve the logarithmic equation algebraically. Approximate the result to three decimal places.

61. $\ln x = -3$

62. $\ln x = 5$

63. $\ln 2x = 2.4$

64. $\ln 4x = 1$

65. $\log_{10} x = 6$

66. $\log_{10} 3z = 2$

67. $6 \log_3(0.5x) = 11$

68. $5 \log_{10}(x - 2) = 11$

69. $3 \ln 5x = 10$

70. $2 \ln x = 7$

71. $\ln \sqrt{x + 2} = 1$

72. $\ln \sqrt{x - 8} = 5$

73. $7 + 3 \ln x = 5$

74. $2 - 6 \ln x = 10$

75. $\ln x - \ln(x + 1) = 2$

76. $\ln x + \ln(x + 1) = 1$

77. $\ln x + \ln(x - 2) = 1$

78. $\ln x + \ln(x + 3) = 1$

79. $\ln(x + 5) = \ln(x - 1) - \ln(x + 1)$

80. $\ln(x + 1) - \ln(x - 2) = \ln x$

81. $\log_2(2x - 3) = \log_2(x + 4)$

82. $\log_{10}(x - 6) = \log_{10}(2x + 1)$

83. $\log_{10}(x + 4) - \log_{10} x = \log_{10}(x + 2)$

84. $\log_2 x + \log_2(x + 2) = \log_2(x + 6)$

85. $\log_4 x - \log_4(x - 1) = \frac{1}{2}$

86. $\log_3 x + \log_3(x - 8) = 2$

87. $\log_{10} 8x - \log_{10}\left(1 + \sqrt{x}\right) = 2$

88. $\log_{10} 4x - \log_{10}\left(12 + \sqrt{x}\right) = 2$

In Exercises 89–92, use a graphing utility to solve the equation. Approximate the result to three decimal places. Verify your result algebraically.

89. $2^x - 7 = 0$

90. $500 - 1500e^{-x/2} = 0$

91. $3 - \ln x = 0$

92. $10 - 4 \ln(x - 2) = 0$

Make a Decision: **Compound Interest** In Exercises 93 and 94, find the time required for a $1000 investment to double at interest rate r, compounded continuously.

93. $r = 0.085$

94. $r = 0.12$

Make a Decision: **Compound Interest** In Exercises 95 and 96, find the time required for a $1000 investment to triple at interest rate r, compounded continuously.

95. $r = 0.0675$

96. $r = 0.075$

97. Suburban Wildlife The variety of suburban nondomesticated wildlife in a community is approximated by the model

$$V = 15 \cdot 10^{0.02x}, \quad 0 \le x \le 36$$

where x is the number of months since the development was completed. Use this model to approximate the number of months since the development was completed when $V = 50$.

98. Native Prairie Grasses The number A of native prairie grasses per acre within a farming region is approximated by the model

$$A = 10.5 \cdot 10^{0.04x}, \quad 0 \le x \le 24$$

where x is the number of months since the acre was plowed. Use this model to approximate the number of months since the field was plowed in a test plot for which $A = 70$.

99. Demand Function The demand function for a limited edition die cast model car is given by

$$p = 4000\left(1 - \frac{6}{6 + e^{-0.003x}}\right).$$

(a) Find the demand x for a price of $p = \$350$.

(b) Find the demand x for a price of $p = \$300$.

(c) Use a graphing utility to confirm graphically the results found in parts (a) and (b).

100. Demand Function The demand function for an entertainment center is given by

$$p = 5000\left(1 - \frac{4}{4 + e^{-0.002x}}\right).$$

(a) Find the demand x for a price of $p = \$600$.

(b) Find the demand x for a price of $p = \$400$.

(c) Use a graphing utility to confirm graphically the results found in parts (a) and (b).

101. Forest Yield The yield V (in millions of cubic feet per acre) for a forest at age t years is given by

$$V = 6.7e^{-48.1/t}, \quad t > 0.$$

(a) Use a graphing utility to find the time necessary to obtain a yield of 1.3 million cubic feet per acre.

(b) Use a graphing utility to find the time necessary to obtain a yield of 2 million cubic feet per acre.

102. Human Memory Model In a group project on learning theory, a mathematical model for the percent P (in decimal form) of correct responses after n trials was found to be

$$P = \frac{0.83}{1 + e^{-0.2n}}, \quad n \ge 0.$$

(a) After how many trials will 60% of the responses be correct? (That is, for what value of n will $P = 0.6$?)

(b) Use a graphing utility to graph the memory model and confirm the result found in part (a).

(c) Write a paragraph describing the memory model.

103. U.S. Currency   The value y (in billions of dollars) of U.S. currency in circulation (outside the U.S. Treasury and not held by banks) from 1996 to 2002 can be approximated by the model

$$y = -218 + 333 \ln t, \quad 6 \leq t \leq 12$$

where t represents the year, with $t = 6$ corresponding to 1996. (Source: Board of Governors of the Federal Reserve System)

(a) Use a graphing utility to graph the model.

(b) Use a graphing utility to estimate the year when the value of U.S. currency in circulation exceeded $500 billion.

(c) Verify your answer to part (b) algebraically.

104. Retail Trade   The average monthly sales y (in billions of dollars) in retail in the United States from 1995 to 2002 can be approximated by the model

$$y = 24 + 99 \ln t, \quad 5 \leq t \leq 12$$

where t represents the year, with $t = 5$ corresponding to 1995. (Source: U.S. Council of Economic Advisors)

(a) Use a graphing utility to graph the model.

(b) Use a graphing utility to estimate the year when the average monthly sales exceeded $250 billion.

(c) Verify your answer to part (b) algebraically.

105. *Make a Decision*: Automobiles   Automobiles are designed with crumple zones that help protect their occupants in crashes. The crumple zones allow the occupants to move short distances when the automobiles come to abrupt stops. The greater the distance moved, the fewer g's the crash victims experience. (One g is equal to the acceleration due to gravity. For very short periods of time, humans have withstood as much as 40 g's.) In crash tests with vehicles moving at 90 kilometers per hour, analysts measured the numbers of g's experienced during deceleration by crash dummies that were permitted to move x meters during impact. The data are shown in the table.

x	0.2	0.4	0.6	0.8	1.0
g's	158	80	53	40	32

A model for these data is given by

$$y = -3.00 + 11.88 \ln x + \frac{36.94}{x}$$

where y is the number of g's.

(a) Complete the table using the model.

x	0.2	0.4	0.6	0.8	1.0
y					

(b) Use a graphing utility to graph the data points and the model in the same viewing window. How do they compare?

(c) Use the model to estimate the distance traveled during impact if the passenger deceleration must not exceed 30 g's.

(d) Do you think it is practical to lower the number of g's experienced during impact to fewer than 23? Explain your reasoning.

106. Average Heights   The percent of American males (between 20 and 29 years old) who are less than x inches tall is approximated by

$$m = -0.016 + \frac{1.037}{1 + e^{-0.5345(x - 69.41)}},$$

$$60 \leq x \leq 75$$

and the percent of American females (between 20 and 29 years old) who are less than x inches tall is approximated by

$$f = -0.013 + \frac{1.027}{1 + e^{-0.5868(x - 64.14)}},$$

$$56 \leq x \leq 72$$

where m and f are the percents (in decimal form) and x is the height (in inches) (see figure). (Source: U.S. National Center for Health Statistics)

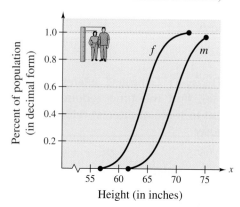

(a) What is the median height for each sex between 20 and 29 years old? (In other words, for what values of x are m and f equal to 0.5?)

(b) Write a paragraph describing each height model.

3.5   Exponential and Logarithmic Models

Objectives

- Construct and use a model for exponential growth or exponential decay.
- Use a Gaussian model to solve an application problem.
- Use a logistic growth model to solve an application problem.
- Use a logarithmic model to solve an application problem.
- Choose an appropriate model involving exponential or logarithmic functions for a real-life situation.

Introduction

The five most common types of mathematical models involving exponential functions and logarithmic functions are as follows.

1. Exponential growth model: $\qquad y = ae^{bx}, \qquad b > 0$

2. Exponential decay model: $\qquad y = ae^{-bx}, \qquad b > 0$

3. Gaussian model: $\qquad y = ae^{-(x-b)^2/c}$

4. Logistic growth model: $\qquad y = \dfrac{a}{1 + be^{-rx}}$

5. Logarithmic models: $\qquad y = a + b \ln x, \qquad y = a + b \log_{10} x$

The basic shape of each of these graphs is shown in Figure 3.21.

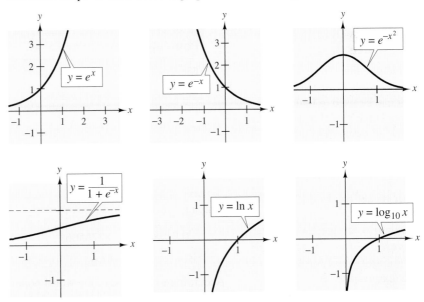

Figure 3.21

Exponential Growth and Decay

EXAMPLE 1 POPULATION INCREASE

The world population (in millions) from 1995 through 2003 is shown in the table. (Source: U.S. Census Bureau)

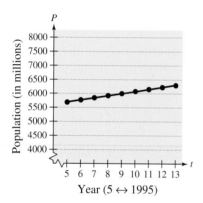

Year	Population
1995	5686
1996	5769
1997	5847
1998	5925
1999	6003

Year	Population
2000	6080
2001	6157
2002	6229
2003	6302

An exponential growth model that approximates these data is

$$P = 5342e^{0.012846t}, \quad 5 \le t \le 13$$

where P is the population (in millions) and $t = 5$ represents 1995. Compare the values given by the model with the estimates given by the U.S. Census Bureau. According to this model, when will the world population reach 6.7 billion?

SOLUTION

The following table compares the two sets of population figures. The graph of the model along with the original data values is shown in Figure 3.22.

Year	1995	1996	1997	1998	1999	2000	2001	2002	2003
Population	5686	5769	5847	5925	6003	6080	6157	6229	6302
Model	5696	5770	5845	5920	5997	6074	6153	6232	6313

To find when the world population will reach 6.7 billion, let $P = 6700$ in the model and solve for t.

$5342e^{0.012846t} = P$	Write original model.
$5342^{0.012846t} = 6700$	Let $P = 6700$.
$e^{0.012846t} \approx 1.25421$	Divide each side by 5342.
$\ln e^{0.012846t} \approx \ln 1.25421$	Take natural log of each side.
$0.012846t \approx 0.226506$	Inverse Property
$t \approx 17.6$	Divide each side by 0.012846.

According to the model, the world population will reach 6.7 billion in 2007.

✓CHECKPOINT Now try Exercise 25.

Figure 3.22

TECHNOLOGY

Some graphing utilities have an *exponential regression* feature that can be used to find an exponential model that represents data. If you have such a graphing utility, try using it to find a model for the data given in Example 1. How does your model compare with the model given in Example 1? For instructions on how to use the *regression* feature, see Appendix A; for specific keystrokes, go to the text website at *college.hmco.com*.

An exponential model increases (or decreases) by the same percent each year. What is the annual percent increase for the exponential model in Example 1?

In Example 1, you were given the exponential growth model. But suppose this model were not given; how could you find such a model? If you are given a set of data, as in Example 1, but you are not given the exponential growth model that fits the data, you can choose any two of the points and substitute them in the general exponential growth model $y = ae^{bx}$. This technique is demonstrated in Example 2.

EXAMPLE 2   FINDING AN EXPONENTIAL GROWTH MODEL

Find an exponential growth model whose graph passes through the points $(0, 4453)$ and $(7, 5024)$, as shown in Figure 3.23(a).

SOLUTION

The general form of the model is

$$y = ae^{bx}.$$

From the fact that the graph passes through the point $(0, 4453)$, you know that $y = 4453$ when $x = 0$. By substituting these values into the general model, you have

$$4453 = ae^0 \quad\Longrightarrow\quad a = 4453.$$

In a similar way, from the fact that the graph passes through the point $(7, 5024)$, you know that $y = 5024$ when $x = 7$. By substituting these values into the model, you obtain

$$5024 = 4453e^{7b} \quad\Longrightarrow\quad b = \frac{1}{7}\ln\frac{5024}{4453} \approx 0.01724.$$

So, the exponential growth model is

$$y = 4453e^{0.01724x}.$$

The graph of the model is shown in Figure 3.23(b).

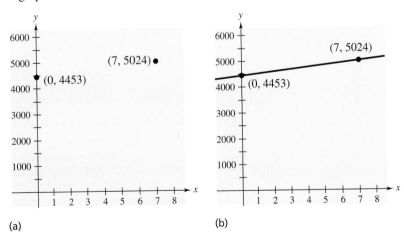

(a)   (b)

Figure 3.23

✓ **CHECKPOINT**   Now try Exercise 21.

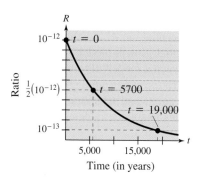

Figure 3.24

© Tom Bean/CORBIS

In 1960, Willard Libby of the University of Chicago won the Nobel Prize for Chemistry for the carbon 14 method, a valuable tool when estimating the age of ancient materials.

In living organic material, the ratio of the number of radioactive carbon isotopes (carbon 14) to the number of nonradioactive carbon isotopes (carbon 12) is about 1 to 10^{12}. When organic material dies, its carbon 12 content remains fixed, whereas its radioactive carbon 14 begins to decay with a half-life of about 5700 years. To estimate the age of dead organic material, scientists use the following formula, which denotes the ratio of carbon 14 to carbon 12 present at any time t (in years).

$$R = \frac{1}{10^{12}} e^{-t/8223} \qquad \text{Carbon dating model}$$

In Figure 3.24, note that R decreases as the time t increases. Any material that is composed of carbon, such as wood, bone, hair, pottery, paper, and water, can be dated.

EXAMPLE 3   CARBON DATING

The ratio of carbon 14 to carbon 12 in a newly discovered fossil is

$$R = \frac{1}{10^{13}}.$$

Estimate the age of the fossil.

SOLUTION

In the carbon dating model, substitute the given value of R to obtain the following.

$$\frac{1}{10^{12}} e^{-t/8223} = R \qquad \text{Write original model.}$$

$$\frac{e^{-t/8223}}{10^{12}} = \frac{1}{10^{13}} \qquad \text{Substitute } \frac{1}{10^{13}} \text{ for } R.$$

$$e^{-t/8223} = \frac{1}{10} \qquad \text{Multiply each side by } 10^{12}.$$

$$\ln e^{-t/8223} = \ln \frac{1}{10} \qquad \text{Take natural log of each side.}$$

$$-\frac{t}{8223} \approx -2.3026 \qquad \text{Inverse Property}$$

$$t \approx 18{,}934 \qquad \text{Multiply each side by } -8223.$$

So, you can estimate the age of the fossil to be about 19,000 years.

✓**CHECKPOINT**   Now try Exercise 29.

An exponential model can be used to determine the *decay* of radioactive isotopes. For instance, to find how much of an initial 10 grams of radioactive radium (^{226}Ra) with a half-life of 1599 years is left after 500 years, you would use the exponential decay model, as follows.

$$y = ae^{-bt} \implies \frac{1}{2}(10) = 10e^{-b(1599)} \implies \ln \frac{1}{2} = -1599b \implies -\frac{\ln \frac{1}{2}}{1599} = b$$

Using the value of b found above, $a = 10$, and $t = 500$, the amount left is

$$y = 10e^{-[-\ln(1/2)/1599](500)} \approx 8.05 \text{ grams.}$$

Gaussian Models

As mentioned at the beginning of this section, Gaussian models are of the form

$$y = ae^{-(x-b)^2/c}.$$

This type of model is commonly used in probability and statistics to represent populations that are **normally distributed.** For *standard* normal distributions, the model takes the form

$$y = \frac{1}{\sqrt{2\pi}} e^{-x^2/2}.$$

The graph of a Gaussian model is called a **bell-shaped curve.** Try to sketch the standard normal distribution curve with a graphing utility. Can you see why it is called a bell-shaped curve?

   The **average value** for a population can be found from the bell-shaped curve by observing where the maximum *y*-value of the function occurs. The *x*-value corresponding to the maximum *y*-value of the function represents the average value of the independent variable—in this case, *x*.

| EXAMPLE 4 | SAT SCORES | |

In 2003, the SAT (Scholastic Aptitude Test) mathematics scores for college-bound seniors in the United States roughly followed a normal distribution given by

$$y = 0.0035e^{-(x-519)^2/26,450}, \quad 200 \leq x \leq 800$$

where *x* is the SAT score for mathematics. Sketch the graph of this function. From the graph, estimate the average SAT score.   (Source: College Board)

SOLUTION

The graph of the function is shown in Figure 3.25. On this bell-shaped curve, the maximum value of the curve represents the average score. From the graph, you can estimate that the average mathematics score for college-bound seniors in 2003 was 519.

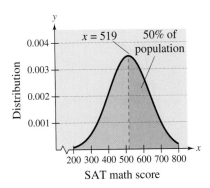

Figure 3.25

✓CHECKPOINT   Now try Exercise 39.

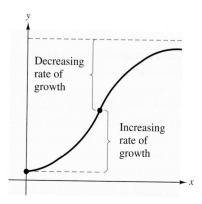

Figure 3.26 Logistic Curve

Logistic Growth Models

Some populations initially have rapid growth, followed by a declining rate of growth, as shown by the graph in Figure 3.26. One model for describing this type of growth pattern is the **logistic curve** given by the function

$$y = \frac{a}{1 + be^{-rx}},$$

where y is the population size and x is the time. An example is a bacteria culture that is initially allowed to grow under ideal conditions, followed by less favorable conditions that inhibit growth. A logistic growth curve is also called a **sigmoidal curve.**

EXAMPLE 5   SPREAD OF A VIRUS

On a college campus of 5000 students, one student returned from vacation with a contagious flu virus. The spread of the virus through the student population is given by

$$y = \frac{5000}{1 + 4999e^{-0.8t}}, \quad t \geq 0$$

where y is the total number of students infected after t days. The college will cancel classes when 40% or more of the students are infected.

a. How many students are infected after 5 days?

b. After how many days will the college cancel classes?

SOLUTION

a. After 5 days, the number of students infected is

$$y = \frac{5000}{1 + 4999e^{-0.8(5)}} = \frac{5000}{1 + 4999e^{-4}} \approx 54.$$

b. Classes are cancelled when the number infected is $(0.40)(5000) = 2000$. So, solve for t in the following equation.

$$2000 = \frac{5000}{1 + 4999e^{-0.8t}}$$

$$1 + 4999e^{-0.8t} = 2.5$$

$$e^{-0.8t} = \frac{1.5}{4999}$$

$$\ln e^{-0.8t} = \ln \frac{1.5}{4999}$$

$$-0.8t = \ln \frac{1.5}{4999}$$

$$t \approx 10.1$$

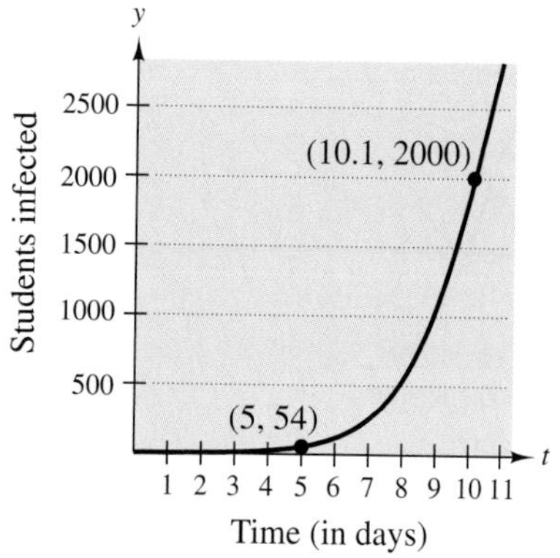

So, after 10 days, at least 40% of the students will be infected, and the college will cancel classes. The graph of the function is shown in Figure 3.27.

✓CHECKPOINT   Now try Exercise 41.

Students infected

2500

(10.1, 2000)
2000

1500

1000

500

(5, 54)

1 2 3 4 5 6 7 8 9 10 11

Time (in days)

Figure 3.27

Logarithmic Models

© Reinhardt/Zeitenspiegel/VISUM/The Image Works

The severity of the destruction caused by an earthquake depends on its magnitude and how long the earthquake occurs. Earthquakes can destroy buildings, and can cause landslides and tsunamis. Here, the Red Cross searches for survivors in an earthquake in Algeria.

EXAMPLE 6   MAGNITUDES OF EARTHQUAKES

On the Richter scale, the magnitude R of an earthquake of intensity I is given by

$$R = \log_{10}\left(\frac{I}{I_0}\right)$$

where $I_0 = 1$ is the minimum intensity used for comparison. Find the intensity per unit of area for each earthquake. (Intensity is a measure of the wave energy of an earthquake.)

a. Prince William Sound, Alaska, in 1964; $R = 9.2$

b. Central Alaska in 2002; $R = 7.9$

SOLUTION

a. Because $I_0 = 1$ and $R = 9.2$,

$$9.2 = \log_{10} I$$
$$I = 10^{9.2} \approx 1{,}584{,}893{,}000.$$

b. For $R = 7.9$,

$$7.9 = \log_{10} I$$
$$I = 10^{7.9} \approx 79{,}433{,}000.$$

Note that an increase of 1.3 units on the Richter scale (from 7.9 to 9.2) represents an intensity change by a factor of

$$\frac{1{,}584{,}893{,}000}{79{,}433{,}000} \approx 20.$$

In other words, the Prince William Sound earthquake of 1964 had a magnitude about 20 times greater than that of Central Alaska's earthquake in 2002.

✓*CHECKPOINT*   Now try Exercise 47.

EXAMPLE 7   pH LEVELS

Acidity, or pH level, is a measure of the hydrogen ion concentration $[H^+]$ (measured in moles of hydrogen per liter) of a solution. Use the model given by $pH = -\log_{10}[H^+]$ to determine the hydrogen ion concentration of milk of magnesia, which has a pH of 10.5.

SOLUTION

$pH = -\log_{10}[H^+]$	Write original model.
$10.5 = -\log_{10}[H^+]$	Substitute 10.5 for pH.
$-10.5 = \log_{10}[H^+]$	Multiply each side by -1.
$10^{-10.5} = 10^{\log_{10}[H^+]}$	Exponentiate each side (base 10).
$3.16 \times 10^{-11} = [H^+]$	Simplify.

So, the hydrogen ion concentration of milk of magnesia is 3.16×10^{-11} mole of hydrogen per liter.

✓*CHECKPOINT*   Now try Exercise 51.

Comparing Models

So far you have been given the type of model to use for a data set. Now you will use the general trends of the graphs of the five models presented in this section to choose appropriate models for real-life situations.

EXAMPLE 8   CHOOSING AN APPROPRIATE MODEL

Make a Decision   Decide whether to use an exponential growth model or a logistic growth model to represent each data set.

a.

b.

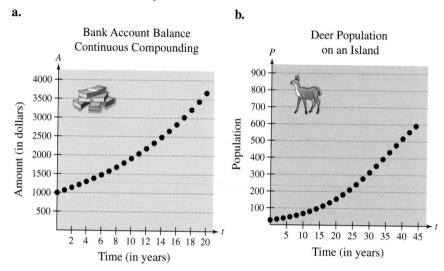

SOLUTION

a. As long as withdrawls and deposits are not made and the interest rate remains constant, the bank account balance will grow exponentially. So, an exponential growth model is an appropriate model.

b. The growth of the deer population will slow as the population approaches the carrying capacity of the island. So, a logistic growth model is an appropriate model.

✔CHECKPOINT   Now try Exercise 57.

Discussing the Concept | Exponential and Logarithmic Models

Use your school's library, the Internet, or some other reference source to find an application that fits one of the five models discussed in this section. After you have collected data for the model, use a graphing utility to create a scatter plot of the data. Then use the *regression* feature of the graphing utility to find a model that describes the data you have collected.

3.5 Warm Up

The following warm-up exercises involve skills that were covered in earlier sections. You will use these skills in the exercise set for this section. For additional help, review Sections 3.1, 3.2, and 3.4.

In Exercises 1–6, sketch the graph of the equation.

1. $y = e^{0.1x}$

2. $y = e^{-0.25x}$

3. $y = e^{-x^2/5}$

4. $y = \dfrac{2}{1 + e^{-x}}$

5. $y = \log_{10} 2x$

6. $y = \ln 4x$

In Exercises 7 and 8, solve the equation algebraically.

7. $3e^{2x} = 7$

8. $4 \ln 5x = 14$

In Exercises 9 and 10, solve the equation graphically.

9. $2e^{-0.2x} = 0.002$

10. $6 \ln 2x = 12$

3.5 Exercises

Compound Interest In Exercises 1–10, complete the table for a savings account in which interest is compounded continuously.

Initial Investment	Annual % Rate	Time to Double	Amount After 10 Years
1. $5000	7%		
2. $1000	$9\frac{1}{4}\%$		
3. $500		10 yr	
4. $10,000		5 yr	
5. $1000			$2281.88
6. $2000			$3000
7.	11%		$19,205
8.	8%		$20,000
9. $5000			$11,127.70
10. $250			$600

Radioactive Decay In Exercises 11–16, complete the table for the radioactive isotope.

Isotope	Half-Life (Years)	Initial Quantity	Amount After 1000 Years
11. ^{226}Ra	1599	4 g	
12. ^{226}Ra	1599		0.15 g
13. ^{14}C	5715		3.5 g
14. ^{14}C	5715	8 g	
15. ^{239}Pu	24,100		1.6 g
16. ^{239}Pu	24,100		0.38 g

In Exercises 17–20, classify the model as an exponential growth model or an exponential decay model.

17. $y = 3e^{0.5t}$

18. $y = 2e^{-0.6t}$

19. $y = 20e^{-1.5t}$

20. $y = 4e^{0.07t}$

In Exercises 21–24, find the constants C and k such that the exponential function $y = Ce^{kt}$ passes through the points on the graph.

21.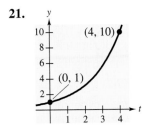

22.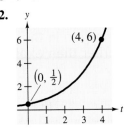

23.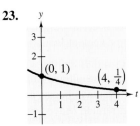

24.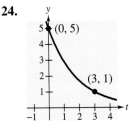

25. **Population** The population P of a city is given by $P = 105,300e^{0.015t}$, where t represents the year, with $t = 0$ corresponding to 2000. Sketch the graph of this equation. According to this model, in what year will the city have a population of 150,000?

26. **Population** The population P of a city is given by

$$P = 10,000e^{kt}$$

where t represents the year, with $t = 0$ corresponding to 2000. In 1955, the population was 3350. Find the value of k and use this result to predict the population in the year 2007.

27. **Bacteria Growth** The number of bacteria N in a culture is given by the model $N = 100e^{kt}$, where t is the time (in hours), with $t = 0$ corresponding to the time when $N = 100$. When $t = 8$, there are 225 bacteria. How long does it take the bacteria population to double in size? To triple in size?

28. **Bacteria Growth** The number of bacteria N in a culture is given by the model $N = 250e^{kt}$, where t is the time (in hours), with $t = 0$ corresponding to the time when $N = 250$. When $t = 10$, there are 280 bacteria. How long does it take the bacteria population to double in size? To triple in size?

29. **Carbon Dating** The ratio of carbon 14 to carbon 12 in a piece of wood discovered in a cave is $R = 1/8^{14}$. Estimate the age of the piece of wood.

30. **Carbon Dating** The ratio of carbon 14 to carbon 12 in a piece of paper buried in a tomb is $R = 1/13^{11}$. Estimate the age of the piece of paper.

IN THE NEWS

First City in the New World?

Peru's Caral suggests civilization emerged in the Americas 1,000 years earlier than experts believed

Six earth-and-rock mounds rise out of the windswept desert of the Supe Valley near the coast of Peru. . . . These are human-made pyramids, and compelling new evidence indicates they are the remains of a city that flourished nearly 5,000 years ago. If true, it would be the oldest urban center in the Americas and among the most ancient in all the world. . . .

In 1996, Shady's team [Peruvian archaeologist Ruth Shady Solís of San Marcos University] began the mammoth task of excavating Pirámide Mayor, the largest of the pyramids. After carefully clearing away several millennia's worth of rubble and sand, they unearthed staircases, circular walls covered with remnants of colored plaster, and squared brickwork. Finally in the foundation, they found the preserved remains of reeds woven into bags, known as *shicras*. The original workers, she surmised, must have filled these bags with stones from a hillside quarry a mile away and laid them atop one another inside retaining walls, gradually giving rise to the city of Caral's immense structures.

Shady knew that the reeds were ideal subjects for radiocarbon dating and could make her case. . . . In December 2000, Shady's suspicions were confirmed: the reeds were 4,600 years old. She took the news calmly, but [Jonathan] Haas [of Chicago's Field Museum] says he "was virtually in hysterics for three days afterward." . . . The news stunned other scientists. "It was almost unbelievable," says Betty Meggers, an archaeologist at the Smithsonian Institution. "This data pushed back the oldest known dates for an urban center in the Americas by more than 1,000 years."

Reprinted by permission of the author, John F. Ross.

31. Before the reeds were carbon dated, Shady suspected that Caral was about 1000 years older than nearby ancient civilizations, which were 3300 years old. Find the suspected ratio of carbon 14 to carbon 12 in the reeds.

32. Find the actual ratio of carbon 14 to carbon 12 in the reeds in December 2000.

33. **Radioactive Decay** What percent of a present amount of radioactive cesium (^{137}Cs) will remain after 100 years? Use the fact that radioactive cesium has a half-life of 30 years.

34. **Radioactive Decay** Find the half-life of radioactive iodine (^{131}I) if, after 20 days, 0.53 kilogram of an initial 3 kilograms remains.

35. **Learning Curve** The management at a factory has found that the maximum number of units a worker can produce in a day is 30. The learning curve for the number of units N produced per day after a new employee has worked t days is given by

$$N = 30(1 - e^{kt}).$$

After 20 days on the job, a particular worker produced 19 units in 1 day.

(a) Find the learning curve for this worker (first find the value of k).

(b) How many days should pass before this worker is producing 25 units per day?

36. Sales   The number of sales units S (in thousands) of a book after it has been on the market for t years is given by $S = 100(1 - e^{kt})$. Find S as a function of t if 15,000 books have been sold after 1 year. How many books will have been sold after 5 years?

37. *Make a Decision*: Cassette Shipments   From 1984 to 2002, manufacturers' shipments y (in millions) of cassettes can be approximated by the function

$$y = \begin{cases} e^{6.10 - 0.154t + 0.0195t^2}, & 4 \le t \le 8 \\ -34.05t + 771.0, & 9 \le t \le 22 \end{cases}$$

where $t = 4$ represents 1984.   (Source: Recording Industry Association of America)

(a) Use a graphing utility to graph the function.

(b) Describe the change in the growth pattern that occurred in 1989.

38. *Make a Decision*: Textbook Sales   The sales of a college textbook that was published in 1997 and revised in 2001 can be approximated by the function

$$y = \begin{cases} e^{0.12 + 2.939t - 0.2199t^2}, & 7 \le t \le 10 \\ e^{-14.7 + 4.69t - 0.22t^2}, & 11 \le t \le 14 \end{cases}$$

where y is the number of copies sold and $t = 7$ represents 1997.

(a) Use a graphing utility to graph the function.

(b) The graph is typical of college textbook sales. How would you explain this sales pattern?

39. Women's Heights   The distribution of heights of American women (between 30 and 39 years of age) can be approximated by the function

$$p = 0.148e^{-(x - 64.9)^2/14.45}, \quad 56 \le x \le 72$$

where x is the height (in inches) and p is the percent (in decimal form). Use a graphing utility to graph the function. Then determine the average height of women in this age bracket.   (Source: U.S. National Center for Health Statistics)

40. Men's Heights   The distribution of heights of American men (between 30 and 39 years of age) can be approximated by the function

$$p = 0.140e^{-(x - 70)^2/16.27}, \quad 62 \le x \le 75$$

where x is the height (in inches) and p is the percent (in decimal form). Use a graphing utility to graph the function. Then determine the average height of men in this age bracket.   (Source: U.S. National Center for Health Statistics)

41. Stocking a Lake with Fish   A lake is stocked with 500 fish, and the fish population P increases according to the logistic curve

$$P = \frac{10,000}{1 + 19e^{-t/5}}, \quad t \ge 0$$

where t is the time (in months).

(a) Use a graphing utility to graph the logistic curve.

(b) Find the fish population after 5 months.

(c) After how many months will the fish population reach 2000?

42. Endangered Species   A conservation organization releases 100 animals of an endangered species into a game preserve. The organization believes that the preserve has a carrying capacity of 1000 animals and that the growth of the herd will be modeled by the logistic curve

$$p = \frac{1000}{1 + 9e^{-kt}}, \quad t \ge 0$$

where p is the number of animals and t is the time (in years). The herd size is 134 after 2 years. Find k. Then find the population after 5 years.

43. Camping Equipment   The sales S (in millions of dollars) of camping equipment in the United States from 1996 to 2001 are shown in the table.   (Source: National Sporting Goods Association)

Year	Sales, S
1996	1127
1997	1153
1998	1204
1999	1265
2000	1354
2001	1370

(a) Use a graphing utility to create a scatter plot of the data. Let t represent the year, with $t = 6$ corresponding to 1996.

(b) Use the *regression* feature of a graphing utility to find an exponential model for the data. Use the Inverse Property $b = e^{\ln b}$ to rewrite the model as an exponential model in base e.

(c) Use a graphing utility to graph the model.

(d) Use the model to estimate camping equipment sales in 2002 and 2003.

44. Research and Development The percents P of research and development expenditures that were defense related in the United States from 1987 to 2001 are shown in the table. (Source: U.S. National Science Foundation)

Year	Percent, P	Year	Percent, P
1987	32	1995	19
1988	30	1996	18
1989	28	1997	17
1990	25	1998	16
1991	22	1999	15
1992	22	2000	13
1993	21	2001	14
1994	20		

(a) Use a graphing utility to create a scatter plot of the data. Let t represent the year, with $t = 7$ corresponding to 1987.

(b) Use the *regression* feature of a graphing utility to find an exponential model for the data. Use the Inverse Property $b = e^{\ln b}$ to rewrite the model as an exponential model in base e.

(c) Use a graphing utility to graph the model.

(d) Use the model to estimate the percent of research and development expenditures that were defense related in 2002 and 2003.

45. Make a Decision: Full-Length CDs The percents P of sound recordings that were full-length CDs in the United States from 1991 to 2002 are shown in the table. (Source: Recording Industry Association of America)

Year	Percent, P	Year	Percent, P
1991	38.9	1997	70.2
1992	46.5	1998	74.8
1993	51.1	1999	83.2
1994	58.4	2000	89.3
1995	65.0	2001	89.2
1996	68.4	2002	90.5

(a) Use a graphing utility to create a scatter plot of the data. Let t represent the year, with $t = 1$ corresponding to 1991.

(b) Use the *regression* feature of a graphing utility to find an exponential model for the data. Use the Inverse Property $b = e^{\ln b}$ to rewrite the model as an exponential model in base e.

(c) Use the *regression* feature of a graphing utility to find a logistic model for the data.

(d) Use a graphing utility to graph the exponential and logistic models with the original data. Use both models to estimate the percents of recordings that were full-length CDs in 2003 and 2004.

(e) Do both models give reasonable estimates? Explain.

46. Make a Decision: High School Graduates The table shows the percent P of the U.S. population (25 years old or older) from 1990 to 2002 who had not completed high school. (Source: U.S. Census Bureau)

Year	Percent, P	Year	Percent, P
1990	22.4	1997	17.9
1991	21.6	1998	17.2
1992	20.6	1999	16.6
1993	19.8	2000	15.9
1994	19.1	2001	15.9
1995	18.3	2002	15.9
1996	18.3		

(a) Use a graphing utility to create a scatter plot of the data. Let t represent the year, with $t = 0$ corresponding to 1990.

(b) Use the *regression* feature of a graphing utility to find a quadratic model and an exponential model for the data. Use the Inverse Property $b = e^{\ln b}$ to rewrite the exponential model in base e.

(c) Use a graphing utility to graph the quadratic model and the exponential model (in base e) with the original data.

(d) Use both models to predict the percents for the years 2005, 2006, and 2007. Do both models give reasonable predictions? Explain.

Earthquake Magnitudes In Exercises 47 and 48, use the Richter scale (see Example 6) for measuring the magnitude of earthquakes.

47. Find the magnitude R (on the Richter scale) of an earthquake of intensity I (let $I_0 = 1$).

(a) $I = 80,500,000$ (b) $I = 48,275,000$

48. Find the intensity I of an earthquake measuring R on the Richter scale (let $I_0 = 1$).

(a) Vanuatu Islands in 2002, $R = 7.3$

(b) Near coast of Peru in 2001, $R = 8.4$

Intensity of Sound In Exercises 49 and 50, find the level of sound using the following information for determining sound intensity. The level of sound L (in decibels) of a sound with an intensity of I is given by

$$L = 10 \log_{10} \frac{I}{I_0}$$

where I_0 is an intensity of 10^{-12} watt per square meter, corresponding roughly to the faintest sound that can be heard by the human ear.

49. (a) $I = 10^{-10}$ watt per m^2 (quiet room)

(b) $I = 10^{-5}$ watt per m^2 (busy street corner)

50. (a) $I = 10^{-3}$ watt per m^2 (loud car horn)

(b) $I \approx 10^0$ watt per m^2 (threshold of pain)

pH Levels In Exercises 51–54, use the acidity model given in Example 7.

51. Compute $[H^+]$ for a solution for which pH $= 5.8$.

52. Compute $[H^+]$ for a solution for which pH $= 7.3$.

53. A grape has a pH of 3.5, and baking soda has a pH of 8.0. The hydrogen ion concentration of the grape is how many times that of the baking soda?

54. The pH of a solution is decreased by one unit. The hydrogen ion concentration is increased by what factor?

55. Estimating the Time of Death At 8:30 A.M., a coroner was called to the home of a person who had died during the night. In order to estimate the time of death, the coroner took the person's temperature twice. At 9:00 A.M. the temperature was 85.7° F, and at 11:00 A.M. the temperature was 82.8° F. From these two temperature readings, the coroner was able to determine that the time elapsed since death and the body temperature are related by the formula

$$t = -10 \ln \frac{T - 70}{98.6 - 70}$$

where t is the time (in hours) elapsed since the person died and T is the temperature (in degrees Fahrenheit) of the person's body. The coroner assumed that the person had a normal body temperature of 98.6° F at death, and that the room temperature was a constant 70° F. Use this formula to estimate the time of death of the person.

56. *Make a Decision*: **Thawing a Package of Steaks** You take a three-pound package of steaks out of the freezer at 11 A.M. and place it in the refrigerator. Will the steaks be thawed in time to be grilled at 6 P.M.? Assume that the refrigerator temperature is 40° F and that the freezer temperature is 0° F. Use the formula for Newton's Law of Cooling

$$t = -5.05 \ln \frac{T - 40}{0 - 40}$$

where t is the time in hours (with $t = 0$ corresponding to 11 A.M.) and T is the temperature of the package of steaks (in degrees Fahrenheit).

57. *Make a Decision*: **Worker's Productivity** The number of units per day n that a new worker can produce after t days on the job is listed in the table. Use a graphing utility to create a scatter plot of the data. Do the data fit an exponential model or a logarithmic model? Use the *regression* feature of the graphing utility to find the model. Graph the model with the original data. Is the model a good fit? Can you think of a better model to use for these data? Explain.

Days, t	5	10	15	20	25
Units, n	6	13	22	34	56

58. *Make a Decision*: **Chemical Reaction** The table shows the yield y (in milligrams) of a chemical reaction after x minutes. Use a graphing utility to create a scatter plot of the data. Do the data fit an exponential model or a logarithmic model? Use the *regression* feature of the graphing utility to find the model. Graph the model with the original data. Is this model a good fit for the data?

Minutes, x	1	2	3	4	5	6	7	8
Yield, y	1.5	7.4	10.2	13.4	15.8	16.3	18.2	18.3

Make a Decision Project: Newton's Law of Cooling

t	T_1	T_2
0	182	176
2	180	170
8	160	158
10	156	156
21	138	140
24	134	136
28	128	132

When a warm object is placed in a cool room, its temperature will gradually change to the temperature of the room. (The same is true for a cool object placed in a warm room.)

According to Newton's Law of Cooling, the rate at which the object's temperature T changes is proportional to the difference between its temperature and the temperature of the room L. This statement can be translated as

$$T = L + Ce^{kt}$$

where k is a constant and $L + C$ is the original temperature of the object.

To confirm Newton's Law of Cooling experimentally, you can place a warm object in a cool room and monitor its temperature. The table at the left shows the temperatures (in degrees Fahrenheit) of 300 milliliters of water in a ceramic cup (T_1) and a Styrofoam cup (T_2) for several times t (in minutes). The room temperature at the time of data collection was 70°F. The data were collected using a cooking thermometer and a digital watch.

1. *Make a Decision*   Enter the time t and the temperature of the water in the ceramic cup T_1 in a graphing utility and create a scatter plot. Enter the time t and the temperature of the water in the Styrofoam cup T_2 in a graphing utility and create a scatter plot. If possible, plot both sets of data in the same viewing window. How are the scatter plots alike? How are the scatter plots different?

2. **Fit a Model to the Data**   Use the *regression* feature of a graphing utility to find an exponential model for the temperature of the water in the ceramic cup. How well does the model fit the data? Create a table that shows the time t, the temperature of the water T_1, and the difference d_1 between T_1 and the room temperature. Now use the *regression* feature to find an exponential model for the temperature difference data (d_1). For the model

$$y = a \cdot b^x$$

what are the values of a and b?

3. **Change a Model's Form and Compare**   Use the difference model from Question 2 and the fact that $b = e^{\ln b}$ to rewrite the model in the form $T - L = Ce^{kt}$. Now write the model for the temperature of a cooling object in the form

$$T = L + Ce^{kt}.$$

Construct a table that compares the values of this model to the actual data.

4. **Fit a Model to the Data**   Repeat Questions 2 and 3 using the data for the Styrofoam cup (T_2).

5. *Make a Decision*   Use a graphing utility to graph the models from Question 3 for both types of cups. What observations can you make? Which cup seems to keep the water warm for a longer period of time? Use the models to predict when the water in each cup will reach room temperature.

Chapter Summary

After studying this chapter, you should have acquired the following skills.
These skills are keyed to the Review Exercises that begin on page 246.
Answers to odd-numbered Review Exercises are given in the back of the book.

3.1 • Sketch the graph of an exponential function.   *Review Exercises 1–4, 9–16*

   • Use the compound interest formula.   *Review Exercises 17–20*

   • Use an exponential model to solve an application problem.   *Review Exercises 21, 22*

3.2 • Recognize and evaluate a logarithmic function.   *Review Exercises 23–36*

   • Sketch the graph of a logarithmic function.   *Review Exercises 5–8, 37–42*

   • Use a logarithmic model to solve an application problem.   *Review Exercises 43–46*

3.3 • Evaluate a logarithm using the change-of-base formula.   *Review Exercises 47–50*

   • Use properties of logarithms to evaluate or rewrite a logarithmic   *Review Exercises 51–58*
     expression.

   • Use properties of logarithms to expand or condense a   *Review Exercises 59–70*
     logarithmic expression.

   • Use logarithmic functions to model and solve real-life applications.   *Review Exercises 71, 72*

3.4 • Solve an exponential equation.   *Review Exercises 73–78*

   • Solve a logarithmic equation.   *Review Exercises 79–86*

   • Use an exponential or a logarithmic model to solve an   *Review Exercises 87, 88*
     application problem.

3.5 • Construct and use a model for exponential growth or   *Review Exercises 89–95*
     exponential decay.

   • Use a Gaussian model to solve an application problem.   *Review Exercise 96*

   • Use a logistic growth model to solve an application problem.   *Review Exercise 97*

   • Use a logarithmic model to solve an application problem.   *Review Exercises 98, 99*

   • Choose an appropriate model involving exponential or logarithmic   *Review Exercise 100*
     functions for a real-life situation.

Review Exercises

In Exercises 1–8, match the function with its graph. [The graphs are labeled (a), (b), (c), (d), (e), (f), (g), and (h).]

(a)
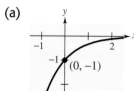

(b)
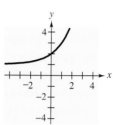

(c)
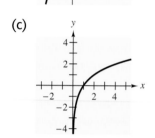

(d)
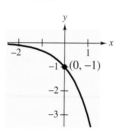

(e)
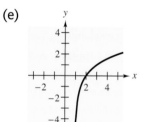

(f)
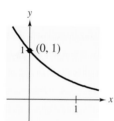

(g)
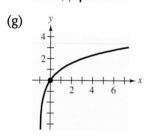

(h)
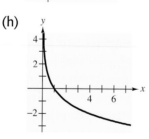

1. $f(x) = -3^x$
2. $f(x) = 3^{-x}$
3. $f(x) = -3^{-x}$
4. $f(x) = 2^x + 1$
5. $f(x) = \log_2 x$
6. $f(x) = \log_2(x - 1)$
7. $f(x) = -\log_2 x$
8. $f(x) = \log_2(x + 1)$

In Exercises 9–16, sketch the graph of the function.

9. $y = 4^x$
10. $y = e^{2x}$
11. $y = \left(\frac{1}{2}\right)^x$
12. $f(x) = \left(\frac{1}{2}\right)^{x+1}$
13. $f(x) = 3e^{0.2x}$
14. $f(x) = 4^{x-2}$
15. $y = 4^{-x^2}$
16. $f(x) = 100e^{-0.2x}$

Compound Interest In Exercises 17 and 18, complete the table to find the balance A for P dollars invested at a rate r for t years, compounded n times per year.

n	1	2	4	12	365	Continuous
A						

17. $P = \$3500$, $r = 10.5\%$, $t = 10$ years
18. $P = \$2000$, $r = 12\%$, $t = 30$ years

Compound Interest In Exercises 19 and 20, complete the table to determine the amount P that should be invested at a rate r to produce a final balance of $A = \$200{,}000$ in t years.

t	1	10	20	30	40	50
P						

19. $r = 8\%$, compounded continuously
20. $r = 10\%$, compounded monthly

21. **Investment Plan** You deposit \$8000 into a fund that yields 4.75% interest, compounded continuously. How much money will be in the fund after 5 years?

22. **Population** The population P of a town increases according to the model $P(t) = 5000e^{0.05t}$, where t is the time in years, with $t = 4$ corresponding to 2004. Use the model to approximate the population in 2006 and 2008.

In Exercises 23–26, use the definition of a logarithm to write the equation in logarithmic form.

23. $4^3 = 64$
24. $25^{3/2} = 125$
25. $e^2 = 7.3890 \ldots$
26. $e^x = 8$

In Exercises 27–30, use the definition of a logarithm to write the equation in exponential form.

27. $\log_3 81 = 4$
28. $\log_5 0.2 = -1$
29. $\ln 1 = 0$
30. $\ln 4 = 1.3862 \ldots$

In Exercises 31–36, evaluate the expression without using a calculator.

31. $\log_2 32$

32. $\log_9 3$

33. $\ln e^7$

34. $\log_4 \frac{1}{4}$

35. $\ln e^{-1/2}$

36. $\ln 1$

In Exercises 37 and 38, use the fact that f and g are inverse functions of each other to sketch their graphs in the same coordinate plane.

37. $f(x) = 10^x$, $g(x) = \log_{10} x$

38. $f(x) = e^x$, $g(x) = \ln x$

In Exercises 39–42, find the domain, vertical asymptote, and x-intercept of the logarithmic function. Then sketch its graph.

39. $f(x) = \log_3(x - 4)$

40. $f(x) = 6 + \log_{10} x$

41. $g(x) = \frac{1}{4}\ln x$

42. $g(x) = \ln(3 - x)$

43. Human Memory Model Students in a sociology class were given an exam and then retested monthly with an equivalent exam. The average score for the class is given by the human memory model

$$f(t) = 85 - 14 \log_{10}(t + 1), \quad 0 \le t \le 4$$

where t is the time (in months). How did the average score change over the four-month period?

44. Investment Time A principal P, invested at 6.75% interest and compounded continuously, increases to an amount that is K times the principal after t years, where t is given by

$$t = \frac{\ln K}{0.0675}.$$

Complete the table and describe the result.

K	1	2	3	4	6	8	10
t							

45. Antler Spread The antler spread a (in inches) and shoulder height h (in inches) of an adult American elk are related by the model

$$h = 116 \log_{10}(a + 40) - 176.$$

(a) Approximate the shoulder height of an elk with an antler spread of 55 inches.

(b) Use a graphing utility to graph the model.

46. Snow Removal The number of miles s of roads cleared of snow is approximated by the model

$$s = 25 - \frac{13 \ln(h/12)}{\ln 3}, \quad 2 \le h \le 15$$

where h is the depth of the snow in inches.

(a) Use the model to find s when $h = 10$ inches.

(b) Use a graphing utility to graph the model.

In Exercises 47–50, evaluate the logarithm using the change-of-base formula. Do each problem twice, once with common logarithms and once with natural logarithms. (Round your answer to three decimal places.)

47. $\log_3 10$

48. $\log_{1/4} 7$

49. $\log_{12} 200$

50. $\log_3 0.28$

In Exercises 51–54, approximate the logarithm using the properties of logarithms, given $\log_b 2 \approx 0.3562$, $\log_b 3 \approx 0.5646$, and $\log_b 5 \approx 0.8271$.

51. $\log_b 6$

52. $\log_b\left(\frac{4}{25}\right)$

53. $\log_b \sqrt{3}$

54. $\log_b 30$

In Exercises 55–58, find the *exact* value of the logarithm.

55. $\log_7 49$

56. $\log_6 \frac{1}{36}$

57. $\ln e^{3.2}$

58. $\ln \sqrt[5]{e^3}$

In Exercises 59–64, use the properties of logarithms to expand the expression as a sum, difference, and/or multiple of logarithms. (Assume all variables are positive.)

59. $\log_{10} \dfrac{x}{y}$

60. $\ln \dfrac{xy^2}{z}$

61. $\ln\left(x\sqrt{x - 3}\right)$

62. $\ln \sqrt[3]{\dfrac{x^3}{y^2}}$

63. $\log_{10}(y - 5)^3$

64. $\ln 2xyz$

In Exercises 65–70, condense the expression to the logarithm of a single quantity.

65. $\log_4 2 + \log_4 3$

66. $\ln y + 2 \ln z$

67. $\frac{1}{2} \ln x$

68. $4 \log_3 x + \log_3 y - 2 \log_3 z$

69. $\ln x - \ln(x - 3) - \ln(x + 1)$

70. $\log_{10}(x + 2) + 2 \log_{10} x - 3 \log_{10}(x + 4)$

71. Curve Fitting   Find a logarithmic equation that relates x and y (see figure).

x	1	2	3	4	5	6
y	1	2.520	4.327	6.350	8.550	10.903

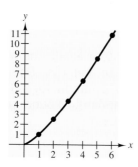

72. Human Memory Model   Students in a learning theory study were given an exam and then retested monthly for 6 months with an equivalent exam. The average scores for the class are shown in the table, with $t = 1$ representing 1 month after the initial exam. Use the table to find a logarithmic equation that relates t and s.

Month, t	Score, s
1	84.2
2	78.4
3	72.1
4	68.5
5	67.1
6	65.3

In Exercises 73–86, solve the equation.

73. $e^x = 8$

74. $4^{2x} = 17$

75. $45^{-3x} = 146$

76. $e^{2x} - 7e^x + 10 = 0$

77. $3^{2x+1} - 3 = 24$

78. $4^{3x-1} + 7 = 71$

79. $\ln 3x = 8.2$

80. $2 \log_3 4x = 15$

81. $-2 + \ln 5x = 0$

82. $\ln 4x^2 = 21$

83. $\ln x - \ln 3 = 2$

84. $\log_3 x - \log_3 4 = 5$

85. $\log_2 \sqrt[3]{x + 1} = 1$

86. $\ln \sqrt{x + 1} = 2$

87. Demand Function   The demand function for a desk is given by

$$p = 6000\left(1 - \frac{5}{5 + e^{-0.004x}}\right).$$

Find the demand x for a price of (a) $p = \$500$ and (b) $p = \$400$.

88. Demand Function   The demand function for a bicycle is given by

$$p = 4000\left(1 - \frac{3}{3 + e^{-0.004x}}\right).$$

Find the demand x for a price of (a) $p = \$700$ and (b) $p = \$400$.

Radioactive Decay   In Exercises 89 and 90, complete the table for the radioactive isotope.

Isotope	Half-Life (Years)	Initial Quantity	Amount After 1000 Years
89. ^{14}C	5715	12 g	
90. ^{239}Pu	24,100		3.1 g

91. Population   The population P of a city is given by

$$P = 270,000e^{0.019t}$$

where t represents the year, with $t = 5$ corresponding to 2005.

(a) Use a graphing utility to graph this equation.

(b) According to this model, in what year will the population of the city reach 320,000?

92. Population   The population P of a city is given by

$$P = 50,000e^{kt}$$

where t represents the year, with $t = 0$ corresponding to 2000. In 1990, the population was 34,500.

(a) Find the value of k and use this result to predict the population in the year 2030.

(b) Use a graphing utility to confirm the result of part (a).

93. Bacteria Growth   The number of bacteria N in a culture is given by the model

$$N = 200e^{kt}$$

where t is the time (in hours), with $t = 0$ corresponding to the time when $N = 200$. When $t = 6$, there are 500 bacteria. How long does it take the bacteria population to double in size? To triple in size?

94. Bacteria Growth The number of bacteria N in a culture is given by the model $N = 200e^{kt}$, where t is the time (in hours), with $t = 0$ corresponding to the time when $N = 200$. When $t = 5$, there are 350 bacteria. How long does it take for the bacteria population to double in size? To triple in size?

95. Learning Curve The management at a factory has found that the maximum number of units a worker can produce in a day is 50. The learning curve for the number of units N produced per day after a new employee has worked t days is given by

$$N = 50(1 - e^{kt}).$$

After 20 days on the job, a particular worker produced 31 units in 1 day.

(a) Find the learning curve for this worker.

(b) How many days should pass before this worker is producing 45 units per day?

96. Test Scores The scores on a general aptitude test roughly follow a normal distribution given by

$$y = 0.0028e^{-[(x-300)^2]/20,000}, \quad 100 \le x \le 500.$$

Sketch the graph of this function. Estimate the average score on this test.

97. Wildlife Management A state parks and wildlife department releases 100 deer into a wilderness area. The department believes that the carrying capacity of the area is 500 deer and that the growth of the herd will be modeled by the logistic curve

$$P = \frac{500}{1 + 4e^{-kt}}, \quad t \ge 0$$

where P is the number of deer and t is the time (in years).

(a) The herd size is 170 after 2 years. Find k.

(b) Find the population after 5 years.

98. Earthquake Magnitudes On the Richter scale, the magnitude R of an earthquake of intensity I is given by

$$R = \log_{10} \frac{I}{I_0}$$

where $I_0 = 1$ is the minimum intensity used for comparison. Find the intensity per unit of area for each value of R.

(a) $R = 8.4$ (b) $R = 6.85$ (c) $R = 9.1$

99. *Make a Decision*: Thawing a Package of Steaks You take a five-pound package of steaks out of a freezer at 10 A.M. and place it in the refrigerator. Will the steaks be thawed in time to be grilled at 6 P.M.? Assume that the refrigerator temperature is 40°F and the freezer temperature is 0° F. Use the formula

$$t = -0.46 \ln \frac{T - 40}{0 - 40}$$

where t is the time in hours (with $t = 0$ corresponding to 10 A.M.) and T is the temperature of the package of steaks (in degrees Fahrenheit).

100. *Make a Decision*: Broadway Shows The gross ticket sales s (in millions of dollars) for Broadway shows from 1994 to 2001 are shown in the table. (Source: The League of American Theaters and Producers, Inc.)

Year	Sales, s
1994	356
1995	406
1996	436
1997	499
1998	558
1999	588
2000	603
2001	666

(a) Use a graphing utility to create a scatter plot of the data. Let t represent the year, with $t = 4$ corresponding to 1994.

(b) Use the *regression* feature of a graphing utility to find an exponential model for the data. Use the Inverse Property $b = e^{\ln b}$ to rewrite the model as an exponential model in base e.

(c) Use the *regression* feature of a graphing utility to find a logarithmic model for the data.

(d) Use a graphing utility to graph the exponential and logarithmic models with the original data. Which model do you think best fits the data?

(e) Use both models to estimate the gross ticket sales in 2004 and 2005.

(f) Using your results from part (e), does each model give a reasonable estimate? Explain your reasoning.

Chapter Test

Take this test as you would take a test in class. After you are done, check your work against the answers given in the back of the book.

In Exercises 1–4, sketch the graph of the function.

1. $y = 2^x$

2. $y = e^{-2x}$

3. $y = \ln x$

4. $y = \log_3(x - 1)$

In Exercises 5 and 6, students in a psychology class were given an exam and then retested monthly with an equivalent exam. The average score for the class is given by the human memory model $f(t) = 87 - 15 \log_{10}(t + 1)$, $0 \le t \le 4$, where t is the time (in months).

5. What was the average score on the original exam? After 2 months? After 4 months?

6. The students in this psychology class participated in a study that required that they continue taking an equivalent exam every 6 months for 2 years. If the model remained valid, what would the average score be after 12 months? After 18 months? What could this indicate about human memory?

In Exercises 7–10, expand the logarithmic expression.

7. $\ln \dfrac{x^2 y^3}{z}$

8. $\log_{10} 3xyz^2$

9. $\log_2\left(x \sqrt[3]{x} - 2\right)$

10. $\log_8 \sqrt[5]{x^2 + 1}$

In Exercises 11 and 12, condense the logarithmic expression.

11. $\ln y + 2 \ln z - 3 \ln x$

12. $\frac{2}{3}(\log_{10} x + \log_{10} y)$

In Exercises 13–16, solve the equation.

13. $2^{4x} = 21$

14. $e^{2x} - 8e^x + 12 = 0$

15. $-3 + \log_2 4x = 0$

16. $\ln \sqrt{x + 2} = 3$

17. You deposit $30,000 into a fund that pays 7.25% interest, compounded continuously. When will the balance be greater than $100,000?

18. The population P of a city is given by $P = 70,000e^{0.023t}$, where t represents the year, with $t = 4$ corresponding to 2004. When will the city have a population of 100,000? Explain.

19. The number of bacteria N in a culture is given by $N = 100e^{kt}$, where t is the time (in hours), with $t = 0$ corresponding to the time when $N = 100$. When $t = 8$, $N = 500$. How long does it take the bacteria population to double?

20. Carbon 14 has a half-life of 5715 years. You have an initial quantity of 10 grams. How many grams will remain after 10,000 years? After 20,000 years?

21. If you are given the annual bear population on a small Alaskan island for the past decade, would you expect the bear population to grow exponentially or logistically? Explain your reasoning.

Cumulative Test: Chapters 1–3

Take this test as you would take a test in class. After you are done, check your work against the answers given in the back of the book.

In Exercises 1–6, use the functions given by $f(x) = x^2 + 1$ and $g(x) = 3x - 5$ to find the indicated function.

1. $(f + g)(x)$ **2.** $(f - g)(x)$ **3.** $(fg)(x)$

4. $\left(\dfrac{f}{g}\right)(x)$ **5.** $(f \circ g)(x)$ **6.** $(g \circ f)(x)$

In Exercises 7–11, sketch the graph of the function. Describe the domain and range of the function.

7. $f(x) = (x - 2)^2 + 3$ **8.** $g(x) = \dfrac{2}{x - 3}$ **9.** $h(x) = 2^{-x}$

10. $f(x) = \log_4 (x - 1)$ **11.** $g(x) = \begin{cases} x + 5, & x < 0 \\ 5, & x = 0 \\ x^2 + 5, & x > 0 \end{cases}$

12. The profit P (in dollars) for a software company is given by

$$P = -0.001x^2 + 150x - 175,000$$

where x is the number of units produced. What production level will yield a maximum profit?

In Exercises 13–15, perform the indicated operation and write the result in standard form.

13. $(10 + 2i)(3 - 4i)$ **14.** $(4 + 5i)^2$

15. Write the quotient in standard form: $\dfrac{1 + 2i}{3 - i}$.

16. Use the Quadratic Formula to solve $3x^2 - 5x + 7 = 0$.

17. Find all the zeros of $f(x) = x^4 + 10x^2 + 9$ given that $3i$ is a zero. Explain your reasoning.

18. Use long division or synthetic division to divide.

(a) $(6x^3 - 4x^2) \div (2x^2 + 1)$ (b) $(2x^4 + 3x^3 - 6x + 5) \div (x + 2)$

In Exercises 19 and 20, solve the equation.

19. $e^{2x} - 11e^x + 24 = 0$ **20.** $\frac{1}{3} \ln(x - 3) = 4$

21. The IQ scores for adults roughly follow the normal distribution given by

$$y = 0.0266e^{-(x - 100)^2/450}, \quad 70 \le x \le 130$$

where x is the IQ score. Use a graphing utility to graph the function. From the graph, estimate the average IQ score.

4 Systems of Equations and Inequalities

WYCOMBE ON A HIGH IN THE FA CUP — ALAN WATKINS HOW HAGUE MISSED

he markets are saying is that the fad for electronic technology is well and truly over — Anatole K.

THE TIME

DIEU ET MON DROIT

Why the Barmy

The Guard

E WALL STREET JOURNAL

INTERNATIONAL BUSINESS DAILY

Chapter 4

Chapter Sections

4.1 Solving Systems Using Substitution

4.2 Solving Systems Using Elimination

4.3 Linear Systems in Three or More Variables

4.4 Systems of Inequalities

4.5 Linear Programming

Make a Decision Are Newspapers on Their Way Out?

As a result of the growth of television news programming in the 1960s, newspapers are no longer the most dominant source of news for people in the United States.

From 1990 to 2002, the number y_1 of evening newspapers published in the United States decreased according to the linear model

$$y_1 = -33.4t + 1064, \quad 0 \le t \le 12.$$

From 1990 to 2002, the number y_2 of morning newspapers published in the United States increased according to the linear model

$$y_2 = 19.6t + 560, \quad 0 \le t \le 12.$$

In both models, t represents the year, with $t = 0$ corresponding to 1990. The graphs of these two models are shown at the right.   (Source: Editor & Publisher Co.)

Make a Decision Because the graphs intersect, what conclusion can you make?

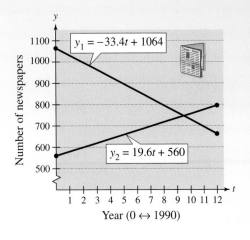

To explore this real-world application further, go to the Make a Decision Project on page 310.

253

4.1　Solving Systems Using Substitution

Objectives

- **Solve a system of equations by the method of substitution.**
- **Solve a system of equations graphically.**
- **Construct and use a system of equations to solve an application problem.**

The Method of Substitution

Up to this point in the text, most problems have involved either a function of one variable or a single equation in two variables. However, many problems in science, business, and engineering involve two or more equations in two or more variables. To solve such problems, you need to find the solutions of a **system of equations.** Here is an example of a system of two equations in x and y.

$$\begin{cases} 2x + y = 5 & \text{Equation 1} \\ 3x - 2y = 4 & \text{Equation 2} \end{cases}$$

A **solution** of this system is an ordered pair that satisfies each equation in the system. For instance, the ordered pair $(2, 1)$ is a solution of this system. To check this, you can substitute 2 for x and 1 for y in *each* equation.

$$2x + y = 5 \qquad \text{Write Equation 1.}$$
$$2(2) + 1 \stackrel{?}{=} 5 \qquad \text{Substitute 2 for } x \text{ and 1 for } y.$$
$$4 + 1 = 5 \qquad \text{Solution checks in Equation 1. } \checkmark$$

$$3x - 2y = 4 \qquad \text{Write Equation 2.}$$
$$3(2) - 2(1) \stackrel{?}{=} 4 \qquad \text{Substitute 2 for } x \text{ and 1 for } y.$$
$$6 - 2 = 4 \qquad \text{Solution checks in Equation 2. } \checkmark$$

Finding the set of all solutions is called **solving the system of equations.** There are several different ways to solve systems of equations. In this chapter, you will study three of the most common techniques, beginning with the **method of substitution.**

Method of Substitution

1. *Solve* one of the equations for one variable in terms of the other.

2. *Substitute* the expression found in Step 1 into the other equation to obtain an equation in one variable.

3. *Solve* the equation obtained in Step 2.

4. *Back-substitute* the solution in Step 3 into the expression obtained in Step 1 to find the value of the other variable.

5. *Check* that the solution satisfies *each* of the original equations.

When using the method of substitution to solve a system of equations, it does not matter which variable you solve for first. You will obtain the same solution regardless. When making your choice, you should choose the variable that is easier to work with. For instance, solve for a variable that has a coefficient of 1 or -1 to avoid working with fractions.

EXAMPLE 1   SOLVING A SYSTEM OF TWO EQUATIONS BY SUBSTITUTION

Solve the system of equations.

$$\begin{cases} x + y = 4 & \text{Equation 1} \\ x - y = 2 & \text{Equation 2} \end{cases}$$

SOLUTION

Begin by solving for y in Equation 1.

$$y = 4 - x \qquad \text{Revised Equation 1}$$

Next, substitute this expression for y into Equation 2 and solve the resulting single-variable equation for x.

$x - y = 2$	Write Equation 2.
$x - (4 - x) = 2$	Substitute $4 - x$ for y.
$x - 4 + x = 2$	Distributive Property
$2x = 6$	Combine like terms.
$x = 3$	Divide each side by 2.

Finally, you can solve for y by *back-substituting* $x = 3$ into the equation $y = 4 - x$, to obtain

$y = 4 - x$	Write revised Equation 1.
$y = 4 - 3$	Substitute 3 for x.
$y = 1$	Solve for y.

The solution is the ordered pair $(3, 1)$. You can check this as follows.

CHECK

$x + y = 4$	Write Equation 1.
$3 + 1 \stackrel{?}{=} 4$	Substitute for x and y.
$4 = 4$	Solution checks in Equation 1. ✓
$x - y = 2$	Write Equation 2.
$3 - 1 \stackrel{?}{=} 2$	Substitute for x and y.
$2 = 2$	Solution checks in Equation 2. ✓

✓CHECKPOINT   Now try Exercise 15.

The term *back-substitution* implies that you work *backwards*. First you solve for one of the variables, and then you substitute that value *back* into one of the equations in the system to find the value of the other variable.

Because many steps are required to solve a system of equations, it is easy to make errors in arithmetic. You should always *check your solution by substituting it into each equation in the original system.*

DISCOVERY

Use a graphing utility to graph $y_1 = -x + 4$ and $y_2 = x - 2$ in the same viewing window. Use the *trace* feature to find the coordinates of the point of intersection. Are the coordinates the same as the solution found in Example 1? Explain.

TECHNOLOGY

For instructions on how to use the *trace* feature, see Appendix A; for specific keystrokes, go to the text website at *college.hmco.com*.

| EXAMPLE 2 | SOLVING A SYSTEM BY SUBSTITUTION | |

You invest $12,000 in two funds paying 9% and 11% simple interest. The total yearly interest is $1180. How much of the $12,000 did you invest at each rate?

SOLUTION

Verbal Model:

$$\boxed{\begin{array}{c}9\% \\ \text{fund}\end{array}} + \boxed{\begin{array}{c}11\% \\ \text{fund}\end{array}} = \boxed{\begin{array}{c}\text{Total} \\ \text{investment}\end{array}}$$

$$\boxed{\begin{array}{c}9\% \\ \text{interest}\end{array}} + \boxed{\begin{array}{c}11\% \\ \text{interest}\end{array}} = \boxed{\begin{array}{c}\text{Total} \\ \text{interest}\end{array}}$$

Labels: Amount in 9% fund $= x$ (dollars)
          Interest for 9% fund $= 0.09x$ (dollars)
          Amount in 11% fund $= y$ (dollars)
          Interest for 11% fund $= 0.11y$ (dollars)
          Total investment $= \$12{,}000$ (dollars)
          Total interest $= \$1180$ (dollars)

System: $\begin{cases} x + \quad y = 12{,}000 & \text{Equation 1} \\ 0.09x + 0.11y = \quad 1180 & \text{Equation 2} \end{cases}$

To begin, it is convenient to multiply each side of Equation 2 by 100. This eliminates the need to work with decimals.

$$100(0.09x + 0.11y) = 100(1180) \qquad \text{Multiply each side by 100.}$$

$$9x + 11y = 118{,}000 \qquad \text{Revised Equation 2}$$

To solve this system, you can solve for x in Equation 1.

$$x = 12{,}000 - y \qquad \text{Revised Equation 1}$$

Then, substitute this expression for x into Equation 2, and solve the resulting equation for y.

$$9x + 11y = 118{,}000 \qquad \text{Write revised Equation 2.}$$

$$9(12{,}000 - y) + 11y = 118{,}000 \qquad \text{Substitute } 12{,}000 - y \text{ for } x.$$

$$108{,}000 - 9y + 11y = 118{,}000 \qquad \text{Distributive Property}$$

$$2y = 10{,}000 \qquad \text{Combine like terms.}$$

$$y = 5000 \qquad \text{Divide each side by 2.}$$

Next, back-substitute the value $y = 5000$ to solve for x.

$$x = 12{,}000 - y \qquad \text{Write revised Equation 1.}$$

$$x = 12{,}000 - 5000 \qquad \text{Substitute 5000 for } y.$$

$$x = 7000 \qquad \text{Solve for } x.$$

The solution is $(7000, 5000)$. So, you invested $7000 in the 9% interest fund and $5000 in the 11% interest fund. Check this in the original statement of the problem.

✓CHECKPOINT   Now try Exercise 21.

The equations in Examples 1 and 2 are linear. The method of substitution can also be used to solve systems in which one or both of the equations are nonlinear.

EXAMPLE 3   SUBSTITUTION: TWO-SOLUTION CASE

Solve the system of equations.

$$\begin{cases} x^2 - x - y = 1 & \text{Equation 1} \\ -x + y = -1 & \text{Equation 2} \end{cases}$$

SOLUTION

Begin by solving for y in Equation 2 to obtain $y = x - 1$. Next, substitute this expression for y into Equation 1 and solve for x.

$x^2 - x - y = 1$	Write Equation 1.
$x^2 - x - (x - 1) = 1$	Substitute $x - 1$ for y.
$x^2 - 2x + 1 = 1$	Simplify.
$x^2 - 2x = 0$	General form
$x(x - 2) = 0$	Factor.
$x = 0, 2$	Solve for x.

Back-substituting these values of x to solve for the corresponding values of y produces the two solutions $(0, -1)$ and $(2, 1)$. Check these solutions in the original system.

✓*CHECKPOINT*   Now try Exercise 27.

EXAMPLE 4   SUBSTITUTION: NO-REAL-SOLUTION CASE

Solve the system of equations.

$$\begin{cases} -x + y = 4 & \text{Equation 1} \\ x^2 + y = 3 & \text{Equation 2} \end{cases}$$

SOLUTION

Begin by solving Equation 1 for y to obtain $y = x + 4$. Next, substitute this expression for y into Equation 2, and solve for x.

$x^2 + y = 3$	Write Equation 2.
$x^2 + (x + 4) = 3$	Substitute $x + 4$ for y.
$x^2 + x + 1 = 0$	Simplify.
$x = \dfrac{-1 \pm \sqrt{1^2 - 4(1)(1)}}{2}$	Use the Quadratic Formula.
$x = \dfrac{-1 \pm \sqrt{-3}}{2}$	Simplify.

Because the discriminant is negative, the equation $x^2 + x + 1 = 0$ has no (real) solution. So, this system has no (real) solution.

✓*CHECKPOINT*   Now try Exercise 31.

Graphical Approach to Finding Solutions

From Examples 2, 3, and 4, you can see that a system of two equations in two unknowns can have exactly one solution, more than one solution, or no solution. In practice, you can gain insight about the location and number of solutions of a system of equations by graphing each of the equations in the same coordinate plane. The solution(s) of the system correspond to the **point(s) of intersection** of the graphs. For instance, in Figure 4.1(a), the two equations graph as two lines with a *single point* of intersection. The two equations in Example 3 graph as a parabola and a line with *two points* of intersection, as shown in Figure 4.1(b). Moreover, the two equations in Example 4 graph as a line and a parabola that happen to have *no points* of intersection, as shown in Figure 4.1(c).

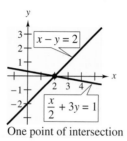
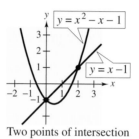
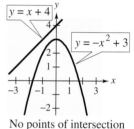

One point of intersection   Two points of intersection   No points of intersection

(a) One solution   (b) Two solutions   (c) No solution

Figure 4.1

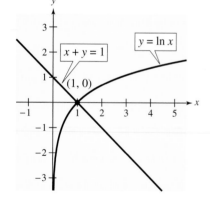

Figure 4.2

EXAMPLE 5   SOLVING A SYSTEM OF EQUATIONS GRAPHICALLY

Solve the system of equations.

$$\begin{cases} y = \ln x & \text{Equation 1} \\ x + y = 1 & \text{Equation 2} \end{cases}$$

SOLUTION

The graph of each equation is shown in Figure 4.2. From the graph, it is clear that there is only one point of intersection. Also, it appears that $(1, 0)$ is the solution point. You can confirm this by substituting 1 for x and 0 for y in *both* equations.

CHECK

Equation 1: $0 = \ln 1$ ✓   Equation 2: $1 + 0 = 1$ ✓

✓CHECKPOINT   Now try Exercise 37.

TECHNOLOGY

Your graphing utility may have an *intersect* feature that approximates the point(s) of intersection of two graphs. Use the *intersect* feature to verify the solution to Example 5. For instructions on how to use the *intersect* feature, see Appendix A; for specific keystrokes, go to the text website at *college.hmco.com*.

Applications

The total cost C of producing x units of a product typically has two components—the initial cost and the cost per unit. When enough units have been sold so that the total revenue R equals the total cost C, the sales are said to have reached the **break-even point.** You will find that the break-even point corresponds to the point of intersection of the cost and revenue curves.

EXAMPLE 6   BREAK-EVEN ANALYSIS

A shoe company invests $300,000 in equipment. Each pair of shoes costs $3 to produce and is sold for $60. How many pairs of shoes must be sold before the business breaks even?

SOLUTION

The total cost of producing x units is

Total cost	=	Cost per unit	·	Number of units	+	Initial cost

$$C = 3x + 300,000. \qquad \text{Equation 1}$$

The revenue obtained by selling x units is

Total revenue	=	Price per unit	·	Number of units

$$R = 60x. \qquad \text{Equation 2}$$

Because the break-even point occurs when $R = C$, you have $C = 60x$, and the system of equations to be solved is

$$\begin{cases} C = 3x + 300,000 & \text{Equation 1} \\ C = 60x & \text{Equation 2} \end{cases}.$$

Now you can solve by substitution.

$60x = 3x + 300,000$	Substitute $60x$ for C in Equation 1.
$57x = 300,000$	Subtract $3x$ from each side.
$x = \dfrac{300,000}{57}$	Divide each side by 57.
$x \approx 5263.$	Use a calculator.

The company must sell about 5263 pairs of shoes to break even. Note in Figure 4.3 that sales less than the break-even point correspond to an overall loss, whereas sales greater than the break-even point correspond to a profit.

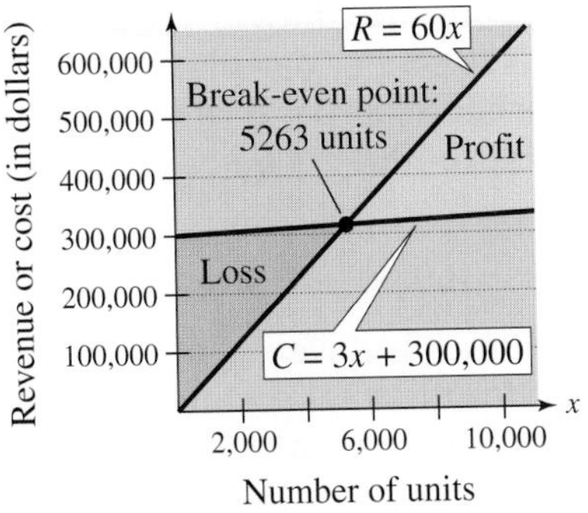

✓CHECKPOINT   Now try Exercise 47.

In 2003, the average price for athletic footwear in the United States was $38.88. (Source: National Sporting Goods Association)

© Royalty-Free/CORBIS

Figure 4.3

Revenue or cost (in dollars) vs. Number of units graph:
$R = 60x$
Break-even point: 5263 units
Profit
Loss
$C = 3x + 300,000$
600,000 / 500,000 / 400,000 / 300,000 / 200,000 / 100,000
2,000 / 6,000 / 10,000

Another way to view the solution in Example 6 is to consider the profit function $P = R - C$. The break-even point occurs when the profit is 0, which is the same as saying that $R = C$.

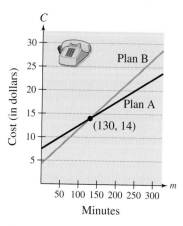

Figure 4.4

EXAMPLE 7   LONG–DISTANCE PHONE PLANS

Make a Decision You are choosing between two long-distance telephone companies. Company A charges $0.05 per minute plus a basic monthly fee of $7.50. Company B charges $0.075 per minute plus a basic monthly fee of $4.25. Which company should you choose if you use 100 long-distance minutes each month?

SOLUTION

Models for each long-distance phone plan are

$$C = 0.05m + 7.5 \qquad \text{Plan A}$$

$$C = 0.075m + 4.25 \qquad \text{Plan B}$$

where C is the monthly phone cost and m is the number of monthly long-distance minutes used. (See Figure 4.4.) Because the first equation has already been solved for C in terms of m, you can substitute this value into the second equation and solve for m, as follows.

$$0.05m + 7.5 = 0.075m + 4.25$$

$$0.05m - 0.075m = 4.25 - 7.5$$

$$-0.025m = -3.25$$

$$m = 130$$

Company B charges less than Company A when you use less than 130 long-distance minutes per month. You should choose Company B.

✓CHECKPOINT   Now try Exercise 51.

Discussing the Concept | Interpreting Points of Intersection

You plan to rent a 14-foot truck for a two-day local move. At truck rental agency A, you can rent a truck for $29.95 per day plus $0.99 per mile. At agency B, you can rent a truck for $50 per day plus $0.49 per mile. The total cost y (in dollars) for the truck from agency A is

$$y = (\$29.95 \text{ per day})(2 \text{ days}) + 0.99x = 59.90 + 0.99x$$

where x is the total number of miles the truck is driven. Write a total cost equation in terms of x and y for the total cost for the truck from agency B. Use a graphing utility to graph the two equations and find the point of intersection. Interpret the meaning of the point of intersection in the context of the problem. Which agency should you choose if you plan to travel a total of 50 miles over the two-day move? Why? How does the situation change if you plan to drive a total of 100 miles instead?

4.1 Warm Up

The following warm-up exercises involve skills that were covered in earlier sections. You will use these skills in the exercise set for this section. For additional help, review Sections R1.5, R2.1, R2.3, and 1.1.

In Exercises 1–4, sketch the graph of the equation.

1. $y = -\frac{1}{3}x + 6$

2. $y = 2(x - 3)$

3. $x^2 + y^2 = 4$

4. $y = 5 - (x - 3)^2$

In Exercises 5–8, perform the indicated operations and simplify.

5. $(3x + 2y) - 2(x + y)$

6. $(-10u + 3v) + 5(2u - 8v)$

7. $x^2 + (x - 3)^2 + 6x$

8. $y^2 - (y + 1)^2 + 2y$

In Exercises 9 and 10, solve the equation.

9. $3x + (x - 5) = 15 + 4$

10. $y^2 + (y - 2)^2 = 2$

4.1 Exercises

In Exercises 1–4, determine whether each ordered pair is a solution of the system of equations.

1. $\begin{cases} x + 4y = -3 \\ 5x - y = 6 \end{cases}$
   (a) $(-1, -1)$
   (b) $(1, -1)$

2. $\begin{cases} 2x - y = 2 \\ x + 3y = 8 \end{cases}$
   (a) $(2, 1)$
   (b) $(2, 2)$

3. $\begin{cases} y = -2e^x \\ 3x - y = 2 \end{cases}$
   (a) $(-2, 0)$
   (b) $(-1, 2)$

4. $\begin{cases} -\log_{10} x + 3 = y \\ \frac{1}{9}x + y = \frac{28}{9} \end{cases}$
   (a) $(1, 3)$
   (b) $\left(9, \frac{37}{9}\right)$

In Exercises 5–14, solve the system by the method of substitution. Then use the graph to confirm your solution.

5. $\begin{cases} x + y = -1 \\ -2x + y = -7 \end{cases}$

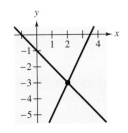

6. $\begin{cases} x - y = -5 \\ x + 2y = 4 \end{cases}$

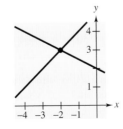

7. $\begin{cases} x - y = -3 \\ x^2 - y = -1 \end{cases}$

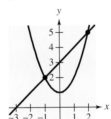

8. $\begin{cases} x^2 - y = 0 \\ x^2 - 4x + y = 0 \end{cases}$

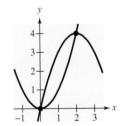

9. $\begin{cases} x - y = 0 \\ x^3 - 5x + y = 0 \end{cases}$

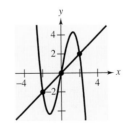

10. $\begin{cases} y = x^3 - 3x^2 + 3 \\ 2x + y = 3 \end{cases}$

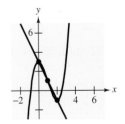

11. $\begin{cases} 3x + y = 4 \\ x^2 + y^2 = 16 \end{cases}$

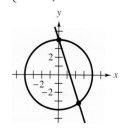

12. $\begin{cases} 3x - 4y = 18 \\ x^2 + y^2 = 36 \end{cases}$

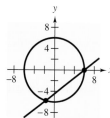

13. $\begin{cases} y = -x^2 + 1 \\ y = x^2 - 1 \end{cases}$

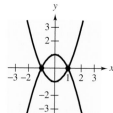

14. $\begin{cases} y = x^2 - 3x - 4 \\ y = -x^2 + 3x + 4 \end{cases}$

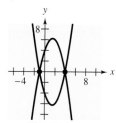

In Exercises 15–36, solve the system by the method of substitution.

15. $\begin{cases} 2x - y = -3 \\ -3x - 4y = -1 \end{cases}$

16. $\begin{cases} x + 2y = 1 \\ 5x - 4y = -23 \end{cases}$

17. $\begin{cases} 2x - y + 2 = 0 \\ 4x + y - 5 = 0 \end{cases}$

18. $\begin{cases} 6x - 3y - 4 = 0 \\ x + 2y - 4 = 0 \end{cases}$

19. $\begin{cases} x - y = 7 \\ 2x + y = 23 \end{cases}$

20. $\begin{cases} x - 2y = -2 \\ 3x - y = 6 \end{cases}$

21. $\begin{cases} 0.3x - 0.4y - 0.33 = 0 \\ 0.1x + 0.2y - 0.21 = 0 \end{cases}$

22. $\begin{cases} 1.5x + 0.8y = 2.3 \\ 0.3x - 0.2y = 0.1 \end{cases}$

23. $\begin{cases} \frac{1}{5}x + \frac{1}{2}y = 8 \\ x + y = 20 \end{cases}$

24. $\begin{cases} \frac{1}{2}x + \frac{3}{4}y = 10 \\ \frac{3}{2}x - y = 4 \end{cases}$

25. $\begin{cases} 6x + 5y = -3 \\ -x - \frac{5}{6}y = -7 \end{cases}$

26. $\begin{cases} -\frac{2}{3}x + y = 2 \\ 2x - 3y = 6 \end{cases}$

27. $\begin{cases} y = 2x \\ y = x^2 - 1 \end{cases}$

28. $\begin{cases} x + y = 4 \\ x^2 - y = 2 \end{cases}$

29. $\begin{cases} 3x - 7y + 6 = 0 \\ x^2 - y^2 = 4 \end{cases}$

30. $\begin{cases} x^2 + y^2 = 25 \\ 2x + y = 10 \end{cases}$

31. $\begin{cases} x - 2y = 4 \\ x^2 - y = 0 \end{cases}$

32. $\begin{cases} x^2 + y^2 = 5 \\ x - y = 1 \end{cases}$

33. $\begin{cases} y = x^4 - 2x^2 + 1 \\ y = 1 - x^2 \end{cases}$

34. $\begin{cases} y = x^3 - 2x^2 + x - 1 \\ y = -x^2 + 3x - 1 \end{cases}$

35. $\begin{cases} xy - 2 = 0 \\ y = \sqrt{x - 1} \end{cases}$

36. $\begin{cases} xy = 3 \\ y = \sqrt{x - 2} \end{cases}$

In Exercises 37–44, use a graphing utility to find the point(s) of intersection of the graphs. Then confirm your solution algebraically.

37. $\begin{cases} y = x^2 + 3x - 1 \\ y = -x^2 - 2x + 2 \end{cases}$

38. $\begin{cases} y = -2x^2 + x - 1 \\ y = x^2 - 2x - 1 \end{cases}$

39. $\begin{cases} x - y + 3 = 0 \\ x^2 - 4x + 7 = y \end{cases}$

40. $\begin{cases} x - y = 3 \\ x - y^2 = 1 \end{cases}$

41. $\begin{cases} y = e^x \\ x - y + 1 = 0 \end{cases}$

42. $\begin{cases} y = \sqrt{x} \\ y = x \end{cases}$

43. $\begin{cases} 2x - y + 3 = 0 \\ x^2 + y^2 - 4x = 0 \end{cases}$

44. $\begin{cases} x^2 + y^2 = 8 \\ y = x^2 + 4 \end{cases}$

Break-Even Analysis In Exercises 45–48, find the sales necessary to break even ($R = C$) for the cost C of producing x units and the revenue R obtained by selling x units. (Round your answer to the nearest whole unit.)

45. $C = 8650x + 250{,}000; R = 9950x$

46. $C = 5.5\sqrt{x} + 10{,}000; R = 3.29x$

47. $C = 2.65x + 350{,}000; R = 4.15x$

48. $C = 0.08x + 50{,}000; R = 0.25x$

49. Break-Even Analysis You invest $15,000 in equipment to make CDs. The CDs can be produced for $3.71 each and will be sold for $16.21 each. Estimate how many CDs you must sell to break even.

50. Break-Even Analysis You invest $5000 in a greenhouse. The planter, potting soil, and seed for each plant costs $6.43 and each plant will be sold for $12.68. Approximately how many plants must you sell to break even?

51. Comparing Populations From 1990 to 2002, the population of Oklahoma grew more slowly than that of Oregon. Two models that represent the populations of the two states are

$$\begin{cases} P = 30.8t + 3144 & \text{Oklahoma} \\ P = 57.0t + 2870 & \text{Oregon} \end{cases}$$

where P is the population (in thousands) and t represents the year, with $t = 0$ corresponding to 1990. Estimate when the population of Oregon overtook the population of Oklahoma. (Source: U.S. Census Bureau)

52. Comparing Populations From 1990 to 2002, the population of Wisconsin grew more slowly than that of Arizona. Two models that represent the populations of the two states are

$$\begin{cases} P = 45.7t + 4920 & \text{Wisconsin} \\ P = 156.0t + 3602 & \text{Arizona} \end{cases}$$

where P is the population (in thousands) and $t = 0$ represents 1990. Estimate when the population of Arizona overtook the population of Wisconsin. (Source: U.S. Census Bureau)

IN THE NEWS
A Wide World of Investment

We sent five professional investors to look for investments from around the world that would prove rewarding well into 2004. . . . They found plenty of opportunities to see profits relative to invested capital, but these were concentrated in a few industries: financials, pharmaceuticals and, interestingly, technology.

Our five virtual travelers are Caesar Bryan, manager of the Gabelli International Growth Fund and Gabelli Gold Fund; David Halpert, managing partner at Prince Street Capital Management, which runs an emerging-markets hedge fund; Ross Margolies, manager of the Salomon Brothers Capital Fund; Ronald Saba, principal at NorthRoad Capital Management; and Wendy Trevisani, an associate portfolio manager at Thornburg Investment Management in Santa Fe. . . .

Also in Europe, Saba likes **ING**, the Dutch financial company, which is trading at nine times earnings with a yield above 5%: "Sustainable ROE [return on equity], probably 13% to 14%. Good quality company. It's had a tremendous move since March, but that's only because it was trading at five, six times earnings with a yield well north of 6%."

If juicy bank dividends appeal, consider Saba's other picks: **Danske Bank,** trading around ten times earnings and yielding 4%, and ING's domestic competitor, **ABN Amro,** which costs ten times its annual earnings and yields 5%. Meanwhile, **Lloyd's TSB Group,** a British High Street bank, yields more than 8% and also costs ten times its earnings. . . .

Reprinted by Permission of Forbes Magazine.
Copyright © 2004 Forbes, Inc.

---

53. A total of $25,000 is invested in two funds paying 8% and 14% simple interest. The total yearly interest is $2900. How much of the $25,000 is invested at each rate?

54. A total of $25,000 is invested in two funds paying 9% and 13% simple interest. The total yearly interest is $2890. How much of the $25,000 is invested at each rate?

---

55. Choice of Two Jobs You are offered two different sales jobs. One company offers an annual salary of $24,500 plus a year-end bonus of 2% of your total sales. The other company offers a salary of $18,000 plus a year-end bonus of 6% of your total sales. How much would you have to sell in a year in order to earn more money from the second company?

56. *Make a Decision*: **Alternative Fuel** From 1995 to 2001, the fuel consumption of electric vehicles increased more slowly than that of vehicles powered by 85% ethanol. Two models that approximate the fuel consumption are

$$\begin{cases} G = 96.03t^2 - 1007.5t + 3154 & \text{Electric} \\ G = 240.79t^2 - 2364.5t + 5873 & \text{85\% Ethanol} \end{cases}$$

where G is fuel consumption (in thousands of gasoline-equivalent gallons) and t represents the year, with $t = 5$ corresponding to 1995. Use a graphing utility to determine whether, according to the models, the fuel consumption of vehicles run by 85% ethanol exceeded the fuel consumption of vehicles run by electricity. If so, in what year did this happen? (Source: Energy Information Association)

57. *Make a Decision*: **SAT or ACT?** The number of participants in SAT and ACT testing from 1990 to 2002 can be approximated by the models

$$\begin{cases} y = 2.25t^2 - 0.8t + 1025 & \text{SAT} \\ y = -0.009t^3 + 0.56t^2 + 20.6t + 800 & \text{ACT} \end{cases}$$

where y is the number of participants (in thousands) and t represents the year, with $t = 0$ corresponding to 1990. Use a graphing utility to determine whether, according to the models, the number of participants in ACT testing will overtake the number of participants in SAT testing. Do you think these models will continue to be accurate? Explain your reasoning. (Source: College Entrance Examination Board; ACT, Inc.)

4.2   Solving Systems Using Elimination

Objectives

- Solve a linear system by the method of elimination.
- Interpret the solution of a linear system graphically.
- Construct and use a linear system to solve an application problem.

The Method of Elimination

In Section 4.1, you studied two methods for solving a system of equations: substitution and graphing. In this section, you will study a third method called the **method of elimination.** The key step in the method of elimination is to obtain, for one of the variables, coefficients that differ only in sign, so that *adding* the two equations eliminates this variable. The following system provides an example.

$$3x + 5y = 7 \qquad \text{Equation 1}$$
$$\underline{-3x - 2y = -1} \qquad \text{Equation 2}$$
$$3y = 6 \qquad \text{Add equations.}$$

Note that by adding the two equations, you eliminate the variable x and obtain a single equation in y. Solving this equation for y produces $y = 2$, which you can then back-substitute into one of the original equations to solve for x.

EXAMPLE 1   THE METHOD OF ELIMINATION

Solve the system of linear equations.

$$\begin{cases} 3x + 2y = 4 & \text{Equation 1} \\ 5x - 2y = 8 & \text{Equation 2} \end{cases}$$

STUDY TIP

The method of substitution can also be used to solve the system in Example 1. Use substitution to solve the system. Which method do you think is easier? Many people find that the method of elimination is more efficient.

SOLUTION

Because the coefficients for y differ only in sign, you can eliminate the y-terms by adding the two equations. This leaves you with an equation in only one variable.

$$3x + 2y = 4 \qquad \text{Write Equation 1.}$$
$$\underline{5x - 2y = 8} \qquad \text{Write Equation 2.}$$
$$8x = 12 \qquad \text{Add equations.}$$

So, $x = \frac{3}{2}$. By back-substituting this value into Equation 1, you can solve for y, as follows.

$$3x + 2y = 4 \qquad \text{Write Equation 1.}$$
$$3\left(\tfrac{3}{2}\right) + 2y = 4 \qquad \text{Substitute } \tfrac{3}{2} \text{ for } x.$$
$$y = -\tfrac{1}{4} \qquad \text{Solve for } y.$$

The solution is $\left(\tfrac{3}{2}, -\tfrac{1}{4}\right)$. Check this in the original system.

✓CHECKPOINT   Now try Exercise 11.

To obtain coefficients (for one of the variables) that differ only in sign, you often need to multiply one or both of the equations by a suitable constant, as demonstrated in Example 2.

EXAMPLE 2 THE METHOD OF ELIMINATION

Solve the system of linear equations.

$$\begin{cases} 2x - 3y = -7 & \text{Equation 1} \\ 3x + y = -5 & \text{Equation 2} \end{cases}$$

SOLUTION

For this system, you can obtain coefficients that differ only in sign by multiplying Equation 2 by 3. Then, by adding the two equations, you can eliminate the y-terms, leaving you with an equation in only one variable.

$$\begin{array}{llll} 2x - 3y = -7 & \Longrightarrow & 2x - 3y = -7 & \text{Write Equation 1.} \\ 3x + y = -5 & \Longrightarrow & 9x + 3y = -15 & \text{Multiply Equation 2 by 3.} \\ & & \overline{11x \quad = -22} & \text{Add equations.} \end{array}$$

By dividing each side by 11, you can see that $x = -2$. By back-substituting this value of x into Equation 1, you can solve for y.

$$\begin{array}{ll} 2x - 3y = -7 & \text{Write Equation 1.} \\ 2(-2) - 3y = -7 & \text{Substitute } -2 \text{ for } x. \\ -3y = -3 & \text{Add 4 to each side.} \\ y = 1 & \text{Solve for } y. \end{array}$$

The solution is $(-2, 1)$. Check this in the original system, as follows.

CHECK

$$\begin{array}{ll} 2(-2) - 3(1) \stackrel{?}{=} -7 & \text{Substitute into Equation 1.} \\ -4 - 3 = -7 & \text{Equation 1 checks. ✓} \\ 3(-2) + 1 \stackrel{?}{=} -5 & \text{Substitute into Equation 2.} \\ -6 + 1 = -5 & \text{Equation 2 checks. ✓} \end{array}$$

The ordered pair $(-2, 1)$ is a solution of both equations. So, $(-2, 1)$ is a solution of the original system.

✓CHECKPOINT Now try Exercise 13.

In Example 2, the two systems of linear equations

$$\begin{cases} 2x - 3y = -7 \\ 3x + y = -5 \end{cases} \quad \text{and} \quad \begin{cases} 2x - 3y = -7 \\ 9x + 3y = -15 \end{cases}$$

are called **equivalent systems** because they have precisely the same solution set. The operations that can be performed on a system of linear equations to produce an equivalent system are (1) interchanging any two equations, (2) multiplying an equation by a nonzero constant, and (3) adding a multiple of one equation to any other equation in the system.

The Method of Elimination

To use the **method of elimination** to solve a system of two linear equations in x and y, use the following steps.

1. Examine the system to determine which variable can most easily be eliminated.

2. Obtain coefficients for x (or y) that differ only in sign by multiplying all terms of one or both equations by suitably chosen constants.

3. Add the equations to eliminate one variable and solve the resulting equation.

4. Back-substitute the value obtained in Step 3 into either of the original equations and solve for the other variable.

5. Check your solution in both of the original equations.

EXAMPLE 3   THE METHOD OF ELIMINATION

Solve the system of linear equations.

$$\begin{cases} 5x + 3y = 9 & \text{Equation 1} \\ 2x - 4y = 14 & \text{Equation 2} \end{cases}$$

SOLUTION

You can obtain coefficients of y that differ only in sign by multiplying Equation 1 by 4 and multiplying Equation 2 by 3.

$$
\begin{aligned}
5x + 3y &= 9 &\Longrightarrow&& 20x + 12y &= 36 &&\text{Multiply Equation 1 by 4.} \\
2x - 4y &= 14 &\Longrightarrow&& 6x - 12y &= 42 &&\text{Multiply Equation 2 by 3.} \\
&&&& \overline{26x} &= 78 &&\text{Add equations.}
\end{aligned}
$$

From this equation, you can see that $x = 3$. By back-substituting this value of x into the second equation, you can solve for y, as follows.

$$
\begin{aligned}
2x - 4y &= 14 &&\text{Write Equation 2.} \\
2(3) - 4y &= 14 &&\text{Substitute 3 for } x. \\
-4y &= 8 &&\text{Subtract 6 from each side.} \\
y &= -2 &&\text{Solve for } y.
\end{aligned}
$$

The solution is $(3, -2)$. Check this in the original system.

✔CHECKPOINT   Now try Exercise 15.

Remember that you can check the solution of a system of equations graphically. For instance, to check the solution found in Example 3, graph both equations in the same viewing window, as shown in Figure 4.5. Notice that the two lines intersect at $(3, -2)$.

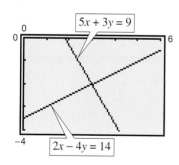

Figure 4.5

Example 4 illustrates a strategy for solving a system of linear equations that has decimal coefficients.

EXAMPLE 4	A LINEAR SYSTEM HAVING DECIMAL COEFFICIENTS

Solve the system of linear equations.

$$\begin{cases} 0.02x - 0.05y = -0.38 \\ 0.03x + 0.04y = 1.04 \end{cases}$$

Equation 1
Equation 2

SOLUTION

Because the coefficients in this system have two decimal places, you can begin by multiplying each equation by 100. (This produces a system in which the coefficients are all integers.)

$$\begin{cases} 2x - 5y = -38 \\ 3x + 4y = 104 \end{cases}$$

Revised Equation 1
Revised Equation 2

Now, to obtain x-coefficients that differ only in sign, multiply Equation 1 by 3 and Equation 2 by -2.

$$2x - 5y = -38 \quad\Longrightarrow\quad 6x - 15y = -114 \quad \text{Multiply Equation 1 by 3.}$$
$$\underline{3x + 4y = 104} \quad\Longrightarrow\quad \underline{-6x - 8y = -208} \quad \text{Multiply Equation 2 by } -2.$$
$$\qquad\qquad\qquad -23y = -322 \quad \text{Add equations.}$$

So, you can conclude that

$$y = \frac{-322}{-23} = 14.$$

Now, back-substitute $y = 14$ into any of the original or revised equations of the system that contain the variable y. Back-substituting this value into revised Equation 2 produces the following.

$$3x + 4y = 104 \qquad\qquad \text{Write revised Equation 2.}$$
$$3x + 4(14) = 104 \qquad\qquad \text{Substitute 14 for } y.$$
$$3x = 48 \qquad\qquad \text{Subtract 56 from each side.}$$
$$x = 16 \qquad\qquad \text{Solve for } x.$$

The solution is $(16, 14)$. Check this in the original system.

✓**CHECKPOINT** Now try Exercise 29.

DISCOVERY

Rewrite each system of equations in slope-intercept form and graph the system using a graphing utility. What is the relationship between the slopes of the two lines and the number of points of intersection?

a. $\begin{cases} 2x + 4y = 8 \\ 4x - 3y = -6 \end{cases}$ **b.** $\begin{cases} -x + 5y = 15 \\ 2x - 10y = -7 \end{cases}$ **c.** $\begin{cases} x - y = 9 \\ 2x - 2y = 18 \end{cases}$

Graphical Interpretation of Solutions

It is possible for a *general* system of equations to have exactly one solution, two or more solutions, or no solution. If a system of *linear* equations has two different solutions, it must have an *infinite* number of solutions. To see why this is true, consider the following graphical interpretations of systems of two linear equations in two variables. (Remember that the graph of a linear equation in two variables is a line.)

Graph			
Graphical Interpretation	The two lines intersect.	The two lines coincide (are identical).	The two lines are parallel.
Intersection	Single point of intersection	Infinitely many points of intersection	No point of intersection
Slopes of Lines	Slopes are not equal.	Slopes are equal.	Slopes are equal.
Number of Solutions	Exactly one solution	Infinitely many solutions	No solution
Type of System	**Independent (consistent) system**	**Dependent (consistent) system**	**Inconsistent system**

Note that the word *consistent* means that the system of linear equations has at least one solution, whereas the word *inconsistent* means that the system of linear equations has no solution. Note that a *consistent* system with exactly one solution is *independent*.

You can see from the graphs above that a comparison of the slopes of two lines gives useful information about the number of solutions of the corresponding system of equations. For instance:

Independent (consistent) systems have lines with different slopes.
Dependent (consistent) systems have lines with equal slopes and equal *y*-intercepts.
Inconsistent systems have lines with equal slopes, but different *y*-intercepts.

So, when solving a system of equations graphically, it is helpful to know the slopes of the lines directly. Writing these linear equations in the slope-intercept form

$$y = mx + b \qquad \text{Slope-intercept form}$$

enables you to identify the slopes quickly.

In Examples 5 and 6, note how you can use the method of elimination to determine that a system of linear equations has no solution or infinitely many solutions.

EXAMPLE 5   THE METHOD OF ELIMINATION: NO-SOLUTION CASE

Solve the system of linear equations.

$$\begin{cases} x - 2y = 3 & \text{Equation 1} \\ -2x + 4y = 1 & \text{Equation 2} \end{cases}$$

SOLUTION

To obtain coefficients that differ only in sign, multiply Equation 1 by 2.

$$\begin{array}{ll} x - 2y = 3 & \Longrightarrow \quad 2x - 4y = 6 & \text{Multiply Equation 1 by 2.} \\ \underline{-2x + 4y = 1} & \Longrightarrow \quad \underline{-2x + 4y = 1} & \text{Write Equation 2.} \\ & \qquad\qquad 0 = 7 & \text{False statement} \end{array}$$

Because there are no values of x and y for which $0 = 7$, you can conclude that the system is inconsistent and has no solution. The lines corresponding to the two equations given in this system are shown in Figure 4.6. Note that the two lines are parallel, and therefore have no point of intersection.

✓CHECKPOINT   Now try Exercise 21.

In Example 5, note that the occurrence of a false statement, such as $0 = 7$, indicates that the system has no solution. In the next example, note that the occurrence of a statement that is true for all values of the variables, such as $0 = 0$, indicates that the system has infinitely many solutions.

EXAMPLE 6   THE METHOD OF ELIMINATION: MANY-SOLUTIONS CASE

Solve the system of linear equations.

$$\begin{cases} 2x - y = 1 & \text{Equation 1} \\ 4x - 2y = 2 & \text{Equation 2} \end{cases}$$

SOLUTION

To obtain coefficients that differ only in sign, multiply Equation 2 by $-\frac{1}{2}$.

$$\begin{array}{ll} 2x - y = 1 & \Longrightarrow \quad 2x - y = 1 & \text{Write Equation 1.} \\ \underline{4x - 2y = 2} & \Longrightarrow \quad \underline{-2x + y = -1} & \text{Multiply Equation 2 by } -\frac{1}{2}. \\ & \qquad\qquad 0 = 0 & \text{Add equations.} \end{array}$$

Because the two equations turn out to be equivalent (have the same solution set), you can conclude that the system has infinitely many solutions. The solution set consists of all points (x, y) lying on the line $2x - y = 1$, as shown in Figure 4.7. To represent the solution set as an ordered pair, let $x = a$, where a is any real number. Then $y = 2a - 1$ and the solution set can be written as $(a, 2a - 1)$.

✓CHECKPOINT   Now try Exercise 27.

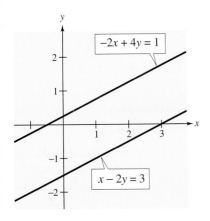

Figure 4.6 No Solution

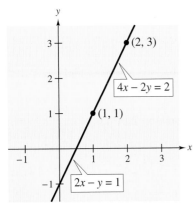

Figure 4.7 Infinite Number of Solutions

Applications

At this point, you may be asking the question, "How can I tell which application problems can be solved using a system of linear equations?" The answer comes from the following considerations.

1. Does the problem involve more than one unknown quantity?

2. Are there two (or more) equations or conditions to be satisfied?

If one or both of these conditions occur, the appropriate mathematical model for the problem may be a system of linear equations. Example 7 shows how to construct such a model.

EXAMPLE 7   AN APPLICATION OF A LINEAR SYSTEM

An airplane flying into a headwind travels the 2000-mile flying distance between Wilmington, Delaware and Tucson, Arizona in 4 hours and 24 minutes. On the return flight, the same distance is traveled in 4 hours. Find the air speed of the plane and the speed of the wind, assuming that both remain constant.

SOLUTION

The two unknown quantities are the speeds of the wind and the plane. If r_1 is the air speed of the plane and r_2 is the speed of the wind, then

$$r_1 - r_2 = \text{speed of the plane } against \text{ the wind}$$

$$r_1 + r_2 = \text{speed of the plane } with \text{ the wind}$$

as shown in Figure 4.8. Using the formula distance = (rate)(time) for these two speeds, you obtain the following equations.

$$2000 = (r_1 - r_2)\left(4 + \frac{24}{60}\right)$$

$$2000 = (r_1 + r_2)(4)$$

These two equations simplify as follows.

$$\begin{cases} 5000 = 11r_1 - 11r_2 & \text{Equation 1} \\ 500 = r_1 + r_2 & \text{Equation 2} \end{cases}$$

To solve this system by elimination, multiply Equation 2 by 11.

$$5000 = 11r_1 - 11r_2 \implies 5000 = 11r_1 - 11r_2 \quad \text{Write Equation 1.}$$

$$\underline{500 = r_1 + r_2} \implies \underline{5500 = 11r_1 + 11r_2} \quad \text{Multiply Equation 2 by 11.}$$

$$10{,}500 = 22r_1 \quad \text{Add equations.}$$

So, the solution is

$$r_1 = \frac{10{,}500}{22} = \frac{5250}{11} \approx 477.27 \text{ miles per hour} \qquad \text{Speed of plane}$$

$$r_2 = 500 - \frac{5250}{11} = \frac{250}{11} \approx 22.73 \text{ miles per hour.} \qquad \text{Speed of wind}$$

Check this solution in the original statement of the problem.

✓CHECKPOINT   Now try Exercise 33.

Original flight

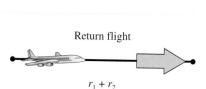

$r_1 - r_2$

Return flight

$r_1 + r_2$

Figure 4.8

In a free market, the demands for many products are related to the prices of the products. As the prices decrease, the demands by consumers increase and the amounts that producers are able or willing to supply decrease.

EXAMPLE 8   FINDING THE POINT OF EQUILIBRIUM

Make a Decision   The demand and supply equations for a DVD are given by

$$\begin{cases} p = 35 - 0.0001x & \text{Demand equation} \\ p = 8 + 0.0001x & \text{Supply equation} \end{cases}$$

where p is the price (in dollars) and x represents the number of DVDs. For how many units will the quantity demanded equal the quantity supplied? What price corresponds to this value?

SOLUTION

To obtain p-coefficients that differ only in sign, multiply the first equation by -1.

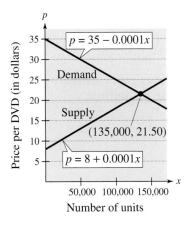

Figure 4.9

$$p = 35 - 0.0001x \quad \Longrightarrow \quad -p = -35 + 0.0001x \qquad \text{Multiply demand equation by } -1.$$

$$\underline{p = 8 + 0.0001x} \quad \Longrightarrow \quad \underline{p = 8 + 0.0001x}$$

$$0 = -27 + 0.0002x \qquad \text{Add equations.}$$

Solving the equation $0 = -27 + 0.0002x$, you get $x = 135{,}000$. So, the quantity demanded equals the quantity supplied for 135,000 units (see Figure 4.9). The price that corresponds to this x-value is obtained by back-substituting $x = 135{,}000$ into either of the original equations. For instance, back-substituting into the demand equation produces

$$p = 35 - 0.0001(135{,}000) = 35 - 13.5 = \$21.50.$$

Try back-substituting $x = 135{,}000$ into the supply equation to see that you obtain the same price. The solution (135,000, 21.50) is called the *point of equilibrium.* The **point of equilibrium** is the price p and the number of units x that satisfy both the demand and supply equations.

✓CHECKPOINT   Now try Exercise 47.

Discussing the Concept I   Creating Consistent and Inconsistent Systems

For each system, find the value of k such that the system has infinitely many solutions.

a. $\begin{cases} 4x + 3y = -8 \\ x + ky = -2 \end{cases}$ b. $\begin{cases} 3x - 12y = 9 \\ x - 4y = k \end{cases}$

Is it possible to find values of k such that the systems have no solutions? Explain why or why not for each system. Is it possible to find values of k such that the systems have unique solutions? Explain why or why not for each system. Summarize the differences between these two systems of equations.

4.2 ⟨ Warm Up ⟩

The following warm-up exercises involve skills that were covered in earlier sections. You will use these skills in the exercise set for this section. For additional help, review Section 1.2.

In Exercises 1 and 2, sketch the graph of the equation.

1. $2x + y = 4$

2. $5x - 2y = 3$

In Exercises 3 and 4, find an equation of the line passing through the two points.

3. $(-1, 3), (4, 8)$

4. $(2, 6), (5, 1)$

In Exercises 5 and 6, determine the slope of the line.

5. $3x + 6y = 4$

6. $7x - 4y = 10$

In Exercises 7–10, determine whether the lines represented by the pair of equations are parallel, perpendicular, or neither.

7. $2x - 3y = -10$
$3x + 2y = 11$

8. $4x - 12y = 5$
$-2x + 6y = 3$

9. $5x + y = 2$
$3x + 2y = 1$

10. $x - 3y = 2$
$6x + 2y = 4$

4.2 ⟨ Exercises ⟩

In Exercises 1–10, solve the system by elimination. Then use the graph to confirm your solution. Copy the graph and label each line with the appropriate equation.

1. $\begin{cases} 3x - 2y = 2 \\ x + 2y = 6 \end{cases}$

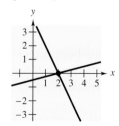

2. $\begin{cases} -x + 3y = 2 \\ x - 4y = -4 \end{cases}$

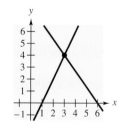

3. $\begin{cases} x - 4y = 2 \\ 2x + y = 4 \end{cases}$

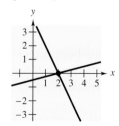

4. $\begin{cases} 2x - y = 2 \\ 4x + 3y = 24 \end{cases}$

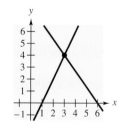

5. $\begin{cases} x - y = 1 \\ -2x + 2y = 5 \end{cases}$

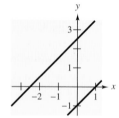

6. $\begin{cases} 3x + 2y = 2 \\ 6x + 4y = 14 \end{cases}$

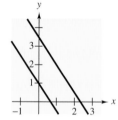

7. $\begin{cases} 3x - 2y = 6 \\ -6x + 4y = -12 \end{cases}$

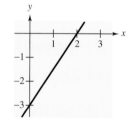

8. $\begin{cases} 2x + 4y = 8 \\ 6x + 12y = 24 \end{cases}$

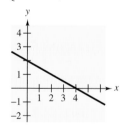

9. $\begin{cases} 9x - 3y = -1 \\ 3x + 6y = -5 \end{cases}$ 10. $\begin{cases} 5x + 3y = 18 \\ 2x - 7y = -1 \end{cases}$

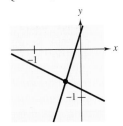

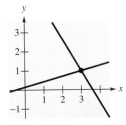

In Exercises 11–30, solve the system by elimination.

11. $\begin{cases} 4x - 3y = 11 \\ -6x + 3y = 3 \end{cases}$ 12. $\begin{cases} 3x - 5y = 2 \\ 2x + 5y = 13 \end{cases}$

13. $\begin{cases} 3x - y = 17 \\ 5x + 5y = -5 \end{cases}$ 14. $\begin{cases} x + 7y = 12 \\ 3x - 5y = 10 \end{cases}$

15. $\begin{cases} 3x + 2y = 10 \\ 2x + 5y = 3 \end{cases}$ 16. $\begin{cases} 8r + 16s = 20 \\ 16r + 50s = 55 \end{cases}$

17. $\begin{cases} 2u + v = 120 \\ u + 2v = 120 \end{cases}$ 18. $\begin{cases} 5u + 6v = 24 \\ 3u + 5v = 18 \end{cases}$

19. $\begin{cases} 4b + 3m = 3 \\ 3b + 11m = 13 \end{cases}$ 20. $\begin{cases} 3b + 3m = 7 \\ 3b + 5m = 3 \end{cases}$

21. $\begin{cases} 6r - 5s = 3 \\ -1.2r + s = 0.5 \end{cases}$ 22. $\begin{cases} 1.8x + 1.2y = 4 \\ 9x + 6y = 3 \end{cases}$

23. $\begin{cases} \dfrac{x}{4} + \dfrac{y}{6} = 1 \\ x - y = 3 \end{cases}$

24. $\begin{cases} \dfrac{2}{3}x + \dfrac{1}{6}y = \dfrac{2}{3} \\ 4x + y = 4 \end{cases}$

25. $\begin{cases} \dfrac{x + 3}{4} + \dfrac{y - 1}{3} = 1 \\ x - y = 3 \end{cases}$

26. $\begin{cases} \dfrac{x - 1}{2} + \dfrac{y + 2}{3} = 4 \\ x - 2y = 5 \end{cases}$

27. $\begin{cases} 2.5x - 3y = 1.5 \\ 10x - 12y = 6 \end{cases}$

28. $\begin{cases} 3.5x - 2y = 2.6 \\ 7x - 2.5y = 4 \end{cases}$

29. $\begin{cases} 0.05x - 0.03y = 0.21 \\ 0.07x + 0.02y = 0.16 \end{cases}$

30. $\begin{cases} 0.02x - 0.05y = -0.19 \\ 0.03x + 0.04y = 0.52 \end{cases}$

Make a Decision In Exercises 31 and 32, the graphs of the two equations appear to be parallel. Are they? Explain your reasoning.

31. $\begin{cases} 200y - x = 200 \\ 199y - x = -198 \end{cases}$ 32. $\begin{cases} 25x - 24y = 0 \\ 13x - 12y = 120 \end{cases}$

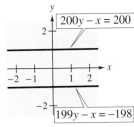

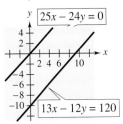

33. **Airplane Speed** An airplane flying into a headwind travels the 1800-mile flying distance between Los Angeles, California and South Bend, Indiana in 3 hours and 36 minutes. On the return flight, the distance is traveled in 3 hours. Find the air speed of the plane and the speed of the wind, assuming that both remain constant.

34. **Airplane Speed** Two planes start from the same airport and fly in opposite directions. The second plane starts $\frac{1}{2}$ hour after the first plane, but its speed is 50 miles per hour faster. Find the air speed of each plane if, 2 hours after the first plane departs, the planes are 2000 miles apart.

35. **Acid Mixture** Ten gallons of a 30% acid solution is obtained by mixing a 20% solution with a 50% solution. How much of each solution is required to obtain the specified concentration of the final mixture?

36. **Fuel Mixture** Five hundred gallons of 89-octane gasoline is obtained by mixing 87-octane gasoline with 92-octane gasoline. How much of each type of gasoline is required to obtain the specified mixture? (Octane ratings can be interpreted as percents. A high octane rating indicates that the gasoline is knock resistant. A low octane rating indicates that the gasoline is knock prone.)

37. **Investment Portfolio** A total of $12,000 is invested in two corporate bonds that pay 10.5% and 12% simple interest. The total annual interest is $1380. How much is invested in each bond?

38. **Investment Portfolio** A total of $32,000 is invested in two municipal bonds that pay 5.75% and 6.25% simple interest. The total annual interest is $1930. How much is invested in each bond?

39. *Make a Decision*: **Ticket Sales** You are the manager of a theater. On Saturday morning you are going over the ticket sales for Friday evening. A total of 740 tickets were sold. The tickets for adults and children sold for $8.50 and $4.00, respectively, and the total receipts for the performance were $4688. However, your assistant manager did not record how many of each type of ticket were sold. From the information you have, can you determine how many of each type were sold? Explain your reasoning.

40. *Make a Decision*: **Shoe Sales** You are the manager of a shoe store. On Sunday morning you are going over the receipts for the previous week's sales. A total of 320 pairs of cross training shoes were sold. One style sold for $56.95 and the other sold for $72.95. The total receipts were $21,024. The cash register that was supposed to keep track of the number of each type of shoe sold malfunctioned. Can you recover the information? If so, how many of each type were sold?

Supply and Demand In Exercises 41–44, find the point of equilibrium for the pair of demand and supply equations.

Demand	Supply
41. $p = 56 - 0.0001x$	$p = 22 + 0.00001x$
42. $p = 60 - 0.00001x$	$p = 15 + 0.00004x$
43. $p = 140 - 0.00002x$	$p = 80 + 0.00001x$
44. $p = 400 - 0.0002x$	$p = 225 + 0.0005x$

45. *Make a Decision*: **Restaurants** The total sales (in billions of dollars) for fast-food and full-service restaurants from 1999 to 2003 are shown in the table. (Source: National Restaurant Association)

Year	Fast-food	Full-service
1999	103.0	125.4
2000	107.1	133.8
2001	111.1	140.4
2002	116.1	146.1
2003	120.9	153.2

(a) Use a graphing utility to create a scatter plot of the data for fast-food sales and use the *regression* feature to find a linear model. Let x represent the year, with $x = 9$ corresponding to 1999. Repeat the procedure for the data for full-service sales.

(b) Assuming that the amounts for the given 5 years are representative of future years, will fast-food sales ever equal full-service sales?

46. *Make a Decision*: **PC and Internet Users** The numbers (in millions) of personal computers and Internet users in the world from 1997 to 2001 are shown in the table. (Source: International Telecommunication Union)

Year	PCs	Internet users
1997	325	117
1998	375	183
1999	435	277
2000	500	399
2001	555	502

(a) Use a graphing utility to create a scatter plot of the data for the number of PCs and use the *regression* feature to find a linear model. Let x represent the year, with $x = 7$ corresponding to 1997. Repeat the procedure for the data for Internet users.

(b) Assuming the amounts for the given 5 years are representative of future years, will the number of Internet users ever exceed the number of PCs?

47. *Make a Decision*: **Supply and Demand** The supply and demand equations for a small television are given by

$$\begin{cases} p + 0.45x = 220 & \text{Demand} \\ p - 0.35x = 60 & \text{Supply} \end{cases}$$

where p is the price (in dollars) and x represents the number of televisions. For how many units will the quantity demanded equal the quantity supplied? What price corresponds to this value?

48. *Make a Decision*: **Supply and Demand** The supply and demand equations for a graphing calculator are given by

$$\begin{cases} p + 3.5x = 299 & \text{Demand} \\ p - 0.15x = 80 & \text{Supply} \end{cases}$$

where p is the price (in dollars) and x represents the number of graphing calculators. For how many units will the quantity demanded equal the quantity supplied? What price corresponds to this value?

Fitting a Line to Data   In Exercises 49–52, find the least squares regression line $y = ax + b$ for the points $(x_1, y_1), (x_2, y_2), \ldots, (x_n, y_n)$ by solving the system for a and b. (If you are unfamiliar with summation notation, look at the discussion in Section 6.1.)

$$\begin{cases} nb + \left(\sum_{i=1}^{n} x_i\right)a = \sum_{i=1}^{n} y \\ \left(\sum_{i=1}^{n} x_i\right)b + \left(\sum_{i=1}^{n} x_i^2\right)a = \sum_{i=1}^{n} x_i y_i \end{cases}$$

49. $\begin{cases} 5b + 10a = 20.2 \\ 10b + 30a = 50.1 \end{cases}$ **50.** $\begin{cases} 5b + 10a = 11.7 \\ 10b + 30a = 25.6 \end{cases}$

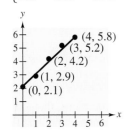

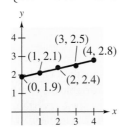

51. $\begin{cases} 7b + 21a = 35.1 \\ 21b + 91a = 114.2 \end{cases}$ **52.** $\begin{cases} 6b + 15a = 23.6 \\ 15b + 55a = 48.8 \end{cases}$

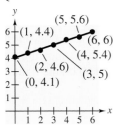

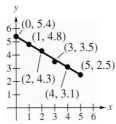

53. Cell Phone Bill   The average monthly bill y (in dollars) for cell phone service in the United States from 1998 to 2002 is shown in the table.   (Source: Cellular Telecommunications & Internet Association)

Year	1998	1999	2000	2001	2002
t	0	1	2	3	4
Monthly bill, y	39.43	41.24	45.27	47.37	48.40

(a) Solve the following system for a and b to find the least squares regression line $y = at + b$ for the data. Let t represent the year, with $t = 0$ corresponding to 1998.

$$\begin{cases} 5b + 10a = 221.71 \\ 10b + 30a = 467.49 \end{cases}$$

(b) Use a graphing utility to graph the regression line and estimate the average monthly cell phone bill in 2004.

(c) Use the *regression* feature of a graphing utility to find a linear model for the data. Compare this model with the one you found in part (a).

(d) Use the linear equation found in part (c) to estimate the average monthly cell phone bill in 2004. Compare this estimate with the one you found in part (b).

54. Revolving Credit   The amount y of revolving credit (in billions of dollars) in the United States from 1998 to 2002 is shown in the table.   (Source: Board of Governors of the Federal Reserve System)

Year	1998	1999	2000	2001	2002
t	0	1	2	3	4
Revolving credit, y	563	598	668	701	713

(a) Solve the following system for a and b to find the least squares regression line $y = at + b$ for the data. Let t represent the year, with $t = 0$ corresponding to 1998.

$$\begin{cases} 5b + 10a = 3243 \\ 10b + 30a = 6889 \end{cases}$$

(b) Use a graphing utility to graph the regression line and estimate the amount of revolving credit in 2004.

(c) Use the *regression* feature of a graphing utility to find a linear model for the data. Compare this model with the one you found in part (a).

(d) Use the linear equation found in part (c) to estimate the amount of revolving credit in 2004. Compare this estimate with the one you found in part (b).

55. Make a Decision   A system of linear equations reduces to $-23 = 0$. What can you conclude?

56. Make a Decision   A system of linear equations reduces to $0 = 0$. What can you conclude?

57. Reasoning   Design a system of two linear equations in two variables that has (a) exactly one solution, (b) infinitely many solutions, and (c) no solution. Solve each system algebraically.

58. Reasoning   Briefly explain whether or not it is possible for a consistent system of linear equations to have exactly two solutions.

4.3 Linear Systems in Three or More Variables

Objectives

- Solve a linear system in row-echelon form using back-substitution.
- Use Gaussian elimination to solve a linear system.
- Solve a nonsquare linear system.
- Construct and use a linear system in three or more variables to solve an application problem.
- Find the equation of a circle or a parabola using a linear system in three or more variables.

Row-Echelon Form and Back-Substitution

The method of elimination can be applied to a system of linear equations in more than two variables. In fact, this method easily adapts to computer use for solving linear systems with dozens of variables.

When elimination is used to solve a system of linear equations, the goal is to rewrite the system in a form to which back-substitution can be applied. To see how this works, consider the following two systems of linear equations.

$$\begin{cases} x - 2y + 3z = 9 \\ -x + 3y = -4 \\ 2x - 5y + 5z = 17 \end{cases} \qquad \begin{cases} x - 2y + 3z = 9 \\ y + 3z = 5 \\ z = 2 \end{cases}$$

The system on the right is said to be in **row-echelon form,** which means that it has a "stair-step" pattern with leading coefficients of 1. After comparing the two systems, it should be clear that it is easier to solve the system on the right.

EXAMPLE 1 USING BACK-SUBSTITUTION

Solve the system of linear equations.

$$\begin{cases} x - 2y + 3z = 9 & \text{Equation 1} \\ y + 3z = 5 & \text{Equation 2} \\ z = 2 & \text{Equation 3} \end{cases}$$

SOLUTION

From Equation 3, you know the value of z. To solve for y, substitute $z = 2$ into Equation 2 to obtain

$$y + 3(2) = 5 \quad \Longrightarrow \quad y = -1.$$

Finally, substitute $y = -1$ and $z = 2$ into Equation 1 to obtain

$$x - 2(-1) + 3(2) = 9 \quad \Longrightarrow \quad x = 1.$$

The solution is $x = 1$, $y = -1$, and $z = 2$, which can be written as the **ordered triple** $(1, -1, 2)$. Check this in the original system of equations.

✓CHECKPOINT   Now try Exercise 1.

Gaussian Elimination

Two systems of equations are **equivalent** if they have the same solution set. To solve a system that is not in row-echelon form, first convert it to an *equivalent* system that is in row-echelon form. To see how this is done, let's take another look at the method of elimination, as applied to a system of two linear equations.

EXAMPLE 2   THE METHOD OF ELIMINATION

Solve the system of linear equations.

$$\begin{cases} 3x - 2y = -1 & \text{Equation 1} \\ x - y = 0 & \text{Equation 2} \end{cases}$$

SOLUTION

An easy way of obtaining a leading coefficient of 1 is to interchange the two equations.

$$\begin{cases} x - y = 0 \\ 3x - 2y = -1 \end{cases} \qquad \text{Interchange two equations in the system.}$$

$$\begin{cases} -3x + 3y = 0 \\ 3x - 2y = -1 \end{cases} \qquad \text{Multiply the first equation by } -3.$$

$$\begin{array}{r} -3x + 3y = 0 \\ 3x - 2y = -1 \\ \hline y = -1 \end{array} \qquad \begin{array}{l} \text{Add the multiple of the first equation to the} \\ \text{second equation to obtain a new equation.} \end{array}$$

$$\begin{cases} x - y = 0 \\ y = -1 \end{cases} \qquad \text{New system in row-echelon form}$$

Now, using back-substitution, you can determine that the solution is $y = -1$ and $x = -1$, which can be written as the ordered pair $(-1, -1)$. Check this in the original system of equations.

✓CHECKPOINT   Now try Exercise 3.

The process of rewriting a system of equations in row-echelon form by using the three basic row operations is called **Gaussian elimination,** after the German mathematician Carl Friedrich Gauss. Example 2 shows the chain of equivalent systems used to solve a linear system in two variables.

Operations That Produce Equivalent Systems

Each of the following **row operations** on a system of linear equations produces an *equivalent* system of linear equations.

1. Interchange two equations.

2. Multiply one of the equations by a nonzero constant.

3. Add a multiple of one of the equations to another equation to replace the latter equation.

EXAMPLE 3 USING ELIMINATION TO SOLVE A SYSTEM

Solve the system of linear equations.

$$\begin{cases} x - 2y + 3z = 9 & \text{Equation 1} \\ -x + 3y = -4 & \text{Equation 2} \\ 2x - 5y + 5z = 17 & \text{Equation 3} \end{cases}$$

SOLUTION

Because the leading coefficient of Equation 1 is 1, you can begin by saving the x in the upper left position and eliminating the other x-terms from the first column.

$$\begin{cases} x - 2y + 3z = 9 \\ y + 3z = 5 \\ 2x - 5y + 5z = 17 \end{cases}$$

> Adding the first equation to the second equation produces a new second equation.

$$\begin{cases} x - 2y + 3z = 9 \\ y + 3z = 5 \\ -y - z = -1 \end{cases}$$

> Adding -2 times the first equation to the third equation produces a new third equation.

Now that all but the first x have been eliminated from the first column, go to work on the second column. (You need to eliminate y from the third equation.)

$$\begin{cases} x - 2y + 3z = 9 \\ y + 3z = 5 \\ 2z = 4 \end{cases}$$

> Adding the second equation to the third equation produces a new third equation.

Finally, you need a coefficient of 1 for z in the third equation.

$$\begin{cases} x - 2y + 3z = 9 \\ y + 3z = 5 \\ z = 2 \end{cases}$$

> Multiplying the third equation by $\frac{1}{2}$ produces a new third equation.

This is the same system that was solved in Example 1, and, as in that example, you can conclude that the solution is

$$x = 1, \quad y = -1, \quad \text{and} \quad z = 2.$$

✓CHECKPOINT   Now try Exercise 5.

In Example 3, you can check the solution by substituting $x = 1$, $y = -1$, and $z = 2$ into each original equation, as follows.

Equation 1: $(1) - 2(-1) + 3(2) = 9$ ✓

Equation 2: $-(1) + 3(-1) = -4$ ✓

Equation 3: $2(1) - 5(-1) + 5(2) = 17$ ✓

The next example involves an inconsistent system—one that has no solution. The key to recognizing an inconsistent system is that at some stage in the elimination process, you obtain an absurdity such as $0 = -2$.

EXAMPLE 4   AN INCONSISTENT SYSTEM

Solve the system of linear equations.

$$\begin{cases} x - 3y + z = 1 & \text{Equation 1} \\ 2x - y - 2z = 2 & \text{Equation 2} \\ x + 2y - 3z = -1 & \text{Equation 3} \end{cases}$$

SOLUTION

$$\begin{cases} x - 3y + z = 1 \\ 5y - 4z = 0 \\ x + 2y - 3z = -1 \end{cases}$$

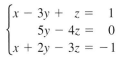 Adding -2 times the first equation to the second equation produces a new second equation.

$$\begin{cases} x - 3y + z = 1 \\ 5y - 4z = 0 \\ 5y - 4z = -2 \end{cases}$$

Adding -1 times the first equation to the third equation produces a new third equation.

$$\begin{cases} x - 3y + z = 1 \\ 5y - 4z = 0 \\ 0 = -2 \end{cases}$$

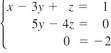 Adding -1 times the second equation to the third equation produces a new third equation.

Because the third "equation" is impossible, you can conclude that this system is inconsistent and therefore has no solution. Moreover, because this system is equivalent to the original system, you can conclude that the original system also has no solution.

✓**CHECKPOINT**   Now try Exercise 7.

As with a system of linear equations in two variables, the solution(s) of a system of linear equations in more than two variables must fall into one of three categories. Because an equation in three variables represents a plane in space, the possible solutions can be shown graphically. See Figure 4.10.

The Number of Solutions of a Linear System

For a system of linear equations, exactly one of the following is true.

1. There is exactly one solution. [See Figure 4.10(a).]

2. There are infinitely many solutions. [See Figures 4.10(b) and (c).]

3. There is no solution. [See Figures 4.10(d) and (e).]

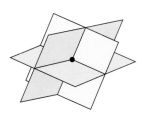

(a) Solution: one point

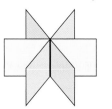

(b) Solution: one line

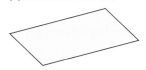

(c) Solution: one plane

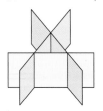

(d) Solution: none

(e) Solution: none

Figure 4.10

DISCOVERY

The total numbers of sides and diagonals of regular polygons with three, four, and five sides are three, six, and ten, respectively, as shown in the figure.

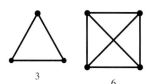

3

6

10

15

Find a quadratic function $y = ax^2 + bx + c$, where y represents the total number of sides and diagonals and x represents the number of sides, that fits these data. Check to see if the quadratic function gives the correct answers for a polygon with six sides.

| EXAMPLE 5 | A SYSTEM WITH INFINITELY MANY SOLUTIONS |

Solve the system of linear equations.

$$\begin{cases} x + y - 3z = -1 & \text{Equation 1} \\ y - z = 0 & \text{Equation 2} \\ -x + 2y = 1 & \text{Equation 3} \end{cases}$$

SOLUTION

$$\begin{cases} x + y - 3z = -1 \\ y - z = 0 \\ 3y - 3z = 0 \end{cases}$$

> Adding the first equation to the third equation produces a new third equation.

$$\begin{cases} x + y - 3z = -1 \\ y - z = 0 \\ 0 = 0 \end{cases}$$

> Adding -3 times the second equation to the third equation produces a new third equation.

This means that Equation 3 depends on Equations 1 and 2 in the sense that it gives us no additional information about the variables. So, the original system is equivalent to the system

$$\begin{cases} x + y - 3z = -1 \\ y - z = 0 \end{cases}.$$

In this last equation, solve for y in terms of z to obtain $y = z$. Back-substituting for y into the previous equation produces $x = 2z - 1$. Finally, letting $z = a$, the solutions to the original system are all of the form

$$x = 2a - 1, \quad y = a, \quad \text{and} \quad z = a$$

where a is a real number. So, every ordered triple of the form

$$(2a - 1, a, a), \quad a \text{ is a real number}$$

is a solution of the system.

✓**CHECKPOINT**   Now try Exercise 11.

In Example 5, there are other ways to write the same infinite set of solutions. For instance, the solutions could have been written as

$$\left(b, \tfrac{1}{2}(b + 1), \tfrac{1}{2}(b + 1)\right), \quad b \text{ is a real number.}$$

To convince yourself that this description produces the same set of solutions, consider the following.

Substitution	*Solution*
$a = 0$	$(2(0) - 1, 0, 0) = (-1, 0, 0)$
$b = -1$	$\left(-1, \tfrac{1}{2}(-1 + 1), \tfrac{1}{2}(-1 + 1)\right) = (-1, 0, 0)$
$a = 1$	$(2(1) - 1, 1, 1) = (1, 1, 1)$
$b = 1$	$\left(1, \tfrac{1}{2}(1 + 1), \tfrac{1}{2}(1 + 1)\right) = (1, 1, 1)$

In both cases, you obtain the same ordered triples. So, when comparing descriptions of an infinite solution set, keep in mind that there is more than one way to describe the set.

Nonsquare Systems

So far, each system of linear equations you have looked at has been **square,** which means that the number of equations is equal to the number of variables. In a **nonsquare** system, the number of equations differs from the number of variables. A system of linear equations cannot have a unique solution unless there are at least as many equations as there are variables in the system.

> **EXAMPLE 6**   A SYSTEM WITH FEWER EQUATIONS THAN VARIABLES

Solve the system of linear equations.

$$\begin{cases} x - 2y + z = 2 & \text{Equation 1} \\ 2x - y - z = 1 & \text{Equation 2} \end{cases}$$

SOLUTION

Begin by rewriting the system in row-echelon form, as follows.

$$\begin{cases} x - 2y + z = 2 \\ \quad\quad 3y - 3z = -3 \end{cases}$$

Adding -2 times the first equation to the second equation produces a new second equation.

$$\begin{cases} x - 2y + z = 2 \\ \quad\quad y - z = -1 \end{cases}$$

Multiplying the second equation by $\frac{1}{3}$ produces a new second equation.

Solving for y in terms of z, you get $y = z - 1$, and back-substitution into Equation 1 yields

$$x - 2(z - 1) + z = 2$$

$$x - 2z + 2 + z = 2$$

$$x = z.$$

Finally, by letting $z = a$, you have the solution

$$x = a, \quad y = a - 1, \quad \text{and} \quad z = a$$

where a is a real number. So, every ordered triple of the form

$$(a, a - 1, a), \quad a \text{ is a real number}$$

is a solution of the system. Because there were originally three variables and only two equations, the system cannot have a unique solution.

✓CHECKPOINT   Now try Exercise 21.

In Example 6, try choosing some values of a to obtain different solutions of the system, such as $(1, 0, 1)$, $(2, 1, 2)$, and $(3, 2, 3)$. Then check each of the solutions in the original system. For example, you can check the solution $(1, 0, 1)$ as follows.

Equation 1: $1 - 2(0) + 1 = 2$ ✓

Equation 2: $2(1) - 0 - 1 = 1$ ✓

Applications

EXAMPLE 7   AN INVESTMENT PORTFOLIO

Make a Decision   You have a portfolio totaling $450,000 and want to invest in (1) certificates of deposit, (2) municipal bonds, (3) blue-chip stocks, and (4) growth or speculative stocks. The certificates pay 9% simple annual interest, and the municipal bonds pay 6% simple annual interest. You expect the blue-chip stocks to return 10% simple annual interest and the growth stocks to return 15% simple annual interest. You want a combined annual return of 8%, and you also want to have only one-third of the portfolio invested in stocks. How much should be allocated to each type of investment?

SOLUTION

To solve this problem, let C, M, B, and G represent the amounts in the four types of investments. Because the total investment is $450,000, you can write the following equation.

$$C + M + B + G = 450{,}000$$

A second equation can be derived from the fact that the combined annual return should be 8%.

$$0.09C + 0.06M + 0.10B + 0.15G = 0.08(450{,}000)$$

Finally, because only one-third of the total investment should be allocated to stocks, you can write

$$B + G = \tfrac{1}{3}(450{,}000).$$

These three equations make up the following system.

$$\begin{cases} C + M + B + G = 450{,}000 & \text{Equation 1} \\ 0.09C + 0.06M + 0.10B + 0.15G = 36{,}000 & \text{Equation 2} \\ B + G = 150{,}000 & \text{Equation 3} \end{cases}$$

Using elimination, you find that the system has infinitely many solutions, which can be written as follows.

$$C = -\tfrac{5}{3}a + 100{,}000$$

$$M = \tfrac{5}{3}a + 200{,}000$$

$$B = -a + 150{,}000$$

$$G = a$$

So, you have many different options. One possible solution is to choose $a = 30{,}000$, which yields the following portfolio.

1. Certificates of deposit: $50,000

2. Municipal bonds: $250,000

3. Blue-chip stocks: $120,000

4. Growth or speculative stocks: $30,000

✓CHECKPOINT   Now try Exercise 43.

EXAMPLE 8   DATA ANALYSIS: CURVE-FITTING

Find a quadratic equation, $y = ax^2 + bx + c$, whose graph passes through the points $(-1, 3)$, $(1, 1)$, and $(2, 6)$.

SOLUTION

Because the graph of $y = ax^2 + bx + c$ passes through the points $(-1, 3)$, $(1, 1)$, and $(2, 6)$, you can write the following.

When $x = -1$, $y = 3$: $a(-1)^2 + b(-1) + c = 3$

When $x = 1$, $y = 1$: $a(1)^2 + b(1) + c = 1$

When $x = 2$, $y = 6$: $a(2)^2 + b(2) + c = 6$

This produces the following system of linear equations.

$$\begin{cases} a - b + c = 3 & \text{Equation 1} \\ a + b + c = 1 & \text{Equation 2} \\ 4a + 2b + c = 6 & \text{Equation 3} \end{cases}$$

The solution of this system is $a = 2$, $b = -1$, and $c = 0$. So, the equation of the parabola is $y = 2x^2 - x$, as shown in Figure 4.11.

✔CHECKPOINT   Now try Exercise 33.

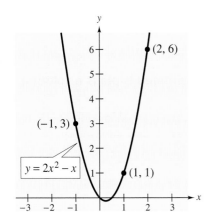

Figure 4.11

© David Young-Wolff/CORBIS

Discussing the Concept | Mathematical Modeling

You work for an outerwear manufacturer, and the marketing department is concerned about sales trends in Arizona. Your manager has asked you to investigate climate data in hopes of explaining sales patterns. Your manager gives you the table below, which gives the average monthly temperature y (in degrees Fahrenheit) for Phoenix, Arizona for the month x, where $x = 1$ corresponds to November.   (Source: National Climactic Data Center)

x	y
1	62
3	54
5	63

Construct a scatter plot of the data. Decide what type of mathematical model might be appropriate for the data, and use the methods you have learned so far to find an appropriate model. Your manager would like to know the average monthly temperatures for December and February. Explain to your manager how you found your model, what it represents, and how it can be used to find the December and February average temperatures. Investigate the usefulness of this model for the rest of the year. Would you recommend using the model to predict monthly average temperatures for the whole year? Explain your reasoning.

4.3 Warm Up

The following warm-up exercises involve skills that were covered in earlier sections. You will use these skills in the exercise set for this section. For additional help, review Sections R2.2, 1.1, 4.1, and 4.2.

In Exercises 1–4, solve the system of linear equations.

1. $\begin{cases} x + y = 25 \\ y = 10 \end{cases}$

2. $\begin{cases} 2x - 3y = 4 \\ 6x = -12 \end{cases}$

3. $\begin{cases} x + y = 32 \\ x - y = 24 \end{cases}$

4. $\begin{cases} 2r - s = 5 \\ r + 2s = 10 \end{cases}$

In Exercises 5–8, determine whether the ordered triple is a solution of the equation.

5. $5x - 3y + 4z = 2$
$(-1, -2, 1)$

6. $x - 2y + 12z = 9$
$(6, 3, 2)$

7. $2x - 5y + 3z = -9$
$(a - 2, a + 1, a)$

8. $-5x + y + z = 21$
$(a - 4, 4a + 1, a)$

In Exercises 9 and 10, solve for x in terms of a.

9. $x + 2y - 3z = 4$
$y = 1 - a, z = a$

10. $x - 3y + 5z = 4$
$y = 2a + 3, z = a$

4.3 Exercises

In Exercises 1 and 2, use back-substitution to solve the system of linear equations.

1. $\begin{cases} x - y + z = 4 \\ 2y + z = -6 \\ z = -2 \end{cases}$

2. $\begin{cases} 4x - 2y + z = 8 \\ -y + z = 4 \\ z = 2 \end{cases}$

In Exercises 3–27, solve the system of equations.

3. $\begin{cases} 4x + y - 3z = 11 \\ 2x - 3y + 2z = 9 \\ x + y + z = -3 \end{cases}$

4. $\begin{cases} 6y + 4z = -12 \\ 3x + 3y = 9 \\ 2x - 3z = 10 \end{cases}$

5. $\begin{cases} 3x + 2z = 13 \\ x + 2y + z = -5 \\ -3y - z = 10 \end{cases}$

6. $\begin{cases} 2x + 3y + z = -4 \\ 2x - 4y + 3z = 18 \\ 3x - 2y + 2z = 9 \end{cases}$

7. $\begin{cases} 3x - 2y + 4z = 1 \\ x + y - 2z = 3 \\ 2x - 3y + 6z = 8 \end{cases}$

8. $\begin{cases} 5x - 3y + 2z = 3 \\ 2x + 4y - z = 7 \\ x - 11y + 4z = 3 \end{cases}$

9. $\begin{cases} 3x + 3y + 5z = 1 \\ 3x + 5y + 9z = 0 \\ 5x + 9y + 17z = 0 \end{cases}$

10. $\begin{cases} 2x + y - z = 13 \\ x + 2y + z = 2 \\ 8x - 3y + 4z = -2 \end{cases}$

11. $\begin{cases} x + 2y - 7z = -4 \\ 2x + y + z = 13 \\ 3x + 9y - 36z = -33 \end{cases}$

12. $\begin{cases} 2x + y - 3z = 4 \\ 4x + 2z = 10 \\ -2x + 3y - 13z = -8 \end{cases}$

13. $\begin{cases} x + 4z = 13 \\ 4x - 2y + z = 7 \\ 2x - 2y - 7z = -19 \end{cases}$

14. $\begin{cases} 4x - y + 5z = 11 \\ x + 2y - z = 5 \\ 5x - 8y + 13z = 7 \end{cases}$

15. $\begin{cases} x + 4z = 1 \\ x + y + 10z = 10 \\ 2x - y + 2z = -5 \end{cases}$

16. $\begin{cases} 3x - 2y - 6z = 4 \\ -3x + 2y + 6z = 1 \\ x - y - 5z = 3 \end{cases}$

17. $\begin{cases} 4x + 3y + 5z = 10 \\ 5x + 2y + 10z = 13 \\ 3x + y - 2z = -9 \end{cases}$

18. $\begin{cases} 2x + 5y = 25 \\ 3x - 2y + 4z = 1 \\ 4x - 3y + z = 9 \end{cases}$

19. $\begin{cases} 5x + 5y - z = 0 \\ 10x + 5y + 2z = 0 \\ 5x + 15y - 9z = 0 \end{cases}$

20. $\begin{cases} 2x + 3y = 0 \\ 4x + 3y - z = 0 \\ 8x + 3y + 3z = 0 \end{cases}$

21. $\begin{cases} 12x + 5y + z = 0 \\ 12x + 4y - z = 0 \end{cases}$

22. $\begin{cases} x - 2y + 5z = 2 \\ 3x + 2y - z = -2 \end{cases}$

23. $\begin{cases} x - 3y + 2z = 18 \\ 5x - 13y + 12z = 80 \end{cases}$

24. $\begin{cases} 2x - 3y + z = -2 \\ -4x + 9y = 7 \end{cases}$

25. $\begin{cases} 2x + 3y + 3z = 7 \\ 4x + 18y + 15z = 44 \end{cases}$

26. $\begin{cases} x + 3w = 4 \\ 2y - z - w = 0 \\ 3y - 2w = 1 \\ 2x - y + 4z = 5 \end{cases}$

27. $\begin{cases} x + y + z + w = 6 \\ 2x + 3y - w = 0 \\ -3x + 4y + z + 2w = 4 \\ x + 2y - z + w = 0 \end{cases}$

28. One solution for Exercise 22 is $(-a, 2a - 1, a)$. A student gives $(b, -2b - 1, -b)$ as a solution to the same exercise. Explain why both solutions are correct.

In Exercises 29–32, write three ordered triples of the given form.

29. $\left(a, a - 5, \frac{2}{3}a + 1\right)$ **30.** $(3a, 5 - a, a)$

31. $\left(\frac{1}{2}a, 3a, 5\right)$ **32.** $\left(-\frac{1}{2}a + 5, -1, a\right)$

In Exercises 33 and 34, find the equation of the parabola $y = ax^2 + bx + c$ that passes through the points.

33.
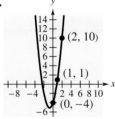

34.
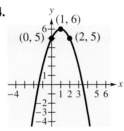

In Exercises 35 and 36, find the equation of the circle $x^2 + y^2 + Dx + Ey + F = 0$ that passes through the points.

35.
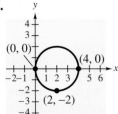

36.
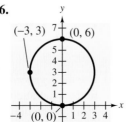

37. Investment A clothing company borrows $775,000. Some of the money is borrowed at 8%, some at 9%, and some at 10% simple annual interest. How much is borrowed at each rate if the total annual interest is $67,500 and the amount borrowed at 8% is four times the amount borrowed at 10%?

38. Investment A computer company borrows $800,000. Some of the money is borrowed at 8%, some at 9%, and some at 10% simple annual interest. How much is borrowed at each rate if the total annual interest is $67,000 and the amount borrowed at 8% is five times the amount borrowed at 10%?

39. Grades of Paper A manufacturer sells a 50-pound package of paper that consists of three grades of computer paper. Grade C costs $4.50 per pound, grade B costs $5.50 per pound, and grade A costs $8.00 per pound. Half of the 50-pound package consists of the two cheaper grades. The cost of the 50-pound package is $322.50. How many pounds of each grade of paper are there in the 50-pound package?

40. Hair Products A hair product company sells three types of hair products for $30, $20, and $10 per unit. In 1 year, the total revenue for the three products was $800,000, which corresponded to the sale of 40,000 units. The company sold half as many units of the $30 product as units of the $20 product. How many units of each product were sold?

41. Crop Spraying A mixture of 6 gallons of chemical A, 8 gallons of chemical B, and 13 gallons of chemical C is required to kill a destructive crop insect. Commercial spray X contains 1, 2, and 2 parts, respectively, of these chemicals. Commercial spray Y contains only chemical C. Commercial spray Z contains chemicals A, B, and C in equal amounts. How much of each type of commercial spray is needed to get the desired mixture?

42. Acid Mixture A chemist needs 10 liters of a 25% acid solution. The solution is to be mixed from three solutions whose acid concentrations are 10%, 20%, and 50%. How many liters of each solution should the chemist use to satisfy the following?

(a) Use as little as possible of the 50% solution.

(b) Use as much as possible of the 50% solution.

(c) Use 2 liters of the 50% solution.

Make a Decision: **Investment Portfolio**   In Exercises 43 and 44, you have a total of $500,000 that is to be invested in (1) certificates of deposit, (2) municipal bonds, (3) blue-chip stocks, and (4) growth or speculative stocks. How much should be put in each type of investment?

43. The certificates of deposit pay 2.5% simple annual interest, and the municipal bonds pay 10% simple annual interest. Over a five-year period, you expect the blue-chip stocks to return 12% simple annual interest and the growth stocks to return 18% simple annual interest. You want a combined annual return of 10% and you also want to have only one-fourth of the portfolio invested in stocks.

44. The certificates of deposit pay 3% simple annual interest, and the municipal bonds pay 10% simple annual interest. Over a five-year period, you expect the blue-chip stocks to return 12% simple annual interest and the growth stocks to return 15% simple annual interest. You want a combined annual return of 10% and you also want to have only one-fourth of the portfolio invested in stocks.

Fitting a Parabola to Data   In Exercises 45–48, find the least squares regression parabola

$y = ax^2 + bx + c$

for the points (x_1, y_1), (x_2, y_2), $\ldots$, (x_n, y_n) by solving the system of linear equations for a, b, and c.

45.
$$\begin{cases} 5c + 10a = 15.5 \\ 10b = 6.3 \\ 10c + 34a = 32.1 \end{cases}$$

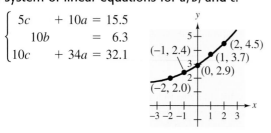

46.
$$\begin{cases} 5c + 10a = 15.0 \\ 10b = 17.3 \\ 10c + 34a = 34.5 \end{cases}$$

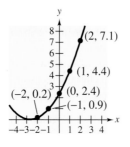

47.
$$\begin{cases} 6c + 3b + 19a = 23.9 \\ 3c + 19b + 27a = -7.2 \\ 19c + 27b + 115a = 48.8 \end{cases}$$

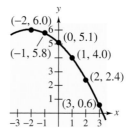

48.
$$\begin{cases} 6c + 3b + 19a = 13.1 \\ 3c + 19b + 27a = -2.6 \\ 19c + 27b + 115a = 29.0 \end{cases}$$

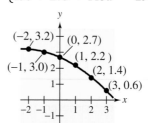

49. Personal Watercraft   The total numbers y (in thousands) of personal watercraft purchased in the United States from 1998 to 2002 are shown in the table. In the table, x represents the year, with $x = 0$ corresponding to 2000.   (Source: National Marine Manufacturers Association)

Year, x	Number, y
-2	130
-1	106
0	92
1	80.9
2	79.3

(a) Find the least squares regression parabola $y = ax^2 + bx + c$ for the data by solving the following system.
$$\begin{cases} 5c + 10a = 488.2 \\ 10b = -126.5 \\ 10c + 34a = 1024.1 \end{cases}$$

(b) Use the *regression* feature of a graphing utility to find a quadratic model for the data. Compare the quadratic model with the model found in part (a).

50. Genetically Modified Cotton The global area y (in millions of hectares) of genetically modified cotton crops planted from 1999 to 2003 is shown in the table. In the table, x represents the year, with $x = 0$ corresponding to 2000. (Source: Clive James, 2003)

Year, x	Area, y
-1	3.7
0	5.3
1	6.8
2	6.8
3	7.2

(a) Find the least squares regression parabola $y = ax^2 + bx + c$ for the data by solving the following system.

$$\begin{cases} 5c + 5b + 15a = 29.8 \\ 5c + 15b + 35a = 38.3 \\ 15c + 35b + 99a = 102.5 \end{cases}$$

(b) Use the *regression* feature of a graphing utility to find a quadratic model for the data. How do the two models compare?

IN THE NEWS

MP3s Are Big Music's Savior, Not Slayer

By Grainger David

. . . The spate of controversial music-business research—from GartnerG2's report on online music service Pressplay 2.0 ("the most significant development in digital music in two years") to the usual harangue from the Recording Industry Association of America (sue the Internet!)—reached its zenith last month with Forrester Research's "Downloads Save the Music Business," which predicts that downloads will generate $2.1 billion for labels by 2007. . . .

Forrester predicts that after one more year of depressing sales, labels will supply more content on the cheap; the story goes that by 2007 this will create a downloading wave of tsunamic proportions that will wash all this additional money onto the music industry.

One indication that it is starting to happen is last month's launch of Pressplay 2.0, a joint venture between Sony and Universal. The updated service offers downloads at lower prices and allows more freedom. (For 99 cents a pop, plus a monthly download fee, you can store a file wherever you'd like.) . . .

Grainger, David. "MP3s Are Big Music's Savior, Not Slayer," Fortune, 16 September 2002, 44.

51. The bar graph shows the predicted revenue (in billions of dollars) from digital music sales for 2002 to 2007. (Source: Forrester Research)

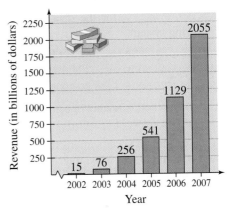

Find the least squares regression parabola $y = ax^2 + bx + c$ for the data by solving the following system. (Let x represent the year, with $x = 2$ corresponding to 2002.) Then use the model to predict the revenue from digital music sales in 2009.

$$\begin{cases} 6c + 27b + 139a = 4072 \\ 27c + 139b + 783a = 25{,}146 \\ 139c + 783b + 4675a = 159{,}704 \end{cases}$$

52. Use the *regression* feature of a graphing utility to verify your answers in Exercise 51.

53. Reasoning Is it possible for a square linear system to have no solution? Explain.

54. Reasoning Is it possible for a square linear system to have infinitely many solutions? Explain.

55. *Make a Decision*: **Stopping Distance** In testing of the new braking system of an automobile, the speed (in miles per hour) and the stopping distance (in feet) were recorded in the table below.

Speed, x	Stopping distance, y
30	54
40	116
50	203
60	315
70	452

(a) Find the least squares regression parabola $y = ax^2 + bx + c$ for the data by solving the following system.

$$\begin{cases} 5c + 250b + 13{,}500a = 1140 \\ 250c + 13{,}500b + 775{,}000a = 66{,}950 \\ 13{,}500c + 775{,}000b + 46{,}590{,}000a = 4{,}090{,}500 \end{cases}$$

(b) Use the *regression* feature of a graphing utility to check your answer to part (a).

(c) If the car design specifications require the car to stop within 520 feet when traveling 75 miles per hour, does the new braking system meet this specification?

56. **Prices of Homes** The median sale prices y (in thousands of dollars) of existing one-family homes in the United States from 1995 to 2002 are shown in the table. In the table, x represents the year, with $x = 0$ corresponding to 1998. (Source: National Association of Realtors)

Year, x	Sale price, y
−3	110.5
−2	115.8
−1	121.8
0	128.4
1	133.3
2	139.0
3	147.8
4	158.1

(a) Find the least squares regression parabola $y = ax^2 + bx + c$ for the data by solving the following system.

$$\begin{cases} 8c + 4b + 44a = 1054.7 \\ 4c + 44b + 64a = 802.2 \\ 44c + 64b + 452a = 6128.6 \end{cases}$$

(b) Use the *regression* feature of a graphing utility to check your answer to part (a).

(c) Estimate the median sale price in 2004 and 2005.

Math MATTERS

Regular Polygons and Regular Polyhedra

A regular polygon is a polygon that has n sides of equal length and n equal angles. For instance, a regular three-sided polygon is called an equilateral triangle, a regular four-sided polygon is called a square, and so on. There are infinitely many different types of regular polygons. It is possible to construct a regular polygon with 100 or even more sides (though it would look very much like a circle).

For solid figures, the story is quite different. A regular polyhedron is a solid figure each of whose sides is a regular polygon (of the same size) and each of whose angles is formed by the same number of sides. At first, one might think that there are infinitely many different types of regular polyhedra, but in fact it can be shown that there are only five. The five

different types are tetrahedron (4 triangular sides), cube (6 square sides), octahedron (8 triangular sides), dodecahedron (12 pentagonal sides), and icosahedron (20 triangular sides), as shown in the figures.

Tetrahedron

Cube (Hexahedron)

Octahedron

Dodecahedron

Icosahedron

Mid-Chapter Quiz

Take this quiz as you would take a quiz in class. After you are done, check your work against the answers given in the back of the book.

In Exercises 1 and 2, solve the system algebraically. Use a graphing utility to verify the solution.

1. $\begin{cases} x + y = 4 \\ y = 2\sqrt{x} + 1 \end{cases}$

2. $\begin{cases} x^2 + y^2 = 9 \\ y = 2x + 1 \end{cases}$

In Exercises 3 and 4, find the number of units x that need to be sold to break even.

3. $C = 12.50x + 10,000$, $R = 18.95x$ 4. $C = 3.79x + 400,000$, $R = 4.59x$

In Exercises 5 and 6, solve the system by substitution or elimination.

5. $\begin{cases} 2.5x - y = 6 \\ 3x + 4y = 2 \end{cases}$

6. $\begin{cases} \frac{1}{2}x + \frac{1}{3}y = 1 \\ x - 2y = -2 \end{cases}$

7. Find the point of equilibrium for the pair of supply and demand equations. Verify the solution graphically.

   Demand: $p = 45 - 0.001x$

   Supply: $p = 23 + 0.0002x$

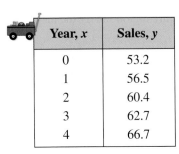

Year, x	Sales, y
0	53.2
1	56.5
2	60.4
3	62.7
4	66.7

Table for 8

8. The total sales y (in billions of dollars) of nondurable toys and sports supplies in the United States from 1997 to 2001 are shown in the table at the left. In the table, x represents the year, with $x = 0$ corresponding to 1997. Solve the following system for a and b to find the least squares regression line $y = ax + b$ for the data.   (Source: U.S. Department of Commerce)

$$\begin{cases} 5b + 10a = 299.5 \\ 10b + 30a = 632.2 \end{cases}$$

In Exercises 9–11, solve the system of equations.

9. $\begin{cases} 2x + 3y - z = -7 \\ x + 3z = 10 \\ 2y + z = -1 \end{cases}$

10. $\begin{cases} x + y - 2z = 12 \\ 2x - y - z = 6 \\ y - z = 6 \end{cases}$

11. $\begin{cases} 3x + 2y + z = 17 \\ -x + y + z = 4 \\ x - y - z = 3 \end{cases}$

Year, x	Teams, y
-2	178
-1	166
0	163
1	155
2	149

Table for 12

12. The number y (in thousands) of adult amateur softball teams in the United States from 1997 to 2001 is shown in the table at the left. In the table, x represents the year, with $x = 0$ corresponding to 1999. Solve the following system for a, b, and c to find the least squares regression parabola $y = ax^2 + bx + c$ for the data.   (Source: Amateur Softball Association)

$$\begin{cases} 5c + 10a = 811 \\ 10b = -69 \\ 10c + 34a = 1629 \end{cases}$$

Systems of Inequalities

Objectives

- Sketch the graph of an inequality in two variables.
- Solve a system of inequalities.
- Construct and use a system of inequalities to solve an application problem.

The Graph of an Inequality

The following statements are inequalities in two variables:

$$3x - 2y < 6 \quad \text{and} \quad 2x^2 + 3y^2 \geq 6.$$

An ordered pair (a, b) is a **solution of an inequality** in x and y if the inequality is true when a and b are substituted for x and y, respectively. The **graph of an inequality** is the collection of all solutions of the inequality. To sketch the graph of an inequality, begin by sketching the graph of the *corresponding equation*. The graph of the equation will normally separate the plane into two or more regions. In each such region, one of the following must be true.

1. *All* points in the region are solutions of the inequality.

2. *No* point in the region is a solution of the inequality.

So, you can determine whether the points in an entire region satisfy the inequality simply by testing *one* point in the region.

Sketching the Graph of an Inequality in Two Variables

1. Replace the inequality sign by an equal sign, and sketch the graph of the resulting equation. (Use a dashed line for < or > and a solid line for ≤ or ≥.)

2. Test one point in each of the regions formed by the graph in Step 1. If the point satisfies the inequality, shade the entire region to denote that every point in the region satisfies the inequality.

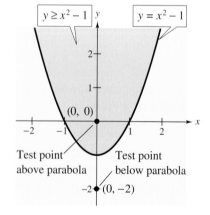

Test point above parabola

Test point below parabola

Figure 4.12

EXAMPLE 1   SKETCHING THE GRAPH OF AN INEQUALITY

Sketch the graph of the inequality $y \geq x^2 - 1$.

SOLUTION

The graph of the corresponding *equation* $y = x^2 - 1$ is a parabola, as shown in Figure 4.12. By testing a point *above* the parabola $(0, 0)$ and a point *below* the parabola $(0, -2)$, you can see that the points that satisfy the inequality are those lying above (or on) the parabola.

✓CHECKPOINT   Now try Exercise 9.

The inequality in Example 1 is a nonlinear inequality in two variables. Most of the following examples involve **linear inequalities** such as $ax + by < c$ (a and b are not both zero). The graph of a linear inequality is a half-plane lying on one side of the line $ax + by = c$. The simplest linear inequalities are those corresponding to horizontal or vertical lines, as shown in Example 2.

EXAMPLE 2   SKETCHING THE GRAPH OF A LINEAR INEQUALITY

Sketch the graph of each linear inequality.

a. $x > -2$ **b.** $y \leq 3$

SOLUTION

a. The graph of the corresponding equation $x = -2$ is a vertical line. The points that satisfy the inequality $x > -2$ are those lying to the right of this line, as shown in Figure 4.13.

b. The graph of the corresponding equation $y = 3$ is a horizontal line. The points that satisfy the inequality $y \leq 3$ are those lying below (or on) this line, as shown in Figure 4.14.

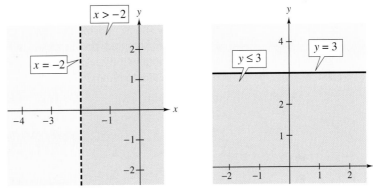

Figure 4.13   **Figure 4.14**

✓CHECKPOINT   Now try Exercise 11.

EXAMPLE 3   SKETCHING THE GRAPH OF A LINEAR INEQUALITY

Sketch the graph of $x - y < 2$.

SOLUTION

The graph of the corresponding equation $x - y = 2$ is a line, as shown in Figure 4.15. Because the origin $(0, 0)$ satisfies the inequality, the graph consists of the half-plane lying above the line. (Try checking a point below the line. Regardless of which point you choose, you will see that it does not satisfy the inequality.)

✓CHECKPOINT   Now try Exercise 15.

STUDY TIP

To graph a linear inequality, it can help to write the inequality in slope-intercept form. For instance, by writing $x - y < 2$ in the form

$$y > x - 2$$

you can see that the solution points lie *above* the line $x - y = 2$ (or $y = x - 2$), as shown in Figure 4.15.

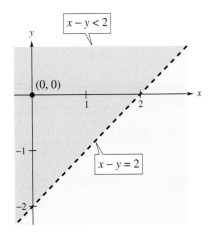

Figure 4.15

Systems of Inequalities

Many practical problems in business, science, and engineering involve *systems of linear inequalities.* A **solution** of a system of inequalities in x and y is a point (x, y) that satisfies each inequality in the system.

To sketch the graph of a system of inequalities in two variables, first sketch the graph of each individual inequality (on the same coordinate system) and then find the region that is *common* to every graph in the system. For systems of *linear* inequalities, it is helpful to find the vertices of the solution region.

EXAMPLE 4   SOLVING A SYSTEM OF INEQUALITIES

Sketch the graph (and label the vertices) of the solution set of the system.

$$\begin{cases} x - y < 2 & \text{Inequality 1} \\ x > -2 & \text{Inequality 2} \\ y \leq 3 & \text{Inequality 3} \end{cases}$$

SOLUTION

The graphs of these inequalities are shown in Figures 4.15, 4.13, and 4.14, respectively, on page 291. The triangular region common to all three graphs can be found by superimposing the graphs on the same coordinate plane, as shown in Figure 4.16. To find the vertices of the region, solve the three systems of equations obtained by taking the *pairs* of equations representing the boundaries of the individual regions.

Vertex A: $(-2, -4)$
Obtained by solving the system

$$\begin{cases} x - y = 2 \\ x = -2 \end{cases}$$

Vertex B: $(5, 3)$
Obtained by solving the system

$$\begin{cases} x - y = 2 \\ y = 3 \end{cases}$$

Vertex C: $(-2, 3)$
Obtained by solving the system

$$\begin{cases} x = -2 \\ y = 3 \end{cases}$$

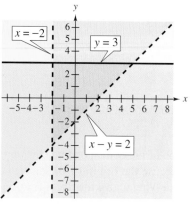 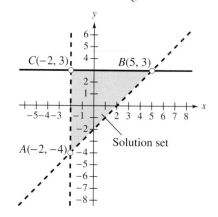

Figure 4.16

✔CHECKPOINT   Now try Exercise 23.

> **STUDY TIP**
>
> Using different colored pencils to shade the solution of each inequality in a system makes identifying the solution of the system of inequalities easier. The region common to every graph in the system is where all the shaded regions overlap. This region represents the solution set of the system.

For the triangular region shown in Figure 4.16, each point of intersection of a pair of boundary lines corresponds to a vertex. With more complicated regions, two border lines can sometimes intersect at a point that is *not* a vertex of the region, as shown in Figure 4.17. In order to keep track of which points of intersection are actually vertices of the region, you should sketch the region and refer to your sketch as you find each point of intersection.

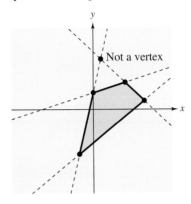

Figure 4.17   Boundary lines can intersect at a point that is not a vertex.

| EXAMPLE 5 | SOLVING A SYSTEM OF INEQUALITIES

Sketch the region containing all points that satisfy the system of inequalities.

$$\begin{cases} x^2 - y \le 1 & \text{Inequality 1} \\ -x + y \le 1 & \text{Inequality 2} \end{cases}$$

SOLUTION

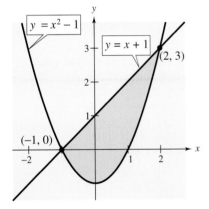

Figure 4.18

As shown in Figure 4.18, the points that satisfy the inequality $x^2 - y \le 1$ are the points lying above (or on) the parabola given by

$$y = x^2 - 1. \qquad \text{Parabola}$$

The points satisfying the inequality $-x + y \le 1$ are the points lying below (or on) the line given by

$$y = x + 1. \qquad \text{Line}$$

To find the points of intersection of the parabola and the line, solve the system of corresponding equations.

$$\begin{cases} x^2 - y = 1 \\ -x + y = 1 \end{cases}$$

Using the method of substitution, you can find the solutions to be $(-1, 0)$ and $(2, 3)$. The graph of the solution set of the system is shown in Figure 4.18.

 CHECKPOINT   Now try Exercise 29.

When solving a system of inequalities, you should be aware that the system might have no solution. For instance, the system

$$\begin{cases} x + y > 3 \\ x + y < -1 \end{cases}$$

has no solution points, because the quantity $(x + y)$ cannot be both less than -1 and greater than 3, as shown in Figure 4.19.

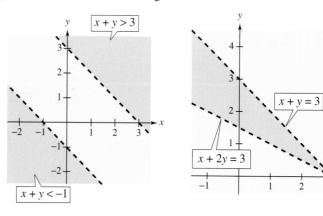

Figure 4.19 No Solution   **Figure 4.20** Unbounded Region

Another possibility is that the solution set of a system of inequalities can be unbounded. For instance, the solution set of

$$\begin{cases} x + y < 3 \\ x + 2y > 3 \end{cases}$$

forms an *infinite wedge,* as shown in Figure 4.20.

TECHNOLOGY

Inequalities and Graphing Utilities   A graphing utility can be used to graph an inequality. The graph of $y \geq x^2 - 2$ is shown below.

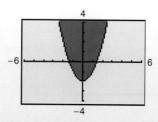

Use your graphing utility to graph each inequality given below. For specific keystrokes on how to graph inequalities and systems of inequalities, go to the text website at *college.hmco.com.*

a. $y \leq 2x + 2$ **b.** $y \geq \frac{1}{2}x^2 - 4$ **c.** $y \leq x^3 - 4x^2 + 4$

Applications

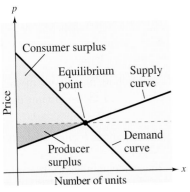

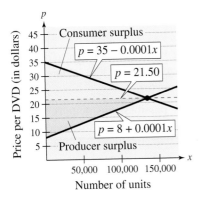

Figure 4.21

Example 8 of Section 4.2 discussed the *point of equilibrium* for a demand function and a supply function. The next example discusses two related concepts that economists call **consumer surplus** and **producer surplus**. As shown in Figure 4.21, the consumer surplus is defined as the area of the region that lies *below* the demand curve, *above* the horizontal line passing through the equilibrium point, and to the right of the *y*-axis. Similarly, the producer surplus is defined as the area of the region that lies *above* the supply curve, *below* the horizontal line passing through the equilibrium point, and to the right of the *y*-axis. The consumer surplus is a measure of the amount that consumers would have been willing to pay *above what they actually paid*, whereas the producer surplus is a measure of the amount that producers would have been willing to receive *below what they actually received*.

EXAMPLE 6   CONSUMER AND PRODUCER SURPLUS

The demand and supply equations for a DVD are given by

$$\begin{cases} p = 35 - 0.0001x & \text{Demand equation} \\ p = \ \ 8 + 0.0001x & \text{Supply equation} \end{cases}$$

where *p* is the price (in dollars) and *x* represents the number of DVDs. Find the consumer surplus and producer surplus for these two equations.

SOLUTION

In Example 8 of Section 4.2, you saw that the point of equilibrium for these equations is

$(135,000, 21.50)$.

So, the horizontal line passing through this point is $p = 21.50$. Now you can determine that the consumer surplus and producer surplus are the areas of the triangular regions given by the following systems of inequalities, respectively.

Consumer Surplus

$$\begin{cases} p \leq 35 - 0.0001x \\ p \geq 21.50 \\ x \geq 0 \end{cases}$$

Producer Surplus

$$\begin{cases} p \geq 8 + 0.0001x \\ p \leq 21.50 \\ x \geq 0 \end{cases}$$

In Figure 4.22, you can see that the consumer and producer surpluses are defined as the areas of the shaded triangles. The base of the triangle representing the consumer surplus is 135,000 because the *x*-value of the point of equilibrium is 135,000. To find the height of this triangle, subtract the *p*-value of the point of equilibrium, 21.50, from the *p*-intercept of the demand equation, 35, to obtain 13.50. You can find the base and height of the triangle representing the producer surplus in a similar manner.

Consumer surplus $= \frac{1}{2}(\text{base})(\text{height}) = \frac{1}{2}(135,000)(13.50) = \$911,250$

Producer surplus $= \frac{1}{2}(\text{base})(\text{height}) = \frac{1}{2}(135,000)(13.50) = \$911,250.$

✓**CHECKPOINT**   Now try Exercise 49.

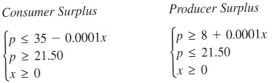

Figure 4.22

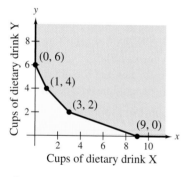

© Michael Keller/CORBIS

EXAMPLE 7 NUTRITION

The liquid portion of a diet is to provide at least 300 calories, 36 units of vitamin A, and 90 units of vitamin C daily. A cup of dietary drink X provides 60 calories, 12 units of vitamin A, and 10 units of vitamin C. A cup of dietary drink Y provides 60 calories, 6 units of vitamin A, and 30 units of vitamin C.

a. Set up a system of linear inequalities that describes how many cups of each drink should be consumed each day to meet or exceed the minimum daily requirements for calories and vitamins.

b. *Make a Decision* A nutritionist normally gives a patient 6 cups of dietary drink X and 1 cup of dietary drink Y per day. Supplies on dietary drink X are running low. Use the graph of the system of linear inequalities to determine other combinations of drinks X and Y that can be given that will meet the minimum daily requirements.

SOLUTION

a. Begin by letting x and y represent the following.

x = number of cups of drink X y = number of cups of drink Y

To meet or exceed the minimum daily requirements, the following inequalities must be satisfied.

$$\begin{cases} 60x + 60y \geq 300 & \text{Calories} \\ 12x + 6y \geq 36 & \text{Vitamin A} \\ 10x + 30y \geq 90 & \text{Vitamin C} \\ x \geq 0 \\ y \geq 0 \end{cases}$$

The last two inequalities are included because x and y cannot be negative. The graph of this system of inequalities is shown in Figure 4.23. (More is said about this application in Example 5 in Section 4.5.)

b. From Figure 4.23, there are many different possible substitutions that the nutritionist can make. Because supplies are running low on dietary drink X, the nutritionist should choose a combination that contains a small amount of drink X. For instance, 1 cup of dietary drink X and 4 cups of dietary drink Y will also meet the minimum requirement.

✓CHECKPOINT   Now try Exercise 55.

Figure 4.23

Discussing the Concept | Writing a System of Inequalities

Write a system of inequalities that includes $x \geq 0$ and $y \geq 0$. Find the solution set by sketching a graph of the system and identifying the vertices. Exchange your system for the system written by another student. Find the solution set and identify the vertices of the system that you received. Does your answer agree with the answer obtained by the student who wrote the problem?

4.4 ⟩ Warm Up

The following warm-up exercises involve skills that were covered in earlier sections. You will use these skills in the exercise set for this section. For additional help, review Sections 1.1, 1.2, 2.1, 4.1, and 4.2.

In Exercises 1–6, classify the graph of the equation as a line, a parabola, or a circle.

1. $x + y = 3$

2. $y = x^2 - 4$

3. $x^2 + y^2 = 9$

4. $y = -x^2 + 1$

5. $4x - y = 8$

6. $y^2 = 16 - x^2$

In Exercises 7–10, solve the system of equations.

7. $\begin{cases} x + 2y = 3 \\ 4x - 7y = -3 \end{cases}$

8. $\begin{cases} 2x - 3y = 4 \\ x + 5y = 2 \end{cases}$

9. $\begin{cases} x^2 + y = 5 \\ 2x - 4y = 0 \end{cases}$

10. $\begin{cases} x^2 + y^2 = 13 \\ x + y = 5 \end{cases}$

4.4 ⟩ Exercises

In Exercises 1–6, match the inequality with its graph. [The graphs are labeled (a), (b), (c), (d), (e), and (f).]

1. $2x + 3y \le 6$

2. $2x - y \ge -2$

3. $x^2 - y \le 2$

4. $y \le 4 - x^2$

5. $y > x^4 - 5x^2 + 4$

6. $3x^4 + y < 6x^2$

(a)
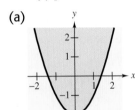

(b)
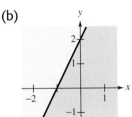

In Exercises 7–20, sketch the graph of the inequality.

7. $x \ge 2$

8. $x < 4$

9. $y + 2x^2 > 0$

10. $y^2 - x < 0$

11. $y > -1$

12. $y \le 3$

13. $y < 2 - x$

14. $y > 2x - 4$

15. $2y - x \ge 4$

16. $5x + 3y \ge -15$

17. $y \le \dfrac{1}{1 + x^2}$

18. $y < \ln x$

19. $x^2 + y^2 \le 4$

20. $x^2 + y^2 > 4$

(c)
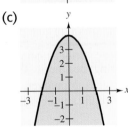

(d)
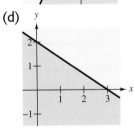

(e)
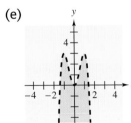

(f)
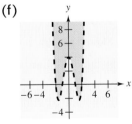

In Exercises 21–40, graph the solution set of the system of inequalities.

21. $\begin{cases} x + y \le 2 \\ -x + y \le 2 \\ y \ge 0 \end{cases}$

22. $\begin{cases} 3x + 2y < 6 \\ x > 1 \\ y > 0 \end{cases}$

23. $\begin{cases} x + y \le 5 \\ x \quad\quad \ge 2 \\ \quad\quad y \ge 0 \end{cases}$

24. $\begin{cases} 2x + y \ge 2 \\ x \quad\quad \le 2 \\ \quad\quad y \le 1 \end{cases}$

25. $\begin{cases} -3x + 2y < 6 \\ x + 4y > -2 \\ 2x + y < 3 \end{cases}$

26. $\begin{cases} x - 7y > -36 \\ 5x + 2y > 5 \\ 6x - 5y > 6 \end{cases}$

27. $\begin{cases} 2x + y < 2 \\ 6x + 3y > 2 \end{cases}$

28. $\begin{cases} 5x - 3y < -6 \\ 5x - 3y > -9 \end{cases}$

29. $\begin{cases} y \ge -3 \\ y \le 1 - x^2 \end{cases}$

30. $\begin{cases} x - y^2 > 0 \\ y > (x - 3)^2 - 4 \end{cases}$

31. $\begin{cases} x^2 + y^2 \le 16 \\ x^2 + y^2 < 1 \end{cases}$

32. $\begin{cases} x^2 + y^2 \le 25 \\ x^2 + y^2 \ge 9 \end{cases}$

33. $\begin{cases} x > y^2 \\ x < y + 2 \end{cases}$

34. $\begin{cases} x < 2y - y^2 \\ 0 < x + y \end{cases}$

35. $\begin{cases} y \le \sqrt{3x} + 1 \\ y \ge x + 1 \end{cases}$

36. $\begin{cases} y < \sqrt{2x} + 3 \\ y > x + 3 \end{cases}$

37. $\begin{cases} y < x^3 - 2x + 1 \\ y > -2x \\ x \le 1 \end{cases}$

38. $\begin{cases} x \ge 1 \\ x - 2y \le 3 \\ 3x + 2y \ge 9 \\ x + y \le 6 \end{cases}$

39. $\begin{cases} y \le e^x \\ y \ge \ln x \\ x \ge \frac{1}{2} \\ x \le 2 \end{cases}$

40. $\begin{cases} y \le e^{-x^2/2} \\ y \ge 0 \\ -1 \le x \le 1 \end{cases}$

In Exercises 41–46, write a system of inequalities that describes the region.

41. Rectangular region with vertices $(1, 1)$, $(8, 1)$, $(8, 5)$, and $(1, 5)$

42. Parallelogram region with vertices $(0, 0)$, $(5, 0)$, $(1, 6)$, and $(6, 6)$

43. Triangular region with vertices $(0, 0)$, $(7, 0)$, and $(3, 5)$

44. Triangular region with vertices $(-2, 0)$, $(6, 0)$, and $(0, 5)$

45. Sector of a circle

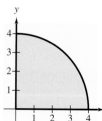

46. Sector of a circle

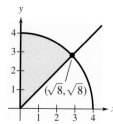

$(\sqrt{8}, \sqrt{8})$

47. Furniture Production   A furniture company can sell all the tables and chairs it produces. Each table requires 2 hours in the assembly center and $1\frac{1}{2}$ hours in the finishing center. Each chair requires $1\frac{1}{2}$ hours in the assembly center and $1\frac{1}{2}$ hours in the finishing center. The company's assembly center is available 18 hours per day, and its finishing center is available 15 hours per day. If x is the number of tables produced per day and y is the number of chairs, find a system of inequalities describing all possible production levels, and sketch the graph of the system.

48. Computer Inventory   A store sells two models of computers. Because of the demand, it is necessary to stock at least twice as many units of model A as units of model B. The costs to the store for the two models are $500 and $700, respectively. The management does not want more than $30,000 in computer inventory at any one time, and it wants at least six model A computers and three model B computers in inventory at all times. Find a system of inequalities describing all possible inventory levels, and sketch the graph of the system.

Consumer and Producer Surplus   In Exercises 49–52, find the consumer surplus and producer surplus for the pair of demand and supply equations.

	Demand	Supply
49.	$p = 56 - 0.0001x$	$p = 22 + 0.00001x$
50.	$p = 60 - 0.00001x$	$p = 15 + 0.00004x$
51.	$p = 140 - 0.00002x$	$p = 80 + 0.00001x$
52.	$p = 600 - 0.0002x$	$p = 125 + 0.0006x$

53. Investment   A person plans to invest up to $30,000 in two different interest-bearing accounts. Each account is to contain at least $6000. Moreover, one account should have at least twice the amount that is in the other account.

(a) Find a system of inequalities that describes the amounts that can be deposited in each account.

(b) Sketch the graph of the system.

54. Concert Ticket Sales   Two types of tickets are to be sold for a concert. One type costs $20 per ticket and the other type costs $30 per ticket. The promoter of the concert must sell at least 20,000 tickets, including at least 8000 of the $20 tickets and at least 5000 of the $30 tickets. Moreover, the gross receipts must total at least $480,000 in order for the concert to be held.

(a) Find a system of inequalities describing the different numbers of tickets that can be sold.

(b) Sketch the graph of the system.

55. *Make a Decision*: **Diet Supplement**   A dietitian is asked to design a special diet supplement using two different foods. Each ounce of food X contains 20 units of calcium, 10 units of iron, and 15 units of vitamin B. Each ounce of food Y contains 15 units of calcium, 20 units of iron, and 20 units of vitamin B. The minimum daily requirements for the diet are 400 units of calcium, 250 units of iron, and 220 units of vitamin B.

(a) Find a system of inequalities describing the different amounts of food X and food Y that can be used in the diet.

(b) Sketch the graph of the system.

(c) A nutritionist normally gives a patient 18 ounces of food X and 3.5 ounces of food Y per day. Supplies of food X are running low. What other combinations of foods X and Y can be given to the patient to meet the minimum daily requirements?

56. *Make a Decision*: **Diet Supplement**   A dietitian is asked to design a special diet supplement using two different foods. Each ounce of food X contains 12 units of calcium, 10 units of iron, and 20 units of vitamin B. Each ounce of food Y contains 15 units of calcium, 20 units of iron, and 12 units of vitamin B. The minimum daily requirements for the diet are 300 units of calcium, 280 units of iron, and 300 units of vitamin B.

(a) Find a system of inequalities describing the different amounts of food X and food Y that can be used in the diet.

(b) Sketch the graph of the system.

(c) A nutritionist normally gives a patient 10 ounces of food X and 12 ounces of food Y per day. Supplies of food Y are running low. What other combinations of foods X and Y can be given to the patient to meet the minimum daily requirements?

57. Federal Student Aid   The amount of federal student aid y (in billions of dollars) awarded in the United States from 1999 to 2002 can be approximated by the linear model

$$y = 4.99t - 5.4, \quad 9 \le t \le 12$$

where t represents the year, with $t = 9$ corresponding to 1999.   (Source: U.S. Department of Education)

(a) The *total* amount of federal student aid awarded during this four-year period can be approximated by finding the area of the trapezoid represented by the following system.

$$\begin{cases} y \le 4.99t - 5.4 \\ y \ge 0 \\ t \ge 8.5 \\ t \le 12.5 \end{cases}$$

Graph this region using a graphing utility.

(b) Use the formula for the area of a trapezoid to approximate the total amount of federal student aid awarded.

58. Snowboarders   The number y (in millions) of people who participated in snowboarding more than once a year in the United States from 1992 to 2002 can be approximated by the linear model

$$y = 0.42t + 0.4, \quad 2 \le t \le 12$$

where t represents the year, with $t = 2$ corresponding to 1992.   (Source: National Sporting Goods Association)

(a) The *total* number of snowboarders during this 11-year period can be approximated by finding the area of the trapezoid represented by the following system.

$$\begin{cases} y \le 0.42t + 0.4 \\ y \ge 0 \\ t \ge 1.5 \\ t \le 12.5 \end{cases}$$

Graph this region using a graphing utility.

(b) Use the formula for the area of a trapezoid to approximate the total number of snowboarders.

4.5   Linear Programming

Objectives
- **Use linear programming to minimize or maximize an objective function.**
- **Use linear programming to optimize an application.**

Linear Programming: A Graphical Approach

Many applications in business and economics involve a process called **optimization,** in which you are asked to find the minimum cost, the maximum profit, or the minimum use of resources. In this section you will study an optimization strategy called **linear programming.**

A two-dimensional linear programming problem consists of a linear **objective function** and a system of linear inequalities called **constraints.** The objective function gives the quantity that is to be maximized (or minimized), and the constraints determine the set of **feasible solutions.** For example, consider a linear programming problem in which you are asked to maximize the value of

$$z = ax + by \qquad \text{Objective function}$$

subject to a set of constraints that determines the region in Figure 4.24. Because every point in the region satisfies each constraint, it is not clear how you should go about finding the point that yields a maximum value of z. Fortunately, it can be shown that if there is an optimal solution, it must occur at one of the vertices of the region. This means that *you can find the maximum value by testing z at each of the vertices.*

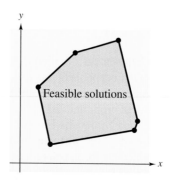

Feasible solutions

Figure 4.24

Optimal Solution of a Linear Programming Problem

If a linear programming problem has a solution, it must occur at a vertex of the set of feasible solutions. If the problem has more than one solution, then at least one solution must occur at a vertex of the set of feasible solutions. In either case, the value of the objective function is unique.

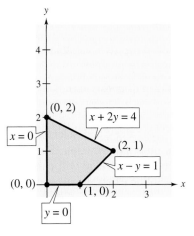

$x + 2y = 4$

$x = 0$

$(2, 1)$

$x - y = 1$

$(0, 0)$

$(1, 0)$

$y = 0$

Figure 4.25

EXAMPLE 1 SOLVING A LINEAR PROGRAMMING PROBLEM

Find the maximum value of

$$z = 3x + 2y \qquad \text{Objective function}$$

subject to the following constraints.

$$\left. \begin{array}{r} x \geq 0 \\ y \geq 0 \\ x + 2y \leq 4 \\ x - y \leq 1 \end{array} \right\} \qquad \text{Constraints}$$

SOLUTION

The constraints form the region shown in Figure 4.25. At the four vertices of this region, the objective function has the following values.

At $(0, 0)$: $z = 3(0) + 2(0) = 0$

At $(1, 0)$: $z = 3(1) + 2(0) = 3$

At $(2, 1)$: $z = 3(2) + 2(1) = 8 \qquad$ Maximum value of z

At $(0, 2)$: $z = 3(0) + 2(2) = 4$

So, the maximum value of z is 8, and this occurs when $x = 2$ and $y = 1$.

✓CHECKPOINT   Now try Exercise 21.

In Example 1, try testing some of the *interior* points of the region. You will see that the corresponding values of z are less than 8. Here are some examples.

At $(1, 1)$: $z = 3(1) + 2(1) = 5$

At $\left(1, \frac{1}{2}\right)$: $z = 3(1) + 2\left(\frac{1}{2}\right) = 4$

To see why the maximum value of the objective function in Example 1 must occur at a vertex, consider writing the objective function in slope-intercept form

$$y = -\frac{3}{2}x + \frac{z}{2} \qquad \text{Family of lines}$$

where $z/2$ is the y-intercept of the objective function. This equation represents a family of lines, each of slope $-\frac{3}{2}$. Of these infinitely many lines, you want the one that has the largest z-value while still intersecting the region determined by the constraints. In other words, of all the lines whose slope is $-\frac{3}{2}$, you want the one that has the largest y-intercept *and* intersects the given region, as shown in Figure 4.26. It should be clear that such a line will pass through one (or more) of the vertices of the region.

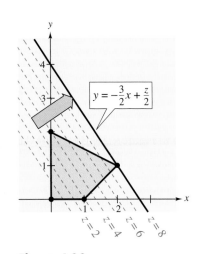

$y = -\frac{3}{2}x + \frac{z}{2}$

Figure 4.26

Remember that a vertex of a region can be found using a system of linear equations. The system will consist of the equations of the lines passing through the vertex.

Solving a Linear Programming Problem

To solve a linear programming problem involving two variables by the graphical method, use the following steps.

1. Sketch the region corresponding to the system of constraints.

2. Find the vertices of the region.

3. Test the objective function at each of the vertices and select the values of the variables that optimize the objective function. For a bounded region, both a minimum and a maximum value will exist. (For an unbounded region, *if* an optimal solution exists, it will occur at a vertex.)

The guidelines above will work whether the objective function is to be maximized or minimized. For instance, the same test used in Example 1 to find the maximum value of z can be used to conclude that the minimum value of z is 0 and that this value occurs at the vertex $(0, 0)$.

EXAMPLE 2   SOLVING A LINEAR PROGRAMMING PROBLEM

Find (a) the maximum value and (b) the minimum value of the objective function

$$z = 4x + 6y \qquad \text{Objective function}$$

subject to the following constraints.

$$\left. \begin{array}{r} x \geq 0 \\ y \geq 0 \\ -x + y \leq 11 \\ x + y \leq 27 \\ 2x + 5y \leq 90 \end{array} \right\} \qquad \text{Constraints}$$

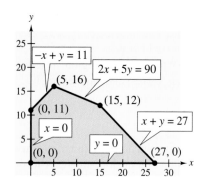

Figure 4.27

SOLUTION

a. The region bounded by the constraints is shown in Figure 4.27. By testing the objective function at each vertex, you obtain the following.

At $(0, 0)$: $z = 4(0) + 6(0) = 0$

At $(0, 11)$: $z = 4(0) + 6(11) = 66$

At $(5, 16)$: $z = 4(5) + 6(16) = 116$

At $(15, 12)$: $z = 4(15) + 6(12) = 132$ Maximum value of z

At $(27, 0)$: $z = 4(27) + 6(0) = 108$

So, the maximum value of z is 132, and this occurs when $x = 15$ and $y = 12$.

b. Using the values of z at the vertices in part (a), you can conclude that the minimum value of z is 0, and that this value occurs when $x = 0$ and $y = 0$.

✓CHECKPOINT   Now try Exercise 11.

The steps used to find the minimum and maximum values of an objective function are precisely the same. In other words, once you have evaluated the objective function at the vertices of the feasible region, you simply choose the largest value as the maximum and the smallest value as the minimum.

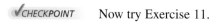

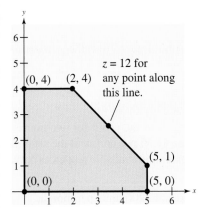

Figure 4.28

It is possible for the maximum (or minimum) value in a linear programming problem to occur at *two* different vertices. For instance, at the vertices of the region shown in Figure 4.28, the objective function

$$z = 2x + 2y \qquad \text{Objective function}$$

has the following values.

At $(0, 0)$: $z = 2(0) + 2(0) = 0$

At $(0, 4)$: $z = 2(0) + 2(4) = 8$

At $(2, 4)$: $z = 2(2) + 2(4) = 12$ Maximum value of z

At $(5, 1)$: $z = 2(5) + 2(1) = 12$ Maximum value of z

At $(5, 0)$: $z = 2(5) + 2(0) = 10$

In this case, you can conclude that the objective function has a maximum value not only at the vertices $(2, 4)$ and $(5, 1)$; it also has a maximum value (of 12) at *any point on the line segment connecting these two vertices*. Note that the objective function

$$y = -x + \tfrac{1}{2}z$$

has the same slope as the line through the vertices $(2, 4)$ and $(5, 1)$.

Some linear programming problems have no optimal solution. This can occur if the region determined by the constraints is *unbounded*. Example 3 illustrates such a problem.

EXAMPLE 3   AN UNBOUNDED REGION

Find the maximum value of

$$z = 4x + 2y \qquad \text{Objective function}$$

where $x \geq 0$ and $y \geq 0$, subject to the following constraints.

$$\left. \begin{array}{r} x + 2y \geq 4 \\ 3x + y \geq 7 \\ -x + 2y \leq 7 \end{array} \right\} \qquad \text{Constraints}$$

SOLUTION

The region determined by the constraints is shown in Figure 4.29. For this unbounded region, there is no maximum value of z. To see this, note that the point $(x, 0)$ lies in the region for all values of $x \geq 4$. By choosing x to be large, you can obtain values of

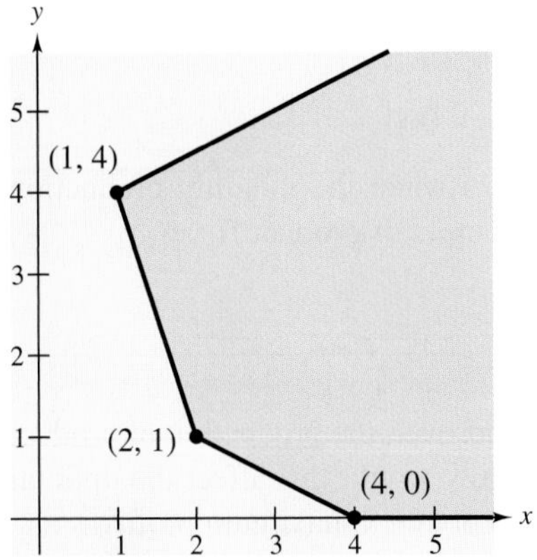

$$z = 4(x) + 2(0) = 4x$$

that are as large as you want. So, there is no maximum value of z. For this problem, there *is* a minimum value of $z = 10$, which occurs at the vertex $(2, 1)$, as shown below.

At $(1, 4)$: $z = 4(1) + 2(4) = 12$

At $(2, 1)$: $z = 4(2) + 2(1) = 10$ Minimum value of z

At $(4, 0)$: $z = 4(4) + 2(0) = 16$

Figure 4.29

✓CHECKPOINT   Now try Exercise 13.

Applications

Example 4 shows how linear programming can be used to find the maximum profit in a business application.

| EXAMPLE 4 | OPTIMAL PROFIT | |

Make a Decision　A manufacturer wants to maximize the profit for two house-ware products. The first product yields a profit of $1.50 per unit, and the second product yields a profit of $2.00 per unit. Market tests and available resources have indicated the following constraints.

1. The combined production level should not exceed 1200 units per month.

2. The demand for product II is no more than half the demand for product I.

3. The production level of product I is less than or equal to 600 units plus three times the production level of product II.

What is the optimal production level for each product?

SOLUTION

If you let x be the number of units of product I and y be the number of units of product II, the objective function (for the combined profit) is given by

$$P = 1.5x + 2y.$$　　　　　　Objective function

The three constraints translate into the following linear inequalities.

1. $x + y \le 1200$		$x + y \le 1200$
2. $y \le \frac{1}{2}x$		$-x + 2y \le 0$
3. $x \le 3y + 600$		$x - 3y \le 600$

Because neither x nor y can be negative, you also have the two additional constraints of $x \ge 0$ and $y \ge 0$. Figure 4.30 shows the region determined by the constraints. To find the maximum profit, test the value of P at the vertices of the region.

At $(0, 0)$:　　　$P = 1.5(0)$　　$+ 2(0)$　$=$　0

At $(800, 400)$:　$P = 1.5(800)$　$+ 2(400) = 2000$　　Maximum profit

At $(1050, 150)$: $P = 1.5(1050) + 2(150) = 1875$

At $(600, 0)$:　　$P = 1.5(600)$　$+ 2(0)$　$=$　900

So, the maximum profit is $2000, and it occurs when the monthly production levels consist of 800 units of product I and 400 units of product II.

✓CHECKPOINT　　Now try Exercise 33.

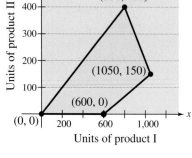

Figure 4.30

　　In Example 4, suppose the manufacturer improved the production of product I so that it yielded a profit of $2.50 per unit. How would this affect the optimal number of units the manufacturer should sell to obtain a maximum profit?

EXAMPLE 5 OPTIMAL COST

Make a Decision The liquid part of a diet is to provide at least 300 calories, 36 units of vitamin A, and 90 units of vitamin C daily. A cup of dietary drink X costs $0.12 and provides 60 calories, 12 units of vitamin A, and 10 units of vitamin C. A cup of dietary drink Y costs $0.15 and provides 60 calories, 6 units of vitamin A, and 30 units of vitamin C. How many cups of each drink should be consumed each day to obtain an optimal cost and still meet the daily requirements?

SOLUTION

As in Example 7 on page 296, let x be the number of cups of dietary drink X and let y be the number of cups of dietary drink Y.

$$\left.\begin{array}{lrcl} \text{For calories:} & 60x + 60y & \geq & 300 \\ \text{For vitamin A:} & 12x + 6y & \geq & 36 \\ \text{For vitamin C:} & 10x + 30y & \geq & 90 \\ & x & \geq & 0 \\ & y & \geq & 0 \end{array}\right\} \quad \text{Constraints}$$

The cost C is given by

$$C = 0.12x + 0.15y. \qquad \text{Objective function}$$

The graph of the region corresponding to the constraints is shown in Figure 4.31. Because you want to incur as little cost as possible, you want to determine the *minimum* cost. To determine the minimum cost, test C at each vertex of the region, as follows.

At $(0, 6)$: $C = 0.12(0) + 0.15(6) = 0.90$

At $(1, 4)$: $C = 0.12(1) + 0.15(4) = 0.72$

At $(3, 2)$: $C = 0.12(3) + 0.15(2) = 0.66$ Minimum value of C

At $(9, 0)$: $C = 0.12(9) + 0.15(0) = 1.08$

So, the minimum cost is $0.66 per day, and this occurs when 3 cups of drink X and 2 cups of drink Y are consumed each day.

✓CHECKPOINT   Now try Exercise 35.

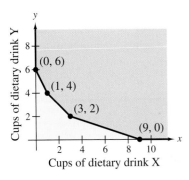

Figure 4.31

Discussing the Concept | Analysis of Constraints

Explain what difficulties you might encounter with each set of linear programming constraints.

a.
$$\begin{array}{rcl} x - y & < & 0 \\ 3x + y & > & 9 \\ -4x + y & > & -2 \end{array}$$

b.
$$\begin{array}{rcl} 2x + y & > & 11 \\ x - y & > & 0 \\ x & < & 4 \\ y & < & 0 \end{array}$$

4.5 ⟩ Warm Up

The following warm-up exercises involve skills that were covered in earlier sections. You will use these skills in the exercise set for this section. For additional help, review Sections 1.2, 4.1, and 4.4.

In Exercises 1–4, sketch the graph of the linear equation.

1. $y + x = 3$

2. $y - x = 12$

3. $x = 0$

4. $y = 4$

In Exercises 5–8, find the point of intersection of the two lines.

5. $x + y = 4, x = 0$

6. $x + 2y = 12, y = 0$

7. $x + y = 4, 2x + 3y = 9$

8. $x + 2y = 12, 2x + y = 9$

In Exercises 9 and 10, sketch the graph of the inequality.

9. $2x + 3y \geq 18$

10. $4x + 3y \geq 12$

4.5 ⟩ Exercises

In Exercises 1–8, find the minimum and maximum values of the objective function and where they occur, subject to the indicated constraints. (For each exercise, the graph of the region determined by the constraints is provided.)

1. *Objective function:*

   $z = 6x + 5y$

   Constraints:

   $$x \geq 0$$
   $$y \geq 0$$
   $$x + y \leq 6$$

   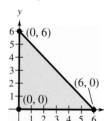

2. *Objective function:*

   $z = 2x + 8y$

   Constraints:

   $$x \geq 0$$
   $$y \geq 0$$
   $$2x + y \leq 4$$

   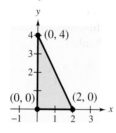

3. *Objective function:*

   $z = 8x + 7y$

   Constraints:

   See Exercise 1.

4. *Objective function:*

   $z = 7x + 3y$

   Constraints:

   See Exercise 2.

5. *Objective function:*

   $z = 3x + 2y$

   Constraints:

   $$x \geq 0$$
   $$y \geq 0$$
   $$x + 3y \leq 15$$
   $$4x + y \leq 16$$

   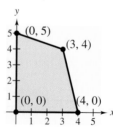

6. *Objective function:*

   $z = 5x + 4y$

   Constraints:

   $$x \geq 0$$
   $$2x + 3y \geq 6$$
   $$3x - 2y \leq 9$$
   $$x + 5y \leq 20$$

   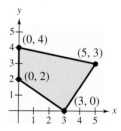

7. *Objective function:*

   $z = 5x + 0.5y$

   Constraints:

   See Exercise 5.

8. *Objective function:*

   $z = x + 6y$

   Constraints:

   See Exercise 6.

In Exercises 9–20, sketch the region determined by the constraints. Then find the minimum and maximum values of the objective function and where they occur, subject to the indicated constraints.

9. *Objective function:*

$z = 6x + 10y$

Constraints:

$x \geq 0$

$y \geq 0$

$3x + 5y \leq 15$

10. *Objective function:*

$z = 7x + 8y$

Constraints:

$x \geq 0$

$y \geq 0$

$x + 2y \leq 8$

11. *Objective function:*

$z = 9x + 4y$

Constraints:

See Exercise 9.

12. *Objective function:*

$z = 7x + 2y$

Constraints:

See Exercise 10.

13. *Objective function:*

$z = 4x + 5y$

Constraints:

$x \geq 0$

$y \geq 0$

$x + y \geq 8$

$3x + 5y \geq 30$

14. *Objective function:*

$z = 4x + 5y$

Constraints:

$x \geq 0$

$y \geq 0$

$x + y \leq 5$

$x + 2y \leq 6$

15. *Objective function:*

$z = 2x + 7y$

Constraints:

See Exercise 13.

16. *Objective function:*

$z = 2x - y$

Constraints:

See Exercise 14.

17. *Objective function:*

$z = 5x + 2y$

Constraints:

$x \geq 0$

$y \geq 0$

$x + 2y \leq 40$

$x + y \geq 30$

$2x + 3y \geq 72$

18. *Objective function:*

$z = x$

Constraints:

$x \geq 0$

$y \geq 0$

$2x + 3y \leq 60$

$2x + y \leq 28$

$4x + y \leq 48$

19. *Objective function:*

$z = 2x + 5y$

Constraints:

See Exercise 17.

20. *Objective function:*

$z = y$

Constraints:

See Exercise 18.

In Exercises 21–24, maximize the objective function subject to the constraints $3x + y \leq 15$, $4x + 3y \leq 30$, $x \geq 0$, and $y \geq 0$.

21. $z = 2x + y$

22. $z = 5x + y$

23. $z = x + y$

24. $z = 3x + y$

In Exercises 25–28, maximize the objective function subject to the constraints $x + 4y \leq 20$, $x + y \leq 8$, $3x + 2y \leq 21$, $x \geq 0$, and $y \geq 0$.

25. $z = 2x + 5y$

26. $z = 3x + 5y$

27. $z = 12x + 5y$

28. $z = 15x + 8y$

Think About It In Exercises 29–32, find an objective function that has a maximum or minimum value at the indicated vertex of the constraint region shown. (There are many correct answers.)

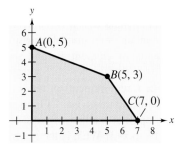

29. The maximum occurs at vertex A.

30. The maximum occurs at vertex B.

31. The maximum occurs at vertex C.

32. The minimum occurs at vertex A.

33. *Make a Decision*: **Optimal Profit** A fruit grower raises two crops, A and B. The profit is $185 per acre for crop A and $245 per acre for crop B. Research and available resources have indicated the following constraints.

- The fruit grower has 150 acres of land for raising the crops.
- It takes 1 day to trim an acre of crop A and 2 days to trim an acre of crop B, and there are 240 days per year available for trimming.
- It takes 0.3 day to pick an acre of crop A and 0.1 day to pick an acre of crop B, and there are 30 days per year available for picking.

What is the optimal acreage for each fruit? What is the optimal profit?

34. *Make a Decision*: **Optimal Profit** The costs to a store for two models of computers are $900 and $1075. The $900 model yields a profit of $100 and the $1075 model yields a profit of $125. Market tests and available resources have indicated the following constraints.
- The merchant estimates that the total monthly demand will not exceed 275 units.
- The merchant does not want to invest more than $260,625 in computer inventory.

What is the optimal inventory level for each model? What is the optimal profit?

35. *Make a Decision*: **Optimal Cost** A farming cooperative mixes two brands of cattle feed. Brand X costs $30 per bag, and brand Y costs $25 per bag. Research and available resources have indicated the following constraints.
- Brand X contains two units of nutritional element A, two units of element B, and two units of element C.
- Brand Y contains one unit of nutritional element A, nine units of element B, and three units of element C.
- The minimum requirements of nutrients A, B, and C are 12 units, 36 units, and 24 units, respectively.

What is the optimal number of bags of each brand that should be mixed? What is the optimal cost?

36. *Make a Decision*: **Optimal Cost** Two gasolines, type A and type B, have octane ratings of 89 and 93, respectively. Type A costs $1.65 per gallon and type B costs $1.89 per gallon. What is the optimal blend of gasoline with an octane rating of at least 91? What is the optimal cost?

37. *Make a Decision*: **Optimal Profit** A manufacturer produces two models of bicycles. The times (in hours) required for assembling, painting, and packaging each model are shown in the table.

Process	Model A	Model B
Assembling	2	2.5
Painting	4	1
Packaging	1	0.75

The total times available for assembling, painting, and packaging are 4000 hours, 4800 hours, and 1500 hours, respectively. The profits per unit are $50 for model A and $75 for model B. What is the optimal production level for each model? What is the optimal profit?

38. *Make a Decision*: **Optimal Profit** A company makes two models of chairs. The times (in hours) required for assembling, painting, and packaging are shown in the table.

Process	Model A	Model B
Assembling	2.5	3
Painting	2	1
Packaging	0.75	1.25

The total times available for assembling, painting, and packaging are 4000 hours, 2500 hours, and 1500 hours, respectively. The profits per unit are $60 for model A and $75 for model B. What is the optimal production level for each model? What is the optimal profit?

39. *Make a Decision*: **Optimal Revenue** An accounting firm charges $2500 for an audit and $350 for a tax return. Research and available resources have indicated the following constraints.
- The firm has 900 hours of staff time available each week.
- The firm has 155 hours of review time available each week.
- Each audit requires 75 hours of staff time and 10 hours of review time.
- Each tax return requires 12.5 hours of staff time and 2.5 hours of review time.

What numbers of audits and tax returns will bring in an optimal revenue?

40. *Make a Decision*: **Optimal Revenue** The accounting firm in Exercise 39 lowers its charge for an audit to $2000. What numbers of audits and tax returns will bring in an optimal revenue?

41. *Make a Decision*: **Media Selection** A company has budgeted a maximum of $1,000,000 for advertising an allergy medication nationally. Each minute of television time costs $100,000 and each one-page newspaper ad costs $20,000. Each television ad is expected to be viewed by 20 million viewers, and each newspaper ad is expected to be seen by 5 million readers. The company's market research department recommends that at most 80% of the advertising budget be spent on television ads. What is the optimal amount that should be spent on advertising for each type of ad? What is the optimal total audience?

42. *Make a Decision*: **Optimal Profit** A fruit juice company makes two drinks by blending apple and pineapple juices. The first uses 30% apple juice and 70% pineapple, and the second uses 60% apple juice and 40% pineapple. There are 1000 liters of apple juice and 1500 liters of pineapple juice available. The profit for the first drink is $0.70 per liter and the profit for the second drink is $0.60 per liter. What is the optimal production level for each type of juice? What is the optimal profit?

43. *Make a Decision*: **Investments** An investor has up to $250,000 to invest in two types of investments. Type A pays 7% annually and type B pays 12% annually. To have a well-balanced portfolio, the investor imposes the following conditions. At least one-fourth of the total portfolio is to be allocated to type A investments and at least one-fourth of the portfolio is to be allocated to type B investments. What is the optimal amount that should be invested in each investment? What is the optimal return?

44. *Make a Decision*: **Investments** An investor has up to $450,000 to invest in two types of investments. Type A pays 8% annually and type B pays 14% annually. To have a well-balanced portfolio, the investor imposes the following conditions. At least one-half of the total portfolio is to be allocated to type A investments and at least one-fourth of the portfolio is to be allocated to type B investments. What is the optimal amount that should be invested in each investment? What is the optimal return?

45. *Make a Decision*: **Optimal Profit** A company makes two models of a patio furniture set. The times for assembling, finishing, and packaging model A are 3 hours, 2.5 hours, and 0.6 hour, respectively. The times for model B are 2.75 hours, 1 hour, and 1.25 hours. The total times available for assembling, finishing, and packaging are 3000 hours, 2400 hours, and 1200 hours, respectively. The profit per unit for model A is $100 and the profit per unit for model B is $85. What is the optimal production level for each model? What is the optimal profit?

46. *Make a Decision*: **Optimal Profit** A manufacturer produces two models of elliptical cross training exercise machines. The times for assembling, finishing, and packaging model A are 3 hours, 2 hours, and 1.2 hours, respectively. The times for model B are 4 hours, 2.5 hours, and 0.9 hour. The total times available for assembling, finishing, and packaging are 6000 hours, 4000 hours, and 900 hours, respectively. The profits per unit are $300 for model A and $375 for model B. What is the optimal production level for each model? What is the optimal profit?

In Exercises 47–52, the given linear programming problem has an unusual characteristic. Sketch a graph of the solution region for the problem and describe the unusual characteristic. Find the maximum value of the objective function and where it occurs.

47. *Objective function:*

$z = 2.5x + y$

Constraints:

$$x \geq 0$$
$$y \geq 0$$
$$3x + 5y \leq 15$$
$$5x + 2y \leq 10$$

48. *Objective function:*

$z = x + y$

Constraints:

$$x \geq 0$$
$$y \geq 0$$
$$-x + y \leq 1$$
$$-x + 2y \leq 4$$

49. *Objective function:*

$z = -x + 2y$

Constraints:

$$x \geq 0$$
$$y \geq 0$$
$$x \leq 10$$
$$x + y \leq 7$$

50. *Objective function:*

$z = x + y$

Constraints:

$$x \geq 0$$
$$y \geq 0$$
$$-x + y \leq 1$$
$$-3x + y \geq 3$$

51. *Objective function:*

$z = 3x + 4y$

Constraints:

$$x \geq 0$$
$$y \geq 0$$
$$x + y \leq 1$$
$$2x + y \leq 4$$

52. *Objective function:*

$z = x + 2y$

Constraints:

$$x \geq 0$$
$$y \geq 0$$
$$x + 2y \leq 4$$
$$2x + y \leq 4$$

Make a Decision Project: Newspapers

t	y_1	y_2
0	1084	559
1	1042	571
2	996	596
3	954	623
4	935	635
5	891	656
6	846	686
7	816	705
8	781	721
9	760	736
10	727	766
11	704	776
12	692	777

With 24-hour television news networks and the Internet, you may think that newspapers are becoming obsolete. This, however, is not the case. Newspapers continue to be a portable and convenient form of news. In 2002, there were 55.2 million daily newspapers in circulation each day in the United States.

The table at the left shows the number y_1 of evening newspapers and the number y_2 of morning newspapers published in the United States from 1990 to 2002. Let t represent the year, with $t = 0$ corresponding to 1990. Use this information to investigate the following questions. (Source: Editor & Publisher Co.)

1. **Fitting a Linear Model to Data** Use the *regression* feature of a graphing utility to verify the linear model for y_1 given on page 253.

2. **Comparing a Model to Actual Data** Make a numerical comparison by constructing a table comparing the values given by the linear model found in Question 1 and the actual data values. Does the numerical comparison confirm that the model you found is valid?

3. **Fitting a Linear Model and Comparing to Actual Data** Repeat Questions 1 and 2 using the data for y_2.

4. *Make a Decision* Use a graphing utility to graph y_1 and y_2 in the same viewing window. Do the graphs intersect? Does that fact agree with the data?

5. *Make a Decision* Algebraically determine the point of intersection of the two graphs. Verify your result by using the *intersect* feature of the graphing utility. Interpret the point of intersection in the context of the problem.

6. **Fitting a Linear Model to Data** The numbers N (in millions) of morning newspapers in circulation each day from 1990 to 2002 are shown in the table below. Use the *regression* feature of a graphing utility to find a linear model for the data. Let t represent the year, with $t = 0$ corresponding to 1990.

t	0	1	2	3	4	5	6
N	41.3	41.5	42.4	43.1	43.4	44.3	44.8

t	7	8	9	10	11	12
N	45.4	45.6	46.0	46.8	46.8	46.6

7. *Make a Decision* From 1990 to 2002, the number of morning newspapers published and circulated increased. Did the average circulation per morning paper (number in circulation ÷ number published) increase or decrease? Explain your reasoning.

8. *Make a Decision* If you were a publisher of a newspaper that published a morning and an evening newspaper, would you continue to publish both? If not, which would you choose to publish? Explain your reasoning.

Chapter Summary

After studying this chapter, you should have acquired the following skills.
These skills are keyed to the Review Exercises that begin on page 312.
Answers to odd-numbered Review Exercises are given in the back of the book.

4.1 • Solve a system of equations by the method of substitution.   *Review Exercises 1–6*

  • Solve a system of equations graphically.   *Review Exercises 7, 8, 14*

  • Construct and use a system of equations to solve an application problem.   *Review Exercises 9–14*

4.2 • Solve a linear system by the method of elimination.   *Review Exercises 15–20, 42*

  • Interpret the solution of a linear system graphically.   *Review Exercises 21, 22*

  • Construct and use a linear system to solve an application problem.   *Review Exercises 23–26*

4.3 • Solve a linear system in row-echelon form using back-substitution.   *Review Exercises 27, 28*

  • Use Gaussian elimination to solve a linear system.   *Review Exercises 29, 30, 33, 34*

  • Solve a nonsquare linear system.   *Review Exercises 31, 32*

  • Construct and use a linear system in three or more variables to solve an application problem.   *Review Exercises 39–41, 44*

  • Find the equation of a circle or a parabola using a linear system in three or more variables.   *Review Exercises 35–38, 43*

4.4 • Sketch the graph of an inequality in two variables.   *Review Exercises 45–48*

  • Solve a system of inequalities.   *Review Exercises 49–54*

  • Construct and use a system of inequalities to solve an application problem.   *Review Exercises 55–60*

4.5 • Use linear programming to minimize or maximize an objective function.   *Review Exercises 61–68*

  • Use linear programming to optimize an application.   *Review Exercises 69–76*

Review Exercises

In Exercises 1–6, solve the system by the method of substitution.

1. $\begin{cases} x + 3y = 10 \\ 4x - 5y = -28 \end{cases}$

2. $\begin{cases} 3x - y - 13 = 0 \\ 4x + 3y - 26 = 0 \end{cases}$

3. $\begin{cases} \frac{1}{2}x + \frac{3}{5}y = -2 \\ 2x + y = 6 \end{cases}$

4. $\begin{cases} 1.3x + 0.9y = 7.5 \\ 0.4x - 0.5y = -0.8 \end{cases}$

5. $\begin{cases} x^2 + y^2 = 100 \\ x + 2y = 20 \end{cases}$

6. $\begin{cases} y = x^3 - 2x^2 - 2x - 3 \\ y = -x^2 + 4x - 3 \end{cases}$

In Exercises 7 and 8, use a graphing utility to find the point(s) of intersection of the graphs of the equations.

7. $\begin{cases} y = x^2 - 3x + 11 \\ y = -x^2 + 2x + 8 \end{cases}$

8. $\begin{cases} y = \sqrt{9 - x^2} \\ y = e^x + 1 \end{cases}$

9. **Break-Even Analysis** You are setting up a basket-weaving business and have made an initial investment of $20,000. The cost of each basket is $3.25 and the selling price is $6.95. How many baskets must you sell to break even? (Round to the nearest whole unit.)

10. **Investment Portfolio** A total of $50,000 is invested in two funds paying 6.75% and 7.5% simple interest. The total yearly interest is $3637.50. How much of the $50,000 is invested at each rate?

11. **Choice of Two Jobs** You are offered two different jobs selling computers. One company offers an annual salary of $30,000 plus a year-end bonus of 2.5% of your total sales. The other company offers a salary of $24,000 plus a year-end bonus of 5.5% of your total sales. How much would you have to sell in order to make the second offer the better offer?

12. **Comparing Populations** The populations for Alaska and North Dakota from 1997 to 2002 can be approximated by the models

$\begin{cases} p = 5.7t + 573 & \text{Alaska} \\ p = -3.3t + 674 & \text{North Dakota} \end{cases}$

where p is the population (in thousands) and t represents the year, with $t = 7$ corresponding to 1997. According to these two models, when did the population of Alaska overtake the population of North Dakota? (Source: U.S. Census Bureau)

13. *Make a Decision*: **6 o'Clock News** Television stations A and B are competing for the 6 P.M. newscast audience. Station A has developed a new format that it believes will increase its 6 P.M. news audience. The numbers of 6 P.M. news viewers each month for the two news stations can be approximated by the models

$\begin{cases} y = 950x + 10,000 & \text{Station A (new format)} \\ y = -875x + 18,000 & \text{Station B} \end{cases}$

where y is the number of viewers and x represents the month, with $x = 1$ corresponding to the first month of the new format. Will the station manager and news director of Station A achieve their goal of having more viewers for the 6 P.M. news than Station B? If so, when will that happen?

14. **Comparing Product Sales** Your company produces three types of CD players. The research and development department has designed a new CD player that is predicted to sell better than the three players combined. The sales equations for the CD players are

$\begin{cases} S = 2150.78 - 156.8t + 0.05t^2 & \text{Three players combined} \\ S = 1121.27 + 15.37t + 4.98t^2 & \text{New player} \end{cases}$

where S is the sales (in thousands) and t represents the year, with $t = 0$ corresponding to 2000. Use a graphing utility to determine when the sales of the new CD player will overtake the combined sales of the other players.

In Exercises 15–20, solve the system by elimination.

15. $\begin{cases} 2x - 3y = 21 \\ 3x + y = 4 \end{cases}$

16. $\begin{cases} 3u + 5v = 9 \\ 12u + 10v = 22 \end{cases}$

17. $\begin{cases} 1.25x - 2y = 3.5 \\ 5x - 8y = 14 \end{cases}$

18. $\begin{cases} 1.5x + 2.5y = 8.5 \\ 6x + 10y = 24 \end{cases}$

19. $\begin{cases} \dfrac{x - 2}{3} + \dfrac{y + 3}{4} = 5 \\ 2x - y = 7 \end{cases}$

20. $\begin{cases} \dfrac{3}{5}x + \dfrac{2}{7}y = 10 \\ x + 2y = 38 \end{cases}$

In Exercises 21 and 22, describe the graph of the solution of the linear system.

21. $\begin{cases} 2x + y = -1 \\ 3x - 2y = -5 \end{cases}$

22. $\begin{cases} x - 2y = -1 \\ -2x + 4y = 2 \end{cases}$

23. Acid Mixture   Twelve gallons of a 25% acid solution are obtained by mixing a 10% solution with a 50% solution.

(a) Write a system of equations that represents the problem and use a graphing utility to graph the equations in the same viewing window.

(b) How much of each solution is required to obtain the specified concentration of the final mixture?

24. *Make a Decision*: Compact Disc Sales   You are the manager of a music store. At the end of the week you are going over receipts for the previous week's sales. Eight hundred and fifty compact discs were sold. One type of CD sold for $10.95 and a second type sold for $17.95. The total CD receipts were $11,407.50. The cash register that was supposed to keep track of the number of each type of CD sold malfunctioned. Can you recover the information? If so, how many of each type of CD were sold?

Supply and Demand   In Exercises 25 and 26, find the point of equilibrium for the pair of demand and supply equations.

Demand	*Supply*
25. $p = 37 - 0.0002x$	$p = 22 + 0.00001x$
26. $p = 120 - 0.0001x$	$p = 45 + 0.0002x$

In Exercises 27–34, solve the system of equations.

27. $\begin{cases} 4x - 3y + 2z = 1 \\ \quad\quad 2y - 4z = 2 \\ \quad\quad\quad\quad z = 2 \end{cases}$ **28.** $\begin{cases} 2x + y - 4z = 6 \\ \quad 3y + z = 2 \\ \quad\quad\quad z = -4 \end{cases}$

29. $\begin{cases} 2x + y + z = 6 \\ x - 4y - z = 3 \\ x + y + z = 4 \end{cases}$ **30.** $\begin{cases} x + 3y - z = 13 \\ 2x \quad\quad - 5z = 23 \\ 4x - y - 2z = 4 \end{cases}$

31. $\begin{cases} x + y + z = 10 \\ -2x + 3y + 4z = 22 \end{cases}$

32. $\begin{cases} 5x - 12y + 7z = 16 \\ 3x - 7y + 4z = 9 \end{cases}$

33. $\begin{cases} 2x + 6y - z = 1 \\ x - 3y + z = 2 \\ \frac{3}{2}x + \frac{3}{2}y \quad = 6 \end{cases}$

34. $\begin{cases} x + y + z + w = 8 \\ \quad 4y + 5z - 2w = 3 \\ 2x + 3y - z \quad = -2 \\ 3x + 2y \quad - 4w = -20 \end{cases}$

In Exercises 35 and 36, find the equation of the parabola $y = ax^2 + bx + c$ that passes through the points.

35.
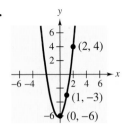

36.
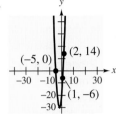

In Exercises 37 and 38, find the equation of the circle $x^2 + y^2 + Dx + Ey + F = 0$ that passes through the points.

37.
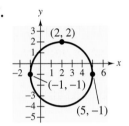

38.
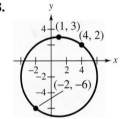

39. *Make a Decision*: Investment Portfolio   An investor has a portfolio totaling $500,000 that is to be allocated among the following types of investments: (1) certificates of deposit, (2) municipal bonds, (3) blue-chip stocks, and (4) growth or speculative stocks. The certificates of deposit pay 5% simple annual interest, and the municipal bonds pay 8% simple annual interest. Over a five-year period, the investor expects the blue-chip stocks to return 10% simple annual interest and the growth stocks to return 15% simple annual interest. The investor wishes a combined return of 9.45% and also wants to have only two-fifths of the portfolio invested in stocks. How much should be allocated to each type of investment if the amount invested in certificates of deposit is twice that invested in municipal bonds?

40. Coffee Maker Types   A company sells three types of coffee makers for $85, $75, and $50 per unit. In 1 year, the total revenue for the three coffee makers was $7,700,000, which corresponded to the sale of 120,000 units. The company sold half as many units of the $85 coffee maker as units of the $75 coffee maker. How many units of each coffee maker were sold?

41. Investment   You receive \$8580 a year in simple annual interest from three investments. The interest rates for the three investments are 6%, 8%, and 10%. The value of the 10% investment is two times that of the 6% investment, and the 8% investment is \$1000 more than the 6% investment. What is the amount of each investment?

42. Fitting a Line to Data   Find the least squares regression line $y = ax + b$ for the points $(0, 1.6)$, $(1, 2.4)$, $(2, 3.6)$, $(3, 4.7)$, and $(4, 5.5)$ by solving the following system of linear equations for a and b.

$$\begin{cases} 5b + 10a = 17.8 \\ 10b + 30a = 45.7 \end{cases}$$

43. Fitting a Parabola to Data   Find the least squares regression parabola $y = ax^2 + bx + c$ for the points

$$(-2, 0.4), (-1, 0.9), (0, 1.9), (1, 2.1), \text{ and } (2, 3.8)$$

by solving the following system of linear equations for a, b, and c.

$$\begin{cases} 5c + 10a = 9.1 \\ 10b = 8.0 \\ 10c + 34a = 19.8 \end{cases}$$

44. Revenue   The revenue y (in billions of dollars) for McDonald's Corporation from 1998 to 2002 is shown in the table, where t represents the year, with $t = 0$ corresponding to 2000. (Source: McDonald's Corporation)

Year, t	Revenue, y
-2	12.4
-1	13.3
0	14.2
1	14.9
2	15.4

(a) Create a scatter plot of the data.

(b) Solve the following system for a and b to find the least squares regression line $y = at + b$ for the data.

$$\begin{cases} 5b = 70.2 \\ 10a = 7.6 \end{cases}$$

(c) Solve the following system for a, b, and c to find the least squares regression parabola $y = at^2 + bt + c$ for the data.

$$\begin{cases} 5c + 10a = 70.2 \\ 10b = 7.6 \\ 10c + 34a = 139.4 \end{cases}$$

(d) Use the *regression* feature of a graphing utility to find linear and quadratic models for the data. Compare them to the least squares regression models found in parts (b) and (c).

(e) Use a graphing utility to graph the linear and quadratic models. Use the *trace* feature with each model to estimate the revenues in 2003 and 2004. How close are the estimates to each other for each year?

In Exercises 45–48, sketch the graph of the inequality.

45. $y \leq 5 - \frac{1}{2}x$

46. $3y - x \geq 7$

47. $y - 4x^2 > -1$

48. $y \leq \dfrac{3}{x^2 + 2}$

In Exercises 49–54, graph the solution set of the system of inequalities.

49. $\begin{cases} 2x + 3y < 9 \\ x > 0 \\ y > 0 \end{cases}$

50. $\begin{cases} 2x - y > 6 \\ x < 5 \\ y \leq 8 \end{cases}$

51. $\begin{cases} 3x - y > -4 \\ 2x + y > -1 \\ 7x + y < 4 \end{cases}$

52. $\begin{cases} x + y > 4 \\ 3x + y < 10 \end{cases}$

53. $\begin{cases} x^2 + y^2 \leq 9 \\ x^2 - x - 2 \leq y \end{cases}$

54. $\begin{cases} \ln x < y \\ y > -1 \\ x < 4 \end{cases}$

In Exercises 55 and 56, write a system of inequalities that describes the region.

55. Parallelogram region with vertices $(3, 6)$, $(9, 7)$, $(1, 1)$, and $(7, 2)$

56. Triangular region with vertices $(4, 7)$, $(2, 1)$, and $(10, 3)$

Consumer and Producer Surplus   In Exercises 57 and 58, find the consumer surplus and producer surplus for the pair of demand and supply equations.

Demand	Supply
57. $p = 160 - 0.0001x$	$p = 70 + 0.0002x$
58. $p = 130 - 0.0002x$	$p = 30 + 0.0003x$

59. DVD Player Inventory   A store sells two models of DVD players. Because of the demand, it is necessary to stock at least twice as many units of model A as units of model B. The costs to the store for the two models are $200 and $300, respectively. The management does not want more than $4000 in DVD player inventory at any one time, and it wants at least four model A DVD players and two model B DVD players in inventory at all times. Find a system of inequalities that describes all possible inventory levels, and sketch the graph of the system.

60. Concert Ticket Sales   Two types of tickets are to be sold for a concert. One type costs $30 per ticket and the other type costs $50 per ticket. The promoter of the concert must sell at least 15,000 tickets, including at least 8000 of the $30 tickets and at least 4000 of the $50 tickets. Moreover, the gross receipts must total at least $550,000 in order for the concert to be held. Find a system of inequalities describing the different numbers of tickets that can be sold, and sketch the graph of the system.

In Exercises 61–64, find the minimum and maximum values of the objective function and where they occur, subject to the indicated constraints. (For each exercise, the graph of the region determined by the constraints is provided.)

61. *Objective function:*

$z = 5x + 6y$

Constraints:

$x \geq 0$

$y \geq 0$

$x + y \leq 8$

62. *Objective function:*

$z = 15x + 12y$

Constraints:

$x \geq 0$

$y \geq 0$

$x + 3y \leq 12$

$3x + 2y \leq 15$

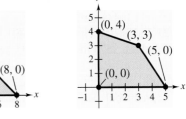

Figure for 61   Figure for 62

63. *Objective function:*

$z = 8x + 3y$

Constraints:

$0 \leq x \leq 50$

$0 \leq y \leq 35$

$4x + 5y \leq 275$

64. *Objective function:*

$z = 50x + 60y$

Constraints:

$x \geq 0$

$y \geq 0$

$3x + 4y \geq 1200$

$5x + 6y \leq 3000$

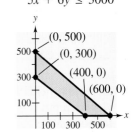

In Exercises 65–68, sketch the region determined by the constraints. Then find the minimum and maximum values of the objective function and where they occur, subject to the indicated constraints.

65. *Objective function:*

$z = 6x + 8y$

Constraints:

$x \geq 0$

$y \geq 0$

$x + 4y \leq 16$

$3x + 2y \leq 18$

66. *Objective function:*

$z = 5x + 8y$

Constraints:

$0 \leq x \leq 5$

$y \geq 0$

$x + 2y \leq 12$

$2x + 3y \leq 19$

67. *Objective function:*

$z = 8x + 3y$

Constraints:

$0 \leq x \leq 5$

$0 \leq y \leq 7$

$x + y \leq 9$

$3x + y \leq 17$

68. *Objective function:*

$z = 10x + 11y$

Constraints:

$x \geq 0$

$y \geq 0$

$2x + 5y \leq 30$

$x + y \geq 3$

$2x + y \leq 14$

69. *Make a Decision*: **Optimal Profit** The costs to a merchant for two models of digital camcorders are $525 and $675. The $525 model yields a profit of $75 and the $675 model yields a profit of $125. The merchant estimates that the total monthly demand will not exceed 350 units. Find the number of units of each model that should be stocked in order to optimize profit. What is the optimal profit? Assume that the merchant does not want to invest more than $206,250 in digital camcorder inventory.

70. *Make a Decision*: **Optimal Profit** The costs to a merchant for two models of color televisions are $270 and $455. The $270 model yields a profit of $30 and the $455 model yields a profit of $45. The merchant estimates that the total monthly demand will not exceed 100 units. Find the number of units of each model that should be stocked in order to optimize profit. What is the optimal profit? Assume that the merchant does not want to invest more than $36,250 in color television inventory.

71. *Make a Decision*: **Optimal Profit** A factory manufactures two color television set models: a basic model that yields $100 profit and a deluxe model that yields a profit of $200. The times (in hours) required for assembling, finishing, and packaging each model are shown in the table.

Process	Basic Model	Deluxe Model
Assembling	2	5
Finishing	1	2
Packaging	1	1

The total times available for assembling, finishing, and packaging are 3000 hours, 1400 hours, and 1000 hours, respectively. What is the optimal production level for each model? What is the optimal profit?

72. *Make a Decision*: **Optimal Profit** A company makes two models of desks. The times (in hours) required for assembling, finishing, and packaging each model are shown in the table.

Process	Model A	Model B
Assembling	3.5	8
Finishing	2.5	2
Packaging	1.3	0.7

The total times available for assembling, finishing, and packaging are 5600 hours, 2000 hours, and 910 hours, respectively. The profits per unit are $100 for model A and $150 for model B. What is the optimal production level for each model? What is the optimal profit?

73. *Make a Decision*: **Optimal Cost** A pet supply company mixes two brands of dry dog food. Brand X costs $15 per bag and contains 8 units of nutritional element A, 1 unit of nutritional element B, and 2 units of nutritional element C. Brand Y costs $30 per bag and contains 2 units of nutritional element A, 1 unit of nutritional element B, and 7 units of nutritional element C. Each bag of mixed dog food must contain at least 16 units of element A, 5 units of element B, and 20 units of element C. Find the number of bags of each brand that should be mixed to produce a mixture having an optimal cost per bag. What is the optimal cost?

74. *Make a Decision*: **Optimal Cost** Two gasolines, type A and type B, have octane ratings of 85 and 89, respectively. Type A costs $1.59 per gallon and type B costs $1.82 per gallon. Determine the blend of optimal cost that has an octane rating of at least 87. What is the optimal cost?

75. *Make a Decision*: **Optimal Revenue** An accounting firm has 800 hours of staff time and 90 hours of review time available each week. The firm charges $2500 for an audit and $200 for a tax return. Each audit requires 100 hours of staff time and 10 hours of review time. Each tax return requires 10 hours of staff time and 2 hours of review time. What numbers of audits and tax returns will bring in an optimal revenue? What is the optimal revenue?

76. *Make a Decision*: **Optimal Profit** The accounting firm in Exercise 75 realizes a profit of $1000 for each audit and $110 for each tax return. What combination of audits and tax returns will yield an optimal profit? What is the optimal profit?

Chapter Test

Take this test as you would take a test in class. After you are done, check your work against the answers given in the back of the book.

In Exercises 1–6, solve the system of equations using the indicated method.

1. *Substitution*

$$\begin{cases} 5x - 7y = -18 \\ 4x + 3y = 20 \end{cases}$$

2. *Substitution*

$$\begin{cases} x + y = 3 \\ x^2 + y = 9 \end{cases}$$

3. *Graphing*

$$\begin{cases} 5x - y = 6 \\ 2x^2 + y = 8 \end{cases}$$

4. *Graphing*

$$\begin{cases} 1.5x - 2.25y = 8 \\ 2.5x + 2y = 5.75 \end{cases}$$

5. *Elimination*

$$\begin{cases} 2x - 4y + z = 11 \\ x + 2y + 3z = 9 \\ 3y + 5z = 12 \end{cases}$$

6. *Elimination*

$$\begin{cases} 3x - 2y + z = 16 \\ 5x - z = 6 \\ 2x - y - z = 3 \end{cases}$$

7. A total of $50,000 is invested in two funds paying 8% and 8.5% simple interest. The total annual interest is $4150. How much is invested in each fund?

8. Find the point of equilibrium for a system that has a demand equation of $p = 45 - 0.0003x$ and a supply equation of $p = 29 + 0.00002x$.

9. The percents y of sound recordings purchased through tape/record clubs from 1998 to 2002 are shown in the table. Find the least squares regression parabola $y = at^2 + bt + c$ for the data by solving the following system. (Source: Recording Industry Association of America)

$$\begin{cases} 5c + 10a = 34.6 \\ 10b = -11.8 \\ 10c + 34a = 66 \end{cases}$$

Estimate the percent of sound recordings purchased through tape/record clubs in 2003.

Year	t	Percent, y
1998	-2	9.0
1999	-1	7.9
2000	0	7.6
2001	1	6.1
2002	2	4.0

Table for 9

In Exercises 10–13, sketch the graph of the inequality.

10. $x \geq 0$ **11.** $y \geq 0$ **12.** $x + 3y \leq 12$ **13.** $3x + 2y \leq 15$

14. Sketch the solution set of the system of inequalities composed of the inequalities in Exercises 10–13.

15. Find the minimum and maximum values of the objective function $z = 6x + 7y$, subject to the constraints given in Exercises 10–13.

16. A manufacturer produces two models of stair climbers. The times required for assembling, painting, and packaging each model are as follows.

- Assembling: 3.5 hours for model A; 8 hours for model B
- Painting: 2.5 hours for model A; 2 hours for model B
- Packaging: 1.3 hours for model A; 0.9 hour for model B

The total times available for assembling, painting, and packaging are 5600 hours, 2000 hours, and 900 hours, respectively. The profits per unit are $200 for model A and $275 for model B. What is the optimal production level for each model? What is the optimal profit? Explain your reasoning.

5

Matrices and Determinants

Chapter 5

Chapter Sections

5.1 Matrices and Linear Systems

5.2 Operations with Matrices

5.3 The Inverse of a Square Matrix

5.4 The Determinant of a Square Matrix

5.5 Applications of Matrices and Determinants

Make a Decision ➤ Who's Switching Service Providers?

Your advertising group has been hired by Company C, a cell phone service provider, to develop an advertising campaign to increase its market share. Currently your client competes with two other service providers, Companies A and B. With this new campaign, you predict that each month (1) 90% of A's subscribers will remain with A, 4% will switch to B, and 6% will switch to C; (2) 92% of B's subscribers will remain with B, 5% will switch to A, and 3% will switch to C; and (3) 94% of C's subscribers will remain with C, 2% will switch to A, and 4% will switch to B. The current shares of the market are given in the following matrix.

Month	Share
0	0.1
1	0.127
2	0.152
3	0.175
4	0.197
5	0.217
6	0.235

$$x_0 = \begin{bmatrix} 0.2 \\ 0.7 \\ 0.1 \end{bmatrix} \begin{matrix} \text{Company A} \\ \text{Company B} \\ \text{Company C} \end{matrix}$$

The predicted market shares for Company C for the first 6 months of the campaign are shown in the table and bar graph.

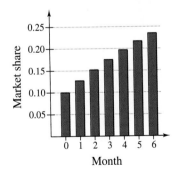

Make a Decision How can you predict the shares of the three companies after 1 month?

To explore this real-world application further, go to the Make a Decision Project on page 379.

Objectives

- **Determine the order of a matrix.**
- **Perform elementary row operations on a matrix to write the matrix in row-echelon form or reduced row-echelon form.**
- **Solve a system of linear equations using Gaussian elimination.**
- **Solve a system of linear equations using Gauss-Jordan elimination.**

Matrices

In this section, you will study a streamlined technique for solving systems of linear equations. This technique involves the use of a rectangular array of real numbers called a **matrix.** The plural of matrix is **matrices.**

Definition of a Matrix

If m and n are positive integers, an $m \times n$ matrix (read "m by n") is a rectangular array

$$\begin{bmatrix} a_{11} & a_{12} & a_{13} & \cdots & a_{1n} \\ a_{21} & a_{22} & a_{23} & \cdots & a_{2n} \\ a_{31} & a_{32} & a_{33} & \cdots & a_{3n} \\ \vdots & \vdots & \vdots & & \vdots \\ a_{m1} & a_{m2} & a_{m3} & \cdots & a_{mn} \end{bmatrix} \quad m \text{ rows}$$

n columns

in which each **entry,** a_{ij}, of the matrix is a number. An $m \times n$ matrix has m **rows** (horizontal lines) and n **columns** (vertical lines).

The entry in the ith row and jth column is denoted by the *double subscript* notation a_{ij}. That is, a_{21} refers to the entry in row 2, column 1. A matrix having m rows and n columns is said to be of **order** $m \times n$. If $m = n$, the matrix is **square** of order n. For a square matrix, the entries $a_{11}, a_{22}, a_{33}, \ldots$ are the **main diagonal** entries.

[STUDY TIP

A matrix that has only one row [such as the matrix in Example 1(a)] is called a **row matrix,** and a matrix that has only one column is called a **column matrix.**

EXAMPLE 1   ORDER OF MATRICES

The following matrices have the indicated orders.

a. *Order:* 1×4

$$\begin{bmatrix} 1 & -3 & 0 & \frac{1}{2} \end{bmatrix}$$

b. *Order:* 2×2

$$\begin{bmatrix} 0 & 0 \\ 0 & 0 \end{bmatrix}$$

c. *Order:* 3×2

$$\begin{bmatrix} 5 & 0 \\ 2 & -2 \\ -7 & 4 \end{bmatrix}$$

✓CHECKPOINT   Now try Exercise 1.

A matrix derived from a system of linear equations (each written in standard form with the constant term on the right) is the **augmented matrix** of the system. Moreover, the matrix derived from the coefficients of the system (but that does not include the constant terms) is the **coefficient matrix** of the system. Note in the matrices below the use of 0 for the missing y-variable in the third equation, and also note the fourth column of constant terms in the augmented matrix. The vertical dots in the augmented matrix separate the coefficients of the linear system from the constant terms.

$$\text{\textit{System}} \qquad\qquad \text{\textit{Augmented Matrix}} \qquad\qquad \text{\textit{Coefficient Matrix}}$$

$$\begin{cases} x - 4y + 3z = 5 \\ -x + 3y - z = -3 \\ 2x - 4z = 6 \end{cases} \quad \left[\begin{array}{ccc:c} 1 & -4 & 3 & 5 \\ -1 & 3 & -1 & -3 \\ 2 & 0 & -4 & 6 \end{array}\right] \quad \left[\begin{array}{ccc} 1 & -4 & 3 \\ -1 & 3 & -1 \\ 2 & 0 & -4 \end{array}\right]$$

When forming either the coefficient matrix or the augmented matrix of a system, you should begin by vertically aligning the variables in the equations and using zeros for the missing variables.

$$\text{\textit{Original System}} \qquad \text{\textit{Line Up Variables}} \qquad \text{\textit{Form Augmented Matrix}}$$

$$\begin{cases} x + 3y = 9 \\ -y + 4z = -2 \\ x - 5z = 0 \end{cases} \quad \begin{cases} x + 3y = 9 \\ -y + 4z = -2 \\ x - 5z = 0 \end{cases} \quad \left[\begin{array}{ccc:c} 1 & 3 & 0 & 9 \\ 0 & -1 & 4 & -2 \\ 1 & 0 & -5 & 0 \end{array}\right]$$

Elementary Row Operations

In Section 4.3, you studied three operations that can be used on a system of linear equations to produce an equivalent system.

1. Interchange two equations.

2. Multiply an equation by a nonzero constant.

3. Add a multiple of an equation to another equation.

In matrix terminology, these three operations correspond to **elementary row operations.** An elementary row operation on an augmented matrix of a given system of linear equations produces a new augmented matrix corresponding to a new (but equivalent) system of linear equations. Two matrices are **row-equivalent** if one can be obtained from the other by a sequence of elementary row operations.

Elementary Row Operations

1. Interchange two rows.

2. Multiply a row by a nonzero constant.

3. Add a multiple of a row to another row.

Although elementary row operations are simple to perform, they involve a lot of arithmetic. Because it is easy to make a mistake, you should get in the habit of noting, next to the row you are changing, the elementary row operation performed in each step so that you can go back and check your work.

The next example demonstrates each of the elementary row operations that can be performed on a matrix to produce an equivalent matrix.

EXAMPLE 2   ELEMENTARY ROW OPERATIONS

a. Interchange the first and second rows.

Original Matrix

$$\begin{bmatrix} 0 & 1 & 3 & 4 \\ -1 & 2 & 0 & 3 \\ 2 & -3 & 4 & 1 \end{bmatrix}$$

New Row-Equivalent Matrix

$$\begin{matrix} R_2 \\ R_1 \end{matrix} \begin{bmatrix} -1 & 2 & 0 & 3 \\ 0 & 1 & 3 & 4 \\ 2 & -3 & 4 & 1 \end{bmatrix}$$

b. Multiply the first row by $\frac{1}{2}$.

Original Matrix

$$\begin{bmatrix} 2 & -4 & 6 & -2 \\ 1 & 3 & -3 & 0 \\ 5 & -2 & 1 & 2 \end{bmatrix}$$

New Row-Equivalent Matrix

$$\frac{1}{2}R_1 \rightarrow \begin{bmatrix} 1 & -2 & 3 & -1 \\ 1 & 3 & -3 & 0 \\ 5 & -2 & 1 & 2 \end{bmatrix}$$

c. Add -2 times the first row to the third row.

Original Matrix

$$\begin{bmatrix} 1 & 2 & -4 & 3 \\ 0 & 3 & -2 & -1 \\ 2 & 1 & 5 & -2 \end{bmatrix}$$

New Row-Equivalent Matrix

$$\begin{bmatrix} 1 & 2 & -4 & 3 \\ 0 & 3 & -2 & -1 \\ 0 & -3 & 13 & -8 \end{bmatrix}$$

$-2R_1 + R_3 \rightarrow$

Note that the elementary row operation is written beside the row that is *changing*.

✓CHECKPOINT   Now try Exercise 9.

TECHNOLOGY

Most graphing utilities can perform elementary row operations on matrices. Consult your user's guide to determine how to enter a matrix and perform elementary row operations. The screens below show how one graphing utility displays each new row-equivalent matrix from Example 2. For specific instructions on how to use the elementary row operations features of a graphing utility, go to the text website at *college.hmco.com*.

a. Interchange the first and second rows.

```
rowSwap([A],1,2)

   [[-1 2  0 3]
    [0  1  3 4]
    [2 -3 4 1]]
```

b. Multiply the first row by $\frac{1}{2}$.

```
*row(.5,[A],1)
   [[1 -2 3 -1]
    [1  3 -3 0 ]
    [5 -2 1  2 ]]
```

c. Add -2 times the first row to the third row.

```
*row+(-2,[A],1,3)
)
   [[1  2 -4  3 ]
    [0  3 -2 -1]
    [0 -3 13 -8]]
```

In Example 3 of Section 4.3, you used Gaussian elimination with back-substitution to solve a system of linear equations. The next example demonstrates the matrix version of Gaussian elimination. The two methods are essentially the same. The basic difference is that with matrices you do not need to keep writing the variables.

EXAMPLE 3   USING ELEMENTARY ROW OPERATIONS

Linear System	*Associated Augmented Matrix*

$$\begin{cases} x - 2y + 3z = 9 \\ -x + 3y = -4 \\ 2x - 5y + 5z = 17 \end{cases}$$

$$\left[\begin{array}{ccc:c} 1 & -2 & 3 & 9 \\ -1 & 3 & 0 & -4 \\ 2 & -5 & 5 & 17 \end{array}\right]$$

Add the first equation to the second equation.

Add the first row to the second row $(R_1 + R_2)$.

$$\begin{cases} x - 2y + 3z = 9 \\ y + 3z = 5 \\ 2x - 5y + 5z = 17 \end{cases}$$

$$R_1 + R_2 \rightarrow \left[\begin{array}{ccc:c} 1 & -2 & 3 & 9 \\ 0 & 1 & 3 & 5 \\ 2 & -5 & 5 & 17 \end{array}\right]$$

Add -2 times the first equation to the third equation.

Add -2 times the first row to the third row $(-2R_1 + R_3)$.

$$\begin{cases} x - 2y + 3z = 9 \\ y + 3z = 5 \\ -y - z = -1 \end{cases}$$

$$-2R_1 + R_3 \rightarrow \left[\begin{array}{ccc:c} 1 & -2 & 3 & 9 \\ 0 & 1 & 3 & 5 \\ 0 & -1 & -1 & -1 \end{array}\right]$$

Add the second equation to the third equation.

Add the second row to the third row $(R_2 + R_3)$.

$$\begin{cases} x - 2y + 3z = 9 \\ y + 3z = 5 \\ 2z = 4 \end{cases}$$

$$R_2 + R_3 \rightarrow \left[\begin{array}{ccc:c} 1 & -2 & 3 & 9 \\ 0 & 1 & 3 & 5 \\ 0 & 0 & 2 & 4 \end{array}\right]$$

Multiply the third equation by $\frac{1}{2}$.

Multiply the third row by $\frac{1}{2}$.

$$\begin{cases} x - 2y + 3z = 9 \\ y + 3z = 5 \\ z = 2 \end{cases}$$

$$\tfrac{1}{2}R_3 \rightarrow \left[\begin{array}{ccc:c} 1 & -2 & 3 & 9 \\ 0 & 1 & 3 & 5 \\ 0 & 0 & 1 & 2 \end{array}\right]$$

At this point, you can use back-substitution to find that the solution is $x = 1$, $y = -1$, and $z = 2$, as was done in Example 1 of Section 4.3.

✓CHECKPOINT   Now try Exercise 41.

Remember that you can check a solution by substituting the values of x, y, and z into each equation in the original system. For example, you can check the solution to Example 3 as follows.

Equation 1: $1 - 2(-1) + 3(2) = 9$ ✓

Equation 2: $-1 + 3(-1) = -4$ ✓

Equation 3: $2(1) - 5(-1) + 5(2) = 17$ ✓

The last matrix in Example 3 is said to be in *row-echelon form*. The term *echelon* refers to the stair-step pattern formed by the nonzero elements of the matrix. To be in this form, a matrix must have the following properties.

Row-Echelon Form and Reduced Row-Echelon Form

A matrix in **row-echelon form** has the following properties.

1. Any rows consisting entirely of zeros occur at the bottom of the matrix.

2. For each row that does not consist entirely of zeros, the first nonzero entry is 1 (called a **leading 1**).

3. For two successive (nonzero) rows, the leading 1 in the higher row is farther to the left than the leading 1 in the lower row.

A matrix in *row-echelon form* is in **reduced row-echelon form** if every column that has a leading 1 has zeros in every position above and below its leading 1.

EXAMPLE 4 ROW-ECHELON FORM

Determine whether each matrix is in row-echelon form. If it is, determine whether the matrix is in reduced row-echelon form.

TECHNOLOGY

Some graphing utilities can automatically transform a matrix to row-echelon form and reduced row-echelon form. For specific keystrokes, go to the text website at *college.hmco.com*.

a. $\begin{bmatrix} 1 & 2 & -1 & 4 \\ 0 & 1 & 0 & 3 \\ 0 & 0 & 1 & -2 \end{bmatrix}$
   b. $\begin{bmatrix} 1 & 2 & -1 & 2 \\ 0 & 0 & 0 & 0 \\ 0 & 1 & 2 & -4 \end{bmatrix}$

c. $\begin{bmatrix} 1 & -5 & 2 & -1 & 3 \\ 0 & 0 & 1 & 3 & -2 \\ 0 & 0 & 0 & 1 & 4 \\ 0 & 0 & 0 & 0 & 1 \end{bmatrix}$
   d. $\begin{bmatrix} 1 & 0 & 0 & -1 \\ 0 & 1 & 0 & 2 \\ 0 & 0 & 1 & 3 \\ 0 & 0 & 0 & 0 \end{bmatrix}$

e. $\begin{bmatrix} 1 & 2 & -3 & 4 \\ 0 & 2 & 1 & -1 \\ 0 & 0 & 1 & -3 \end{bmatrix}$
   f. $\begin{bmatrix} 0 & 1 & 0 & 5 \\ 0 & 0 & 1 & 3 \\ 0 & 0 & 0 & 0 \end{bmatrix}$

SOLUTION

The matrices in (a), (c), (d), and (f) are in row-echelon form. The matrices in (d) and (f) are in reduced row-echelon form because every column that has a leading 1 has zeros in every position above and below its leading 1. The matrix in (b) is not in row-echelon form because the row of all zeros does not occur at the bottom of the matrix. The matrix in (e) is not in row-echelon form because the first nonzero entry in row 2 is not a leading 1.

✓CHECKPOINT   Now try Exercise 13.

Every matrix is row-equivalent to a matrix in row-echelon form. For instance, in Example 4, you can change the matrix in part (e) to row-echelon form by multiplying its second row by $\frac{1}{2}$. What elementary row operation could you perform on the matrix in part (b) so that it would be in row-echelon form?

Gaussian Elimination with Back-Substitution

Gaussian elimination with back-substitution works well for solving systems of linear equations by hand or with a computer. For this algorithm, the order in which the elementary row operations are performed is important. You should operate from *left to right* by columns, using elementary row operations to obtain zeros in all entries directly below the leading 1's.

EXAMPLE 5   GAUSSIAN ELIMINATION WITH BACK-SUBSTITUTION

Solve the system.

$$\begin{cases} y + z - 2w = -3 \\ x + 2y - z = 2 \\ 2x + 4y + z - 3w = -2 \\ x - 4y - 7z - w = -19 \end{cases}$$

SOLUTION

$$\underset{R_1}{\overset{R_2}{\curvearrowleft}}\ \begin{bmatrix} 1 & 2 & -1 & 0 & \vdots & 2 \\ 0 & 1 & 1 & -2 & \vdots & -3 \\ 2 & 4 & 1 & -3 & \vdots & -2 \\ 1 & -4 & -7 & -1 & \vdots & -19 \end{bmatrix}$$

Interchange R_1 and R_2 so that the first column has a leading 1 in the upper left corner.

$$\begin{matrix} \\ \\ -2R_1 + R_3 \rightarrow \\ -R_1 + R_4 \rightarrow \end{matrix}\ \begin{bmatrix} 1 & 2 & -1 & 0 & \vdots & 2 \\ 0 & 1 & 1 & -2 & \vdots & -3 \\ 0 & 0 & 3 & -3 & \vdots & -6 \\ 0 & -6 & -6 & -1 & \vdots & -21 \end{bmatrix}$$

Perform operations on R_3 and R_4 so that the first column has zeros below its leading 1.

$$\begin{matrix} \\ \\ \\ 6R_2 + R_4 \rightarrow \end{matrix}\ \begin{bmatrix} 1 & 2 & -1 & 0 & \vdots & 2 \\ 0 & 1 & 1 & -2 & \vdots & -3 \\ 0 & 0 & 3 & -3 & \vdots & -6 \\ 0 & 0 & 0 & -13 & \vdots & -39 \end{bmatrix}$$

Perform operations on R_4 so that the second column has zeros below its leading 1.

$$\begin{matrix} \\ \\ \frac{1}{3}R_3 \rightarrow \\ \\ \end{matrix}\ \begin{bmatrix} 1 & 2 & -1 & 0 & \vdots & 2 \\ 0 & 1 & 1 & -2 & \vdots & -3 \\ 0 & 0 & 1 & -1 & \vdots & -2 \\ 0 & 0 & 0 & -13 & \vdots & -39 \end{bmatrix}$$

Multiply R_3 by $\frac{1}{3}$ so that the third row has a leading 1.

$$\begin{matrix} \\ \\ \\ -\frac{1}{13}R_4 \rightarrow \end{matrix}\ \begin{bmatrix} 1 & 2 & -1 & 0 & \vdots & 2 \\ 0 & 1 & 1 & -2 & \vdots & -3 \\ 0 & 0 & 1 & -1 & \vdots & -2 \\ 0 & 0 & 0 & 1 & \vdots & 3 \end{bmatrix}$$

Multiply R_4 by $-\frac{1}{13}$ so that the fourth row has a leading 1.

The matrix is now in row-echelon form, and the corresponding system is

$$\begin{cases} x + 2y - z = 2 \\ y + z - 2w = -3 \\ z - w = -2 \\ w = 3 \end{cases}.$$

Using back-substitution, you can determine that the solution is $x = -1$, $y = 2$, $z = 1$, and $w = 3$. Check this in the original system of equations.

✔CHECKPOINT   Now try Exercise 49.

Gaussian Elimination with Back-Substitution

1. Write the augmented matrix of the system of linear equations.

2. Use elementary row operations to rewrite the augmented matrix in row-echelon form.

3. Write the system of linear equations corresponding to the matrix in row-echelon form, and use back-substitution to find the solution.

When solving a system of linear equations, remember that it is possible for the system to have no solution. If, in the elimination process, you obtain a row with zeros except for the last entry, it is unnecessary to continue the elimination process. You can simply conclude that the system has no solution, or is inconsistent.

EXAMPLE 6   A SYSTEM WITH NO SOLUTION

Solve the system.

$$\begin{cases} 3x + 2y - z = 1 \\ x \quad\quad + z = 6 \\ 2x - 3y + 5z = 4 \\ x - y + 2z = 4 \end{cases}$$

SOLUTION

$$\begin{bmatrix} 3 & 2 & -1 & \vdots & 1 \\ 1 & 0 & 1 & \vdots & 6 \\ 2 & -3 & 5 & \vdots & 4 \\ 1 & -1 & 2 & \vdots & 4 \end{bmatrix} \quad \begin{array}{c} R_4 \\ \\ \\ R_1 \end{array} \begin{bmatrix} 1 & -1 & 2 & \vdots & 4 \\ 1 & 0 & 1 & \vdots & 6 \\ 2 & -3 & 5 & \vdots & 4 \\ 3 & 2 & -1 & \vdots & 1 \end{bmatrix}$$

$$\begin{array}{c} \\ -R_1 + R_2 \rightarrow \\ -2R_1 + R_3 \rightarrow \\ -3R_1 + R_4 \rightarrow \end{array} \begin{bmatrix} 1 & -1 & 2 & \vdots & 4 \\ 0 & 1 & -1 & \vdots & 2 \\ 0 & -1 & 1 & \vdots & -4 \\ 0 & 5 & -7 & \vdots & -11 \end{bmatrix}$$

$$\begin{array}{c} \\ \\ R_2 + R_3 \rightarrow \\ \end{array} \begin{bmatrix} 1 & -1 & 2 & \vdots & 4 \\ 0 & 1 & -1 & \vdots & 2 \\ 0 & 0 & 0 & \vdots & -2 \\ 0 & 5 & -7 & \vdots & -11 \end{bmatrix}$$

Note that the third row of this matrix consists of zeros except for the last entry. This means that the original system of linear equations is *inconsistent*. You can see why this is true by converting back to a system of linear equations.

$$\begin{cases} x - y + 2z = 4 \\ y - z = 2 \\ 0 = -2 \\ 5y - 7z = -11 \end{cases}$$

Because the third equation is not possible, the system has no solution.

✓**CHECKPOINT**   Now try Exercise 53.

Gauss-Jordan Elimination

With Gaussian elimination, elementary row operations are applied to a matrix to obtain a (row-equivalent) row-echelon form of the matrix. A second method of elimination, called **Gauss-Jordan elimination** after Carl Friedrich Gauss and Wilhelm Jordan (1842–1899), continues the reduction process until a *reduced* row-echelon form is obtained. This procedure is demonstrated in Example 7.

EXAMPLE 7 GAUSS-JORDAN ELIMINATION

Use Gauss-Jordan elimination to solve the system.

$$\begin{cases} x - 2y + 3z = 9 \\ -x + 3y = -4 \\ 2x - 5y + 5z = 17 \end{cases}$$

SOLUTION

In Example 3, Gaussian elimination was used to obtain the row-echelon form

$$\begin{bmatrix} 1 & -2 & 3 & \vdots & 9 \\ 0 & 1 & 3 & \vdots & 5 \\ 0 & 0 & 1 & \vdots & 2 \end{bmatrix}.$$

Now, apply elementary row operations until you obtain a matrix in reduced row-echelon form. To do this, you must produce zeros above each of the leading 1's, as follows.

$$2R_2 + R_1 \longrightarrow \begin{bmatrix} 1 & 0 & 9 & \vdots & 19 \\ 0 & 1 & 3 & \vdots & 5 \\ 0 & 0 & 1 & \vdots & 2 \end{bmatrix}$$

Perform operations on R_1 so that the second column has a zero above its leading 1.

$$\begin{matrix} -9R_3 + R_1 \longrightarrow \\ -3R_3 + R_2 \longrightarrow \end{matrix} \begin{bmatrix} 1 & 0 & 0 & \vdots & 1 \\ 0 & 1 & 0 & \vdots & -1 \\ 0 & 0 & 1 & \vdots & 2 \end{bmatrix}$$

Perform operations on R_1 and R_2 so that the third column has zeros above its leading 1.

Now, converting back to a system of linear equations, you have

$$\begin{cases} x = 1 \\ y = -1. \\ z = 2 \end{cases}$$

The beauty of Gauss-Jordan elimination is that, from the reduced row-echelon form, you can simply read the solution.

✓CHECKPOINT Now try Exercise 55.

The elimination procedure described in this section sometimes results in fractional coefficients. For instance, in the elimination procedure for the system

$$\begin{cases} 2x - 5y + 5z = 17 \\ 3x - 2y + 3z = 11 \\ -3x + 3y = 6 \end{cases}$$

you may be inclined to multiply the first row by $\frac{1}{2}$ to produce a leading 1, which will result in working with fractional coefficients. You can sometimes avoid fractions by judiciously choosing the order in which you apply elementary row operations.

EXAMPLE 8   COMPARING ROW-ECHELON FORMS

Make a Decision   Compare the following row-echelon form with the one found in Example 3. Is it the same? Does it yield the same solution?

$$\begin{cases} x - 2y + 3z = \ \ 9 \\ -x + 3y \qquad\ \ = -4 \\ 2x - 5y + 5z = \ 17 \end{cases}$$

$$\begin{bmatrix} 1 & -2 & 3 & \vdots & 9 \\ -1 & 3 & 0 & \vdots & -4 \\ 2 & -5 & 5 & \vdots & 17 \end{bmatrix}$$

$$\begin{matrix} R_2 \\ R_1 \end{matrix} \begin{bmatrix} -1 & 3 & 0 & \vdots & -4 \\ 1 & -2 & 3 & \vdots & 9 \\ 2 & -5 & 5 & \vdots & 17 \end{bmatrix}$$

$$-R_1 \rightarrow \begin{bmatrix} 1 & -3 & 0 & \vdots & 4 \\ 1 & -2 & 3 & \vdots & 9 \\ 2 & -5 & 5 & \vdots & 17 \end{bmatrix}$$

$$\begin{matrix} -R_1 + R_2 \rightarrow \\ -2R_1 + R_3 \rightarrow \end{matrix} \begin{bmatrix} 1 & -3 & 0 & \vdots & 4 \\ 0 & 1 & 3 & \vdots & 5 \\ 0 & 1 & 5 & \vdots & 9 \end{bmatrix}$$

$$-R_2 + R_3 \rightarrow \begin{bmatrix} 1 & -3 & 0 & \vdots & 4 \\ 0 & 1 & 3 & \vdots & 5 \\ 0 & 0 & 2 & \vdots & 4 \end{bmatrix}$$

$$\tfrac{1}{2}R_3 \rightarrow \begin{bmatrix} 1 & -3 & 0 & \vdots & 4 \\ 0 & 1 & 3 & \vdots & 5 \\ 0 & 0 & 1 & \vdots & 2 \end{bmatrix}$$

SOLUTION

This row-echelon form is different from that obtained in Example 3. The corresponding system of linear equations for this matrix is

$$\begin{cases} x - 3y \qquad\ = 4 \\ \quad\ \ y + 3z = 5. \\ \qquad\qquad z = 2 \end{cases}$$

Using back-substitution on this system, you obtain the solution

$$x = 1, y = -1, \text{ and } z = 2$$

which is the same solution that was obtained in Example 3. This row-echelon form is not the same as the one found in Example 3, but both forms yield the same solution.

✓CHECKPOINT   Now try Exercise 77.

In Example 8, you discovered that the row-echelon form of a matrix is *not* unique. That is, two different sequences of elementary row operations may yield different row-echelon forms.

| EXAMPLE 9 | A SYSTEM WITH AN INFINITE NUMBER OF SOLUTIONS |

Solve the system.

$$\begin{cases} 2x + 4y - 2z = 0 \\ 3x + 5y \quad\quad = 1 \end{cases}$$

SOLUTION

$$\begin{bmatrix} 2 & 4 & -2 & \vdots & 0 \\ 3 & 5 & 0 & \vdots & 1 \end{bmatrix} \qquad \tfrac{1}{2}R_1 \rightarrow \begin{bmatrix} 1 & 2 & -1 & \vdots & 0 \\ 3 & 5 & 0 & \vdots & 1 \end{bmatrix}$$

$$-3R_1 + R_2 \rightarrow \begin{bmatrix} 1 & 2 & -1 & \vdots & 0 \\ 0 & -1 & 3 & \vdots & 1 \end{bmatrix}$$

$$-R_2 \rightarrow \begin{bmatrix} 1 & 2 & -1 & \vdots & 0 \\ 0 & 1 & -3 & \vdots & -1 \end{bmatrix}$$

$$-2R_2 + R_1 \rightarrow \begin{bmatrix} 1 & 0 & 5 & \vdots & 2 \\ 0 & 1 & -3 & \vdots & -1 \end{bmatrix}$$

STUDY TIP

Remember that the solution set of a system with an infinite number of solutions can be written in several ways. For example, the solution set in Example 9 could have been written as

$$\left(\frac{1 - 5b}{3}, b, \frac{b + 1}{3} \right)$$

where b is a real number.

The corresponding system of equations is

$$\begin{cases} x \quad\quad + 5z = 2 \\ \quad y - 3z = -1 \end{cases}.$$

Solving for x and y in terms of z, you have $x = -5z + 2$ and $y = 3z - 1$. To write a solution of the system that does not use any of the three variables of the system, let a represent any real number and let $z = a$. Now, substitute a for z in the equations for x and y.

$$x = -5z + 2 = -5a + 2$$

$$y = 3z - 1 = 3a - 1$$

So, the solution set has the form

$$(-5a + 2, 3a - 1, a)$$

where a is a real number. Try substituting values for a to obtain a few solutions. Then check each solution in the original system of equations.

✓CHECKPOINT   Now try Exercise 61.

Discussing the Concept | Unique Solution Versus Infinitely Many Solutions

Construct one linear system in three variables that has a unique solution and another linear system in three variables that has an infinite number of solutions. Exchange systems with another student. Solve the systems you receive using the methods of this section. Compare and discuss your solutions with each other. Did you both write the infinite solution set in the same form? If not, verify that your answers represent the same solution set.

5.1 Warm Up

The following warm-up exercises involve skills that were covered in earlier sections. You will use these skills in the exercise set for this section. For additional help, review Sections R1.2, 4.1, and 4.3.

In Exercises 1–4, evaluate the expression.

1. $2(-1) - 3(5) + 7(2)$

2. $-4(-3) + 6(7) + 8(-3)$

3. $11\left(\frac{1}{2}\right) - 7\left(-\frac{3}{2}\right) - 5(2)$

4. $\frac{2}{3}\left(\frac{1}{2}\right) + \frac{4}{3}\left(-\frac{1}{3}\right)$

In Exercises 5 and 6, decide whether $x = 1$, $y = 3$, and $z = -1$ is a solution of the system.

5.
$$\begin{cases} 4x - 2y + 3z = -5 \\ x + 3y - z = 11 \\ -x + 2y = 5 \end{cases}$$

6.
$$\begin{cases} -x + 2y + z = 4 \\ 2x - 3z = 5 \\ 3x + 5y - 2z = 21 \end{cases}$$

In Exercises 7–10, use back-substitution to solve the system of linear equations.

7.
$$\begin{cases} 2x - 3y = 4 \\ y = 2 \end{cases}$$

8.
$$\begin{cases} 5x + 4y = 0 \\ y = -3 \end{cases}$$

9.
$$\begin{cases} x - 3y + z = 0 \\ y - 3z = 8 \\ z = 2 \end{cases}$$

10.
$$\begin{cases} 2x - 5y + 3z = -2 \\ y - 4z = 0 \\ z = 1 \end{cases}$$

5.1 Exercises

In Exercises 1–8, determine the order of the matrix.

1. $\begin{bmatrix} 5 & -1 & 6 \\ 3 & 1 & -2 \end{bmatrix}$

2. $\begin{bmatrix} 4 & -1 \end{bmatrix}$

3. $\begin{bmatrix} 6 & 4 & 1 \\ 8 & 3 & 0 \\ -1 & 2 & 1 \\ 1 & 5 & 4 \end{bmatrix}$

4. $\begin{bmatrix} 1 \\ 0 \\ 3 \\ 5 \\ 6 \end{bmatrix}$

5. $\begin{bmatrix} 33 & 45 \\ -9 & 20 \\ 12 & 15 \\ 16 & -2 \end{bmatrix}$

6. $\begin{bmatrix} 12 & -2 & 4 \\ -3 & 4 & 0 \\ -8 & 12 & 2 \end{bmatrix}$

7. $\begin{bmatrix} 2 & 7 & 11 & -3 \\ -1 & 10 & -5 & 0 \end{bmatrix}$

8. $\begin{bmatrix} -11 \end{bmatrix}$

In Exercises 9–12, identify the elementary row operation(s) being performed to obtain the new row-equivalent matrix.

Original Matrix	New Row-Equivalent Matrix

9. $\begin{bmatrix} -2 & 5 & 1 \\ 3 & -1 & -8 \end{bmatrix}$ $\qquad$ $\begin{bmatrix} 13 & 0 & -39 \\ 3 & -1 & -8 \end{bmatrix}$

Original Matrix	New Row-Equivalent Matrix

10. $\begin{bmatrix} 3 & -1 & -4 \\ -4 & 3 & 7 \end{bmatrix}$ $\qquad$ $\begin{bmatrix} 3 & -1 & -4 \\ 5 & 0 & -5 \end{bmatrix}$

Original Matrix	New Row-Equivalent Matrix

11. $\begin{bmatrix} 0 & -1 & -5 & 5 \\ -1 & 3 & -7 & 6 \\ 4 & -5 & 1 & 3 \end{bmatrix}$ $\qquad$ $\begin{bmatrix} -1 & 3 & -7 & 6 \\ 0 & -1 & -5 & 5 \\ 0 & 7 & -27 & 27 \end{bmatrix}$

Original Matrix	New Row-Equivalent Matrix

12. $\begin{bmatrix} -1 & -2 & 3 & -2 \\ 2 & -5 & 1 & -7 \\ 5 & 4 & -7 & 6 \end{bmatrix}$ $\qquad$ $\begin{bmatrix} -1 & -2 & 3 & -2 \\ 0 & -9 & 7 & -11 \\ 0 & -6 & 8 & -4 \end{bmatrix}$

In Exercises 13–18, determine whether the matrix is in row-echelon form. If it is, determine if it is also in reduced row-echelon form.

13. $\begin{bmatrix} 1 & 0 & 0 & 0 \\ 0 & 1 & 1 & 5 \\ 0 & 0 & 0 & 0 \end{bmatrix}$

14. $\begin{bmatrix} 1 & 0 & 2 & 1 \\ 0 & 1 & -3 & 10 \\ 0 & 0 & 1 & 0 \end{bmatrix}$

15. $\begin{bmatrix} 2 & 0 & 4 & 0 \\ 0 & -1 & 3 & 6 \\ 0 & 0 & 1 & 5 \end{bmatrix}$ **16.** $\begin{bmatrix} 0 & 0 & 0 & 0 \\ 0 & 1 & 0 & 5 \\ 0 & 0 & 1 & 3 \end{bmatrix}$

17. $\begin{bmatrix} 1 & 3 & 0 & 0 & 0 & 0 \\ 0 & 0 & 1 & 8 & 1 & 0 \\ 0 & 0 & 0 & 0 & 1 & 1 \\ 0 & 0 & 0 & 0 & 1 & 1 \end{bmatrix}$

18. $\begin{bmatrix} 1 & 0 & 0 & 10 \\ 0 & 1 & 3 & 9 \\ 0 & 0 & 0 & 1 \\ 0 & 0 & 0 & 0 \end{bmatrix}$

19. Use a graphing utility to perform the sequence of row operations to reduce the matrix to row-echelon form.

$$\begin{bmatrix} 1 & 1 & 2 \\ 3 & 4 & -3 \\ 2 & -1 & 1 \end{bmatrix}$$

(a) Add -3 times R_1 to R_2.

(b) Add -2 times R_1 to R_3.

(c) Add 3 times R_2 to R_3.

(d) Multiply R_3 by $-\frac{1}{30}$.

20. Use a graphing utility to perform the sequence of row operations to reduce the matrix to *reduced* row-echelon form.

$$\begin{bmatrix} 7 & 1 \\ 0 & 2 \\ -3 & 4 \\ 4 & 1 \end{bmatrix}$$

(a) Add R_3 to R_4.

(b) Interchange R_1 and R_4.

(c) Add 3 times R_1 to R_3.

(d) Add -7 times R_1 to R_4.

(e) Multiply R_2 by $\frac{1}{2}$.

(f) Add the appropriate multiple of R_2 to R_1, R_3, and R_4.

In Exercises 21–24, write the matrix in row-echelon form. (*Note:* Row-echelon forms are not unique.)

21. $\begin{bmatrix} 1 & 2 & -1 & 5 \\ 3 & 2 & 1 & 11 \\ 4 & 8 & 1 & 10 \end{bmatrix}$

22. $\begin{bmatrix} 1 & 2 & -1 & 3 \\ 3 & 7 & -5 & 14 \\ -2 & -1 & -3 & 8 \end{bmatrix}$

23. $\begin{bmatrix} 1 & -1 & -1 & 1 \\ 5 & -4 & 1 & 8 \\ -6 & 8 & 18 & 0 \end{bmatrix}$

24. $\begin{bmatrix} 1 & -3 & 0 & -7 \\ -3 & 10 & 1 & 23 \\ 1 & 0 & 1 & 12 \\ 4 & -10 & 2 & -24 \end{bmatrix}$

In Exercises 25–28, write the matrix in *reduced* row-echelon form.

25. $\begin{bmatrix} 4 & 4 & 8 \\ 1 & 2 & 2 \\ -3 & 6 & -9 \end{bmatrix}$ **26.** $\begin{bmatrix} 1 & 3 & 2 \\ 5 & 15 & 9 \\ 2 & 6 & 10 \end{bmatrix}$

27. $\begin{bmatrix} 1 & 2 & 3 & -5 \\ 1 & 2 & 4 & -9 \\ -2 & -4 & -4 & 3 \\ 4 & 8 & 11 & -14 \end{bmatrix}$

28. $\begin{bmatrix} 1 & -3 \\ -1 & 8 \\ 0 & 4 \\ -2 & 10 \end{bmatrix}$

In Exercises 29–32, write the system of linear equations represented by the augmented matrix. (Use the variables w, x, y, and z.)

29. $\left[\begin{array}{cc:c} 2 & 4 & 6 \\ -1 & 3 & -8 \end{array}\right]$ **30.** $\left[\begin{array}{cc:c} 7 & -2 & 7 \\ -8 & 3 & -3 \end{array}\right]$

31. $\left[\begin{array}{ccc:c} 1 & 0 & 2 & -10 \\ 0 & 3 & -1 & 5 \\ 4 & 2 & 0 & 3 \end{array}\right]$

32. $\left[\begin{array}{cccc:c} 5 & 8 & 2 & 0 & -1 \\ -2 & 15 & 5 & 1 & 9 \\ 1 & 6 & -7 & 0 & -3 \end{array}\right]$

In Exercises 33–38, write the augmented matrix for the system of linear equations.

33. $\begin{cases} 2x - y = 3 \\ 5x + 7y = 12 \end{cases}$ **34.** $\begin{cases} 8x + 3y = 25 \\ 3x - 9y = 12 \end{cases}$

35. $\begin{cases} x + 10y - 3z = 2 \\ 5x - 3y + 4z = 0 \\ 2x + 4y = 6 \end{cases}$ **36.** $\begin{cases} 2x + 3y - z = 8 \\ y + 2z = -10 \\ x - 2y - 3z = 21 \end{cases}$

37. $\begin{cases} 9w - 3x + 20y + z = 13 \\ 12w - 8y = 5 \\ w + 2x + 3y - 4z = -2 \\ -w - x + y + z = 1 \end{cases}$

38. $\begin{cases} w + 2x - 3y + z = 18 \\ 3w \quad\;\; - 5y \quad\quad\; = 8 \\ w + x + y + 2z = 15 \\ -w - x + 2y + z = -3 \end{cases}$

In Exercises 39–42, write the system of equations represented by the augmented matrix. Use back-substitution to find the solution. (Use w, x, y, and z.)

39. $\begin{bmatrix} 1 & -5 & \vdots & 6 \\ 0 & 1 & \vdots & -2 \end{bmatrix}$

40. $\begin{bmatrix} 1 & 2 & -1 & \vdots & 3 \\ 0 & 1 & -2 & \vdots & -3 \\ 0 & 0 & 1 & \vdots & 4 \end{bmatrix}$

41. $\begin{bmatrix} 1 & 3 & -1 & \vdots & 15 \\ 0 & 1 & 4 & \vdots & -12 \\ 0 & 0 & 1 & \vdots & -5 \end{bmatrix}$

42. $\begin{bmatrix} 1 & 2 & -2 & 0 & \vdots & -1 \\ 0 & 1 & 1 & 2 & \vdots & 9 \\ 0 & 0 & 1 & 0 & \vdots & 2 \\ 0 & 0 & 0 & 1 & \vdots & -3 \end{bmatrix}$

In Exercises 43–46, an augmented matrix that represents a system of linear equations (in variables x, y, and z) has been reduced using Gauss-Jordan elimination. Write the solution represented by the augmented matrix.

43. $\begin{bmatrix} 1 & 0 & \vdots & -4 \\ 0 & 1 & \vdots & 6 \end{bmatrix}$

44. $\begin{bmatrix} 1 & 0 & \vdots & 9 \\ 0 & 1 & \vdots & -3 \end{bmatrix}$

45. $\begin{bmatrix} 1 & 0 & 0 & \vdots & -4 \\ 0 & 1 & 0 & \vdots & -8 \\ 0 & 0 & 1 & \vdots & 2 \end{bmatrix}$

46. $\begin{bmatrix} 1 & 0 & 0 & \vdots & 3 \\ 0 & 1 & 0 & \vdots & -1 \\ 0 & 0 & 1 & \vdots & 0 \end{bmatrix}$

In Exercises 47–70, use matrices to solve the system of equations (if possible). Use Gaussian elimination with back-substitution or Gauss-Jordan elimination.

47. $\begin{cases} x + 2y = 7 \\ 2x + y = 8 \end{cases}$

48. $\begin{cases} 2x + 6y = 16 \\ 2x + 3y = 7 \end{cases}$

49. $\begin{cases} -3x + 5y = -22 \\ 3x + 4y = 4 \\ 4x - 8y = 32 \end{cases}$

50. $\begin{cases} x + 2y = 0 \\ x + y = 6 \\ 3x - 2y = 8 \end{cases}$

51. $\begin{cases} 8x - 4y = 7 \\ 5x + 2y = 1 \end{cases}$

52. $\begin{cases} x - 3y = 5 \\ -2x + 6y = -10 \end{cases}$

53. $\begin{cases} -x + 2y = 1.5 \\ 2x - 4y = 3 \end{cases}$

54. $\begin{cases} 2x - y = -0.1 \\ 3x + 2y = 1.6 \end{cases}$

55. $\begin{cases} 2x + 2y - z = 2 \\ x - 3y + z = -28 \\ -x + y = 14 \end{cases}$

56. $\begin{cases} -x + y - z = -14 \\ 2x - y + z = 21 \\ 3x + 2y + z = 19 \end{cases}$

57. $\begin{cases} 2x \quad\;\; + 3z = 3 \\ 4x - 3y + 7z = 5 \\ 8x - 9y + 15z = 9 \end{cases}$

58. $\begin{cases} 2x - y + 3z = 24 \\ 2y - z = 14 \\ 7x - 5y = 6 \end{cases}$

59. $\begin{cases} x + y - 5z = 3 \\ x - 2z = 1 \\ 2x - y - z = 1 \end{cases}$

60. $\begin{cases} x - 3z = -2 \\ 3x + y - 2z = 5 \\ 2x + 2y + z = 4 \end{cases}$

61. $\begin{cases} x + 2y + z = 8 \\ 3x + 7y + 6z = 26 \end{cases}$

62. $\begin{cases} 4x + 12y - 7z - 20w = 22 \\ 3x + 9y - 5z - 28w = 30 \end{cases}$

63. $\begin{cases} 3x + 3y + 12z = 6 \\ x + y + 4z = 2 \\ 2x + 5y + 20z = 10 \\ -x + 2y + 8z = 4 \end{cases}$

64. $\begin{cases} 2x + 10y + 2z = 6 \\ x + 5y + 2z = 6 \\ x + 5y + z = 3 \\ -3x + 15y - 3z = -9 \end{cases}$

65. $\begin{cases} 2x + y - z + 2w = -16 \\ 3x + 4y + w = 1 \\ x + 5y + 2z + 6w = -3 \\ 5x + 2y - z + w = 3 \end{cases}$

66. $\begin{cases} x + 2y + 2z + 4w = 11 \\ 3x + 6y + 5z + 12w = 30 \end{cases}$

67. $\begin{cases} x + 2y = 0 \\ -x - y = 0 \end{cases}$

68. $\begin{cases} x + 2y = 0 \\ 2x + 4y = 0 \end{cases}$

69. $\begin{cases} x + y + z = 0 \\ 2x + 3y + z = 0 \\ 3x + 5y + z = 0 \end{cases}$

70. $\begin{cases} x - 2y + z + 3w = 0 \\ x - y + w = 0 \\ y - z + 2w = 0 \end{cases}$

71. Borrowing Money A small clothing corporation borrowed $800,000 at simple annual interest to expand its product line. Some of the money was borrowed at 8%, some at 9%, and some at 12%. Use a system of equations to determine how much was borrowed at each rate if the total annual interest was $73,000 and the amount borrowed at 8% was twice the amount borrowed at 12%. Solve the system using matrices.

72. Borrowing Money   A shoe company borrowed $800,000 at simple annual interest to expand its product line. Some of the money was borrowed at 7%, some at 8.5%, and some at 9.5%. Use a system of equations to determine how much was borrowed at each rate if the total annual interest was $62,375 and the amount borrowed at 8.5% was four times the amount borrowed at 9.5%. Solve the system using matrices.

73. You and a friend solve the following system of equations independently.

$$\begin{cases} 2x - 4y - 3z = 3 \\ x + 3y + z = -1 \\ 5x + y - 2z = 2 \end{cases}$$

You write your solution set as $(a, -a, 2a - 1)$, where a is any real number. Your friend's solution set is $\left(\frac{1}{2}b + \frac{1}{2}, -\frac{1}{2}b - \frac{1}{2}, b\right)$, where b is any real number. Are you both correct? Explain. If you let $a = 3$, what value of b should be selected so that you both have the same ordered triple?

74. Describe how you would explain to another student that the augmented matrix below represents a dependent system of equations. Describe a way to write the infinitely many solutions of this system.

$$\begin{bmatrix} 1 & -2 & 3 & \vdots & -6 \\ 0 & 1 & 2 & \vdots & 5 \\ 0 & 0 & 0 & \vdots & 0 \end{bmatrix}$$

75. *Make a Decision*: **Energy Imports**   From 1990 to 2001, the total energy imports y (in quadrillion Btu) to the United States increased in a pattern that was approximately linear (see figure). Find the least squares regression line $y = at + b$ for the data shown in the figure by solving the following system using matrices. Let t represent the year, with $t = 0$ corresponding to 1990.

$$\begin{cases} 12b + 66a = 284.31 \\ 66b + 506a = 1719.72 \end{cases}$$

Use the result to estimate the total energy imports in 2006. Is the estimate reasonable? Explain.   (Source: Energy Information Administration)

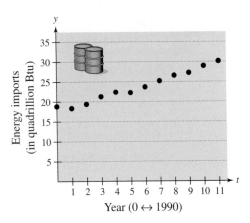

Year (0 ↔ 1990)

Figure for 75

76. *Make a Decision*: **Fossil Fuels**   From 1990 to 2001, the total fossil fuel consumption y (in quadrillion Btu) in the United States increased in a pattern that was approximately linear (see figure). Find the least squares regression line $y = at + b$ for the data shown in the figure by solving the following system using matrices. Let t represent the year, with $t = 0$ corresponding to 1990.

$$\begin{cases} 12b + 66a = 940.436 \\ 66b + 506a = 5346.066 \end{cases}$$

Use the result to estimate the total fossil fuel consumption in 2007. Is the estimate reasonable? Explain. (Source: Energy Information Administration)

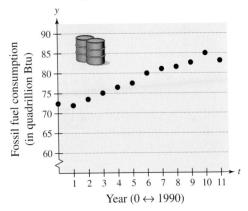

Year (0 ↔ 1990)

77. *Make a Decision*   Compare the following reduced row-echelon form with the one found in Exercise 55. Is it the same? Does it yield the same solution? Explain.

$$\begin{bmatrix} 1 & 0 & 0 & \vdots & -6 \\ 0 & 1 & 0 & \vdots & 8 \\ 0 & 0 & 1 & \vdots & 2 \end{bmatrix}$$

Objectives

- **Determine whether two matrices are equal.**
- **Add or subtract two matrices and multiply a matrix by a scalar.**
- **Find the product of two matrices.**
- **Solve a matrix equation.**
- **Use matrix multiplication to solve an application problem.**

Equality of Matrices

In Section 5.1, you used matrices to solve systems of linear equations. Matrices, however, can do much more than this. There is a rich mathematical theory of matrices, and its applications are numerous. This section and the next introduce some fundamentals of matrix theory. It is standard mathematical convention to represent matrices in any of the following three ways.

1. A matrix can be denoted by an uppercase letter such as A, B, or C.

2. A matrix can be denoted by a representative element enclosed in brackets, such as $[a_{ij}]$, $[b_{ij}]$, or $[c_{ij}]$.

3. A matrix can be denoted by a rectangular array of numbers such as

$$A = [a_{ij}] = \begin{bmatrix} a_{11} & a_{12} & a_{13} & \cdots & a_{1n} \\ a_{21} & a_{22} & a_{23} & \cdots & a_{2n} \\ a_{31} & a_{32} & a_{33} & \cdots & a_{3n} \\ \vdots & \vdots & \vdots & & \vdots \\ a_{m1} & a_{m2} & a_{m3} & \cdots & a_{mn} \end{bmatrix}.$$

Two matrices $A = [a_{ij}]$ and $B = [b_{ij}]$ are **equal** if they have the same order $(m \times n)$ and $a_{ij} = b_{ij}$ for $1 \le i \le m$ and $1 \le j \le n$. In other words, two matrices are equal if their corresponding entries are equal.

EXAMPLE 1   EQUALITY OF MATRICES

Solve for a_{11}, a_{12}, a_{21}, and a_{22} in the matrix equation.

$$\begin{bmatrix} a_{11} & a_{12} \\ a_{21} & a_{22} \end{bmatrix} = \begin{bmatrix} 2 & -1 \\ -3 & 0 \end{bmatrix}$$

SOLUTION

Because two matrices are equal only if their corresponding entries are equal, you can conclude that

$$a_{11} = 2, \quad a_{12} = -1, \quad a_{21} = -3, \quad \text{and} \quad a_{22} = 0.$$

✓**CHECKPOINT**   Now try Exercise 1.

Matrix Addition and Scalar Multiplication

You can **add** two matrices (of the same order) by adding their corresponding entries.

Definition of Matrix Addition

If $A = [a_{ij}]$ and $B = [b_{ij}]$ are matrices of order $m \times n$, their **sum** is the $m \times n$ matrix given by

$$A + B = [a_{ij} + b_{ij}].$$

The sum of two matrices of different orders is undefined.

EXAMPLE 2   ADDITION OF MATRICES

a. $\begin{bmatrix} -1 & 2 \\ 0 & 1 \end{bmatrix} + \begin{bmatrix} 1 & 3 \\ -1 & 2 \end{bmatrix} = \begin{bmatrix} -1+1 & 2+3 \\ 0-1 & 1+2 \end{bmatrix} = \begin{bmatrix} 0 & 5 \\ -1 & 3 \end{bmatrix}$

b. $\begin{bmatrix} 0 & 1 & -2 \\ 1 & 2 & 3 \end{bmatrix} + \begin{bmatrix} 0 & 0 & 0 \\ 0 & 0 & 0 \end{bmatrix} = \begin{bmatrix} 0 & 1 & -2 \\ 1 & 2 & 3 \end{bmatrix}$

c. $\begin{bmatrix} 1 \\ -3 \\ -2 \end{bmatrix} + \begin{bmatrix} -1 \\ 3 \\ 2 \end{bmatrix} = \begin{bmatrix} 0 \\ 0 \\ 0 \end{bmatrix}$

d. The sum of

$$A = \begin{bmatrix} 2 & 1 & 0 \\ 4 & 0 & -1 \\ 3 & -2 & 2 \end{bmatrix} \quad \text{and} \quad B = \begin{bmatrix} 0 & 1 \\ -1 & 3 \\ 2 & 4 \end{bmatrix}$$

is undefined because A is of order 3×3 and B is of order 3×2.

✓CHECKPOINT   Now try Exercise 5(a).

When working with matrices, numbers are usually referred to as **scalars.** In this text, scalars will always be real numbers. You can multiply a matrix A by a scalar c by multiplying each entry in A by c, as shown below.

Scalar   Matrix

$$3\begin{bmatrix} -1 & 2 \\ 6 & 5 \end{bmatrix} = \begin{bmatrix} 3(-1) & 3(2) \\ 3(6) & 3(5) \end{bmatrix} = \begin{bmatrix} -3 & 6 \\ 18 & 15 \end{bmatrix}$$

Definition of Scalar Multiplication

If $A = [a_{ij}]$ is an $m \times n$ matrix and c is a scalar, the **scalar multiple** of A by c is the $m \times n$ matrix given by

$$cA = [ca_{ij}].$$

The symbol $-A$ represents the negation of A, or the scalar product $(-1)A$. Moreover, if A and B are of the same order, then $A - B$ represents the sum of A and $(-1)B$. That is,

$$A - B = A + (-1)B.$$ Subtraction of matrices

EXAMPLE 3   SCALAR MULTIPLICATION AND MATRIX SUBTRACTION

For the following matrices, find (a) $3A$, (b) $-B$, and (c) $3A - B$.

$$A = \begin{bmatrix} 2 & 2 & 4 \\ -3 & 0 & -1 \\ 2 & 1 & 2 \end{bmatrix} \quad \text{and} \quad B = \begin{bmatrix} 2 & 0 & 0 \\ 1 & -4 & 3 \\ -1 & 3 & 2 \end{bmatrix}$$

SOLUTION

a. $3A = 3\begin{bmatrix} 2 & 2 & 4 \\ -3 & 0 & -1 \\ 2 & 1 & 2 \end{bmatrix}$ Scalar multiplication

$$= \begin{bmatrix} 3(2) & 3(2) & 3(4) \\ 3(-3) & 3(0) & 3(-1) \\ 3(2) & 3(1) & 3(2) \end{bmatrix}$$ Multiply each entry by 3.

$$= \begin{bmatrix} 6 & 6 & 12 \\ -9 & 0 & -3 \\ 6 & 3 & 6 \end{bmatrix}$$ Simplify.

b. $-B = (-1)\begin{bmatrix} 2 & 0 & 0 \\ 1 & -4 & 3 \\ -1 & 3 & 2 \end{bmatrix}$ Definition of negation

$$= \begin{bmatrix} -2 & 0 & 0 \\ -1 & 4 & -3 \\ 1 & -3 & -2 \end{bmatrix}$$ Multiply each entry by -1.

c. $3A - B = \begin{bmatrix} 6 & 6 & 12 \\ -9 & 0 & -3 \\ 6 & 3 & 6 \end{bmatrix} - \begin{bmatrix} 2 & 0 & 0 \\ 1 & -4 & 3 \\ -1 & 3 & 2 \end{bmatrix}$ Matrix subtraction

$$= \begin{bmatrix} 4 & 6 & 12 \\ -10 & 4 & -6 \\ 7 & 0 & 4 \end{bmatrix}$$ Subtract corresponding entries.

✓**CHECKPOINT**   Now try Exercises 5(b)–(d).

It is often convenient to rewrite the scalar multiple cA by factoring c out of every entry in the matrix. For instance, in the first matrix below the scalar $\frac{1}{2}$ has been factored out of the matrix, and in the second matrix the scalar -2 has been factored out of the matrix.

$$\begin{bmatrix} \frac{1}{2} & -\frac{3}{2} \\ \frac{5}{2} & \frac{1}{2} \end{bmatrix} = \frac{1}{2}\begin{bmatrix} 1 & -3 \\ 5 & 1 \end{bmatrix}$$

$$\begin{bmatrix} -4 & -20 \\ -10 & -2 \end{bmatrix} = -2\begin{bmatrix} 2 & 10 \\ 5 & 1 \end{bmatrix}$$

The properties of matrix addition and scalar multiplication are similar to those of addition and multiplication of real numbers.

Properties of Matrix Addition and Scalar Multiplication

If A, B, and C are $m \times n$ matrices and c and d are scalars, then the following properties are true.

1. $A + B = B + A$ Commutative Property of Matrix Addition

2. $A + (B + C) = (A + B) + C$ Associative Property of Matrix Addition

3. $(cd)A = c(dA)$ Associative Property of Scalar Multiplication

4. $1A = A$ Scalar Identity

5. $c(A + B) = cA + cB$ Distributive Property

6. $(c + d)A = cA + dA$ Distributive Property

Note that the Associative Property of Matrix Addition allows you to write expressions such as $A + B + C$ without ambiguity, because you obtain the same sum no matter how the matrices are grouped. In other words, you obtain the same sum whether you group $A + B + C$ as $(A + B) + C$ or as $A + (B + C)$. This same reasoning applies to sums of four or more matrices.

EXAMPLE 4   ADDITION OF MORE THAN TWO MATRICES

Add the following four matrices.

$$\begin{bmatrix} 1 \\ 2 \\ -3 \end{bmatrix}, \begin{bmatrix} -1 \\ -1 \\ 2 \end{bmatrix}, \begin{bmatrix} 0 \\ 1 \\ 4 \end{bmatrix}, \begin{bmatrix} 2 \\ -3 \\ -2 \end{bmatrix}$$

SOLUTION

By adding corresponding entries, you obtain the following sum of four matrices.

$$\begin{bmatrix} 1 \\ 2 \\ -3 \end{bmatrix} + \begin{bmatrix} -1 \\ -1 \\ 2 \end{bmatrix} + \begin{bmatrix} 0 \\ 1 \\ 4 \end{bmatrix} + \begin{bmatrix} 2 \\ -3 \\ -2 \end{bmatrix} = \begin{bmatrix} 2 \\ -1 \\ 1 \end{bmatrix}$$

✓CHECKPOINT   Now try Exercise 11.

TECHNOLOGY

Most graphing utilities can add and subtract matrices and multiply matrices by scalars. Use your graphing utility to find (a) $A + B$, (b) $A - B$, (c) $4A$, and (d) $4A + B$. For specific keystrokes on how to perform matrix operations using a graphing utility, go to the text website at *college.hmco.com*.

$$A = \begin{bmatrix} 2 & -3 \\ -1 & 0 \end{bmatrix} \quad \text{and} \quad B = \begin{bmatrix} -1 & 4 \\ 2 & -5 \end{bmatrix}$$

One important property of addition of real numbers is that the number 0 is the additive identity. That is, $c + 0 = c$ for any real number c. For matrices, a similar property holds. That is, if A is an $m \times n$ matrix and O is the $m \times n$ **zero matrix** consisting entirely of zeros, then $A + O = A$.

In other words, O is the **additive identity** for the set of all $m \times n$ matrices. For example, the following matrices are the additive identities for the sets of all 2×3 and 2×2 matrices, respectively.

$$O = \begin{bmatrix} 0 & 0 & 0 \\ 0 & 0 & 0 \end{bmatrix} \quad \text{and} \quad O = \begin{bmatrix} 0 & 0 \\ 0 & 0 \end{bmatrix}$$

$\underbrace{}_{2 \times 3 \text{ zero matrix}}$ $\underbrace{}_{2 \times 2 \text{ zero matrix}}$

The algebra of real numbers and the algebra of matrices have many similarities. For example, compare the following solutions.

Real Numbers	*$m \times n$ Matrices*
(Solve for x.)	*(Solve for X.)*
$x + a = b$	$X + A = B$
$x + a + (-a) = b + (-a)$	$X + A + (-A) = B + (-A)$
$x + 0 = b - a$	$X + O = B - A$
$x = b - a$	$X = B - A$

This means that you can apply some of your knowledge of solving real number equations to solving matrix equations. It is often easier to complete the algebraic steps first, and then substitute the matrices into the equation, as illustrated in Example 5.

EXAMPLE 5 SOLVING A MATRIX EQUATION

Solve for X in the equation $3X + A = B$, where

$$A = \begin{bmatrix} 1 & -2 \\ 0 & 3 \end{bmatrix} \quad \text{and} \quad B = \begin{bmatrix} -3 & 4 \\ 2 & 1 \end{bmatrix}.$$

SOLUTION

Begin by solving the equation for X to obtain

$$3X = B - A \quad \Longrightarrow \quad X = \frac{1}{3}(B - A).$$

Now, using the matrices A and B, you have

$$X = \frac{1}{3}\left(\begin{bmatrix} -3 & 4 \\ 2 & 1 \end{bmatrix} - \begin{bmatrix} 1 & -2 \\ 0 & 3 \end{bmatrix} \right) \qquad \text{Substitute the matrices.}$$

$$= \frac{1}{3}\begin{bmatrix} -4 & 6 \\ 2 & -2 \end{bmatrix} \qquad \text{Subtract matrix } A \text{ from matrix } B.$$

$$= \begin{bmatrix} -\frac{4}{3} & 2 \\ \frac{2}{3} & -\frac{2}{3} \end{bmatrix}. \qquad \text{Multiply the resulting matrix by } \frac{1}{3}.$$

✓*CHECKPOINT* Now try Exercise 19.

Matrix Multiplication

The third basic matrix operation is **matrix multiplication.** At first glance the definition may seem unusual. You will see later, however, that this definition of the product of two matrices has many practical applications.

Definition of Matrix Multiplication

If $A = [a_{ij}]$ is an $m \times n$ matrix and $B = [b_{ij}]$ is an $n \times p$ matrix, the **product** AB is an $m \times p$ matrix

$$AB = [c_{ij}]$$

where $c_{ij} = a_{i1}b_{1j} + a_{i2}b_{2j} + a_{i3}b_{3j} + \cdots + a_{in}b_{nj}.$

The definition of matrix multiplication indicates a *row-by-column* multiplication, where the entry in the *i*th row and *j*th column of the product AB is obtained by multiplying the entries in the *i*th row of A by the corresponding entries in the *j*th column of B and then adding the results. The general pattern for matrix multiplication is as follows.

$$\begin{bmatrix} a_{11} & a_{12} & a_{13} & \cdots & a_{1n} \\ a_{21} & a_{22} & a_{23} & \cdots & a_{2n} \\ a_{31} & a_{32} & a_{33} & \cdots & a_{3n} \\ \vdots & \vdots & \vdots & & \vdots \\ a_{i1} & a_{i2} & a_{i3} & \cdots & a_{in} \\ \vdots & \vdots & \vdots & & \vdots \\ a_{m1} & a_{m2} & a_{m3} & \cdots & a_{mn} \end{bmatrix} \begin{bmatrix} b_{11} & b_{12} & \cdots & b_{1j} & \cdots & b_{1p} \\ b_{21} & b_{22} & \cdots & b_{2j} & \cdots & b_{2p} \\ b_{31} & b_{32} & \cdots & b_{3j} & \cdots & b_{3p} \\ \vdots & \vdots & & \vdots & & \vdots \\ b_{n1} & b_{n2} & \cdots & b_{nj} & \cdots & b_{np} \end{bmatrix} = \begin{bmatrix} c_{11} & c_{12} & \cdots & c_{1j} & \cdots & c_{1p} \\ c_{21} & c_{22} & \cdots & c_{2j} & \cdots & c_{2p} \\ \vdots & \vdots & & & & \vdots \\ c_{i1} & c_{i2} & \cdots & c_{ij} & \cdots & c_{ip} \\ \vdots & \vdots & & & & \vdots \\ c_{m1} & c_{m2} & \cdots & c_{mj} & \cdots & c_{mp} \end{bmatrix}$$

$$a_{i1}b_{1j} + a_{i2}b_{2j} + a_{i3}b_{3j} + \cdots + a_{in}b_{nj} = c_{ij}$$

TECHNOLOGY

Some graphing utilities can multiply two matrices. Use your graphing utility to find the product AB.

$$A = \begin{bmatrix} 1 & 2 & 3 \\ 2 & -5 & 1 \end{bmatrix}$$

$$B = \begin{bmatrix} -3 & 2 & 1 \\ 4 & -2 & 0 \\ 1 & 2 & 3 \end{bmatrix}$$

Now use your graphing utility to find the product BA. What is the result of this operation? For specific keystrokes, go to the text website at *college.hmco.com*.

EXAMPLE 6   FINDING THE PRODUCT OF TWO MATRICES

Find the product AB using $A = \begin{bmatrix} -1 & 3 \\ 4 & -2 \\ 5 & 0 \end{bmatrix}$ and $B = \begin{bmatrix} -3 & 2 \\ -4 & 1 \end{bmatrix}$.

SOLUTION

First, note that the product AB is defined because the number of columns of A is equal to the number of rows of B. Moreover, the product AB has order 3×2. To find the entries of the product, multiply each row of A by each column of B.

$$AB = \begin{bmatrix} -1 & 3 \\ 4 & -2 \\ 5 & 0 \end{bmatrix} \begin{bmatrix} -3 & 2 \\ -4 & 1 \end{bmatrix}$$

$$= \begin{bmatrix} (-1)(-3) + (3)(-4) & (-1)(2) + (3)(1) \\ (4)(-3) + (-2)(-4) & (4)(2) + (-2)(1) \\ (5)(-3) + (0)(-4) & (5)(2) + (0)(1) \end{bmatrix} = \begin{bmatrix} -9 & 1 \\ -4 & 6 \\ -15 & 10 \end{bmatrix}$$

✓CHECKPOINT   Now try Exercise 21.

Be sure you understand that for the product of two matrices to be defined, the number of *columns* of the first matrix must equal the number of *rows* of the second matrix. That is, the middle two indices must be the same and the outside two indices give the order of the product, as shown in the following diagram.

$$\underset{m \times n}{A} \qquad \underset{n \times p}{B} \qquad = \qquad \underset{m \times p}{AB}$$

Equal
Order of AB

DISCOVERY

Use a graphing utility to multiply the matrices

$$A = \begin{bmatrix} 1 & 2 \\ 3 & 4 \end{bmatrix} \text{ and}$$

$$B = \begin{bmatrix} 0 & 1 \\ 2 & 3 \end{bmatrix}.$$

Do you obtain the same result for the product AB as for the product BA? What does this tell you about matrix multiplication and commutativity?

EXAMPLE 7   MATRIX MULTIPLICATION

Find the product AB using $A = \begin{bmatrix} 6 & 2 & 0 \\ 3 & -1 & 2 \\ 1 & 4 & 6 \end{bmatrix}$ and $B = \begin{bmatrix} 1 & 0 \\ 2 & 7 \\ -3 & 5 \end{bmatrix}$.

SOLUTION

Note that the order of A is 3×3 and the order of B is 3×2. So, the product AB has the order 3×2.

$$AB = \begin{bmatrix} 6 & 2 & 0 \\ 3 & -1 & 2 \\ 1 & 4 & 6 \end{bmatrix} \begin{bmatrix} 1 & 0 \\ 2 & 7 \\ -3 & 5 \end{bmatrix}$$

$$= \begin{bmatrix} 6(1) + 2(2) + 0(-3) & 6(0) + 2(7) + 0(5) \\ 3(1) + (-1)(2) + 2(-3) & 3(0) + (-1)(7) + 2(5) \\ 1(1) + 4(2) + 6(-3) & 1(0) + 4(7) + 6(5) \end{bmatrix}$$

$$= \begin{bmatrix} 10 & 14 \\ -5 & 3 \\ -9 & 58 \end{bmatrix}$$

✓**CHECKPOINT**   Now try Exercise 23.

EXAMPLE 8   PATTERNS IN MATRIX MULTIPLICATION

a. $\underset{2 \times 3}{\begin{bmatrix} 1 & 0 & 3 \\ 2 & -1 & -2 \end{bmatrix}} \underset{3 \times 3}{\begin{bmatrix} -2 & 4 & 2 \\ 1 & 0 & 0 \\ -1 & 1 & -1 \end{bmatrix}} = \underset{2 \times 3}{\begin{bmatrix} -5 & 7 & -1 \\ -3 & 6 & 6 \end{bmatrix}}$

b. $\underset{2 \times 2}{\begin{bmatrix} 3 & 4 \\ -2 & 5 \end{bmatrix}} \underset{2 \times 2}{\begin{bmatrix} 1 & 0 \\ 0 & 1 \end{bmatrix}} = \underset{2 \times 2}{\begin{bmatrix} 3 & 4 \\ -2 & 5 \end{bmatrix}}$

c. The product AB for the following matrices is not defined.

$$A = \underset{3 \times 2}{\begin{bmatrix} -2 & 1 \\ 1 & -3 \\ 1 & 4 \end{bmatrix}} \text{ and } B = \underset{3 \times 4}{\begin{bmatrix} -2 & 3 & 1 & 4 \\ 0 & 1 & -1 & 2 \\ 2 & -1 & 0 & 1 \end{bmatrix}}$$

✓**CHECKPOINT**   Now try Exercise 25.

EXAMPLE 9   PATTERNS IN MATRIX MULTIPLICATION

a. $\begin{bmatrix} 1 & -2 \end{bmatrix} \begin{bmatrix} 2 \\ -1 \end{bmatrix} = \begin{bmatrix} 4 \end{bmatrix}$ **b.** $\begin{bmatrix} 2 \\ -1 \end{bmatrix} \begin{bmatrix} 1 & -2 \end{bmatrix} = \begin{bmatrix} 2 & -4 \\ -1 & 2 \end{bmatrix}$

  1×2 2×1 1×1 2×1 1×2 2×2

✓*CHECKPOINT*   Now try Exercise 27.

In Example 9, note that the two products are different. Even if AB and BA are defined, matrix multiplication is not, in general, commutative. That is, for most matrices, $AB \neq BA$.

Properties of Matrix Multiplication

Let A, B, and C be matrices and let c be a scalar.

1. $A(BC) = (AB)C$ Associative Property of Matrix Multiplication

2. $A(B + C) = AB + AC$ Left Distributive Property

3. $(A + B)C = AC + BC$ Right Distributive Property

4. $c(AB) = (cA)B = A(cB)$ Associative Property of Scalar Multiplication

Definition of Identity Matrix

The $n \times n$ matrix that consists of 1's on its main diagonal and 0's elsewhere is called the **identity matrix of order n** and is denoted by

$$I_n = \begin{bmatrix} 1 & 0 & 0 & \cdots & 0 \\ 0 & 1 & 0 & \cdots & 0 \\ 0 & 0 & 1 & \cdots & 0 \\ \vdots & \vdots & \vdots & & \vdots \\ 0 & 0 & 0 & \cdots & 1 \end{bmatrix}.$$ Identity matrix

Note that an identity matrix must be *square*. When the order is understood to be n, you can denote I_n simply by I.

If A is an $n \times n$ matrix, the identity matrix has the property that $AI_n = A$ and $I_nA = A$. For example,

$$\begin{bmatrix} 3 & -2 & 5 \\ 1 & 0 & 4 \\ -1 & 2 & -3 \end{bmatrix} \begin{bmatrix} 1 & 0 & 0 \\ 0 & 1 & 0 \\ 0 & 0 & 1 \end{bmatrix} = \begin{bmatrix} 3 & -2 & 5 \\ 1 & 0 & 4 \\ -1 & 2 & -3 \end{bmatrix}$$

and

$$\begin{bmatrix} 1 & 0 & 0 \\ 0 & 1 & 0 \\ 0 & 0 & 1 \end{bmatrix} \begin{bmatrix} 3 & -2 & 5 \\ 1 & 0 & 4 \\ -1 & 2 & -3 \end{bmatrix} = \begin{bmatrix} 3 & -2 & 5 \\ 1 & 0 & 4 \\ -1 & 2 & -3 \end{bmatrix}.$$

Applications

One application of matrix multiplication is representation of a system of linear equations. Note how the system

$$\begin{cases} a_{11}x_1 + a_{12}x_2 + a_{13}x_3 = b_1 \\ a_{21}x_1 + a_{22}x_2 + a_{23}x_3 = b_2 \\ a_{31}x_1 + a_{32}x_2 + a_{33}x_3 = b_3 \end{cases}$$

STUDY TIP

The column matrix B is also called a *constant matrix*. Its entries are the constant terms in the system of equations.

can be written as the matrix equation $AX = B$, where A is the *coefficient matrix* of the system, and X and B are column matrices.

$$\underbrace{\begin{bmatrix} a_{11} & a_{12} & a_{13} \\ a_{21} & a_{22} & a_{23} \\ a_{31} & a_{32} & a_{33} \end{bmatrix}}_{A} \underbrace{\begin{bmatrix} x_1 \\ x_2 \\ x_3 \end{bmatrix}}_{\times \; X} = \underbrace{\begin{bmatrix} b_1 \\ b_2 \\ b_3 \end{bmatrix}}_{B}$$

EXAMPLE 10   SOLVING A SYSTEM OF LINEAR EQUATIONS

Consider the system of linear equations.

$$\begin{cases} x_1 - 2x_2 + x_3 = -4 \\ \quad\quad x_2 + 2x_3 = 4 \\ 2x_1 + 3x_2 - 2x_3 = 2 \end{cases}$$

a. Write this system as a matrix equation $AX = B$.

b. Use Gauss-Jordan elimination on $[A \,\vdots\, B]$ to solve for the matrix X.

SOLUTION

a. In matrix form $AX = B$, the system is written as follows.

$$\underbrace{\begin{bmatrix} 1 & -2 & 1 \\ 0 & 1 & 2 \\ 2 & 3 & -2 \end{bmatrix}}_{\text{Coefficient matrix}} \begin{bmatrix} x_1 \\ x_2 \\ x_3 \end{bmatrix} = \underbrace{\begin{bmatrix} -4 \\ 4 \\ 2 \end{bmatrix}}_{\text{Constant matrix}}$$

STUDY TIP

The notation $[A \,\vdots\, B]$ represents the augmented matrix formed when matrix B is adjoined to matrix A. The notation $[I \,\vdots\, X]$ represents the reduced row-echelon form of the augmented matrix that yields the solution of the system.

b. The augmented matrix is

$$[A \,\vdots\, B] = \begin{bmatrix} 1 & -2 & 1 & \vdots & -4 \\ 0 & 1 & 2 & \vdots & 4 \\ 2 & 3 & -2 & \vdots & 2 \end{bmatrix}.$$

Using Gauss-Jordan elimination, you can rewrite this equation as

$$[I \,\vdots\, X] = \begin{bmatrix} 1 & 0 & 0 & \vdots & -1 \\ 0 & 1 & 0 & \vdots & 2 \\ 0 & 0 & 1 & \vdots & 1 \end{bmatrix}.$$

So, the solution of the matrix equation is

$$X = \begin{bmatrix} x_1 \\ x_2 \\ x_3 \end{bmatrix} = \begin{bmatrix} -1 \\ 2 \\ 1 \end{bmatrix}.$$

✓CHECKPOINT   Now try Exercise 35.

EXAMPLE 11   LONG-DISTANCE PHONE PLANS

Make a Decision The charges (in dollars per minute) of two long-distance telephone companies are shown in the table.

	Company A	Company B
In-state	0.07	0.095
State-to-state	0.10	0.08
International	0.28	0.25

You plan to use 120 minutes on in-state long-distance calls, 80 minutes on state-to-state calls, and 20 minutes on international calls. Use matrices to determine which company you should choose to be your long-distance carrier.

SOLUTION

The charges C and amounts of time T spent on the phone can be written in matrix form as

$$C = \begin{bmatrix} 0.07 & 0.095 \\ 0.10 & 0.08 \\ 0.28 & 0.25 \end{bmatrix} \quad \text{and} \quad T = \begin{bmatrix} 120 & 80 & 20 \end{bmatrix}.$$

The total amount that each company charges is given by the product

$$TC = \begin{bmatrix} 120 & 80 & 20 \end{bmatrix} \begin{bmatrix} 0.07 & 0.095 \\ 0.10 & 0.08 \\ 0.28 & 0.25 \end{bmatrix} = \begin{bmatrix} 22 & 22.8 \end{bmatrix}.$$

Company A charges $22 for the calls and Company B charges $22.80. Company A charges less for the calling pattern, so you should choose Company A.

✓CHECKPOINT   Now try Exercise 57.

Discussing the Concept | Matrix Multiplication

Discuss the matrix order requirements for multiplication of two matrices. Determine which of the following matrix multiplications AB is (are) defined. For each case in which AB is defined, what is the order of the resulting matrix?

a. A is of order 1×3
   B is of order 2×1
c. A is of order 3×4
   B is of order 4×2

b. A is of order 2×3
   B is of order 2×3
d. A is of order 3×1
   B is of order 3×3

Discuss why matrix multiplication is not, in general, commutative. Give an example of two 2×2 matrices such that $AB \neq BA$. Find an example of two 2×2 matrices such that $AB = BA$.

5.2 Warm Up

The following warm-up exercises involve skills that were covered in earlier sections. You will use these skills in the exercise set for this section. For additional help, review Sections R1.2 and 5.1.

In Exercises 1 and 2, evaluate the expression.

1. $-3\left(-\frac{5}{6}\right) + 10\left(-\frac{3}{4}\right)$

2. $-22\left(\frac{5}{2}\right) + 6(8)$

In Exercises 3 and 4, determine whether the matrix is in *reduced* row-echelon form.

3. $\begin{bmatrix} 0 & 1 & 0 & -5 \\ 1 & 0 & 3 & 2 \\ 0 & 0 & 1 & 0 \end{bmatrix}$

4. $\begin{bmatrix} 1 & 0 & 0 & 2 & 3 \\ 0 & 0 & 0 & 0 & 0 \\ 0 & 1 & 1 & 3 & 10 \end{bmatrix}$

In Exercises 5 and 6, write the augmented matrix for the system of linear equations.

5. $\begin{cases} -5x + 10y = 12 \\ 7x - 3y = 0 \end{cases}$

6. $\begin{cases} 10x + 15y - 9z = 42 \\ 6x - 5y = 0 \end{cases}$

In Exercises 7–10, solve the system of linear equations represented by the augmented matrix.

7. $\begin{bmatrix} 1 & 0 & \vdots & 0 \\ 0 & 1 & \vdots & 2 \end{bmatrix}$

8. $\begin{bmatrix} 1 & 0 & -1 & \vdots & 2 \\ 0 & 1 & 1 & \vdots & 3 \end{bmatrix}$

9. $\begin{bmatrix} 1 & 2 & 1 & \vdots & 0 \\ 0 & 0 & 1 & \vdots & -1 \\ 0 & 0 & 0 & \vdots & 0 \end{bmatrix}$

10. $\begin{bmatrix} 1 & -1 & 0 & \vdots & 3 \\ 0 & 1 & -2 & \vdots & 1 \\ 0 & 0 & 1 & \vdots & -1 \end{bmatrix}$

5.2 Exercises

In Exercises 1–4, find x and y.

1. $\begin{bmatrix} 4 & x \\ -1 & y \end{bmatrix} = \begin{bmatrix} 4 & -3 \\ -1 & 2 \end{bmatrix}$

2. $\begin{bmatrix} x & -7 \\ 9 & y \end{bmatrix} = \begin{bmatrix} 5 & -7 \\ 9 & -8 \end{bmatrix}$

3. $\begin{bmatrix} -4 & 3 \\ 6 & -1 \\ 8 & 2 \\ 5 & 9 \end{bmatrix} = \begin{bmatrix} x-2 & 3 \\ 6 & -1 \\ 8 & -x \\ 5 & 2y-1 \end{bmatrix}$

4. $\begin{bmatrix} x+2 & 8 & -3 \\ 1 & 2y & 2x \\ 7 & -2 & y+2 \end{bmatrix} = \begin{bmatrix} 2x+6 & 8 & -3 \\ 1 & 18 & -8 \\ 7 & -2 & 11 \end{bmatrix}$

In Exercises 5–10, find (a) $A + B$, (b) $A - B$, (c) $3A$, and (d) $3A - 2B$.

5. $A = \begin{bmatrix} 5 & -2 \\ 3 & 1 \end{bmatrix}, B = \begin{bmatrix} 3 & 1 \\ -2 & 6 \end{bmatrix}$

6. $A = \begin{bmatrix} 7 & 4 \\ -4 & 5 \end{bmatrix}, B = \begin{bmatrix} -3 & 1 \\ 8 & -4 \end{bmatrix}$

7. $A = \begin{bmatrix} 6 & -1 \\ 2 & 4 \\ -3 & 5 \end{bmatrix}, B = \begin{bmatrix} 1 & 4 \\ -1 & 5 \\ 1 & 10 \end{bmatrix}$

8. $A = \begin{bmatrix} 6 & 8 & -3 & 2 & 1 \\ -4 & 2 & 1 & 5 & -2 \end{bmatrix}$,

$B = \begin{bmatrix} 6 & 0 & 4 & -1 & 3 \\ 4 & 5 & -2 & 1 & 2 \end{bmatrix}$

9. $A = \begin{bmatrix} 2 & 2 & -1 \\ 1 & 1 & -2 \\ 1 & -1 & 3 \end{bmatrix}, B = \begin{bmatrix} 1 & 1 & -1 \\ -3 & 4 & 9 \\ 0 & -7 & 8 \end{bmatrix}$

10. $A = \begin{bmatrix} 3 \\ 2 \\ -1 \end{bmatrix}, B = \begin{bmatrix} -4 \\ 6 \\ 2 \end{bmatrix}$

In Exercises 11–16, evaluate the expression.

11. $\begin{bmatrix} -5 & 0 \\ 3 & -6 \end{bmatrix} + \begin{bmatrix} 7 & 1 \\ -2 & -1 \end{bmatrix} + \begin{bmatrix} -10 & -8 \\ 14 & 6 \end{bmatrix}$

12. $\begin{bmatrix} 6 & 8 \\ -1 & 0 \end{bmatrix} + \begin{bmatrix} 0 & 5 \\ -3 & -1 \end{bmatrix} + \begin{bmatrix} -11 & -7 \\ 2 & -1 \end{bmatrix}$

13. $4\left(\begin{bmatrix} -4 & 0 & 1 \\ 0 & 2 & 3 \end{bmatrix} - \begin{bmatrix} 2 & 1 & -2 \\ 3 & -6 & 0 \end{bmatrix} \right)$

14. $\frac{1}{2}([5 \quad -2 \quad 4 \quad 0] + [14 \quad 6 \quad -18 \quad 9])$

15. $-3\left(\begin{bmatrix} 0 & -3 \\ 7 & 2 \end{bmatrix} + \begin{bmatrix} -6 & 3 \\ 8 & 1 \end{bmatrix} \right) - 2\begin{bmatrix} 4 & -4 \\ 7 & -9 \end{bmatrix}$

16. $-1\begin{bmatrix} 4 & 11 \\ -2 & -1 \\ 9 & 3 \end{bmatrix} + \frac{1}{6}\left(\begin{bmatrix} -5 & -1 \\ 3 & 4 \\ 0 & 13 \end{bmatrix} + \begin{bmatrix} 7 & 5 \\ -9 & -1 \\ 6 & -1 \end{bmatrix} \right)$

In Exercises 17–20, solve for X when

$$A = \begin{bmatrix} -2 & -1 \\ 1 & 0 \\ 3 & -4 \end{bmatrix} \quad \text{and} \quad B = \begin{bmatrix} 0 & 3 \\ 2 & 0 \\ -4 & -1 \end{bmatrix}.$$

17. $X = 3A - 2B$

18. $2X = 2A - B$

19. $2X + 3A = B$

20. $2A + 4B = -2X$

In Exercises 21–28, find AB, if possible.

21. $A = \begin{bmatrix} 3 & -2 \\ 4 & 5 \\ 1 & -1 \end{bmatrix}$, $B = \begin{bmatrix} -1 & 4 & -2 & 5 \\ 2 & 1 & 3 & -1 \end{bmatrix}$

22. $A = \begin{bmatrix} 0 & -1 & 0 \\ 4 & 0 & 2 \\ 8 & -1 & 7 \end{bmatrix}$, $B = \begin{bmatrix} 2 & 1 \\ -3 & 4 \\ 1 & 6 \end{bmatrix}$

23. $A = \begin{bmatrix} -1 & 3 \\ 4 & -5 \\ 0 & 2 \end{bmatrix}$, $B = \begin{bmatrix} 1 & 2 \\ 0 & 7 \end{bmatrix}$

24. $A = \begin{bmatrix} 1 & 0 & 0 \\ 0 & 4 & 0 \\ 0 & 0 & -2 \end{bmatrix}$, $B = \begin{bmatrix} 3 & 0 & 0 \\ 0 & -1 & 0 \\ 0 & 0 & 5 \end{bmatrix}$

25. $A = \begin{bmatrix} 5 & 0 & 0 \\ 0 & -8 & 0 \\ 0 & 0 & 7 \end{bmatrix}$, $B = \begin{bmatrix} \frac{1}{5} & 0 & 0 \\ 0 & -\frac{1}{8} & 0 \\ 0 & 0 & \frac{1}{2} \end{bmatrix}$

26. $A = \begin{bmatrix} 0 & 1 & 0 \\ 3 & 0 & 2 \\ 5 & 0 & 0 \end{bmatrix}$, $B = \begin{bmatrix} 4 \\ -2 \\ 0 \\ 1 \end{bmatrix}$

27. $A = \begin{bmatrix} 6 \\ -2 \\ 1 \\ 6 \end{bmatrix}$, $B = [10 \quad 12]$

28. $A = \begin{bmatrix} 1 & 0 & 3 & -2 & 4 \\ 6 & 13 & 8 & -17 & 10 \end{bmatrix}$, $B = \begin{bmatrix} 1 & 6 \\ 4 & 2 \end{bmatrix}$

In Exercises 29–34, find (a) AB, (b) BA, and, if possible, (c) A^2. (*Note:* $A^2 = AA$.)

29. $A = \begin{bmatrix} 1 & 2 \\ 4 & 2 \end{bmatrix}$, $B = \begin{bmatrix} 2 & -1 \\ -1 & 8 \end{bmatrix}$

30. $A = \begin{bmatrix} 2 & -1 \\ 1 & 4 \end{bmatrix}$, $B = \begin{bmatrix} 0 & 0 \\ 3 & -3 \end{bmatrix}$

31. $A = \begin{bmatrix} -1 & 2 & 3 \\ 4 & 1 & -1 \end{bmatrix}$, $B = \begin{bmatrix} 1 & 3 \\ -1 & -2 \\ 2 & 4 \end{bmatrix}$

32. $A = \begin{bmatrix} 1 & -1 & 7 \\ 2 & -1 & 8 \\ 3 & 1 & -1 \end{bmatrix}$, $B = \begin{bmatrix} 1 & 1 & 2 \\ 2 & 1 & 1 \\ 1 & -3 & 2 \end{bmatrix}$

33. $A = [-4 \quad 2 \quad 3]$, $B = \begin{bmatrix} 1 \\ 0 \\ 5 \end{bmatrix}$

34. $A = [3 \quad 2 \quad 1 \quad 0]$, $B = \begin{bmatrix} 2 \\ 3 \\ 1 \\ 0 \end{bmatrix}$

In Exercises 35–40, (a) write the system of linear equations as a matrix equation $AX = B$, and (b) use Gauss-Jordan elimination on the augmented matrix $[A \ \vdots \ B]$ to solve for the matrix X.

35. $\begin{cases} -x + y = 4 \\ -2x + y = 0 \end{cases}$ **36.** $\begin{cases} 2x + 3y = 5 \\ x + 4y = 10 \end{cases}$

37. $\begin{cases} x + 2y = 3 \\ 3x - y = 2 \end{cases}$

38. $\begin{cases} 2x - 4y + z = 0 \\ -x + 3y + z = 1 \\ x + y = 3 \end{cases}$

39. $\begin{cases} x - 2y + 3z = 9 \\ -x + 3y - z = -6 \\ 2x - 5y + 5z = 17 \end{cases}$

40. $\begin{cases} x + y - 3z = -1 \\ -x + 2y = 1 \\ -y + z = 0 \end{cases}$

41. **Factory Production** A sunglass corporation has four factories, each of which manufactures two products. The number of units of product i produced at factory j in one day is represented by a_{ij} in the matrix

$$A = \begin{bmatrix} 100 & 120 & 60 & 40 \\ 140 & 160 & 200 & 80 \end{bmatrix}.$$

Find the production levels if production is increased by 15%. (*Hint:* Because an increase of 15% corresponds to $100\% + 15\%$, multiply the matrix by 1.15.)

42. **Factory Production** A tire corporation has three factories, each of which manufactures two products. The number of units of product i produced at factory j in one day is represented by a_{ij} in the matrix

$$A = \begin{bmatrix} 80 & 120 & 140 \\ 40 & 100 & 80 \end{bmatrix}.$$

Find the production levels if production is decreased by 5%. (*Hint:* Because a decrease of 5% corresponds to $100\% - 5\%$, multiply the matrix by 0.95.)

43. **Hotel Pricing** A convention planning service has identified three suitable hotels for a convention. The quoted room rates are for single, double, triple, and quadruple occupancy. The current cost for each type of room by hotel is represented by the matrix A.

$$
\begin{array}{c}
\text{Hotel} \quad \text{Hotel} \quad \text{Hotel} \\
x \qquad\; y \qquad\; z \\
A = \begin{bmatrix} 85 & 92 & 110 \\ 100 & 120 & 130 \\ 110 & 130 & 140 \\ 110 & 140 & 155 \end{bmatrix}
\begin{array}{l} \text{Single} \\ \text{Double} \\ \text{Triple} \\ \text{Quadruple} \end{array}
\end{array} \text{Occupancy}
$$

If room rates are guaranteed not to increase by more than 15% by the time of the convention, what is the maximum rate per room per hotel?

44. **Vacation Packages** A vacation service has identified four resort hotels with a special all-inclusive package (room and meals included) at a popular travel destination. The quoted room rates are for double or family (maximum of four people) occupancy for 5 days and 4 nights. The current cost for each type of room by hotel is represented by the matrix A.

$$
\begin{array}{c}
\text{Hotel} \quad \text{Hotel} \quad \text{Hotel} \quad \text{Hotel} \\
w \qquad\; x \qquad\; y \qquad\; z \\
A = \begin{bmatrix} 615 & 670 & 740 & 990 \\ 995 & 1030 & 1180 & 1105 \end{bmatrix}
\begin{array}{l} \text{Double} \\ \text{Family} \end{array}
\end{array} \text{Occupancy}
$$

If room rates are guaranteed not to increase by more than 12% by next season, what is the maximum rate per package per hotel?

45. **Inventory Levels** A company sells five different models of computers through three retail outlets. The inventories of the five models at the three outlets are given by the matrix S.

$$
\begin{array}{c}
\text{Model} \\
\begin{array}{ccccc} A & B & C & D & E \end{array} \\
S = \begin{bmatrix} 3 & 2 & 2 & 3 & 0 \\ 0 & 2 & 3 & 4 & 3 \\ 4 & 2 & 1 & 3 & 2 \end{bmatrix} \begin{array}{l} 1 \\ 2 \\ 3 \end{array} \text{Outlet}
\end{array}
$$

The wholesale and retail prices for each model are given by the matrix T.

$$
\begin{array}{c}
\text{Price} \\
\begin{array}{cc} \text{Wholesale} & \text{Retail} \end{array} \\
T = \begin{bmatrix} \$900 & \$1200 \\ \$1200 & \$1450 \\ \$1400 & \$1650 \\ \$2650 & \$3250 \\ \$3050 & \$3375 \end{bmatrix} \begin{array}{l} A \\ B \\ C \\ D \\ E \end{array} \text{Model}
\end{array}
$$

(a) What is the total retail price of the inventory at Outlet 1?

(b) What is the total wholesale price of the inventory at Outlet 3?

(c) Compute the product ST and interpret the result in the context of the problem.

46. **Labor/Wage Requirements** A company that manufactures boats has the following labor-hour and wage requirements.

Labor-Hour Requirements (per boat)

$$
\begin{array}{c}
\text{Department} \\
\begin{array}{ccc} \text{Cutting} & \text{Assembly} & \text{Packaging} \end{array} \\
S = \begin{bmatrix} 1.0 \text{ hour} & 0.5 \text{ hour} & 0.2 \text{ hour} \\ 1.6 \text{ hours} & 1.0 \text{ hour} & 0.2 \text{ hour} \\ 2.5 \text{ hours} & 2.0 \text{ hours} & 0.4 \text{ hour} \end{bmatrix} \begin{array}{l} \text{Small} \\ \text{Medium} \\ \text{Large} \end{array}
\begin{array}{l} \text{Boat} \\ \text{size} \end{array}
\end{array}
$$

Wage Requirements (per hour)

$$
\begin{array}{c}
\text{Plant} \\
\begin{array}{cc} A & B \end{array} \\
T = \begin{bmatrix} \$5 & \$12 \\ \$10 & \$9 \\ \$7 & \$8 \end{bmatrix} \begin{array}{l} \text{Cutting} \\ \text{Assembly} \\ \text{Packaging} \end{array} \text{Department}
\end{array}
$$

(a) What is the labor cost for a medium boat at Plant B?

(b) What is the labor cost for a large boat at Plant A?

(c) Compute ST and interpret the result.

Think About It   In Exercises 47–54, let matrices A, B, C, and D be of orders 2×3, 2×3, 3×2, and 2×2, respectively. Determine whether the matrices are of proper order to perform the operation(s). If so, give the order of the answer.

47. $A + 2C$

48. $B - 3C$

49. AB

50. BC

51. $BC - D$

52. $CB - D$

53. $D(A - 3B)$

54. $(BC - D)A$

IN THE NEWS

Not Sure, Think I Can or Know I Can: Which One Are You?

One of the more interesting findings of NAIC's [National Association of Investors Corporation] first Voice of the American Shareholder survey is that although American investors aren't a homogeneous group—they have differing levels of enjoyment, confidence, knowledge, perspectives on the economy, online experiences and reliance on financial professionals—they tend to fall somewhat neatly into three distinct groups. Members of these groups, which for now we'll call the Not Sure I Cans, Think I Cans and Know I Cans, share many behaviors, experiences and attitudes that distinguish themselves from other investors. . . .

The largest group is the Not Sures, accounting for 38 percent of the shareholder population. "Investing for this group does not hold much appeal," Harris Interactive says. "These are the shareholders who do not enjoy investing, are not active regarding investing either online or offline, and feel less knowledgeable and confident about investing and their abilities."

The Know I Cans are the second-largest group at 33 percent. "These are the active and enthusiastic shareholders," Harris says, "These shareholders truly enjoy investing, are highly active both online and offline, and feel very knowledgeable and confident about their abilities."

The Think I Cans constituted the remaining 29 percent. "These are shareholders who hold a solid 'middle ground' when it comes to investing," the market research and consulting firm says. "Investing is somewhat enjoyable for them, and they feel somewhat knowledgeable and are moderately active online and offline."

There were some striking differences among the three groups in several categories. The average amount invested in individual stocks was much higher for the Know I Cans ($102,931) than the Think I Cans ($44,541) and the Not Sures ($25,897). But the Not Sures had more invested in stock mutual funds (an average of $117,543) than the Know I Cans ($84,171) and Think I Cans ($57,154). . . .

"Not sure, think I can or know I can: Which one are you?" Better Investing, *February 2004, 32.*

The average amounts invested in individual stocks and mutual funds by the three types of investors are given by the matrix A.

$$A = \begin{bmatrix} 102{,}931 & 84{,}171 \\ 44{,}541 & 57{,}154 \\ 25{,}897 & 117{,}543 \end{bmatrix} \begin{matrix} \text{"Know I Can" investors} \\ \text{"Think I Can" investors} \\ \text{"Not Sure I Can" investors} \end{matrix}$$

with columns labeled *Individual stocks* and *Mutual funds*.

Assume the interest rate for the individual stocks is 9.3% simple annual interest and the interest rate for the mutual funds is 8.5% simple annual interest. The matrix P gives the interest rates for each stock category.

$$P = \begin{bmatrix} 0.093 \\ 0.085 \end{bmatrix} \begin{matrix} \text{Individual stocks} \\ \text{Mutual funds} \end{matrix}$$

with column labeled *Interest rate*.

55. Compute AP.

56. Interpret the product AP.

57. *Make a Decision*: **Long-Distance Plans**   You are choosing between two monthly long-distance phone plans offered by two different companies. Company A charges $0.05 per minute for in-state calls, $0.12 per minute for state-to-state calls, and $0.30 per minute for international calls. Company B charges $0.085 per minute for in-state calls, $0.10 per minute for state-to-state calls, and $0.25 per minute for international calls. In a month, you normally use 20 minutes on in-state calls, 60 minutes on state-to-state calls, and 30 minutes on international calls.

(a) Write a matrix C that represents the charges for each type of call i by each company j. State what each entry c_{ij} of the matrix represents.

(b) Write a matrix T that represents the times spent on the phone for each type of call. State what each entry t_{ij} of the matrix represents.

(c) Find the product TC and state what each entry of the matrix represents.

(d) Which company should you choose? Explain.

58. **Contract Bonuses** Professional athletes frequently have bonus or incentive clauses in their contracts. For example, a defensive football player might receive a bonus for a sack, an interception, and/or a key tackle. In one contract, a sack is worth $2000, an interception is worth $1000, and a key tackle is worth $800. Use matrices to calculate the bonuses for defensive players A, B, and C under this contract if the following table shows the numbers of sacks, interceptions, and key tackles in a game.

Player	Sacks	Interceptions	Key tackles
Player A	3	0	4
Player B	1	2	5
Player C	2	3	3

59. **Voting Preference** The matrix

From

$$
\begin{array}{cc}
 & \begin{array}{ccc} R & D & I \end{array} \\
P = & \begin{bmatrix} 0.6 & 0.1 & 0.1 \\ 0.2 & 0.7 & 0.1 \\ 0.2 & 0.2 & 0.8 \end{bmatrix} \begin{array}{c} R \\ D \\ I \end{array}
\end{array} \right\} \text{To}
$$

is called a *stochastic matrix*. Each entry p_{ij} $(i \ne j)$ represents the proportion of the voting population that changes from party i to party j, and p_{ii} represents the proportion that remains loyal to the party from one election to the next. Use a graphing utility to find P^2. (This matrix gives the transition probabilities from the first election to the third.)

60. *Make a Decision*: **Voting Preference** Use a graphing utility to find P^3, P^4, P^5, P^6, P^7, and P^8 for the matrix given in Exercise 59. Can you detect a pattern as P is raised to higher and higher powers?

Make a Decision In Exercises 61 and 62, find a matrix B such that AB is the identity matrix. Is there more than one correct result?

61. $A = \begin{bmatrix} 1 & 3 \\ 1 & 2 \end{bmatrix}$ 62. $A = \begin{bmatrix} 2 & 1 \\ 5 & 2 \end{bmatrix}$

63. If a, b, and c are real numbers such that $c \ne 0$ and $ac = bc$, then $a = b$. However, if A, B, and C are matrices such that $AC = BC$, then A is *not* necessarily equal to B. Illustrate this using the following matrices.

$$
A = \begin{bmatrix} 1 & 2 & 3 \\ 0 & 5 & 4 \\ 3 & -2 & 1 \end{bmatrix}, \ B = \begin{bmatrix} 4 & -6 & 3 \\ 5 & 4 & 4 \\ -1 & 0 & 1 \end{bmatrix},
$$

and $C = \begin{bmatrix} 0 & 0 & 0 \\ 0 & 0 & 0 \\ 4 & -2 & 3 \end{bmatrix}$

64. If a and b are real numbers such that $ab = 0$, then $a = 0$ or $b = 0$. However, if A and B are matrices such that $AB = O$, then it is *not* necessarily true that $A = O$ or $B = O$. Illustrate this using the following matrices.

$$
A = \begin{bmatrix} 3 & 3 \\ 4 & 4 \end{bmatrix} \quad \text{and} \quad B = \begin{bmatrix} 1 & -1 \\ -1 & 1 \end{bmatrix}
$$

Find another example of two nonzero matrices whose product is the zero matrix.

65. *Make a Decision*: **Cable Television** Two competing companies offer cable television to a city with 100,000 households. Gold Cable Company has 25,000 subscribers and Galaxy Cable Company has 30,000 subscribers. (The other 45,000 households do not subscribe.) The percent changes in cable subscriptions each year are shown in the matrix below.

Percent Changes

	From Gold	From Galaxy	From Non-subscriber
To Gold	0.70	0.15	0.15
To Galaxy	0.20	0.80	0.15
To Nonsubscriber	0.10	0.05	0.70

Percent Changes

(a) Find the number of subscribers each company will have in 1 year using matrix multiplication. Explain how you obtained your answer.

(b) Find the number of subscribers each company will have in 2 years using matrix multiplication. Explain how you obtained your answer.

(c) Find the number of subscribers each company will have in 3 years using matrix multiplication. Explain how you obtained your answer.

(d) What is happening to the number of subscribers to each company? What is happening to the number of nonsubscribers?

5.3 The Inverse of a Square Matrix

Objectives

- Verify that a matrix is the inverse of a given matrix.
- Find the inverse of a matrix.
- Find the inverse of a 2 × 2 matrix using a formula.
- Use an inverse matrix to solve a system of linear equations.

The Inverse of a Matrix

This section further develops the algebra of matrices. To begin, consider the real number equation $ax = b$. To solve this equation for x, multiply each side of the equation by a^{-1} (provided $a \neq 0$).

$$ax = b$$
$$(a^{-1}a)x = a^{-1}b$$
$$(1)x = a^{-1}b$$
$$x = a^{-1}b$$

The number a^{-1} is called the *multiplicative inverse of a* because it has the property that $a^{-1}a = 1$. The definition of the multiplicative inverse of a matrix is similar.

Definition of the Inverse of a Square Matrix

Let A be an $n \times n$ matrix and let I_n be the $n \times n$ identity matrix. If there exists a matrix A^{-1} such that

$$AA^{-1} = I_n = A^{-1}A$$

then A^{-1} is called the **inverse** of A. (The symbol A^{-1} is read "A inverse.")

EXAMPLE 1 THE INVERSE OF A MATRIX

Show that B is the inverse of A, where

$$A = \begin{bmatrix} -1 & 2 \\ -1 & 1 \end{bmatrix} \quad \text{and} \quad B = \begin{bmatrix} 1 & -2 \\ 1 & -1 \end{bmatrix}.$$

SOLUTION

To show that B is the inverse of A, show that $AB = I = BA$, as follows.

$$AB = \begin{bmatrix} -1 & 2 \\ -1 & 1 \end{bmatrix}\begin{bmatrix} 1 & -2 \\ 1 & -1 \end{bmatrix} = \begin{bmatrix} -1+2 & 2-2 \\ -1+1 & 2-1 \end{bmatrix} = \begin{bmatrix} 1 & 0 \\ 0 & 1 \end{bmatrix}$$

$$BA = \begin{bmatrix} 1 & -2 \\ 1 & -1 \end{bmatrix}\begin{bmatrix} -1 & 2 \\ -1 & 1 \end{bmatrix} = \begin{bmatrix} -1+2 & 2-2 \\ -1+1 & 2-1 \end{bmatrix} = \begin{bmatrix} 1 & 0 \\ 0 & 1 \end{bmatrix}$$

✓**CHECKPOINT** Now try Exercise 1.

STUDY TIP

Recall that it is not always true that $AB = BA$, even if both products are defined. However, if A and B are both square matrices and $AB = I_n$, it can be shown that $BA = I_n$. So, in Example 1, you need only check that $AB = I_2$.

If a matrix A has an inverse, A is called **invertible** (or **nonsingular**); otherwise, A is called **singular.** A nonsquare matrix cannot have an inverse. To see this, note that if A is of order $m \times n$ and B is of order $n \times m$ (where $m \neq n$), the products AB and BA are of different orders and therefore cannot be equal to each other. Not all square matrices have inverses (see the matrix at the bottom of page 352). If, however, a matrix does have an inverse, that inverse is unique. The following example shows how to use a system of equations to find an inverse.

EXAMPLE 2 FINDING THE INVERSE OF A MATRIX

Find the inverse of the matrix

$$A = \begin{bmatrix} 1 & 4 \\ -1 & -3 \end{bmatrix}.$$

SOLUTION

To find the inverse of A, try to solve the matrix equation $AX = I$ for X.

$$\underset{A}{\begin{bmatrix} 1 & 4 \\ -1 & -3 \end{bmatrix}} \underset{X}{\begin{bmatrix} x_{11} & x_{12} \\ x_{21} & x_{22} \end{bmatrix}} = \underset{I}{\begin{bmatrix} 1 & 0 \\ 0 & 1 \end{bmatrix}} \qquad \text{Write matrix equation.}$$

$$\begin{bmatrix} x_{11} + 4x_{21} & x_{12} + 4x_{22} \\ -x_{11} - 3x_{21} & -x_{12} - 3x_{22} \end{bmatrix} = \begin{bmatrix} 1 & 0 \\ 0 & 1 \end{bmatrix} \qquad \text{Multiply } A \text{ times } X.$$

Equating corresponding entries, you obtain the following two systems of linear equations.

$$\begin{cases} x_{11} + 4x_{21} = 1 \\ -x_{11} - 3x_{21} = 0 \end{cases} \qquad \begin{cases} x_{12} + 4x_{22} = 0 \\ -x_{12} - 3x_{22} = 1 \end{cases}$$

You can solve these systems using the methods learned in Chapter 4. From the first system you can determine that $x_{11} = -3$ and $x_{21} = 1$, and from the second system you can determine that $x_{12} = -4$ and $x_{22} = 1$. So, the inverse of A is

$$X = A^{-1}$$

$$= \begin{bmatrix} -3 & -4 \\ 1 & 1 \end{bmatrix}.$$

You can use matrix multiplication to check this result.

CHECK

$$AA^{-1} = \begin{bmatrix} 1 & 4 \\ -1 & -3 \end{bmatrix} \begin{bmatrix} -3 & -4 \\ 1 & 1 \end{bmatrix}$$

$$= \begin{bmatrix} 1 & 0 \\ 0 & 1 \end{bmatrix} \checkmark$$

$$A^{-1}A = \begin{bmatrix} -3 & -4 \\ 1 & 1 \end{bmatrix} \begin{bmatrix} 1 & 4 \\ -1 & -3 \end{bmatrix}$$

$$= \begin{bmatrix} 1 & 0 \\ 0 & 1 \end{bmatrix} \checkmark$$

✓CHECKPOINT   Now try Exercise 11.

Finding Inverse Matrices

DISCOVERY

Select two 2×2 matrices A and B that have inverses. Calculate $(AB)^{-1}$ and then calculate $B^{-1}A^{-1}$ and $A^{-1}B^{-1}$. Make a conjecture about the inverse of the product of two invertible matrices.

In Example 2, note that the two systems of linear equations have the *same coefficient matrix A*. Rather than solve the two systems represented by

$$\begin{bmatrix} 1 & 4 & \vdots & 1 \\ -1 & -3 & \vdots & 0 \end{bmatrix} \quad \text{and} \quad \begin{bmatrix} 1 & 4 & \vdots & 0 \\ -1 & -3 & \vdots & 1 \end{bmatrix}$$

separately, you can solve them simultaneously by adjoining the identity matrix to the coefficient matrix to obtain

$$\begin{array}{cc} A & I \end{array}$$
$$\begin{bmatrix} 1 & 4 & \vdots & 1 & 0 \\ -1 & -3 & \vdots & 0 & 1 \end{bmatrix}.$$

Then, applying Gauss-Jordan elimination to this matrix, you can solve *both* systems with a single elimination process, as follows.

$$\begin{bmatrix} 1 & 4 & \vdots & 1 & 0 \\ -1 & -3 & \vdots & 0 & 1 \end{bmatrix}$$

$$R_1 + R_2 \rightarrow \begin{bmatrix} 1 & 4 & \vdots & 1 & 0 \\ 0 & 1 & \vdots & 1 & 1 \end{bmatrix}$$

$$-4R_2 + R_1 \rightarrow \begin{bmatrix} 1 & 0 & \vdots & -3 & -4 \\ 0 & 1 & \vdots & 1 & 1 \end{bmatrix}$$

So, from the "doubly augmented" matrix $[A : I]$, you obtained the matrix $[I : A^{-1}]$.

$$\begin{array}{cccc} A & & I & \end{array} \qquad \begin{array}{cccc} I & & A^{-1} \end{array}$$

$$\begin{bmatrix} 1 & 4 & \vdots & 1 & 0 \\ -1 & -3 & \vdots & 0 & 1 \end{bmatrix} \implies \begin{bmatrix} 1 & 0 & \vdots & -3 & -4 \\ 0 & 1 & \vdots & 1 & 1 \end{bmatrix}$$

This procedure (or algorithm) works for an arbitrary square matrix that has an inverse.

Finding an Inverse Matrix

Let A be a square matrix of order n.

1. Write the $n \times 2n$ matrix that consists of the given matrix A on the left and the $n \times n$ identity matrix I on the right to obtain $[A : I]$. Note that the matrices A and I are separated by a dotted line. This process is called **adjoining** the matrices A and I.

2. If possible, row reduce A to I using elementary row operations on the *entire* matrix $[A : I]$. The result will be the matrix $[I : A^{-1}]$. If this is not possible, A is not invertible.

3. Check your work by multiplying to see that $AA^{-1} = I = A^{-1}A$.

EXAMPLE 3   FINDING THE INVERSE OF A MATRIX

Find the inverse of the matrix $A = \begin{bmatrix} 1 & -1 & 0 \\ 1 & 0 & -1 \\ 6 & -2 & -3 \end{bmatrix}$.

SOLUTION

Begin by adjoining the identity matrix to A to form the matrix

$$[A \;\vdots\; I] = \begin{bmatrix} 1 & -1 & 0 & \vdots & 1 & 0 & 0 \\ 1 & 0 & -1 & \vdots & 0 & 1 & 0 \\ 6 & -2 & -3 & \vdots & 0 & 0 & 1 \end{bmatrix}.$$

Use elementary row operations to obtain the form $[I \;\vdots\; A^{-1}]$, as follows.

$$\begin{matrix} \\ -R_1 + R_2 \rightarrow \\ -6R_1 + R_3 \rightarrow \end{matrix} \begin{bmatrix} 1 & -1 & 0 & \vdots & 1 & 0 & 0 \\ 0 & 1 & -1 & \vdots & -1 & 1 & 0 \\ 0 & 4 & -3 & \vdots & -6 & 0 & 1 \end{bmatrix}$$

$$\begin{matrix} R_2 + R_1 \rightarrow \\ \\ -4R_2 + R_3 \rightarrow \end{matrix} \begin{bmatrix} 1 & 0 & -1 & \vdots & 0 & 1 & 0 \\ 0 & 1 & -1 & \vdots & -1 & 1 & 0 \\ 0 & 0 & 1 & \vdots & -2 & -4 & 1 \end{bmatrix}$$

$$\begin{matrix} R_3 + R_1 \rightarrow \\ R_3 + R_2 \rightarrow \\ \\ \end{matrix} \begin{bmatrix} 1 & 0 & 0 & \vdots & -2 & -3 & 1 \\ 0 & 1 & 0 & \vdots & -3 & -3 & 1 \\ 0 & 0 & 1 & \vdots & -2 & -4 & 1 \end{bmatrix}$$

So, the matrix A is invertible and its inverse is

$$A^{-1} = \begin{bmatrix} -2 & -3 & 1 \\ -3 & -3 & 1 \\ -2 & -4 & 1 \end{bmatrix}.$$

Try confirming this result by multiplying A and A^{-1} to obtain I.

✓CHECKPOINT   Now try Exercise 21.

The process shown in Example 3 applies to any $n \times n$ matrix A. If A has an inverse, this process will find it. When using this process, if the matrix A does not reduce to the identity matrix, then A does not have an inverse.

To confirm that matrix A shown below has no inverse, begin by adjoining the identity matrix to A to form the following.

$$A = \begin{bmatrix} 1 & 2 & 0 \\ 3 & -1 & 2 \\ -2 & 3 & -2 \end{bmatrix} \implies [A \;\vdots\; I] = \begin{bmatrix} 1 & 2 & 0 & \vdots & 1 & 0 & 0 \\ 3 & -1 & 2 & \vdots & 0 & 1 & 0 \\ -2 & 3 & -2 & \vdots & 0 & 0 & 1 \end{bmatrix}$$

Then use elementary row operations to obtain

$$\begin{bmatrix} 1 & 2 & 0 & \vdots & 1 & 0 & 0 \\ 0 & -7 & 2 & \vdots & -3 & 1 & 0 \\ 0 & 0 & 0 & \vdots & -2 & 1 & 1 \end{bmatrix}.$$

At this point in the elimination process, you can see that it is impossible to obtain the identity matrix I on the left. So, A is not invertible.

The Inverse of a 2 × 2 Matrix (Quick Method)

DISCOVERY

Use a graphing utility with matrix operations to find the inverse of the matrix

$$A = \begin{bmatrix} 1 & -3 \\ -2 & 6 \end{bmatrix}.$$

What message appears on the screen? Why does the graphing utility display an error message?

Using Gauss-Jordan elimination to find the inverse of a matrix works well (even as a computer technique) for matrices of order 3 × 3 or greater. For 2 × 2 matrices, however, many people prefer to use a formula for the inverse rather than Gauss-Jordan elimination. This simple formula, which works *only* for 2 × 2 matrices, is explained as follows. If A is a 2 × 2 matrix given by

$$A = \begin{bmatrix} a & b \\ c & d \end{bmatrix}$$

then A is invertible if and only if $ad - bc \neq 0$. Moreover, if $ad - bc \neq 0$, the inverse is given by

$$A^{-1} = \frac{1}{ad - bc} \begin{bmatrix} d & -b \\ -c & a \end{bmatrix}.$$

Try verifying this inverse by multiplication.

The denominator $ad - bc$ is called the *determinant* of the 2 × 2 matrix A. You will study determinants in the next section.

EXAMPLE 4   FINDING THE INVERSE OF A 2 × 2 MATRIX

Make a Decision Decide whether each matrix is invertible. If it is invertible, find the inverse of the matrix.

a. $A = \begin{bmatrix} 3 & -1 \\ -2 & 2 \end{bmatrix}$ **b.** $B = \begin{bmatrix} 3 & -1 \\ -6 & 2 \end{bmatrix}$

SOLUTION

a. For the matrix A, apply the formula for the determinant of a 2 × 2 matrix to obtain

$$ad - bc = 3(2) - (-1)(-2) = 4.$$

Because this quantity is not zero, the matrix is invertible. The inverse is formed by interchanging the entries on the main diagonal, changing the signs of the other two entries, and multiplying by the scalar $\frac{1}{4}$, as follows.

$$A^{-1} = \frac{1}{ad - bc} \begin{bmatrix} d & -b \\ -c & a \end{bmatrix} \qquad \text{Formula for inverse of a 2 × 2 matrix}$$

$$= \frac{1}{4} \begin{bmatrix} 2 & 1 \\ 2 & 3 \end{bmatrix} \qquad \text{Substitute for } a, b, c, d, \text{ and the determinant.}$$

$$= \begin{bmatrix} \frac{1}{4}(2) & \frac{1}{4}(1) \\ \frac{1}{4}(2) & \frac{1}{4}(3) \end{bmatrix} \qquad \text{Multiply by the scalar } \frac{1}{4}.$$

$$= \begin{bmatrix} \frac{1}{2} & \frac{1}{4} \\ \frac{1}{2} & \frac{3}{4} \end{bmatrix} \qquad \text{Simplify.}$$

b. For the matrix B, you have

$$ad - bc = 3(2) - (-1)(-6) = 0.$$

Because $ad - bc = 0$, B is not invertible.

✓**CHECKPOINT**   Now try Exercise 37.

Systems of Linear Equations

We know that a system of linear equations can have exactly one solution, infinitely many solutions, or no solution. If the coefficient matrix A of a *square* system (a system that has the same number of equations as variables) is invertible, then the system has a unique solution, which is defined as follows.

A System of Equations with a Unique Solution

If A is an invertible matrix, then the system of linear equations represented by $AX = B$ has a unique solution given by $X = A^{-1}B$.

EXAMPLE 5   SOLVING A SYSTEM OF EQUATIONS USING AN INVERSE MATRIX

Use an inverse matrix to solve the system.

$$\begin{cases} 2x + 3y + z = -1 \\ 3x + 3y + z = 1 \\ 2x + 4y + z = -2 \end{cases}$$

SOLUTION

Begin by writing the system in the matrix form $AX = B$.

$$\begin{bmatrix} 2 & 3 & 1 \\ 3 & 3 & 1 \\ 2 & 4 & 1 \end{bmatrix} \begin{bmatrix} x \\ y \\ z \end{bmatrix} = \begin{bmatrix} -1 \\ 1 \\ -2 \end{bmatrix}$$

Next, use Gauss-Jordan elimination to find A^{-1}.

$$A^{-1} = \begin{bmatrix} -1 & 1 & 0 \\ -1 & 0 & 1 \\ 6 & -2 & -3 \end{bmatrix}$$

Finally, multiply B by A^{-1} on the left to obtain the solution.

$$X = A^{-1}B = \begin{bmatrix} -1 & 1 & 0 \\ -1 & 0 & 1 \\ 6 & -2 & -3 \end{bmatrix} \begin{bmatrix} -1 \\ 1 \\ -2 \end{bmatrix} = \begin{bmatrix} 2 \\ -1 \\ -2 \end{bmatrix}$$

So, the solution is $x = 2$, $y = -1$, and $z = -2$.

✔CHECKPOINT   Now try Exercise 47.

Discussing the Concept ǀ Methods of Problem Solving

Describe how the method used to solve Example 5 is similar to the method used to solve a simple equation such as

$2x = 10$.

How are the two methods different?

5.3 Warm Up

The following warm-up exercises involve skills that were covered in earlier sections. You will use these skills in the exercise set for this section. For additional help, review Sections 5.1 and 5.2.

In Exercises 1–8, perform the indicated matrix operations.

1. $4\begin{bmatrix} 1 & 6 \\ 0 & -4 \\ 12 & 2 \end{bmatrix}$

2. $\dfrac{1}{2}\begin{bmatrix} 11 & 10 & 48 \\ 1 & 0 & 16 \\ 0 & 2 & 8 \end{bmatrix}$

3. $\begin{bmatrix} 1 & -10 & 3 \\ 4 & 1 & 0 \end{bmatrix} - 2\begin{bmatrix} 3 & -4 & 8 \\ 0 & 7 & 1 \end{bmatrix}$

4. $\begin{bmatrix} 5 & 20 \\ -7 & 15 \end{bmatrix} - 3\begin{bmatrix} 6 & 3 \\ 4 & -2 \end{bmatrix}$

5. $\begin{bmatrix} 1 & -2 \\ -1 & 3 \end{bmatrix}\begin{bmatrix} 3 & 2 \\ 1 & 1 \end{bmatrix}$

6. $\begin{bmatrix} 1 & 0 \\ 0 & 1 \end{bmatrix}\begin{bmatrix} 6 & 5 \\ 3 & -2 \end{bmatrix}$

7. $\begin{bmatrix} 2 & 0 & 0 \\ 0 & -1 & 0 \\ 0 & 0 & 3 \end{bmatrix}\begin{bmatrix} \frac{1}{2} & 0 & 0 \\ 0 & -1 & 0 \\ 0 & 0 & \frac{1}{3} \end{bmatrix}$

8. $\begin{bmatrix} 1 & -1 & 0 \\ 1 & 0 & -1 \\ 6 & -2 & -3 \end{bmatrix}\begin{bmatrix} -2 & -3 & 1 \\ -3 & -3 & 1 \\ -2 & -4 & 1 \end{bmatrix}$

In Exercises 9 and 10, rewrite the matrix in reduced row-echelon form.

9. $\begin{bmatrix} 3 & -2 & 1 & 0 \\ 4 & -3 & 0 & 1 \end{bmatrix}$

10. $\begin{bmatrix} 1 & 1 & 2 & 1 & 0 & 0 \\ -1 & 0 & 3 & 0 & 1 & 0 \\ 1 & 2 & 8 & 0 & 0 & 1 \end{bmatrix}$

5.3 Exercises

In Exercises 1–10, show that B is the inverse of A.

1. $A = \begin{bmatrix} 7 & 4 \\ 5 & 3 \end{bmatrix}$, $B = \begin{bmatrix} 3 & -4 \\ -5 & 7 \end{bmatrix}$

2. $A = \begin{bmatrix} -4 & 1 \\ -9 & 2 \end{bmatrix}$, $B = \begin{bmatrix} 2 & -1 \\ 9 & -4 \end{bmatrix}$

3. $A = \begin{bmatrix} 2 & -1 \\ 5 & -4 \end{bmatrix}$, $B = \begin{bmatrix} \frac{4}{3} & -\frac{1}{3} \\ \frac{5}{3} & -\frac{2}{3} \end{bmatrix}$

4. $A = \begin{bmatrix} 1 & -2 \\ 3 & -10 \end{bmatrix}$, $B = \begin{bmatrix} \frac{5}{2} & -\frac{1}{2} \\ \frac{3}{4} & -\frac{1}{4} \end{bmatrix}$

5. $A = \begin{bmatrix} -2 & 2 & 3 \\ 1 & -1 & 0 \\ 0 & 1 & 4 \end{bmatrix}$, $B = \dfrac{1}{3}\begin{bmatrix} -4 & -5 & 3 \\ -4 & -8 & 3 \\ 1 & 2 & 0 \end{bmatrix}$

6. $A = \begin{bmatrix} 2 & -17 & 11 \\ -1 & 11 & -7 \\ 0 & 3 & -2 \end{bmatrix}$, $B = \begin{bmatrix} 1 & 1 & 2 \\ 2 & 4 & -3 \\ 3 & 6 & -5 \end{bmatrix}$

7. $A = \begin{bmatrix} -1 & 0 & 2 \\ 1 & -2 & 0 \\ 1 & 0 & 3 \end{bmatrix}$, $B = \dfrac{1}{10}\begin{bmatrix} -6 & 0 & 4 \\ -3 & -5 & 2 \\ 2 & 0 & 2 \end{bmatrix}$

8. $A = \begin{bmatrix} -1 & 1 & -3 \\ 2 & -1 & 4 \\ -1 & 1 & -2 \end{bmatrix}$, $B = \begin{bmatrix} 2 & 1 & -1 \\ 0 & 1 & 2 \\ -1 & 0 & 1 \end{bmatrix}$

9. $A = \begin{bmatrix} 2 & 0 & 1 & 1 \\ 3 & 0 & 0 & 1 \\ -1 & 1 & -2 & 1 \\ 4 & -1 & 1 & 0 \end{bmatrix}$,

$B = \begin{bmatrix} -1 & 2 & -1 & -1 \\ -4 & 9 & -5 & -6 \\ 0 & 1 & -1 & -1 \\ 3 & -5 & 3 & 3 \end{bmatrix}$

10. $A = \begin{bmatrix} -1 & 1 & 0 & -1 \\ 1 & -1 & 2 & 0 \\ -1 & 1 & 2 & 0 \\ 0 & -1 & 1 & 1 \end{bmatrix}$,

$B = \dfrac{1}{4}\begin{bmatrix} -4 & 1 & 1 & -4 \\ -4 & -1 & 3 & -4 \\ 0 & 1 & 1 & 0 \\ -4 & -2 & 2 & 0 \end{bmatrix}$

In Exercises 11–36, find the inverse of the matrix (if it exists).

11. $\begin{bmatrix} 8 & 4 \\ -2 & -2 \end{bmatrix}$

12. $\begin{bmatrix} 1 & 2 \\ 3 & 7 \end{bmatrix}$

13. $\begin{bmatrix} 2 & 3 \\ 6 & 9 \end{bmatrix}$

14. $\begin{bmatrix} -7 & 33 \\ 4 & -19 \end{bmatrix}$

15. $\begin{bmatrix} -1 & 1 \\ -2 & 1 \end{bmatrix}$

16. $\begin{bmatrix} 2 & 3 \\ 1 & 4 \end{bmatrix}$

17. $\begin{bmatrix} 0 & 4 \\ -3 & 6 \end{bmatrix}$

18. $\begin{bmatrix} 11 & 1 \\ -1 & 0 \end{bmatrix}$

19. $\begin{bmatrix} 2 & 7 & 1 \\ -3 & -9 & 2 \end{bmatrix}$

20. $\begin{bmatrix} -2 & 5 \\ 6 & -15 \\ 0 & 1 \end{bmatrix}$

21. $\begin{bmatrix} 1 & 1 & 1 \\ 3 & 5 & 4 \\ 3 & 6 & 5 \end{bmatrix}$

22. $\begin{bmatrix} 1 & 2 & 2 \\ 3 & 7 & 9 \\ -1 & -4 & -7 \end{bmatrix}$

23. $\begin{bmatrix} 1 & 2 & -1 \\ 3 & 7 & -10 \\ -5 & -7 & -15 \end{bmatrix}$

24. $\begin{bmatrix} 10 & 5 & -7 \\ -5 & 1 & 4 \\ 3 & 2 & -2 \end{bmatrix}$

25. $\begin{bmatrix} 1 & 1 & 2 \\ 3 & 1 & 0 \\ -2 & 0 & 3 \end{bmatrix}$

26. $\begin{bmatrix} 3 & 2 & 2 \\ 2 & 2 & 2 \\ -4 & 4 & 3 \end{bmatrix}$

27. $\begin{bmatrix} 3 & 0 & 0 \\ 0 & -2 & 0 \\ 0 & 0 & 4 \end{bmatrix}$

28. $\begin{bmatrix} 2 & 0 & 0 \\ 0 & 3 & 0 \\ 0 & 0 & 5 \end{bmatrix}$

29. $\begin{bmatrix} 1 & 0 & 0 \\ 3 & 4 & 0 \\ 2 & 5 & 5 \end{bmatrix}$

30. $\begin{bmatrix} 1 & 0 & 0 \\ 3 & 0 & 0 \\ 2 & 5 & 5 \end{bmatrix}$

31. $\begin{bmatrix} 1 & 0 & 3 & 0 \\ 0 & 2 & 0 & 4 \\ 1 & 0 & 3 & 0 \\ 0 & 2 & 0 & 4 \end{bmatrix}$

32. $\begin{bmatrix} 1 & 3 & -2 & 0 \\ 0 & 2 & 4 & 6 \\ 0 & 0 & -2 & 1 \\ 0 & 0 & 0 & 5 \end{bmatrix}$

33. $\begin{bmatrix} -8 & 0 & 0 & 0 \\ 0 & 1 & 0 & 0 \\ 0 & 0 & 4 & 0 \\ 0 & 0 & 0 & -5 \end{bmatrix}$

34. $\begin{bmatrix} -1 & 0 & 1 & 0 \\ 0 & 2 & 0 & -1 \\ 2 & 0 & -1 & 0 \\ 0 & -1 & 0 & 1 \end{bmatrix}$

35. $\begin{bmatrix} 1 & -2 & -1 & -2 \\ 3 & -5 & -2 & -3 \\ 2 & -5 & -2 & -5 \\ -1 & 4 & 4 & 11 \end{bmatrix}$

36. $\begin{bmatrix} 4 & 8 & -7 & 14 \\ 2 & 5 & -4 & 6 \\ 0 & 2 & 1 & -7 \\ 3 & 6 & -5 & 10 \end{bmatrix}$

In Exercises 37–42, use the formula on page 353 to find the inverse of the matrix (if it exists).

37. $\begin{bmatrix} 5 & -2 \\ 2 & 3 \end{bmatrix}$

38. $\begin{bmatrix} 7 & 12 \\ -8 & -5 \end{bmatrix}$

39. $\begin{bmatrix} -4 & -6 \\ 2 & 3 \end{bmatrix}$

40. $\begin{bmatrix} -12 & 3 \\ 5 & -2 \end{bmatrix}$

41. $\begin{bmatrix} \frac{7}{2} & -\frac{3}{4} \\ \frac{1}{5} & \frac{4}{5} \end{bmatrix}$

42. $\begin{bmatrix} -\frac{1}{4} & \frac{9}{4} \\ \frac{5}{3} & \frac{8}{9} \end{bmatrix}$

In Exercises 43–46, use the inverse matrix found in Exercise 15 to solve the system of linear equations.

43. $\begin{cases} -x + y = 4 \\ -2x + y = 0 \end{cases}$

44. $\begin{cases} -x + y = -3 \\ -2x + y = 5 \end{cases}$

45. $\begin{cases} -x + y = 20 \\ -2x + y = 10 \end{cases}$

46. $\begin{cases} -x + y = 0 \\ -2x + y = 7 \end{cases}$

In Exercises 47–50, use the inverse matrix found in Exercise 16 to solve the system of linear equations.

47. $\begin{cases} 2x + 3y = 5 \\ x + 4y = 10 \end{cases}$

48. $\begin{cases} 2x + 3y = 0 \\ x + 4y = 3 \end{cases}$

49. $\begin{cases} 2x + 3y = 4 \\ x + 4y = 2 \end{cases}$

50. $\begin{cases} 2x + 3y = 1 \\ x + 4y = -2 \end{cases}$

In Exercises 51 and 52, use the inverse matrix found in Exercise 35 to solve the system of linear equations.

51. $\begin{cases} x_1 - 2x_2 - x_3 - 2x_4 = 0 \\ 3x_1 - 5x_2 - 2x_3 - 3x_4 = 1 \\ 2x_1 - 5x_2 - 2x_3 - 5x_4 = -1 \\ -x_1 + 4x_2 + 4x_3 + 11x_4 = 2 \end{cases}$

52. $\begin{cases} x_1 - 2x_2 - x_3 - 2x_4 = 1 \\ 3x_1 - 5x_2 - 2x_3 - 3x_4 = -2 \\ 2x_1 - 5x_2 - 2x_3 - 5x_4 = 0 \\ -x_1 + 4x_2 + 4x_3 + 11x_4 = -3 \end{cases}$

In Exercises 53–60, use an inverse matrix to solve (if possible) the system of linear equations.

53. $\begin{cases} 3x + 4y = -2 \\ 5x + 3y = 4 \end{cases}$ **54.** $\begin{cases} 18x + 12y = 13 \\ 30x + 24y = 23 \end{cases}$

55. $\begin{cases} -0.4x + 0.8y = 1.6 \\ 2x - 4y = 5 \end{cases}$ **56.** $\begin{cases} 0.2x - 0.6y = 2.4 \\ -x + 1.4y = -8.8 \end{cases}$

57. $\begin{cases} -\frac{1}{4}x + \frac{3}{8}y = -2 \\ \frac{3}{2}x + \frac{3}{4}y = -12 \end{cases}$ **58.** $\begin{cases} \frac{5}{6}x - y = -20 \\ \frac{4}{3}x - \frac{7}{2}y = -51 \end{cases}$

59. $\begin{cases} 4x - y + z = -5 \\ 2x + 2y + 3z = 10 \\ 5x - 2y + 6z = 1 \end{cases}$ **60.** $\begin{cases} 4x - 2y + 3z = -2 \\ 2x + 2y + 5z = 16 \\ 8x - 5y - 2z = 4 \end{cases}$

In Exercises 61 and 62, develop for the given matrix a system of equations that will have the given solution. Use an inverse matrix to verify that the system of equations gives the desired solution.

61. $\begin{bmatrix} 2 & 1 & 3 \\ 4 & 0 & -2 \\ 0 & 3 & 2 \end{bmatrix}$ $\begin{matrix} x = 2 \\ y = -3 \\ z = 5 \end{matrix}$

62. $\begin{bmatrix} 1 & 0 & 2 \\ 1 & 1 & 1 \\ 2 & -1 & 0 \end{bmatrix}$ $\begin{matrix} x = 5 \\ y = -2 \\ z = 1 \end{matrix}$

Bond Investment In Exercises 63–66, you are investing in AAA-rated bonds, A-rated bonds, and B-rated bonds. Your average yield is 9% on AAA bonds, 7% on A bonds, and 8% on B bonds. Twice as much is invested in B bonds as in A bonds. The desired system of linear equations (where x, y, and z represent the amounts invested in AAA, A, and B bonds, respectively) is as follows.

$$\begin{cases} x + y + z = \text{(total investment)} \\ 0.09x + 0.07y + 0.08z = \text{(annual return)} \\ 2y - z = 0 \end{cases}$$

Use the inverse of the coefficient matrix of this system to find the amount invested in each type of bond for the given total investments and annual returns.

63. Total investment = \$35,000; annual return = \$2950

64. Total investment = \$50,000; annual return = \$4180

65. Total investment = \$36,000; annual return = \$3040

66. Total investment = \$45,000; annual return = \$3770

Acquisition of Raw Materials In Exercises 67–70, consider a company that produces computer chips, resistors, and transistors. Each computer chip requires 2 units of copper, 2 units of zinc, and 1 unit of glass. Each resistor requires 1 unit of copper, 3 units of zinc, and 2 units of glass. Each transistor requires 3 units of copper, 2 units of zinc, and 2 units of glass. The desired system of linear equations (where x, y, and z represent the numbers of computer chips, resistors, and transistors, respectively) is as follows.

$$\begin{cases} 2x + y + 3z = \text{(units of copper)} \\ 2x + 3y + 2z = \text{(units of zinc)} \\ x + 2y + 2z = \text{(units of glass)} \end{cases}$$

Use the inverse of the coefficient matrix of this system to find the numbers of computer chips, resistors, and transistors that the company can produce with the given amounts of raw materials.

67. 80 units of copper   **68.** 100 units of copper
    90 units of zinc   110 units of zinc
    55 units of glass   80 units of glass

69. 200 units of copper   **70.** 350 units of copper
    260 units of zinc   400 units of zinc
    180 units of glass   325 units of glass

Raw Materials In Exercises 71–74, consider a company that specializes in gourmet chocolate baked goods—chocolate muffins, cookies, and brownies. Each chocolate muffin requires 2 units of chocolate, 3 units of flour, and 2 units of sugar. Each chocolate cookie requires 1 unit of chocolate, 1 unit of flour, and 1 unit of sugar. Each chocolate brownie requires 2 units of chocolate, 1 unit of flour, and 1.5 units of sugar. Find the numbers of muffins, cookies, and brownies that the company can produce with the given amounts of raw materials.

71. 700 units of chocolate   **72.** 525 units of chocolate
    500 units of flour   480 units of flour
    600 units of sugar   500 units of sugar

73. 800 units of chocolate   **74.** 1000 units of chocolate
    750 units of flour   950 units of flour
    725 units of sugar   900 units of sugar

75. *Make a Decision*: **Child Support** The total values y (in billions of dollars) of child support collections from 1994 to 2001 are shown in the figure. The least squares regression parabola $y = at^2 + bt + c$ for these data is found by solving the system

$$\begin{cases} 8c + 60b + 492a = 112.7 \\ 60c + 492b + 4320a = 902.4. \\ 492c + 4320b + 39{,}876a = 7796.6 \end{cases}$$

Let t represent the year, with $t = 4$ corresponding to 1994. (Source: U.S. Department of Health and Human Services)

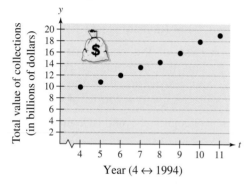

(a) Use a graphing utility to find an inverse matrix to solve this system, and find the equation of the least squares regression parabola.

(b) Use the result of part (a) to estimate the value of child support collections in 2002.

(c) The U.S. Department of Health and Human Services predicted that the value of child support collections in 2002 would be $20.1 billion. How does this value compare with your estimate in part (b)? Do both estimates seem reasonable?

76. *Make a Decision*: **Hotel Room Rates** The average room rates y (in dollars) for hotels from 1992 to 2001 are shown in the figure. The least squares regression parabola $y = at^2 + bt + c$ for these data is found by solving the system

$$\begin{cases} 10c + 65b + 505a = 729.3 \\ 65c + 505b + 4355a = 5027.7. \\ 505c + 4355b + 39{,}973a = 40{,}585.2 \end{cases}$$

Let t represent the year, with $t = 2$ corresponding to 1992. (Source: American Hotel & Lodging Association)

(a) Use a graphing utility to find an inverse matrix to solve this system, and write the equation of the least squares regression parabola.

(b) Use the result of part (a) to estimate the average room rate for hotels in 2002.

(c) The actual average room rate in 2002 was $83.54. How does this value compare with your estimate in part (b)?

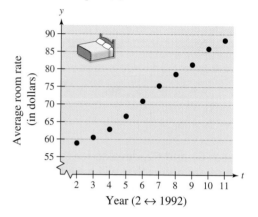

Figure for 76

In Exercises 77 and 78, use the following matrices.

$$A = \begin{bmatrix} 4 & 3 \\ -2 & 1 \end{bmatrix}, B = \begin{bmatrix} 1 & -2 \\ 3 & 4 \end{bmatrix}, C = \begin{bmatrix} 13 & 4 \\ 1 & 8 \end{bmatrix}$$

77. Find AB and BA. What do you observe about the two products?

78. Find C^{-1}, $A^{-1} \cdot B^{-1}$, and $B^{-1} \cdot A^{-1}$. What do you observe about the three resulting matrices?

In Exercises 79 and 80, find a value for k that will make the matrix invertible and then find a value for k that will make the matrix singular. (There are many correct answers.)

79. $\begin{bmatrix} 4 & 3 \\ -2 & k \end{bmatrix}$ **80.** $\begin{bmatrix} 2k + 1 & 3 \\ -7 & 1 \end{bmatrix}$

81. Exploration Consider matrices of the form

$$A = \begin{bmatrix} a_{11} & 0 & 0 & 0 & \cdots & 0 \\ 0 & a_{22} & 0 & 0 & \cdots & 0 \\ 0 & 0 & a_{33} & 0 & \cdots & 0 \\ \vdots & \vdots & \vdots & \vdots & \cdots & \vdots \\ 0 & 0 & 0 & 0 & \cdots & a_{nn} \end{bmatrix}$$

(a) Write a 2×2 matrix and a 3×3 matrix of the form of A. Find the inverse of each.

(b) Use the result of part (a) to make a conjecture about the inverses of matrices of the form of A.

Mid-Chapter Quiz

Take this quiz as you would take a quiz in class. After you are done, check your work against the answers given in the back of the book.

In Exercises 1 and 2, write a matrix of the given order.

1. 3×4

2. 1×3

In Exercises 3 and 4, write the augmented matrix for the system of equations.

3. $\begin{cases} 3x + 2y = -2 \\ 5x - y = 19 \end{cases}$

4. $\begin{cases} x \qquad + 3z = -5 \\ x + 2y - z = 3 \\ 3x \qquad + 4z = 0 \end{cases}$

5. Use Gaussian elimination with back-substitution to solve the augmented matrix found in Exercise 3.

6. Use Gauss-Jordan elimination to solve the augmented matrix found in Exercise 4.

In Exercises 7–12, use the following matrices to find the indicated matrix (if possible).

$$A = \begin{bmatrix} 1 & -2 \\ 3 & 4 \end{bmatrix}, \quad B = \begin{bmatrix} -1 & 2 & -3 \\ 2 & 0 & 5 \end{bmatrix}, \quad C = \begin{bmatrix} 0 & -2 \\ 3 & 1 \end{bmatrix}$$

7. $2A + 3C$

8. AB

9. $A - 2C$

10. C^2

11. A^{-1}

12. B^{-1}

In Exercises 13 and 14, solve for X using matrices A and C from Exercises 7–12.

13. $X = 3A - 2C$

14. $2X + 4C = 2A$

In Exercises 15 and 16, find matrices A, X, and B such that the system can be written as $AX = B$. Then solve for X.

15. $\begin{cases} x - 3y = 10 \\ -2x + y = -10 \end{cases}$

16. $\begin{cases} 2x - y + z = 3 \\ 3x \qquad - z = 15 \\ \qquad 4y + 3z = -1 \end{cases}$

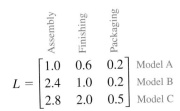

$$L = \begin{bmatrix} 1.0 & 0.6 & 0.2 \\ 2.4 & 1.0 & 0.2 \\ 2.8 & 2.0 & 0.5 \end{bmatrix} \begin{matrix} \text{Model A} \\ \text{Model B} \\ \text{Model C} \end{matrix}$$

Labor-Hour Requirements
(in hours per hang glider)

$$W = \begin{bmatrix} 15 & 12 \\ 8 & 9 \\ 7 & 6 \end{bmatrix} \begin{matrix} \text{Assembly} \\ \text{Finishing} \\ \text{Packaging} \end{matrix}$$

Wage Requirements
(in dollars per hour)

Matrices for 17–20

In Exercises 17–20, a hang glider manufacturer has the labor-hour and wage requirements indicated at the left.

17. What is the labor cost for model A?

18. What is the labor cost for model B?

19. What is the labor cost for model C?

20. Find LW and interpret the result.

The Determinant of a Square Matrix

Objectives

- Evaluate the determinant of a 2 × 2 matrix.
- Find the minors and cofactors of a matrix.
- Find the determinant of a square matrix.
- Find the determinant of a triangular matrix.

The Determinant of a 2 × 2 Matrix

Every *square* matrix can be associated with a real number called its **determinant.** Determinants have many uses, and several will be discussed in this and the next section. Historically, the use of determinants arose from special number patterns that occur when systems of linear equations are solved. For instance, the system

$$\begin{cases} a_1x + b_1y = c_1 \\ a_2x + b_2y = c_2 \end{cases}$$

has a solution given by

$$x = \frac{c_1b_2 - c_2b_1}{a_1b_2 - a_2b_1} \quad \text{and} \quad y = \frac{a_1c_2 - a_2c_1}{a_1b_2 - a_2b_1}$$

provided that $a_1b_2 - a_2b_1 \neq 0$. Note that the denominator of each fraction is the same. This denominator is called the **determinant** of the coefficient matrix of the system.

Coefficient Matrix	Determinant
$A = \begin{bmatrix} a_1 & b_1 \\ a_2 & b_2 \end{bmatrix}$	$\det(A) = a_1b_2 - a_2b_1$

The determinant of the matrix A can also be denoted by vertical bars on both sides of the matrix, as indicated in the following definition.

Definition of the Determinant of a 2 × 2 Matrix

The **determinant** of the matrix

$$A = \begin{bmatrix} a_1 & b_1 \\ a_2 & b_2 \end{bmatrix}$$

is given by

$$\det(A) = |A| = \begin{vmatrix} a_1 & b_1 \\ a_2 & b_2 \end{vmatrix} = a_1b_2 - a_2b_1.$$

In this text, $\det(A)$ and $|A|$ are used interchangeably to represent the determinant of A. Although vertical bars are also used to denote the absolute value of a real number, the context will show which use is intended.

A convenient method for remembering the formula for the determinant of a 2×2 matrix is shown in the following diagram.

$$\det(A) = \begin{vmatrix} a_1 & b_1 \\ a_2 & b_2 \end{vmatrix} = a_1 b_2 - a_2 b_1$$

Note that the determinant is given by the difference of the products of the two diagonals of the matrix. In Example 1 you will see that the determinant of a matrix can be positive, zero, or negative.

EXAMPLE 1   THE DETERMINANT OF A 2×2 MATRIX

Find the determinant of each matrix.

a. $A = \begin{bmatrix} 2 & -3 \\ 1 & 2 \end{bmatrix}$ **b.** $B = \begin{bmatrix} 2 & 1 \\ 4 & 2 \end{bmatrix}$ **c.** $C = \begin{bmatrix} 0 & 3 \\ 2 & 4 \end{bmatrix}$

SOLUTION

Use the formula $\det(A) = \begin{vmatrix} a_1 & b_1 \\ a_2 & b_2 \end{vmatrix} = a_1 b_2 - a_2 b_1$.

a. $\det(A) = \begin{vmatrix} 2 & -3 \\ 1 & 2 \end{vmatrix} = 2(2) - 1(-3) = 4 + 3 = 7$

b. $\det(B) = \begin{vmatrix} 2 & 1 \\ 4 & 2 \end{vmatrix} = 2(2) - 4(1) = 4 - 4 = 0$

c. $\det(C) = \begin{vmatrix} 0 & 3 \\ 2 & 4 \end{vmatrix} = 0(4) - 2(3) = 0 - 6 = -6$

✓CHECKPOINT   Now try Exercise 3.

STUDY TIP

The determinant of a matrix of order 1×1 is defined simply as the entry of the matrix. For instance, if $A = [-2]$, then $\det(A) = -2$.

TECHNOLOGY

Most graphing utilities can evaluate the determinant of a matrix. Use a graphing utility to find the determinant of matrix A from Example 1. The result should be 7, as shown below. For specific keystrokes on how to use a graphing utility to evaluate the determinant of a matrix, go to the text website at *college.hmco.com*.

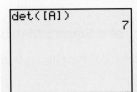

```
det([A])
                    7
```

Try evaluating the determinant of B with your graphing utility.

$$B = \begin{bmatrix} 3 & -1 & 1 \\ 0 & 2 & 1 \end{bmatrix}$$

What happens when you try to evaluate the determinant of a nonsquare matrix?

Minors and Cofactors

To define the determinant of a square matrix of order 3×3 or higher, it is convenient to introduce the concepts of **minors** and **cofactors**.

Minors of a Square Matrix

If A is a square matrix, the **minor** M_{ij} of the entry a_{ij} is the determinant of the matrix obtained by deleting the ith row and jth column of A.

EXAMPLE 2   FINDING THE MINORS OF A MATRIX

Find all the minors of $A = \begin{bmatrix} 0 & 2 & 1 \\ 3 & -1 & 2 \\ 4 & 0 & 1 \end{bmatrix}$.

SOLUTION

To find the minor M_{11}, delete the first row and first column of A and evaluate the determinant of the resulting matrix.

$$\begin{bmatrix} 0 & 2 & 1 \\ 3 & -1 & 2 \\ 4 & 0 & 1 \end{bmatrix}, \quad M_{11} = \begin{vmatrix} -1 & 2 \\ 0 & 1 \end{vmatrix} = -1(1) - 0(2) = -1$$

Sign Pattern for Cofactors

$$\begin{bmatrix} + & - & + \\ - & + & - \\ + & - & + \end{bmatrix}$$

3×3 matrix

$$\begin{bmatrix} + & - & + & - \\ - & + & - & + \\ + & - & + & - \\ - & + & - & + \end{bmatrix}$$

4×4 matrix

$$\begin{bmatrix} + & - & + & - & + & \cdots \\ - & + & - & + & - & \cdots \\ + & - & + & - & + & \cdots \\ - & + & - & + & - & \cdots \\ + & - & + & - & + & \cdots \\ \vdots & \vdots & \vdots & \vdots & \vdots & \end{bmatrix}$$

$n \times n$ matrix

Notice that *odd* positions (where $i + j$ is odd) have negative signs and *even* positions (where $i + j$ is even) have positive signs.

Similarly, to find M_{12}, delete the first row and second column.

$$\begin{bmatrix} 0 & 2 & 1 \\ 3 & -1 & 2 \\ 4 & 0 & 1 \end{bmatrix}, \quad M_{12} = \begin{vmatrix} 3 & 2 \\ 4 & 1 \end{vmatrix} = 3(1) - 4(2) = -5$$

Continuing this pattern, you obtain the following minors.

$$\begin{array}{lll} M_{11} = -1 & M_{12} = -5 & M_{13} = 4 \\ M_{21} = 2 & M_{22} = -4 & M_{23} = -8 \\ M_{31} = 5 & M_{32} = -3 & M_{33} = -6 \end{array}$$

✓CHECKPOINT   Now try Exercise 15(a).

Cofactors of a Square Matrix

The **cofactor** C_{ij} of the entry a_{ij} is given by $C_{ij} = (-1)^{i+j} M_{ij}$.

To find the cofactors of the matrix A in Example 2, combine the checkerboard pattern of signs for a 3×3 matrix (at the left) with the minors.

$$\begin{array}{lll} C_{11} = -1 & C_{12} = 5 & C_{13} = 4 \\ C_{21} = -2 & C_{22} = -4 & C_{23} = 8 \\ C_{31} = 5 & C_{32} = 3 & C_{33} = -6 \end{array}$$

The Determinant of a Square Matrix

The definition below is called **inductive** because it uses determinants of matrices of order $n - 1$ to define the determinant of a matrix of order n.

Determinant of a Square Matrix

If A is a square matrix (of order 2×2 or greater), then the determinant of A is the sum of the entries in any row (or column) of A multiplied by their respective cofactors. For instance, expanding along the first row of an $n \times n$ matrix yields

$$|A| = a_{11}M_{11} - a_{12}M_{12} + \cdots - a_{1n}M_{1n}. \qquad n \text{ is even.}$$

or

$$|A| = a_{11}M_{11} - a_{12}M_{12} + \cdots + a_{1n}M_{1n} \qquad n \text{ is odd.}$$

Applying this definition to find a determinant is called **expanding by minors,** or **expanding by cofactors.**

Try checking that for a 2×2 matrix this definition yields

$$|A| = a_{11}a_{22} - a_{12}a_{21},$$

as previously defined.

EXAMPLE 3   THE DETERMINANT OF A MATRIX OF ORDER 3×3

Find the determinant of

$$A = \begin{bmatrix} 0 & 2 & 1 \\ 3 & -1 & 2 \\ 4 & 0 & 1 \end{bmatrix}.$$

SOLUTION

Note that this is the same matrix that was given in Example 2. There you found the minors of the entries in the first row to be

$$M_{11} = -1, \quad M_{12} = -5, \quad \text{and} \quad M_{13} = 4.$$

So, by the definition of a determinant, you have the following.

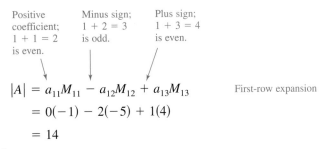

$$|A| = a_{11}M_{11} - a_{12}M_{12} + a_{13}M_{13} \qquad \text{First-row expansion}$$

$$= 0(-1) - 2(-5) + 1(4)$$

$$= 14$$

✓ CHECKPOINT   Now try Exercise 25.

In Example 3 the determinant was found by expanding by the minors in the first row. You could have used any row or column. For instance, you could have expanded along the second column to obtain

$$|A| = -a_{12}M_{12} + a_{22}M_{22} - a_{32}M_{32} \qquad \text{Second-column expansion}$$

$$= -2(-5) + (-1)(-4) - 0(-3)$$

$$= 14.$$

When expanding by minors, you do not need to find the minors of zero entries, because zero times its minor is zero.

$$a_{ij}M_{ij} = (0)M_{ij} = 0$$

So, the row (or column) containing the most zeros is usually the best choice for expansion by minors. This is demonstrated in the next example.

EXAMPLE 4 THE DETERMINANT OF A MATRIX OF ORDER 4 × 4

Make a Decision Find the determinant of

$$A = \begin{bmatrix} 1 & -2 & 3 & 0 \\ -1 & 1 & 0 & 2 \\ 0 & 2 & 0 & 3 \\ 3 & 4 & 0 & 2 \end{bmatrix}.$$

Expand by minors on the row or column that appears to make the computations easiest.

SOLUTION

After inspecting this matrix, you can see that three of the entries in the third column are zeros. Expanding on the third column appears to make the computations easiest.

$$|A| = 3(M_{13}) - 0(M_{23}) + 0(M_{33}) - 0(M_{43})$$

Because M_{23}, M_{33}, and M_{43} have zero coefficients, you need only find the minor M_{13}. To do this, delete the first row and third column of A and evaluate the determinant of the resulting matrix.

$$M_{13} = \begin{vmatrix} -1 & 1 & 2 \\ 0 & 2 & 3 \\ 3 & 4 & 2 \end{vmatrix}$$

Expanding by minors in the second row yields the following.

$$M_{13} = -0\begin{vmatrix} 1 & 2 \\ 4 & 2 \end{vmatrix} + 2\begin{vmatrix} -1 & 2 \\ 3 & 2 \end{vmatrix} - 3\begin{vmatrix} -1 & 1 \\ 3 & 4 \end{vmatrix}$$

$$= 0 + 2(-2 - 6) - 3(-4 - 3)$$

$$= 5$$

So, you obtain $|A| = 3M_{13} = 3(5) = 15$.

✓**CHECKPOINT** Now try Exercise 37.

Try using a graphing utility to confirm the result of Example 4.

There is an alternative method that is commonly used to evaluate the determinant of a 3×3 matrix A. This method works *only* for 3×3 matrices. To apply this method, copy the first and second columns of A to form fourth and fifth columns. The determinant of A is then obtained by adding the products of the three "downward diagonals" and subtracting the products of the three "upward diagonals," as shown in the following diagram.

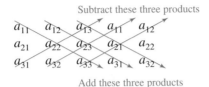

Subtract these three products

Add these three products

So, the determinant of the 3×3 matrix A is given by

$$|A| = a_{11}a_{22}a_{33} + a_{12}a_{23}a_{31} + a_{13}a_{21}a_{32}$$
$$- a_{31}a_{22}a_{13} - a_{32}a_{23}a_{11} - a_{33}a_{21}a_{12}.$$

EXAMPLE 5   THE DETERMINANT OF A 3×3 MATRIX

Find the determinant of

$$A = \begin{bmatrix} 0 & 2 & 1 \\ 3 & -1 & 2 \\ 4 & -4 & 1 \end{bmatrix}.$$

SOLUTION

Because A is a 3×3 matrix, you can use the alternative procedure for finding $|A|$. Begin by recopying the first two columns and then computing the six diagonal products, as follows.

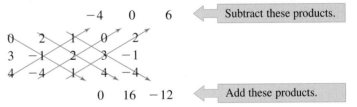

Now, by adding the lower three products and subtracting the upper three products, you find the determinant of A to be

$$|A| = 0 + 16 - 12 - (-4) - 0 - 6$$
$$= 2.$$

✓**CHECKPOINT**   Now try Exercise 53.

Be sure you understand that the diagonal process illustrated in Example 5 is valid *only* for matrices of order 3×3. For matrices of higher orders, another method must be used, such as expansion by minors or a graphing utility.

Triangular Matrices

Evaluating determinants of matrices of order 4 or higher can be tedious. There is, however, an important exception: the determinant of a **triangular** matrix. A triangular matrix is a square matrix with all *zero* entries either below or above its main diagonal. A square matrix is **upper triangular** if it has all zero entries below its main diagonal and **lower triangular** if it has all zero entries above its main diagonal. A matrix that is both upper and lower triangular is called **diagonal.** That is, a diagonal matrix is one in which all entries above and below the main diagonal are zero.

Upper Triangular Matrix

$$\begin{bmatrix} a_{11} & a_{12} & a_{13} & \cdots & a_{1n} \\ 0 & a_{22} & a_{23} & \cdots & a_{2n} \\ 0 & 0 & a_{33} & \cdots & a_{3n} \\ \vdots & \vdots & \vdots & & \vdots \\ 0 & 0 & 0 & \cdots & a_{nn} \end{bmatrix}$$

Lower Triangular Matrix

$$\begin{bmatrix} a_{11} & 0 & 0 & \cdots & 0 \\ a_{21} & a_{22} & 0 & \cdots & 0 \\ a_{31} & a_{32} & a_{33} & \cdots & 0 \\ \vdots & \vdots & \vdots & & \vdots \\ a_{n1} & a_{n2} & a_{n3} & \cdots & a_{nn} \end{bmatrix}$$

To find the determinant of a triangular matrix of any order, simply form the product of the entries on the main diagonal.

EXAMPLE 6 THE DETERMINANT OF A TRIANGULAR MATRIX

a.
$$\begin{vmatrix} 2 & 0 & 0 & 0 \\ 4 & -2 & 0 & 0 \\ -5 & 6 & 1 & 0 \\ 1 & 5 & 3 & 3 \end{vmatrix} = 2(-2)(1)(3) = -12$$

b.
$$\begin{vmatrix} -1 & 0 & 0 & 0 & 0 \\ 0 & 3 & 0 & 0 & 0 \\ 0 & 0 & 2 & 0 & 0 \\ 0 & 0 & 0 & 4 & 0 \\ 0 & 0 & 0 & 0 & -2 \end{vmatrix} = -1(3)(2)(4)(-2) = 48$$

✓*CHECKPOINT* Now try Exercise 49.

Discussing the Concept | The Determinant of a Triangular Matrix

Explain why the determinant of a 3 × 3 triangular matrix is the product of its main-diagonal entries.

$$\begin{vmatrix} a_{11} & a_{12} & a_{13} \\ 0 & a_{22} & a_{23} \\ 0 & 0 & a_{33} \end{vmatrix} = a_{11}a_{22}a_{33}$$

Is this true of higher-order triangular matrices? Explain.

5.4 Warm Up

The following warm-up exercises involve skills that were covered in earlier sections. You will use these skills in the exercise set for this section. For additional help, review Sections R1.2, R1.3, and 5.2.

In Exercises 1–4, perform the indicated matrix operations.

1. $\begin{bmatrix} 1 & -2 \\ 0 & 3 \end{bmatrix} + \begin{bmatrix} 2 & 7 \\ 4 & -3 \end{bmatrix}$

2. $\begin{bmatrix} -2 & 5 \\ 3 & -2 \end{bmatrix} - \begin{bmatrix} 0 & -3 \\ 1 & 2 \end{bmatrix}$

3. $3\begin{bmatrix} 3 & -4 & 2 \\ 1 & 0 & -1 \\ 0 & 1 & -2 \end{bmatrix}$

4. $4\begin{bmatrix} 0 & 2 & 3 \\ -1 & 2 & 3 \\ -2 & 1 & -2 \end{bmatrix}$

In Exercises 5–10, perform the indicated arithmetic operations.

5. $[(1)(3) + (-3)(2)] - [(1)(4) + (3)(5)]$

6. $[(4)(4) + (-1)(-3)] - [(-1)(2) + (-2)(7)]$

7. $\dfrac{4(7) - 1(-2)}{(-5)(-2) - 3(4)}$

8. $\dfrac{3(6) - 2(7)}{6(-5) - 2(1)}$

9. $-5(-1)^2[6(-2) - 7(-3)]$

10. $4(-1)^3[3(6) - 2(7)]$

5.4 Exercises

In Exercises 1–14, find the determinant of the matrix.

1. $[-5]$

2. $[6]$

3. $\begin{bmatrix} 1 & 3 \\ 2 & 7 \end{bmatrix}$

4. $\begin{bmatrix} -3 & 4 \\ -2 & 1 \end{bmatrix}$

5. $\begin{bmatrix} 5 & 6 \\ 2 & 3 \end{bmatrix}$

6. $\begin{bmatrix} -7 & -4 \\ 8 & 7 \end{bmatrix}$

7. $\begin{bmatrix} 9 & 3 \\ 12 & 4 \end{bmatrix}$

8. $\begin{bmatrix} -5 & -2 \\ 10 & 4 \end{bmatrix}$

9. $\begin{bmatrix} 2 & 1 \\ 3 & 4 \end{bmatrix}$

10. $\begin{bmatrix} -3 & 1 \\ 5 & 2 \end{bmatrix}$

11. $\begin{bmatrix} -\frac{1}{2} & \frac{1}{3} \\ -6 & \frac{1}{3} \end{bmatrix}$

12. $\begin{bmatrix} \frac{2}{3} & \frac{4}{3} \\ -1 & -\frac{1}{3} \end{bmatrix}$

13. $\begin{bmatrix} \frac{2}{3} & 0 \\ -1 & 6 \end{bmatrix}$

14. $\begin{bmatrix} 9 & -\frac{1}{4} \\ 8 & 0 \end{bmatrix}$

In Exercises 15–18, find all (a) minors and (b) cofactors of the matrix.

15. $\begin{bmatrix} 3 & 4 \\ 2 & -5 \end{bmatrix}$

16. $\begin{bmatrix} 11 & 0 \\ -3 & 2 \end{bmatrix}$

17. $\begin{bmatrix} 3 & -2 & 8 \\ 3 & 2 & -6 \\ -1 & 3 & 6 \end{bmatrix}$

18. $\begin{bmatrix} -2 & 9 & 4 \\ 7 & -6 & 0 \\ 6 & 7 & -6 \end{bmatrix}$

In Exercises 19–24, find the determinant of the matrix by the method of expansion by minors. Expand using the indicated row or column.

19. $\begin{bmatrix} 4 & 1 & -3 \\ 6 & 5 & -2 \\ -1 & 3 & -4 \end{bmatrix}$

   (a) Row 3

   (b) Column 2

20. $\begin{bmatrix} -3 & 4 & 2 \\ 6 & 3 & 1 \\ 4 & -7 & -8 \end{bmatrix}$

   (a) Row 2

   (b) Column 3

21. $\begin{bmatrix} 7 & 0 & -4 \\ 2 & -3 & 0 \\ 5 & 8 & 1 \end{bmatrix}$

   (a) Row 1

   (b) Column 3

22. $\begin{bmatrix} 10 & -5 & 5 \\ 30 & 0 & 10 \\ 0 & 10 & 1 \end{bmatrix}$

   (a) Row 3

   (b) Column 1

23. $\begin{bmatrix} 6 & 0 & -3 & 5 \\ 4 & 13 & 6 & -8 \\ -1 & 0 & 7 & 4 \\ 8 & 6 & 0 & 2 \end{bmatrix}$

   (a) Row 2   (b) Column 2

24. $\begin{bmatrix} 10 & 8 & 3 & -7 \\ 4 & 0 & 5 & -6 \\ 0 & 3 & 2 & 7 \\ 1 & 0 & -3 & 2 \end{bmatrix}$

   (a) Row 3   (b) Column 1

In Exercises 25–42, find the determinant of the matrix. Expand by minors on the row or column that appears to make the computations easiest. Use a graphing utility to confirm your result.

25. $\begin{bmatrix} 1 & 4 & -2 \\ 3 & 2 & 0 \\ -1 & 4 & 3 \end{bmatrix}$ **26.** $\begin{bmatrix} 2 & -1 & 3 \\ 1 & 4 & 4 \\ 1 & 0 & 2 \end{bmatrix}$

27. $\begin{bmatrix} 2 & 4 & 6 \\ 0 & 3 & 1 \\ 0 & 0 & -5 \end{bmatrix}$ **28.** $\begin{bmatrix} -3 & 0 & 0 \\ 7 & 11 & 0 \\ 1 & 2 & 2 \end{bmatrix}$

29. $\begin{bmatrix} 2 & -1 & 0 \\ 4 & 2 & 1 \\ 4 & 2 & 1 \end{bmatrix}$ **30.** $\begin{bmatrix} -2 & 2 & 3 \\ 1 & -1 & 0 \\ 0 & 1 & 4 \end{bmatrix}$

31. $\begin{bmatrix} 0.3 & 0.2 & 0.2 \\ 0.2 & 0.2 & 0.2 \\ -0.4 & 0.4 & 0.3 \end{bmatrix}$ **32.** $\begin{bmatrix} 0.1 & 0.2 & 0.3 \\ -0.3 & 0.2 & 0.2 \\ 0.5 & 0.4 & 0.4 \end{bmatrix}$

33. $\begin{bmatrix} 1 & 4 & -2 \\ 3 & 6 & -6 \\ -2 & 1 & 4 \end{bmatrix}$ **34.** $\begin{bmatrix} -1 & 3 & 1 \\ 4 & 2 & 5 \\ -2 & 1 & 6 \end{bmatrix}$

35. $\begin{bmatrix} 6 & 3 & -7 \\ 0 & 0 & 0 \\ 4 & -6 & 3 \end{bmatrix}$ **36.** $\begin{bmatrix} 5 & 0 & 3 \\ -4 & 0 & 8 \\ 3 & 0 & -6 \end{bmatrix}$

37. $\begin{bmatrix} 3 & 6 & -5 & 4 \\ -2 & 0 & 6 & 0 \\ 1 & 1 & 2 & 2 \\ 0 & 3 & -1 & -1 \end{bmatrix}$

38. $\begin{bmatrix} 2 & 6 & 6 & 2 \\ 2 & 7 & 3 & 6 \\ 1 & 5 & 0 & 1 \\ 3 & 7 & 0 & 7 \end{bmatrix}$

39. $\begin{bmatrix} 5 & 3 & 0 & 6 \\ 4 & 6 & 4 & 12 \\ 0 & 2 & -3 & 4 \\ 0 & 1 & -2 & 2 \end{bmatrix}$ **40.** $\begin{bmatrix} 1 & 4 & 3 & 2 \\ -5 & 6 & 2 & 1 \\ 0 & 0 & 0 & 0 \\ 3 & -2 & 1 & 5 \end{bmatrix}$

41. $\begin{bmatrix} 3 & 2 & 4 & -1 & 5 \\ -2 & 0 & 1 & 3 & 2 \\ 1 & 0 & 0 & 4 & 0 \\ 6 & 0 & 2 & -1 & 0 \\ 3 & 0 & 5 & 1 & 0 \end{bmatrix}$

42. $\begin{bmatrix} 5 & 2 & 0 & 0 & -2 \\ 0 & 1 & 4 & 3 & 2 \\ 0 & 0 & 2 & 6 & 3 \\ 0 & 0 & 3 & 4 & 1 \\ 0 & 0 & 0 & 0 & 2 \end{bmatrix}$

In Exercises 43–52, evaluate the determinant of the matrix. Do not use a graphing utility.

43. $\begin{bmatrix} 2 & 0 & 0 \\ 4 & -3 & 0 \\ 6 & 5 & 1 \end{bmatrix}$ **44.** $\begin{bmatrix} 1 & 0 & 0 \\ -4 & -1 & 0 \\ 5 & 1 & 5 \end{bmatrix}$

45. $\begin{bmatrix} 2 & 3 & -1 & -1 \\ 0 & -1 & -3 & 5 \\ 0 & 0 & -2 & 7 \\ 0 & 0 & 0 & -4 \end{bmatrix}$ **46.** $\begin{bmatrix} 4 & 0 & 0 & 0 \\ 1 & -4 & 0 & 0 \\ 2 & 1 & -1 & 0 \\ 6 & -2 & 3 & -1 \end{bmatrix}$

47. $\begin{bmatrix} 1 & 0 & 0 & 0 & 0 \\ 0 & 2 & 0 & 0 & 0 \\ 0 & 0 & 3 & 0 & 0 \\ 0 & 0 & 0 & 4 & 0 \\ 0 & 0 & 0 & 0 & 5 \end{bmatrix}$

48. $\begin{bmatrix} -2 & 0 & 0 & 0 & 0 \\ 0 & 3 & 0 & 0 & 0 \\ 0 & 0 & -1 & 0 & 0 \\ 0 & 0 & 0 & 2 & 0 \\ 0 & 0 & 0 & 0 & -4 \end{bmatrix}$

49. $\begin{bmatrix} 4 & 0 & 0 & 0 \\ 6 & -5 & 0 & 0 \\ 1 & 3 & 2 & 0 \\ 1 & 2 & 7 & -1 \end{bmatrix}$ **50.** $\begin{bmatrix} 5 & 3 & 6 & 1 \\ 0 & -10 & 4 & 3 \\ 0 & 0 & 5 & 2 \\ 0 & 0 & 0 & 8 \end{bmatrix}$

51. $\begin{bmatrix} -6 & 7 & 2 & 0 & 5 \\ 0 & -1 & 3 & 4 & -3 \\ 0 & 0 & -7 & 0 & 4 \\ 0 & 0 & 0 & -2 & 1 \\ 0 & 0 & 0 & 0 & -2 \end{bmatrix}$

52. $\begin{bmatrix} -3 & 0 & 0 & 0 & 0 \\ 4 & 1 & 0 & 0 & 0 \\ 7 & -8 & 7 & 0 & 0 \\ 6 & 4 & 0 & -2 & 0 \\ 1 & 5 & 1 & -10 & 6 \end{bmatrix}$

Make a Decision In Exercises 53–56, find the determinant of the matrix. Tell whether you used expansion by minors, the product of the entries on the main diagonal, or upward and downward diagonals.

53. $\begin{bmatrix} 2 & 1 & 3 \\ 7 & 3 & -2 \\ 4 & 1 & 1 \end{bmatrix}$ **54.** $\begin{bmatrix} 6 & -5 & 2 \\ 0 & 5 & -3 \\ 0 & 0 & 2 \end{bmatrix}$

55. $\begin{bmatrix} 3 & 0 & 0 \\ 4 & -2 & 0 \\ 5 & 4 & 3 \end{bmatrix}$ **56.** $\begin{bmatrix} 3 & 2 & -4 \\ -1 & 5 & -3 \\ 0 & 1 & 0 \end{bmatrix}$

In Exercises 57–60, find a 4 × 4 *upper* triangular matrix whose determinant is equal to the given value and a 4 × 4 *lower* triangular matrix whose determinant is equal to the given value. Use a graphing utility to confirm your results.

57. −18

58. −40

59. 28

60. 36

In Exercises 61–64, explain why the determinant of the matrix is equal to zero.

61. $\begin{bmatrix} 3 & 4 & -2 & 7 \\ 1 & 3 & -1 & 2 \\ 0 & 5 & 7 & 1 \\ 1 & 3 & -1 & 2 \end{bmatrix}$

62. $\begin{bmatrix} 3 & 2 & -1 \\ -6 & -4 & 2 \\ 5 & -7 & 9 \end{bmatrix}$

63. $\begin{bmatrix} 3 & 0 & 1 & 7 \\ 2 & -1 & 4 & 3 \\ 11 & 5 & -7 & 8 \\ -6 & 3 & -12 & -9 \end{bmatrix}$

64. $\begin{bmatrix} -1 & 3 & 2 \\ 5 & 7 & 0 \\ -1 & 3 & 2 \end{bmatrix}$

Math *MATTERS*

Guess the Number

Here is a guessing game that can be made using the four cards shown in the figure (note that the fourth card has numbers on the front *and* back). To play the game, ask someone to think of a number between 1 and 15. Ask the person if the number is on the first card. If it is, place the card face up with the "YES" on top. If it isn't, place the card face up with the "NO" on top. Repeat this with each of the four cards (using the same number), stacking the cards one on top of another. Be sure that the fourth card is used last. After all four cards are in a stack, turn the stack over. The person's number will be the number that is showing through a window in the cards. Can you explain why this card game works?

Card 1   Card 2   Card 3

Card 4 (front)   Card 4 (back)

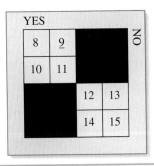

Applications of Matrices and Determinants

Objectives

- **Find the area of a triangle using a determinant.**
- **Determine whether three points are collinear using a determinant.**
- **Use a determinant to find an equation of a line.**
- **Encode and decode a cryptogram using a matrix.**

Area of a Triangle

In this section, you will study some additional applications of matrices and determinants. The first involves a formula for finding the area of a triangle whose vertices are given by three points on a rectangular coordinate system.

Area of a Triangle

The area of a triangle with vertices (x_1, y_1), (x_2, y_2), and (x_3, y_3) is given by

$$\text{Area} = \pm \frac{1}{2} \begin{vmatrix} x_1 & y_1 & 1 \\ x_2 & y_2 & 1 \\ x_3 & y_3 & 1 \end{vmatrix}$$

where the symbol $(\pm)$ indicates that the appropriate sign should be chosen to yield a positive area.

EXAMPLE 1   FINDING THE AREA OF A TRIANGLE

Find the area of the triangle whose vertices are $(1, 0)$, $(2, 2)$, and $(4, 3)$, as shown in Figure 5.1.

SOLUTION

Let $(x_1, y_1) = (1, 0)$, $(x_2, y_2) = (2, 2)$, and $(x_3, y_3) = (4, 3)$. Then, to find the area of the triangle, evaluate the determinant

$$\begin{vmatrix} x_1 & y_1 & 1 \\ x_2 & y_2 & 1 \\ x_3 & y_3 & 1 \end{vmatrix} = \begin{vmatrix} 1 & 0 & 1 \\ 2 & 2 & 1 \\ 4 & 3 & 1 \end{vmatrix} = 1 \begin{vmatrix} 2 & 1 \\ 3 & 1 \end{vmatrix} - 0 \begin{vmatrix} 2 & 1 \\ 4 & 1 \end{vmatrix} + 1 \begin{vmatrix} 2 & 2 \\ 4 & 3 \end{vmatrix}$$

$$= 1(-1) - 0(-2) + 1(-2) = -3.$$

Using this value, you can conclude that the area of the triangle is

$$\text{Area} = -\frac{1}{2} \begin{vmatrix} 1 & 0 & 1 \\ 2 & 2 & 1 \\ 4 & 3 & 1 \end{vmatrix} = -\frac{1}{2}(-3) = \frac{3}{2}.$$ Choose $(-)$ so that the area is positive.

✔CHECKPOINT   Now try Exercise 1.

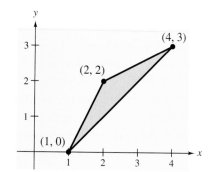

Figure 5.1

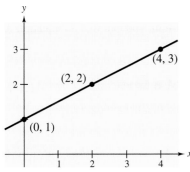

Figure 5.2

Lines in the Plane

Suppose the three points in Example 1 had been on the same line. What would have happened had the area formula been applied to three such points? The answer is that the determinant would have been zero. Consider, for instance, the three collinear points $(0, 1)$, $(2, 2)$, and $(4, 3)$, as shown in Figure 5.2. The area of the "triangle" that has these three points as vertices is

$$\frac{1}{2}\begin{vmatrix} 0 & 1 & 1 \\ 2 & 2 & 1 \\ 4 & 3 & 1 \end{vmatrix} = \frac{1}{2}\left(0\begin{vmatrix} 2 & 1 \\ 3 & 1 \end{vmatrix} - 1\begin{vmatrix} 2 & 1 \\ 4 & 1 \end{vmatrix} + 1\begin{vmatrix} 2 & 2 \\ 4 & 3 \end{vmatrix}\right)$$

$$= \frac{1}{2}[0(-1) - 1(-2) + 1(-2)] = 0.$$

This result is generalized as follows.

Test for Collinear Points

Three points (x_1, y_1), (x_2, y_2), and (x_3, y_3) are collinear (lie on the same line) if and only if

$$\begin{vmatrix} x_1 & y_1 & 1 \\ x_2 & y_2 & 1 \\ x_3 & y_3 & 1 \end{vmatrix} = 0.$$

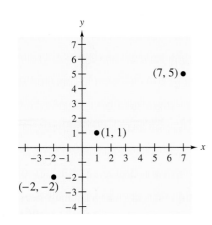

Figure 5.3

EXAMPLE 2	TESTING FOR COLLINEAR POINTS

Make a Decision Decide whether the points $(-2, -2)$, $(1, 1)$, and $(7, 5)$ are collinear. (See Figure 5.3.)

SOLUTION

Letting $(x_1, y_1) = (-2, -2)$, $(x_2, y_2) = (1, 1)$, and $(x_3, y_3) = (7, 5)$, you have

$$\begin{vmatrix} x_1 & y_1 & 1 \\ x_2 & y_2 & 1 \\ x_3 & y_3 & 1 \end{vmatrix} = \begin{vmatrix} -2 & -2 & 1 \\ 1 & 1 & 1 \\ 7 & 5 & 1 \end{vmatrix}$$

$$= -2\begin{vmatrix} 1 & 1 \\ 5 & 1 \end{vmatrix} - (-2)\begin{vmatrix} 1 & 1 \\ 7 & 1 \end{vmatrix} + 1\begin{vmatrix} 1 & 1 \\ 7 & 5 \end{vmatrix}$$

$$= -2(-4) - (-2)(-6) + 1(-2)$$

$$= -6.$$

Because the value of this determinant is *not* zero, you can conclude that the three points *do not* lie on the same line and are not collinear.

✓CHECKPOINT   Now try Exercise 17.

Another way to test for collinear points in Example 2 is to find the slope of the line between $(-2, -2)$ and $(1, 1)$ and the slope of the line between $(-2, -2)$ and $(7, 5)$. Try doing this. If the slopes are equal, then the points are collinear. If the slopes are not equal, then the points are not collinear.

The test for collinear points can be adapted to another use. That is, if you are given two points on a rectangular coordinate system, you can find an equation of the line passing through the two points, as follows.

Two-Point Form of the Equation of a Line

An equation of the line passing through the distinct points (x_1, y_1) and (x_2, y_2) is given by

$$\begin{vmatrix} x & y & 1 \\ x_1 & y_1 & 1 \\ x_2 & y_2 & 1 \end{vmatrix} = 0.$$

EXAMPLE 3 FINDING AN EQUATION OF A LINE

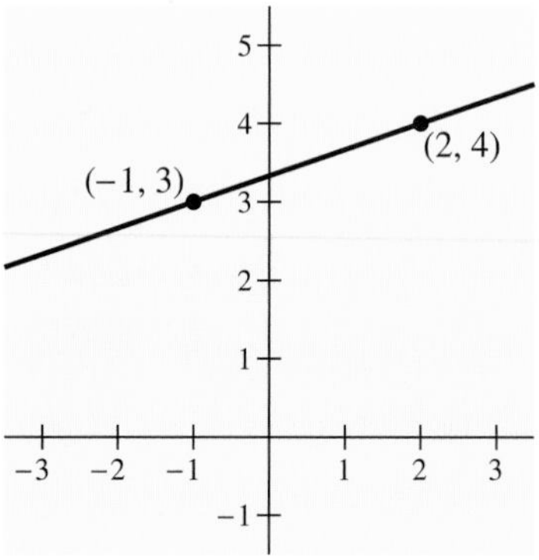

Find an equation of the line passing through the two points $(2, 4)$ and $(-1, 3)$, as shown in Figure 5.4.

SOLUTION

Applying the determinant formula for the equation of a line produces

$$\begin{vmatrix} x & y & 1 \\ 2 & 4 & 1 \\ -1 & 3 & 1 \end{vmatrix} = 0.$$

To evaluate this determinant, you can expand by minors along the first row to obtain the following.

$$x\begin{vmatrix} 4 & 1 \\ 3 & 1 \end{vmatrix} - y\begin{vmatrix} 2 & 1 \\ -1 & 1 \end{vmatrix} + 1\begin{vmatrix} 2 & 4 \\ -1 & 3 \end{vmatrix} = 0$$

$$x(1) - y(3) + (1)(10) = 0$$

$$x - 3y + 10 = 0$$

So, an equation of the line is

$$x - 3y + 10 = 0.$$

✓CHECKPOINT   Now try Exercise 25.

Figure 5.4

Note that this method of finding the equation of a line works for all lines, including horizontal and vertical lines. For instance, the equation of the vertical line through $(2, 0)$ and $(2, 2)$ is

$$\begin{vmatrix} x & y & 1 \\ 2 & 0 & 1 \\ 2 & 2 & 1 \end{vmatrix} = 0$$

$$4 - 2x = 0$$

$$x = 2.$$

Cryptography

© CORBIS

During World War II, Navajo soldiers created a code using their native language to send messages between batallions. Native words were assigned to represent characters in the English alphabet, and they created a number of expressions for important military terms, like iron-fish to mean submarine. Without the Navajo Code Talkers, the Second World War might have had a very different outcome.

A **cryptogram** is a message written according to a secret code. (The Greek word *kryptos* means "hidden.") Matrix multiplication can be used to **encode** and **decode** messages. To begin, you need to assign a number to each letter in the alphabet (with 0 assigned to a blank space), as follows.

0 = __	9 = I	18 = R
1 = A	10 = J	19 = S
2 = B	11 = K	20 = T
3 = C	12 = L	21 = U
4 = D	13 = M	22 = V
5 = E	14 = N	23 = W
6 = F	15 = O	24 = X
7 = G	16 = P	25 = Y
8 = H	17 = Q	26 = Z

The message is then converted to numbers and partitioned into **uncoded row matrices,** each having n entries, as demonstrated in Example 4.

EXAMPLE 4   FORMING UNCODED ROW MATRICES

Write the uncoded row matrices of order 1×3 for the message

MEET ME MONDAY.

SOLUTION

Partitioning the message (including blank spaces, but ignoring punctuation) into groups of three produces the following uncoded row matrices.

$$\begin{bmatrix} 13 & 5 & 5 \end{bmatrix} \quad \begin{bmatrix} 20 & 0 & 13 \end{bmatrix} \quad \begin{bmatrix} 5 & 0 & 13 \end{bmatrix} \quad \begin{bmatrix} 15 & 14 & 4 \end{bmatrix} \quad \begin{bmatrix} 1 & 25 & 0 \end{bmatrix}$$
$$\quad\text{M} \;\;\; \text{E} \;\;\; \text{E} \qquad \text{T} \qquad\;\;\; \text{M} \;\;\; \text{E} \qquad\;\;\; \text{M} \;\;\; \text{O} \;\;\; \text{N} \qquad \text{D} \;\;\; \text{A} \;\;\; \text{Y}$$

Note that a blank space is used to fill out the last uncoded row matrix.

✓CHECKPOINT   Now try Exercise 31.

To **encode** a message, choose an $n \times n$ invertible matrix A and multiply the uncoded row matrices by A to obtain **coded row matrices.** The uncoded matrix should be on the left, whereas the encoding matrix A should be on the right. Here is an example.

Uncoded Matrix   *Encoding Matrix A*   *Coded Matrix*

$$\begin{bmatrix} 13 & 5 & 5 \end{bmatrix} \quad \begin{bmatrix} 1 & -2 & 2 \\ -1 & 1 & 3 \\ 1 & -1 & -4 \end{bmatrix} = \quad \begin{bmatrix} 13 & -26 & 21 \end{bmatrix}$$

This technique is further illustrated in Example 5.

EXAMPLE 5   ENCODING A MESSAGE

Use the following matrix to encode the message MEET ME MONDAY.

$$A = \begin{bmatrix} 1 & -2 & 2 \\ -1 & 1 & 3 \\ 1 & -1 & -4 \end{bmatrix}$$

SOLUTION

The coded row matrices are obtained by multiplying each of the uncoded row matrices found in Example 4 by the matrix A, as follows.

Uncoded Matrix	*Encoding Matrix A*		*Coded Matrix*
$[13 \quad 5 \quad 5]$	$\begin{bmatrix} 1 & -2 & 2 \\ -1 & 1 & 3 \\ 1 & -1 & -4 \end{bmatrix}$	$=$	$[13 \quad -26 \quad 21]$
$[20 \quad 0 \quad 13]$	$\begin{bmatrix} 1 & -2 & 2 \\ -1 & 1 & 3 \\ 1 & -1 & -4 \end{bmatrix}$	$=$	$[33 \quad -53 \quad -12]$
$[5 \quad 0 \quad 13]$	$\begin{bmatrix} 1 & -2 & 2 \\ -1 & 1 & 3 \\ 1 & -1 & -4 \end{bmatrix}$	$=$	$[18 \quad -23 \quad -42]$
$[15 \quad 14 \quad 4]$	$\begin{bmatrix} 1 & -2 & 2 \\ -1 & 1 & 3 \\ 1 & -1 & -4 \end{bmatrix}$	$=$	$[5 \quad -20 \quad 56]$
$[1 \quad 25 \quad 0]$	$\begin{bmatrix} 1 & -2 & 2 \\ -1 & 1 & 3 \\ 1 & -1 & -4 \end{bmatrix}$	$=$	$[-24 \quad 23 \quad 77]$

So, the sequence of coded row matrices is

$$[13 -26 \quad 21][33 -53 -12][18 -23 -42][5 -20 \quad 56][-24 \quad 23 \quad 77].$$

Finally, removing the matrix notation produces the following cryptogram.

$$13 \quad -26 \quad 21 \quad 33 \quad -53 \quad -12 \quad 18 \quad -23 \quad -42 \quad 5 \quad -20 \quad 56 \quad -24 \quad 23 \quad 77$$

✔**CHECKPOINT**   Now try Exercise 35.

For those who do not know the encoding matrix A, decoding the cryptogram found in Example 5 is difficult. But for an authorized receiver who knows the encoding matrix A, decoding is simple. The receiver need only multiply the coded row matrices by A^{-1} (on the right) to retrieve the uncoded row matrices. Here is an example.

$$\underbrace{[13 \quad -26 \quad 21]}_{\text{Coded}} A^{-1} = \underbrace{[13 \quad 5 \quad 5]}_{\text{Uncoded}}$$

The receiver could then easily refer to the number code chart on page 373 and translate $[13 \quad 5 \quad 5]$ into the letters M E E.

EXAMPLE 6   DECODING A MESSAGE

Use the inverse of the matrix $A = \begin{bmatrix} 1 & -2 & 2 \\ -1 & 1 & 3 \\ 1 & -1 & -4 \end{bmatrix}$ to decode the cryptogram.

$$13 \ \ -26 \ \ 21 \ \ 33 \ \ -53 \ \ -12 \ \ 18 \ \ -23 \ \ -42 \ \ 5 \ \ -20 \ \ 56 \ \ -24 \ \ 23 \ \ 77$$

SOLUTION

First find A^{-1} by using the techniques demonstrated in Section 5.3. A^{-1} is the decoding matrix. Then partition the message into groups of three to form the coded row matrices. Multiply each coded row matrix on the right by A^{-1} to obtain the decoded row matrices.

Coded Matrix   *Decoding Matrix* A^{-1} *Decoded Matrix*

$$\begin{bmatrix} 13 & -26 & 21 \end{bmatrix} \begin{bmatrix} -1 & -10 & -8 \\ -1 & -6 & -5 \\ 0 & -1 & -1 \end{bmatrix} = \begin{bmatrix} 13 & 5 & 5 \end{bmatrix}$$

$$\begin{bmatrix} 33 & -53 & -12 \end{bmatrix} \begin{bmatrix} -1 & -10 & -8 \\ -1 & -6 & -5 \\ 0 & -1 & -1 \end{bmatrix} = \begin{bmatrix} 20 & 0 & 13 \end{bmatrix}$$

$$\begin{bmatrix} 18 & -23 & -42 \end{bmatrix} \begin{bmatrix} -1 & -10 & -8 \\ -1 & -6 & -5 \\ 0 & -1 & -1 \end{bmatrix} = \begin{bmatrix} 5 & 0 & 13 \end{bmatrix}$$

$$\begin{bmatrix} 5 & -20 & 56 \end{bmatrix} \begin{bmatrix} -1 & -10 & -8 \\ -1 & -6 & -5 \\ 0 & -1 & -1 \end{bmatrix} = \begin{bmatrix} 15 & 14 & 4 \end{bmatrix}$$

$$\begin{bmatrix} -24 & 23 & 77 \end{bmatrix} \begin{bmatrix} -1 & -10 & -8 \\ -1 & -6 & -5 \\ 0 & -1 & -1 \end{bmatrix} = \begin{bmatrix} 1 & 25 & 0 \end{bmatrix}$$

So, the message is as follows.

$$\begin{array}{ccccc} \begin{bmatrix} 13 & 5 & 5 \end{bmatrix} & \begin{bmatrix} 20 & 0 & 13 \end{bmatrix} & \begin{bmatrix} 5 & 0 & 13 \end{bmatrix} & \begin{bmatrix} 15 & 14 & 4 \end{bmatrix} & \begin{bmatrix} 1 & 25 & 0 \end{bmatrix} \\ \text{M E E} & \text{T M E} & \text{M O} & \text{N D} & \text{A Y} \end{array}$$

✓CHECKPOINT   Now try Exercise 41.

Discussing the Concept | Decoding

Show how to use a graphing utility with matrix operations to decode the following cryptogram. (Use the matrix given in Example 6.)

$$12 \ \ -25 \ \ 15 \ \ \ \ 28 \ \ -32 \ \ -89 \ \ \ \ 10 \ \ -10 \ \ -49 \ \ \ 12$$

$$-12 \ \ -51 \ \ 17 \ \ -31 \ \ \ \ 10 \ \ \ \ 10 \ \ -28 \ \ \ \ 55 \ \ \ \ 4 \ \ -8 \ \ 8$$

5.5 Warm Up

The following warm-up exercises involve skills that were covered in earlier sections. You will use these skills in the exercise set for this section. For additional help, review Sections 5.2, 5.3, and 5.4.

In Exercises 1–6, evaluate the determinant.

1. $\begin{vmatrix} 4 & 3 \\ -3 & -2 \end{vmatrix}$

2. $\begin{vmatrix} 10 & -20 \\ -1 & 2 \end{vmatrix}$

3. $\begin{vmatrix} 4 & 0 \\ -3 & -2 \end{vmatrix}$

4. $\begin{vmatrix} x & x^2 \\ 1 & 2x \end{vmatrix}$

5. $\begin{vmatrix} 4 & 0 & -2 \\ 3 & 1 & 2 \\ -8 & 0 & 6 \end{vmatrix}$

6. $\begin{vmatrix} 3 & 2 & 5 \\ 0 & 0 & -4 \\ -6 & 1 & 1 \end{vmatrix}$

In Exercises 7 and 8, find the inverse of the matrix.

7. $A = \begin{bmatrix} 1 & 3 \\ 2 & 7 \end{bmatrix}$

8. $A = \begin{bmatrix} 10 & 5 & -2 \\ -4 & -2 & 1 \\ 1 & 1 & 0 \end{bmatrix}$

In Exercises 9 and 10, perform the indicated matrix multiplication.

9. $\begin{bmatrix} 0.1 & 0.2 \\ 0.4 & 0.3 \end{bmatrix}\begin{bmatrix} 0.4 \\ 0.5 \end{bmatrix}$

10. $\begin{bmatrix} 2 & 5 \end{bmatrix}\begin{bmatrix} 1 & 2 \\ 1 & 2 \end{bmatrix}$

5.5 Exercises

In Exercises 1–10, use a determinant to find the area of the triangle with the given vertices.

1.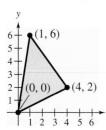

2.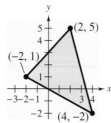

3.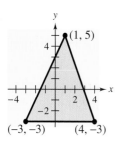

4.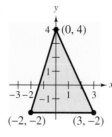

5.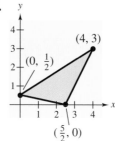

6.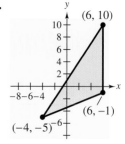

7. $(-2, 4), (2, 3), (-1, 5)$ 8. $(0, -2), (-1, 4), (3, 5)$
9. $(-3, 5), (2, 6), (3, -5)$ 10. $(-2, 4), (1, 5), (3, -2)$

In Exercises 11 and 12, find a value of y such that the triangle with the given vertices has an area of 4 square units.

11. $(-5, 1), (0, 2), (-2, y)$
12. $(-4, 2), (-3, 5), (-1, y)$

In Exercises 13 and 14, find a value of y such that the triangle with the given vertices has an area of 6 square units.

13. $(-2, -3), (1, -1), (-8, y)$

14. $(1, 0), (5, -3), (-3, y)$

15. Area of a Region   A large region of forest has been infected with gypsy moths. The region is roughly triangular, as shown in the figure. From the northernmost vertex A of the region, the distances to the other vertices are 25 miles south and 10 miles east (for vertex B), and 20 miles south and 28 miles east (for vertex C). Use a graphing utility to approximate the number of square miles in this region.

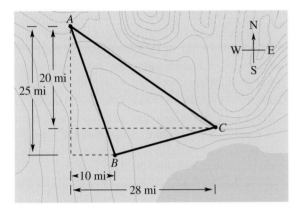

16. Area of a Region   You own a triangular tract of land, as shown in the figure. To estimate the number of square feet in the tract, you start at one vertex, walk 65 feet east and 50 feet north to the second vertex, and then walk 85 feet west and 30 feet north to the third vertex. Use a graphing utility to determine how many square feet there are in the tract of land.

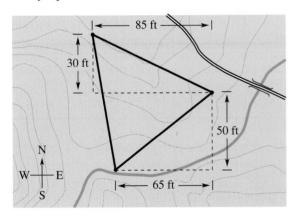

Make a Decision   In Exercises 17–22, use a determinant to determine whether the points are collinear.

17. $(-4, -7), (0, -4), (4, -1)$

18. $(2, 4), (4, 5), (-2, 2)$

19. $(-1, -7), (0, -3), (1, 2)$

20. $(-2, -11), (4, 13), (2, 5)$

21. $(1, 7), (0, 4), (-1, 2)$

22. $(4, 3), (3, 1), (2, -1)$

In Exercises 23 and 24, find y such that the points are collinear.

23. $(2, -5), (3, y), (5, -2)$

24. $(-6, 2), (-4, y), (-3, 5)$

In Exercises 25–30, use a determinant to find an equation of the line passing through the points.

25. $(-1, 2), (5, 3)$ **26.** $(3, 1), (-2, -5)$

27. $(-4, 3), (2, 1)$ **28.** $(10, 7), (-2, -7)$

29. $\left(-\frac{1}{2}, 3\right), \left(\frac{5}{2}, 1\right)$ **30.** $\left(\frac{2}{3}, 4\right), (6, 12)$

In Exercises 31 and 32, find the uncoded 1×2 row matrices for the message. Then encode the message using the encoding matrix.

Message	*Encoding Matrix*
31. COME HOME SOON	$\begin{bmatrix} 1 & 2 \\ 3 & 5 \end{bmatrix}$
32. HELP IS ON THE WAY	$\begin{bmatrix} -2 & 3 \\ -1 & 1 \end{bmatrix}$

In Exercises 33 and 34, find the uncoded 1×3 row matrices for the message. Then encode the message using the encoding matrix.

Message	*Encoding Matrix*
33. CALL ME TOMORROW	$\begin{bmatrix} 1 & -1 & 0 \\ 1 & 0 & -1 \\ -6 & 2 & 3 \end{bmatrix}$
34. PLEASE SEND MONEY	$\begin{bmatrix} 4 & 2 & 1 \\ -3 & -3 & -1 \\ 3 & 2 & 1 \end{bmatrix}$

In Exercises 35–40, write a cryptogram for the message using the matrix

$$A = \begin{bmatrix} 1 & 2 & 2 \\ 3 & 7 & 9 \\ -1 & -4 & -7 \end{bmatrix}.$$

35. LANDING SUCCESSFUL

36. BEAM ME UP SCOTTY

37. HAPPY BIRTHDAY

38. OPERATION OVERLORD

39. CONTACT AT DAWN

40. HEAD DUE WEST

In Exercises 41–44, use A^{-1} to decode the cryptogram.

41. $A = \begin{bmatrix} 1 & 2 \\ 3 & 5 \end{bmatrix}$

11, 21, 64, 112, 25, 50, 29, 53, 23, 46, 40, 75, 55, 92

42. $A = \begin{bmatrix} 2 & 3 \\ 3 & 4 \end{bmatrix}$

19, 26, 41, 57, 28, 42, 78, 109, 64, 87, 62, 83, 63, 87, 28, 42, 73, 102, 46, 69

43. $A = \begin{bmatrix} 4 & 2 & 1 \\ -3 & -3 & -1 \\ 3 & 2 & 1 \end{bmatrix}$

94, 35, 25, 44, 16, 10, 4, −10, 1, 27, 15, 9, 71, 43, 22

44. $A = \begin{bmatrix} 1 & -1 & 0 \\ 1 & 0 & -1 \\ -6 & 2 & 3 \end{bmatrix}$

9, −1, −9, 38, −19, −19, 28, −9, −19, −80, 25, 41, −64, 21, 31, −7, −4, 7

In Exercises 45 and 46, decode the cryptogram by using the inverse of the matrix A.

$A = \begin{bmatrix} 1 & 2 & 2 \\ 3 & 7 & 9 \\ -1 & -4 & -7 \end{bmatrix}$

45. 20, 17, −15, −9, −44, −83, 64, 136, 157, 24, 31, 12, 4, −37, −102

46. −10, −57, −111, 74, 168, 209, 35, 75, 85, 16, 35, 42, 34, 55, 43

47. The following cryptogram was encoded with a 2 × 2 matrix.

8, 21, −15, −10, −13, −13, 5, 10, 5, 25, 5, 19, −1, 6, 20, 40, −18, −18, 1, 16

The last word of the message is __RON. What is the message?

48. The following cryptogram was encoded with a 2 × 2 matrix.

5, 2, 25, 11, −2, −7, −15, −15, 32, 14, −8, −13, 38, 19, −19, −19, 37, 16

The last word of the message is __SUE. What is the message?

IN THE NEWS
Spying on Us

. . . Echelon is perhaps the best known and least understood spy tool. Although it is run by the U.S. National Security Agency (NSA), and paid for almost entirely by American taxpayers, it is a multinational spying effort . . . It wasn't until 1957, five years after NSA was created, that the federal government would admit that it even existed.

Simply put, the agency's job is to eavesdrop and share its notes. On a day-to-day basis, this means intercepting radio signals, unscrambling encrypted messages, and distributing the resulting information to a host of espionage organizations. Its chief "customer" is the Central Intelligence Agency.

The electronic signals that Echelon satellites and listening posts capture are separated into two streams, depending upon whether the communications are sent with or without encryption. . . .

Wilson, Jim. *"Spying on Us,"* Popular Mechanics, *April 2001, 70.*

49. Write a cryptogram for the message "NEED ASSISTANCE" using the matrix

$A = \begin{bmatrix} -5 & 2 & 0 \\ 3 & -6 & 8 \\ 0 & 1 & 8 \end{bmatrix}$.

50. Decode the cryptogram 33, 60, 16, 62, 112, 31, 61, 108, 29, 103, 161, 47, 50, 75, 25 by using the inverse of the matrix A.

$A = \begin{bmatrix} 2 & 3 & 1 \\ 3 & 3 & 1 \\ 2 & 4 & 1 \end{bmatrix}$

51. *Make a Decision*: **Cryptography** A code breaker intercepted the encoded message below.

45, −35, 38, −30, 18, −18, 35, −30, 81, −60, 42, −28, 75, −55, 2, −2, 22, −21, 15, −10

Let $A^{-1} = \begin{bmatrix} w & x \\ y & z \end{bmatrix}$.

(a) You know that $[45 \ -35]A^{-1} = [10 \ 15]$ and that $[38 \ -30]A^{-1} = [8 \ 14]$, where A^{-1} is the inverse of the encoding matrix A. How can you find the values of w, x, y, and z?

(b) Decode the message.

Make a Decision ➤ Project: Market Share

Your advertising group has been hired by Company C, a cell phone service provider, to develop an advertising campaign that will increase its market share from 10% to at least 30% in 1 year. Currently, your client competes with two other cell phone service providers, Company A and Company B. Your group has designed a campaign that you believe will meet the 30% objective. With this new campaign, you predict that each month

(1) 90% of A's subscribers will remain with A, 4% will switch to B, and 6% will switch to C;

(2) 92% of B's subscribers will remain with B, 5% will switch to A, and 3% will switch to C; and

(3) 94% of C's subscribers will remain with C, 2% will switch to A, and 4% will switch to B.

$$x_0 = \begin{bmatrix} 0.2 \\ 0.7 \\ 0.1 \end{bmatrix} \begin{array}{l} \text{Company A} \\ \text{Company B} \\ \text{Company C} \end{array}$$

The current shares of the market are given in the matrix at the left. After 1 month, the shares of the three companies are predicted to be

$$x_1 = px_0 = \begin{bmatrix} 0.90 & 0.05 & 0.02 \\ 0.04 & 0.92 & 0.04 \\ 0.06 & 0.03 & 0.94 \end{bmatrix} \begin{bmatrix} 0.2 \\ 0.7 \\ 0.1 \end{bmatrix} = \begin{bmatrix} 0.217 \\ 0.656 \\ 0.127 \end{bmatrix}. \quad \begin{array}{l} \text{Company A} \\ \text{Company B} \\ \text{Company C} \end{array}$$

Use this information to investigate the following questions.

1. **Making a Table**   Construct a table that shows the predicted market share for each of the three companies during each of the first 12 months. The matrix operations on your graphing utility can help you with the calculations.

2. **Fitting a Model to Data**   Enter the results for Company C from Question 1 using the statistical features of your graphing utility.

   (a) Create a scatter plot of the data.

   (b) What type of model do you think best fits the data? Select a regression program on your graphing utility to model the data.

   (c) Graph the model you found in part (b) for the data for Company C.

   (d) Repeat (a), (b), and (c) using the data for Company A and Company B from Question 1.

3. *Make a Decision*   Graph the models for Companies A, B, and C in the same viewing window. Will Company C's market share reach 30% within 12 months? Based on the model for Company C's market share, will Company C maintain a 30% or better market share?

4. *Make a Decision*   Are the models you found good predictors over long periods of time? Would you expect this type of market to fluctuate or be static? As Company C begins to take more of the market share, what action do you predict Companies A and B will take? What possible effect on this system could their actions have?

Chapter Summary

After studying this chapter, you should have acquired the following skills.
These skills are keyed to the Review Exercises that begin on page 381.
Answers to odd-numbered Review Exercises are given in the back of the book.

5.1 • Determine the order of a matrix.   *Review Exercises 1, 2*

• Perform elementary row operations on a matrix to write the matrix   *Review Exercises 3–6*
in row-echelon form or reduced row-echelon form.

• Solve a system of linear equations using Gaussian elimination or   *Review Exercises 7–16*
Gauss-Jordan elimination.

5.2 • Add or subtract two matrices and multiply a matrix by a scalar.   *Review Exercises 17–20, 31, 32*

• Find the product of two matrices.   *Review Exercises 21–28*

• Solve a matrix equation.   *Review Exercises 29, 30*

• Use matrix multiplication to solve an application problem.   *Review Exercises 33, 34*

5.3 • Verify that a matrix is the inverse of a given matrix.   *Review Exercises 35, 36*

• Find the inverse of a matrix.   *Review Exercises 37, 38*

• Find the inverse of a 2×2 matrix using a formula.   *Review Exercises 39, 40*

• Use an inverse matrix to solve a system of linear equations.   *Review Exercises 41–52*

5.4 • Evaluate the determinant of a 2×2 matrix.   *Review Exercises 53–56*

• Find the minors and cofactors of a matrix.   *Review Exercises 57–60*

• Find the determinant of a square matrix.   *Review Exercises 61, 62, 65, 66*

• Find the determinant of a triangular matrix.   *Review Exercises 63, 64*

5.5 • Find the area of a triangle using a determinant.   *Review Exercises 67, 68*

• Determine whether three points are collinear using a determinant.   *Review Exercises 69–72*

• Use a determinant to find an equation of a line.   *Review Exercises 73, 74*

• Encode and decode a cryptogram using a matrix.   *Review Exercises 75–78*

Review Exercises

In Exercises 1 and 2, determine the order of the matrix.

1. $\begin{bmatrix} 3 & 7 & 4 & -2 \\ 1 & 8 & 6 & 1 \end{bmatrix}$ **2.** $\begin{bmatrix} 5 \\ -1 \\ 2 \\ 4 \end{bmatrix}$

In Exercises 3 and 4, write the matrix in row-echelon form.

3. $\begin{bmatrix} 1 & 3 & 0 & 2 \\ 3 & 10 & 1 & 8 \\ 2 & 3 & 3 & 10 \end{bmatrix}$

4. $\begin{bmatrix} 1 & 2 & -1 & 0 \\ -2 & -3 & 3 & 4 \\ 4 & 0 & 1 & 3 \end{bmatrix}$

In Exercises 5 and 6, write the matrix in reduced row-echelon form.

5. $\begin{bmatrix} 1 & 2 & 3 \\ -2 & 0 & 2 \\ 2 & 1 & 2 \end{bmatrix}$

6. $\begin{bmatrix} 2 & 3 & 1 & -5 \\ 1 & 0 & 5 & 2 \\ -1 & 4 & 3 & 6 \\ 0 & -2 & 6 & -8 \end{bmatrix}$

In Exercises 7–14, use matrices to solve the system (if possible). Use Gaussian elimination with back-substitution or Gauss-Jordan elimination.

7. $\begin{cases} 4x - 3y = 18 \\ x + y = 1 \end{cases}$ **8.** $\begin{cases} 2x + 4y = 16 \\ -x + 3y = 17 \end{cases}$

9. $\begin{cases} 2x + 3y - z = 13 \\ 3x + z = 8 \\ x - 2y + 3z = -4 \end{cases}$ **10.** $\begin{cases} 3x + 4y + 2z = 5 \\ 2x + 3y = 7 \\ 2y - 3z = 12 \end{cases}$

11. $\begin{cases} x + 2y + 2z = 10 \\ 2x + 3y + 5z = 20 \end{cases}$ **12.** $\begin{cases} 3x + 10y + 4z = 20 \\ x + 3y - 2z = 8 \end{cases}$

13. $\begin{cases} 2x + y - 3z = 4 \\ x + 2y + 2z = 10 \\ x - 2z = 12 \\ x + y + z = 6 \end{cases}$ **14.** $\begin{cases} 2x + 4y + 2z = 10 \\ x + 3z = 9 \\ 3x - 2y = 4 \\ x + y + z = 8 \end{cases}$

15. Borrowing Money A recreational vehicle company borrowed $900,000 at simple annual interest to expand its product line. Some of the money was borrowed at 8%, some at 10%, and some at 12%.Use a system of equations to determine how much was borrowed at each rate if the total annual interest was $88,000 and the amount borrowed at 10% was three times the amount borrowed at 8%. Solve the system using matrices.

16. Borrowing Money A restaurant owner borrowed $800,000 at simple annual interest to renovate a restaurant. Some of the money was borrowed at 8.5%, some at 9.5%, and some at 10%. Use a system of equations to determine how much was borrowed at each rate if the total annual interest was $72,500 and the amount borrowed at 8.5% was four times the amount borrowed at 10%. Solve the system using matrices.

In Exercises 17–20, find (a) $A + B$, (b) $A - B$, (c) $4A$, and (d) $4A - 3B$.

17. $A = \begin{bmatrix} -1 & 5 \\ 2 & 1 \end{bmatrix}$, $B = \begin{bmatrix} 4 & 2 \\ -6 & 3 \end{bmatrix}$

18. $A = \begin{bmatrix} 1 & 0 & 2 \\ -1 & 3 & 5 \\ 2 & -2 & 3 \end{bmatrix}$, $B = \begin{bmatrix} 2 & 0 & 1 \\ 3 & -4 & 6 \\ 1 & 2 & -3 \end{bmatrix}$

19. $A = \begin{bmatrix} 1 & 3 & -2 & 6 \\ 0 & 1 & 3 & 2 \end{bmatrix}$,

$B = \begin{bmatrix} 2 & 1 & 4 & -5 \\ 3 & -6 & 3 & -2 \end{bmatrix}$

20. $A = \begin{bmatrix} 3 \\ -2 \\ 3 \end{bmatrix}$, $B = \begin{bmatrix} -1 \\ 4 \\ 5 \end{bmatrix}$

In Exercises 21–26, find AB, if possible.

21. $A = \begin{bmatrix} 1 & 4 \\ -2 & -1 \\ 3 & 2 \end{bmatrix}$, $B = \begin{bmatrix} -4 \\ 3 \end{bmatrix}$

22. $A = \begin{bmatrix} 3 \\ 2 \\ 4 \\ 6 \end{bmatrix}$, $B = \begin{bmatrix} 2 & 0 & -1 \end{bmatrix}$

23. $A = \begin{bmatrix} 4 & 0 & 0 \\ 0 & 3 & 0 \\ 0 & 0 & -2 \end{bmatrix}$, $B = \begin{bmatrix} \frac{1}{4} & 0 & 0 \\ 0 & \frac{1}{3} & 0 \\ 0 & 0 & -\frac{1}{2} \end{bmatrix}$

24. $A = \begin{bmatrix} 3 & 1 \\ 4 & 7 \\ 1 & 1 \end{bmatrix}$, $B = \begin{bmatrix} 1 & 2 & -2 \\ 3 & 4 & 0 \\ 0 & 1 & 0 \end{bmatrix}$

25. $A = \begin{bmatrix} 1 & 2 & 3 & 6 & -1 \\ 2 & 8 & 0 & 0 & 2 \end{bmatrix}$, $B = \begin{bmatrix} 3 & 2 \\ 4 & -1 \end{bmatrix}$

26. $A = \begin{bmatrix} 0 & 0 & 2 \\ 1 & 0 & 6 \\ 0 & 2 & 2 \end{bmatrix}$, $B = \begin{bmatrix} 3 & 4 & 0 & 1 \\ 2 & 1 & 0 & 0 \\ 0 & 0 & 1 & 1 \end{bmatrix}$

In Exercises 27 and 28, find (a) AB, (b) BA, and, if possible, (c) A^2. (*Note:* $A^2 = AA$.)

27. $A = \begin{bmatrix} 1 & -3 & 4 \end{bmatrix}$, $B = \begin{bmatrix} 2 \\ -2 \\ -1 \end{bmatrix}$

28. $A = \begin{bmatrix} 1 & 0 & 2 \\ 3 & 1 & -2 \\ 1 & 1 & 1 \end{bmatrix}$, $B = \begin{bmatrix} 2 & 0 & 0 \\ 1 & -2 & 1 \\ 5 & 4 & -2 \end{bmatrix}$

In Exercises 29 and 30, solve for X when

$A = \begin{bmatrix} 1 & -2 \\ 0 & 1 \\ 2 & 3 \end{bmatrix}$ and $B = \begin{bmatrix} 0 & 1 \\ 1 & 1 \\ 3 & 5 \end{bmatrix}$.

29. $X = 4A - 3B$

30. $2X - 3A = B$

31. Factory Production A window corporation has four factories, each of which manufactures three products. The number of units of product i produced at factory j in 1 day is represented by a_{ij} in the matrix

$A = \begin{bmatrix} 80 & 120 & 20 & 40 \\ 40 & 60 & 80 & 20 \\ 140 & 60 & 100 & 80 \end{bmatrix}$.

Find the production levels if production is increased by 25%.

32. Factory Production An electronics manufacturer has three factories, each of which manufactures four products. The number of units of product i produced at factory j in 1 day is represented by a_{ij} in the matrix

$A = \begin{bmatrix} 120 & 140 & 60 \\ 80 & 100 & 40 \\ 40 & 160 & 80 \\ 20 & 120 & 100 \end{bmatrix}$.

Find the production levels if production is decreased by 15%.

33. Inventory Levels A company sells four different models of car sound systems through three retail outlets. The inventories of the four models at the three outlets are given by matrix S.

Model

	A	B	C	D	
	3	2	1	4	1
$S =$	1	3	4	3	2 } Outlet
	5	3	2	2	3

The wholesale and retail prices of the four models are given by matrix T.

Price

	Wholesale	Retail	
	350	600	A
$T =$	425	705	B } Model
	300	455	C
	750	1150	D

Use a graphing utility to compute ST and interpret the result.

34. Labor/Wage Requirements A company that manufactures racing bicycles has the following labor-hour and wage requirements.

Labor-Hour Requirements (per bicycle)

Department

	Cutting	Assembly	Packaging	
	0.9 hour	0.8 hour	0.2 hour	Basic
$S =$	1.5 hours	1.0 hour	0.4 hour	Light } Models
	3.5 hours	3.0 hours	0.5 hour	Ultra-Light

Wage Requirements (per hour)

Plant

	A	B	
	\$15	\$13	Cutting
$T =$	\$9	\$10	Assembly } Department
	\$8	\$7	Packaging

(a) What is the labor cost for a light racing bicycle at Plant A?

(b) What is the labor cost for an ultra-light racing bicycle at Plant B?

(c) Use a graphing utility to compute ST and interpret the result.

In Exercises 35 and 36, show that B is the inverse of A.

35. $A = \begin{bmatrix} 1 & 2 & 1 \\ 3 & 6 & 4 \\ 0 & 1 & 3 \end{bmatrix}$, $B = \begin{bmatrix} -14 & 5 & -2 \\ 9 & -3 & 1 \\ -3 & 1 & 0 \end{bmatrix}$

36. $A = \begin{bmatrix} 2 & 0 & 1 & 2 \\ 3 & 0 & 0 & 1 \\ -1 & 1 & 2 & 0 \\ 0 & -1 & 2 & 2 \end{bmatrix}$,

$B = \dfrac{1}{9}\begin{bmatrix} -4 & 6 & 1 & 1 \\ 10 & -6 & 2 & -7 \\ -7 & 6 & 4 & 4 \\ 12 & -9 & -3 & -3 \end{bmatrix}$

In Exercises 37 and 38, find the inverse of the matrix (if it exists).

37. $\begin{bmatrix} -1 & 0 & 0 \\ 0 & 2 & 0 \\ 0 & 0 & 4 \end{bmatrix}$ **38.** $\begin{bmatrix} 3 & 2 & 2 \\ 0 & 2 & 1 \\ 1 & 0 & 1 \end{bmatrix}$

In Exercises 39 and 40, use the formula on page 353 to find the inverse of the matrix.

39. $\begin{bmatrix} 1 & 3 \\ 2 & 5 \end{bmatrix}$ **40.** $\begin{bmatrix} -2 & 1 \\ 4 & 3 \end{bmatrix}$

In Exercises 41 and 42, use the inverse matrix found in Exercise 39 to solve the system of linear equations.

41. $\begin{cases} x + 3y = 15 \\ 2x + 5y = 26 \end{cases}$ **42.** $\begin{cases} x + 3y = 7 \\ 2x + 5y = 11 \end{cases}$

In Exercises 43 and 44, use the inverse matrix found in Exercise 38 to solve the system of linear equations.

43. $\begin{cases} 3x + 2y + 2z = 13 \\ 2y + z = 4 \\ x + z = 5 \end{cases}$ **44.** $\begin{cases} 3x + 2y + 2z = 12 \\ 2y + z = 13 \\ x + z = 3 \end{cases}$

In Exercises 45–48, use an inverse matrix to solve (if possible) the system of linear equations.

45. $\begin{cases} -3x + 10y = 8 \\ 5x - 17y = -13 \end{cases}$ **46.** $\begin{cases} 5x - y = 13 \\ -9x + 2y = -24 \end{cases}$

47. $\begin{cases} 3x + 2y - z = 6 \\ x - y + 2z = -1 \\ 5x + y + z = 7 \end{cases}$

48. $\begin{cases} -x + 4y - 2z = 12 \\ 2x - 9y + 5z = -25 \\ -x + 5y - 4z = 10 \end{cases}$

Computer Models In Exercises 49 and 50, consider a company that produces three computer models. Model A requires 2 units of plastic, 2 units of computer chips, and 1 unit of computer "cards." Model B requires 1 unit of plastic, 3 units of computer chips, and 2 units of computer cards. Model C requires 3 units of plastic, 2 units of computer chips, and 2 units of computer cards. A system of linear equations (where $x, y,$ and z represent models A, B, and C, respectively) is as follows.

$\begin{cases} 2x + y + 3z = \text{(units of plastic)} \\ 2x + 3y + 2z = \text{(units of computer chips)} \\ x + 2y + 2z = \text{(units of computer cards)} \end{cases}$

Use the inverse of the coefficient matrix of this system to find the numbers of models A, B, and C that the company can produce with the given amounts of components.

49. 800 units of plastic
   1000 units of computer chips
   700 units of computer cards

50. 750 units of plastic
   875 units of computer chips
   600 units of computer cards

51. College Applications The percent y of U.S. college freshmen who applied to three or more colleges from 1997 to 2001 increased in a pattern that was approximately parabolic. The least squares regression parabola $y = at^2 + bt + c$ for the data is found by solving the system

$\begin{cases} 5c + 45b + 415a = 239.8 \\ 45c + 415b + 3915a = 2176.9. \\ 415c + 3915b + 37{,}699a = 20{,}238.7 \end{cases}$

Let t represent the year, with $t = 7$ corresponding to 1997. (Source: The Higher Education Research Institute)

(a) Use a graphing utility to find an inverse matrix to solve this system, and find the equation of the least squares regression parabola.

(b) Use the result of part (a) to approximate the percent of college freshmen applying to three or more colleges in 2006.

52. **Social Science** The percents y of U.S. college freshmen who identified social science as their probable field of study from 1997 to 2001 increased in a pattern that was approximately linear. The least squares regression line $y = at + b$ for the data is found by solving the system

$$\begin{cases} 5b + 45a = 48.8 \\ 45b + 415a = 441.7 \end{cases}$$

Let t represent the year, with $t = 7$ corresponding to 1997. (Source: The Higher Education Research Institute)

(a) Use a graphing utility to find an inverse matrix to solve this system, and find the equation of the least squares regression line.

(b) Use the result of part (a) to approximate the percent of college freshmen identifying social science as their probable field of study in 2007.

In Exercises 53–56, find the determinant of the 2 × 2 matrix.

53. $\begin{bmatrix} 8 & 4 \\ 3 & 2 \end{bmatrix}$

54. $\begin{bmatrix} 7 & 2 \\ 9 & -3 \end{bmatrix}$

55. $\begin{bmatrix} 5 & 2 \\ 0 & 0 \end{bmatrix}$

56. $\begin{bmatrix} 3 & 0 \\ 0 & -7 \end{bmatrix}$

In Exercises 57–60, find all (a) minors and (b) cofactors of the matrix.

57. $\begin{bmatrix} 2 & -1 \\ 7 & 4 \end{bmatrix}$

58. $\begin{bmatrix} 3 & 6 \\ 5 & -4 \end{bmatrix}$

59. $\begin{bmatrix} 3 & 2 & -1 \\ -2 & 5 & 0 \\ 1 & 8 & 6 \end{bmatrix}$

60. $\begin{bmatrix} 8 & 3 & 4 \\ 6 & 5 & -9 \\ -4 & 1 & 2 \end{bmatrix}$

Make a Decision In Exercises 61–66, find the determinant of the matrix. Tell whether you used expansion by minors, the product of the entries on the main diagonal, or upward and downward diagonals.

61. $\begin{bmatrix} 1 & 2 & 3 \\ 8 & 6 & 7 \\ 0 & 2 & -1 \end{bmatrix}$

62. $\begin{bmatrix} -2 & 3 & 3 \\ -1 & 0 & 5 \\ 1 & 2 & -1 \end{bmatrix}$

63. $\begin{bmatrix} 1 & 3 & 2 & 4 \\ 0 & -1 & 2 & 2 \\ 0 & 0 & 3 & 0 \\ 0 & 0 & 0 & 4 \end{bmatrix}$

64. $\begin{bmatrix} 2 & 0 & 0 & 0 \\ 3 & 4 & 0 & 0 \\ 5 & 1 & -2 & 0 \\ 6 & 3 & 1 & 1 \end{bmatrix}$

65. $\begin{bmatrix} -2 & 4 & 1 \\ 6 & 1 & 2 \\ 5 & 3 & 4 \end{bmatrix}$

66. $\begin{bmatrix} 4 & 7 & -1 \\ 2 & -3 & 4 \\ -5 & 1 & -1 \end{bmatrix}$

In Exercises 67 and 68, use a determinant to find the area of the triangle with the given vertices.

67. $(3, 4), (2, -3), (-1, -4)$

68. $(1, -4), (-2, 3), (0, 6)$

Make a Decision In Exercises 69–72, use a determinant to determine whether the points are collinear.

69. $(0, 3), (1, 5), (2, 8)$

70. $(2, 6), (-2, 3), (0, 5)$

71. $(-4, 1), (6, 6), (0, 3)$

72. $(-3, -1), (0, 5), (-4, -3)$

In Exercises 73 and 74, use a determinant to find an equation of the line passing through the points.

73. $(-7, 3), (8, 2)$ 74. $(5, -4), (-3, 2)$

In Exercises 75 and 76, find the uncoded row matrices for the message. Then encode the message using the encoding matrix.

Message	Encoding Matrix
75. TRANSMIT NOW	$\begin{bmatrix} 2 & 3 \\ 3 & 4 \end{bmatrix}$
76. CALL AT MIDNIGHT	$\begin{bmatrix} 1 & 2 & 2 \\ 3 & 7 & 9 \\ -1 & -4 & -7 \end{bmatrix}$

In Exercises 77 and 78, use A^{-1} to decode the cryptogram.

77. $A = \begin{bmatrix} 1 & 2 \\ -1 & 3 \end{bmatrix}$

14, 53, -17, 96, 5, 10, 12, 64, 5, 10, 3, 11, 25, 50

78. $A = \begin{bmatrix} 1 & -1 & 0 \\ 1 & 0 & -1 \\ -6 & 2 & 3 \end{bmatrix}$

$-14, -1, 10, -38, 2, 27, -94, 18, 57, 7, -11, -1,$
$-96, 20, 57, -74, 23, 35, 17, -12, -5$

Chapter Test

Take this test as you would take a test in class. After you are done, check your work against the answers given in the back of the book.

In Exercises 1 and 2, write an augmented matrix that represents the system.

1. $\begin{cases} 3x + y = 1 \\ x - y = 7 \end{cases}$

2. $\begin{cases} 3x + 4y + 2z = 4 \\ 2x + 3y = -2 \\ 2y - 3z = -13 \end{cases}$

In Exercises 3–5, use matrices to solve the system.

3. $\begin{cases} x + 2y + 3z = 16 \\ 5x + 4y - z = 22 \end{cases}$

4. $\begin{cases} x - 2y + z = 14 \\ y - 3z = 2 \\ z = -6 \end{cases}$

5. $\begin{cases} 2x - 3y + z = 14 \\ x + 2y = -4 \\ y - z = -4 \end{cases}$

In Exercises 6–9, use the matrices to find the indicated matrix.

$$A = \begin{bmatrix} 1 & 3 \\ 2 & 4 \end{bmatrix}, \quad B = \begin{bmatrix} 2 & -1 & 3 \\ 4 & 0 & 1 \end{bmatrix}, \quad C = \begin{bmatrix} 0 & -2 \\ 3 & 5 \end{bmatrix}, \quad D = \begin{bmatrix} 3 \\ 2 \\ -1 \end{bmatrix}$$

6. $2A + C$

7. CA

8. BD

9. A^2

In Exercises 10–12, find the inverse of the matrix.

10. $A = \begin{bmatrix} 2 & -1 \\ -3 & 4 \end{bmatrix}$

11. $A = \begin{bmatrix} 1 & 0 \\ 0 & 1 \end{bmatrix}$

12. $A = \begin{bmatrix} 3 & 4 & 2 \\ 2 & 3 & 0 \\ 0 & 2 & -3 \end{bmatrix}$

In Exercises 13–15, find the determinant of the matrix.

13. $\begin{bmatrix} 3 & -1 \\ 4 & 7 \end{bmatrix}$

14. $\begin{bmatrix} 3 & 2 & -1 \\ 1 & 0 & 2 \\ 4 & 5 & 2 \end{bmatrix}$

15. $\begin{bmatrix} 2 & 0 & 0 \\ 0 & 5 & 0 \\ 0 & 0 & -2 \end{bmatrix}$

16. Use the inverse matrix found in Exercise 12 to solve the system in Exercise 2.

17. Find two nonzero matrices whose product is a zero matrix.

18. Find the area of the triangle whose vertices are $(-3, 1)$, $(0, 4)$, and $(5, 2)$.

19. Use a determinant to decide whether $(2, 1)$, $(-3, -14)$, and $(4, 7)$ are collinear.

20. A manufacturer produces three models of a product, which are shipped to two warehouses. The number of units i that are shipped to warehouse j is represented by a_{ij} in matrix A at the left. The price per unit is represented by matrix B. Find the product BA and interpret the result.

$$A = \begin{bmatrix} 1000 & 3000 \\ 2000 & 4000 \\ 5000 & 8000 \end{bmatrix}$$

$$B = \begin{bmatrix} \$50 & \$35 & \$28 \end{bmatrix}$$

Matrices for 20

6

Sequences, Series, and Probability

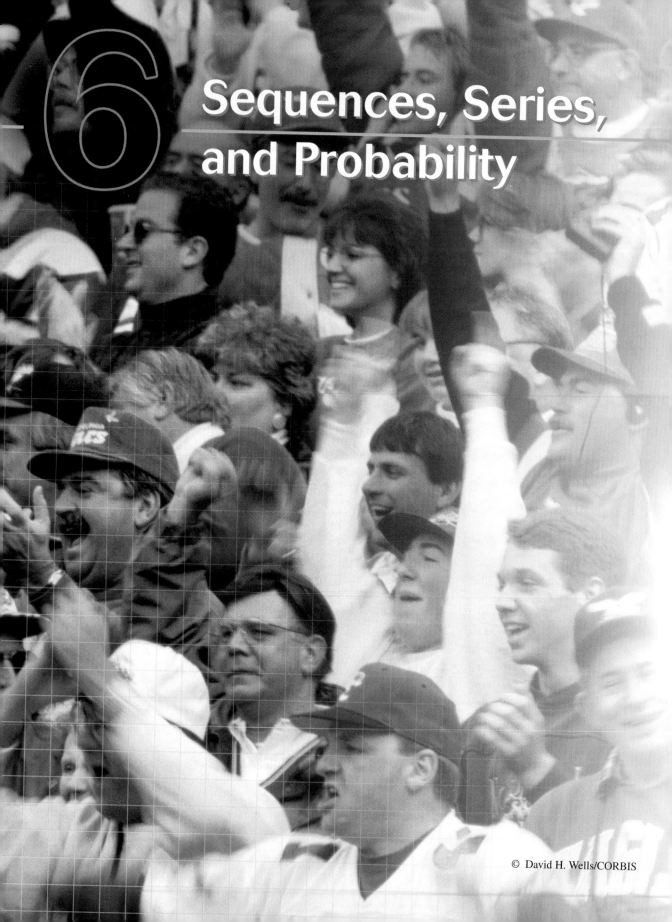

© David H. Wells/CORBIS

Chapter 6

Chapter Sections

6.1 Sequences and Summation Notation

6.2 Arithmetic Sequences and Partial Sums

6.3 Geometric Sequences and Series

6.4 The Binomial Theorem

6.5 Counting Principles

6.6 Probability

6.7 Mathematical Induction

Make a Decision ➤ Where Is Your Money Going?

In an economy, one person's expenditures become another person's income. If individuals in an economy spend 75% of their income, then for an initial 1 dollar the first person spends $0.75, the second person spends $0.75 × 75% = $0.56, the third person spends $0.56 × 75% = $0.42, and so on.

When a city hosts a large special event, such as the Super Bowl, the event can have a huge economic impact on the host city and surrounding areas. For instance, the table and bar graph show the cycled income in a city that hosted a large football event. In the table, x represents the cycle number of a dollar and y represents the total spending (in millions of dollars) for that cycle.

Make a Decision Suppose that the table at the right just listed the total spending amounts for the second and fourth cycles. How would you determine the spending amounts for the first and third cycles?

Cycle, x	Total spending, y
1	76.000
2	60.800
3	48.640
4	38.912
5	31.130
6	24.904

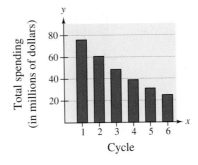

To explore this real-world application further, go to the Make a Decision Project on page 458.

6.1   Sequences and Summation Notation

Objectives

- **Use sequence notation to find the terms of a sequence.**
- **Evaluate a factorial expression.**
- **Find the sum of a finite sequence and use summation notation to write the sum of a sequence.**
- **Use a sequence to solve an application problem.**

Sequences

In mathematics, the word *sequence* is used in much the same way as in ordinary English. Saying that a collection is listed *in sequence* means that it is ordered so that it has a first member, a second member, a third member, and so on.

Mathematically, you can think of a sequence as a *function* whose domain is the set of positive integers. Rather than using function notation, however, sequences are usually written using subscript notation, as indicated in the definition below.

Definition of Sequence

An **infinite sequence** is a function whose domain is the set of positive integers. The function values

$$a_1, a_2, a_3, a_4, \ldots, a_n, \ldots$$

are the **terms** of the sequence. If the domain of the function consists of the first n positive integers only, the sequence is called a **finite sequence.**

On occasion it is convenient to begin subscripting a sequence with 0 instead of 1 so that the terms of the sequence become $a_0, a_1, a_2, a_3, \ldots$.

EXAMPLE 1   FINDING TERMS OF A SEQUENCE

a. The first four terms of the sequence given by $a_n = 3n - 2$ are

$$a_1 = 3(1) - 2 = 1 \qquad \text{1st term}$$
$$a_2 = 3(2) - 2 = 4 \qquad \text{2nd term}$$
$$a_3 = 3(3) - 2 = 7 \qquad \text{3rd term}$$
$$a_4 = 3(4) - 2 = 10. \qquad \text{4th term}$$

b. The first four terms of the sequence given by $a_n = 3 + (-1)^n$ are

$$a_1 = 3 + (-1)^1 = 3 - 1 = 2 \qquad a_2 = 3 + (-1)^2 = 3 + 1 = 4$$
$$a_3 = 3 + (-1)^3 = 3 - 1 = 2 \qquad a_4 = 3 + (-1)^4 = 3 + 1 = 4.$$

✓CHECKPOINT   Now try Exercise 1.

When working with sequences, note that any variable can be used as a subscript. The most commonly used variable subscripts in sequence and series notation are i, j, k, and n. Also note that the terms of a sequence need not all be positive, as shown in Example 2.

EXAMPLE 2   FINDING TERMS OF A SEQUENCE

The first four terms of the sequence given by $a_n = \dfrac{(-1)^n}{2n-1}$ are

$$a_1 = \frac{(-1)^1}{2(1)-1} = \frac{-1}{2-1} = -1 \qquad \text{1st term}$$

$$a_2 = \frac{(-1)^2}{2(2)-1} = \frac{1}{4-1} = \frac{1}{3} \qquad \text{2nd term}$$

$$a_3 = \frac{(-1)^3}{2(3)-1} = \frac{-1}{6-1} = -\frac{1}{5} \qquad \text{3rd term}$$

$$a_4 = \frac{(-1)^4}{2(4)-1} = \frac{1}{8-1} = \frac{1}{7}. \qquad \text{4th term}$$

✓**CHECKPOINT**   Now try Exercise 9.

Try finding the first four terms of the sequence whose nth term is

$$a_n = \frac{(-1)^{n+1}}{2n-1}.$$

How do they differ from the first four terms of the sequence in Example 2?

It is important to realize that simply listing the first few terms is not sufficient to define a unique sequence—the nth term *must be given*. To see this, consider the following sequences, both of which have the same first three terms.

$$\frac{1}{2}, \frac{1}{4}, \frac{1}{8}, \frac{1}{16}, \ldots, \frac{1}{2^n}, \ldots$$

$$\frac{1}{2}, \frac{1}{4}, \frac{1}{8}, \frac{1}{15}, \ldots, \frac{6}{(n+1)(n^2-n+6)}, \ldots$$

TECHNOLOGY

Use the *table* feature of a graphing utility to find the first four terms of each sequence. Be sure to start the table with $n = 1$ and set the table to increment n by 1.

a. $a_n = 2^n + 1$ **b.** $a_n = (-1)^n(2^n + 1)$ **c.** $a_n = (-1)^{n+1}(2^n + 1)$

Describe the differences in the terms of the sequences.

For instructions on how to use the *table* feature, see Appendix A; for specific keystrokes, go to the text website at *college.hmco.com*.

Factorial Notation

Some very important sequences in mathematics involve terms that are defined with special types of products called **factorials.**

Definition of Factorial

If n is a positive integer, n **factorial** is defined by

$$n! = 1 \cdot 2 \cdot 3 \cdot 4 \cdots (n - 1) \cdot n.$$

As a special case, zero factorial is defined as $0! = 1$.

Here are some values of $n!$ for the first several nonnegative integers. Notice that $0!$ is 1 by definition.

$$0! = 1$$

$$1! = 1$$

$$2! = 1 \cdot 2 = 2$$

$$3! = 1 \cdot 2 \cdot 3 = 6$$

$$4! = 1 \cdot 2 \cdot 3 \cdot 4 = 24$$

$$5! = 1 \cdot 2 \cdot 3 \cdot 4 \cdot 5 = 120$$

The value of n does not have to be very large before the value of $n!$ becomes huge. For instance, $10! = 3,628,800$.

TECHNOLOGY

Most scientific and graphing calculators have a *factorial* feature built in. Use the *factorial* feature to verify the factorials at the right. Try finding such values as $8!$, $15!$, and $20!$. How does the calculator's display change for increasingly large values of n? For specific keystrokes on how to use the *factorial* feature, go to the text website at *college.hmco.com.*

EXAMPLE 3   FINDING TERMS OF A SEQUENCE INVOLVING FACTORIALS

List the first five terms of the sequence given by

$$a_n = \frac{2}{n!}.$$

Begin with $n = 0$.

SOLUTION

$$a_0 = \frac{2}{0!} = \frac{2}{1} = 2 \qquad \text{0th term}$$

$$a_1 = \frac{2}{1!} = \frac{2}{1} = 2 \qquad \text{1st term}$$

$$a_2 = \frac{2}{2!} = \frac{2}{2} = 1 \qquad \text{2nd term}$$

$$a_3 = \frac{2}{3!} = \frac{2}{6} = \frac{1}{3} \qquad \text{3rd term}$$

$$a_4 = \frac{2}{4!} = \frac{2}{24} = \frac{1}{12} \qquad \text{4th term}$$

✓CHECKPOINT   Now try Exercise 13.

Factorials follow the same conventions for order of operations as do exponents. For instance,

$$2n! = 2(n!) = 2(1 \cdot 2 \cdot 3 \cdot 4 \cdots n)$$

whereas $(2n)! = 1 \cdot 2 \cdot 3 \cdot 4 \cdots 2n$.

EXAMPLE 4   FINDING TERMS OF A SEQUENCE INVOLVING FACTORIALS

List the first five terms of the sequence given by

$$a_n = \frac{2^n}{n!}.$$

Begin with $n = 0$.

SOLUTION

$$a_0 = \frac{2^0}{0!} = \frac{1}{1} = 1 \qquad \text{0th term}$$

$$a_1 = \frac{2^1}{1!} = \frac{2}{1} = 2 \qquad \text{1st term}$$

$$a_2 = \frac{2^2}{2!} = \frac{4}{2} = 2 \qquad \text{2nd term}$$

$$a_3 = \frac{2^3}{3!} = \frac{8}{6} = \frac{4}{3} \qquad \text{3rd term}$$

$$a_4 = \frac{2^4}{4!} = \frac{16}{24} = \frac{2}{3} \qquad \text{4th term}$$

✓**CHECKPOINT**   Now try Exercise 19.

When working with fractions involving factorials, you will often find that the fractions can be reduced.

EXAMPLE 5   SIMPLIFYING FACTORIAL EXPRESSIONS

Simplify each factorial expression.

a. $\dfrac{8!}{2! \cdot 6!}$ **b.** $\dfrac{2! \cdot 6!}{3! \cdot 5!}$ **c.** $\dfrac{n!}{(n - 1)!}$

SOLUTION

a. $\dfrac{8!}{2! \cdot 6!} = \dfrac{1 \cdot 2 \cdot 3 \cdot 4 \cdot 5 \cdot 6 \cdot 7 \cdot 8}{1 \cdot 2 \cdot 1 \cdot 2 \cdot 3 \cdot 4 \cdot 5 \cdot 6} = \dfrac{7 \cdot 8}{2} = 28$

b. $\dfrac{2! \cdot 6!}{3! \cdot 5!} = \dfrac{1 \cdot 2 \cdot 1 \cdot 2 \cdot 3 \cdot 4 \cdot 5 \cdot 6}{1 \cdot 2 \cdot 3 \cdot 1 \cdot 2 \cdot 3 \cdot 4 \cdot 5} = \dfrac{6}{3} = 2$

c. $\dfrac{n!}{(n - 1)!} = \dfrac{1 \cdot 2 \cdot 3 \cdots (n - 1) \cdot n}{1 \cdot 2 \cdot 3 \cdots (n - 1)} = n$

✓**CHECKPOINT**   Now try Exercise 33.

Series and Summation Notation

The sum of the terms of an infinite sequence is called an **infinite series.** The sum of the first n terms of a sequence is called a **finite series** or the **nth partial sum.** There is a convenient notation for the sum of the terms of a finite sequence. It is called **summation notation** or **sigma notation** because it involves the use of the uppercase Greek letter sigma, written as Σ.

Definition of Summation Notation

The sum of the first n terms of a sequence is represented by

$$\sum_{i=1}^{n} a_i = a_1 + a_2 + a_3 + a_4 + \cdots + a_n$$

where i is the **index of summation,** n is the **upper limit of summation,** and 1 is the **lower limit of summation.**

Summation notation is an instruction to add the terms of a sequence. From the above definition, the lower limit of the summation tells you where to begin the sum and the upper limit of the summation tells you where to end the sum.

EXAMPLE 6 SUMMATION NOTATION FOR SUMS

a. $\displaystyle\sum_{i=1}^{5} 3i = 3(1) + 3(2) + 3(3) + 3(4) + 3(5)$

$$= 3(1 + 2 + 3 + 4 + 5)$$

$$= 3(15) = 45$$

b. $\displaystyle\sum_{k=3}^{6} (1 + k^2) = (1 + 3^2) + (1 + 4^2) + (1 + 5^2) + (1 + 6^2)$

$$= 10 + 17 + 26 + 37 = 90$$

c. $\displaystyle\sum_{i=0}^{8} \frac{1}{i!} = \frac{1}{0!} + \frac{1}{1!} + \frac{1}{2!} + \frac{1}{3!} + \frac{1}{4!} + \frac{1}{5!} + \frac{1}{6!} + \frac{1}{7!} + \frac{1}{8!}$

$$= 1 + 1 + \frac{1}{2} + \frac{1}{6} + \frac{1}{24} + \frac{1}{120} + \frac{1}{720} + \frac{1}{5040} + \frac{1}{40,320}$$

$$\approx 2.71828$$

Note that the sum in part (c) is very close to the irrational number $e \approx 2.718281828$. It can be shown that as more terms of the sequence whose nth term is $1/n!$ are added, the sum becomes closer and closer to e.

✓CHECKPOINT   Now try Exercise 51.

In Example 6, note that the lower index of a summation does not have to be 1. Also note that the index does not have to be the letter i. For instance, in part (b), the letter k is the index.

DISCOVERY

Consider the sequence

$$a_n = \left(1 + \frac{1}{n}\right)^n.$$

Use a graphing utility to find a_{100}, a_{1000}, $a_{10,000}$, and $a_{100,000}$. Compare your results to the approximation of e given by a graphing utility. What do you observe?

| EXAMPLE 7 | WRITING A SUM IN SUMMATION NOTATION |

Write each partial sum in summation notation. Decide whether the notation is unique.

a. $\dfrac{3}{1+2} + \dfrac{3}{2+2} + \dfrac{3}{3+2} + \dfrac{3}{4+2} + \dfrac{3}{5+2}$ b. $2 - 4 + 8 - 16$

SOLUTION

a. Begin by looking for similarities and differences in each of the terms. Each term has 3 in the numerator. In the denominator, 2 is added to a number that is increasing by 1 for each term. Let i be the index of summation, let 1 be the lower limit of summation, and let 5 be the upper limit of summation. Then, using summation notation, you can write

$$\frac{3}{1+2} + \frac{3}{2+2} + \frac{3}{3+2} + \frac{3}{4+2} + \frac{3}{5+2} = \sum_{i=1}^{5} \frac{3}{i+2}.$$

The sum can also be written in other ways, such as

$$\sum_{i=3}^{7} \frac{3}{i}, \quad \sum_{i=0}^{4} \frac{3}{i+3}, \quad \text{and} \quad \sum_{i=4}^{8} \frac{3}{i-1}.$$

The summation notation for the sum is not unique.

b. This series has terms with alternating signs, and each term can be written as a power of 2. Let n be the index of summation, let 1 be the lower limit of summation, and let 4 be the upper limit of summation. Then, using summation notation, you can write

$$2 - 4 + 8 - 16 = (-1)^{1+1}(2)^1 + (-1)^{2+1}(2)^2 + (-1)^{3+1}(2)^3 + (-1)^{4+1}(2)^4$$

$$= \sum_{n=1}^{4} (-1)^{n+1}(2)^n.$$

The sum can also be written in other ways, such as

$$\sum_{n=0}^{3} (-1)^n(2)^{n+1} \quad \text{and} \quad \sum_{n=2}^{5} (-1)^n(2)^{n-1}.$$

The summation notation for the sum is not unique.

✓CHECKPOINT   Now try Exercise 69.

STUDY TIP

The sum of the terms of any *finite* sequence must be a finite number. Variations in the upper and lower limits of summation can produce quite different-looking summation notations for *the same sum*, as you discovered in Example 7. For example, consider the following two sums.

$$\sum_{i=1}^{3} 3(2^i) = 3 \sum_{i=1}^{3} 2^i$$
$$= 3(2^1 + 2^2 + 2^3)$$

$$\sum_{i=0}^{2} 3(2^{i+1}) = 3 \sum_{i=0}^{2} 2^{i+1}$$
$$= 3(2^1 + 2^2 + 2^3)$$

The following properties of sums are helpful for adding the terms of a sequence.

Properties of Sums

1. $\displaystyle\sum_{i=1}^{n} c = cn,$ c is a constant.

2. $\displaystyle\sum_{i=1}^{n} ca_i = c \sum_{i=1}^{n} a_i,$ c is a constant.

3. $\displaystyle\sum_{i=1}^{n} (a_i + b_i) = \sum_{i=1}^{n} a_i + \sum_{i=1}^{n} b_i$ 4. $\displaystyle\sum_{i=1}^{n} (a_i - b_i) = \sum_{i=1}^{n} a_i - \sum_{i=1}^{n} b_i$

Application

Sequences have many applications in business and science. One such application is illustrated in Example 8.

EXAMPLE 8   POPULATION OF THE UNITED STATES

Make a Decision Figure 6.1 shows the total population of the United States from 1960 to 2002. These data can be approximated by the model

$$a_n = 0.0008n^3 - 0.036n^2 + 2.67n + 181.5, \qquad n = 0, 1, \ldots, 42$$

where a_n is the population (in thousands) and n represents the year, with $n = 0$ corresponding to 1960. Find the last three terms of this finite sequence. The projected population of the United States in the year 2010 is 308.9 million. Use the model to predict the population in 2010. Do you think this model should be used to predict the U.S. population in future years?   (Source: U.S. Census Bureau)

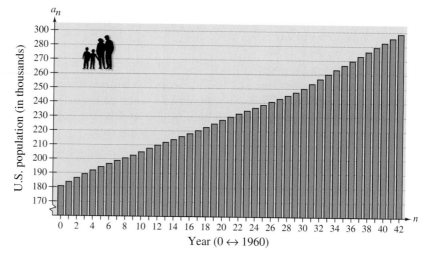

Figure 6.1

SOLUTION

The last three terms of this finite sequence are as follows.

$$a_{40} = 0.0008(40)^3 - 0.036(40)^2 + 2.67(40) + 181.5 \approx 282 \quad \text{2000 population}$$

$$a_{41} = 0.0008(41)^3 - 0.036(41)^2 + 2.67(41) + 181.5 \approx 286 \quad \text{2001 population}$$

$$a_{42} = 0.0008(42)^3 - 0.036(42)^2 + 2.67(42) + 181.5 \approx 289 \quad \text{2002 population}$$

To predict the population in 2010, find the 50th term of the sequence as follows.

$$a_{50} = 0.0008(50)^3 - 0.036(50)^2 + 2.67(50) + 181.5 = 325$$

This estimate seems a little too high when compared with the Census Bureau's estimate. This model should not be used to predict the U.S. population in future years. From Figure 6.1, you can see that the population grew in a linear or quadratic pattern from 1960 to 2002. So, a linear or quadratic model would be better for predicting the population in future years.

✓CHECKPOINT   Now try Exercise 83.

6.1 Warm Up

The following warm-up exercises involve skills that were covered in earlier sections. You will use these skills in the exercise set for this section. For additional help, review Sections R1.3, R1.6, and 1.4.

1. Find $f(2)$ for $f(n) = \dfrac{2n}{n^2 + 1}$.

2. Find $f(3)$ for $f(n) = \dfrac{4}{3(n + 1)}$.

In Exercises 3–6, factor the expression.

3. $4n^2 - 1$

4. $4n^2 - 8n + 3$

5. $n^2 - 3n + 2$

6. $n^2 + 3n + 2$

In Exercises 7–10, perform the indicated operations and simplify.

7. $\left(\dfrac{2}{3}\right)\left(\dfrac{3}{4}\right)\left(\dfrac{4}{5}\right)\left(\dfrac{5}{6}\right)$

8. $\dfrac{2 \cdot 4 \cdot 6 \cdot 8}{2^4}$

9. $\dfrac{1}{2 \cdot 2} + \dfrac{1}{2 \cdot 3} + \dfrac{1}{2 \cdot 4}$

10. $\dfrac{1}{1 \cdot 2} + \dfrac{1}{2 \cdot 3} + \dfrac{1}{3 \cdot 4}$

6.1 Exercises

In Exercises 1–22, write the first five terms of the sequence. (Assume that n begins with 1.)

1. $a_n = 2n - 7$

2. $a_n = 4 - 2n$

3. $a_n = 2^n$

4. $a_n = (-2)^n$

5. $a_n = \left(\dfrac{1}{2}\right)^n$

6. $a_n = \left(-\dfrac{1}{2}\right)^n$

7. $a_n = \dfrac{2}{n + 1}$

8. $a_n = \dfrac{n + 1}{3}$

9. $a_n = \dfrac{(-1)^n}{n}$

10. $a_n = \dfrac{(-1)^{n+1}}{n}$

11. $a_n = 3 - \dfrac{1}{2^n}$

12. $a_n = \dfrac{3^n}{4^n}$

13. $a_n = \dfrac{3^n}{n!}$

14. $a_n = \dfrac{n!}{n}$

15. $a_n = \dfrac{n^2 - 1}{n + 1}$

16. $a_n = \dfrac{n^2 - 1}{n^2 + 2}$

17. $a_n = \dfrac{(-1)^n}{n^2}$

18. $a_n = \dfrac{(-1)^n n}{n + 1}$

19. $a_n = \dfrac{3n!}{(n - 1)!}$

20. $a_n = \dfrac{(n + 1)!}{n!}$

21. $a_n = \dfrac{5n^2 - n + 1}{n^2}$

22. $a_n = \dfrac{3n^2 - n + 4}{2n^2 + 1}$

In Exercises 23–28, find the indicated term of the sequence.

23. $a_n = (-1)^n(3n - 2)$
$a_{25} = $

24. $a_n = (-1)^{n-1}[n(n - 1)]$
$a_{16} = $

25. $a_n = \dfrac{2^n}{n!}$
$a_{10} = $

26. $a_n = \dfrac{n!}{2n}$
$a_8 = $

27. $a_n = \dfrac{4n}{2n^2 - 3}$
$a_{11} = $

28. $a_n = \dfrac{4n^2 - n + 3}{n(n - 1)(n + 2)}$
$a_{13} = $

In Exercises 29–32, evaluate the expression.

29. $\dfrac{8!}{6!}$

30. $\dfrac{10!}{9!}$

31. $\dfrac{28!}{25!}$

32. $\dfrac{49!}{50!}$

In Exercises 33–36, simplify the factorial expression. Then evaluate the expression when $n = 0$ and $n = 1$.

33. $\dfrac{(n + 1)!}{n!}$

34. $\dfrac{(n + 2)!}{n!}$

35. $\dfrac{(2n - 1)!}{(2n + 1)!}$

36. $\dfrac{(2n + 2)!}{(2n)!}$

In Exercises 37–50, write an expression for the *most apparent* nth term of the sequence. (Assume that n begins with 1.)

37. $1, 3, 5, 7, 9, \ldots$

38. $5, 8, 11, 14, 17, \ldots$

39. $2, -4, 6, -8, 10, \ldots$

40. $1, -1, 1, -1, 1, \ldots$

41. $2, \frac{1}{2}, \frac{2}{9}, \frac{1}{8}, \frac{2}{25}, \ldots$

42. $1, \frac{1}{4}, \frac{1}{9}, \frac{1}{16}, \frac{1}{25}, \ldots$

43. $\frac{2}{3}, \frac{3}{4}, \frac{4}{5}, \frac{5}{6}, \frac{6}{7}, \ldots$

44. $\frac{2}{1}, \frac{3}{3}, \frac{4}{5}, \frac{5}{7}, \frac{6}{9}, \ldots$

45. $\frac{1}{2}, -\frac{1}{4}, \frac{1}{8}, -\frac{1}{16}, \frac{1}{32}, \ldots$

46. $\frac{1}{3}, \frac{2}{9}, \frac{4}{27}, \frac{8}{81}, \frac{16}{243}, \ldots$

47. $1 + \frac{1}{1}, 1 + \frac{1}{2}, 1 + \frac{1}{3}, 1 + \frac{1}{4}, 1 + \frac{1}{5}, \ldots$

48. $1 + \frac{1}{2}, 1 + \frac{3}{4}, 1 + \frac{7}{8}, 1 + \frac{15}{16}, 1 + \frac{31}{32}, \ldots$

49. $1, \frac{1}{2}, \frac{1}{6}, \frac{1}{24}, \frac{1}{120}, \ldots$

50. $1, 2, \dfrac{2^2}{2}, \dfrac{2^3}{6}, \dfrac{2^4}{24}, \dfrac{2^5}{120}, \ldots$

In Exercises 51–68, find the sum.

51. $\displaystyle\sum_{i=1}^{4} (6i + 3)$

52. $\displaystyle\sum_{i=1}^{6} (4i - 2)$

53. $\displaystyle\sum_{i=1}^{7} -3i$

54. $\displaystyle\sum_{i=1}^{5} -3(i - 2)$

55. $\displaystyle\sum_{k=1}^{4} 10$

56. $\displaystyle\sum_{k=1}^{5} 4$

57. $\displaystyle\sum_{i=0}^{4} i^2$

58. $\displaystyle\sum_{i=0}^{5} 3i^2$

59. $\displaystyle\sum_{k=2}^{6} \frac{1}{2k}$

60. $\displaystyle\sum_{j=3}^{5} \frac{1}{j}$

61. $\displaystyle\sum_{i=1}^{4} (i - 1)^2$

62. $\displaystyle\sum_{k=2}^{5} (k + 1)(k - 3)$

63. $\displaystyle\sum_{i=1}^{4} (-1)^i (2i + 4)$

64. $\displaystyle\sum_{i=1}^{4} (-2)^i$

65. $\displaystyle\sum_{k=0}^{4} \frac{1}{k!}$

66. $\displaystyle\sum_{j=0}^{5} \frac{2}{(j + 1)!}$

67. $\displaystyle\sum_{j=0}^{6} \frac{(j + 2)!}{(2j)!}$

68. $\displaystyle\sum_{i=1}^{4} \frac{i}{(i - 1)!}$

In Exercises 69–78, use summation notation to write the sum.

69. $\dfrac{1}{3(1)} + \dfrac{1}{3(2)} + \dfrac{1}{3(3)} + \cdots + \dfrac{1}{3(9)}$

70. $\dfrac{5}{1 + 1} + \dfrac{5}{1 + 2} + \dfrac{5}{1 + 3} + \cdots + \dfrac{5}{1 + 15}$

71. $\left[2\left(\frac{1}{8}\right) + 3\right] + \left[2\left(\frac{2}{8}\right) + 3\right] + \cdots + \left[2\left(\frac{8}{8}\right) + 3\right]$

72. $\left[1 - \left(\frac{1}{6}\right)^2\right] + \left[1 - \left(\frac{2}{6}\right)^2\right] + \cdots + \left[1 - \left(\frac{6}{6}\right)^2\right]$

73. $3 - 9 + 27 - 81 + 243 - 729$

74. $1 - \frac{1}{2} + \frac{1}{4} - \frac{1}{8} + \cdots - \frac{1}{128}$

75. $\dfrac{1}{1^2} - \dfrac{1}{2^2} + \dfrac{1}{3^2} - \dfrac{1}{4^2} + \cdots - \dfrac{1}{20^2}$

76. $\dfrac{1}{1 \cdot 3} + \dfrac{1}{2 \cdot 4} + \dfrac{1}{3 \cdot 5} + \cdots + \dfrac{1}{10 \cdot 12}$

77. $\frac{1}{4} + \frac{3}{8} + \frac{7}{16} + \frac{15}{32} + \frac{31}{64}$

78. $\frac{1}{2} + \frac{2}{4} + \frac{6}{8} + \frac{24}{16} + \frac{120}{32} + \frac{720}{64}$

IN THE NEWS

A new formula for picking off pieces of pi

. . . Mathematicians have discovered a surprisingly simple formula for computing digits of the number pi (π). Unlike previously known expressions, this one allows them to calculate isolated digits—say, the billionth digit of pi—without computing and keeping track of all the preceding numbers.

The only catch is that the formula works for binary, but not for decimal, digits. Thus, it's possible to determine that the forty billionth binary digit of pi is 1, followed by 00100100001110 . . . . . But there's no way to convert these numbers into decimal form without knowing all the binary digits that come before the string.

$$\pi = \sum_{i=0}^{\pi} \frac{1}{16^i}\left(\frac{4}{8i + 1} - \frac{2}{8i + 4} - \frac{1}{8i + 5} - \frac{1}{8i + 6}\right)$$

New formula that serves as the basis for computing isolated binary (or hexadecimal) digits of pi.

However, the answer that comes out of the expression for pi gives only hexadecimal (base 16) digits, which can be readily converted to binary. "The frustrating thing is that it doesn't work in base 10 (for decimal digits)," Borwein remarks. . . .

Reprinted with permission from Science News, the weekly magazine of science. Copyright © 1995 by Science Service.

---

79. Evaluate the above sum from $i = 0$ to $i = 2$. Is the result accurate to within five decimal places of pi? ($\pi \approx 3.14159$)

80. Evaluate the above sum from $i = 0$ to $i = 3$. Is the result accurate to within five decimal places of pi?

81. *Make a Decision* Evaluate $\displaystyle\sum_{j=0}^{4} j^2$ and $\displaystyle\sum_{k=2}^{6} (k - 2)^2$. What do you observe about the two sums? Is the summation notation for a sum unique? Explain.

82. *Make a Decision* Evaluate $\displaystyle\sum_{j=0}^{6} \frac{(j + 1)!}{j!}$ and $\displaystyle\sum_{k=1}^{7} \frac{k!}{(k - 1)!}$. What do you observe about the two sums? Is the summation notation for a sum unique? Explain.

83. *Make a Decision*: **Compound Interest** A deposit of $10,000 is made in an account that earns 8.5% interest compounded quarterly. The balance in the account after n quarters is given by

$$A_n = 10,000\left(1 + \frac{0.085}{4}\right)^n, \quad n = 1, 2, 3, \ldots$$

(a) Compute the first eight terms of the sequence.

(b) Find the balance in the account after 10 years by computing the 40th term of the sequence.

(c) Is the balance after 20 years twice the balance after 10 years? Explain.

84. *Make a Decision*: **Ratio of Males to Females** The ratio a_n of males to females in the United States from 1940 to 2010 is approximated by the model

$$a_n = 1.04 - 0.034n + 0.0032n^2,$$
$$n = 1, 2, \ldots, 8$$

where n is the year, with $n = 1, 2, 3, \ldots, 8$ corresponding to 1940, 1950, 1960, . . . , 2010. (Source: U.S. Census Bureau)

(a) Use a graphing utility to find the terms of this finite sequence. Interpret the meaning of the terms in the context of the real-life situation.

(b) Construct a bar graph that represents the sequence. Describe any trends you see from your graph.

(c) In 2000, the population of the United States was approximately 281 million. In that year, how many people were females? How many were males?

85. **Hospital Costs** The average daily cost a_n (in dollars) to a patient in a hospital in the United States from 1995 to 2001 is approximated by the model

$$a_n = 974 + 21.6n + 2.96n^2, \quad n = 0, 1, \ldots, 6$$

where n represents the year, with $n = 0$ corresponding to 1995. Use a graphing utility to find the terms of this finite sequence. Construct a bar graph that represents the sequence. Predict the average cost of a day in a hospital in 2007. (Source: Health Forum)

86. **Annual Payroll** The annual payroll a_n (in billions of dollars) for new car dealerships in the United States from 1997 to 2002 can be approximated by the model

$$a_n = 37.0 + 3.43n - 0.200n^2, \quad n = 0, 1, 2, 3, 4, 5$$

where n represents the year, with $n = 0$ corresponding to 1997 (see figure). Find the total payroll from 1997 to 2002 by evaluating the sum

$$\sum_{n=0}^{5} (37.0 + 3.43n - 0.200n^2).$$

(Source: National Automobile Dealers Association)

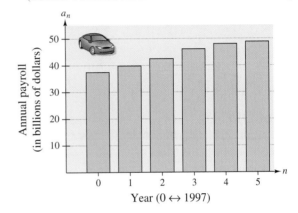

87. **Total Revenue** The total revenue a_n (in billions of dollars) for United Parcel Service, Inc., from 1999 to 2003 can be approximated by the model

$$a_n = 0.2833n^3 - 1.779n^2 + 4.18n + 27.1,$$
$$n = 0, 1, 2, 3, 4$$

where n represents the year, with $n = 0$ corresponding to 1999. Find the total revenue from 1999 to 2003 by evaluating the sum

$$\sum_{n=0}^{4} (0.2833n^3 - 1.779n^2 + 4.18n + 27.1).$$

(Source: United Parcel Service, Inc.)

6.2 Arithmetic Sequences and Partial Sums

Objectives

- Recognize and find the nth term of an arithmetic sequence.
- Find a formula for an arithmetic sequence.
- Find the nth partial sum of an arithmetic sequence.
- Use an arithmetic sequence to solve an application problem.

Arithmetic Sequences

A sequence whose consecutive terms have a common difference is called an **arithmetic sequence**.

DISCOVERY

Determine the arithmetic rule used to generate the terms of each sequence.

a. 1, 4, 7, 10, 13, . . .

b. 6, 12, 18, 24, . . .

c. 100, 90, 80, 70, . . .

Definition of Arithmetic Sequence

A sequence is **arithmetic** if the differences between consecutive terms are the same. So, the sequence

$$a_1, a_2, a_3, a_4, \ldots, a_n, \ldots$$

is arithmetic if there is a number d such that

$$a_2 - a_1 = d, \quad a_3 - a_2 = d, \quad a_4 - a_3 = d, \text{ and so on.}$$

The number d is the **common difference** of the arithmetic sequence.

EXAMPLE 1 EXAMPLES OF ARITHMETIC SEQUENCES

a. The sequence whose nth term is $4n + 3$ is arithmetic. For this sequence, the common difference between consecutive terms is 4.

$$\underbrace{7, 11}, 15, 19, \ldots, 4n + 3, \ldots \qquad \text{Begin with } n = 1.$$

$$11 - 7 = 4$$

b. The sequence whose nth term is $7 - 5n$ is arithmetic. For this sequence, the common difference between consecutive terms is -5.

$$\underbrace{2, -3}, -8, -13, \ldots, 7 - 5n, \ldots \qquad \text{Begin with } n = 1.$$

$$-3 - 2 = -5$$

c. The sequence whose nth term is $\frac{1}{4}(n + 3)$ is arithmetic. For this sequence, the common difference between consecutive terms is $\frac{1}{4}$.

$$\underbrace{1, \frac{5}{4}}, \frac{3}{2}, \frac{7}{4}, \ldots, \frac{n + 3}{4}, \ldots \qquad \text{Begin with } n = 1.$$

$$\frac{5}{4} - 1 = \frac{1}{4}$$

✓CHECKPOINT   Now try Exercise 1.

In Example 1, notice that each of the arithmetic sequences has an nth term that is of the form $dn + c$, where the common difference of the sequence is d. This result is summarized as follows.

The nth Term of an Arithmetic Sequence

The nth term of an arithmetic sequence has the form

$$a_n = dn + c$$

where d is the common difference between consecutive terms of the sequence and $c = a_1 - d$.

EXAMPLE 2   FINDING THE NTH TERM OF AN ARITHMETIC SEQUENCE

Find a formula for the nth term of the arithmetic sequence whose common difference is 3 and whose first term is 2.

SOLUTION

Because the sequence is arithmetic, you know that the formula for the nth term is of the form $a_n = dn + c$. Moreover, because the common difference is $d = 3$, the formula must have the form

$$a_n = 3n + c.$$

Because $a_1 = 2$, it follows that

$$c = a_1 - d = 2 - 3 = -1.$$

So, the formula for the nth term is

$$a_n = 3n - 1.$$

The sequence has the form 2, 5, 8, 11, 14, . . . , $3n - 1$, ...

✓CHECKPOINT   Now try Exercise 21.

EXAMPLE 3   FINDING THE NTH TERM OF AN ARITHMETIC SEQUENCE

Find the ninth term of the arithmetic sequence whose first two terms are 2 and 9.

SOLUTION

For this sequence, the common difference is $d = 9 - 2 = 7$. There are two ways to find the ninth term. One way is simply to write out the first nine terms (by repeatedly adding 7).

2, 9, 16, 23, 30, 37, 44, 51, 58

Another way to find the ninth term is first to find a formula for the nth term. Because the first term is 2, it follows that

$$c = a_1 - d = 2 - 7 = -5.$$

So, a formula for the nth term is $a_n = 7n - 5$, which implies that the ninth term is

$$a_9 = 7(9) - 5 = 58.$$

✓CHECKPOINT   Now try Exercise 29.

If you know the nth term of an arithmetic sequence *and* you know the common difference of the sequence, you can find the $(n + 1)$th term by using the **recursion formula**

$$a_{n+1} = a_n + d. \qquad \text{Recursion formula}$$

With this formula, you can find any term of an arithmetic sequence, *provided* that you know the previous term. For instance, if you know the first term, you can find the second term. Then, knowing the second term, you can find the third term, and so on.

If you substitute $a_1 - d$ for c in the formula $a_n = dn + c$, then the nth term of an arithmetic sequence has the alternative recursion formula

$$a_n = a_1 + (n - 1)d. \qquad \text{Alternative recursion formula}$$

EXAMPLE 4   USING A RECURSION FORMULA

Find the 15th term of the arithmetic sequence whose common difference is 11 and whose 14th term is 136.

SOLUTION

Because you know the 14th term and want to find the 15th term, use the recursion formula.

$$a_{n+1} = a_n + d$$
$$a_{14+1} = a_{14} + d$$
$$a_{15} = 136 + 11$$
$$= 147$$

So, the 15th term of the sequence is $a_{15} = 147$.

✓CHECKPOINT   Now try Exercise 33.

EXAMPLE 5   FINDING THE FIRST SEVERAL TERMS OF AN ARITHMETIC SEQUENCE

The fourth term of an arithmetic sequence is 20, and the 13th term is 65. Write the first several terms of this sequence.

SOLUTION

You know that $a_4 = 20$ and $a_{13} = 65$. So, you must add the common difference d nine times to the fourth term to obtain the 13th term. Therefore, the fourth and 13th terms of the sequence are related by

$$a_{13} = a_4 + 9d. \qquad a_4 \text{ and } a_{13} \text{ are nine terms apart.}$$

Using $a_4 = 20$ and $a_{13} = 65$, you can conclude that $d = 5$, which implies that the sequence is as follows.

a_1	a_2	a_3	a_4	a_5	a_6	a_7	a_8	a_9	a_{10}	a_{11}
5,	10,	15,	20,	25,	30,	35,	40,	45,	50,	55, . . .

✓CHECKPOINT   Now try Exercise 37.

The Sum of a Finite Arithmetic Sequence

There is a simple formula for the sum of a finite arithmetic sequence. Be sure you see that this formula works only for *arithmetic* sequences.

The Sum of a Finite Arithmetic Sequence

The sum of a finite arithmetic sequence with n terms is given by

$$S_n = \frac{n}{2}(a_1 + a_n).$$

EXAMPLE 6   FINDING THE SUM OF A FINITE ARITHMETIC SEQUENCE

Find the sum: $1 + 3 + 5 + 7 + 9 + 11 + 13 + 15 + 17 + 19$.

SOLUTION

To begin, notice that the sequence is arithmetic (with a common difference of 2). Moreover, the sequence has 10 terms. So, the sum of the sequence is

$$S_n = 1 + 3 + 5 + 7 + 9 + 11 + 13 + 15 + 17 + 19$$

$$= \frac{n}{2}(a_1 + a_n) \qquad \text{Formula for sum of a finite arithmetic sequence}$$

$$= \frac{10}{2}(1 + 19) \qquad \text{Substitute 10 for } n, 1 \text{ for } a_1, \text{ and } 19 \text{ for } a_n.$$

$$= 5(20) \qquad \text{Simplify.}$$

$$= 100. \qquad \text{Multiply.}$$

✔CHECKPOINT   Now try Exercise 47.

EXAMPLE 7   FINDING THE SUM OF A FINITE ARITHMETIC SEQUENCE

Find the sum of the integers from 1 to 100.

SOLUTION

The integers from 1 to 100 form an arithmetic sequence that has 100 terms. So, you can use the formula for the sum of an arithmetic sequence, as follows.

$$S_n = 1 + 2 + 3 + 4 + 5 + 6 + \cdots + 99 + 100$$

$$= \frac{n}{2}(a_1 + a_n) \qquad \text{Formula for sum of a finite arithmetic sequence}$$

$$= \frac{100}{2}(1 + 100) \qquad \text{Substitute 100 for } n, 1 \text{ for } a_1, \text{ and } 100 \text{ for } a_n.$$

$$= 50(101) \qquad \text{Simplify.}$$

$$= 5050 \qquad \text{Multiply.}$$

✔CHECKPOINT   Now try Exercise 55.

DISCOVERY

To *develop* the formula for the sum of a finite arithmetic sequence, consider writing the sum in two different ways. One way is to write the sum as

$$S_n = a_1 + (a_1 + d)$$
$$+ (a_1 + 2d) + \cdots$$
$$+ [a_1 + (n - 1)d].$$

In the second way, you repeatedly subtract d from the nth term to obtain

$$S_n = a_n + (a_n - d)$$
$$+ (a_n - 2d) + \cdots$$
$$+ [a_n - (n - 1)d].$$

Can you discover a way to combine these two versions of S to obtain the following formula?

$$S_n = \frac{n}{2}(a_1 + a_n)$$

The sum of the first n terms of an infinite sequence is the **nth partial sum.** The nth partial sum can be found by using the formula for the sum of a finite arithmetic sequence.

EXAMPLE 8   FINDING A PARTIAL SUM OF AN ARITHMETIC SEQUENCE

Find the 150th partial sum of the arithmetic sequence

$$5, 16, 27, 38, 49, \ldots .$$

SOLUTION

For this arithmetic sequence, you have $a_1 = 5$ and $d = 16 - 5 = 11$. So, $c = a_1 - d = 5 - 11 = -6$, and the nth term is

$$a_n = 11n - 6.$$

So, $a_{150} = 11(150) - 6 = 1644$, and the sum of the first 150 terms is as follows.

$$S_n = \frac{n}{2}(a_1 + a_n) \qquad \text{nth partial sum formula}$$

$$= \frac{150}{2}(5 + 1644) \qquad \text{Substitute 150 for } n, 5 \text{ for } a_1, \text{ and 1644 for } a_n.$$

$$= 75(1649) = 123{,}675 \qquad \text{Simplify.}$$

✔**CHECKPOINT**   Now try Exercise 57.

Applications

EXAMPLE 9   SEATING CAPACITY

An auditorium has 20 rows of seats. There are 20 seats in the first row, 22 seats in the second row, 24 seats in the third row, and so on (see Figure 6.2). How many seats are there in all 20 rows?

SOLUTION

The number of seats in the rows forms an arithmetic sequence in which the common difference is $d = 2$. Because $c = a_1 - d = 20 - 2 = 18$, you can determine that the formula for the nth term in the sequence is $a_n = 2n + 18$. So, the 20th term in the sequence is $a_{20} = 2(20) + 18 = 58$, and the total number of seats is

$$S_{20} = 20 + 22 + 24 + \cdots + 58$$

$$= \frac{n}{2}(a_1 + a_n) \qquad \text{nth partial sum formula}$$

$$= \frac{20}{2}(20 + 58) \qquad \text{Substitute 20 for } n, 20 \text{ for } a_1, \text{ and 58 for } a_n.$$

$$= 10(78) = 780. \qquad \text{Simplify.}$$

✔**CHECKPOINT**   Now try Exercise 69.

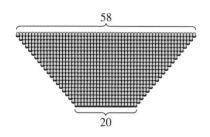
58

20

Figure 6.2

| EXAMPLE 10 | TOTAL SALES | |

Make a Decision A small coffee shop sells $10,000 worth of beverages during its first year. The owner of the business has set a goal of increasing annual sales by $7500 each year for 9 years. Assuming that this goal is met, find the total sales during the first 10 years this business is in operation. What kinds of economic factors could prevent the shop owner from meeting the sales goal?

SOLUTION

Starbucks reported a net revenue of over 4 billion dollars in 2003.
(Source: *www.starbucks.com*)

The annual sales form an arithmetic sequence in which $a_1 = 10,000$ and $d = 7500$. So,

$$c = a_1 - d = 10,000 - 7500 = 2500$$

and the nth term of the sequence is

$$a_n = 7500n + 2500.$$

This implies that the 10th term of the sequence is

$$a_{10} = (7500)(10) + 2500 = 77,500.$$

So, the total sales for the first 10 years are as follows.

$$S_n = \frac{n}{2}(a_1 + a_n) \qquad \text{nth partial sum formula}$$

$$= \frac{10}{2}(10,000 + 77,500) \qquad \text{Substitute 10 for n, 10,000 for a_1, and 77,500 for a_n.}$$

$$= 5(87,500) \qquad \text{Simplify.}$$

$$= \$437,500 \qquad \text{Simplify.}$$

Many economic factors, such as location of the shop, demand by consumers, and marketing techniques, could prevent the shop owner from meeting the sales goal.

✓CHECKPOINT   Now try Exercise 77.

Discussing the Concept | Numerical Relationships

Decide whether it is possible to fill in the blanks in each of the following such that the resulting sequence is arithmetic. If so, find a recursion formula for the sequence.

a. -7, ___, ___, ___, ___, ___, 11

b. 17, ___, ___, ___, ___, ___, ___, ___, ___, 71

c. 2, 6, ___, ___, 162

d. 4, 7.5, ___, ___, ___, ___, ___, ___, ___, ___, 39

e. 8, 12, ___, ___, ___, 60.75

6.2 Warm Up

The following warm-up exercises involve skills that were covered in earlier sections. You will use these skills in the exercise set for this section. For additional help, review Sections R1.1, R1.2, 1.4, and 6.1.

In Exercises 1 and 2, find the sum.

1. $\sum_{i=1}^{6} (2i - 1)$

2. $\sum_{i=1}^{10} (4i + 2)$

In Exercises 3 and 4, find the distance between the two real numbers.

3. $\frac{5}{2}, 8$

4. $\frac{4}{3}, \frac{14}{3}$

In Exercises 5 and 6, evaluate the function as indicated.

5. Find $f(3)$ for $f(n) = 10 + (n - 1)4$.

6. Find $f(10)$ for $f(n) = 1 + (n - 1)\frac{1}{3}$.

In Exercises 7–10, evaluate the expression.

7. $\frac{11}{2}(1 + 25)$

8. $\frac{16}{2}(4 + 16)$

9. $\frac{20}{2}[2(5) + (12 - 1)3]$

10. $\frac{8}{2}[2(-3) + (15 - 1)5]$

6.2 Exercises

In Exercises 1–10, determine whether the sequence is arithmetic. If it is, find the common difference.

1. $3, 7, 11, 15, 19, \ldots$

2. $20, 17, 14, 11, 8, \ldots$

3. $-3, -2, 0, 3, 7, \ldots$

4. $-12, -8, -4, 0, 4, 8, \ldots$

5. $3, \frac{5}{2}, 2, \frac{3}{2}, 1, \frac{1}{2}, \ldots$

6. $\frac{9}{4}, 2, \frac{7}{4}, \frac{3}{2}, \frac{5}{4}, 1, \ldots$

7. $\frac{1}{3}, \frac{2}{3}, \frac{4}{3}, \frac{8}{3}, \frac{16}{3}, \frac{32}{3}, \ldots$

8. $\ln 1, \ln 2, \ln 3, \ln 4, \ln 5, \ldots$

9. $5.3, 5.7, 6.1, 6.5, 6.9, \ldots$

10. $1^2, 2^2, 3^2, 4^2, 5^2, \ldots$

In Exercises 11–20, write the first five terms of the sequence. Determine whether the sequence is arithmetic. If it is, find the common difference.

11. $a_n = 5 + 3n$

12. $a_n = 100 - 3n$

13. $a_n = (2^n)n$

14. $a_n = 2^{n-1}$

15. $a_n = (2 + n) - (1 + n)$

16. $a_n = 1 + (n - 1)4$

17. $a_n = \dfrac{1}{n + 1}$

18. $a_n = \dfrac{n + 1}{n}$

19. $a_n = (-1)^n$

20. $a_n = (-1)^{2n+1}$

In Exercises 21–30, find a formula for a_n for the arithmetic sequence.

21. $a_1 = 4, d = 5$

22. $a_1 = -2, d = 3$

23. $a_1 = 50, d = -12$

24. $a_1 = 120, d = -15$

25. $a_1 = 7, a_5 = 27$

26. $a_1 = -5, a_7 = -17$

27. $a_3 = 8, a_9 = 32$

28. $a_6 = 29, a_{18} = 95$

29. $2, 8, 14, 20, 26, \ldots$

30. $3, -7, -17, -27, -37, \ldots$

In Exercises 31–38, write the first five terms of the arithmetic sequence.

31. $a_1 = 10, d = -4$

32. $a_1 = 2, d = \frac{1}{3}$

33. $a_1 = 5, a_{n+1} = a_n + 9$

34. $a_1 = 7, a_{n+1} = a_n - 2$

35. $a_1 = 2, a_{12} = 46$

36. $a_5 = 28, a_{10} = 3$

37. $a_8 = 26, a_{12} = 10$

38. $a_4 = 16, a_{10} = 46$

In Exercises 39–42, match the arithmetic sequence with its graph. [The graphs are labeled (a), (b), (c), and (d).]

(a)
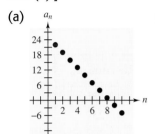

(b)
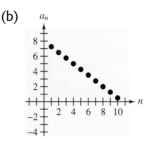

(c)
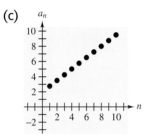

(d)
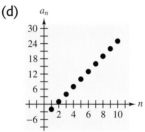

39. $a_n = -\frac{3}{4}n + 8$ **40.** $a_n = 3n - 5$

41. $a_n = 2 + \frac{3}{4}n$ **42.** $a_n = 25 - 3n$

In Exercises 43–46, use a graphing utility to graph the first 10 terms of the sequence.

43. $a_n = 15 - \frac{3}{2}n$ **44.** $a_n = -5 + 2n$

45. $a_n = 0.2n + 3$ **46.** $a_n = -0.3n + 8$

In Exercises 47–54, find the indicated nth partial sum of the arithmetic sequence.

47. $4, 8, 12, 16, \ldots, n = 6$

48. $3, 11, 19, 27, \ldots, n = 10$

49. $-10, -8, -6, -4, \ldots, n = 12$

50. $-2, -9, -16, -23, \ldots, n = 10$

51. $0.5, 0.9, 1.3, 1.7, \ldots, n = 10$

52. $1.50, 1.45, 1.40, 1.35, \ldots, n = 20$

53. $a_1 = 100, a_{25} = 220, n = 25$

54. $a_1 = 15, a_{100} = 307, n = 100$

55. Find the sum of the first 100 odd integers.

56. Find the sum of the integers from -10 to 50.

In Exercises 57–66, find the partial sum.

57. $\displaystyle\sum_{n=1}^{50} n$ **58.** $\displaystyle\sum_{n=1}^{100} 2n$

59. $\displaystyle\sum_{n=32}^{100} 5n$ **60.** $\displaystyle\sum_{n=51}^{100} 7n$

61. $\displaystyle\sum_{n=0}^{50} (1000 - 5n)$ **62.** $\displaystyle\sum_{n=1}^{250} (1000 - n)$

63. $\displaystyle\sum_{n=1}^{20} (2n + 5)$ **64.** $\displaystyle\sum_{n=1}^{500} (n + 3)$

65. $\displaystyle\sum_{n=1}^{100} \frac{n + 4}{2}$ **66.** $\displaystyle\sum_{n=0}^{100} \frac{8 - 3n}{16}$

Job Offer   In Exercises 67 and 68, consider a job offer with the given starting salary and the given annual raise. (a) Determine the salary during the sixth year of employment. (b) Determine the total compensation from the company through six full years of employment.

	Starting Salary	*Annual Raise*
67.	$32,500	$1500
68.	$36,800	$1750

69. Seating Capacity   Determine the seating capacity of an auditorium with 30 rows of seats if there are 15 seats in the first row, 20 seats in the second row, 25 seats in the third row, and so on.

70. Seating Capacity   Determine the seating capacity of an auditorium with 36 rows of seats if there are 15 seats in the first row, 18 seats in the second row, 21 seats in the third row, and so on.

71. Total Sales   The annual sales a_n (in millions of dollars) for The Cheesecake Factory, Inc., from 1998 to 2003 can be approximated by the model

$$a_n = 248.6 + 101.64n, \quad n = 0, 1, \ldots, 5$$

where n represents the year, with $n = 0$ corresponding to 1998.   (Source: The Cheesecake Factory, Inc.)

(a) Construct a bar graph showing the annual sales for The Cheesecake Factory, Inc., from 1998 to 2003.

(b) Find the total sales from 1998 to 2003.

72. Total Sales   The annual sales a_n (in millions of dollars) for Starbucks Corporation from 1998 to 2003 can be modeled by

$$a_n = 1161.4 + 546.86n, \quad n = 0, 1, \ldots, 5$$

where n represents the year, with $n = 0$ corresponding to 1998.   (Source: Starbucks Corporation)

(a) Construct a bar graph showing the annual sales for Starbucks Corporation from 1998 to 2003.

(b) Find the total sales from 1998 to 2003.

73. Falling Object A heavy object (with negligible air resistance) is dropped from a plane. During the first second of fall, the object falls 16 feet; during the second second, it falls 48 feet; during the third second, it falls 80 feet; and during the fourth second, it falls 112 feet. If this pattern continues, how many feet will the object fall in 8 seconds?

74. Falling Object A heavy object (with negligible air resistance) is dropped from a plane. During the first second of fall, the object falls 4.9 meters; during the second second, it falls 14.7 meters; during the third second, it falls 24.5 meters; and during the fourth second, it falls 34.3 meters. If this pattern continues, how many meters will the object fall in 10 seconds?

75. Brick Pattern A brick patio is roughly the shape of a trapezoid (see figure). The patio has 20 rows of bricks. The first row has 14 bricks and the 20th row has 33 bricks. How many bricks are in the patio?

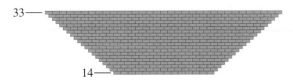

76. *Make a Decision*: **Total Profit** A small landscaping company makes a profit of $50,000 during its first year. The company president sets a goal of increasing profit by $10,000 each year for 4 years. Assuming that this goal is met, find the total profit during the first 5 years of this business. What kinds of economic factors could prevent the company from meeting the profit goal?

77. *Make a Decision*: **Total Sales** An entrepreneur sells $20,000 worth of gumballs in the first year of operation. The entrepreneur has set a goal of increasing annual sales by $5000 each year for 9 years. Assuming that this goal is met, find the total sales during the first 10 years of this gumball business. What kinds of economic factors could prevent the entrepreneur from meeting the sales goal?

78. *Make a Decision* Given an arithmetic sequence for which only the first few terms are known, is it possible to find the nth term? Explain your reasoning.

79. *Make a Decision* If the only known information about a finite arithmetic sequence is its first term and its last term, is it possible to find the sum of the sequence? Explain your reasoning.

80. Total Revenue The annual revenue a_n (in billions of dollars) for McDonald's Corporation from 1998 to 2003 is shown in the table, where n represents the year, with $n = 0$ corresponding to 1998.   (Source: McDonald's Corporation)

Year, n	Revenue, a_n
0	12.421
1	13.259
2	14.243
3	14.870
4	15.406
5	17.140

(a) Construct a bar graph showing the annual revenue from 1998 to 2003.

(b) Use the *linear regression* feature of a graphing utility to find an arithmetic sequence that approximates the annual revenue from 1998 to 2003.

(c) Use summation notation to represent the *total* revenue from 1998 to 2003. Find the total revenue.

81. Total Revenue The annual revenue a_n (in millions of dollars) for Outback Steakhouse, Inc., from 1998 to 2003 is shown in the table, where n represents the year, with $n = 0$ corresponding to 1998.   (Source: Outback Steakhouse, Inc.)

Year, n	Revenue, a_n
0	1358.9
1	1646.0
2	1906.0
3	2127.1
4	2362.1
5	2744.4

(a) Construct a bar graph showing the annual revenue from 1998 to 2003.

(b) Use the *linear regression* feature of a graphing utility to find an arithmetic sequence that approximates the annual revenue from 1998 to 2003.

(c) Use summation notation to represent the *total* revenue from 1998 to 2003. Find the total revenue.

6.3 Geometric Sequences and Series

Objectives

- Recognize and find the *n*th term of a geometric sequence.
- Find the sum of a finite geometric sequence.
- Find the sum of an infinite geometric series.
- Use a geometric sequence to solve an application problem.

Geometric Sequences

In Section 6.2, you learned that a sequence whose consecutive terms have a common *difference* is an arithmetic sequence. In this section, you will study another important type of sequence called a **geometric sequence**. Consecutive terms of a geometric sequence have a common *ratio,* as indicated in the following definition.

DISCOVERY

Determine the rule used to generate the terms of each sequence.

a. $1, \frac{1}{2}, \frac{1}{4}, \frac{1}{8}, \ldots$

b. $2, 8, 32, 128, \ldots$

c. $\frac{1}{2}, \frac{3}{4}, \frac{9}{8}, \frac{27}{16}, \ldots$

Definition of Geometric Sequence

A sequence is **geometric** if the ratios of consecutive terms are the same. So, the sequence $a_1, a_2, a_3, a_4, \ldots, a_n, \ldots$ is geometric if there is a number r such that

$$\frac{a_2}{a_1} = r, \quad \frac{a_3}{a_2} = r, \quad \frac{a_4}{a_3} = r, \quad r \neq 0$$

and so on. The number r is the **common ratio** of the geometric sequence.

EXAMPLE 1   **EXAMPLES OF GEOMETRIC SEQUENCES**

a. The sequence whose *n*th term is 2^n is geometric. For this sequence, the common ratio between consecutive terms is 2.

$$2, 4, 8, 16, \ldots, 2^n, \ldots \qquad \qquad \text{Begin with } n = 1.$$

$$\underbrace{\quad}$$
$$\frac{4}{2} = 2$$

b. The sequence whose *n*th term is $4(3^n)$ is geometric. For this sequence, the common ratio between consecutive terms is 3.

$$12, 36, 108, 324, \ldots, 4(3^n), \ldots \qquad \text{Begin with } n = 1.$$

$$\frac{36}{12} = 3$$

c. The sequence whose *n*th term is $\left(-\frac{1}{3}\right)^n$ is geometric. For this sequence, the common ratio between consecutive terms is $-\frac{1}{3}$.

$$-\frac{1}{3}, \frac{1}{9}, -\frac{1}{27}, \frac{1}{81}, \ldots, \left(-\frac{1}{3}\right)^n, \ldots \qquad \text{Begin with } n = 1.$$

$$\frac{1/9}{-1/3} = -\frac{1}{3}$$

✔**CHECKPOINT**   Now try Exercise 1.

In Example 1, notice that each of the geometric sequences has an nth term that is of the form ar^n, where the common ratio of the sequence is r.

The nth Term of a Geometric Sequence

The nth term of a geometric sequence has the form

$$a_n = a_1 r^{n-1}$$

where r is the common ratio of consecutive terms of the sequence. So, every geometric sequence can be written in the following form.

$$a_1, \quad a_2, \quad a_3, \quad a_4, \quad a_5, \quad \ldots, \quad a_n, \quad \ldots$$
$$\downarrow \quad \downarrow \quad \downarrow \quad \downarrow \quad \downarrow \quad \quad \downarrow$$
$$a_1, \; a_1 r, \; a_1 r^2, \; a_1 r^3, \; a_1 r^4, \; \ldots, \; a_1 r^{n-1}, \; \ldots$$

If you know the nth term of a geometric sequence, you can find the $(n + 1)$th term by multiplying by r. That is, $a_{n+1} = r a_n$.

EXAMPLE 2   FINDING THE TERMS OF A GEOMETRIC SEQUENCE

Write the first five terms of the geometric sequence whose first term is $a_1 = 3$ and whose common ratio is $r = 2$.

SOLUTION

Starting with 3, repeatedly multiply by 2 to obtain the following.

$a_1 = 3$	1st term
$a_2 = 3(2^1) = 6$	2nd term
$a_3 = 3(2^2) = 12$	3rd term
$a_4 = 3(2^3) = 24$	4th term
$a_5 = 3(2^4) = 48$	5th term

✓CHECKPOINT   Now try Exercise 11.

EXAMPLE 3   FINDING A TERM OF A GEOMETRIC SEQUENCE

Find the 15th term of the geometric sequence whose first term is 20 and whose common ratio is 1.05.

SOLUTION

$a_n = a_1 r^{n-1}$	Formula for geometric sequence
$a_{15} = 20(1.05)^{15-1}$	Substitute 20 for a_1, 1.05 for r, and 15 for n.
≈ 39.599	Use a calculator.

✓CHECKPOINT   Now try Exercise 19.

TECHNOLOGY

You can generate the geometric sequence in Example 2 with a graphing utility using the following steps.

3 [ENTER]

2 [×] [2nd] [ANS]

Now press [ENTER] repeatedly to generate the terms of the sequence.

EXAMPLE 4   FINDING A TERM OF A GEOMETRIC SEQUENCE

Find the 12th term of the geometric sequence

$$5, 15, 45, \ldots$$

SOLUTION

The common ratio of this sequence is $r = \frac{15}{5} = 3$. Because the first term is $a_1 = 5$, you can determine the 12th term ($n = 12$) to be

$a_n = a_1 r^{n-1}$	Formula for geometric sequence
$a_{12} = 5(3)^{12-1}$	Substitute 5 for a_1, 3 for r, and 12 for n.
$= 5(177,147)$	Use a calculator.
$= 885,735.$	Simplify.

✓CHECKPOINT   Now try Exercise 21.

If you know any two terms of a geometric sequence, you can use that information to find a formula for the nth term of the sequence.

EXAMPLE 5   FINDING A TERM OF A GEOMETRIC SEQUENCE

The fourth term of a geometric sequence is 125, and the 10th term is $\frac{125}{64}$. Find the 14th term. (Assume that the terms of the sequence are positive.)

SOLUTION

Because r is the common ratio of *consecutive* terms of a sequence, you have to find the relationship between the 10th term and the fourth term.

$$a_{10} = a_1 r^9$$
$$= a_1 \cdot r \cdot r \cdot r \cdot r^6$$
$$= a_1 \cdot \frac{a_2}{a_1} \cdot \frac{a_3}{a_2} \cdot \frac{a_4}{a_3} \cdot r^6$$
$$a_{10} = a_4 r^6.$$

Because $a_{10} = 125/64$ and $a_4 = 125$, you can solve for r as follows.

$$\frac{125}{64} = 125r^6$$

$$\frac{1}{64} = r^6$$

$$\pm\frac{1}{2} = r$$

Because the terms of the sequence are assumed to be positive, choose the positive solution, $r = \frac{1}{2}$. You can obtain the 14th term by multiplying the 10th term by r^4.

$$a_{14} = a_{10}r^4 = \frac{125}{64}\left(\frac{1}{2}\right)^4 = \frac{125}{1024}$$

✓CHECKPOINT   Now try Exercise 29.

The Sum of a Finite Geometric Sequence

The formula for the sum of a *finite* geometric sequence is as follows.

The Sum of a Finite Geometric Sequence

The sum of the finite geometric sequence

$$a_1, a_1 r, a_1 r^2, a_1 r^3, a_1 r^4, \ldots, a_1 r^{n-1}$$

with common ratio $r \neq 1$ is given by

$$S_n = \sum_{i=1}^{n} a_1 r^{i-1} = a_1 \left(\frac{1 - r^n}{1 - r} \right).$$

EXAMPLE 6 FINDING THE SUM OF A FINITE GEOMETRIC SEQUENCE

Find the sum $\sum_{n=1}^{12} 4(0.3)^{n-1}$.

SOLUTION

By writing out a few terms, you have

$$\sum_{n=1}^{12} 4(0.3)^{n-1} = 4(0.3)^0 + 4(0.3)^1 + 4(0.3)^2 + 4(0.3)^3 + \cdots + 4(0.3)^{11}.$$

Now, because $a_1 = 4$, $r = 0.3$, and $n = 12$, you can apply the formula for the sum of a finite geometric sequence to obtain

$$\sum_{n=1}^{12} 4(0.3)^{n-1} = a_1 \left(\frac{1 - r^n}{1 - r} \right) = 4 \left[\frac{1 - (0.3)^{12}}{1 - 0.3} \right] \approx 5.714.$$

✓**CHECKPOINT**   Now try Exercise 41.

When using the formula for the sum of a finite geometric sequence, be careful to check that the sum is of the form

$$\sum_{i=1}^{n} a_1 r^{i-1}.$$

If the sum is not of this form, you must adjust the formula for the nth partial sum. For instance, if the sum in Example 6 were $\sum_{n=1}^{12} 4(0.3)^n$, then you would evaluate the sum as follows.

$$\sum_{n=1}^{12} 4(0.3)^n = 4(0.3) + 4(0.3)^2 + 4(0.3)^3 + \cdots + 4(0.3)^{12}$$

$$= 4(0.3) + [4(0.3)](0.3) + [4(0.3)](0.3)^2 + \cdots + [4(0.3)](0.3)^{11}.$$

So, $a_1 = 4(0.3)$, $r = 0.3$, and $n = 12$, and the sum is

$$S_{12} = 4(0.3) \left(\frac{1 - (0.3)^{12}}{1 - 0.3} \right) \approx 1.714.$$

Geometric Series

The summation of the terms of an infinite geometric *sequence* is called an **infinite geometric series,** or simply a **geometric series.**

The formula for the sum of a *finite* geometric *sequence* can, depending on the value of r, be extended to produce a formula for the sum of an *infinite* geometric series. Specifically, if the common ratio r has the property that $|r| < 1$, it can be shown that r^n becomes arbitrarily close to zero as n increases without bound. Consequently,

$$a_1\left(\frac{1 - r^n}{1 - r}\right) \to a_1\left(\frac{1 - 0}{1 - r}\right) \quad \text{as} \quad n \to \infty.$$

This result is summarized as follows.

The Sum of an Infinite Geometric Series

If $|r| < 1$, the infinite geometric series

$$a_1 + a_1 r + a_1 r^2 + a_1 r^3 + \cdots + a_1 r^{n-1} + \cdots$$

has the sum

$$S = \sum_{i=0}^{\infty} a_1 r^i = \frac{a_1}{1 - r}.$$

EXAMPLE 7   FINDING THE SUM OF AN INFINITE GEOMETRIC SERIES

Find each sum.

a. $\displaystyle\sum_{n=1}^{\infty} 4(0.6)^{n-1}$ **b.** $\displaystyle\sum_{n=1}^{\infty} 3(0.1)^{n-1}$

SOLUTION

a. $\displaystyle\sum_{n=1}^{\infty} 4(0.6)^{n-1} = 4 + 4(0.6) + 4(0.6)^2 + 4(0.6)^3 + \cdots + 4(0.6)^{n-1} + \cdots$

$$= \frac{4}{1 - 0.6} \qquad \frac{a_1}{1 - r}$$

$$= 10$$

b. $\displaystyle\sum_{n=1}^{\infty} 3(0.1)^{n-1} = 3 + 3(0.1) + 3(0.1)^2 + 3(0.1)^3 + \cdots + 3(0.1)^{n-1} + \cdots$

$$= \frac{3}{1 - 0.1} \qquad \frac{a_1}{1 - r}$$

$$= \frac{10}{3}$$

$$\approx 3.33$$

✓*CHECKPOINT*   Now try Exercise 51.

Application

Digital televisions will eventually replace analog TVs because they deliver better sound and picture quality to viewers, and they are a more efficient way for broadcasters to communicate information.

> **EXAMPLE 8**   COMPOUND INTEREST

Make a Decision   You make a deposit of $50 on the first day of each month in a savings account that pays 6% compounded monthly. Will there be enough money in the account in 2 years to buy a $1500 digital TV?

SOLUTION

The first deposit will gain interest for 24 months, and its balance will be

$$A_{24} = 50\left(1 + \frac{0.06}{12}\right)^{24}$$

$$= 50(1.005)^{24}.$$

The second deposit will gain interest for 23 months, and its balance will be

$$A_{23} = 50\left(1 + \frac{0.06}{12}\right)^{23}$$

$$= 50(1.005)^{23}.$$

The last deposit will gain interest for only 1 month, and its balance will be

$$A_{1} = 50\left(1 + \frac{0.06}{12}\right)^{1}$$

$$= 50(1.005).$$

The total balance in the account will be the sum of the balances of the 24 deposits. Using the formula for the sum of a finite geometric sequence, with $A_1 = 50(1.005)$, $r = 1.005$, and $n = 24$, you have

$$S_{24} = 50(1.005)\left[\frac{1 - (1.005)^{24}}{1 - 1.005}\right]$$

$$= \$1277.96.$$

You will not have enough money in your account in 2 years to pay for the TV.

✓CHECKPOINT   Now try Exercise 57.

Discussing the Concept | An Experiment

You will need a piece of string or yarn, a pair of scissors, and a tape measure. Measure out any length of string at least 5 feet long. Double over the string and cut it in half. Take one of the resulting halves, double it over, and cut it in half. Continue this process until you are no longer able to cut a length of string in half. How many cuts were you able to make? Construct a sequence of the resulting string lengths after each cut, starting with the original length of the string. Find a formula for the *n*th term of this sequence. How many cuts could you theoretically make? Discuss why you were not able to make that many cuts.

6.3 Warm Up

The following warm-up exercises involve skills that were covered in earlier sections. You will use these skills in the exercise set for this section. For additional help, review Sections R1.3 and 6.1.

In Exercises 1–4, evaluate the expression.

1. $\left(\frac{4}{5}\right)^3$

2. $\left(\frac{3}{4}\right)^2$

3. 2^{-4}

4. $4(3^4)$

In Exercises 5–8, simplify the expression.

5. $(2n)(3n^2)$

6. $n(3n)^3$

7. $\dfrac{4n^5}{n^2}$

8. $\dfrac{(2n)^3}{8n}$

In Exercises 9 and 10, use summation notation to write the sum.

9. $2 + 2(3) + 2(3^2) + 2(3^3)$

10. $3 + 3(2) + 3(2^2) + 3(2^3)$

6.3 Exercises

In Exercises 1–10, determine whether the sequence is geometric. If it is, find its common ratio and write a formula for a_n.

1. $1, 4, 16, 64, \ldots$

2. $2, 10, 50, 250, \ldots$

3. $4, -12, 36, -108, \ldots$

4. $1, -2, 4, -8, \ldots$

5. $1, -\frac{1}{2}, \frac{1}{4}, -\frac{1}{8}, \ldots$

6. $\frac{3}{2}, 1, \frac{2}{3}, \frac{4}{9}, \ldots$

7. $\frac{1}{2}, \frac{2}{3}, \frac{3}{4}, \frac{4}{5}, \ldots$

8. $9, -6, 4, -\frac{8}{3}, \ldots$

9. $1, \frac{1}{2}, \frac{1}{3}, \frac{1}{4}, \ldots$

10. $\frac{1}{5}, \frac{2}{3}, \frac{3}{9}, \frac{4}{11}, \ldots$

In Exercises 11–18, write the first five terms of the geometric sequence.

11. $a_1 = 4, r = 2$

12. $a_1 = 6, r = 3$

13. $a_1 = 1, r = \frac{1}{3}$

14. $a_1 = 1, r = \frac{2}{3}$

15. $a_1 = 7, r = -\frac{1}{5}$

16. $a_1 = \frac{1}{2}, r = -\frac{1}{3}$

17. $a_1 = 1, r = e$

18. $a_1 = 2, r = e^{0.1}$

In Exercises 19–30, find the indicated term of the geometric sequence.

19. $a_1 = 16, r = \frac{1}{4}$, 5th term

20. $a_1 = 9, r = \frac{2}{3}$, 7th term

21. $a_1 = 6, r = -\frac{1}{3}$, 12th term

22. $a_1 = 1, r = -\frac{2}{3}$, 10th term

23. $a_1 = 100, r = e$, 9th term

24. $a_1 = 6, r = e^{0.1}$, 9th term

25. $a_1 = 500, r = 1.02$, 40th term

26. $a_1 = 1000, r = 1.005$, 60th term

27. $a_1 = 16, a_4 = \frac{27}{4}$, 3rd term

28. $a_2 = 3, a_5 = \frac{3}{64}$, 3rd term

29. $a_2 = -18, a_5 = \frac{2}{3}$, 6th term

30. $a_3 = \frac{16}{3}, a_5 = \frac{64}{27}$, 7th term

In Exercises 31–34, match the geometric sequence with its graph. [The graphs are labeled (a), (b), (c), and (d).]

(a)

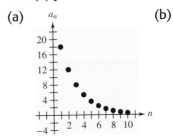

(b)

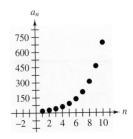

(c)

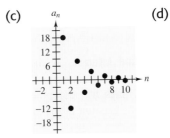

(d)

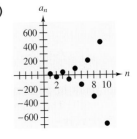

31. $a_n = 18\left(\frac{2}{3}\right)^{n-1}$

32. $a_n = 18\left(-\frac{2}{3}\right)^{n-1}$

33. $a_n = 18\left(\frac{3}{2}\right)^{n-1}$

34. $a_n = 18\left(-\frac{3}{2}\right)^{n-1}$

In Exercises 35–40, use a graphing utility to graph the first 10 terms of the sequence.

35. $a_n = 12(-0.75)^{n-1}$ **36.** $a_n = 10(1.5)^{n-1}$

37. $a_n = 12(-0.4)^{n-1}$ **38.** $a_n = 20(-1.25)^{n-1}$

39. $a_n = 2(1.3)^{n-1}$ **40.** $a_n = 10(1.2)^{n-1}$

In Exercises 41–48, find the sum of the finite geometric sequence.

41. $\displaystyle\sum_{n=1}^{10} 8(2^n)$ **42.** $\displaystyle\sum_{n=1}^{6} 3(4^n)$

43. $\displaystyle\sum_{i=1}^{10} 8\left(\tfrac{1}{4}\right)^{i-1}$ **44.** $\displaystyle\sum_{i=1}^{10} 5\left(\tfrac{1}{3}\right)^{i-1}$

45. $\displaystyle\sum_{n=0}^{8} 2^n$ **46.** $\displaystyle\sum_{n=0}^{8} (-2)^n$

47. $\displaystyle\sum_{n=0}^{20} 3\left(\tfrac{3}{2}\right)^n$ **48.** $\displaystyle\sum_{n=0}^{15} 2\left(\tfrac{4}{3}\right)^n$

In Exercises 49–56, find the sum of the infinite geometric series, if it exists.

49. $\displaystyle\sum_{n=0}^{\infty} \left(\tfrac{1}{2}\right)^n = 1 + \tfrac{1}{2} + \tfrac{1}{4} + \tfrac{1}{8} + \cdots$

50. $\displaystyle\sum_{n=0}^{\infty} 2\left(\tfrac{2}{3}\right)^n = 2 + \tfrac{4}{3} + \tfrac{8}{9} + \tfrac{16}{27} + \cdots$

51. $\displaystyle\sum_{n=1}^{\infty} \left(-\tfrac{1}{2}\right)^{n-1} = 1 - \tfrac{1}{2} + \tfrac{1}{4} - \tfrac{1}{8} + \cdots$

52. $\displaystyle\sum_{n=1}^{\infty} 2\left(-\tfrac{2}{3}\right)^{n-1} = 2 - \tfrac{4}{3} + \tfrac{8}{9} - \tfrac{16}{27} + \cdots$

53. $\displaystyle\sum_{n=0}^{\infty} \left(\tfrac{3}{2}\right)^n = 1 + \tfrac{3}{2} + \tfrac{9}{4} + \tfrac{27}{8} + \cdots$

54. $\displaystyle\sum_{n=0}^{\infty} 4\left(\tfrac{1}{4}\right)^n = 4 + 1 + \tfrac{1}{4} + \tfrac{1}{16} + \cdots$

55. $\displaystyle\sum_{n=0}^{\infty} 3\left(\tfrac{1}{10}\right)^n = 3 + 0.3 + 0.03 + 0.003 + \cdots$

56. $\displaystyle\sum_{n=0}^{\infty} 5\left(\tfrac{4}{3}\right)^n = 5 + \tfrac{20}{3} + \tfrac{80}{9} + \tfrac{320}{27} + \cdots$

57. *Make a Decision*: **Compound Interest**   A deposit of $250 is made at the beginning of each month for 5 years in an account that pays 6% compounded monthly. The balance A in the account at the end of 5 years is

$$A = 250\left(1 + \frac{0.06}{12}\right)^1 + \cdots + 250\left(1 + \frac{0.06}{12}\right)^{60}.$$

(a) Find the balance after 5 years. Is there enough money in the account to buy an $18,000 motorcycle?

(b) How much would the balance increase if the interest rate were raised to 8%? Is this enough money to buy the motorcycle?

58. *Make a Decision*: **Compound Interest**   A deposit of $100 is made at the beginning of each month for 4 years in an account that pays 5% compounded monthly. Use a graphing utility to find the balance A in the account at the end of 4 years. Is there enough money in the account to pay for the construction of a $5000 addition to your house?

$$A = 100\left(1 + \frac{0.05}{12}\right)^1 + \cdots + 100\left(1 + \frac{0.05}{12}\right)^{48}$$

59. Compound Interest   You deposit $100 in an account at the beginning of each month for 10 years. The account pays 8% compounded monthly. Use a graphing utility to find your balance at the end of 10 years. If the interest were compounded continuously, what would be the balance?

60. Compound Interest   You deposit $150 in an account at the beginning of each month for 20 years. The account pays 6% compounded monthly. What is your balance at the end of 20 years? If the interest were compounded continuously, what would be the balance?

61. Profit   The annual profit a_n (in millions of dollars) for Harley-Davidson, Inc., from 1998 to 2003 can be approximated by the model

$$a_n = 210.66e^{0.243n}, \quad n = 0, 1, \ldots, 5$$

where n represents the year, with $n = 0$ corresponding to 1998 (see figure). Use the formula for the sum of a finite geometric sequence to approximate the total profit earned during this six-year period. (Source: Harley-Davidson, Inc.)

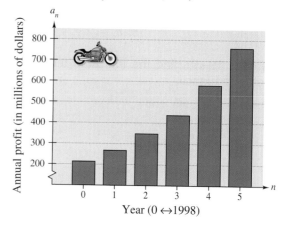

Year (0 ↔ 1998)

62. Profit The annual profit a_n (in millions of dollars) for The Home Depot, Inc., from 1993 to 2002 can be approximated by the model

$$a_n = 468.70e^{0.239n}, \quad n = 0, 1, 2, 3, \ldots, 9$$

where n represents the year, with $n = 0$ corresponding to 1993 (see figure). Use the formula for the sum of a finite geometric sequence to approximate the total profit earned during this 10-year period. (Source: The Home Depot, Inc.)

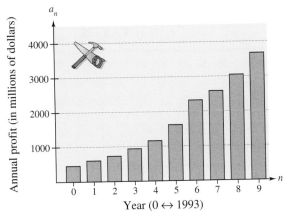

Year (0 ↔ 1993)

63. Revenue The annual revenue a_n (in millions of dollars) for California Pizza Kitchen, Inc., from 1999 to 2003 can be approximated by the model

$$a_n = 177.69e^{0.177n}, \quad n = 0, 1, 2, 3, 4$$

where n represents the year, with $n = 0$ corresponding to 1999. (Source: California Pizza Kitchen, Inc.)

(a) Construct a bar graph showing the annual revenue from 1999 to 2003.

(b) Use the formula for the sum of a finite geometric sequence to approximate the total revenue during this five-year period.

64. Revenue The annual revenue a_n (in millions of dollars) for Krispy Kreme Doughnuts, Inc., from 1997 to 2002 can be approximated by the model

$$a_n = 147.98e^{0.237n}, \quad n = 0, 1, 2, 3, 4, 5$$

where n represents the year, with $n = 0$ corresponding to 1997. (Source: Krispy Kreme Doughnuts, Inc.)

(a) Construct a bar graph showing the annual revenue from 1997 to 2002.

(b) Use the formula for the sum of a finite geometric sequence to approximate the total revenue during this six-year period.

IN THE NEWS

Aging Drives Demand

Biomet, Inc.

. . .A leader in the burgeoning market for replacement hips, artificial knees and related treatments is Biomet, Inc.

This Indiana-based manufacturer of advanced orthopedic devices and supplies has turned in strong, rocksteady growth in sales and earnings for many years. . . .

Performance Results Biomet's net income, excluding special items, totaled $286.7 million in fiscal 2003 (ended May 31, 2003), up 19.6 percent from $239.7 million the previous fiscal year. Diluted earnings per share totaled $1.10 versus 2002's $.88. Net sales rose 15.6 percent to $1.4 billion from $1.2 billion in fiscal 2002.

"Biomet, Inc." Better Investing, September 2003, 40.

65. The annual sales a_n (in millions of dollars) for Biomet, Inc., from 1993 to 2003 can be modeled by

$$a_n = 333.58e^{0.141n}, \quad n = 0, 1, 2, \ldots, 10$$

where n represents the year, with $n = 0$ corresponding to 1993. Use the formula for the sum of a finite geometric sequence to approximate the total sales for Biomet, Inc., during this 11-year period. (Source: Biomet, Inc.)

66. The annual profits a_n (in millions of dollars) for Biomet, Inc., from 1993 to 2003 can be modeled by

$$a_n = 60.02e^{0.152n}, \quad n = 0, 1, 2, \ldots, 10$$

where n represents the year, with $n = 0$ corresponding to 1993. Use the formula for the sum of a finite geometric sequence to approximate the total profit earned by Biomet, Inc., during this 11-year period. (Source: Biomet, Inc.)

67. *Make a Decision* You go to work at a company that pays $0.01 for the first day, $0.02 for the second day, $0.04 for the third day, and so on. If the daily wage keeps doubling, what will be your total income for working 29 days? 30 days? 31 days? Which job pays more the first year: the job that pays $0.01 the first day and twice the previous day's wage for every day thereafter, or a job that pays $30,000 per year?

68. Geometry The sides of a square are 16 inches in length. A new square is formed by connecting the midpoints of the sides of the original square, and two of the resulting triangles are shaded (see figure). If this process is repeated five more times, what will be the total area of the shaded region?

69. Geometry The sides of a square are 27 inches in length. New squares are formed by dividing the original square into nine squares. The center square is then shaded (see figure). Each of the eight unshaded squares is then divided into nine smaller squares and the center square of each is shaded. If this process is repeated four more times, what will be the total area of the shaded region?

Math *MATTERS*

The Snowflake Curve

To create a **snowflake curve,** begin with an equilateral triangle (called Stage 1). To form Stage 2, trisect each side of the triangle and at each of the middle sections construct an equilateral triangle pointing outward. To form each additional stage, repeat this process. The diagram shows the first six stages of the snowflake curve, which is the curve obtained by continuing this process infinitely many times.

In this figure, note that the nth stage of the snowflake curve has $3(4^{n-1})$ sides. If each side in Stage 1 has a length of 1, then each side in Stage 2 will have a length of $\frac{1}{3}$, each side in Stage 3 will have a length of $\left(\frac{1}{3}\right)^2 = \frac{1}{9}$, and so on. So, the total perimeter of the nth stage is

$$3(4^{n-1}) \cdot \left(\frac{1}{3}\right)^{n-1} = 3\left(\frac{4}{3}\right)^{n-1}.$$

Because the value of this expression is unbounded as n approaches infinity, you can conclude that the perimeter of the snowflake curve is infinite. This seems to be a paradox because it is difficult to conceive of a curve that is infinitely long being drawn on a piece of paper that has a finite area.

Stage 1 (3 sides)

Stage 2 (12 sides)

Stage 3 (48 sides)

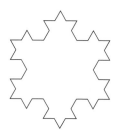

Stage 4 (192 sides)

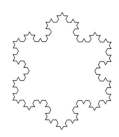

Stage 5 (768 sides)

Stage 6 (3072 sides)

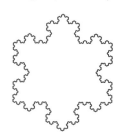

6.4   The Binomial Theorem

Objectives

- Find a binomial coefficient.
- Use Pascal's Triangle to determine a binomial coefficient.
- Expand a binomial using binomial coefficients and find a specific term in a binomial expansion.

Binomial Coefficients

Recall that a *binomial* is a polynomial that has two terms. In this section, you will study a formula that gives a quick method of raising a binomial to a power. To begin, look at the expansion of $(x + y)^n$ for several values of n.

$$(x + y)^0 = 1$$
$$(x + y)^1 = x + y$$
$$(x + y)^2 = x^2 + 2xy + y^2$$
$$(x + y)^3 = x^3 + 3x^2y + 3xy^2 + y^3$$
$$(x + y)^4 = x^4 + 4x^3y + 6x^2y^2 + 4xy^3 + y^4$$
$$(x + y)^5 = x^5 + 5x^4y + 10x^3y^2 + 10x^2y^3 + 5xy^4 + y^5$$

There are several observations you can make about these expansions.

1. In each expansion, there are $n + 1$ terms.

2. In each expansion, x and y have symmetrical roles. The powers of x decrease by 1 in successive terms, whereas the powers of y increase by 1.

3. The degree of each term is n. For instance, in the expansion of $(x + y)^5$, the degree of each term is 5.

$$\overset{4 + 1 = 5 \quad 3 + 2 = 5}{(x + y)^5 = x^5 + 5x^4y^1 + 10x^3y^2 + 10x^2y^3 + 5xy^4 + y^5}$$

4. The coefficients increase and then decrease in a symmetric pattern.

The coefficients of a binomial expansion are called **binomial coefficients.** To find them, you can use the **Binomial Theorem.**

The Binomial Theorem

In the expansion of $(x + y)^n$

$$(x + y)^n = x^n + nx^{n-1}y + \cdots + {}_nC_r x^{n-r}y^r + \cdots + nxy^{n-1} + y^n$$

the coefficient of $x^{n-r}y^r$, the $(r + 1)$th term, is given by

$$_nC_r = \frac{n!}{(n - r)!r!}.$$

STUDY TIP

The notation $\binom{n}{r}$ is often used in place of ${}_nC_r$ to denote a binomial coefficient.

EXAMPLE 1   FINDING BINOMIAL COEFFICIENTS

Find each binomial coefficient.

a. $_8C_2$ **b.** $_{10}C_3$ **c.** $_7C_0$ **d.** $_8C_8$

SOLUTION

a. $_8C_2 = \dfrac{8!}{6! \cdot 2!} = \dfrac{(8 \cdot 7) \cdot 6!}{6! \cdot 2!} = \dfrac{8 \cdot 7}{2 \cdot 1} = 28$ Coefficient of third term in expansion of $(x + y)^8$.

b. $_{10}C_3 = \dfrac{10!}{7! \cdot 3!} = \dfrac{(10 \cdot 9 \cdot 8) \cdot 7!}{7! \cdot 3!} = \dfrac{10 \cdot 9 \cdot 8}{3 \cdot 2 \cdot 1} = 120$ Coefficient of fourth term in expansion of $(x + y)^{10}$.

c. $_7C_0 = \dfrac{7!}{7! \cdot 0!} = 1$ Coefficient of first term in expansion of $(x + y)^7$.

d. $_8C_8 = \dfrac{8!}{0! \cdot 8!} = 1$ Coefficient of ninth term in expansion of $(x + y)^8$.

✓CHECKPOINT   Now try Exercise 1.

When $r \neq 0$ and $r \neq n$, as in parts (a) and (b) of Example 1, there is a simple pattern for evaluating binomial coefficients.

$$_8C_2 = \overbrace{\dfrac{8 \cdot 7}{2 \cdot 1}}^{2 \text{ factors}} \quad \text{and} \quad _{10}C_3 = \overbrace{\dfrac{10 \cdot 9 \cdot 8}{3 \cdot 2 \cdot 1}}^{3 \text{ factors}}$$

2 factorial   3 factorial

EXAMPLE 2   FINDING BINOMIAL COEFFICIENTS

Find each binomial coefficient using the pattern shown above.

a. $_7C_3$ **b.** $\binom{7}{4}$ **c.** $_{12}C_1$ **d.** $\binom{12}{11}$

SOLUTION

a. $_7C_3 = \dfrac{7 \cdot 6 \cdot 5}{3 \cdot 2 \cdot 1} = 35$ Coefficient of fourth term in expansion of $(x + y)^7$.

b. $\binom{7}{4} = \dfrac{7 \cdot 6 \cdot 5 \cdot 4}{4 \cdot 3 \cdot 2 \cdot 1} = 35$ Coefficient of fifth term in expansion of $(x + y)^7$.

c. $_{12}C_1 = \dfrac{12}{1} = 12$ Coefficient of second term in expansion of $(x + y)^{12}$.

d. $\binom{12}{11} = \dfrac{12!}{1! \cdot 11!} = \dfrac{(12) \cdot 11!}{1! \cdot 11!} = \dfrac{12}{1} = 12$ Coefficient of twelfth term in expansion of $(x + y)^{12}$.

✓CHECKPOINT   Now try Exercise 5.

It is not a coincidence that the results in parts (a) and (b) of Example 2 are the same and that the results in parts (c) and (d) are the same. In general, it is true that

$$_nC_r = {_nC_{n-r}}.$$

Pascal's Triangle

There is a convenient way to remember a pattern for binomial coefficients. By arranging the coefficients in a triangular pattern, you obtain the following array, which is called **Pascal's Triangle.** This triangle is named after the famous French mathematician Blaise Pascal (1623–1662).

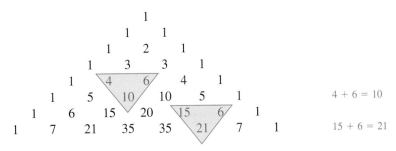

4 + 6 = 10

15 + 6 = 21

The first and last numbers in each row of Pascal's Triangle are 1. Every other number in each row is formed by adding the two numbers immediately above the number. Pascal noticed that the numbers in this triangle are precisely the same numbers that are the coefficients of binomial expansions, as follows.

$$(x + y)^0 = 1 \qquad \text{0th row}$$

$$(x + y)^1 = 1x + 1y \qquad \text{1st row}$$

$$(x + y)^2 = 1x^2 + 2xy + 1y^2 \qquad \text{2nd row}$$

$$(x + y)^3 = 1x^3 + 3x^2y + 3xy^2 + 1y^3 \qquad \text{3rd row}$$

$$(x + y)^4 = 1x^4 + 4x^3y + 6x^2y^2 + 4xy^3 + 1y^4$$

$$(x + y)^5 = 1x^5 + 5x^4y + 10x^3y^2 + 10x^2y^3 + 5xy^4 + 1y^5$$

$$(x + y)^6 = 1x^6 + 6x^5y + 15x^4y^2 + 20x^3y^3 + 15x^2y^4 + 6xy^5 + 1y^6$$

$$(x + y)^7 = 1x^7 + 7x^6y + 21x^5y^2 + 35x^4y^3 + 35x^3y^4 + 21x^2y^5 + 7xy^6 + 1y^7$$

The top row in Pascal's Triangle is called the *zero row* because it corresponds to the binomial expansion $(x + y)^0 = 1$. Similarly, the next row is called the *first row* because it corresponds to the binomial expansion $(x + y)^1 = 1x + 1y$. In general, the *nth row* in Pascal's Triangle gives the coefficients of $(x + y)^n$.

EXAMPLE 3   USING PASCAL'S TRIANGLE

Use the seventh row of Pascal's Triangle to find the binomial coefficients.

$${}_8C_0, \; {}_8C_1, \; {}_8C_2, \; {}_8C_3, \; {}_8C_4, \; {}_8C_5, \; {}_8C_6, \; {}_8C_7, \; {}_8C_8$$

SOLUTION

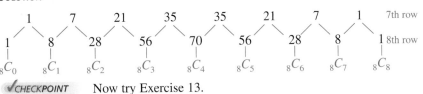

✔**CHECKPOINT**   Now try Exercise 13.

Binomial Expansions

As mentioned at the beginning of this section, when you write out the coefficients for a binomial that is raised to a power, you are **expanding a binomial.** The formulas for binomial coefficients give you an easy way to expand binomials, as demonstrated in the next three examples.

EXAMPLE 4   EXPANDING A BINOMIAL

Write the expansion for the expression

$(x + 1)^3$.

SOLUTION

The binomial coefficients from the third row of Pascal's Triangle are

1, 3, 3, 1.

So, the expansion is as follows.

$$(x + 1)^3 = (1)x^3 + (3)x^2(1) + (3)x(1^2) + (1)(1^3)$$

$$= x^3 + 3x^2 + 3x + 1$$

✓CHECKPOINT   Now try Exercise 15.

To expand binomials representing *differences,* rather than sums, you alternate signs. Here are two examples.

$$(x - 1)^3 = [x + (-1)]^3$$

$$= x^3 - 3x^2 + 3x - 1$$

$$(x - 1)^4 = [x + (-1)]^4$$

$$= x^4 - 4x^3 + 6x^2 - 4x + 1$$

EXAMPLE 5   EXPANDING A BINOMIAL

Write the expansion for each expression.

a. $(2x - 3)^4$ **b.** $(x - 2y)^4$

SOLUTION

The binomial coefficients from the fourth row of Pascal's Triangle are

1, 4, 6, 4, 1.

So, the expansions are as follows.

a. $(2x - 3)^4 = (1)(2x)^4 - (4)(2x)^3(3) + (6)(2x)^2(3^2) - (4)(2x)(3^3) + (1)(3^4)$

$$= 16x^4 - 96x^3 + 216x^2 - 216x + 81$$

b. $(x - 2y)^4 = (1)x^4 - (4)x^3(2y) + (6)x^2(2y)^2 - (4)x(2y)^3 + (1)(2y)^4$

$$= x^4 - 8x^3y + 24x^2y^2 - 32xy^3 + 16y^4$$

✓CHECKPOINT   Now try Exercise 43.

EXAMPLE 6	FINDING A TERM IN A BINOMIAL EXPANSION

a. Find the sixth term of $(a + 2b)^8$.

b. *Make a Decision* Suppose that

$$(a + 2b)^8 = C_1a^8 + C_2a^7b + C_3a^6b^2 + \cdots + C_9b^8$$

where C_i is the coefficient of the ith term. Decide whether all C_i's must be even numbers.

SOLUTION

a. From the Binomial Theorem, you can see that the $(r + 1)$th term is $_nC_r x^{n-r}y^r$. In this case, $6 = r + 1$ and therefore $r = 5$. Because $n = 8$, $x = a$, and $y = 2b$, the sixth term in the binomial expansion is

$$_8C_5 a^{8-5}(2b)^5 = 56 \cdot a^3 \cdot (2b)^5$$
$$= 56(2^5)a^3b^5$$
$$= 1792a^3b^5.$$

b. Because each term except the first contains a power of $2b$, the coefficients of the second through the ninth terms are even. The coefficient of the first term is $_8C_0 = 1$, which is odd. So, the coefficients of the expansion of $(a + 2b)^8$ are not all even.

✓CHECKPOINT   Now try Exercise 47.

Discussing the Concept | Finding a Pattern

By adding the terms in each row of Pascal's Triangle, you obtain the following.

Row 0: $1 = 1$

Row 1: $1 + 1 = 2$

Row 2: $1 + 2 + 1 = 4$

Row 3: $1 + 3 + 3 + 1 = 8$

Row 4: $1 + 4 + 6 + 4 + 1 = 16$

Find a pattern for this sequence. Then use the pattern to find the sum of the terms in the 10th row of Pascal's Triangle. Check your answer by actually adding the terms of the 10th row.

6.4 **Warm Up**

The following warm-up exercises involve skills that were covered in earlier sections. You will use these skills in the exercise set for this section. For additional help, review Sections R1.5 and 6.1.

In Exercises 1–6, perform the indicated operations and simplify.

1. $5x^2(x^3 + 3)$

2. $(x + 5)(x^2 - 3)$

3. $(x + 4)^2$

4. $(2x - 3)^2$

5. $x^2y(3xy^{-2})$

6. $(-2z)^5$

In Exercises 7–10, evaluate the expression.

7. $5!$

8. $\dfrac{8!}{5!}$

9. $\dfrac{10!}{7!}$

10. $\dfrac{6!}{3!3!}$

6.4 **Exercises**

In Exercises 1–10, find the binomial coefficient.

1. $_6C_2$

2. $_9C_4$

3. $_{12}C_0$

4. $_8C_8$

5. $_{20}C_{15}$

6. $_{20}C_5$

7. $_{100}C_{98}$

8. $_{10}C_4$

9. $_{100}C_2$

10. $_{10}C_6$

In Exercises 11–14, evaluate using Pascal's Triangle.

11. $\binom{8}{5}$

12. $\binom{8}{7}$

13. $_7C_4$

14. $_6C_3$

In Exercises 15–34, use the Binomial Theorem to expand the expression. Simplify your answer.

15. $(x + 1)^4$

16. $(x + 1)^6$

17. $(x + 2)^3$

18. $(x + 2)^4$

19. $(x - 2)^4$

20. $(x - 2)^5$

21. $(x + y)^5$

22. $(x + y)^6$

23. $(3s + 2)^4$

24. $(2s + 3)^5$

25. $(3 - 2s)^6$

26. $(2 - 3s)^4$

27. $(x - y)^5$

28. $(2x - y)^5$

29. $(1 - 2x)^3$

30. $(1 - 3x)^4$

31. $(x^2 - y^2)^5$

32. $(x^2 + y^2)^6$

33. $\left(\dfrac{1}{x} + y\right)^5$

34. $\left(x - \dfrac{3}{y}\right)^3$

35. How are the expansions of the binomials $(x + y)^n$ and $(x - y)^n$ alike? How are they different?

36. How many terms are in the expanded form of $(x + y)^5$? $(x + y)^6$? $(x + y)^7$? $(x + y)^8$? How many terms are in $(x + y)^n$?

In Exercises 37–42, use the Binomial Theorem to expand the complex number. Simplify your answer by using the fact that $i^2 = -1$.

37. $(1 + i)^4$

38. $(2 - i)^5$

39. $\left(3 - \sqrt{-4}\right)^6$

40. $\left(5 + \sqrt{-16}\right)^3$

41. $\left(\dfrac{-1}{2} + \dfrac{\sqrt{3}}{2}i\right)^3$

42. $\left(\dfrac{3}{4} - \dfrac{\sqrt{2}}{4}i\right)^4$

In Exercises 43–46, expand the expression by using Pascal's Triangle to determine the coefficients.

43. $(2t - 1)^5$

44. $(x + 4)^5$

45. $(3 + 2z)^4$

46. $(3y + 2)^5$

In Exercises 47–54, find the specified nth term in the expansion of the binomial.

47. $(x + y)^{10}$, $n = 4$

48. $(x - y)^6$, $n = 7$

49. $(x - 6y)^5$, $n = 3$

50. $(x - 10z)^7$, $n = 4$

51. $(4x + 3y)^9$, $n = 8$

52. $(5a + 6b)^5$, $n = 5$

53. $(10x - 3y)^{12}$, $n = 9$

54. $(7x + 2y)^{15}$, $n = 7$

In Exercises 55–62, find the coefficient a of the indicated term in the expansion of the binomial.

55. $(x + 2)^{10}$, ax^6

56. $(x^2 + 3)^{12}$, ax^8

57. $(x - 2y)^{10}$, ax^8y^2

58. $(4x - y)^{10}$, ax^3y^7

59. $(3x - 2y)^9$, ax^4y^5 **60.** $(2x - 3y)^{11}$, ax^3y^8

61. $(x^2 - 1)^8$, ax^8 **62.** $(x^2 + 2)^{12}$, ax^{10}

In Exercises 63–66, use the Binomial Theorem to expand the expression.

63. $\left(\frac{1}{2} + \frac{1}{2}\right)^7$ **64.** $\left(\frac{1}{4} + \frac{3}{4}\right)^{10}$

65. $(0.6 + 0.4)^5$ **66.** $(0.35 + 0.65)^6$

67. Finding a Pattern   Describe the pattern formed by the sums of the numbers along the diagonal segments of Pascal's Triangle (see figure).

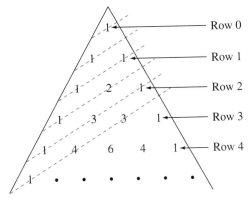

68. Finding a Pattern   Use each of the encircled groups of numbers in the figure to form a 2×2 matrix. Find the determinant of each matrix. Describe the pattern.

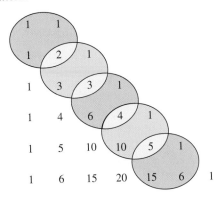

69. Non-Farm Employees   The total number $f(t)$ of non-farm employees (in millions) in the United States from 1980 to 2002 can be approximated by

$$f(t) = 0.015t^2 + 1.74t + 88.3, \quad 0 \le t \le 22$$

where t represents the year, with $t = 0$ corresponding to 1980 (see figure). (Source: U.S. Bureau of Labor Statistics)

(a) You want to adjust this model so that $t = 0$ corresponds to 1990 rather than 1980. To do this, you shift the graph of f 10 units *to the left* and obtain $g(t) = f(t + 10)$. Write $g(t)$ in standard form.

(b) Use a graphing utility to graph f and g in the same viewing window.

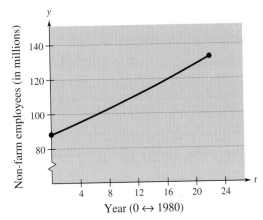

Figure for 69

70. Life Insurance   The average amount $f(t)$ (in thousands of dollars) of life insurance per insured household in the United States from 1980 through 2001 can be approximated by

$$f(t) = 0.008t^2 + 6.95t + 49.3, \quad 0 \le t \le 21$$

where $t = 0$ represents 1980 (see figure). (Source: American Council of Life Insurers)

(a) You want to adjust this model so that $t = 0$ corresponds to 1990 rather than 1980. To do this, you shift the graph of f 10 units *to the left* and obtain $g(t) = f(t + 10)$. Write $g(t)$ in standard form.

(b) Use a graphing utility to graph f and g in the same viewing window.

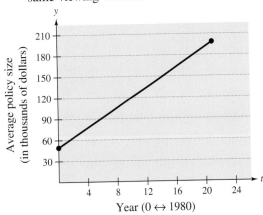

Mid-Chapter Quiz

Take this quiz as you would take a quiz in class. After you are done, check your work against the answers given in the back of the book.

In Exercises 1 and 2, write the first five terms of the sequence. (Assume that n begins with 1.)

1. $a_n = 2n + 1$

2. $a_n = \dfrac{n}{(n+1)^2}$

In Exercises 3 and 4, evaluate the expression.

3. $\dfrac{5!}{3!}$

4. $\dfrac{30!}{27!}$

In Exercises 5–10, write an expression for a_n. (Assume that n begins with 1.)

5. $2, \frac{1}{2}, \frac{2}{9}, \frac{1}{8}, \frac{2}{25}, \ldots$

6. $\frac{1}{2}, \frac{1}{3}, \frac{1}{4}, \frac{1}{5}, \frac{1}{6}, \ldots$

7. Arithmetic: $a_1 = 2, d = 5$

8. Arithmetic: $a_1 = 7, d = 3$

9. Geometric: $a_1 = 10, r = 3$

10. Geometric: $a_1 = 4, r = -\frac{1}{2}$

In Exercises 11–14, determine whether the sequence is arithmetic, geometric, or neither. Explain your reasoning.

11. $a_n = 2 + 3(n - 1)$

12. $1, 1, 2, 3, 5, 8, 13, \ldots$

13. $1, \frac{1}{2}, \frac{1}{4}, \frac{1}{8}, \ldots$

14. $a_n = 3(2)^n$

In Exercises 15–20, find the sum.

15. $\displaystyle\sum_{i=1}^{5} 7$

16. $\displaystyle\sum_{i=1}^{4} 2i^2$

17. $\displaystyle\sum_{n=1}^{25} (4n - 3)$

18. $\displaystyle\sum_{n=1}^{6} 2^n$

19. $\displaystyle\sum_{n=0}^{\infty} \left(-\frac{1}{3}\right)^n$

20. $\displaystyle\sum_{n=0}^{\infty} 5\left(\frac{1}{5}\right)^n$

21. Evaluate (a) $_9C_4$ and (b) $_{12}C_{12}$.

22. Use the Binomial Theorem to expand each expression.
  (a) $(x - 5)^4$ (b) $(2x + 1)^5$

23. Find the ninth term in the expansion of $(x^2 + y)^{10}$.

24. A person accepts a position with a company and will receive a salary of $30,500 for the first year. The person is guaranteed a raise of $1700 per year.
  (a) What will be the salary during the sixth year of employment?
  (b) How much will the company have paid the person by the end of the sixth year?

25. You deposit $100 in an account at the beginning of each month for 20 years. The account pays 8% compounded monthly. What will be your balance at the end of the 20 years?

6.5 Counting Principles

Objectives

- Solve simple counting problems.
- Use the Fundamental Counting Principle to solve counting problems.
- Use permutations to solve counting problems.
- Use combinations to solve counting problems.

Simple Counting Problems

The next two sections of this chapter contain a brief introduction to some of the basic counting principles and their application to probability.

EXAMPLE 1 SELECTING PAIRS OF NUMBERS AT RANDOM

Eight pieces of paper are numbered from 1 to 8 and placed in a box. One piece of paper is drawn from the box, its number is written down, and the piece of paper is *replaced in the box*. Then, a second piece of paper is drawn from the box, and its number is written down. Finally, the two numbers are added together. In how many different ways can a sum of 12 be obtained?

SOLUTION

To solve this problem, count the different ways that a sum of 12 can be obtained using two numbers from 1 to 8.

First number	4	5	6	7	8
Second number	8	7	6	5	4

From this list, you can see that a sum of 12 can occur in five different ways.

✔CHECKPOINT   Now try Exercise 7.

EXAMPLE 2 SELECTING PAIRS OF NUMBERS AT RANDOM

Eight pieces of paper are numbered from 1 to 8 and placed in a box. One piece of paper is drawn from the box and, *without replacing the paper*, a second piece of paper is drawn. The numbers on the pieces of paper are written down and totaled. In how many different ways can a sum of 12 be obtained?

SOLUTION

To solve this problem, you can count the different ways that a sum of 12 can be obtained using two *different* numbers from 1 to 8.

First number	4	5	7	8
Second number	8	7	5	4

So, a sum of 12 can be obtained in four different ways.

✔CHECKPOINT   Now try Exercise 8.

STUDY TIP

The counting problems in Examples 1 and 2 can be distinguished by saying that the random selection in Example 1 occurs **with replacement,** whereas the random selection in Example 2 occurs **without replacement,** which eliminates the possibility of choosing two 6's.

Counting Principles

Examples 1 and 2 describe simple counting problems in which you can *list* each possible way that an event can occur. When it is possible, this is always the best way to solve a counting problem. However, some events can occur in so many different ways that it is not feasible to write out the entire list. In such cases, you must rely on formulas and counting principles. The most important of these is the **Fundamental Counting Principle.**

Fundamental Counting Principle

Let E_1 and E_2 be two events. The first event E_1 can occur in m_1 different ways. After E_1 has occurred, E_2 can occur in m_2 different ways. The number of ways that the two events can occur is $m_1 \cdot m_2$.

The Fundamental Counting Principle can be extended to three or more events. For instance, the number of ways that three events E_1, E_2, and E_3 can occur is $m_1 \cdot m_2 \cdot m_3$.

EXAMPLE 3   USING THE FUNDAMENTAL COUNTING PRINCIPLE

How many different pairs of letters from the English alphabet are possible?

SOLUTION

This experiment has two events. The first event is the choice of the first letter, and the second event is the choice of the second letter. Because the English alphabet contains 26 letters, it follows that the number of pairs of letters is $26 \cdot 26 = 676$.

✓CHECKPOINT   Now try Exercise 11.

EXAMPLE 4   USING THE FUNDAMENTAL COUNTING PRINCIPLE

Telephone numbers in the United States currently have 10 digits. The first three are the *area code* and the next seven are the *local telephone number.* How many different telephone numbers are possible within each area code? (Note that at this time, a local telephone number cannot begin with 0 or 1.)

SOLUTION

Because the first digit cannot be 0 or 1, there are only eight choices for the first digit. For each of the other six digits, there are 10 choices.

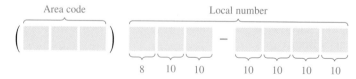

So, the number of local telephone numbers that are possible *within* each area code is $8 \cdot 10 \cdot 10 \cdot 10 \cdot 10 \cdot 10 \cdot 10 = 8{,}000{,}000$.

✓CHECKPOINT   Now try Exercise 13.

Permutations

One important application of the Fundamental Counting Principle is in determining the number of ways that n elements can be arranged (in order). An ordering of n elements is called a **permutation** of the elements.

Definition of Permutation

A **permutation** of n different elements is an ordering of the elements such that one element is first, one is second, one is third, and so on.

EXAMPLE 5   FINDING THE NUMBER OF PERMUTATIONS OF N ELEMENTS

How many permutations are possible of the letters A, B, C, D, E, and F?

SOLUTION

Consider the following reasoning.

First position:	Any of the *six* letters.
Second position:	Any of the remaining *five* letters.
Third position:	Any of the remaining *four* letters.
Fourth position:	Any of the remaining *three* letters.
Fifth position:	Any of the remaining *two* letters.
Sixth position:	The *one* remaining letter.

So, the numbers of choices for the six positions are as follows.

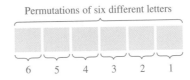

Permutations of six different letters

6   5   4   3   2   1

The total number of permutations of the six letters is

$$6! = 6 \cdot 5 \cdot 4 \cdot 3 \cdot 2 \cdot 1$$
$$= 720.$$

✓CHECKPOINT   Now try Exercise 23.

Number of Permutations of n Elements

The number of permutations of n distinct elements is given by

$$n \cdot (n - 1) \cdots 4 \cdot 3 \cdot 2 \cdot 1 = n!.$$

In other words, there are $n!$ different ways that n distinct elements can be ordered.

© Rick Rickman/NewSport/CORBIS

Occasionally, you are interested in ordering a *subset* of a collection of elements rather than the entire collection. For example, you might want to choose (and order) *r* elements out of a collection of *n* elements. Such an ordering is called a **permutation of *n* elements taken *r* at a time.**

EXAMPLE 6   COUNTING HORSE RACE FINISHES

Eight horses are running in a race. In how many different ways can these horses come in first, second, and third? (Assume that there are no ties.)

SOLUTION

Here are the different possibilities.

Win (first position):	*Eight* choices
Place (second position):	*Seven* choices
Show (third position):	*Six* choices

Using the Fundamental Counting Principle, multiply these three numbers together to obtain the following.

Different orders of horses

8   7   6

So, there are $8 \cdot 7 \cdot 6 = 336$ different orders.

✓CHECKPOINT   Now try Exercise 37.

Permutations of *n* Elements Taken *r* at a Time

The number of permutations of *n* distinct elements taken *r* at a time is

$$_nP_r = \frac{n!}{(n-r)!}$$

$$= n(n-1)(n-2) \cdots (n-r+1).$$

Using this formula, you can rework Example 6 to find that the number of permutations of eight horses taken three at a time is

$$_8P_3 = \frac{8!}{(8-3)!}$$

$$= \frac{8!}{5!}$$

$$= \frac{(8 \cdot 7 \cdot 6) \cdot 5!}{5!}$$

$$= 336$$

which is the same answer obtained in Example 6.

Remember that for permutations, order is important. So, if you were looking at the possible permutations of the letters A, B, C, and D taken three at a time, the permutations (A, B, D) and (B, A, D) would be different because the *order* of the elements is different.

Suppose, however, that you were asked to find the possible permutations of the letters A, A, B, and C. The total number of permutations of the four letters would be $_4P_4 = 4!$. However, not all of these arrangements would be *distinguishable* because there are two A's in the list. To find the number of distinguishable permutations, you can use the following formula.

Distinguishable Permutations

Suppose a set of n objects has n_1 of one kind of object, n_2 of a second kind, n_3 of a third kind, and so on, with $n = n_1 + n_2 + n_3 + \cdots + n_k$. Then the number of **distinguishable permutations** of the n objects is

$$\frac{n!}{n_1! \cdot n_2! \cdot n_3! \cdot \cdots \cdot n_k!}.$$

EXAMPLE 7 DISTINGUISHABLE PERMUTATIONS

In how many distinguishable ways can the letters in BANANA be written?

SOLUTION

This word has six letters, so $n = 6$, but you have to take into account that there are three A's, two N's, and one B. So, the number of distinguishable ways the letters can be written is

$$\frac{n!}{n_1! \cdot n_2! \cdot n_3!} = \frac{6!}{3! \cdot 2! \cdot 1!} = \frac{6 \cdot 5 \cdot 4 \cdot 3!}{3! \cdot 2!} = 60.$$

The 60 different distinguishable permutations are as follows.

AAABNN	AAANBN	AAANNB	AABANN	AABNAN	AABNNA
AANABN	AANANB	AANBAN	AANBNA	AANNAB	AANNBA
ABAANN	ABANAN	ABANNA	ABNAAN	ABNANA	ABNNAA
ANAABN	ANAANB	ANABAN	ANABNA	ANANAB	ANANBA
ANBAAN	ANBANA	ANBNAA	ANNAAB	ANNABA	ANNBAA
BAAANN	BAANAN	BAANNA	BANAAN	BANANA	BANNAA
BNAAAN	BNAANA	BNANAA	BNNAAA	NAAABN	NAAANB
NAABAN	NAABNA	NAANAB	NAANBA	NABAAN	NABANA
NABNAA	NANAAB	NANABA	NANBAA	NBAAAN	NBAANA
NBANAA	NBNAAA	NNAAAB	NNAABA	NNABAA	NNBAAA

✓CHECKPOINT   Now try Exercise 43.

Combinations

When you count the number of possible permutations of a set of elements, *order* is important. As a final topic in this section, you will look at a method of selecting subsets of a larger set in which order is *not important.* Such subsets are called **combinations of *n* elements taken *r* at a time.** For instance, the combinations

$$\{A, B, C\} \quad \text{and} \quad \{B, A, C\}$$

are equivalent because both sets contain the same three elements, and the order in which the elements are listed is *not important.* So, you would count only one of the two sets. A common example of how a combination occurs is a card game in which the player is free to reorder the cards after they have been dealt.

EXAMPLE 8   COMBINATIONS OF *N* ELEMENTS TAKEN *R* AT A TIME

In how many different ways can three letters be chosen from the letters A, B, C, D, and E? (The order of the three letters is not important.)

SOLUTION

The following subsets represent the different combinations of three letters that can be chosen from five letters.

$\{A, B, C\}$	$\{A, B, D\}$
$\{A, B, E\}$	$\{A, C, D\}$
$\{A, C, E\}$	$\{A, D, E\}$
$\{B, C, D\}$	$\{B, C, E\}$
$\{B, D, E\}$	$\{C, D, E\}$

From this list, you can conclude that there are 10 different ways that three letters can be chosen from five letters.

✓CHECKPOINT   Now try Exercise 49.

Combinations of *n* Elements Taken *r* at a Time

The number of combinations of *n* elements taken *r* at a time is

$$_nC_r = \frac{n!}{(n-r)!r!}.$$

Note that the formula for $_nC_r$ is the same one given for binomial coefficients. To see how this formula is used, solve the counting problem in Example 8. In that problem, you are asked to find the number of combinations of five elements taken three at a time. So, $n = 5$, $r = 3$, and the number of combinations is

$$_5C_3 = \frac{5!}{2!3!} = \frac{5 \cdot 4 \cdot \overset{2}{3 \cdot 2!}}{2! \cdot 3 \cdot 2 \cdot 1} = 10$$

which is the same answer obtained in Example 8.

EXAMPLE 9 COUNTING CARD HANDS

A standard poker hand consists of five cards dealt from a deck of 52. How many different poker hands are possible? (After the cards are dealt, the player may reorder them, and therefore order is not important.)

SOLUTION

You can find the number of different poker hands by using the formula for the number of combinations of 52 elements taken five at a time, as follows.

$$_{52}C_5 = \frac{52!}{47!5!} = \frac{52 \cdot 51 \cdot 50 \cdot 49 \cdot 48}{5 \cdot 4 \cdot 3 \cdot 2 \cdot 1} = 2{,}598{,}960 \text{ different hands}$$

✓CHECKPOINT   Now try Exercise 53.

EXAMPLE 10 CHOOSING BETWEEN PERMUTATIONS AND COMBINATIONS

Make a Decision Decide whether each scenario should be counted using permutations or combinations.

a. Number of different arrangements of three types of flowers from an array of 20 types

b. Number of different three-digit pin numbers for a debit card

SOLUTION

To determine which counting principle is necessary to solve each problem, ask yourself two questions: (1) Is the order of the elements important? If yes, then you should solve the problem using permutations. (2) Are the chosen elements a subset of a larger set in which order is not important? If yes, then you should solve the problem using combinations.

a. The order of the flowers is not important. So, the number of possible flower arrangements should be counted using combinations.

b. The order of the digits in the pin number matters. So, the number of possible pin numbers should be counted using permutations.

✓CHECKPOINT   Now try Exercise 55.

Discussing the Concept | Problem Posing

According to NASA, each space shuttle astronaut consumes an average of 2700 calories per day. An evening meal normally consists of 2 main dishes, a vegetable dish, a bread item, and a dessert or fruit. The space shuttle food and beverage list contains 30 items classified as main dishes, 11 vegetable dishes, 4 breads, and 22 desserts/fruits. How many different evening meal menus are possible? Create and solve two other problems that could be asked about the evening meal menus. (*Hint:* You may want to think of variety versus repetition and length of mission.)
(Source: NASA)

6.5 ⟨ Warm Up ⟩

The following warm-up exercises involve skills that were covered in earlier sections. You will use these skills in the exercise set for this section. For additional help, review Sections R1.3, 6.1, and 6.4.

In Exercises 1–4, evaluate the expression.

1. $13 \cdot 8^2 \cdot 2^3$
2. $10^2 \cdot 9^3 \cdot 4$
3. $\dfrac{12!}{2!(7!)(3!)}$
4. $\dfrac{25!}{22!}$

In Exercises 5 and 6, find the binomial coefficient.

5. $_{12}C_7$
6. $_{25}C_{22}$

In Exercises 7–10, simplify the expression.

7. $\dfrac{n!}{(n-4)!}$
8. $\dfrac{(2n)!}{4(2n-3)!}$
9. $\dfrac{2 \cdot 4 \cdot 6 \cdot 8 \, \cdots \, (2n)}{2^n}$
10. $\dfrac{3 \cdot 6 \cdot 9 \cdot 12 \, \cdots \, (3n)}{3^n}$

6.5 ⟨ Exercises ⟩

Random Selection In Exercises 1–8, determine the number of ways a computer can randomly generate the specified integer(s) from 1 through 12.

1. An odd integer
2. An even integer
3. A prime integer
4. An integer that is greater than 9
5. An integer that is divisible by 4
6. An integer that is divisible by 3
7. Two integers whose sum is 8
8. Two *distinct* integers whose sum is 8

9. **Job Applicants** A small college needs two additional faculty members: a chemist and a statistician. There are five applicants for the chemistry position and six applicants for the statistics position. In how many ways can these positions be filled?

10. **Computer Systems** A customer in a computer store can choose one of six monitors, one of three keyboards, and one of eight computers. If all the choices are compatible, how many different systems can be chosen?

11. **Toboggan Ride** Six people are lining up for a ride on a toboggan, but only two of the six are willing to take the first position. With that constraint, in how many ways can the six people be seated on the toboggan? Draw a diagram to illustrate the number of ways that the people can sit.

12. **Course Schedule** A college student is preparing her course schedule for the next semester. She may select one of six mathematics courses, one of five science courses, and one of eight courses from the social sciences and humanities. In how many ways can she select her schedule?

13. **License Plate Numbers** In a certain state, each automobile license plate number consists of two letters followed by a four-digit number. How many distinct license plate numbers can be formed?

14. **License Plate Numbers** In a certain state, each automobile license plate number consists of two letters followed by a four-digit number. To avoid confusion between "O" and "zero" and "I" and "one," the letters "O" and "I" are not used. How many distinct license plate numbers can be formed?

15. **True-False Exam** In how many ways can a 10-question true-false exam be answered? (Assume that no questions are omitted.)

16. **True-False Exam** In how many ways can a 15-question true-false exam be answered? (Assume that no questions are omitted.)

17. **Three-Digit Numbers** How many three-digit numbers can be formed under each condition?
    (a) The leading digit cannot be 0.
    (b) The leading digit cannot be 0 and no repetition of digits is allowed.
    (c) The leading digit cannot be 0 and the number must be a multiple of 5.

18. Four-Digit Numbers How many four-digit numbers can be formed under each condition?

(a) The leading digit cannot be 0.

(b) The leading digit cannot be 0 and no repetition of digits is allowed.

(c) The leading digit cannot be 0 and the number must be a multiple of 5.

19. *Make a Decision*: **Competing Restaurants** Two hamburger restaurants are competing for customers. Restaurant A offers one of six cheeses, one of three types of meat patties, one of four toppings, one of five sauces, and one of five buns. Restaurant B offers one of seven cheeses, one of three meats, one of five toppings, one of five sauces, and one of three buns. How many different burgers can be made at each restaurant? Which restaurant can truthfully advertise that it has more burger choices?

20. *Make a Decision*: **Competing Restaurants** Two Italian restaurants are featuring a "Design Your Own Pasta Dish" sales promotion. Restaurant A offers one of six pastas, one of four sauces, one of five cheeses, and one of eight toppings. Restaurant B offers one of five pastas, one of five sauces, one of six cheeses, and one of seven toppings. How many different pasta dishes can be made at each restaurant? Which restaurant can truthfully advertise that it has more pasta dish choices?

21. Combination Lock A combination lock will open when the right choice of three numbers (from 1 to 40, inclusive) is selected. How many different lock combinations are possible?

22. Combination Lock A combination lock will open when the right choice of three numbers (from 1 to 50, inclusive) is selected. How many different lock combinations are possible?

23. Posing for a Photograph In how many ways can six children line up in one row to have their picture taken?

24. Riding in a Car In how many ways can eight people sit in an eight-passenger van?

25. Concert Seats Four couples have reserved seats in a given row for a concert. In how many different ways can they be seated, given the following conditions?

(a) There are no restrictions.

(b) The two members of each couple wish to sit together.

26. Single File In how many orders can five girls and three boys walk through a doorway single-file, given the following conditions?

(a) There are no restrictions.

(b) The boys go before the girls.

(c) The girls go before the boys.

In Exercises 27–36, evaluate $_nP_r$.

27. $_4P_3$

28. $_5P_4$

29. $_8P_3$

30. $_{10}P_5$

31. $_{12}P_4$

32. $_{20}P_2$

33. $_{100}P_2$

34. $_{48}P_3$

35. $_6P_6$

36. $_7P_7$

37. Choosing Officers From a pool of 15 candidates, the offices of president, vice-president, secretary, and treasurer will be filled. In how many different ways can the offices be filled if each of the 15 candidates can hold any office?

38. Bike Race There are 10 bicyclists entered in a race. In how many different ways can the top three places be decided?

39. Forming an Experimental Group In order to conduct an experiment, five students are randomly selected from a class of 30. How many different groups of five students are possible?

40. Test Questions A student may answer any 12 questions from a total of 15 questions on an exam. How many different ways can the student select the questions?

41. Lottery In Connecticut's Classic Lotto game, a player chooses six distinct numbers from 1 to 44. In how many ways can a player select the six numbers? (The order of selection is not important.)

42. Lottery In South Dakota's Dakota Cash game, a player chooses five distinct numbers from 1 to 35. In how many ways can a player select the five numbers? (The order of selection is not important.)

In Exercises 43–48, find the number of distinguishable permutations of the group of letters.

43. A, S, S, E, S, S

44. P, A, P, A, Y, A

45. R, E, F, E, R, E, E

46. P, A, R, A, L, L, E, L

47. A, L, G, E, B, R, A

48. M, I, S, S, I, S, S, I, P, P, I

49. Number of Subsets How many subsets of four elements can be formed from a set of 100 elements?

50. Number of Subsets How many subsets of five elements can be formed from a set of 80 elements?

51. Forming a Committee A committee composed of three graduate students and two undergraduate students is to be selected from a group of 10 graduates and six undergraduates. How many different committees can be formed?

52. Job Applicants An employer interviews 10 people for four openings in the research and development department of the company. Four of the 10 people are women. If all 10 are qualified, in how many ways can the employer fill the four positions if (a) the selection is random and (b) exactly two selections are women?

53. Poker Hand Five cards are selected from an ordinary deck of 52 playing cards. In how many ways can you get a full house? (A full house consists of three of one kind and two of another. For example, A-A-A-5-5 and K-K-K-10-10 are full houses.)

54. Poker Hand Five cards are selected from an ordinary deck of 52 playing cards. In how many ways can you get a straight flush? (A straight flush consists of five cards that are in order and of the same suit. For example, A♥, 2♥, 3♥, 4♥, 5♥ and 10♠, J♠, Q♠, K♠, A♠ are straight flushes.)

Make a Decision: In Exercises 55 and 56, decide whether the scenario should be counted using permutations or combinations. Explain your reasoning.

55. Number of ways 10 people can line up in a row for concert tickets

56. Number of different three-topping pizzas that can be made from an assortment of 15 different toppings

Geometry In Exercises 57–60, find the number of diagonals of the polygon. (A line segment connecting any two nonadjacent vertices is called a *diagonal* of a polygon.)

57. Pentagon

58. Hexagon

59. Octagon

60. Decagon (10 sides)

In Exercises 61–63, use the example of the tree diagram to help you answer the question.

A college theater company is planning a tour with performances in Phoenix, Dallas, and New Orleans. If there are no restrictions on the order of the performances, in how many ways can the tour be arranged?

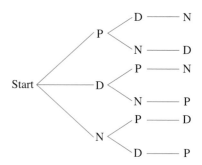

To determine the possible arrangements, count the paths. A path moves from the start to the end of the last branch. In this case, there are six paths, so there are six possible ways to arrange the itinerary of the tour. The tree diagram verifies visually what you know by using counting principles. That is, the number of possible arrangements is $3 \cdot 2 \cdot 1 = 6$.

61. Use a tree diagram to describe visually the possible tour arrangements for the college theater company if the city of Houston is added to the itinerary.

62. Use a tree diagram to describe visually the possible birth order arrangements for the genders of three children within a family.

63. Use a tree diagram to describe visually the possible birth order arrangements for the genders of four children within a family.

64. *Make a Decision* Without calculating the numbers, determine which of the following is greater. Explain.

   (a) The combinations of 10 elements taken six at a time

   (b) The permutations of 10 elements taken six at a time

65. *Make a Decision* Is the value of $_nP_r$ always greater than the value of $_nC_r$? Explain.

6.6 **Probability**

Objectives

- **Determine the sample space of an experiment and find the probability of an event.**
- **Find the probability of mutually exclusive events.**
- **Find the probability of independent events.**
- **Find the probability of the complement of an event.**

The Probability of an Event

In measuring the uncertainties of everyday life, ambiguous terminology, such as "fairly certain" and "highly unlikely," is often used. In mathematics we attempt to assign a number to the likelihood of the occurrence of an event. This measurement is called the **probability** that the event will occur. For example, if you toss a fair coin, the probability that it will land face up is $\frac{1}{2}$.

Any happening whose result is uncertain is called an **experiment.** The possible results of an experiment are **outcomes,** the set of all possible outcomes of an experiment is the **sample space** of the experiment, and any subcollection of a sample space is an **event.** For instance, when a six-sided die is tossed, the sample space can be represented by the numbers from 1 through 6, each being an *equally likely* outcome.

To describe a sample space in such a way that each outcome is equally likely, you must sometimes distinguish between various outcomes in ways that appear artificial. Example 1 illustrates such a situation.

EXAMPLE 1 FINDING THE SAMPLE SPACE

Find the sample space for each event.

a. One coin is tossed.   **b.** Two coins are tossed.   **c.** Three coins are tossed.

SOLUTION

a. Because the coin will land either heads up (denoted by H) or tails up (denoted by T), the sample space is $S = \{H, T\}$.

b. Because either coin can land heads up or tails up, the possible outcomes are

$HH =$ heads up on both coins

$HT =$ heads up on first coin and tails up on second coin

$TH =$ tails up on first coin and heads up on second coin

$TT =$ tails up on both coins.

The sample space is $S = \{HH, HT, TH, TT\}$. This list distinguishes between the two cases HT and TH, even though these two outcomes appear to be similar.

c. Following the notation of part (b), the sample space is

$S = \{HHH, HHT, HTH, HTT, THH, THT, TTH, TTT\}$.

✓CHECKPOINT   Now try Exercise 1.

To calculate the probability of an event, count the number of outcomes in the event and in the sample space. The *number of outcomes* in event E is denoted by $n(E)$, and the number of outcomes in the sample space S is denoted by $n(S)$. The probability that event E will occur is given by $n(E)/n(S)$.

The Probability of an Event

If an event E has $n(E)$ equally likely outcomes and its sample space S has $n(S)$ equally likely outcomes, the **probability** of event E is

$$P(E) = \frac{n(E)}{n(S)}.$$

Because the number of outcomes in an event must be less than or equal to the number of outcomes in the sample space, the probability of an event must be a number between 0 and 1. That is, for any event E, it must be true that $0 \le P(E) \le 1$, as shown in Figure 6.3.

Properties of the Probability of an Event

Let E be an event that is a subset of a finite sample space S.

1. $0 \le P(E) \le 1$

2. If $P(E) = 0$, E *cannot occur* and is called an **impossible event.**

3. If $P(E) = 1$, E *must occur* and is called a **certain event.**

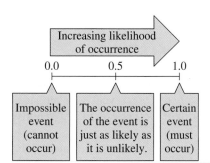

Figure 6.3

Figure 6.4   Standard deck of playing cards

| EXAMPLE 2 | FINDING THE PROBABILITY OF AN EVENT | |

a. Two coins are tossed. What is the probability that both land heads up?

b. A card is drawn from a standard deck of playing cards (see Figure 6.4). What is the probability that it is an ace?

SOLUTION

a. Following the procedure in Example 1(b), let $E = \{HH\}$ and $S = \{HH, HT, TH, TT\}$. The probability of getting two heads is

$$P(E) = \frac{n(E)}{n(S)} = \frac{1}{4}.$$

b. Because there are 52 cards in a standard deck of playing cards and there are four aces (one in each suit), the probability of drawing an ace is

$$P(E) = \frac{n(E)}{n(S)} = \frac{4}{52} = \frac{1}{13}.$$

✓CHECKPOINT   Now try Exercise 7.

Figure 6.5

TECHNOLOGY

You can use the *random integer* feature of a graphing utility to simulate the tossing of a die or set of dice. For instance, the result of rolling a pair of dice five times and summing the outcomes is shown below.

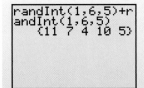

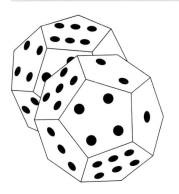

Figure 6.6

EXAMPLE 3   FINDING THE PROBABILITY OF AN EVENT

Two six-sided dice are tossed. What is the probability that the total of the two dice is 7? (See Figure 6.5.)

SOLUTION

Because there are six possible outcomes on each die, you can use the Fundamental Counting Principle to conclude that there are $6 \cdot 6$ or 36 different outcomes when two dice are tossed. To find the probability of rolling a total of 7, you must first count the number of ways this can occur.

First Die	1	2	3	4	5	6
Second Die	6	5	4	3	2	1

So, a total of 7 can be rolled in six ways, which means that the probability of rolling a 7 is

$$P(E) = \frac{n(E)}{n(S)} = \frac{6}{36} = \frac{1}{6}.$$

✓CHECKPOINT   Now try Exercise 11.

EXAMPLE 4   FINDING THE PROBABILITY OF AN EVENT

Make a Decision   Twelve-sided dice can be constructed (in the shape of regular dodecahedrons) such that each of the numbers from 1 to 6 appears twice on each die, as shown in Figure 6.6. Can these dice be used in any game requiring ordinary six-sided dice without changing the probabilities of different outcomes?

SOLUTION

For an ordinary six-sided die, each of the numbers 1, 2, 3, 4, 5, and 6 occurs only once, so the probability of any particular number coming up is

$$P(E) = \frac{n(E)}{n(S)} = \frac{1}{6}.$$

For a twelve-sided die, each number occurs twice, so the probability of any particular number coming up is

$$P(E) = \frac{n(E)}{n(S)} = \frac{2}{12} = \frac{1}{6}.$$

So, the twelve-sided dice can be used in any game requiring ordinary six-sided dice without changing the probabilities of different outcomes.

✓CHECKPOINT   Now try Exercise 13.

In Examples 2, 3, and 4, you simply counted the outcomes in the desired events. For larger sample spaces, however, you should use the counting principles discussed in Section 6.5.

EXAMPLE 5   THE PROBABILITY OF WINNING A LOTTERY

In Louisiana's Lotto game, a player chooses six numbers from 1 to 40. If these six numbers match the six numbers drawn (in any order) by the lottery commission, the player wins (or shares) the jackpot. What is the probability of winning the jackpot?

SOLUTION

Because the order of the six numbers does not matter, use the formula for the number of combinations of 40 elements taken six at a time.

$$n(S) = {}_{40}C_6 = \frac{40 \cdot 39 \cdot 38 \cdot 37 \cdot 36 \cdot 35}{6 \cdot 5 \cdot 4 \cdot 3 \cdot 2 \cdot 1} = 3{,}838{,}380.$$

If a person buys only one ticket, the probability of winning the jackpot is

$$P(E) = \frac{n(E)}{n(S)} = \frac{1}{3{,}838{,}380}.$$

✓CHECKPOINT   Now try Exercise 21.

EXAMPLE 6   RANDOM SELECTION

The total number of colleges and universities in the United States in 2002 is shown in Figure 6.7. One institution is selected at random. What is the probability that the institution is in one of the three southern regions?   (Source: U.S. Department of Education)

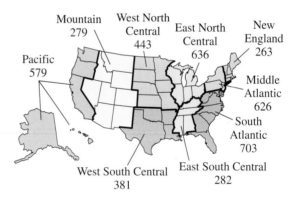

Figure 6.7

SOLUTION

From the figure, the total number of colleges and universities is 4192. Because there are $703 + 282 + 381 = 1366$ colleges and universities in the three southern regions, the probability that the institution is in one of these regions is

$$P(E) = \frac{n(E)}{n(S)} = \frac{1366}{4192} \approx 0.326.$$

✓CHECKPOINT   Now try Exercise 31.

Mutually Exclusive Events

Two events A and B (from the same sample space) are **mutually exclusive** if A and B have no outcomes in common. In the terminology of sets, the **intersection of A and B** is the empty set and

$$P(A \cap B) = 0.$$

For instance, if two dice are tossed, the event A of rolling a total of 6 and the event B of rolling a total of 9 are mutually exclusive. To find the probability that one or the other of two mutually exclusive events will occur, you can *add* their individual probabilities.

Probability of the Union of Two Events

If A and B are events in the same sample space, the probability of A *or* B occurring is given by

$$P(A \cup B) = P(A) + P(B) - P(A \cap B).$$

If A and B are mutually exclusive, then

$$P(A \cup B) = P(A) + P(B).$$

EXAMPLE 7   THE PROBABILITY OF A UNION

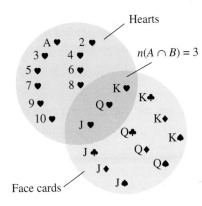

Figure 6.8

One card is selected from a standard deck of 52 playing cards. What is the probability that the card is either a heart or a face card?

SOLUTION

Because the deck has 13 hearts, the probability of selecting a heart (event A) is

$$P(A) = \tfrac{13}{52}.$$

Similarly, because the deck has 12 face cards, the probability of selecting a face card (event B) is

$$P(B) = \tfrac{12}{52}.$$

Because three of the cards are hearts and face cards (see Figure 6.8), it follows that the events A and B are *not* mutually exclusive, and therefore

$$P(A \cap B) = \tfrac{3}{52}.$$

Finally, applying the formula for the probability of the union of two events, you can conclude that the probability of selecting a heart or a face card is

$$P(A \cup B) = P(A) + P(B) - P(A \cap B)$$

$$= \frac{13}{52} + \frac{12}{52} - \frac{3}{52}$$

$$= \frac{22}{52} \approx 0.423.$$

 CHECKPOINT   Now try Exercise 37.

| EXAMPLE 8 | PROBABILITY OF MUTUALLY EXCLUSIVE EVENTS | |

The personnel department of a company has compiled data showing the number of years of service for each employee. The results are shown in the table.

Years of Service	Number of Employees
0–4	157
5–9	89
10–14	74
15–19	63
20–24	42
25–29	38
30–34	37
35–39	21
40–44	8

If an employee is chosen at random, what is the probability that the employee has had 9 or fewer years of service?

SOLUTION

To begin, add the number of employees and find that the total is 529. Next, let event A represent choosing an employee with 0 to 4 years of service and let event B represent choosing an employee with 5 to 9 years of service. Then

$$P(A) = \frac{157}{529} \quad \text{and} \quad P(B) = \frac{89}{529}.$$

Because A and B have no outcomes in common, you can conclude that these two events are mutually exclusive and that

$$P(A \cup B) = P(A) + P(B) = \frac{157}{529} + \frac{89}{529}$$

$$= \frac{246}{529}$$

$$\approx 0.465.$$

So, the probability of choosing an employee who has 9 or fewer years of service is about 0.465.

✓CHECKPOINT   Now try Exercise 47.

Independent Events

Two events are **independent** if the occurrence of one has no effect on the occurrence of the other. For instance, rolling a total of 12 with two six-sided dice has no effect on the outcome of future rolls of the dice. To find the probability that two independent events will occur, *multiply* the probabilities of each.

Probability of Independent Events

If A and B are independent events, the probability that both A and B will occur is

$$P(A \text{ and } B) = P(A) \cdot P(B).$$

EXAMPLE 9 PROBABILITY OF INDEPENDENT EVENTS

A random number generator on a computer selects three integers from 1 to 20. What is the probability that all three numbers are less than or equal to 5?

SOLUTION

The probability of selecting an integer from 1 to 5 is

$$P(A) = \frac{5}{20} = \frac{1}{4}.$$

So, the probability that all three integers are less than or equal to 5 is

$$P(A) \cdot P(A) \cdot P(A) = \left(\frac{1}{4}\right)\left(\frac{1}{4}\right)\left(\frac{1}{4}\right)$$

$$= \frac{1}{64}.$$

✓CHECKPOINT   Now try Exercise 33.

EXAMPLE 10 PROBABILITY OF INDEPENDENT EVENTS

In 2002, approximately 58.3% of the population of the United States was 30 years old or older. In a survey, 10 people were chosen at random from the population. What is the probability that all 10 were 30 years old or older?   (Source: U.S. Census Bureau)

SOLUTION

Let A represent choosing a person who was 30 years old or older. The probability of choosing a person who was 30 years old or older is 0.583, the probability of choosing a second person who was 30 years old or older is 0.583, and so on. Because these events are independent, you can conclude that the probability that all 10 people were 30 years old or older is

$$\underbrace{P(A) \cdot P(A) \cdot \ldots \cdot P(A)}_{10 \text{ occurrences}} = [P(A)]^{10}$$

$$= (0.583)^{10}$$

$$\approx 0.0045.$$

✓CHECKPOINT   Now try Exercise 59.

The Complement of an Event

The **complement of an event** A is the collection of all outcomes in the sample space that are *not* in A. The complement of event A is denoted by A'. Because $P(A \text{ or } A') = 1$ and because A and A' are mutually exclusive, it follows that $P(A) + P(A') = 1$. So, the probability of A' is given by

$$P(A') = 1 - P(A).$$

For instance, if the probability of *winning* a certain game is $P(A) = \frac{1}{4}$, then the probability of *losing* the game is $P(A') = 1 - \frac{1}{4} = \frac{3}{4}$.

Probability of a Complement

Let A be an event and let A' be its complement. If the probability of A is $P(A)$, then the probability of the complement is

$$P(A') = 1 - P(A).$$

EXAMPLE 11   FINDING THE PROBABILITY OF A COMPLEMENT

A digital piano manufacturer has determined that a certain machine averages one faulty piano for every 1000 it produces. What is the probability that an order of 200 pianos will have one or more faulty pianos?

SOLUTION

To solve this problem as stated, you would need to find the probabilities of having exactly one faulty piano, exactly two faulty pianos, exactly three faulty pianos, and so on. However, using complements, you can simply find the probability that all pianos are perfect and then subtract this value from 1. Because the probability that any given piano is perfect is $999/1000$, the probability that all 200 pianos are perfect is

$$P(A) = \left(\frac{999}{1000}\right)^{200} \approx 0.8186.$$

So, the probability that at least one piano is faulty is

$$P(A') = 1 - P(A) \approx 0.1814.$$

✔CHECKPOINT   Now try Exercise 45.

Discussing the Concept | Suspicious Probabilities

When comparing the market shares of two products, a marketing director states that the probability that a customer will pick Brand A is 0.43, the probability that a customer will pick Brand B is 0.46, and the probability that a customer will pick neither brand is 0.14. A person at the meeting questions these probabilities. Why?

6.6 (Warm Up

The following warm-up exercises involve skills that were covered in earlier sections. You will use these skills in the exercise set for this section. For additional help, review Sections R1.2, R1.3, 6.1, and 6.4.

In Exercises 1–8, evaluate the expression.

1. $\frac{1}{4} + \frac{5}{8} - \frac{5}{16}$

2. $\frac{4}{15} + \frac{3}{5} - \frac{1}{3}$

3. $\frac{5 \cdot 4}{5!}$

4. $\frac{5!22!}{27!}$

5. $\frac{4!}{8!12!}$

6. $\frac{9 \cdot 8 \cdot 7 \cdot 6 \cdot 5}{9!}$

7. $\frac{_5C_3}{_{10}C_3}$

8. $\frac{_{10}C_2 \cdot _{10}C_2}{_{20}C_4}$

In Exercises 9 and 10, evaluate the expression. (Round your answer to three decimal places.)

9. $\left(\frac{99}{100}\right)^{100}$

10. $1 - \left(\frac{89}{100}\right)^{50}$

6.6 (Exercises

In Exercises 1–6, determine the sample space for the experiment.

1. A coin and a six-sided die are tossed.

2. A six-sided die is tossed twice and the sum of the points is recorded.

3. A taste tester has to rank three varieties of yogurt (A, B, and C) according to preference.

4. Two marbles are selected from a sack containing two red marbles, two blue marbles, and one black marble. The color of each marble is recorded.

5. Two county supervisors are selected from five supervisors (A, B, C, D and E) to study a recycling plan.

6. A sales representative makes presentations of a product in three homes per day. In each home, there may be a sale (denote by S) or there may be no sale (denote by F).

Heads or Tails?   In Exercises 7–10, a coin is tossed three times. Find the probability of the event.

7. Getting exactly one head

8. Getting a tail on the second toss

9. Getting at least one tail

10. Getting at least two heads

Tossing a Die   In Exercises 11–16, two six-sided dice are tossed. Find the probability of the event.

11. The sum is 5.

12. The sum is less than 10.

13. The sum is at least 7.   **14.** The sum is 2, 3, or 8.

15. The sum is odd and no more than 7.

16. The sum is odd or a prime number.

Drawing a Card   In Exercises 17–20, a card is selected from a standard deck of 52 cards. Find the probability of the event.

17. Getting a face card   **18.** Not getting a face card

19. Getting a red card that is not a face card

20. Getting a card that is a 6 or less (aces are low)

Drawing Marbles   In Exercises 21–24, two marbles are drawn (without replacement) from a bag containing two green, three yellow, and four red marbles. Find the probability of the event.

21. Drawing exactly one red marble

22. Drawing two green marbles

23. Drawing none of the yellow marbles

24. Drawing marbles of different colors

In Exercises 25 and 26, you are given the probability that an event *will* happen. Find the probability that the event *will not* happen.

25. $p = 0.3$ **26.** $p = 0.72$

In Exercises 27 and 28, you are given the probability that an event *will not* happen. Find the probability that the event *will* happen.

27. $p = 0.1$ **28.** $p = 0.68$

29. Winning an Election Three people have been nominated for president of a college class. From a small poll, it is estimated that the probability of Jane winning the election is 0.46, and the probability of Larry winning the election is 0.32. What is the probability of the third candidate winning the election?

30. Winning an Election Taylor, Moore, and Jenkins are candidates for a public office. It is estimated that Moore and Jenkins have about the same probability of winning, and Taylor is believed to be twice as likely to win as either of the others. Find each candidate's probability of winning the election.

31. College Bound In a high school graduating class of 198 students, 43 are on the honor roll. Of these, 37 are going on to college, and of the other 155 students, 102 are going on to college. A student is selected at random from the class. What is the probability that the person chosen is (a) going to college, (b) not going to college, and (c) on the honor roll, but not going to college?

32. Alumni Association The alumni office of a college sends a survey to selected members of the class of 2004. Of the 1254 people who graduated that year, 672 are women, and of those, 124 went to graduate school. Of the 582 male graduates, 198 went to graduate school. An alumnus is selected at random. What is the probability that the person is (a) female, (b) male, and (c) female and did not attend graduate school?

33. Random Number Generator Two integers (from 1 to 30, inclusive) are chosen by a random number generator on a computer. What is the probability that (a) both numbers are even, (b) one number is even and one is odd, (c) both numbers are less than 10, and (d) the same number is chosen twice?

34. Random Number Generator Two integers (from 1 to 40, inclusive) are chosen by a random number generator on a computer. What is the probability that (a) both numbers are even, (b) one number is even and one is odd, (c) both numbers are less than 30, and (d) the same number is chosen twice?

35. Preparing for a Test An instructor gives her class a list of eight study problems, from which she will select five to be answered on an exam. A student knows how to solve six of the problems. Find the probability that the student will be able to answer all five questions on the exam.

36. Preparing for a Test An instructor gives his class a list of 20 study problems, from which he will select 10 to be answered on an exam. A student knows how to solve 15 of the problems. Find the probability that the student will be able to answer all 10 questions on the exam.

37. Drawing Cards from a Deck One card is selected at random from a standard deck of 52 playing cards. Find (a) the probability that the card is an even-numbered card, (b) the probability that the card is a heart or a diamond, and (c) the probability that the card is a nine or a face card.

38. Drawing Cards from a Deck Two cards are selected at random from a standard deck of 52 playing cards. Find the probability that two hearts are selected under each condition.

(a) The cards are drawn in sequence, with the first card being replaced and the deck reshuffled prior to the second drawing.

(b) The two cards are drawn consecutively, without replacement.

39. Game Show On a game show, you are given five different digits to arrange in the proper order to represent the price of a car. If you are correct, you win the car. Find the probability of winning under each condition.

(a) You must guess the position of each digit.

(b) You know the first digit, but must guess the remaining four.

(c) You know the first and last digits, but must guess the remaining three.

40. Game Show On a game show, you are given six different digits to arrange in the proper order to represent the price of a house. If you are correct, you win the house. Find the probability of winning under each condition.

(a) You must guess the position of each digit.

(b) You know the first digit, but must guess the remaining five.

(c) You know the first and last digits, but must guess the remaining four.

41. Letter Mix-Up Five letters and envelopes are addressed to five different people. The letters are randomly inserted into the envelopes. What is the probability that (a) exactly one will be inserted in the correct envelope and (b) at least one will be inserted in the correct envelope?

42. Payroll Mix-Up Three paychecks and envelopes are addressed to three different people. The paychecks get mixed up and are randomly inserted into the envelopes. What is the probability that (a) exactly one will be inserted in the correct envelope and (b) at least one will be inserted in the correct envelope?

43. Poker Hand Five cards are drawn from a standard deck of 52 cards. What is the probability of getting a full house? (See Exercise 53 in Section 6.5.)

44. Poker Hand Five cards are drawn from a standard deck of 52 cards. What is the probability of getting a straight flush? (See Exercise 54 in Section 6.5.)

45. Defective Units A shipment of 1000 compact disc players contains four defective units. A retail outlet has ordered 20 units.

(a) What is the probability that all 20 units are good?

(b) What is the probability that at least one unit is defective?

46. Defective Units A shipment of 12 stereos contains three defective units. Four of the units are shipped to a retail store. What is the probability that (a) all four units are good, (b) exactly two units are good, and (c) at least two units are good?

47. Cash Scholarship A senior high school is composed of 31% freshmen, 26% sophomores, 25% juniors, and 18% seniors. A single student is picked randomly by lottery for a cash scholarship. What is the probability that the scholarship winner is either a freshman, a sophomore, or a junior?

48. Voting-Age Population In 2000, about 22.5% of the voting-age population of the United States lived in the west, about 35.4% lived in the south, about 22.9% lived in the midwest, and about 19.2% lived in the northeast. A person is selected at random from the voting-age population. What is the probability that the person lives in the midwest or the south? (Source: U.S. Census Bureau)

49. Making a Sale A sales representative makes a sale at approximately one-third of the offices she calls on. On a given day, she goes to four offices. What is the probability that she will make a sale at (a) all four offices, (b) none of the offices, and (c) at least one office?

50. Making a Sale A sales representative makes a sale at approximately one-fourth of the businesses he calls on. On a given day, he goes to five businesses. What is the probability that he will make a sale at (a) all five businesses, (b) none of the businesses, and (c) at least one of the businesses?

51. A Boy or a Girl? Assume that the probability of the birth of a child of a particular gender is 50%. In a family with six children, what is the probability that (a) all six children are girls, (b) all six children are of the same gender, and (c) there is at least one girl?

52. A Boy or a Girl? Assume that the probability of the birth of a child of a particular gender is 50%. In a family with four children, what is the probability that (a) all four children are boys, (b) all four children are of the same gender, and (c) there is at least one boy?

Probability In Exercises 53–56, consider n independent trials of an experiment in which each trial has two possible outcomes, called success and failure. The probability of a success on each trial is p, and the probability of a failure is $q = 1 - p$. In this context, the term $_nC_k p^k q^{n-k}$ in the expansion of $(p + q)^n$ gives the probability of k successes in the n trials of the experiment.

53. A fair coin is tossed eight times. To find the probability of obtaining five heads, evaluate the term

$$_8C_5\left(\frac{1}{2}\right)^5\left(\frac{1}{2}\right)^3$$

in the expansion of $\left(\frac{1}{2} + \frac{1}{2}\right)^8$.

54. The probability of a baseball player getting a hit during any given time at bat is $\frac{1}{5}$. To find the probability that the player gets four hits during the next 10 times at bat, evaluate the term

$$_{10}C_4\left(\frac{1}{5}\right)^4\left(\frac{4}{5}\right)^6$$

in the expansion of $\left(\frac{1}{5} + \frac{4}{5}\right)^{10}$.

55. The probability of a sales representative making a sale with any one customer is $\frac{1}{4}$. The sales representative makes 10 contacts a day. To find the probability of making four sales, evaluate the term

$$_{10}C_4\left(\frac{1}{4}\right)^4\left(\frac{3}{4}\right)^6$$

in the expansion of $\left(\frac{1}{4} + \frac{3}{4}\right)^{10}$.

56. To find the probability that the baseball player in Exercise 54 makes five hits during the next 10 times at bat if the probability of a hit is $\frac{1}{3}$, evaluate the term

$$_{10}C_5\left(\frac{1}{3}\right)^5\left(\frac{2}{3}\right)^5$$

in the expansion of $\left(\frac{1}{3} + \frac{2}{3}\right)^{10}$.

57. Use information from news sources, the Internet, or an original experiment to write an example of two mutually exclusive events.

58. Use information from news sources, the Internet, or an original experiment to write an example of two independent events.

59. Emails Received According to a survey, the numbers of emails users receive on a typical day are as shown in the pie graph. Two email users from the survey are chosen at random. What is the probability that both users receive more than 30 emails on a typical day? (Source: Pew Internet & American Life Project)

Number of Emails
Received per Day

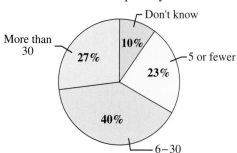

60. Spam According to a survey, the percentages of spam email received on a typical day are shown in the pie graph. Two email users from the survey are chosen at random. What is the probability that 60% or more of both users' emails are spam? (Source: Pew Internet & American Life Project)

Percent of Email
That Is Spam

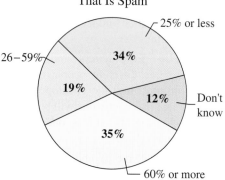

61. Research Habits According to a survey, the relative use of the Internet and the library for research by college students is shown in the pie graph. Three college students from the survey are chosen at random. What is the probability that all three students use the Internet more than the library? (Source: Pew Internet & American Life Project)

Research Habits

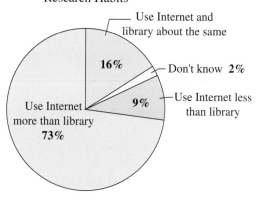

62. *Make a Decision* For a given n and r, is it possible to order the following from least to greatest? Explain.

$$n!, \ _nP_r, \ \frac{n!}{r!}, \ _nC_r$$

6.7 Mathematical Induction

Objectives

- **Use mathematical induction to prove a formula.**
- **Find the sums of powers of integers.**
- **Find a formula for a finite sum.**
- **Use finite differences to find a linear or quadratic model.**

Introduction

In this section, you will study a form of mathematical proof called **mathematical induction**. It is important that you clearly see the logical need for it, so take a closer look at the following problem.

$$S_1 = 1 = 1^2$$
$$S_2 = 1 + 3 = 2^2$$
$$S_3 = 1 + 3 + 5 = 3^2$$
$$S_4 = 1 + 3 + 5 + 7 = 4^2$$
$$S_5 = 1 + 3 + 5 + 7 + 9 = 5^2$$

Judging from the pattern formed by these first five sums, it appears that the sum of the first n integers is

$$S_n = 1 + 3 + 5 + 7 + 9 + \cdots + (2n - 1) = n^2.$$

Although this particular formula *is* valid, it is important for you to see that recognizing a pattern and then simply *jumping to the conclusion* that the pattern must be true for all values of n is *not* a logically valid method of proof. There are many examples in which a pattern appears to be developing for small values of n and then at some point the pattern fails. One of the most famous cases of this was the conjecture by the French mathematician Pierre de Fermat (1601–1665), who speculated that all numbers of the form

$$F_n = 2^{2^n} + 1, \quad n = 0, 1, 2, \ldots$$

are prime. For $n = 0, 1, 2, 3$, and 4, the conjecture is true.

$$F_0 = 3, F_1 = 5, F_2 = 17, F_3 = 257, F_4 = 65{,}537$$

The size of the next Fermat number ($F_5 = 4{,}294{,}967{,}297$) is so great that it was difficult for Fermat to determine whether it was prime or not. However, another well-known mathematician, Leonhard Euler (1707–1783), later found the factorization

$$F_5 = 4{,}294{,}967{,}297 = 641(6{,}700{,}417)$$

which proved that F_5 is not prime and therefore Fermat's conjecture was false.

Just because a rule, pattern, or formula seems to work for several values of n, you cannot simply decide that it is valid for all values of n without going through a *legitimate proof.* Mathematical induction is one method of proof.

The Principle of Mathematical Induction

Let P_n be a statement involving the positive integer n. If

1. P_1 is true, and

2. the truth of P_k implies the truth of P_{k+1} for every positive integer k,

then P_n must be true for all positive integers n.

It is important to recognize that both parts of the Principle of Mathematical Induction are necessary. To apply the Principle of Mathematical Induction, you need to be able to determine the statement P_{k+1} for a given statement P_k.

EXAMPLE 1   A PRELIMINARY EXAMPLE

Find P_{k+1} for the following.

a. P_k: $S_k = \dfrac{k^2(k+1)^2}{4}$

b. P_k: $S_k = 1 + 5 + 9 + \cdots + [4(k-1) - 3] + (4k - 3)$

c. P_k: $3^k \geq 2k + 1$

SOLUTION

a. P_{k+1}: $S_{k+1} = \dfrac{(k+1)^2(k+1+1)^2}{4}$ Replace k by $k+1$.

$= \dfrac{(k+1)^2(k+2)^2}{4}$ Simplify.

b. P_{k+1}: $S_{k+1} = 1 + 5 + 9 + \cdots + \{4[(k+1) - 1] - 3\} + [4(k+1) - 3]$

$= 1 + 5 + 9 + \cdots + (4k - 3) + (4k + 1)$

c. P_{k+1}: $3^{k+1} \geq 2(k+1) + 1$

$3^{k+1} \geq 2k + 3$

CHECKPOINT   Now try Exercise 1.

Figure 6.9

A well-known illustration used to explain why the Principle of Mathematical Induction works is the unending line of dominoes shown in Figure 6.9. If the line actually contains infinitely many dominoes, it is clear that you could not knock the entire line down by knocking down only *one domino* at a time. However, suppose it were true that each domino would knock down the next one as it fell. Then you could knock them all down simply by pushing the first one and starting a chain reaction. Mathematical induction works in the same way. If the truth of P_k implies the truth of P_{k+1} and if P_1 is true, the chain reaction proceeds as follows: P_1 implies P_2, P_2 implies P_3, P_3 implies P_4, and so on.

Using Mathematical Induction

EXAMPLE 2   USING MATHEMATICAL INDUCTION

Use mathematical induction to prove the following formula.

$$S_n = 1 + 3 + 5 + 7 + \cdots + (2n - 1)$$
$$= n^2$$

SOLUTION

Mathematical induction consists of two distinct parts. First, you must show that the formula is true when $n = 1$.

1. When $n = 1$, the formula is valid, because

$$S_1 = 1 = 1^2.$$

The second part of mathematical induction has two steps. The first step is to assume that the formula is valid for *some* integer k. The second step is to use this assumption to prove that the formula is valid for the next integer, $k + 1$.

2. Assuming that the formula

$$S_k = 1 + 3 + 5 + 7 + \cdots + (2k - 1)$$
$$= k^2$$

is true, you must show that the formula $S_{k+1} = (k + 1)^2$ is true.

$$S_{k+1} = 1 + 3 + 5 + 7 + \cdots + (2k - 1) + [2(k + 1) - 1]$$
$$= [1 + 3 + 5 + 7 + \cdots + (2k - 1)] + (2k + 2 - 1)$$
$$= S_k + (2k + 1) \qquad \text{Group terms to form } S_k.$$
$$= k^2 + 2k + 1 \qquad \text{Replace } S_k \text{ by } k^2.$$
$$= (k + 1)^2 \qquad \text{Factor.}$$

Combining the results of parts (1) and (2), you can conclude by mathematical induction that the formula is valid for *all* positive integer values of n.

✔CHECKPOINT   Now try Exercise 5.

STUDY TIP

When using mathematical induction to prove a *summation* formula (such as the one in Example 2), it is helpful to think of S_{k+1} as $S_{k+1} = S_k + a_{k+1}$, where a_{k+1} is the $(k + 1)$th term of the original sum.

It occasionally happens that a statement involving natural numbers is not true for the first $k - 1$ positive integers but is true for all values of $n \geq k$. In these instances, you use a slight variation of the Principle of Mathematical Induction in which you verify P_k rather than P_1. This variation is called the **Extended Principle of Mathematical Induction.** To see the validity of this variation, note from Figure 6.9 that all but the first $k - 1$ dominoes can be knocked down by knocking over the kth domino. This suggests that you can prove a statement P_n to be true for $n \geq k$ by showing that P_k is true and that P_k implies P_{k+1}. In Exercises 35–37 of this section, you are asked to apply this extension of mathematical induction.

EXAMPLE 3	USING MATHEMATICAL INDUCTION

Use mathematical induction to prove the formula

$$S_n = 1^2 + 2^2 + 3^2 + 4^2 + \cdots + n^2$$

$$= \frac{n(n + 1)(2n + 1)}{6}$$

for all integers $n \geq 1$.

SOLUTION

1. When $n = 1$, the formula is valid, because

$$S_1 = 1^2 = \frac{1(1 + 1)(2 \cdot 1 + 1)}{6} = \frac{1(2)(3)}{6}.$$

2. Assuming that

$$S_k = 1^2 + 2^2 + 3^2 + 4^2 + \cdots + k^2$$

$$= \frac{k(k + 1)(2k + 1)}{6}$$

you must show that

$$S_{k+1} = \frac{(k + 1)(k + 1 + 1)[2(k + 1) + 1]}{6} = \frac{(k + 1)(k + 2)(2k + 3)}{6}.$$

To do this, write the following.

$$S_{k+1} = S_k + a_{k+1}$$

$$= (1^2 + 2^2 + 3^2 + 4^2 + \cdots + k^2) + (k + 1)^2$$

$$= \frac{k(k + 1)(2k + 1)}{6} + (k + 1)^2 \qquad \text{By assumption}$$

$$= \frac{k(k + 1)(2k + 1) + 6(k + 1)^2}{6} \qquad \text{Add expressions.}$$

$$= \frac{(k + 1)[k(2k + 1) + 6(k + 1)]}{6} \qquad \text{Factor out } (k + 1).$$

$$= \frac{(k + 1)(2k^2 + 7k + 6)}{6} \qquad \text{Simplify numerator.}$$

$$= \frac{(k + 1)(k + 2)(2k + 3)}{6} \qquad \text{Completely factor numerator.}$$

Combining the results of parts (1) and (2), you can conclude by mathematical induction that the formula is valid for *all* integers $n \geq 1$.

✓**CHECKPOINT** Now try Exercise 9.

When proving a formula by mathematical induction, the only statement that you *need* to verify is P_1. For the inequality in Example 4, you need to show that the inequality holds for $n = 1$ and $n = 2$ because in Step 2 of the mathematical induction, $k > 1$ (as opposed to $k \geq 1$) is the restriction on k.

| EXAMPLE 4 | PROVING AN INEQUALITY BY MATHEMATICAL INDUCTION |

Prove that $n < 2^n$ for all positive integers n.

SOLUTION

1. For $n = 1$ and $n = 2$, the formula is true, because

$$1 < 2^1 \text{ and } 2 < 2^2.$$

2. Assuming that

$$k < 2^k$$

you need to show that $k + 1 < 2^{k+1}$. For $n = k$, you have

$$2^{k+1} = 2(2^k) > 2(k) = 2k. \qquad \text{By assumption}$$

Because $2k = k + k > k + 1$ for all $k > 1$, it follows that

$$2^{k+1} > 2k > k + 1$$

or

$$k + 1 < 2^{k+1}.$$

Combining the results of parts (1) and (2), you can conclude by mathematical induction that $n < 2^n$ for all integers $n \geq 1$.

✓**CHECKPOINT**   Now try Exercise 35.

Sums of Powers of Integers

The formula in Example 3 is one of a collection of useful summation formulas. This and other formulas dealing with the sums of various powers of the first n positive integers are as follows.

Sums of Powers of Integers

1. $1 + 2 + 3 + 4 + \cdots + n = \dfrac{n(n + 1)}{2}$

2. $1^2 + 2^2 + 3^2 + 4^2 + \cdots + n^2 = \dfrac{n(n + 1)(2n + 1)}{6}$

3. $1^3 + 2^3 + 3^3 + 4^3 + \cdots + n^3 = \dfrac{n^2(n + 1)^2}{4}$

4. $1^4 + 2^4 + 3^4 + 4^4 + \cdots + n^4 = \dfrac{n(n + 1)(2n + 1)(3n^2 + 3n - 1)}{30}$

5. $1^5 + 2^5 + 3^5 + 4^5 + \cdots + n^5 = \dfrac{n^2(n + 1)^2(2n^2 + 2n - 1)}{12}$

Each of these formulas for sums can be proven by mathematical induction.

EXAMPLE 5	FINDING A SUM OF POWERS OF INTEGERS

Find $\displaystyle\sum_{n=1}^{7} n^3 = 1^3 + 2^3 + 3^3 + 4^3 + 5^3 + 6^3 + 7^3$.

SOLUTION

Using the formula for the sum of the cubes of the first n positive integers, you obtain the following.

$$\sum_{n=1}^{7} n^3 = 1^3 + 2^3 + 3^3 + 4^3 + 5^3 + 6^3 + 7^3$$

$$= \frac{7^2(7 + 1)^2}{4}$$

$$= \frac{49(64)}{4}$$

$$= 784$$

Check this sum by adding the numbers 1, 8, 27, 64, 125, 216, and 343.

✓**CHECKPOINT**   Now try Exercise 21.

Pattern Recognition

Although choosing a formula on the basis of a few observations does *not* guarantee the validity of the formula, pattern recognition *is* important. Once you have a pattern that you think works, you can try using mathematical induction to prove your formula.

Finding a Formula for the *n*th Term of a Sequence

To find a formula for the *n*th term of a sequence, consider the following guidelines.

1. Calculate the first several terms of the sequence. It is often a good idea to write the terms in both simplified and factored forms.

2. Try to find a recognizable pattern for the terms and write a formula for the *n*th term of the sequence. This is your *hypothesis* or *conjecture*. You might try computing one or two more terms in the sequence to test your hypothesis.

3. Use mathematical induction to prove your hypothesis.

EXAMPLE 6 FINDING A FORMULA FOR A FINITE SUM

Find a formula for the following finite sum.

$$\frac{1}{1 \cdot 2} + \frac{1}{2 \cdot 3} + \frac{1}{3 \cdot 4} + \frac{1}{4 \cdot 5} + \cdots + \frac{1}{n(n+1)}$$

SOLUTION

Begin by writing out the first few sums.

$$S_1 = \frac{1}{1 \cdot 2} = \frac{1}{2} = \frac{1}{1+1}$$

$$S_2 = \frac{1}{1 \cdot 2} + \frac{1}{2 \cdot 3} = \frac{4}{6} = \frac{2}{3} = \frac{2}{2+1}$$

$$S_3 = \frac{1}{1 \cdot 2} + \frac{1}{2 \cdot 3} + \frac{1}{3 \cdot 4} = \frac{9}{12} = \frac{3}{4} = \frac{3}{3+1}$$

$$S_4 = \frac{1}{1 \cdot 2} + \frac{1}{2 \cdot 3} + \frac{1}{3 \cdot 4} + \frac{1}{4 \cdot 5} = \frac{48}{60} = \frac{4}{5} = \frac{4}{4+1}$$

From this sequence, it appears that the formula for the kth sum is

$$S_k = \frac{1}{1 \cdot 2} + \frac{1}{2 \cdot 3} + \frac{1}{3 \cdot 4} + \frac{1}{4 \cdot 5} + \cdots + \frac{1}{k(k+1)}$$

$$= \frac{k}{k+1}.$$

To prove the validity of this hypothesis, use mathematical induction, as follows. Note that you have already verified the formula for $n = 1$, so you can begin by assuming that the formula is valid for $n = k$ and trying to show that it is valid for $n = k + 1$.

$$S_{k+1} = \left[\frac{1}{1 \cdot 2} + \frac{1}{2 \cdot 3} + \frac{1}{3 \cdot 4} + \frac{1}{4 \cdot 5} + \cdots + \frac{1}{k(k+1)} \right] + \frac{1}{(k+1)(k+2)}$$

$$= \frac{k}{k+1} + \frac{1}{(k+1)(k+2)} \qquad \text{By assumption}$$

$$= \frac{k(k+2) + 1}{(k+1)(k+2)} \qquad \text{Add fractions.}$$

$$= \frac{k^2 + 2k + 1}{(k+1)(k+2)} \qquad \text{Distributive Property}$$

$$= \frac{(k+1)^2}{(k+1)(k+2)} \qquad \text{Factor numerator.}$$

$$= \frac{k+1}{k+2} \qquad \text{Simplify.}$$

So, the hypothesis is valid.

✓CHECKPOINT Now try Exercise 33.

Finite Differences

The **first differences** of a sequence are found by subtracting consecutive terms. The **second differences** are found by subtracting consecutive first differences. The first and second differences of the sequence 3, 5, 8, 12, 17, 23, . . . are as follows.

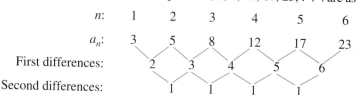

n:	1	2	3	4	5	6
a_n:	3	5	8	12	17	23

First differences: 2   3   4   5   6

Second differences: 1   1   1   1

For this sequence, the second differences are all the same nonzero number. When this happens, the sequence has a perfect *quadratic* model. If the first differences are all the same nonzero number, the sequence has a *linear* model. That is, it is arithmetic.

EXAMPLE 7   FINDING AN APPROPRIATE MODEL

Make a Decision   Decide whether the sequence 6, 12, 26, 48, 78, 116, . . . can be represented perfectly by a linear or a quadratic model. If so, find the model.

SOLUTION

The first and second differences are as follows.

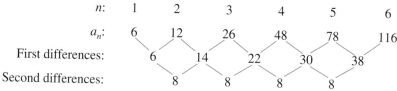

n:	1	2	3	4	5	6
a_n:	6	12	26	48	78	116

First differences: 6   14   22   30   38

Second differences: 8   8   8   8

For this sequence, the second differences are all the same. So, the sequence follows a quadratic model perfectly and you know that the model has the form

$$a_n = an^2 + bn + c.$$

By substituting 1, 2, and 3 for n, you can obtain a system of three linear equations in three variables.

$a_1 = a(1)^2 + b(1) + c = 6$ Substitute 1 for n.

$a_2 = a(2)^2 + b(2) + c = 12$ Substitute 2 for n.

$a_3 = a(3)^2 + b(3) + c = 26$ Substitute 3 for n.

You now have a system of three equations in a, b, and c.

$$\begin{cases} a + b + c = 6 & \text{Equation 1} \\ 4a + 2b + c = 12 & \text{Equation 2} \\ 9a + 3b + c = 26 & \text{Equation 3} \end{cases}$$

Using the techniques discussed in Chapter 4, you can find that the solution of this system is $a = 4$, $b = -6$, and $c = 8$. So, the quadratic model is

$$a_n = 4n^2 - 6n + 8.$$

✓CHECKPOINT   Now try Exercise 51.

EXAMPLE 8   FINDING AN APPROPRIATE MODEL

Make a Decision   Decide whether the sequence 2, 9, 16, 23, 30, 37, . . . can be represented perfectly by a linear or a quadratic model. If so, find the model.

SOLUTION

The first differences are as follows.

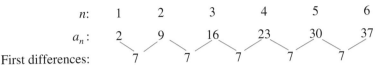

For this sequence, the first differences are all the same. So, the sequence follows a linear model perfectly and you know that the model has the form $a_n = an + b$. By substituting 1 and 2 for n, you can obtain a system of two linear equations in two variables.

$a_1 = a(1) + b = 2$ Substitute 1 for n.

$a_2 = a(2) + b = 9$ Substitute 2 for n.

You now have a system of two equations in a and b.

$$\begin{cases} a + b = 2 & \text{Equation 1} \\ 2a + b = 9 & \text{Equation 2} \end{cases}$$

Using the techniques discussed in Chapter 4, you can find that the solution of this system is $a = 7$ and $b = -5$. So, the linear model is $a_n = 7n - 5$.

✔CHECKPOINT   Now try Exercise 55.

Discussing the Concept | Mathematical Modeling

Use finite differences to determine whether each sequence in the table can be represented by a linear or a quadratic model.

n	1	2	3	4	5	6
a_n	12	14	22	36	56	82
b_n	-23.5	-20.0	-16.5	-13.0	-9.5	-6.0
c_n	7	13	20	26	33	39
d_n	0.8	4.2	9.2	15.8	24.0	33.8

If the sequence can be represented by a linear or a quadratic model, find the appropriate model and determine the value of the 10th term. Discuss how the finite differences technique complements other modeling techniques you have learned thus far.

6.7 Warm Up

The following warm-up exercises involve skills that were covered in earlier sections. You will use these skills in the exercise set for this section. For additional help, review Sections R1.2, R1.3, R1.4, R1.6, R1.7, and 6.1.

In Exercises 1–4, find the sum.

1. $\displaystyle\sum_{k=3}^{6} (2k - 3)$

2. $\displaystyle\sum_{j=1}^{5} (j^2 - j)$

3. $\displaystyle\sum_{k=2}^{5} \frac{1}{k}$

4. $\displaystyle\sum_{i=1}^{2} \left(1 + \frac{1}{i}\right)$

In Exercises 5–10, simplify the expression.

5. $\dfrac{2(k + 1) + 3}{5}$

6. $\dfrac{3(k + 1) - 2}{6}$

7. $2 \cdot 2^{2(k+1)}$

8. $\dfrac{3^{2k}}{3^{2(k+1)}}$

9. $\dfrac{k + 1}{k^2 + k}$

10. $\dfrac{\sqrt{32}}{\sqrt{50}}$

6.7 Exercises

In Exercises 1–4, find P_{k+1} for the given P_k.

1. $P_k = \dfrac{5}{k(k + 1)}$

2. $P_k = \dfrac{3}{(k + 2)(k + 3)}$

3. $P_k = \dfrac{k^2(k - 1)^2}{4}$

4. $P_k = \dfrac{k^2}{2(k + 1)^2}$

In Exercises 5–18, use mathematical induction to prove the formula for every positive integer n.

5. $2 + 4 + 6 + 8 + \cdots + 2n = n(n + 1)$

6. $3 + 7 + 11 + 15 + \cdots + (4n - 1) = n(2n + 1)$

7. $2 + 7 + 12 + 17 + \cdots + (5n - 3) = \dfrac{n}{2}(5n - 1)$

8. $1 + 4 + 7 + 10 + \cdots + (3n - 2) = \dfrac{n}{2}(3n - 1)$

9. $1 + 2 + 2^2 + 2^3 + \cdots + 2^{n-1} = 2^n - 1$

10. $2(1 + 3 + 3^2 + 3^3 + \cdots + 3^{n-1}) = 3^n - 1$

11. $1 + 2 + 3 + 4 + \cdots + n = \dfrac{n(n + 1)}{2}$

12. $1^2 + 2^2 + 3^2 + 4^2 + \cdots + n^2 = \dfrac{n(n + 1)(2n + 1)}{6}$

13. $1^3 + 2^3 + 3^3 + 4^3 + \cdots + n^3 = \dfrac{n^2(n + 1)^2}{4}$

14. $\left(1 + \dfrac{1}{1}\right)\left(1 + \dfrac{1}{2}\right)\left(1 + \dfrac{1}{3}\right) \cdots \left(1 + \dfrac{1}{n}\right) = n + 1$

15. $\displaystyle\sum_{i=1}^{n} i^5 = \dfrac{n^2(n + 1)^2(2n^2 + 2n - 1)}{12}$

16. $\displaystyle\sum_{i=1}^{n} i^4 = \dfrac{n(n + 1)(2n + 1)(3n^2 + 3n - 1)}{30}$

17. $\displaystyle\sum_{i=1}^{n} i(i + 1) = \dfrac{n(n + 1)(n + 2)}{3}$

18. $\displaystyle\sum_{i=1}^{n} \dfrac{1}{(2i - 1)(2i + 1)} = \dfrac{n}{2n + 1}$

In Exercises 19–28, find the sum using the formulas for the sums of powers of integers.

19. $\displaystyle\sum_{n=1}^{20} n$

20. $\displaystyle\sum_{n=1}^{50} n$

21. $\displaystyle\sum_{n=1}^{6} n^2$

22. $\displaystyle\sum_{n=1}^{10} n^3$

23. $\displaystyle\sum_{n=1}^{5} n^4$

24. $\displaystyle\sum_{n=1}^{8} n^5$

25. $\displaystyle\sum_{n=1}^{6} (n^2 - n)$

26. $\displaystyle\sum_{n=1}^{10} (n^3 - n^2)$

27. $\displaystyle\sum_{i=1}^{6} (6i - 8i^3)$

28. $\displaystyle\sum_{j=1}^{5} (2j^3 - j^2)$

In Exercises 29–34, find a formula for the sum of the first n terms of the sequence.

29. $1, 5, 9, 13, \ldots$

30. $25, 22, 19, 16, \ldots$

31. $1, \frac{9}{10}, \frac{81}{100}, \frac{729}{1000}, \ldots$

32. $3, -\frac{9}{2}, \frac{27}{4}, -\frac{81}{8}, \ldots$

33. $\frac{1}{4}, \frac{1}{12}, \frac{1}{24}, \frac{1}{40}, \ldots, \frac{1}{2n(n+1)}, \ldots$

34. $\frac{1}{2 \cdot 3}, \frac{1}{3 \cdot 4}, \frac{1}{4 \cdot 5}, \frac{1}{5 \cdot 6}, \ldots, \frac{1}{(n+1)(n+2)}, \ldots$

In Exercises 35–38, prove the inequality for the indicated integer values of n.

35. $n! > 2^n$, $n \geq 4$

36. $\left(\frac{4}{3}\right)^n > n$, $n \geq 7$

37. $\frac{1}{\sqrt{1}} + \frac{1}{\sqrt{2}} + \frac{1}{\sqrt{3}} + \cdots + \frac{1}{\sqrt{n}} > \sqrt{n}$, $n \geq 2$

38. $\left(\frac{x}{y}\right)^{n+1} < \left(\frac{x}{y}\right)^n$, $n \geq 1$ and $0 < x < y$

In Exercises 39–46, use mathematical induction to prove the given property for all positive integers n.

39. $(ab)^n = a^n b^n$

40. $\left(\frac{a}{b}\right)^n = \frac{a^n}{b^n}$

41. If $x_1 \neq 0, x_2 \neq 0, \ldots, x_n \neq 0$, then
$(x_1 x_2 x_3 \cdots x_n)^{-1} = x_1^{-1} x_2^{-1} x_3^{-1} \cdots x_n^{-1}$.

42. If $x_1 > 0, x_2 > 0, \ldots, x_n > 0$, then
$\ln(x_1 x_2 x_3 \cdots x_n) =$
$\ln x_1 + \ln x_2 + \ln x_3 + \cdots + \ln x_n$.

43. Generalized Distributive Law:
$x(y_1 + y_2 + \cdots + y_n) = xy_1 + xy_2 + \cdots + xy_n$

44. $(a + bi)^n$ and $(a - bi)^n$ are complex conjugates for all $n \geq 1$.

45. A factor of $(n^3 + 3n^2 + 2n)$ is 3.

46. A factor of $(2^{2n-1} + 3^{2n-1})$ is 5.

47. Writing In your own words, explain what is meant by a proof by mathematical induction.

48. Think About It What conclusion can be drawn from the given information about the sequence of statements P_n?

(a) P_3 is true and P_k implies P_{k+1}.

(b) $P_1, P_2, P_3, \ldots, P_{50}$ are all true.

(c) $P_1, P_2,$ and P_3 are all true, but the truth of P_k does not imply the truth of P_{k+1}.

(d) P_2 is true and P_{2k} implies P_{2k+2}.

Make a Decision In Exercises 49–58, write the first five terms of the sequence, where $a_1 = f(1)$. Then calculate the first and second differences of the sequence. Does the sequence have a linear model or a quadratic model? If so, find the model.

49. $f(1) = 0$
$a_n = a_{n-1} + 3$

50. $f(1) = 2$
$a_n = n - a_{n-1}$

51. $f(1) = 3$
$a_n = a_{n-1} - n$

52. $f(1) = -3$
$a_n = -2a_{n-1}$

53. $a_0 = 0$
$a_n = a_{n-1} + n$

54. $a_0 = 2$
$a_n = (a_{n-1})^2$

55. $f(1) = 2$
$a_n = a_{n-1} + 2$

56. $f(1) = 0$
$a_n = a_{n-1} + 2n$

57. $a_0 = 1$
$a_n = a_{n-1} + n^2$

58. $a_0 = 0$
$a_n = a_{n-1} - 1$

In Exercises 59–62, find a quadratic model for the sequence with the indicated terms.

59. $a_0 = 3, a_1 = 3, a_4 = 15$

60. $a_0 = 7, a_1 = 6, a_3 = 10$

61. $a_1 = 0, a_2 = 8, a_4 = 30$

62. $a_0 = -3, a_2 = -5, a_6 = -57$

63. *Make a Decision:* **Population of Vermont** The table shows the population a_n of Vermont (in thousands) from 1998 to 2002. In the table, n represents the year, with $n = 8$ corresponding to 1998. (Source: U.S. Census Bureau)

Year, n	Population, a_n
8	600
9	605
10	609
11	613
12	617

(a) Find the first differences of the data. Are the first differences approximately equal?

(b) Use your results from part (a) to determine whether a linear model can be used to approximate the data. If so, find a model algebraically.

(c) Use the *regression* feature of a graphing utility to find a linear model for the data. Is the model a good fit for the data?

Make a Decision **Project: The Multiplier Effect**

Cycle, x	Total spending, y
1	76.000
2	60.800
3	48.640
4	38.912
5	31.130
6	24.904
7	19.923
8	15.938
9	12.751
10	10.201
11	8.160
12	6.528
13	5.223
14	4.178
15	3.343
16	2.674
17	2.139
18	1.711
19	1.369
20	1.095

The ideas of economist John Maynard Keynes have been influential in the development of economic policy at the national, state, and local levels. Keynesian economics brought the term "multiplier effect" into popular use. The multiplier effect is based on the idea that the expenditures of one person become the income of another. As the second person spends his or her income, then his/her expenditures become a third person's income. This pattern continues as one person's expenditures become another person's income. If individuals in an economy spend 75% of their income (some amount will go to income taxes, forced retirement contributions, voluntary savings/investments, and so forth), then a person spends $0.75 of every dollar earned. For one dollar earned the first person would spend $0.75. The next person spends only $0.56 of the original dollar ($0.75 × 75%). Each subsequent person would spend progressively smaller and smaller amounts of the original dollar, so the income "leaks" out of the spending system. The greater the leakage, the smaller the multiplier effect will be.

The table shows the cycle of income in a city that hosted a large football event. In the table, x represents the cycle number of a dollar and y represents the total spending (in millions of dollars) for that cycle. Use this information to investigate the following questions.

1. **Creating a Scatter Plot**   Use a graphing utility to create a scatter plot of the data shown in the table.

2. *Make a Decision*   What type or types of graphs does the scatter plot in Question 1 suggest might fit the data?

3. **Fitting Models to Data**   Use the *regression* feature of a graphing utility to find the following models for the data.

   (a) Quadratic model, $y = ax^2 + bx + c$

   (b) Power model, $y = ax^b$

   (c) Exponential model, $y = ab^x$

4. **Comparing a Model to Actual Data**   Numerically compare the values given by the models found in Question 3 with the actual data in the table by constructing a table. Graphically compare the models with the data by using a graphing utility to graph each model with the scatter plot.

5. *Make a Decision*   Which model fits the data best? Explain. Find the total economic impact of this event on a community by summing the total dollars spent after 20 cycles and the total according to the model you selected. Did the model fit the data well? Explain.

6. **Further Explorations**   Use your school's library, the Internet, or some other reference source to further investigate the multiplier effect. Can you find examples of situations to which the multiplier effect could apply?

Chapter Summary

After studying this chapter, you should have acquired the following skills.
These skills are keyed to the Review Exercises that begin on page 460.
Answers to odd-numbered Review Exercises are given in the back of the book.

6.1 • Use sequence notation to find the terms of a sequence. — *Review Exercises 1–12*

• Evaluate a factorial expression. — *Review Exercises 13, 14*

• Find the sum of a finite sequence and use summation notation to write the sum of a sequence. — *Review Exercises 15–22*

• Use a sequence to solve an application problem. — *Review Exercises 23, 24*

6.2 • Recognize and find the nth term of an arithmetic sequence. — *Review Exercises 25–32*

• Find a formula for an arithmetic sequence. — *Review Exercises 33–36*

• Find the nth partial sum of an arithmetic sequence. — *Review Exercises 37–40*

• Use an arithmetic sequence to solve an application problem. — *Review Exercises 41, 42*

6.3 • Recognize and find the nth term of a geometric sequence. — *Review Exercises 43–54*

• Find the sum of a finite geometric sequence. — *Review Exercises 55, 56*

• Find the sum of an infinite geometric series. — *Review Exercises 57–60*

• Use a geometric sequence to solve an application problem. — *Review Exercises 61–64*

6.4 • Find a binomial coefficient. — *Review Exercises 65–68*

• Use Pascal's Triangle to determine a binomial coefficient. — *Review Exercises 69, 70*

• Expand a binomial using binomial coefficients and find a specific term in a binomial expansion. — *Review Exercises 71–80*

6.5 • Use the Fundamental Counting Principle to solve counting problems. — *Review Exercises 81–86*

• Use permutations to solve counting problems. — *Review Exercises 87–92*

• Use combinations to solve counting problems. — *Review Exercises 93, 94*

6.6 • Determine the sample space of an experiment and find the probability of an event. — *Review Exercises 95–103, 106*

• Find the probability of mutually exclusive events. — *Review Exercises 100, 102, 104, 105*

• Find the probability of independent events. — *Review Exercises 104, 105*

• Find the probability of the complement of an event. — *Review Exercises 99, 104–108*

6.7 • Use mathematical induction to prove a formula. — *Review Exercises 109–112*

• Find the sums of powers of integers. — *Review Exercises 113–116*

• Find a formula for a finite sum. — *Review Exercises 117, 118*

• Use finite differences to find a linear or quadratic model. — *Review Exercises 119, 120*

Review Exercises

In Exercises 1–6, write the first five terms of the sequence. (Assume that n begins with 1.)

1. $a_n = 4n + 1$

2. $a_n = 3n - 2$

3. $a_n = \dfrac{n}{n + 1}$

4. $a_n = \dfrac{3n - 1}{n}$

5. $a_n = (-1)^{n+1} \dfrac{2^n}{n + 1}$

6. $a_n = (-1)^n \dfrac{3^n - 1}{3^n + 1}$

In Exercises 7–12, write an expression for the *most apparent* nth term of the sequence. (Assume that n begins with 1.)

7. $1, 2, 4, 8, \ldots$

8. $4, -4, 4, -4, \ldots$

9. $1, \frac{1}{2}, \frac{1}{3}, \frac{1}{4}, \frac{1}{5}, \ldots$

10. $-\frac{2}{3}, \frac{3}{4}, -\frac{4}{5}, \frac{5}{6}, -\frac{6}{7}, \ldots$

11. $3, -9, 27, -81, 243, \ldots$

12. $-2, 4, -8, 16, -32, \ldots$

In Exercises 13 and 14, evaluate the expression.

13. $\dfrac{16!}{12!}$

14. $\dfrac{24!}{22!}$

In Exercises 15–18, find the sum.

15. $\displaystyle\sum_{i=1}^{6} (3i - 2)$

16. $\displaystyle\sum_{i=1}^{6} (4i + 5)$

17. $\displaystyle\sum_{k=1}^{6} (k - 1)(k + 2)$

18. $\displaystyle\sum_{k=1}^{8} k^2 + 1$

In Exercises 19–22, use summation notation to write the sum.

19. $\dfrac{3}{1 + 1} + \dfrac{3}{1 + 2} + \dfrac{3}{1 + 3} + \dfrac{3}{1 + 4} + \cdots + \dfrac{3}{1 + 12}$

20. $1 + \frac{1}{2} + \frac{1}{3} + \frac{1}{4} + \frac{1}{5} + \cdots + \frac{1}{60}$

21. $2 - 4 + 8 - 16 + 32 - 64 + \cdots + 8192$

22. $2 + 8 + 18 + 32 + \cdots + 98$

23. **Compound Interest** On your next birthday, you plan to deposit $4000 in an account that earns 6% compounded quarterly. The balance in the account after n quarters is given by

$$A_n = 4000\left(1 + \dfrac{0.06}{4}\right)^n, \qquad n = 1, 2, 3, \ldots$$

(a) Compute the first eight terms of the sequence.

(b) Find the balance in the account after 10 years by computing the 40th term of the sequence.

(c) Assuming you do not withdraw money from the account, find the balance in the account after 30 years by computing the 120th term of the sequence.

24. **Compound Interest** A deposit of $10,000 is made in an account that earns 8% compounded monthly. The balance in the account after n months is given by

$$A_n = 10,000\left(1 + \dfrac{0.08}{12}\right)^n, \qquad n = 1, 2, 3, \ldots$$

(a) Compute the first 10 terms of the sequence.

(b) Find the balance in the account after 10 years by computing the 120th term of the sequence.

In Exercises 25–28, determine whether the sequence is arithmetic. If it is, find the common difference.

25. $3, 8, 13, 18, \ldots$

26. $1, 3, 9, 27, \ldots$

27. $4, \frac{7}{2}, 3, \frac{5}{2}, 2, \ldots$

28. $6.4, 5.6, 4.8, 4.0, 3.2, \ldots$

In Exercises 29–32, write the first five terms of the arithmetic sequence.

29. $a_n = 4 + 5n$

30. $a_n = 2 + (3 - n)5$

31. $a_1 = 4, a_8 = 25$

32. $a_6 = 28, a_{14} = 16$

In Exercises 33–36, find a formula for a_n for the arithmetic sequence.

33. $a_1 = 2, d = 3$

34. $a_1 = 15, d = -3$

35. $a_2 = 93, a_6 = 65$

36. $a_7 = 8, a_{13} = 6$

In Exercises 37–40, find the indicated nth partial sum of the arithmetic sequence.

37. $3, 9, 15, 21, 27, \ldots, n = 8$

38. $-8, -4, 0, 4, 8, \ldots, n = 12$

39. $a_1 = 9, a_{15} = 93, n = 15$

40. $a_1 = 27, a_{40} = 300, n = 20$

41. *Make a Decision*: **Job Offers** You are offered a job by two companies. The position with Company A has a salary of $32,500 for the first year with a guaranteed annual raise of $1600 per year. The position with Company B has a salary of $28,000 for the first year with a guaranteed annual raise of $2500 per year.

(a) What will your salary be during the sixth year of employment at Company A? at Company B?

(b) How much will Company A have paid you at the end of 6 years?

(c) How much will Company B have paid you at the end of 6 years?

(d) Which job should you accept, and why?

42. Seating Capacity Determine the seating capacity of an auditorium with 20 rows of seats if there are 25 seats in the first row, 28 seats in the second row, 31 seats in the third row, and so on.

In Exercises 43–46, determine whether the sequence is geometric. If it is, find its common ratio and write a formula for a_n.

43. $1, -3, 9, -27, \ldots$

44. $3, 9, 15, 21, \ldots$

45. $16, 8, 4, 2, 1, \frac{1}{2}, \frac{1}{4}, \frac{1}{8}, \ldots$

46. $1, -\frac{1}{3}, \frac{1}{9}, -\frac{1}{27}, \frac{1}{81}, -\frac{1}{243}, \ldots$

In Exercises 47–50, write the first five terms of the geometric sequence.

47. $a_1 = 3, r = 2$

48. $a_1 = 2, r = e$

49. $a_1 = 10, r = -\frac{1}{5}$

50. $a_1 = 1, r = \frac{3}{4}$

In Exercises 51–54, find the indicated term of the geometric sequence.

51. $a_1 = 8, r = \frac{1}{2}$, 40th term

52. $a_1 = 100, r = -\frac{1}{10}$, 5th term

53. $a_2 = -100, a_5 = 12.5$, 10th term

54. $a_1 = 10, a_3 = 11.025$, 19th term

In Exercises 55 and 56, find the sum of the finite geometric sequence.

55. $\sum_{n=1}^{10} 4(2^n)$

56. $\sum_{n=0}^{20} 2\left(\frac{2}{3}\right)^n$

In Exercises 57–60, find the sum of the infinite geomtric series, if it exists.

57. $\sum_{n=0}^{\infty} \left(\frac{1}{3}\right)^n$

58. $\sum_{n=0}^{\infty} \left(\frac{3}{2}\right)^n$

59. $\sum_{n=0}^{\infty} \left(\frac{5}{4}\right)^n$

60. $\sum_{n=0}^{\infty} 3\left(\frac{1}{4}\right)^n$

61. *Make a Decision*: **Compound Interest** A deposit of $100 is made at the beginning of each month for 10 years in an account that pays 6% compounded monthly. The balance A at the end of 10 years is as follows.

$$A = 100\left(1 + \frac{0.06}{12}\right)^1 + \cdots + 100\left(1 + \frac{0.06}{12}\right)^{120}$$

(a) Find the balance after 10 years.

(b) Is the balance after 20 years twice what it is after 10 years? Explain your reasoning.

62. Compound Interest You deposit $200 in an account at the beginning of each month for 10 years. The account pays 6% compounded monthly. What will your balance be at the end of 10 years? If the interest were compounded continuously, what would the balance be?

63. Profit The annual profit a_n (in millions of dollars) for a pharmaceuticals company from 1990 to 2005 can be approximated by the model

$$a_n = 165.3e^{0.15n}, \quad n = 0, 1, 2, 3, \ldots, 15$$

where n represents the year, with $n = 0$ corresponding to 1990.

(a) Construct a bar graph that represents the company's profits during the 16-year period.

(b) Find the total profit earned during the 16-year period.

64. Sales The annual sales a_n (in millions of dollars) for a furniture company from 1990 to 2005 can be approximated by the model

$$a_n = 225e^{0.135n}, \quad n = 0, 1, 2, 3, \ldots, 15$$

where n represents the year, with $n = 0$ corresponding to 1990. Use the formula for the sum of a finite geometric sequence to approximate the total sales during the 16-year period.

In Exercises 65–68, find the binomial coefficient.

65. $_7C_4$

66. $_{12}C_8$

67. $_{30}C_{30}$

68. $_{20}C_9$

In Exercises 69 and 70, evaluate using Pascal's Triangle.

69. $_{12}C_5$

70. $_{21}C_6$

In Exercises 71–76, use the Binomial Theorem to expand the expression. Simplify your answer.

71. $(x + 1)^6$

72. $(x - 1)^7$

73. $(3x + y)^5$

74. $(2x - y)^6$

75. $(x - 2y)^8$

76. $(x^2 + 4)^5$

In Exercises 77 and 78, find the specified nth term in the expansion of the binomial.

77. $(a - 3b)^5$, $n = 3$

78. $(2x + y)^{10}$, $n = 8$

In Exercises 79 and 80, find the coefficient a of the indicated term in the expansion of the binomial.

	Binomial	*Term*
79.	$(x + 2)^{12}$	ax^7
80.	$(2x - 5y)^9$	ax^4y^5

81. Computer Systems A customer in a computer store can choose one of six monitors, one of five keyboards, and one of seven computers. If all of the choices are compatible, how many different systems can be chosen?

82. Home Theater System A customer in an electronics store can choose one of six speaker systems, one of five DVD players, and one of six big-screen televisions to design a home theater system. How many different systems can be designed?

83. Roller Coaster Ride Six people are lining up for a ride on a roller coaster. Each coaster car has six seats. All six people want to sit in the same car, but only three of the six are willing to sit in the front two seats of the car. With that constraint, in how many ways can the six people be seated in the car?

84. Aircraft Boarding Eight people are boarding an aircraft. Two have tickets for first class and board before those in economy class. In how many ways can the eight people board the aircraft?

85. Four-Digit Numbers How many four-digit numbers can be formed under each condition?

(a) The leading digit cannot be 0.

(b) The leading digit cannot be 0 and no repetition of digits is allowed.

(c) The leading digit cannot be 0 and the number must be divisible by 2.

86. Five-Digit Numbers How many five-digit numbers can be formed under each condition?

(a) The leading digit cannot be 0.

(b) The leading digit cannot be 0 and no repetition of digits is allowed.

(c) The leading digit cannot be 0 and the number must be odd.

In Exercises 87–90, evaluate $_nP_r$.

87. $_9P_3$

88. $_{30}P_1$

89. $_8P_8$

90. $_{15}P_2$

In Exercises 91 and 92, find the number of distinguishable permutations of the group of letters.

91. I, N, T, E, R, N, E, T

92. S, W, E, A, T, S, H, I, R, T

93. Defective Units A shipment of 30 VCRs contains three defective units. In how many ways can a vending company purchase five of these units and receive (a) all good units, (b) two good units, and (c) at least two good units?

94. Job Applicants An employer interviews 10 people for five openings in the marketing department of a company. Six of the 10 people are women. If all 10 are qualified, in how many ways can the employer fill the five positions if (a) the selection is random, (b) exactly three of those hired are women, and (c) all five of those hired are women?

In Exercises 95 and 96, determine the sample space for the experiment.

95. Guessing the last digit in a telephone number

96. Four coins are tossed.

Drawing a Card In Exercises 97 and 98, a card is selected from a standard deck of 52 cards. Find the probability of the event.

97. Getting a 10, jack, or queen

98. Getting a black card that is not a face card

99. Education In a high school graduating class of 72 students, 28 are on the honor roll. Of these, 18 are going on to college, and of the other 44 students, 12 are going on to college. A student is selected at random from the class. What is the probability that the person chosen is (a) going to college, (b) not going to college, and (c) on the honor roll, but not going to college?

100. Preparing for a Test An instructor gives a class a list of 12 study problems, from which eight will be selected to construct an exam. A student knows how to solve 10 of the problems. Find the probability that the student will be able to answer (a) all eight questions on the exam, (b) exactly seven questions on the exam, and (c) at least seven questions on the exam.

101. Game Show On a game show, you are given five different digits to arrange in the proper order to represent the price of a car. If you are correct, you win the car. Find the probability of winning under each condition.

(a) You must guess the position of each digit.

(b) You know the first two digits, but must guess the remaining three.

102. Payroll Mix-Up Four paychecks and envelopes are addressed to four different people. The paychecks get mixed up and are randomly inserted into the envelopes. What is the probability that (a) exactly one will be inserted in the correct envelope and (b) at least one will be inserted in the correct envelope?

103. Poker Hand Five cards are drawn from a standard deck of 52 playing cards. What is the probability of getting two pair? (Two pair consists of two of one kind, two of another kind, and one of a third kind. For example, K-K-9-9-5 is two pair.)

104. A Boy or a Girl? Assume that the probability of the birth of a child of a particular gender is 50%. In a family with five children, what is the probability that (a) all the children are boys, (b) all the children are of the same gender, and (c) there is at least one boy?

105. A Boy or a Girl? Assume that the probability of the birth of a child of a particular gender is 50%. In a family with seven children, what is the probability that (a) all the children are girls, (b) all the children are of the same gender, and (c) there is at least one girl?

106. Defective Units A shipment of 1000 radar detectors contains five defective units. A retail outlet has ordered 50 units.

(a) What is the probability that all 50 units are good?

(b) What is the probability that at least one unit is defective?

In Exercises 107 and 108, you are given the probability that an event will happen. Find the probability that the event *will not* happen.

107. $p = 0.57$ **108.** $p = 0.23$

In Exercises 109–112, use mathematical induction to prove the formula for every positive integer *n*.

109. $1 + 4 + \cdots + (3n - 2) = \dfrac{n}{2}(3n - 1)$

110. $1 + \dfrac{3}{2} + 2 + \dfrac{5}{2} + \cdots + \dfrac{1}{2}(n + 1) = \dfrac{n}{4}(n + 3)$

111. $\displaystyle\sum_{i=0}^{n-1} ar^i = \dfrac{a(1 - r^n)}{1 - r}$

112. $\displaystyle\sum_{k=0}^{n-1} (a + kd) = \dfrac{n}{2}[2a + (n - 1)d]$

In Exercises 113–116, find the sum using the formulas for the sums of powers of integers.

113. $\displaystyle\sum_{n=1}^{30} n$ **114.** $\displaystyle\sum_{n=1}^{5} n^2$

115. $\displaystyle\sum_{n=1}^{8} n^3$ **116.** $\displaystyle\sum_{n=1}^{6} (n^2 + n)$

In Exercises 117 and 118, find a formula for the sum of the first *n* terms of the sequence.

117. $1, \dfrac{3}{5}, \dfrac{9}{25}, \dfrac{27}{125}, \ldots$ **118.** $5, -\dfrac{5}{2}, \dfrac{5}{4}, -\dfrac{5}{8}, \ldots$

Make a Decision In Exercises 119 and 120, write the first five terms of the sequence, where $a_1 = f(1)$. Then calculate the first and second differences of the sequence. Does the sequence have a linear model or a quadratic model? If so, find the model.

119. $a_0 = 2$; $a_n = a_{n-1} + n$

120. $f(1) = 1$; $a_n = a_{n-1} + 2n$

Chapter Test

Take this test as you would take a test in class. After you are done, check your work against the answers given in the back of the book.

In Exercises 1–4, write the first five terms of the sequence. (Assume that n begins with 1.)

1. $a_n = 2n + 1$ **2.** $a_n = (-1)^n n^2$ **3.** $a_n = (n + 1)!$ **4.** $a_n = \left(\frac{1}{2}\right)^n$

In Exercises 5–8, determine whether the sequence is arithmetic, geometric, or neither. If possible, find its common difference or common ratio.

5. $a_n = 5n - 2$ **6.** $a_n = 2^n + 2$ **7.** $a_n = 5(2^n)$ **8.** $a_n = 3n^2 - 2$

In Exercises 9–11, find the sum.

9. $\displaystyle\sum_{n=1}^{50} (2n + 1)$ **10.** $\displaystyle\sum_{n=1}^{20} 3\left(\frac{3}{2}\right)^n$ **11.** $\displaystyle\sum_{n=1}^{\infty} \left(\frac{1}{3}\right)^n$

12. A deposit of $5000 is made in an account that pays 6% compounded monthly. The balance in the account after n months is given by

$$A = 5{,}000\left(1 + \frac{0.06}{12}\right)^n, \quad n = 1, 2, 3, \ldots$$

Find the balance in the account after 10 years.

13. Use the Binomial Theorem to expand $(2x - 3)^5$.

14. Determine the coefficient of the x^7 term in the expansion of $(x - 3)^{12}$.

In Exercises 15 and 16, evaluate each expression.

15. (a) $_9P_2$ (b) $_{70}P_3$

16. (a) $_{11}C_4$ (b) $_{66}C_5$

17. A customer in an electronics store can choose one of five CD players, one of six speaker systems, and one of four radio/cassette players to design a sound system. How many different systems can be designed?

18. In how many ways can a 10-question true-false exam be answered? (Assume that no question is omitted.) Explain your reasoning.

19. How many different five-digit numbers can be formed if the leading digit cannot be 0 and the number must be odd?

20. Four coins are tossed. What is the probability that all are heads?

21. The weather report calls for 75% chance of rain. According to this report, what is the probability that it will not rain?

22. Use mathematical induction to prove the formula.

$$5 + 10 + 15 + \cdots + 5n = \frac{5n(n + 1)}{2}$$

Cumulative Test: Chapters 4–6

Take this test as you would take a test in class. After you are done, check your work against the answers given in the back of the book.

In Exercises 1 and 2, solve the system of equations using the indicated method.

1. *Elimination*

$$\begin{cases} 2x - 3y + z = 18 \\ 3x \qquad - 2z = -4 \\ x - y + 3z = 20 \end{cases}$$

2. *Matrices*

$$\begin{cases} 3x - 4y + 2z = -32 \\ 2x + 3y \qquad = 8 \\ y - 3z = 19 \end{cases}$$

In Exercises 3 and 4, sketch the graph of the solution set of the system of inequalities.

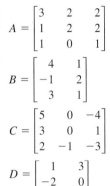

$$A = \begin{bmatrix} 3 & 2 & 2 \\ 1 & 2 & 2 \\ 1 & 0 & 1 \end{bmatrix}$$

$$B = \begin{bmatrix} 4 & 1 \\ -1 & 2 \\ 3 & 1 \end{bmatrix}$$

$$C = \begin{bmatrix} 5 & 0 & -4 \\ 3 & 0 & 1 \\ 2 & -1 & -3 \end{bmatrix}$$

$$D = \begin{bmatrix} 1 & 3 \\ -2 & 0 \end{bmatrix}$$

Matrices for 5–9

3. $\begin{cases} 2x + y \geq -3 \\ x - 3y \leq 2 \end{cases}$

4. $\begin{cases} x - y > 6 \\ 5x + 2y < 10 \end{cases}$

In Exercises 5–8, use the matrices at the left to perform the indicated matrix operation.

5. $2A - C$ 6. AB 7. BD 8. A^{-1}

9. Find the determinants (if possible) of the matrices A, B, C, and D at the left.

10. Find the point of equilibrium for a system with demand equation $p = 75 - 0.0005x$ and supply equation $p = 30 + 0.002x$.

$$A = \begin{bmatrix} 1000 & 3000 & 2000 \\ 2000 & 4000 & 5000 \\ 3000 & 1000 & 1000 \end{bmatrix}$$

$$B = \begin{bmatrix} \$25 & \$32 & \$27 \end{bmatrix}$$

Matrices for 11

11. A factory produces three different models of a product, which are shipped to three different warehouses. The number of units of model i that are shipped to warehouse j is represented by a_{ij} in matrix A at the left. The price per unit for each model is represented by matrix B. Find the product BA and state what each entry of the product represents.

12. Write the first five terms of the sequence given by $a_n = 5 + 2n$.

13. Write the first five terms of the geometric sequence for which $a_1 = 2$ and $r = \frac{1}{3}$.

14. Find the sum $\displaystyle\sum_{i=0}^{\infty} 1.3\left(\frac{1}{10}\right)^{i-1}$.

15. Use the Binomial Theorem to expand $(2x + 3y)^6$.

16. In how many ways can the letters in MATRIX be arranged?

17. A shipment of 2000 microwave ovens contains 15 defective units. You order three units from the shipment. What is the probability that all three are good?

18. On a game show, the digits 7, 8, and 9 must be arranged in the proper order to form the price of an appliance. If the digits are arranged correctly, the contestant wins the appliance. What is the probability of winning if the contestant knows that the price is at least $800?

19. Write the first five terms of the sequence for which $a_1 = 2$ and $a_n = a_{n-1} + n$. Then calculate the first and second differences of the sequence. Does the sequence have a linear model, a quadratic model, or neither?

Appendix A   An Introduction to Graphing Utilities

Graphing utilities such as graphing calculators and computers with graphing software are very valuable tools for visualizing mathematical principles, verifying solutions to equations, exploring mathematical ideas, and developing mathematical models. Although graphing utilities are extremely helpful in learning mathematics, their use does not mean that learning algebra is any less important. In fact, the combination of knowledge of mathematics and the use of graphing utilities enables you to explore mathematics more easily and to a greater depth. If you are using a graphing utility in this course, it is up to you to learn its capabilities and to practice using this tool to enhance your mathematical learning.

In this text, there are many opportunities to use a graphing utility, some of which are described below.

Uses of a Graphing Utility

A graphing utility can be used to

• check or validate answers to problems obtained using algebraic methods.

• discover and explore algebraic properties, rules, and concepts.

• graph functions and approximate solutions of equations involving functions.

• efficiently perform complicated mathematical procedures such as those found in many real-life applications.

• find mathematical models for sets of data.

In this appendix, the features of graphing utilities are discussed from a generic perspective. To learn how to use the features of a specific graphing utility, consult your user's manual or the website for this text, found at *college.hmco.com*, where you will find specific keystrokes for most graphing utilities. Your college library may also have a videotape on how to use your graphing utility.

The Equation Editor

Many graphing utilities are designed to act as "function graphers." In this course, functions and their graphs are studied in detail. A function can be thought of as a rule that describes the relationship between two variables. These rules are frequently written in terms of x and y. For example, the equation $y = 3x + 5$ represents y as a function of x.

Many graphing utilities have an *equation editor* that requires that an equation be written in "$y =$" form in order to be entered, as shown in Figure A.1. (You should note that your *equation editor* screen may not look like the screen shown in Figure A.1.)

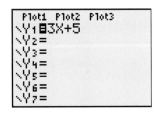

Figure A.1

Figure A.2

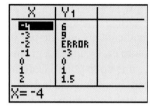

Figure A.3

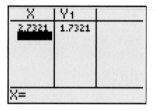

Figure A.4

Table Feature

Most graphing utilities are capable of displaying a table of values with x-values and one or more corresponding y-values. These tables can be used to check solutions of an equation and to generate ordered pairs to assist in graphing an equation by hand.

To use the *table* feature, enter an equation into the *equation editor* in "y =" form. The table may have a setup screen, which allows you to select the starting x-value and the table step or x-increment. You may then have the option of automatically generating values for x and y or building your own table using *ask* mode. In *ask* mode, you enter a value for x and the graphing utility displays the y-value.

For example, enter the equation

$$y = \frac{3x}{x + 2}$$

into the *equation editor* as shown in Figure A.2. In the table setup screen, set the table to start at $x = -4$ and set the table step to 1. When you view the table, notice that the first x-value is -4 and each value after it increases by 1. Also notice that the Y1 column gives the resulting y-value for each x-value, as shown in Figure A.3. The table shows that the y-value when $x = -2$ is ERROR. This means that the equation is undefined when $x = -2$.

With the same equation in the *equation editor*, set the independent variable in the table to *ask* mode. In this mode, you do not need to set the starting x-value or the table step, because you are entering any value you choose for x. You may enter any real value for x—integers, fractions, decimals, irrational numbers, and so forth. If you enter $x = 1 + \sqrt{3}$, the graphing utility may rewrite the number as a decimal approximation, as shown in Figure A.4. You can continue to build your own table by entering additional x-values in order to generate y-values.

If you have several equations in the *equation editor*, the table may generate y-values for each equation.

Creating a Viewing Window

A **viewing window** for a graph is a rectangular portion of the coordinate plane. A viewing window is determined by the following six values.

Xmin = the smallest value of x
Xmax = the largest value of x
Xscl = the number of units per tick mark on the x-axis
Ymin = the smallest value of y
Ymax = the largest value of y
Yscl = the number of units per tick mark on the y-axis

When you enter these six values into a graphing utility, you are setting the viewing window. On some graphing utilities, there is a seventh value on the viewing window labeled Xres. This sets the pixel resolution. For instance, when Xres = 1, functions are evaluated and graphed at each pixel on the x-axis. Some graphing utilities have a standard viewing window, as shown in Figure A.5.

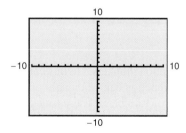

Figure A.5

By choosing different viewing windows for a graph, it is possible to obtain very different impressions of the graph's shape. For instance, Figure A.6 shows four different viewing windows for the graph of

$$y = 0.1x^4 - x^3 + 2x^2.$$

Of these, the view shown in part (a) is the most complete.

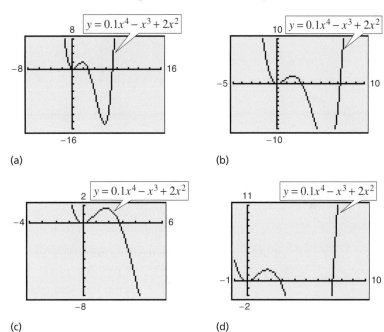

(a)   (b)

(c)   (d)

Figure A.6

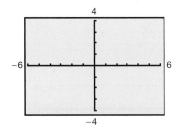

Figure A.7

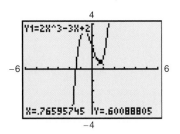

Figure A.8

On most graphing utilities, the display screen is two-thirds as high as it is wide. On such screens, you can obtain a graph with a true geometric perspective by using a **square setting**—one in which

$$\frac{\text{Ymax} - \text{Ymin}}{\text{Xmax} - \text{Xmin}} = \frac{2}{3}.$$

One such setting is shown in Figure A.7. Notice that the x and y tick marks are equally spaced on a square setting, but not on a standard setting.

To see how the viewing window affects the geometric perspective, graph the semicircles $y_1 = \sqrt{9 - x^2}$ and $y_2 = -\sqrt{9 - x^2}$ using a standard viewing window. Then graph y_1 and y_2 using a square viewing window. Note the difference in the shapes of the circles.

Zoom and Trace Features

When you graph an equation, you can move from point to point along its graph using the *trace* feature. As you trace the graph, the coordinates of each point are displayed, as shown in Figure A.8. The *trace* feature combined with the *zoom* feature enables you to obtain better and better approximations of desired points on a graph. For instance, you can use the *zoom* feature of a graphing utility to approximate the x-intercept(s) of a graph. Suppose you want to approximate the x-intercept(s) of the graph of $y = 2x^3 - 3x + 2$.

Begin by graphing the equation, as shown in Figure A.9(a). From the viewing window shown, the graph appears to have only one x-intercept. This intercept lies between -2 and -1. By zooming in on the intercept, you can improve the approximation, as shown in Figure A.9(b). To three decimal places, the solution is $x \approx -1.476$.

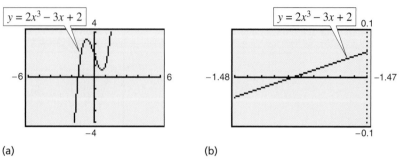

(a)   (b)

Figure A.9

Here are some suggestions for using the *zoom* feature.

1. With each successive zoom-in, adjust the x-scale so that the viewing window shows at least one tick mark on each side of the x-intercept.

2. The error in your approximation will be less than the distance between two tick marks.

3. The *trace* feature can usually be used to add one more decimal place of accuracy without changing the viewing window.

Figure A.10(a) shows the graph of $y = x^2 - 5x + 3$. Figures A.10(b) and A.10(c) show "zoom-in views" of the two x-intercepts. Using these views and the *trace* feature, you can approximate the x-intercepts to be $x \approx 0.697$ and $x \approx 4.303$.

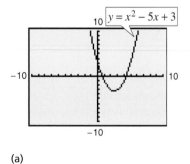

(a)

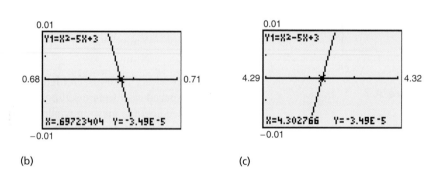

(b)   (c)

Figure A.10

Zero or Root Feature

Using the *zero* or *root* feature, you can find the real zeros of functions of the various types studied in this text—polynomial, exponential, and logarithmic. To find the zeros of a function such as $f(x) = \frac{3}{4}x - 2$, first enter the function as $y_1 = \frac{3}{4}x - 2$ in the *equation editor*. Then use the *zero* or *root* feature, which may require entering lower and upper bound estimates of the zero, as shown in Figures A.11(a) and A.11(b).

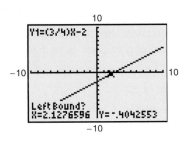

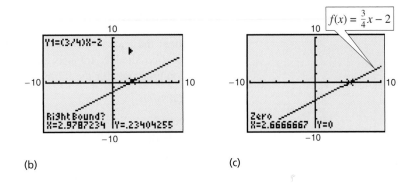

(a)

(b)

(c)

Figure A.11

In Figure A.11(c), you can see that the zero is $x = 2.6666667 \approx 2\frac{2}{3}$.

Intersect Feature

To find the points of intersection of two graphs, you can use the *intersect* feature. For instance, to find the points of intersection of the graphs of

$$y_1 = -x + 2 \quad \text{and} \quad y_2 = x + 4$$

enter these two functions in the *equation editor* and use the *intersect* feature, as shown in Figure A.12.

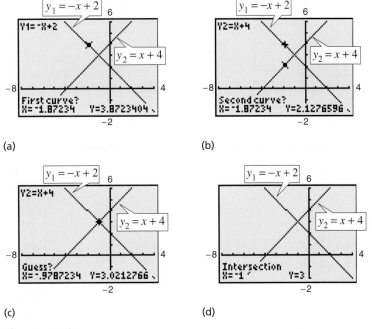

Figure A.12

From Figure A.12(d), you can see that the point of intersection is $(-1, 3)$.

Regression Feature

Throughout the text, you are asked to use the *regression* feature of a graphing utility to find models for sets of data. Most graphing utilities have built-in regression programs for the following.

Regression	Form of Model
Linear	$y = ax + b$ or $y = a + bx$
Quadratic	$y = ax^2 + bx + c$
Cubic	$y = ax^3 + bx^2 + cx + d$
Quartic	$y = ax^4 + bx^3 + cx^2 + dx + e$
Logarithmic	$y = a + b \ln(x)$
Exponential	$y = ab^x$
Power	$y = ax^b$
Logistic	$y = \dfrac{c}{1 + ae^{-bx}}$
Sine	$y = a \sin(bx + c) + d$

For example, you can find a linear model for the number y of daily-fee golf facilities in the United States from 1994 to 2001 using the data shown in the table. (Source: National Golf Foundation)

Year	1994	1995	1996	1997	1998	1999	2000	2001
Number, y	7126	7491	7729	7984	8247	8470	8761	8972

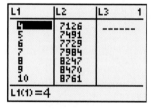

Figure A.13

First let x represent the year, with $x = 4$ corresponding to 1994, and enter the data into the *list editor*, as shown in Figure A.13. Note that the list in the first column contains the years and the list in the second column contains the numbers of daily-fee golf facilities that correspond to the years. Now use your graphing utility's *linear regression* feature to obtain the coefficients a and b for the model $y = ax + b$, as shown in Figure A.14. So, a linear model for the data is given by

$$y = 259.0x + 6155.$$

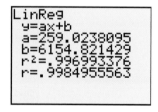

Figure A.14

STUDY TIP

In this text, when regression models are found, the number of decimal places in the constant term of the model is the same as the number of decimal places in the data, and the number of decimal places increases by one for terms of increasing powers of the independent variable.

When you run some regression programs, you may obtain an "r-value," which gives a measure of how well the model fits the data. The closer the value of $|r|$ is to 1, the better the fit. For the data in the table above, you can see from Figure A.14 that $r \approx 0.998$, which implies that the model is a very good fit for the data.

Maximum and Minimum Features

The *maximum* and *minimum* features of a graphing utility find the relative extrema of a function. To find the relative maximum of a function such as

$$f(x) = x^3 - 3x$$

first enter the function as $y_1 = x^3 - 3x$ in the *equation editor*. Then use the *maximum* feature, which may require entering lower and upper bound estimates for the maximum, as shown in Figures A.15(a) and A.15(b).

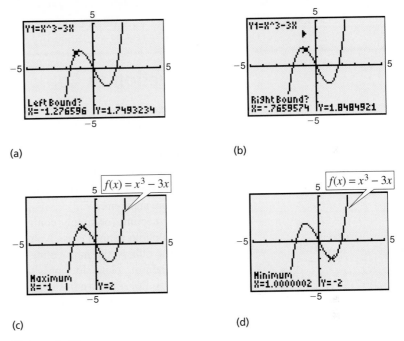

(a)

(b)

(c)

(d)

Figure A.15

In Figure A.15(c), you can see that the relative maximum occurs at $(-1, 2)$. The relative minimum of the function can be found in a similar manner. From Figure A.15(d), you can see that the relative minimum occurs at $(1, -2)$.

Appendix B   Conic Sections

B.1   Conic Sections

Objectives
- Recognize the four basic conics: circles, parabolas, ellipses, and hyperbolas.
- Recognize, graph, and write equations of parabolas (vertex at origin).
- Recognize, graph, and write equations of ellipses (centered at origin).
- Recognize, graph, and write equations of hyperbolas (centered at origin).

Introduction to Conic Sections

Conic sections were discovered during the classical Greek period, which lasted from 600 to 300 B.C. By the beginning of the Alexandrian period, enough was known of conics for Apollonius (262–190 B.C.) to produce an eight-volume work on the subject.

This early Greek study was largely concerned with the geometric properties of conics. It was not until the early seventeenth century that the broad applicability of conics became apparent.

A **conic section** (or simply **conic**) can be described as the intersection of a plane and a double-napped cone. Notice from Figure B.1 that in the formation of the four basic conics, the intersecting plane does not pass through the vertex of the cone. When the plane does pass through the vertex, the resulting figure is a **degenerate conic,** as shown in Figure B.2.

Figure B.1   Conic Sections

Figure B.2   Degenerate Conics

There are several ways to approach the study of conics. You could begin by defining conics in terms of the intersections of planes and cones, as the Greeks did, or you could define them algebraically, in terms of the general second-degree equation

$$Ax^2 + Bxy + Cy^2 + Dx + Ey + F = 0.$$

However, you will study a third approach, in which each of the conics is defined as a **locus,** or collection of points satisfying a certain geometric property. For example, in Section 1.1 you saw how the definition of a circle as *the collection of all points (x, y) that are equidistant from a fixed point (h, k)* led easily to the standard equation of a circle, $(x - h)^2 + (y - k)^2 = r^2$.

You will restrict your study of conics in this section to parabolas with vertices at the origin and ellipses and hyperbolas with centers at the origin. In the following section, you will look at the more general cases.

Parabolas

In Section 2.1 you determined that the graph of the quadratic function given by $f(x) = ax^2 + bx + c$ is a parabola that opens upward or downward. The following definition of a parabola is more general in the sense that it is independent of the orientation of the parabola.

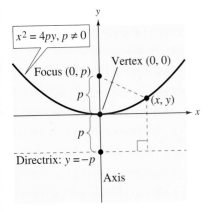

Parabola with vertical axis

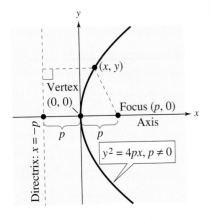

Parabola with horizontal axis

Figure B.3

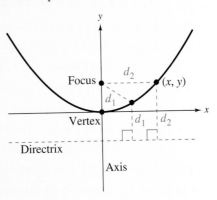

Definition of a Parabola

A **parabola** is the set of all points (x, y) in a plane that are equidistant from a fixed line called the **directrix** and a fixed point called the **focus** (not on the line). The midpoint between the focus and the directrix is called the **vertex,** and the line passing through the focus and the vertex is called the **axis** of the parabola.

Using this definition, you can derive the following standard form of the equation of a parabola.

Standard Equation of a Parabola (Vertex at Origin)

The **standard form of the equation of a parabola** with vertex at $(0, 0)$ and directrix $y = -p$ is given by

$$x^2 = 4py, \quad p \neq 0. \qquad \text{Vertical axis}$$

For directrix $x = -p$, the equation is given by

$$y^2 = 4px, \quad p \neq 0. \qquad \text{Horizontal axis}$$

The focus is on the axis p units (directed distance) from the vertex. See Figure B.3.

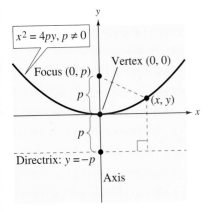

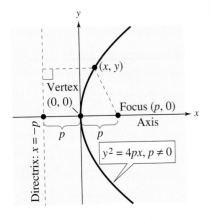

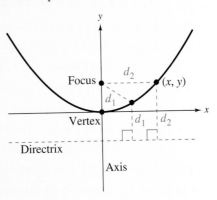

STUDY TIP

Note that the term *parabola* is a technical term used in mathematics and does not simply refer to *any* U-shaped curve.

| EXAMPLE 1 | FINDING THE FOCUS OF A PARABOLA |

Find the focus of the parabola whose equation is $y = -2x^2$.

SOLUTION

Because the squared term in the equation involves x, you know that the axis is vertical, and the equation is of the form

$$x^2 = 4py. \qquad \text{Standard form, vertical axis}$$

You can write the original equation in this form as follows.

$$-2x^2 = y \qquad \text{Write original equation.}$$

$$x^2 = -\frac{1}{2}y \qquad \text{Divide each side by } -2.$$

$$x^2 = 4\left(-\frac{1}{8}\right)y. \qquad \text{Write in standard form.}$$

So, $p = -\frac{1}{8}$. Because p is negative, the parabola opens downward and the focus of the parabola is $(0, p) = \left(0, -\frac{1}{8}\right)$, as shown in Figure B.4.

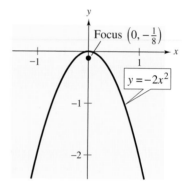

Figure B.4

✓CHECKPOINT   Now try Exercise 9.

| EXAMPLE 2 | FINDING THE STANDARD EQUATION OF A PARABOLA |

Write the standard form of the equation of the parabola with vertex at the origin and focus at $(2, 0)$.

SOLUTION

The axis of the parabola is horizontal, passing through $(0, 0)$ and $(2, 0)$, as shown in Figure B.5. So, the standard form is

$$y^2 = 4px. \qquad \text{Standard form, horizontal axis}$$

Because the focus is $p = 2$ units from the vertex, the equation is

$$y^2 = 4(2)x \qquad \text{Standard form}$$

$$y^2 = 8x.$$

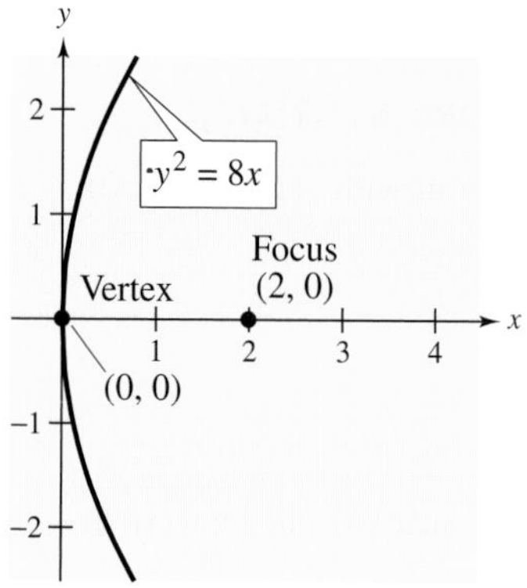

✓CHECKPOINT   Now try Exercise 19.

Figure B.5

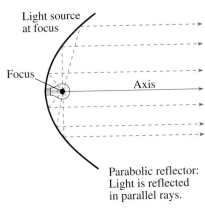

Light source
at focus

Focus

Axis

Parabolic reflector:
Light is reflected
in parallel rays.

Figure B.6

Parabolas occur in a wide variety of applications. For instance, a parabolic reflector can be formed by revolving a parabola about its axis. The resulting surface has the property that all incoming rays parallel to the axis are reflected through the focus of the parabola. This is the principle behind the construction of the parabolic mirrors used in reflecting telescopes. Conversely, the light rays emanating from the focus of the parabolic reflector used in a flashlight are all reflected parallel to one another, as shown in Figure B.6.

Ellipses

The second basic type of conic is called an **ellipse,** and it is defined as follows.

Definition of an Ellipse

An **ellipse** is the set of all points (x, y) in a plane the sum of whose distances from two distinct fixed points, called **foci,** is constant.

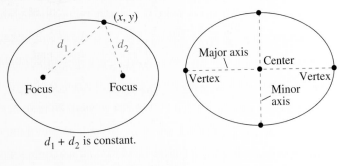

(x, y)

d_1 d_2

Focus   Focus

Major axis   Center

Vertex   Vertex

Minor axis

$d_1 + d_2$ is constant.

The line through the foci intersects the ellipse at two points, called the **vertices.** The chord joining the vertices is called the **major axis,** and its midpoint is called the **center** of the ellipse. The chord perpendicular to the major axis at the center is called the **minor axis** of the ellipse.

You can visualize the definition of an ellipse by imagining two thumbtacks placed at the foci, as shown in Figure B.7. If the ends of a fixed length of string are fastened to the thumbtacks and the string is drawn taut with a pencil, the path traced by the pencil will be an ellipse.

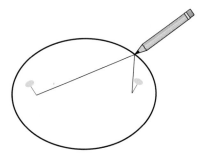

Figure B.7

The standard form of the equation of an ellipse takes one of two forms, depending on whether the major axis is horizontal or vertical.

Standard Equation of an Ellipse (Center at Origin)

The **standard form of the equation of an ellipse** with the center at the origin and major and minor axes of lengths $2a$ and $2b$ (where $0 < b < a$) is

$$\frac{x^2}{a^2} + \frac{y^2}{b^2} = 1 \qquad \text{or} \qquad \frac{x^2}{b^2} + \frac{y^2}{a^2} = 1.$$

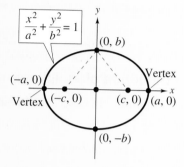

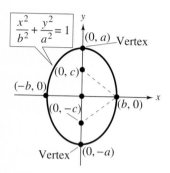

Major axis is horizontal.
Minor axis is vertical.

Major axis is vertical.
Minor axis is horizontal.

The vertices and foci lie on the major axis, a and c units, respectively, from the center. Moreover, a, b, and c are related by the equation $c^2 = a^2 - b^2$.

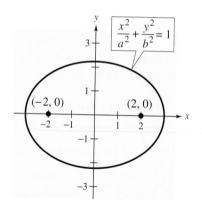

Figure B.8

EXAMPLE 3   FINDING THE STANDARD EQUATION OF AN ELLIPSE

Find the standard form of the equation of the ellipse that has a major axis of length 6 and foci at $(-2, 0)$ and $(2, 0)$, as shown in Figure B.8.

SOLUTION

Because the foci occur at $(-2, 0)$ and $(2, 0)$, the center of the ellipse is $(0, 0)$, and the major axis is horizontal. So, the ellipse has an equation of the form

$$\frac{x^2}{a^2} + \frac{y^2}{b^2} = 1. \qquad \text{Standard form, horizontal major axis}$$

Because the length of the major axis is 6, you have $2a = 6$, which implies that $a = 3$. Moreover, the distance from the center to either focus is $c = 2$. Finally, you have

$$b^2 = a^2 - c^2 = 3^2 - 2^2 = 9 - 4 = 5.$$

Substituting $a^2 = 3^2$ and $b^2 = \left(\sqrt{5}\right)^2$ yields the following equation in standard form.

$$\frac{x^2}{3^2} + \frac{y^2}{\left(\sqrt{5}\right)^2} = 1 \qquad \text{Standard form}$$

This equation simplifies to $\dfrac{x^2}{9} + \dfrac{y^2}{5} = 1.$

✓CHECKPOINT   Now try Exercise 35.

EXAMPLE 4   SKETCHING AN ELLIPSE

Sketch the ellipse given by $4x^2 + y^2 = 36$, and identify the vertices.

SOLUTION

Begin by writing the equation in standard form.

$$4x^2 + y^2 = 36 \qquad \text{Write original equation.}$$

$$\frac{4x^2}{36} + \frac{y^2}{36} = \frac{36}{36} \qquad \text{Divide each side by 36.}$$

$$\frac{x^2}{3^2} + \frac{y^2}{6^2} = 1 \qquad \text{Write in standard form.}$$

$$\frac{x^2}{9} + \frac{y^2}{36} = 1 \qquad \text{Simplify.}$$

Because the denominator of the y^2-term is larger than the denominator of the x^2-term, you can conclude that the major axis is vertical. Moreover, because $a = 6$, the vertices are $(0, -6)$ and $(0, 6)$. Finally, because $b = 3$, the endpoints of the minor axis are $(-3, 0)$ and $(3, 0)$, as shown in Figure B.9.

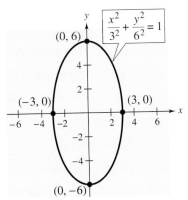

STUDY TIP

The endpoints of the minor axis of an ellipse are commonly referred to as the *co-vertices*. In Figure B.9, the co-vertices are $(-3, 0)$ and $(3, 0)$.

Figure B.9

Note that from the standard form of the equation you can sketch the ellipse by locating the endpoints of the two axes. Because 3^2 is the denominator of the x^2-term, move three units to the *right and left* of the center to locate the endpoints of the horizontal axis. Similarly, because 6^2 is the denominator of the y^2-term, move six units *upward and downward* from the center to locate the endpoints of the vertical axis.

✓**CHECKPOINT**   Now try Exercise 33.

Hyperbolas

The definition of a **hyperbola** is similar to that of an ellipse. The distinction is that, for an ellipse, the *sum* of the distances between the foci and a point on the ellipse is constant, whereas for a hyperbola, the *difference* of the distances between the foci and a point on the hyperbola is constant.

Definition of a Hyperbola

A **hyperbola** is the set of all points (x, y) in a plane the difference of whose distances from two distinct fixed points, called **foci,** is constant.

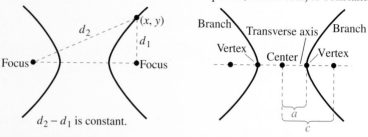

$d_2 - d_1$ is constant.

The graph of a hyperbola has two disconnected parts, called **branches.** The line through the two foci intersects the hyperbola at two points, called **vertices.** The line segment connecting the vertices is called the **transverse axis,** and the midpoint of the transverse axis is called the **center** of the hyperbola.

Standard Equation of a Hyperbola (Center at Origin)

The **standard form of the equation of a hyperbola** with the center at the origin (where $a \neq 0$ and $b \neq 0$) is $\dfrac{x^2}{a^2} - \dfrac{y^2}{b^2} = 1$ or $\dfrac{y^2}{a^2} - \dfrac{x^2}{b^2} = 1.$

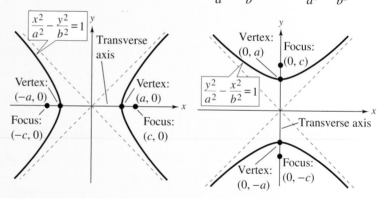

The vertices and foci are, respectively, a and c units from the center. Moreover, a, b, and c are related by the equation $b^2 = c^2 - a^2.$

EXAMPLE 5   FINDING THE STANDARD EQUATION OF A HYPERBOLA

Find the standard form of the equation of the hyperbola with foci at $(-3, 0)$ and $(3, 0)$ and vertices at $(-2, 0)$ and $(2, 0)$, as shown in Figure B.10.

SOLUTION

From the graph, $c = 3$ because the foci are three units from the center. Moreover, $a = 2$ because the vertices are two units from the center. So, it follows that

$$b^2 = c^2 - a^2 = 3^2 - 2^2 = 9 - 4 = 5.$$

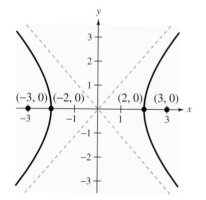

Figure B.10

Because the transverse axis is horizontal, the standard form of the equation is

$$\frac{x^2}{a^2} - \frac{y^2}{b^2} = 1.$$ Standard form, horizontal transverse axis

Finally, substituting $a^2 = 2^2$ and $b^2 = \left(\sqrt{5}\right)^2$ you have

$$\frac{x^2}{2^2} - \frac{y^2}{\left(\sqrt{5}\right)^2} = 1.$$ Write in standard form.

✓CHECKPOINT   Now try Exercise 51.

An important aid in sketching the graph of a hyperbola is the determination of its *asymptotes*, as shown in Figure B.11. Each hyperbola has two asymptotes that intersect at the center of the hyperbola. Furthermore, the asymptotes pass through the corners of a rectangle of dimensions $2a$ by $2b$. The line segment of length $2b$, joining $(0, b)$ and $(0, -b)$ [or $(-b, 0)$ and $(b, 0)$], is referred to as the **conjugate axis** of the hyperbola.

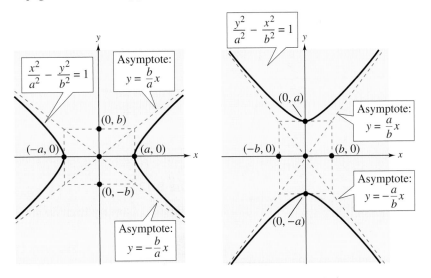

Transverse axis is horizontal.   Transverse axis is vertical.

Figure B.11

Asymptotes of a Hyperbola (Center at Origin)

The **asymptotes of a hyperbola** with center at $(0, 0)$ are

$$y = \frac{b}{a}x \quad \text{and} \quad y = -\frac{b}{a}x$$ Transverse axis is horizontal.

or

$$y = \frac{a}{b}x \quad \text{and} \quad y = -\frac{a}{b}x.$$ Transverse axis is vertical.

| EXAMPLE 6 | SKETCHING A HYPERBOLA |

Sketch the hyperbola whose equation is $4x^2 - y^2 = 16$.

SOLUTION

$$4x^2 - y^2 = 16 \qquad \text{Write original equation.}$$

$$\frac{4x^2}{16} - \frac{y^2}{16} = \frac{16}{16} \qquad \text{Divide each side by 16.}$$

$$\frac{x^2}{2^2} - \frac{y^2}{4^2} = 1 \qquad \text{Write in standard form.}$$

Because the x^2-term is positive, you can conclude that the transverse axis is horizontal and the vertices occur at $(-2, 0)$ and $(2, 0)$. Moreover, the endpoints of the conjugate axis occur at $(0, -4)$ and $(0, 4)$, and you can sketch the rectangle shown in Figure B.12. Finally, by drawing the asymptotes through the corners of this rectangle, you can complete the sketch shown in Figure B.13.

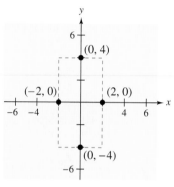

Figure B.12

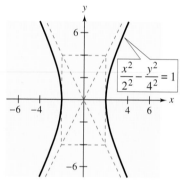

Figure B.13

✓CHECKPOINT   Now try Exercise 49.

| EXAMPLE 7 | FINDING THE STANDARD EQUATION OF A HYPERBOLA |

Find the standard form of the equation of the hyperbola that has vertices at $(0, -3)$ and $(0, 3)$ and asymptotes $y = -2x$ and $y = 2x$, as shown in Figure B.14.

SOLUTION

Because the transverse axis is vertical, the asymptotes are of the form

$$y = \frac{a}{b}x \qquad \text{and} \qquad y = -\frac{a}{b}x. \qquad \text{Transverse axis is vertical.}$$

Using the fact that $y = 2x$ and $y = -2x$, you can determine that $a/b = 2$. Because $a = 3$, you can determine that $b = \frac{3}{2}$. Finally, you can conclude that the hyperbola has the following equation.

$$\frac{y^2}{3^2} - \frac{x^2}{(3/2)^2} = 1 \qquad \text{Write in standard form.}$$

✓CHECKPOINT   Now try Exercise 53.

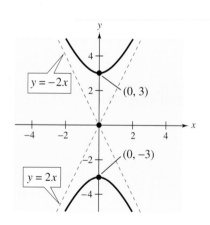

Figure B.14

B.1 Warm Up

The following warm-up exercises involve skills that were covered in earlier sections. You will use these skills in the exercise set for this section. For additional help, review Sections R1.7, R2.3, and 1.1.

In Exercises 1–4, rewrite the equation so that it has no fractions.

1. $\dfrac{x^2}{16} + \dfrac{y^2}{9} = 1$

2. $\dfrac{x^2}{32} + \dfrac{4y^2}{32} = \dfrac{32}{32}$

3. $\dfrac{x^2}{1/4} - \dfrac{y^2}{4} = 1$

4. $\dfrac{3x^2}{1/9} + \dfrac{4y^2}{9} = 1$

In Exercises 5–8, solve for c. (Assume c > 0.)

5. $c^2 = 3^2 - 1^2$

6. $c^2 = 2^2 + 3^2$

7. $c^2 + 2^2 = 4^2$

8. $c^2 - 1^2 = 2^2$

In Exercises 9 and 10, find the distance between the point and the origin.

9. $(0, -4)$

10. $(-2, 0)$

B.1 Exercises

In Exercises 1–8, match the equation with its graph. [The graphs are labeled (a), (b), (c), (d), (e), (f), (g), and (h).]

(a)
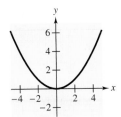

(b)
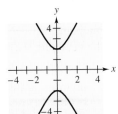

(c)
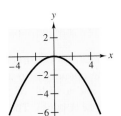

(d)
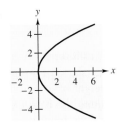

(e)
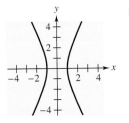

(f)
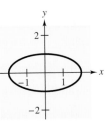

(g)
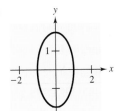

(h)
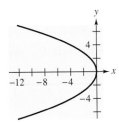

1. $x^2 = 4y$

2. $x^2 = -4y$

3. $y^2 = 4x$

4. $y^2 = -4x$

5. $\dfrac{x^2}{1} + \dfrac{y^2}{4} = 1$

6. $\dfrac{x^2}{4} + \dfrac{y^2}{1} = 1$

7. $\dfrac{x^2}{1} - \dfrac{y^2}{4} = 1$

8. $\dfrac{y^2}{4} - \dfrac{x^2}{1} = 1$

In Exercises 9–16, find the vertex and focus of the parabola and sketch its graph.

9. $y = 4x^2$

10. $y = \tfrac{1}{2}x^2$

11. $y^2 = -6x$

12. $y^2 = 3x$

13. $x^2 + 8y = 0$

14. $x + y^2 = 0$

15. $y^2 - 8x = 0$

16. $x^2 + 12y = 0$

In Exercises 17–26, find the standard form of the equation of the parabola with vertex at the origin.

17. Focus: $\left(0, -\tfrac{3}{2}\right)$

18. Focus: $\left(\tfrac{5}{2}, 0\right)$

19. Focus: $(-2, 0)$

20. Focus: $(0, -2)$

21. Directrix: $y = -1$

22. Directrix: $x = 3$

23. Directrix: $y = 2$

24. Directrix: $x = -2$

25. Passes through the point $(4, 6)$; horizontal axis

26. Passes through the point $(-2, -2)$; vertical axis

In Exercises 27–34, find the center and vertices of the ellipse and sketch its graph.

27. $\dfrac{x^2}{25} + \dfrac{y^2}{16} = 1$

28. $\dfrac{x^2}{144} + \dfrac{y^2}{169} = 1$

29. $\dfrac{x^2}{\frac{25}{9}} + \dfrac{y^2}{\frac{16}{9}} = 1$

30. $\dfrac{x^2}{4} + \dfrac{y^2}{\frac{1}{4}} = 1$

31. $\dfrac{x^2}{9} + \dfrac{y^2}{5} = 1$

32. $\dfrac{x^2}{28} + \dfrac{y^2}{64} = 1$

33. $5x^2 + 3y^2 = 15$

34. $x^2 + 4y^2 = 4$

In Exercises 35–42, find the standard form of the equation of the ellipse with center at the origin.

35. Vertices: $(0, \pm 2)$; minor axis of length 2

36. Vertices: $(\pm 2, 0)$; minor axis of length 3

37. Vertices: $(\pm 5, 0)$; foci: $(\pm 2, 0)$

38. Vertices: $(0, \pm 10)$; foci: $(0, \pm 4)$

39. Foci: $(\pm 5, 0)$; major axis of length 12

40. Foci: $(\pm 2, 0)$; major axis of length 8

41. Vertices: $(0, \pm 5)$; passes through the point $(4, 2)$

42. Major axis vertical; passes through the points $(0, 4)$ and $(2, 0)$

In Exercises 43–50, find the center and vertices of the hyperbola and sketch its graph.

43. $x^2 - y^2 = 1$

44. $\dfrac{x^2}{9} - \dfrac{y^2}{16} = 1$

45. $\dfrac{y^2}{1} - \dfrac{x^2}{4} = 1$

46. $\dfrac{y^2}{9} - \dfrac{x^2}{1} = 1$

47. $\dfrac{y^2}{25} - \dfrac{x^2}{144} = 1$

48. $\dfrac{x^2}{36} - \dfrac{y^2}{4} = 1$

49. $2x^2 - 3y^2 = 6$

50. $3y^2 - 5x^2 = 15$

In Exercises 51–58, find the standard form of the equation of the hyperbola with center at the origin.

51. Vertices: $(0, \pm 2)$; foci: $(0, \pm 4)$

52. Vertices: $(\pm 3, 0)$; foci: $(\pm 5, 0)$

53. Vertices: $(\pm 1, 0)$; asymptotes: $y = \pm 3x$

54. Vertices: $(0, \pm 3)$; asymptotes: $y = \pm 3x$

55. Foci: $(0, \pm 4)$; asymptotes: $y = \pm \frac{1}{2}x$

56. Foci: $(\pm 10, 0)$; asymptotes: $y = \pm \frac{3}{4}x$

57. Vertices: $(0, \pm 3)$; passes through the point $(-2, 5)$

58. Vertices: $(\pm 2, 0)$; passes through the point $\left(3, \sqrt{3}\right)$

59. Satellite Antenna The receiver in a parabolic television dish antenna is 3 feet from the vertex and is located at the focus (see figure). Write an equation for a cross section of the reflector. (Assume that the dish is directed upward and the vertex is at the origin.)

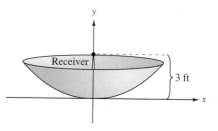

60. Suspension Bridge Each cable of the Golden Gate Bridge is suspended (in the shape of a parabola) between two towers that are 1280 meters apart. The top of each tower is 152 meters above the roadway (see figure). The cables touch the roadway at the midpoint between the towers. Write an equation for the parabolic shape of each cable.

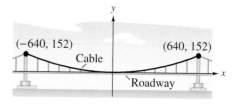

61. Architecture A fireplace arch is to be constructed in the shape of a semiellipse. The opening is to have a height of 2 feet at the center and a width of 5 feet along the base (see figure on next page). The contractor draws the outline of the ellipse by the method shown in Figure B.7. Where should the tacks be placed and what should be the length of the piece of string?

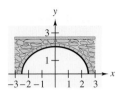

Figure for 61

62. Mountain Tunnel A semielliptical arch over a tunnel for a road through a mountain has a major axis of 100 feet, and its height at the center is 30 feet (see figure). Determine the height of the arch 5 feet from the edge of the tunnel.

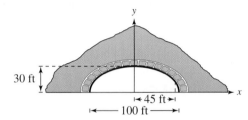

63. Sketch a graph of the ellipse that consists of all points (x, y) such that the sum of the distances between (x, y) and two fixed points is 15 units and the foci are located at the centers of the two sets of concentric circles, as shown in the figure.

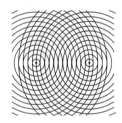

64. Think About It A line segment through a focus of an ellipse with endpoints on the ellipse and perpendicular to its major axis is called a **latus rectum** of the ellipse. Therefore, an ellipse has two latera recta. Knowing the length of the latera recta is helpful in sketching an ellipse because this information yields other points on the curve (see figure). Show that the length of each latus rectum is $2b^2/a$.

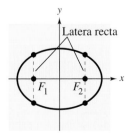

In Exercises 65–68, sketch the ellipse using the latera recta (see Exercise 64).

65. $\dfrac{x^2}{4} + \dfrac{y^2}{1} = 1$ **66.** $\dfrac{x^2}{9} + \dfrac{y^2}{16} = 1$

67. $9x^2 + 4y^2 = 36$ **68.** $5x^2 + 3y^2 = 15$

69. Navigation Long-range navigation for aircraft and ships is accomplished by synchronized pulses transmitted by widely separated transmitting stations. These pulses travel at the speed of light (186,000 miles per second). The difference in the arrival times of these pulses at an aircraft or ship is constant on a hyperbola having the transmitting stations as foci. Assume that two stations 300 miles apart are positioned on a rectangular coordinate system at points with coordinates $(-150, 0)$ and $(150, 0)$ and that a ship is traveling on a path with coordinates $(x, 75)$ (see figure). Find the x-coordinate of the position of the ship if the time difference between the pulses from the transmitting stations is 1000 microseconds (0.001 second).

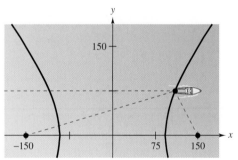

70. Hyperbolic Mirror A hyperbolic mirror (used in some telescopes) has the property that a light ray directed at one focus will be reflected to the other focus (see figure). The focus of the hyperbolic mirror has coordinates $(12, 0)$. Find the vertex of the mirror if its mount at the top edge of the mirror has coordinates $(12, 12)$.

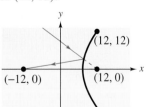

B.2   Conic Sections and Translations

Objectives

- Recognize equations of conics that have been shifted vertically or horizontally in the plane.
- Write and graph equations of conics that have been shifted vertically or horizontally in the plane.

Vertical and Horizontal Shifts of Conics

In Section B.1, you looked at conic sections whose graphs were in **standard position.** In this section, you will study the equations of conic sections that have been shifted vertically or horizontally in the plane. The following summary lists the standard forms of the equations of the four basic conics.

Standard Forms of Equations of Conics

Circle: Center $= (h, k)$; Radius $= r$

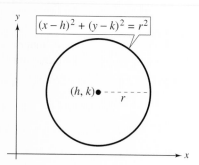

$$(x - h)^2 + (y - k)^2 = r^2$$

(h, k)

Parabola: Vertex $= (h, k)$

Directed distance from vertex to focus $= p$

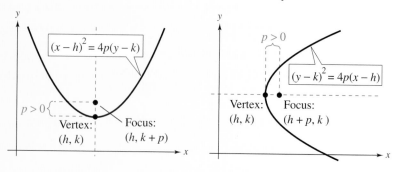

$$(x - h)^2 = 4p(y - k)$$

$p > 0$

Vertex:
(h, k)

Focus:
$(h, k + p)$

$$(y - k)^2 = 4p(x - h)$$

$p > 0$

Vertex:
(h, k)

Focus:
$(h + p, k)$

Standard Forms of Equations of Conics   (continued)

Ellipse: Center $= (h, k)$
            Major axis length $= 2a$; Minor axis length $= 2b$

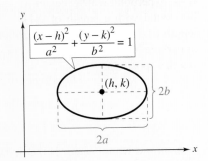

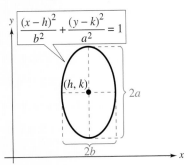

Hyperbola: Center $= (h, k)$
              Transverse axis length $= 2a$; Conjugate axis length $= 2b$

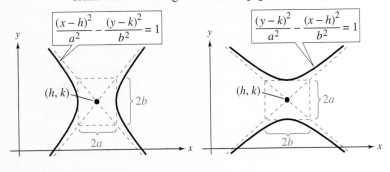

EXAMPLE 1   EQUATIONS OF CONIC SECTIONS

Describe the translation of the graph of each conic.

a. $(x - 1)^2 + (y + 2)^2 = 3^2$ **b.** $(x - 2)^2 = 4(-1)(y - 3)$

c. $\dfrac{(x - 3)^2}{1^2} - \dfrac{(y - 2)^2}{3^2} = 1$ **d.** $\dfrac{(x - 2)^2}{3^2} + \dfrac{(y - 1)^2}{2^2} = 1$

SOLUTION

a. The graph of $(x - 1)^2 + (y + 2)^2 = 3^2$ is a circle whose center is the point $(1, -2)$ and whose radius is 3, as shown in Figure B.15(a) on the following page. Note that the graph of the circle has been shifted one unit to the right and two units downward from standard position.

b. The graph of $(x - 2)^2 = 4(-1)(y - 3)$ is a parabola whose vertex is the point $(2, 3)$. The axis of the parabola is vertical. The focus is one unit above or below the vertex. Moreover, because $p = -1$, it follows that the focus lies *below* the vertex, as shown in Figure B.15(b). Note that the graph of the parabola has been shifted two units to the right and three units upward from standard position.

c. The graph of

$$\frac{(x-3)^2}{1^2} - \frac{(y-2)^2}{3^2} = 1$$

is a hyperbola whose center is the point $(3, 2)$. The transverse axis is horizontal with a length of $2(1) = 2$. The conjugate axis is vertical with a length of $2(3) = 6$, as shown in Figure B.15(c). Note that the graph of the hyperbola has been shifted three units to the right and two units upward from standard position.

d. The graph of

$$\frac{(x-2)^2}{3^2} + \frac{(y-1)^2}{2^2} = 1$$

is an ellipse whose center is the point $(2, 1)$. The major axis of the ellipse is horizontal with a length of $2(3) = 6$. The minor axis of the ellipse is vertical with a length of $2(2) = 4$, as shown in Figure B.15(d). Note that the graph of the ellipse has been shifted two units to the right and one unit upward from standard position.

(a)
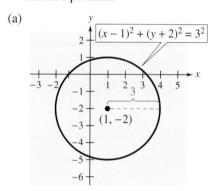

(b)
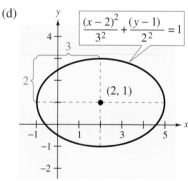

(c)
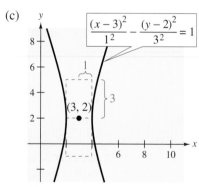

(d)
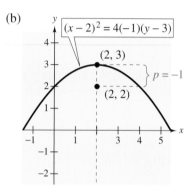

Figure B.15

✓*CHECKPOINT*   Now try Exercise 1.

Writing Equations of Conics in Standard Form

To write the equation of a conic in standard form, complete the square, as demonstrated in Examples 2, 3, and 4.

EXAMPLE 2   FINDING THE STANDARD FORM OF A PARABOLA

Find the vertex and focus of the parabola given by $x^2 - 2x + 4y - 3 = 0$.

SOLUTION

$x^2 - 2x + 4y - 3 = 0$	Write original equation.
$x^2 - 2x = -4y + 3$	Group terms.
$x^2 - 2x + 1 = -4y + 3 + 1$	Add 1 to each side.
$(x - 1)^2 = -4y + 4$	Write in completed square form.
$(x - 1)^2 = 4(-1)(y - 1)$	Write in standard form.

From this standard form, it follows that $h = 1$, $k = 1$, and $p = -1$. Because the axis is vertical and p is negative, the parabola opens downward. The vertex and focus are $(h, k) = (1, 1)$ and $(h, k + p) = (1, 0)$. The graph of this parabola is shown in Figure B.16.

✔CHECKPOINT   Now try Exercise 13.

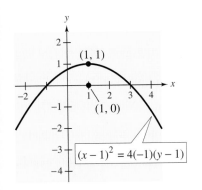

Figure B.16

In Examples 1(b) and 2, p is the *directed distance* from the vertex to the focus. Because the axis of the parabola is vertical and $p = -1$, the focus is one unit *below* the vertex, and the parabola opens downward.

EXAMPLE 3   SKETCHING AN ELLIPSE

Sketch the ellipse given by the equation $x^2 + 4y^2 + 6x - 8y + 9 = 0$.

SOLUTION

$x^2 + 4y^2 + 6x - 8y + 9 = 0$	Write original equation.
$(x^2 + 6x + \quad) + (4y^2 - 8y + \quad) = -9$	Group terms.
$(x^2 + 6x + \quad) + 4(y^2 - 2y + \quad) = -9$	Factor 4 out of y-terms.
$(x^2 + 6x + 9) + 4(y^2 - 2y + 1) = -9 + 9 + 4(1)$	Add 9 and $4(1) = 4$ to each side.
$(x + 3)^2 + 4(y - 1)^2 = 4$	Write in completed square form.
$\dfrac{(x + 3)^2}{4} + \dfrac{(y - 1)^2}{1} = 1$	Divide each side by 4.
$\dfrac{(x + 3)^2}{2^2} + \dfrac{(y - 1)^2}{1^2} = 1$	Write in standard form.

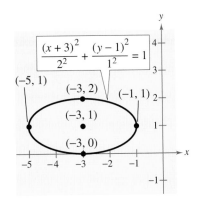

From this standard form, it follows that the center is $(h, k) = (-3, 1)$. Because the denominator of the x-term is $a^2 = 2^2$, the endpoints of the major axis lie two units to the right and left of the center. Similarly, because the denominator of the y-term is $b^2 = 1^2$, the endpoints of the minor axis lie one unit up and down from the center. The ellipse is shown in Figure B.17.

✔CHECKPOINT   Now try Exercise 27.

Figure B.17

| EXAMPLE 4 | SKETCHING A HYPERBOLA |

Sketch the hyperbola given by the equation

$$y^2 - 4x^2 + 4y + 24x - 41 = 0.$$

SOLUTION

$y^2 - 4x^2 + 4y + 24x - 41 = 0$	Write original equation.
$(y^2 + 4y + \quad) - (4x^2 - 24x + \quad) = 41$	Group terms.
$(y^2 + 4y + \quad) - 4(x^2 - 6x + \quad) = 41$	Factor 4 out of x-terms.
$(y^2 + 4y + 4) - 4(x^2 - 6x + 9) = 41 + 4 - 4(9)$	Add 4 and subtract $4(9) = 36$ from each side.
$(y + 2)^2 - 4(x - 3)^2 = 9$	Write in completed square form.
$\dfrac{(y + 2)^2}{9} - \dfrac{4(x - 3)^2}{9} = 1$	Divide each side by 9.
$\dfrac{(y + 2)^2}{9} - \dfrac{(x - 3)^2}{9/4} = 1$	Change 4 to $1/(1/4)$.
$\dfrac{(y + 2)^2}{3^2} - \dfrac{(x - 3)^2}{(3/2)^2} = 1$	Write in standard form.

From the standard form, it follows that the transverse axis is vertical and the center lies at $(h, k) = (3, -2)$. Because the denominator of the y-term is $a^2 = 3^2$, you know that the vertices lie three units above and below the center.

Vertices: $(3, -5)$ and $(3, 1)$

To sketch the hyperbola, draw a rectangle whose top and bottom pass through the vertices. Because the denominator of the x-term is $b^2 = \left(\frac{3}{2}\right)^2$, locate the sides of the rectangle $\frac{3}{2}$ units to the right and left of the center, as shown in Figure B.18. Finally, sketch the asymptotes by drawing lines through the opposite corners of the rectangle. Using these asymptotes, you can complete the graph of the hyperbola, as shown in Figure B.18.

✔CHECKPOINT   Now try Exercise 45.

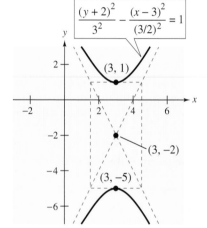

Figure B.18

To find the foci in Example 4, first find c.

$$c^2 = a^2 + b^2$$

$$c^2 = 9 + \frac{9}{4}$$

$$c^2 = \frac{45}{4}$$

$$c = \frac{3\sqrt{5}}{2}$$

Because the transverse axis is vertical, the foci lie c units above and below the center.

Foci: $\left(3, -2 + \dfrac{3\sqrt{5}}{2}\right)$ and $\left(3, -2 - \dfrac{3\sqrt{5}}{2}\right)$

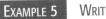

EXAMPLE 5   WRITING THE EQUATION OF AN ELLIPSE

Write the standard form of the equation of the ellipse whose vertices are $(2, -2)$ and $(2, 4)$. The length of the minor axis of the ellipse is 4, as shown in Figure B.19.

SOLUTION

The center of the ellipse lies at the midpoint of its vertices. So, the center is

$$(h, k) = \left(\frac{2 + 2}{2}, \frac{4 + (-2)}{2} \right) = (2, 1).$$ Center is midpoint.

Because the vertices lie on a vertical line and are six units apart, it follows that the major axis is vertical and has a length of $2a = 6$. So, $a = 3$. Moreover, because the minor axis has a length of 4, it follows that $2b = 4$, which implies that $b = 2$. Therefore, you can conclude that the standard form of the equation of the ellipse is as follows.

$$\frac{(x - h)^2}{b^2} + \frac{(y - k)^2}{a^2} = 1$$ Major axis is vertical.

$$\frac{(x - 2)^2}{2^2} + \frac{(y - 1)^2}{3^2} = 1$$ Write in standard form.

✓CHECKPOINT   Now try Exercise 33.

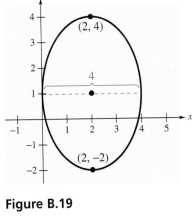

Figure B.19

An interesting application of conic sections involves the orbits of comets in our solar system. Of the 610 comets identified prior to 1970, 245 have elliptical orbits, 295 have parabolic orbits, and 70 have hyperbolic orbits. For example, Halley's comet has an elliptical orbit, and reappearance of this comet can be predicted every 76 years. The center of the sun is a focus of each of these orbits, and each orbit has a vertex at the point where the comet is closest to the sun, as shown in Figure B.20.

If p is the distance between the vertex and the focus (in meters), and v is the speed of the comet at the vertex (in meters per second), then the type of orbit is determined as follows.

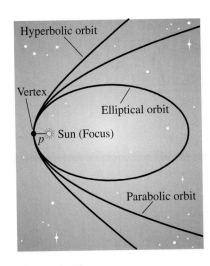

Figure B.20

1. Ellipse: $v < \sqrt{\dfrac{2GM}{p}}$

2. Parabola: $v = \sqrt{\dfrac{2GM}{p}}$

3. Hyperbola: $v > \sqrt{\dfrac{2GM}{p}}$

In each of these equations, $M = 1.989 \times 10^{30}$ kilograms (the mass of the sun) and $G \approx 6.67 \times 10^{-11}$ cubic meter per kilogram-second squared (the universal gravitational constant).

B.2 Warm Up

The following warm-up exercises involve skills that were covered in earlier sections. You will use these skills in the exercise set for this section. For additional help, review Section B.1.

In Exercises 1–10, identify the conic represented by the equation.

1. $\dfrac{x^2}{4} - \dfrac{y^2}{4} = 1$

2. $\dfrac{x^2}{9} + \dfrac{y^2}{1} = 1$

3. $2x + y^2 = 0$

4. $\dfrac{x^2}{9} - \dfrac{y^2}{4} = 1$

5. $\dfrac{x^2}{4} + \dfrac{y^2}{16} = 1$

6. $4x^2 + 4y^2 = 25$

7. $\dfrac{y^2}{4} - \dfrac{x^2}{2} = 1$

8. $x^2 - 6y = 0$

9. $3x - y^2 = 0$

10. $\dfrac{x^2}{9/4} + \dfrac{y^2}{4} = 1$

B.2 Exercises

In Exercises 1–6, describe the translation of the graph of the conic.

1.

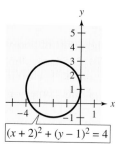

$(x + 2)^2 + (y - 1)^2 = 4$

2.

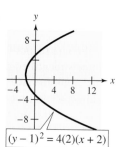

$(y - 1)^2 = 4(2)(x + 2)$

3.

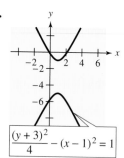

$\dfrac{(y + 3)^2}{4} - (x - 1)^2 = 1$

4.

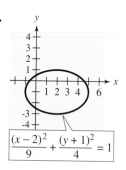

$\dfrac{(x - 2)^2}{9} + \dfrac{(y + 1)^2}{4} = 1$

5.

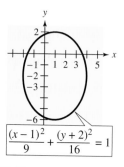

$\dfrac{(x - 1)^2}{9} + \dfrac{(y + 2)^2}{16} = 1$

6.

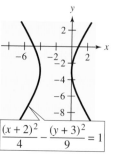

$\dfrac{(x + 2)^2}{4} - \dfrac{(y + 3)^2}{9} = 1$

In Exercises 7–16, find the vertex, focus, and directrix of the parabola and sketch its graph.

7. $(x - 1)^2 + 8(y + 2) = 0$

8. $(x + 3) + (y - 2)^2 = 0$

9. $\left(y + \tfrac{1}{2}\right)^2 = 2(x - 5)$ **10.** $\left(x + \tfrac{1}{2}\right)^2 = 4(y - 3)$

11. $y = \tfrac{1}{4}(x^2 - 2x + 5)$ **12.** $y = -\tfrac{1}{6}(x^2 + 4x - 2)$

13. $4x - y^2 - 2y - 33 = 0$

14. $y^2 + x + y = 0$

15. $y^2 + 6y + 8x + 25 = 0$

16. $x^2 - 2x + 8y + 9 = 0$

In Exercises 17–24, find the standard form of the equation of the parabola.

17. Vertex: $(3, 2)$; focus: $(1, 2)$

18. Vertex: $(-1, 2)$; focus: $(-1, 0)$

19. Vertex: $(0, 4)$; directrix: $y = 2$

20. Vertex: $(-2, 1)$; directrix: $x = 1$

21. Focus: $(2, 2)$; directrix: $x = -2$

22. Focus: $(0, 0)$; directrix: $y = 4$

23. Vertex: $(0, 4)$; passes through $(-2, 0)$ and $(2, 0)$

24. Vertex: $(2, 4)$; passes through $(0, 0)$ and $(4, 0)$

In Exercises 25–32, find the center, foci, and vertices of the ellipse, and sketch its graph.

25. $\dfrac{(x - 1)^2}{9} + \dfrac{(y - 5)^2}{25} = 1$

26. $(x + 2)^2 + \dfrac{(y + 4)^2}{1/4} = 1$

27. $9x^2 + 4y^2 + 36x - 24y + 36 = 0$

28. $9x^2 + 4y^2 - 36x + 8y + 31 = 0$

29. $16x^2 + 25y^2 - 32x + 50y + 16 = 0$

30. $9x^2 + 25y^2 - 36x - 50y + 61 = 0$

31. $12x^2 + 20y^2 - 12x + 40y - 37 = 0$

32. $36x^2 + 9y^2 + 48x - 36y + 43 = 0$

In Exercises 33–40, find the standard form of the equation of the ellipse.

33. Vertices: $(0, 2)$, $(4, 2)$; minor axis of length 2

34. Foci: $(0, 0)$, $(4, 0)$; major axis of length 8

35. Foci: $(0, 0)$, $(0, 8)$; major axis of length 16

36. Center: $(2, -1)$; vertex: $\left(2, \tfrac{1}{2}\right)$;
minor axis of length 2

37. Vertices: $(3, 1)$, $(3, 9)$; minor axis of length 6

38. Center: $(3, 2)$; $a = 3c$; foci: $(1, 2)$, $(5, 2)$

39. Center: $(0, 4)$; $a = 2c$; vertices $(-4, 4)$, $(4, 4)$

40. Vertices: $(5, 0)$, $(5, 12)$; endpoints of the minor axis:
$(0, 6)$, $(10, 6)$

In Exercises 41–50, find the center, vertices, and foci of the hyperbola and sketch its graph, using asymptotes as a sketching aid.

41. $\dfrac{(x - 1)^2}{4} - \dfrac{(y + 2)^2}{1} = 1$

42. $\dfrac{(x + 1)^2}{144} - \dfrac{(y - 4)^2}{25} = 1$

43. $(y + 6)^2 - (x - 2)^2 = 1$

44. $\dfrac{(y - 1)^2}{1/4} - \dfrac{(x + 3)^2}{1/9} = 1$

45. $9x^2 - y^2 - 36x - 6y + 18 = 0$

46. $x^2 - 9y^2 + 36y - 72 = 0$

47. $9y^2 - x^2 + 2x + 54y + 62 = 0$

48. $16y^2 - x^2 + 2x + 64y + 63 = 0$

49. $x^2 - 9y^2 + 2x - 54y - 107 = 0$

50. $9x^2 - y^2 + 54x + 10y + 55 = 0$

In Exercises 51–58, find the standard form of the equation of the hyperbola.

51. Vertices: $(2, 0)$, $(6, 0)$; foci: $(0, 0)$, $(8, 0)$

52. Vertices: $(2, 3)$, $(2, -3)$; foci: $(2, 5)$, $(2, -5)$

53. Vertices: $(4, 1)$, $(4, 9)$; foci: $(4, 0)$, $(4, 10)$

54. Vertices: $(-2, 1)$, $(2, 1)$; foci: $(-3, 1)$, $(3, 1)$

55. Vertices: $(2, 3)$, $(2, -3)$; passes through $(0, 5)$

56. Vertices: $(-2, 1)$, $(2, 1)$; passes through $(4, 3)$

57. Vertices: $(0, 2)$, $(6, 2)$;
asymptotes: $y = \tfrac{2}{3}x$, $y = 4 - \tfrac{2}{3}x$

58. Vertices: $(3, 0)$, $(3, 4)$;
asymptotes: $y = \tfrac{2}{3}x$, $y = 4 - \tfrac{2}{3}x$

In Exercises 59–66, identify the conic by writing the equation in standard form. Then sketch its graph.

59. $x^2 + y^2 - 6x + 4y + 9 = 0$

60. $x^2 + 4y^2 - 6x + 16y + 21 = 0$

61. $4x^2 - y^2 - 4x - 3 = 0$

62. $y^2 - 4y - 4x = 0$

63. $4x^2 + 3y^2 + 8x - 24y + 51 = 0$

64. $4y^2 - 2x^2 - 4y + 8x - 15 = 0$

65. $25x^2 - 10x - 200y - 119 = 0$

66. $4x^2 + 4y^2 - 16y + 15 = 0$

67. Satellite Orbit A satellite in a 100-mile-high circular orbit around Earth has a velocity of approximately 17,500 miles per hour. If this velocity is multiplied by $\sqrt{2}$, then the satellite will have the minimum velocity necessary to escape Earth's gravity, and it will follow a parabolic path with the center of Earth as the focus (see figure).

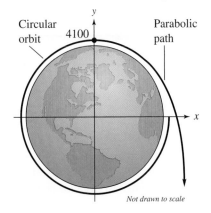

Not drawn to scale

(a) Find the escape velocity of the satellite.

(b) Find an equation of its path (assume the radius of Earth is 4000 miles).

68. Fluid Flow   Water is flowing from a horizontal pipe 48 feet above the ground. The falling stream of water has the shape of a parabola whose vertex $(0, 48)$ is at the end of the pipe (see figure). The stream of water strikes the ground at the point $\left(10\sqrt{3}, 0\right)$. Find the equation of the path taken by the water.

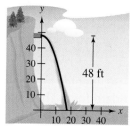

In Exercises 69–75 and 77, e is called the **eccentricity** of an ellipse and is defined by $e = c/a$. It measures the flatness of the ellipse.

69. Find an equation of the ellipse with vertices $(\pm 5, 0)$ and eccentricity $e = \frac{3}{5}$.

70. Find an equation of the ellipse with vertices $(0, \pm 8)$ and eccentricity $e = \frac{1}{2}$.

71. Planetary Motion   Earth moves in an elliptical orbit with the sun at one of the foci (see figure). The length of half of the major axis is $a = 92.956 \times 10^6$ miles and the eccentricity is 0.017. Find the shortest distance (perihelion) and the greatest distance (aphelion) between Earth and the sun.

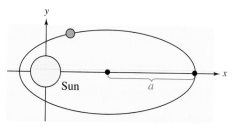

Figure for 71–73

72. Planetary Motion   Pluto moves in an elliptical orbit with the sun at one of the foci (see figure). The length of half of the major axis is 3.670×10^9 miles and the eccentricity is 0.249. Find the shortest distance and the greatest distance between Pluto and the sun.

73. Planetary Motion   Saturn moves in an elliptical orbit with the sun at one of the foci (see figure). The shortest distance and the greatest distance between Saturn and the sun are 1.3495×10^9 kilometers and 1.5040×10^9 kilometers, respectively. Find the eccentricity of the orbit.

74. Satellite Orbit   The first artificial satellite to orbit Earth was Sputnik I (launched by the former Soviet Union in 1957). Its highest point above Earth's surface was 588 miles, and its lowest point was 142 miles (see figure). Assume that the center of Earth is the focus of the elliptical orbit and that the radius of Earth is 4000 miles. Find the eccentricity of the orbit.

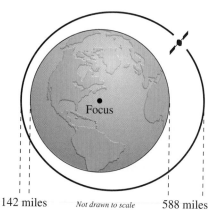

142 miles   *Not drawn to scale*   588 miles

75. Show that the equation of an ellipse can be written as

$$\frac{(x - h)^2}{a^2} + \frac{(y - k)^2}{a^2(1 - e^2)} = 1.$$

Note that as e approaches zero, the ellipse approaches a circle of radius a.

76. Australian Football   In Australia, football by *Australian Rules* (or rugby) is played on elliptical fields. The fields can be a maximum of 170 yards wide and a maximum of 200 yards long. Let the center of a field of maximum size be represented by the point $(0, 85)$. Find the standard form of the equation of the ellipse that represents this field. (Source: Australian Football League)

77. Comet Orbit   Halley's comet has an elliptical orbit with the sun at one focus. The eccentricity of the orbit is approximately 0.97. The length of the major axis of the orbit is approximately 35.88 astronomical units. (An astronomical unit is about 93 million miles.) Find the standard form of the equation of the orbit. Place the center of the orbit at the origin and place the major axis on the x-axis.

78. Comet Orbit   The comet Encke has an elliptical orbit with the sun at one focus. Encke's path ranges from 0.34 to 4.08 astronomical units from the sun. Find the standard form of the equation of the orbit. Place the center of the orbit at the origin and place the major axis on the x-axis.

Answers to Warm Ups, Odd-Numbered Exercises, Quizzes, and Tests

Chapter R1

Section R1.1   *(page R8)*

1. (a) Natural: $\{5\}$

   (b) Integer: $\{-9, 5\}$

   (c) Rational: $\{-9, -\frac{7}{2}, 5, \frac{2}{3}, 0.1\}$

   (d) Irrational: $\{\sqrt{2}\}$

3. (a) Natural: $\{12, 1, \sqrt{4}\}$ *(Note:* $\sqrt{4} = 2$)

   (b) Integer: $\{12, -13, 1, \sqrt{4}\}$

   (c) Rational: $\{12, -13, 1, \sqrt{4}, \frac{3}{2}\}$

   (d) Irrational: $\{\sqrt{6}\}$

5. (a) Natural: $\{\frac{8}{2}, 9\}$ *(Note:* $\frac{8}{2} = 4$)

   (b) Integer: $\{\frac{8}{2}, -4, 9\}$

   (c) Rational: $\{\frac{8}{2}, -\frac{8}{3}, -4, 9, 14.2\}$

   (d) Irrational: $\{\sqrt{10}\}$

7. 0.625 9. $0.\overline{54}$ 11. $-1 < 2.5$

13. $\frac{3}{2} < 7$

15. $-3.5 < 1$

17. $\frac{5}{6} > \frac{2}{3}$

19. $\frac{127}{90}, \frac{584}{413}, \frac{7071}{5000}, \sqrt{2}, \frac{47}{33}$

21. $x < 0$ denotes all negative real numbers.

23. $x \le 5$ denotes all real numbers less than or equal to 5.

25. $x > 3$ denotes all real numbers greater than 3.

27. $-2 < x < 2$ denotes all real numbers greater than -2 and less than 2.

29. $-1 \le x < 0$ denotes all real numbers greater than or equal to -1 and less than 0.

31. $x < 0$ 33. $A \ge 35$ years   35. $3.5\% \le r \le 6\%$

37. 10 39. -6 41. -9 43. -1 45. $\pi - 3$

47. $|-7| = |7|$ 49. $|-3| > -|-3|$

51. $-|-2| = -|2|$ 53. 4 55. $\frac{5}{2}$ 57. $\frac{7}{2}$

59. 51 61. $\frac{128}{75}$ 63. $\left|z - \frac{3}{2}\right| > 1$

65. $|x + 10| \ge 6$ 67. $|y - 0| \ge 6 \Rightarrow |y| \ge 6$

69. 179 miles   71. $37°$

	$\|a - b\|$	$0.05b$	Passes Budget Variance Test
73.	$\$127.88$	$\$1250$	Yes
75.	$\$572.59$	$\$470$	No
77.	$\$671.75$	$\$1882$	No

	$\|a - b\|$	$0.002b$	Passes Quality Control Test
79.	0.02	0.03	Yes
81.	0.045	0.033	No
83.	0.035	0.036	No

85. (a) No. If $u > 0$ and $v < 0$ or $u < 0$ and $v > 0$, then $|u + v| \ne |u| + |v|$.

   (b) Yes. If the signs of u and v are different, then $|u + v| < |u| + |v|$.

87. Answers will vary. Example: The set of natural numbers includes only the integers greater than zero. The set of integers include all numbers that have no fractional or decimal parts. The set of rational numbers includes all numbers that can be written as the quotient of two integers. Any real number that is not a rational number is in the set of irrational numbers.

Section R1.2   *(page R18)*

Warm Up   *(page R18)*

1. $-4 < -2$ 2. $0 > -3$ 3. $\sqrt{3} > 1.73$

4. $-\pi < -3$ 5. $|6 - 4| = 2$ 6. $|2 - (-2)| = 4$

7. $|0 - (-5)| = 5$ 8. $|3 - (-1)| = 4$

9. $|-7| + |7| = 7 + 7 = 14$

10. $-|8 - 10| = -|-2| = -2$

1. $7x, 4$ 3. $x^2, -4x, 8$ 5. $2x^2, -9x, 13$ 7. -6

9. 6 11. (a) -10 (b) -6 13. (a) 14 (b) 2

15. (a) 0 (b) 0

17. (a) Undefined. You cannot divide by zero.   (b) $\frac{1}{2}$

19. Commutative (addition)   21. Inverse (addition)

23. Distributive Property   **25.** Inverse (multiplication)

27. Identity (addition)   **29.** Identity (multiplication)

31. Associative (addition)

33. $x(3y) = (x \cdot 3)y$ Associative (multiplication)
$= (3x)y$ Commutative (multiplication)

35. $2^2 \cdot 3^2$ **37.** $2^3 \cdot 5^2$ **39.** -14 **41.** $\frac{1}{24}$

43. $\frac{7}{20}$ **45.** $\frac{1}{12}$ **47.** -0.13 **49.** 1.56 **51.** 18.81

53. 41.14 **55.** 16.6% **57.** ≈ 2 meals

59. (a) 34.6%

(b) Social Security: $\approx \$456$ billion

Veteran's Benefits: $\approx \$51$ billion

Education: $\approx \$71$ billion

Health: $\approx \$196$ billion

Medicare: $\approx \$231$ billion

Income Security: $\approx \$312$ billion

(c) $\approx \$260$ billion; Answers will vary. Example: No. Because the total human resources expenses will change, the amounts of money spent on Social Security and healthcare will change, and so the difference between these amounts will change.

61. ≈ 5237 students

63. (a) Scientific: 5 $\boxed{\times}$ $\boxed{(}$ 18 $\boxed{-}$ 2 $\boxed{y^x}$ 3 $\boxed{)}$
$\boxed{\div}$ 10 $\boxed{=}$

Graphing: 5 $\boxed{\times}$ $\boxed{(}$ 18 $\boxed{-}$ 2 $\boxed{\wedge}$ 3 $\boxed{)}$
$\boxed{\div}$ 10 $\boxed{\text{ENTER}}$

(b) Scientific: 6 $\boxed{x^2}$ $\boxed{+/-}$ $\boxed{-}$ $\boxed{(}$ 7 $\boxed{+}$ $\boxed{(}$ 2 $\boxed{+/-}$ $\boxed{)}$
$\boxed{y^x}$ 3 $\boxed{)}$ $\boxed{=}$

Graphing: $\boxed{(-)}$ 6 $\boxed{x^2}$ $\boxed{-}$ $\boxed{(}$ 7 $\boxed{+}$ $\boxed{(}$ $\boxed{-}$ 2 $\boxed{)}$ $\boxed{)}$ $\boxed{\wedge}$
3 $\boxed{)}$ $\boxed{\text{ENTER}}$

Section R1.3   *(page R27)*

Warm Up   *(page R27)*

1. 1   **2.** 5   **3.** 4   **4.** 4   **5.** $\frac{1}{4}$ **6.** 1

7. $\frac{3}{7}$ **8.** 0   **9.** $-\frac{1}{8}$ **10.** 1

1. 64   **3.** 8   **5.** 729   **7.** -81 **9.** $\frac{1}{2}$ **11.** 8

13. $-\frac{3}{10}$ **15.** 5184   **17.** $-\frac{3}{5}$ **19.** 1   **21.** 18

23. $\frac{7}{16}$ **25.** $-125z^3$ **27.** $16x^7$ **29.** $10x^4$

31. $-3z^7$ **33.** $\frac{5y^4}{2}$ **35.** $\frac{5184}{y^7}$ **37.** $\frac{7}{x}$ **39.** $\frac{1}{x}$

41. 3^{3n} **43.** $1, x \ne 0$ **45.** $\dfrac{1}{(y+2)^3}$ **47.** $32y^2$

49. $\dfrac{10}{x}$ **51.** $\dfrac{125x^9}{y^{12}}$ **53.** 5.73×10^7 square miles

55. 9.461×10^{12} kilometers   **57.** $350,000,000$ air sacs

59. $0.00000000000000000001602$ coulomb

61. 1×10^{18} attoseconds

63. (a) 6.0×10^4 (b) 2.0×10^{11}

65. (a) 3.071×10^6 (b) 3.077×10^{10}

67. (a) 4.907×10^{17} (b) 1.479

69. (a) $(5.1 - 3.6)^5$ (b) $[1 + 3(2)]^{-2}$

71. (a) $\$19,154.30$ (b) $\$19,147.63$
(c) $\$19,121.84$ (d) $\$19,055.59$

As the number of compoundings per year increases, the balance in the account also increases.

73. $\approx 4.46\%$

Section R1.4   *(page R36)*

Warm Up   *(page R36)*

1. $\frac{4}{27}$ **2.** 48   **3.** $-8x^3$ **4.** $6x^7$

5. $28x^6$ **6.** $\frac{1}{5}x^2$ **7.** $3z^4$ **8.** $\dfrac{25}{4x^2}$ **9.** 1

10. $(x + 2)^{10}$

1. $9^{1/2} = 3$ **3.** $\sqrt[5]{32} = 2$ **5.** $\sqrt{196} = 14$

7. $(-216)^{1/3} = -6$ **9.** $81^{3/4} = 27$ **11.** $\sqrt[3]{27^2} = 9$

13. 3   **15.** 3   **17.** $\frac{1}{2}$ **19.** -125 **21.** 4

23. 216   **25.** $\sqrt{6}$ **27.** $\frac{27}{8}$ **29.** -4

31. $2x\sqrt[3]{2x^2}$ **33.** $\dfrac{5|x|\sqrt{3}}{y^2}$ **35.** $\dfrac{2\sqrt[5]{2}}{y}$ **37.** $\dfrac{\sqrt{3}}{3}$

39. $4\sqrt[3]{4}$ **41.** $\dfrac{x(5 + \sqrt{3})}{11}$ **43.** $3(\sqrt{6} - \sqrt{5})$

45. 25   **47.** $2^{1/2}$ **49.** $x^{3/2}, x \ne 0$ **51.** $2\sqrt[4]{2}$

53. $3^{1/2} = \sqrt{3}$ **55.** $\sqrt[3]{x}$ **57.** $2\sqrt{x}$

59. $31\sqrt{2}$ **61.** $-2\sqrt{y}$ **63.** 3.557 **65.** 2.006

67. 2.938 **69.** 0.382 **71.** $\sqrt{5} + \sqrt{3} > \sqrt{5 + 3}$

73. $5 > \sqrt{3^2 + 2^2}$ **75.** $\sqrt{3} \cdot \sqrt[4]{3} > \sqrt[8]{3}$

77. 24 inches $\times$ 24 inches $\times$ 24 inches   **79.** $\approx 12.8\%$

81. No. The escape velocity is equal to approximately 2375 meters per second, which is greater than the velocity of the rocket.

83. ≈ 1.57 seconds   **85.** ≈ 0.026 inch

87. ≈ 494 vibrations per second

89. a; Higher notes have higher frequencies.   **91.** 1

93. $a^0 = a^{n-n} = \dfrac{a^n}{a^n} = 1$ **95.** Answers will vary.

Mid-Chapter Quiz *(page R39)*

1. $-|-7| < |7|$ **2.** $-(-3) = |-3|$

3. $x \geq 0$ **4.** $r \geq 96.5\%$

5. $0 \leq x < 3$ denotes all numbers greater than or equal to 0 and less than 3.

6. $3x^2, -7x, 2$ **7.** 3   **8.** -4 **9.** $\frac{5}{14}$ **10.** $\frac{11}{9}$

11. $-2x^7$ **12.** $\frac{1}{3}y^4$ **13.** $\frac{27x^6}{y^6}$ **14.** \$2219.64

15. -1 **16.** -64 **17.** 9   **18.** $-\sqrt[3]{3}$

19. $2\sqrt{3}$ **20.** 26 cm × 26 cm × 26 cm

Section R1.5 *(page R46)*

Warm Up *(page R46)*

1. $42x^3$ **2.** $-20z^2$ **3.** $-27x^6$ **4.** $-3x^6$

5. $\frac{9}{4}z^3, z \neq 0$ **6.** $4\sqrt{3}$ **7.** $\frac{9}{4x^2}$ **8.** 8

9. $\sqrt{2}$ **10.** $-3x$

1. Degree: 2
Leading coefficient: 2

3. Degree: 5
Leading coefficient: 1

5. Degree: 5
Leading coefficient: 3

7. Polynomial, $-3x^3 + 2x + 8$, degree 3

9. Not a polynomial

11. Polynomial, $-w^4 + 2w^3 + w^2$, degree 4

13. (a) -1 (b) 3   (c) 7   (d) 11

15. (a) -10 (b) -1 (c) 4   (d) 5

17. $-2x - 10$ **19.** $2x^2 - 4x - 5$

21. $8x^3 + 29x^2 + 11$ **23.** $3x^3 - 6x^2 + 3x$

25. $4x^4 - 12x$ **27.** $30x^3 + 12x^2$ **29.** $x^2 + 7x + 12$

31. $6x^2 - 7x - 5$ **33.** $x^2 - 25$ **35.** $x^2 + 12x + 36$

37. $4x^2 - 20xy + 25y^2$

39. $x^2 + 2xy + y^2 - 6x - 6y + 9$

41. $x^3 + 3x^2 + 3x + 1$ **43.** $8x^3 - 12x^2y + 6xy^2 - y^3$

45. $9y^4 - 1$ **47.** $m^2 - n^2 - 6m + 9$ **49.** $x - y$

51. $x^4 + x^2 + 1$ **53.** $2x^2 + 2x$

55. $500r^2 + 1000r + 500$ **57.** No

59. 9860.37; 11,399.88; In the years 2001 and 2002, the total amounts of Federal Pell Grants were approximately \$9,860,370,000 and \$11,399,880,000, respectively.

61. Volume $= 4x^3 - 184x^2 + 2052x$
$x = 5$ inches: $V = 6160$ cubic inches
$x = 7$ inches: $V = 6720$ cubic inches
$x = 9$ inches: $V = 6480$ cubic inches
$x = 7$ inches produces the greatest volume.

63. $3x^2 + 7x$

Section R1.6 *(page R53)*

Warm Up *(page R53)*

1. $15x^2 - 6x$ **2.** $-2y^2 - 2y$ **3.** $4x^2 + 12x + 9$

4. $9x^2 - 48x + 64$ **5.** $2x^2 + 13x - 24$

6. $-5z^2 - z + 4$ **7.** $4y^2 - 1$ **8.** $x^2 - a^2$

9. $x^3 + 12x^2 + 48x + 64$

10. $8x^3 - 36x^2 + 54x - 27$

1. $3(x + 2)$ **3.** $4x(x^2 - 2)$ **5.** $(x - 1)(x + 5)$

7. $(x + 6)(x - 6)$ **9.** $(4x + 3y)(4x - 3y)$

11. $(x + 1)(x - 3)$ **13.** $(x - 2)^2$ **15.** $(2y + 3)^2$

17. $(5y - 1)^2$ **19.** $(x - 2)(x^2 + 2x + 4)$

21. $(y + 5)(y^2 - 5y + 25)$ **23.** $(2t - 1)(4t^2 + 2t + 1)$

25. $(x + 2)(x - 1)$ **27.** $(w - 2)(w - 3)$

29. $(y + 5)(y - 4)$ **31.** $(x - 20)(x - 10)$

33. $(3x - 2)(x - 1)$ **35.** $(3x + 1)(3x - 2)$

37. $(5x + 1)(x + 5)$ **39.** $(x - 1)(x^2 + 2)$

41. $(2x - 1)(x^2 - 3)$ **43.** $(2 - y^3)(3 + y)$

45. $4x(x - 2)$ **47.** $y(y - 3)(y + 3)$ **49.** $x^2(x - 4)$

51. $(x - 1)^2$ **53.** $(1 - 2x)^2$ **55.** $y(2y + 3)(y - 5)$

57. $2x(x - 2)(x + 1)$ **59.** $(3x + 1)(x^2 + 5)$

61. $x(x - 4)(x^2 + 1)$ **63.** $-x(x + 10)$

65. $(x + 1)^2(x - 1)^2$ **67.** $2(t - 2)(t^2 + 2t + 4)$

69.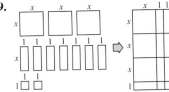

71.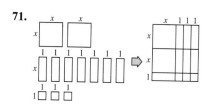

73. $(2x + 1)$ feet

75. (a) $(x - 2)(3x + 4)$　(b) $(x + 5)(x + 6)$

77. $c = \{9, 16, 21, 24, 25\}$; Answers will vary.

79. Answers will vary.

81.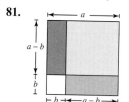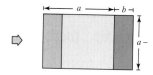

Section R1.7　*(page R60)*

1. All real numbers　　**3.** All real numbers except $x = 2$

5. All real numbers except $x = 0$ and $x = 4$

7. All real numbers greater than or equal to -1

9. $3x$　　**11.** $x - 2$, $x \neq 2$　　**13.** $x + 2$, $x \neq -2$

15. $\dfrac{3x}{2}$, $x \neq 0$　　**17.** $\dfrac{x}{2(x + 1)}$　　**19.** $-\dfrac{1}{2}$, $x \neq 5$

21. $-(x + 5)$, $x \neq 5$　　**23.** $\dfrac{x(x + 3)}{x - 2}$, $x \neq -2$

25. $\dfrac{y - 4}{y + 6}$, $y \neq 3$　　**27.** $-1 - x^2$, $x \neq 2$　　**29.** $z - 2$

31. $\dfrac{1}{5(x - 2)}$, $x \neq 1$　　**33.** $-\dfrac{x(x + 7)}{x + 1}$, $x \neq 9$

35. $\dfrac{r + 1}{r}$, $r \neq 1$　　**37.** $\dfrac{t - 3}{(t + 3)(t - 2)}$, $t \neq -2$

39. $\dfrac{x - 1}{x(x + 1)^2}$, $x \neq -2$　　**41.** $\dfrac{3}{2}$, $x \neq -y$

43. $x(x + 1)$, $x \neq -1, 0$　　**45.** $\dfrac{x + 5}{x - 1}$　　**47.** $\dfrac{2x + 5}{x - 5}$

49. $\dfrac{6x + 13}{x + 3}$　**51.** $\dfrac{x - 4}{(x + 2)(x - 2)(x - 1)}$

53. $\dfrac{2 - x}{x^2 + 1}$, $x \neq 0$　　**55.** $\dfrac{1}{2}$, $x \neq 2$　　**57.** $\dfrac{1}{x}$, $x \neq -1$

59. $\dfrac{2x - 1}{2x}$, $x > 0$　　**61.** (a) 12.65%　(b) $\dfrac{288(NM - P)}{N(12P + NM)}$

63. 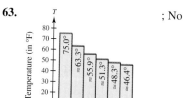 ; No

Review Exercises　*(page R64)*

1. (a) Natural: $\{11\}$

(b) Integer: $\{11, -14\}$

(c) Rational: $\left\{11, -14, -\dfrac{8}{9}, \dfrac{5}{2}, 0.4\right\}$

(d) Irrational: $\left\{\sqrt{6}\right\}$

3. $-4 < -3$

5. $x \leq -6$ denotes all real numbers less than or equal to -6.

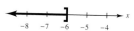

7. $x \geq 0$　　**9.** -14　　**11.** $|-12| > -|12|$　　**13.** 3

15. $|x - 7| \geq 4$　　**17.** $2x^2, -3x, 4$　　**19.** (a) 2　(b) 0

21. Distributive Property　　**23.** -1　　**25.** $\dfrac{2}{3}$　　**27.** 5

29. 0.10　　**31.** -2　　**33.** $8x$　　**35.** 3.5116×10^7

37. $483,400,000$ miles

39. (a) $11,414.125$　(b) $18,380.160$

41.

Year	5	10	15
Balance	\$2697.70	\$3638.79	\$4908.19

Year	20	25
Balance	\$6620.41	\$8929.94

43. $16^{1/2} = 4$　　**45.** 13　　**47.** $2x^2$　　**49.** $2 + \sqrt{3}$

51. $-3\sqrt{x}$　　**53.** $\sqrt{10}$　　**55.** $\sqrt{5}$　　**57.** 11.269

59. $-6x + 26$　　**61.** $x^2 - 8x - 14$　　**63.** $x^2 - x - 2$

65. $x^3 - x^2 + 2x - 2$

67. 260.74

In 2002, the median sales price for a new one-family home in the northeastern United States was $260,740.

69. $x = 5$: 27,132.5

$x = 12$: 139,620.4

There were 27,132,500 and 139,620,400 cell phone subscribers in the United States in 1995 and 2002, respectively.

71. $4(x + 3)(x - 3)$ **73.** $3x(x + 1)(x - 2)$

75. $(x^2 - 2)(x - 4)$ **77.** All real numbers except $x = 3$

79. All real numbers   **81.** $3x$

83. $\dfrac{x - 2}{2}, x \ne -2$ **85.** $x + 2, x \ne 0$

87. $\dfrac{x - 1}{x - 3}, x \ne -1, \dfrac{1}{2}$ **89.** $\dfrac{3x^2 - 4x}{(x - 1)(x - 2)}$

91. $\dfrac{x + 1}{x - 1}, x \ne 0$ **93.** $\dfrac{4x}{2x - 3}, x \ne 0, -\dfrac{3}{2}$

Chapter Test (page R67)

1. -12

2.

Year	5	10	15
Balance	$4470	$6659	$9921

Year	20	25
Balance	$14,780	$22,021

The longer you leave the $3000 in the account, the more money you earn.

3. $-64x^6$ **4.** $-4\sqrt{x}$ **5.** 25   **6.** $4(\sqrt{3} - \sqrt{5})$

7. $2x\sqrt{3x}$ **8.** $\dfrac{5 + \sqrt{7}}{9}$ **9.** $9x^2 + 42x + 49$

10. $-5x^2 + 29x$ **11.** $5(x + 4)(x - 4)$

12. $(2x + 3)^2$ **13.** $(x^2 - 3)(x - 6)$

14. $(x + 2)^2(x - 2)$ **15.** $\frac{1}{3}(x - 4), x \ne -4$

16. $\dfrac{x + 4}{3x + 5}, x \ne -3, \dfrac{5}{3}$ **17.** $\dfrac{4x^2 - 13x}{(x - 3)(x - 4)}$

18. $-\dfrac{x + 26}{(x + 5)(x - 2)}$

19. All real numbers greater than or equal to -3

20. All real numbers except $x = 2$

21. $\dfrac{2x^2 - 5x - 18}{5 + 5x - x^2}, x \ne 1, -2$

22. $x = 5$: 236.75, $x = 12$: 294.36

In 1995 and 2002, the total expenditures for U.S. colleges and universities were about $236.75 billion and $294.36 billion, respectively.

Chapter R2

Section R2.1 (page R76)

Warm Up (page R76)

1. $-3x - 10$ **2.** $5x - 12$ **3.** x **4.** $x + 26$

5. $\dfrac{8x}{15}$ **6.** $\dfrac{3x}{4}$ **7.** $-\dfrac{1}{x(x + 1)}$ **8.** $\dfrac{5}{x}$

9. $\dfrac{7x - 8}{x(x - 2)}$ **10.** $-\dfrac{2}{x^2 - 1}$

1. Identity   **3.** Conditional equation

5. Conditional equation

7. (a) No   (b) No   (c) Yes   (d) No

9. (a) Yes   (b) Yes   (c) No   (d) No

11. (a) Yes   (b) No   (c) No   (d) No

13. (a) No   (b) No   (c) No   (d) Yes

15. (a) No   (b) No   (c) No   (d) No

17. 5   **19.** -4 **21.** 3   **23.** 9   **25.** -26

27. -4 **29.** $-\frac{6}{5}$ **31.** 9   **33.** No solution

35. 10   **37.** 4   **39.** 3   **41.** 5   **43.** No solution

45. $\frac{11}{6}$ **47.** No solution   **49.** 0

51. All real numbers   **53.** No solution

55. Because substituting 2 for x in the equation produces division by zero, $x = 2$ cannot be a solution to the equation.

57. Check by substituting in the original equation, by using the *table* feature of a graphing utility, or by evaluating the solution in the original equation using a graphing utility.

59. $x \approx 138.889$ **61.** $x \approx 62.372$ **63.** $x \approx 19.993$

65. Use the *table* feature in ASK mode or use the scientific calculator part of the graphing utility.

67. (a) 6.46   (b) 6.41; Yes

69. (a) 56.09   (b) 56.13; Yes

71. 2004 ($t \approx 14.02$)   **73.** 61.2 inches

75. 1998 ($t \approx 8.35$)   **77.** 1999 ($t \approx 9.29$)

Section R2.2 (page R87)

Warm Up (page R87)

1. 14   **2.** 4   **3.** -3 **4.** 4   **5.** -2

6. 1   **7.** $\frac{2}{5}$ **8.** $\frac{10}{3}$ **9.** 6   **10.** $-\frac{11}{5}$

1. $x + (x + 1) = 2x + 1$ **3.** $50t$ **5.** $0.2x$ **7.** $6x$

9. $1200 + 25x$ **11.** $525 = n + (n + 1)$; 262, 263

13. $5x - x = 148$; 37, 185

15. $n^2 - 5 = n(n + 1)$; $-5, -4$

17. Coworker's check: $300

Your check: $345

19. Coworker's check: $348.65

Your check: $296.35

21. $\approx 37.03\%$ decrease   **23.** $\approx 39.42\%$ increase

25. (a) ≈ 498.96 million users

(b) ≈ 543.87 million users

(c) ≈ 580.85 million users

27. Two TVs: ≈ 36.278 million households

Three or more TVs: ≈ 43.747 million households

VCR: ≈ 97.097 million households

Basic cable: ≈ 74.690 million households

Premium cable: ≈ 51.216 million households

29. 15 feet $\times$ 22.5 feet   **31.** ≈ 5.7 years

33. 97 or greater   **35.** $22,316.98

37. $1411.76   **39.** $\approx 20.13\%$ **41.** $361.25

43. 3 hours   **45.** $\frac{1}{3}$ hour

47. Family 1 (42 miles per hour): ≈ 3.8 hours

Family 2 (50 miles per hour): 3.2 hours

49. ≈ 1.28 seconds   **51.** 62.5 feet   **53.** $563,952

55. $4500 at 5.5%

$7500 at 7%   **57.** 11.43%

59. 8823 units per month   **61.** ≈ 48 feet

63. ≈ 32.1 gallons   **65.** ≈ 12.31 miles per hour

67. $h = \dfrac{2A}{b}$ **69.** $l = \dfrac{V}{wh}$ **71.** $C = \dfrac{S}{1 + R}$

73. $r = \dfrac{A - P}{Pt}$ **75.** $b = \dfrac{2A - ah}{h}$

77. $n = \dfrac{L + d - a}{d}$ **79.** $h = \dfrac{A}{2\pi r}$

81. A mathematical model should be accurate and easy to use. If the model is very complicated, the user may make errors or choose not to use the model. So, sometimes a reasonably accurate model that is easy to use is better than a very complicated model that is more accurate.

83. ≈ 192.27 cubic inches

Section R2.3   *(page R100)*

Warm Up   *(page R100)*

1. $\dfrac{\sqrt{14}}{10}$ **2.** $4\sqrt{2}$ **3.** 14   **4.** $\dfrac{\sqrt{10}}{4}$

5. $x(3x + 7)$ **6.** $(2x - 5)(2x + 5)$

7. $-(x - 7)(x - 15)$ **8.** $(x - 2)(x + 9)$

9. $(5x - 1)(2x + 3)$ **10.** $(6x - 1)(x - 12)$

---

1. $2x^2 + 5x - 3 = 0$ **3.** $x^2 - 25x = 0$

5. $x^2 - 6x + 7 = 0$ **7.** $2x^2 - 2x + 1 = 0$

9. $3x^2 - 60x - 10 = 0$ **11.** $4, -2$ **13.** $0, -\frac{1}{2}$

15. -5 **17.** $3, -\frac{1}{2}$ **19.** $2, -6$ **21.** $-2, -5$

23. ± 4 **25.** $\pm\sqrt{7} \approx \pm 2.65$ **27.** $\pm 2\sqrt{3} \approx \pm 3.46$

29. $12 + 3\sqrt{2} \approx 16.24$ **31.** $-2 + 2\sqrt{3} \approx 1.46$

$12 - 3\sqrt{2} \approx 7.76$ $-2 - 2\sqrt{3} \approx -5.46$

33. ± 5 **35.** $\pm\dfrac{\sqrt{115}}{5} \approx \pm 2.14$ **37.** ± 8 **39.** 1

41. $\pm\frac{3}{4}$ **43.** $\frac{3}{2}$ **45.** $6, -12$ **47.** $\frac{3}{2}, -\frac{1}{2}$ **49.** $5, -\frac{10}{3}$

51. 9, 3   **53.** $\frac{1}{5}, 1$ **55.** $-1, -5$ **57.** $-\frac{1}{2}$

59. Algebra argument:

$(x + 2)^2 = (x + 2)(x + 2)$ — Definition of exponent

$= x^2 + 2x + 2x + 4$ — FOIL

$= x^2 + 4x + 4$ — Combine like terms.

So, $(x + 2)^2 \ne x^2 + 4$.

Graphing utility argument:

(1) Let $y_1 = (x + 2)^2$ and $y_2 = x^2 + 4$. Use the *table* feature with an arbitrary value of x (but not $x = 0$). The table will show that y_1 is not the same as y_2.

(2) Use the scientific calculator portion of the graphing utility to show that if $x = 5$, $(5 + 2)^2 = 49$ and $5^2 + 4 = 29$. So, $(x + 2)^2$ is not equal to $x^2 + 4$.

61. 34 feet $\times$ 48 feet

63. Base: $2\sqrt{2}$ feet

Height: $2\sqrt{2}$ feet

65. ≈ 3.54 seconds   **67.** ≈ 1.43 seconds

69. ≈ 24.37 seconds   **71.** ≈ 3.54 centimeters

73. 976 miles   **75.** ≈ 1414 feet

77. 50,000 units   **79.** 2015 ($t \approx 15.4$)

81. 1987 ($t \approx 18.74$); The model is a good representation through 2002.

83. The model in Exercise 82 is *not* valid for the population in 2050 because it predicts 536,526,000 people (not 419,854,000).

85. 2002 ($t \approx 11.62$); No; for 1980 ($t = -10$) the model yields a value of $E = 865$, or 865,000 students. From 1995–2001, the enrollment steadily increased from 708,000 to 897,970 students. It is unlikely that in 1980 the enrollment was significantly higher than it was 15 years later in 1995.

Section R2.4   *(page R110)*

Warm Up   *(page R110)*

1. $3\sqrt{17}$ **2.** $2\sqrt{3}$ **3.** $4\sqrt{6}$ **4.** $3\sqrt{73}$

5. $2, -1$ **6.** $\frac{3}{2}, -3$ **7.** $5, -1$ **8.** $\frac{1}{2}, -7$

9. $3, 2$ **10.** $4, -1$

1. One real solution   **3.** Two real solutions

5. No real solutions   **7.** Two real solutions

9. $\frac{1}{2}, -1$ **11.** $\frac{1}{4}, -\frac{3}{4}$ **13.** $1 \pm \sqrt{3}$

15. $-7 \pm \sqrt{5}$ **17.** $-4 \pm 2\sqrt{5}$ **19.** $\frac{2}{3} \pm \frac{\sqrt{7}}{3}$

21. $-\frac{1}{3} \pm \frac{\sqrt{11}}{6}$ **23.** $-\frac{1}{2} \pm \sqrt{2}$ **25.** $\frac{2}{7}$

27. $2 \pm \frac{\sqrt{6}}{2}$ **29.** $6 \pm \sqrt{11}$ **31.** $x \approx 0.976, -0.643$

33. $x \approx 0.561, 0.126$ **35.** $x \approx 1.687, -0.488$

37. -11 **39.** $\pm\sqrt{10}$ **41.** $-\frac{3}{2} \pm \frac{\sqrt{5}}{2}$ **43.** $-2, 4$

45. ± 2 **47.** $50, 50$ **49.** $7, 8$ or $-8, -7$

51. 200 units   **53.** 653 units   **55.** 9 seats per row

57. 14 inches $\times$ 14 inches

59. Moon: ≈ 14.9 seconds
Earth: ≈ 2.6 seconds

61. Shorter period of time on Earth

63. ≈ 259 miles; ≈ 541 miles

65. (a) 1999 ($t \approx 9.28$)

(b) 2005 ($t \approx 14.85$)

(c) No. The model's prediction of $899.66 exceeds the expected consumer spending.

67. 2010 ($t = 10$); Yes. The model in Exercise 66 predicts a spending amount of $378 billion, which is close to the industry's projection of $374 billion.

69. Southbound: ≈ 550 miles per hour

Eastbound: ≈ 600 miles per hour

71. 3761 units or 146,239 units

73. In an application, one of the solutions may not make sense in the context. In Example 5, the other possible solution is $t \approx -29.77$. Because the number of alternative fuel vehicles has been steadily growing since 1993, it is not likely that there were 1,000,000 of them in 1960. So, this solution ($t \approx -29.77$) can be rejected.

Mid-Chapter Quiz   *(page R114)*

1. $x = -6$ **2.** $x = 6$ **3.** $x = -2$ **4.** No solution

5. 328.954   **6.** 431.398

7. Use the table feature in ASK mode or the scientific calculator portion of the graphing utility.

8. $7.50x + 20{,}000 = 80{,}000$; 8000 units

9. 3499 units ($x \approx 3499.214$) or 321,501 units
($x \approx 321{,}500.786$)

10. $x = \frac{2}{3}, -5$ **11.** $x = \pm\sqrt{5}; x \approx \pm 2.24$

12. $x = -3 \pm \sqrt{17}; x \approx -7.12, 1.12$

13. $x = -1 \pm \sqrt{6}$ **14.** $x = \dfrac{-7 \pm \sqrt{73}}{6}$

15. $x \approx 1.568, -0.068$ **16.** No real solutions

17. One real solution

18. Answers will vary. Sample answer: Use the FOIL method $[(x + 3)^2 = (x + 3)(x + 3) = x^2 + 6x + 9]$, use the *table* feature of your graphing utility, or use the scientific calculator portion of your graphing utility to evaluate the solution.

19. ≈ 3.95 seconds   **20.** 6 inches $\times$ 6 inches

Section R2.5   *(page R123)*

Warm Up   *(page R123)*

1. 11   **2.** $20, -3$ **3.** $5, -45$ **4.** $0, -\frac{1}{5}$

5. $\frac{2}{3}, -2$ **6.** $\frac{11}{6}, -\frac{5}{2}$ **7.** $1, -5$ **8.** $\frac{3}{2}, -\frac{5}{2}$

9. $\dfrac{3 \pm \sqrt{5}}{2}$ **10.** $2 \pm \sqrt{2}$

1. $3, -1, 0$ **3.** $0, \pm\dfrac{3\sqrt{2}}{2}$ **5.** ± 3 **7.** $-3, 0$

9. $\pm 2, 7$ **11.** ± 1 **13.** $\pm\sqrt{11}, \pm 1$ **15.** ± 2

17. $\pm\frac{1}{2}, \pm 4$ **19.** $1, -2$ **21.** 50   **23.** 26

25. -16 **27.** $\frac{1}{4}$ **29.** $6, 5$ **31.** $2, -5$ **33.** 0

35. $-59, 69$ **37.** 1   **39.** $\pm\sqrt{69}$ **41.** $\dfrac{-3 \pm \sqrt{21}}{6}$

43. $4, -5$ **45.** -1 **47.** $1, -3$ **49.** $1, -3$

51. $3, -2$　　**53.** $\sqrt{3}, -3$　　**55.** $10, -1$

57. The quadratic equation was not written in standard form before the values for $a, b,$ and c were substituted in the Quadratic Formula. The standard form for this equation is $3x^2 - 7x - 4 = 0$ ($a = 3, b = -7,$ and $c = -4$), and the correct solution is

$$x = \frac{-(-7) \pm \sqrt{(-7)^2 - 4(3)(-4)}}{2(3)}.$$

59. $x \approx \pm 1.038$　　**61.** $x \approx 16.756$　　**63.** 34　　**65.** 7%

67. $\approx 19.2\%$　　**69.** 26,250 passengers

71. 62 years old; This model is not used for people over the age of 65 because, as x increases past $x = 65$, the y-values are not low enough to produce realistic life expectancies.

73. 2,566,025 units; It does not make sense for demand x or price p to be less than zero.

75. ≈ 12.12 feet　　**77.** $13\frac{1}{3}$ minutes　　**79.** $11\frac{1}{9}$ hours

Section R2.6　*(page R134)*

Warm Up　*(page R134)*

1. $-\frac{1}{2}$　　**2.** $-\frac{1}{6}$　　**3.** -3　　**4.** -6　　**5.** $x \geq 0$

6. $-3 < z < 10$　　**7.** $P \leq 2$　　**8.** $W \geq 200$

9. $2, 7$　　**10.** $0, 1$

1. $-1 \leq x \leq 5$; Bounded　　**3.** $x > 11$; Unbounded

5. $x < -2$; Unbounded　　**7.** c　　**8.** h　　**9.** f

10. e　　**11.** g　　**12.** a　　**13.** b　　**14.** d

15. (a) Yes　(b) No　(c) Yes　(d) No

17. (a) Yes　(b) No　(c) No　(d) Yes

19. (a) Yes　(b) Yes　(c) Yes　(d) No

21. $x \geq 6$　　　　　　**23.** $x > -4$

25. $x < 25$　　　　　　**27.** $x > 2$

29. $x \leq -\frac{1}{3}$　　　　**31.** $x < -18$

33. $x > \frac{2}{5}$　　　　　**35.** $2 \leq x < 4$

37. $-1 < x < 3$　　　　　**39.** $-\frac{9}{2} < x < \frac{15}{2}$

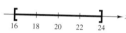

41. $-\frac{3}{4} < x < -\frac{1}{4}$　　　**43.** $-6 < x < 6$

45. $x < -6, x > 6$　　　　**47.** $-8 < x < 2$

49. $16 \leq x \leq 24$　　　　**51.** $x < -\frac{1}{2}, x > \frac{11}{2}$

53. $x \leq -7, x \geq 13$　　　　**55.** $4 < x < 5$

57. $x \leq -\frac{29}{2}, x \geq -\frac{11}{2}$　　　**59.** No solution

61. $|x| \leq 2$　　**63.** $|x - 9| \geq 3$　　**65.** $|x - 12| \leq 10$

67. $|x + 3| > 5$　　**69.** More than 400 miles

71. Greater than 12.5%　　**73.** 24 weeks

75. (a)

x	10	20	30	40	50
R	\$1159.50	\$2319.00	\$3478.50	\$4638.00	\$5797.50
C	\$1700.00	\$2650.00	\$3600.00	\$4550.00	\$5500.00

(b) $x \geq 36$ units

77. Less than 24,062.5 miles　　**79.** $x \geq 128.93$

81. 2005 ($t < 15.07$)　　**83.** $[\approx 106.864, \approx 109.464]$

85. Overcharged or undercharged up to \$0.47

87. $[65.8, 71.2]$　　**89.** $[20, 80]$

Math Matters　*(page R137)*

Cube	Ratio of $\dfrac{\text{surface area}}{\text{weight}}$
1	6
2	3
3	2
4	1.5

Section R2.7 *(page R145)*

Warm Up *(page R145)*

1. $y < -6$ **2.** $z > -\frac{9}{2}$ **3.** $-3 \le x < 1$

4. $x \le -5$ **5.** $-3 < x$ **6.** $5 < x < 7$

7. $-\frac{7}{2} \le x \le \frac{7}{2}$ **8.** $x < 2, x > 4$

9. $x < -6, x > -2$ **10.** $-2 \le x \le 6$

1. $-3 \le x \le 3$

3. $x < -2, x > 2$

5. $-7 < x < 3$

7. $x \le -5, x \ge 1$

9. $-3 < x < 2$

11. $x < -1, x > 1$

13. $-3 < x < 1$

15. $x < 0, 0 < x < \frac{3}{2}$

17. $-2 \le x \le 0, x \ge 2$

19. $-1 \le x \le 1, x \ge 2$

21. $x < -1, 0 < x < 1$

23. $x < -1, x > 4$

25. $5 < x < 15$

27. $-5 < x < -\frac{3}{2}, x > -1$

29. $-\frac{3}{4} < x < 3, x \ge 6$

31. $[-2, 2]$

33. $-\frac{9}{2} \le x \le \frac{9}{2}$ **35.** $(-\infty, 3], [4, \infty)$

37. All real numbers   **39.** All real numbers

41. The cube root of any real number is a real number.

43. $-3.51 < x < 3.51$ **45.** $-0.13 < x < 25.13$

47. $2.26 < x < 2.39$ **49.** Between 4 and 6 seconds

51. ≈ 13.8 meters $\le l \le \approx 36.2$ meters

53. (a) $90{,}000 \le x < 100{,}000$ (b) $\$30 \le p \le \$32$

   (c) 185,968 units

55. 9.5% **57.** 2006 $(t > 15.71)$

59. 2006/2007 $(t > 17.28)$

Review Exercises *(page R150)*

1. Conditional equation

3. (a) No   (b) Yes   (c) Yes   (d) No

5. $-\frac{1}{2}$ **7.** -10 **9.** $-\frac{2}{3}$ **11.** 377.778   **13.** 12

15. $130 - x = 100$; 30 pounds   **17.** 29.5 feet $\times$ 59 feet

19. $\$12$ **21.** $\$161.25$ **23.** 2 hours

25. ≈ 2.9 quarts   **27.** $-\frac{1}{2}, \frac{4}{3}$ **29.** 3, 8

31. $\pm\sqrt{11}, \approx \pm 3.32$

33. $-4 + 3\sqrt{2} \approx 0.24$

   $-4 - 3\sqrt{2} \approx -8.24$

35. (1) Use the *table* feature in ASK mode with the variable equal to a solution.

   (2) Use the scientific calculator portion of the graphing utility to evaluate the quadratic equation at a particular solution.

37. 15 feet $\times$ 27 feet   **39.** 200,000 units or 300,000 units

41. Two real solutions   **43.** $6 \pm \sqrt{6}$

45. $\dfrac{-19 \pm \sqrt{165}}{2}$ **47.** $-3 \pm 2\sqrt{3}$

49. $1.866, -0.283$ **51.** Moon: ≈ 6.09 seconds

   Earth: 2.5 seconds

53. $0, -1, 4$ **55.** $-3, \sqrt[3]{5}$ **57.** $\frac{25}{4}$ **59.** No solution

61. $\pm 4\sqrt{2}$ **63.** $-3, \frac{7}{5}$

65. $2 \pm \sqrt{19}$ **67.** $\$600$ **69.** $\approx 21.2\%$

71. $x < 11$

73. $-\frac{13}{2} < x < \frac{11}{2}$

75. $-12 < x < -8$

77. $x \ge 36$ units

79. $-1 < x < 3$ **81.** $x < -3, 0 < x < 3$

83. $x < \frac{6}{5}, x > 4$ **85.** $-1.69 < x < 1.69$

87. $1.65 < x < 1.74$ **89.** $x \ge 10$

91. All real numbers   **93.** $x \le 6$ or $x \ge 9$

95. Between 3.65 and 4.72 seconds

97. Greater than 6.96%   **99.** $25,359 \le x \le 94,641$

101. (a)

t	0	5	10	11
R	188,175	207,047	225,919	229,693

(b) 2009 ($t \ge 19.03$)

Chapter Test   *(page R154)*

1. $\frac{17}{23}$ **2.** (a) All real numbers   (b) $-3 \le x \le 3$

3. April: $175,364.00   **4.** $-\frac{5}{3}, \frac{1}{2}$ **5.** $4, -\frac{3}{2}$
May: $140,291.20

6. $\pm\sqrt{15}$ **7.** $\dfrac{-13 \pm \sqrt{69}}{2}$ **8.** $\dfrac{11 \pm \sqrt{145}}{6}$

9. $1.038, -0.446$ **10.** $2, -\frac{10}{3}$ **11.** 4

12. $-1, 1, -3, 3$ **13.** $-6, 6$

14. Selling either 341,421 units or 58,579 units will produce a revenue of $2,000,000.

15. $x < 3$ **16.** $x \le -4, x \ge \frac{28}{5}$

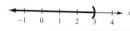

17. $(-11, -7)$ **18.** $x \le -2, 0 \le x \le 2$

19. More than 10,839 units but less than 129,161 units

20. 2004 ($t > 23.8$)

Cumulative Test: Chapters R1–R2   *(page R155)*

1. $-32x^6$ **2.** $3x^2\sqrt{2x}$ **3.** $\dfrac{3 + \sqrt{5}}{2}$

4. $\left(x + \sqrt{3}\right)\left(x - \sqrt{3}\right)(x - 6)$ **5.** $\dfrac{x + 4}{5}, x \ne 4$

6. $\dfrac{y - x}{x + y}, xy \ne 0$

7. (a) 172,000,000   (b) 2006 ($t = 15.84$)

8. $5, \frac{1}{2}$ **9.** $0.734, -1.022$ **10.** $\frac{8}{3}, -\frac{10}{3}$

11. $5 - 2\sqrt{2}$ **12.** $\pm 1, \pm 4$ **13.** $\pm 3\sqrt{2}$

14. $-3 < x < \frac{11}{3}$ **15.** $-2\sqrt{2} \le x \le 0, x \ge 2\sqrt{2}$

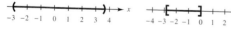

16. $-\frac{16}{3} \le x \le \frac{26}{3}$

17. At least 10,470 units but no more than 222,864 units

18. 2007 ($t > 16.97$)

Chapter 1
Section 1.1   *(page 12)*

Warm Up   *(page 12)*

1. 5   **2.** $3\sqrt{2}$ **3.** 1   **4.** -2
5. $3\left(\sqrt{2} + \sqrt{5}\right)$ **6.** $2\left(\sqrt{3} + \sqrt{11}\right)$ **7.** $-3, 11$
8. 9, 1   **9.** $0, \pm 3$ **10.** ± 2

1. (a)   **3.** (a)

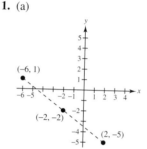
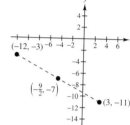

(b) 10   (c) $(-2, -2)$ (b) 17   (c) $\left(-\frac{9}{2}, -7\right)$

5. (a)   **7.** (a)

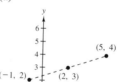
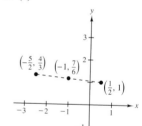

(b) $2\sqrt{10}$ (c) $(2, 3)$ (b) $\dfrac{\sqrt{82}}{3}$ (c) $\left(-1, \frac{7}{6}\right)$

9. (a)   **11.** (a)

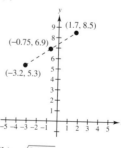

(b) $\sqrt{34.25}$ (b) $6\sqrt{277}$
(c) $(-0.75, 6.9)$ (c) $(6, -45)$

13. 5   **15.** $\sqrt{109}$ **17.** $x = 15, -9$ **19.** $y = 9, -23$

21. (a) Yes   (b) Yes   **23.** (a) Yes   (b) Yes

25.

x	y
-2	-2.5
0	-1
1	-0.25
$\frac{4}{3}$	0
2	0.5

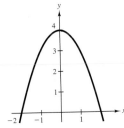

27. $\left(\frac{1}{2}, 0\right), (0, -1)$ **29.** $(-2, 0), (1, 0), (0, -2)$

31. $(0, 0), (-2, 0)$ **33.** $\left(\frac{4}{3}, 0\right), (0, 2)$

35. Every ordered pair on the x-axis has a y-coordinate of zero $[(x, 0)]$, so to find an x-intercept we let $y = 0$. Similarly, every ordered pair on the y-axis has an x-coordinate of zero $[(0, y)]$, so to find a y-intercept we let $x = 0$.

37. y-axis symmetry   **39.** x-axis symmetry

41. x-axis symmetry   **43.** y-axis symmetry

45. Origin symmetry

47.

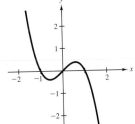

49.

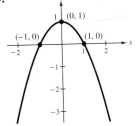

51. c   **52.** d   **53.** f   **54.** a   **55.** e   **56.** b

57.

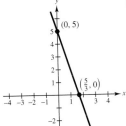

Intercepts: $\left(\frac{5}{3}, 0\right), (0, 5)$
Symmetry: none

59.

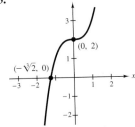

Intercepts:
$(-1, 0), (1, 0), (0, 1)$
Symmetry: y-axis

61.

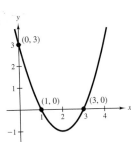

Intercepts: $(3, 0), (1, 0), (0, 3)$
Symmetry: none

63.

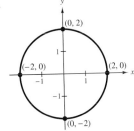

Intercepts: $\left(-\sqrt[3]{2}, 0\right), (0, 2)$
Symmetry: none

65.

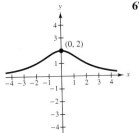

Intercepts: no x-intercepts; $(0, 2)$
Symmetry: y-axis

67.

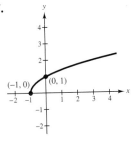

Intercepts: $(-1, 0), (0, 1)$
Symmetry: none

69.

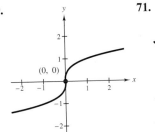

Intercept: $(0, 0)$
Symmetry: origin

71.

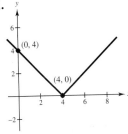

Intercepts: $(4, 0), (0, 4)$
Symmetry: none

73.

Intercepts: $(-1, 0), (0, 1), (0, -1)$
Symmetry: x-axis

75.

Intercepts: $(-2, 0), (2, 0), (0, 2), (0, -2)$
Symmetry: x-axis, y-axis, origin

77. $x^2 + y^2 = 9$ **79.** $(x + 4)^2 + (y - 1)^2 = 2$

81. $(x + 1)^2 + (y - 2)^2 = 5$

83. $(x - 1)^2 + (y - 1)^2 = 25$

85. $(x - 3)^2 + (y + 2)^2 = 16$

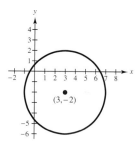

(3,−2)

87. $(x - 1)^2 + (y + 3)^2 = 0$

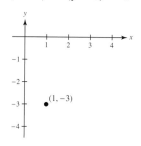

(1, −3)

89. $\left(x - \frac{1}{2}\right)^2 + \left(y - \frac{1}{2}\right)^2 = 2$ **91.** $\left(x + \frac{1}{2}\right)^2 + \left(y + \frac{5}{4}\right)^2 = \frac{9}{4}$

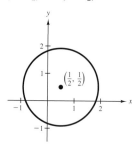

$\left(\frac{1}{2}, \frac{1}{2}\right)$

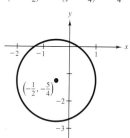

$\left(-\frac{1}{2}, -\frac{5}{4}\right)$

93. Center: $(2, -3)$; Radius: 4

$x^2 + y^2 - 4x + 6y - 3 = 0$

95. 610 dollars per fine ounce; 1980   **97.** 27%

99. (a)

Year	Data	Model
1970	70.8	71.1
1975	72.6	72.3
1980	73.7	73.3
1985	74.7	74.2
1990	75.4	75.0
1995	75.8	75.7
2000	77.0	76.2

(b)

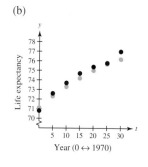

Year $(0 \leftrightarrow 1970)$

(c) ≈ 76.8 years

101.

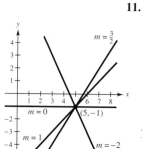

Year $(0 \leftrightarrow 1990)$

$y(13) = \$1.81, y(14) = \$2.09$; Yes

103. $\approx \$6.92$ billion

Section 1.2   *(page 24)*

Warm Up   *(page 24)*

1. $-\frac{9}{2}$ **2.** $-\frac{13}{3}$ **3.** $-\frac{5}{4}$ **4.** $\frac{1}{2}$

5. $y = \frac{2}{3}x - \frac{5}{3}$ **6.** $y = -2x$

7. $y = 3x - 1$ **8.** $y = \frac{2}{3}x + 5$

9. $y = -2x + 7$ **10.** $y = x + 3$

1. 1   **3.** 0   **5.** −3   **7.** Positive

9.   **11.**

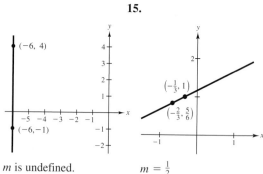

$m = 1$

13.   **15.**

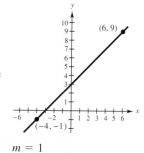

m is undefined. $m = \frac{1}{2}$

17. Answers will vary.

Sample answer: $(3, -2), (-1, -2), (0, -2)$

19. Answers will vary.

Sample answer: $(2, -3), (2, -7), (2, 9)$

21. Answers will vary.

Sample answer: $(6, -5), (7, -4), (8, -3)$

23. Answers will vary.

Sample answer: $(2, 3), (-4, 0), (4, 4)$

25. $x - y - 7 = 0$

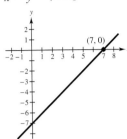

27. $2x + y = 0$

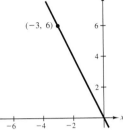

29. $x + 3y - 4 = 0$

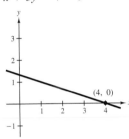

31. $x - 6 = 0$

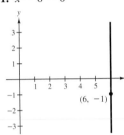

33. $y + 7 = 0$

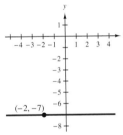

35. $8x - 6y - 17 = 0$

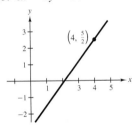

37. $m = 4, (0, -6)$

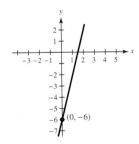

39. m is undefined;
no y-intercept

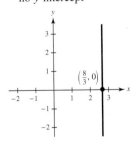

41. $m = -\frac{7}{6}, (0, 5)$

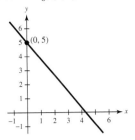

43. $x + 2y + 1 = 0$ **45.** $x + 9 = 0$

47. $x + 2y - 3 = 0$ **49.** $2x - 5y + 1 = 0$

51. Answers will vary. Sample answer: You could graph a vertical line and pick two convenient points on the vertical line to find the slope. Regardless of the vertical line selected and the points selected, the slope will have zero in the denominator. Division by zero is not possible, so the slope does not exist.

53. $4x - y - 4 = 0$ **55.** $12x + 3y + 2 = 0$

57. Perpendicular   **59.** Neither   **61.** Parallel

63. (a) $2x - y - 10 = 0$ **65.** (a) $4x - 6y - 5 = 0$

(b) $x + 2y - 10 = 0$ (b) $36x + 24y + 7 = 0$

67. (a) $y = 0$ (b) $x + 1 = 0$ **69.** $F = \frac{9}{5}C + 32$

71. $A = 4.5t + 750$ **73.** No $\left(m = \frac{11}{144}\right)$

75. No; $166,000   **77.** $34,450

79. $23.732 billion; Yes   **81.** $8.64 billion; No

83. $p = \frac{1}{33}d + 1; \frac{1}{33}$ atmosphere per foot

Section 1.3 *(page 35)*

Warm Up *(page 35)*

1.

2.

3. 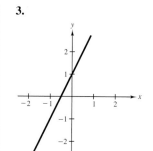 **4.**

5. $y = x + 2$ **6.** $y = \frac{3}{2}x + 3$ **7.** $y = x + 2$

8. $y = \frac{6}{7}x + 4$ **9.** $5x + 40y - 213 = 0$

10. $29x + 60y - 448 = 0$

1.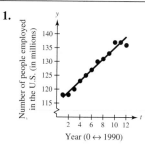

The model is a good fit for the actual data.

3.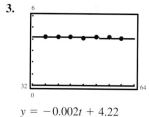

$y = -0.002t + 4.22$

Predicted time: 4.09 minutes; Yes

5. $y = \frac{3}{8}x$ **7.** $y = 20x$ **9.** $y = \frac{3.2}{7}x$ or $y = \frac{16}{35}x$

11. $I = 0.075P$ **13.** (a) $y = 0.0368x$ (b) \$6808

15. (a) $C = \frac{33}{13}I$

(b)

Inches	5	10	20	25	30
Centimeters	12.69	25.38	50.77	63.46	76.15

17. $K = \frac{2000}{4653}M$ **19.** $V = 125t + 1915,\ 5 \le t \le 10$

21. $V = 30{,}400 - 2000t,\ 5 \le t \le 10$

23. $V = 12{,}500t + 91{,}500,\ 5 \le t \le 10$

25. (a) $h = 7000 - 20t$ (b) 2:13:50 P.M.

27. $V = 875 - 175t,\ 0 \le t \le 5$ **29.** $S = 0.85L$

31. $W = 0.75x + 11.50$

33. b; Slope $= -10$; The amount owed *decreases* by \$10 per week.

35. a; Slope $= 0.25$; The amount received *increases* by \$0.25 per mile driven.

37. (a) $N = 950 + 50t$ (b) 1950 students

39. Yes; Answers will vary. Sample answer:

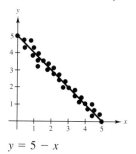

$y = 5 - x$

41. No

43. Yes; Answers will vary. Sample answer:

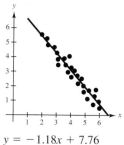

$y = -1.18x + 7.76$

45. (a) 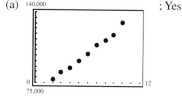 ; Yes

(b) $R = 6340.2t + 64437$

(c) 6340.2; Revenue increased by \$6340.20 million each year.

(d) 2001: \$134,179 million
2003: \$146,860 million; Yes

47. (a)

(b) Answers will vary.

Sample answer: $y = 1.07t + 124.6$

(c) 1.07; The length of the winning discus throw increased by 1.07 feet every year.

(d) 235.9 feet

(e) and (f) Answers will vary.

49. (a) $x = 78.36 - \frac{3}{55}p$ (b) 44 units

51. (a)

; Yes

(b) $C = -0.37t + 63.1$

(c) 2003: 58.29 million newspapers

2005: 57.55 million newspapers

(d) Answers will vary. Sample answer: Circulation is decreasing due to competition from the Internet and other sources of information. One strategy is to publish the paper online, charging for online memberships.

Section 1.4   *(page 48)*

Warm Up   *(page 48)*

1. -73 **2.** 13   **3.** $2(x + 2)$ **4.** $-8(x - 2)$

5. $y = \frac{7}{5} - \frac{2}{5}x$ **6.** $y = \pm x$ **7.** $x \le -2, x \ge 2$

8. $-3 \le x \le 3$ **9.** All real numbers

10. $x \le 1, x \ge 2$

1. This is a function from A to B, because each element of A is matched with an element of B.

3. Not a function; The relationship does not match an element of A with an element of B.

5. This is a function from A to B, because each element of A is matched with an element of B.

7. This is a function from A to B, because each element of A is matched with an element of B.

9. Not a function; The relationship assigns two elements of B to the element c of A.

11. Not a function from A to B; The relationship defines a function from B to A.

13. $\{(-2, 4), (-1, 1), (0, 0), (1, 1), (2, 4)\}$

15. $\{(-2, 0), (-1, 1), (0, \sqrt{2}), (1, \sqrt{3}), (2, 2)\}$

17. Not a function   **19.** Function   **21.** Function

23. Not a function   **25.** Function

27. (a) -6 (b) 34   (c) $6 - 4t$ (d) $2 - 4c$

29. (a) -1 (b) $\frac{1}{15}$ (c) $\frac{1}{t^2 - 2t}$ (d) $\frac{1}{t^2 - 1}$

31. (a) -1 (b) -9 (c) $2x - 5$ (d) $-\frac{5}{2}$

33. (a) 0   (b) 3   (c) $x^2 + 2x$ (d) -0.75

35. (a) 1   (b) -7 (c) $3 - 2|x|$ (d) 2.5

37. (a) $\frac{1}{7}$ (b) $-\frac{1}{9}$ (c) Undefined   (d) $\frac{1}{y^2 + 6y}$

39. (a) 1   (b) -1 (c) 1   (d) $\frac{|x - 1|}{x - 1}$

41. (a) -1 (b) 2   (c) 4   (d) 6

43. 5   **45.** ± 3 **47.** $0, 1, -1$ **49.** $\frac{10}{7}$

51. All real numbers   **53.** All real numbers except $t = 0$

55. All real numbers   **57.** $-1 \le x \le 1$

59. All real numbers except $x = 0, -2$

61. All real numbers $x \ge -1$ except $x = 2$ **63.** $x > 0$

65. The domain of $f(x) = \sqrt{x - 2}$ is all real numbers $x \ge 2$, because an even root of a negative number is not a real number. The domain of $g(x) = \sqrt[3]{x - 2}$ is all real numbers. f and g have different domains because an odd root of a negative number is a real number but an even root of a negative number is not a real number.

67. (a) $V = 4x(6 - x)^2$ (b) Domain: $0 < x < 6$

(c) 20 cubic inches

69. $h = \sqrt{d^2 - 2000^2}$;   Domain: $d \ge 2000$

71. Yes $[y(30) = 6]$ **73.** 3,099,200 mobile homes

75. (a)

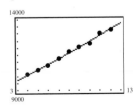

(b) $y = 404.9t + 8423$

$y = 4.11t^2 + 339.2t + 8659$

(c) Linear   Quadratic

Answers will vary. Sample answer: The quadratic model is a better fit. The linear model is easier to use but the quadratic model will give more accurate results.

77. (a) $C = 35,000 + 1.15x$ (b) $\overline{C} = \dfrac{35,000}{x} + 1.15$

(c)

x	100	1000	10,000	100,000
$\overline{C}$	351.15	36.15	4.65	1.5

(d) Answers will vary. Sample answer: The average cost per unit decreases as x gets larger.

79. (a) $r(t) = 0.8t$

(b) $A = \pi(0.8t)^2$

Time, t	1	2	3	4	5
Radius, r (feet)	0.8	1.6	2.4	3.2	4
Area, A (square feet)	2.011	8.042	18.096	32.170	50.265

(c) $\dfrac{A(2)}{A(1)} = \dfrac{8.042}{2.011} \approx 3.999,\ \dfrac{A(4)}{A(2)} = \dfrac{32.170}{8.042} \approx 4.000$

Predicted area when $t = 8$: 128.68 square feet

Calculated area when $t = 8$: 128.680 square feet

81. (a) Incorrect   (b) Correct

Mid-Chapter Quiz   *(page 53)*

1. (a)

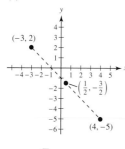

(b) $7\sqrt{2}$

(c) $\left(\tfrac{1}{2}, -\tfrac{3}{2}\right)$

2. (a)

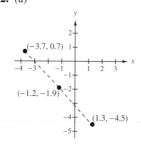

(b) $\sqrt{52.04}$

(c) $(-1.2, -1.9)$

3. (a)

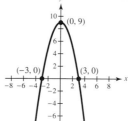

(b) $\dfrac{\sqrt{101}}{2}$

(c) $\left(\dfrac{3}{2}, -\dfrac{9}{4}\right)$

4. If the sales follow a linear growth pattern, then the sales will be $2,170,000 in 2007.

5. $2x - 3y + 9 = 0$

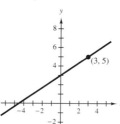

6. $y - 4 = 0$

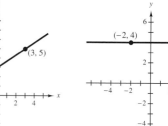

7. $x - 2 = 0$

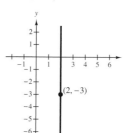

8. $2x + y + 9 = 0$

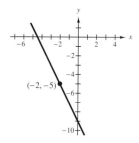

9.

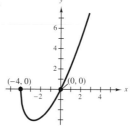

Intercepts: $(-3, 0), (3, 0), (0, 9)$

Symmetry: y-axis

10.

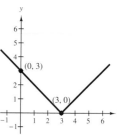

Intercepts: $(-4, 0), (0, 0)$

Symmetry: none

11.

Intercepts: $(3, 0), (0, 3)$

Symmetry: none

12. $(x - 2)^2 + (y + 3)^2 = 16$ **13.** $x^2 + \left(y + \tfrac{1}{2}\right)^2 = 4$

14. $(x - 1)^2 + (y + 2)^2 = 9$

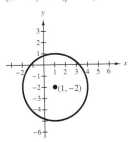

15. (a) 2   (b) −7   **16.** (a) 1   (b) −20

17. $x \geq -2$ **18.** All real numbers

19.

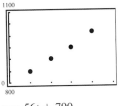

$y = 56t + 790$

20. \$1126   **21.** $A = \dfrac{C^2}{4\pi}$

Section 1.5   *(page 61)*

Warm Up   *(page 61)*

1. 2   **2.** 0   **3.** $-\dfrac{3}{x}$ **4.** $x^2 + 3$ **5.** $0, \pm 4$

6. $\frac{1}{2}, 1$ **7.** All real numbers except $x = 4$

8. All real numbers except $x = 4, 5$ **9.** $t \leq \frac{5}{3}$

10. All real numbers

1. Domain: $[1, \infty)$; Range: $[0, \infty)$; 0

3. Domain: $(-\infty, \infty)$; Range: $(-\infty, 4]$; 4

5. Domain: $(-\infty, \infty)$; Range: $(-\infty, \infty)$; −1

7. Domain: $[-5, 5]$; Range: $[0, 5]$; 5

9. Function   **11.** Not a function

13. Increasing on $(-\infty, \infty)$; No change

15. Increasing on $(-\infty, 0)$ and $(2, \infty)$, decreasing on $(0, 2)$; Behavior changes at $(0, 0)$ and $(2, -4)$.

17. Increasing on $(-1, 0)$ and $(1, \infty)$, decreasing on $(-\infty, -1)$ and $(0, 1)$; Behavior changes at $(-1, -3)$, $(0, 0)$, and $(1, -3)$.

19. Increasing on $(-2, \infty)$, decreasing on $(-3, -2)$; Behavior changes at $(-2, -2)$.

21. Minimum: $(2, -3)$
Increasing: $(2, \infty)$
Decreasing: $(-\infty, 2)$

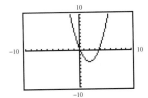

23. Relative maximum: $(0, 0)$
Relative minimum: $(2, -4)$
Increasing: $(-\infty, 0), (2, \infty)$
Decreasing: $(0, 2)$

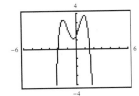

25. Relative maxima: $(-1.54, 3.29), (0.95, 3.77)$
Relative minimum: $(-0.34, 1.14)$
Increasing: $(-\infty, -1.54), (-0.34, 0.95)$
Decreasing: $(-1.54, -0.34), (0.95, \infty)$

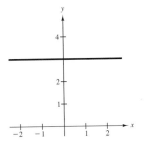

27. Even   **29.** Odd   **31.** Neither even nor odd

33. Even   **35.** Neither even nor odd

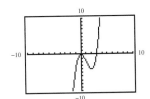

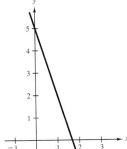

37. Odd

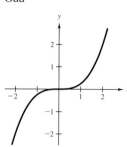

39. Neither even nor odd

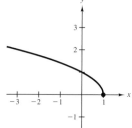

41. Neither even nor odd   **43.** Neither even nor odd

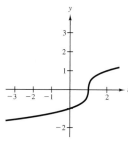

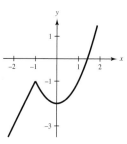

45. Neither even nor odd

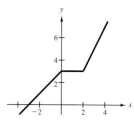

47.

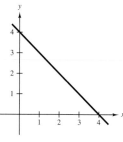

49.

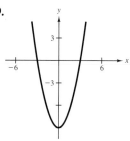

51.

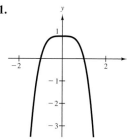

53.

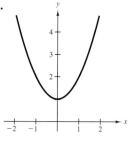

55.

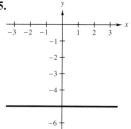

57.

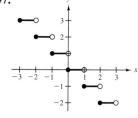

59.

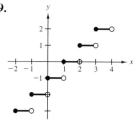

61. Maximum: 3,455,520

Minimum: 3,277,719

Decreasing: 1991–1992

Increasing: 1992–2002

It is not realistic to assume that the population will continue to follow this model, because the function decreases after the year 2005, and eventually yields negative values.

63.

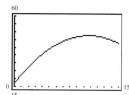

Increasing: 1990–2001

Decreasing: 2001–2005

Maximum revenue: $4,754,000

Minimum revenue: $1,700,000

65. Maximum or minimum values may occur at endpoints.

67.

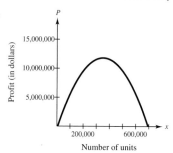

Approximately 350,000 units

69.

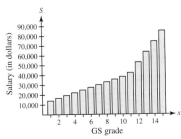

71.

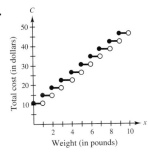

73. (a) $y = 0.0083t^3 - 0.125t^2 + 0.82t + 19.0, \quad 3 \le t \le 7$

(b) $y = 0.20t + 19.9, \quad 8 \le t \le 12$

(c)

$$y = \begin{cases} 0.0083t^3 - 0.125t^2 + 0.82t + 19.0, & 3 \le t \le 7 \\ 0.20t + 19.9, & 8 \le t \le 11 \end{cases}$$

Section 1.6   *(page 71)*

Warm Up   *(page 71)*

1. 12   **2.** $\dfrac{-2x}{-x - 3}$ **3.** $0, \pm\sqrt{10}$ **4.** $\dfrac{4}{3}, -2$

5. $f(x) = -2$

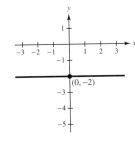

6. $f(x) = -x$

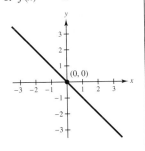

7. $f(x) = x + 5$

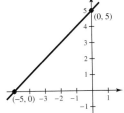

8. $f(x) = 2 - x$

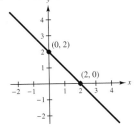

9. $f(x) = 3x - 4$

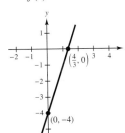

10. $f(x) = 9x + 10$

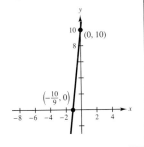

1. Shifted three units up

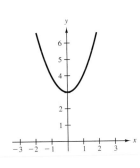

3. Shifted four units to the left

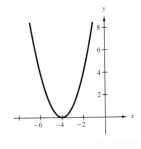

5. Shifted two units up and two units to the right

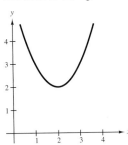

7. Reflected about the *x*-axis and shifted one unit up

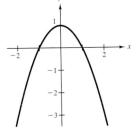

9. Shifted two units up

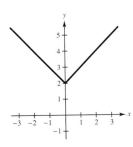

11. Shifted one unit to the right

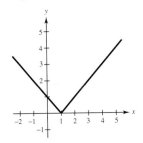

13. Reflected about the x-axis and shifted three units up

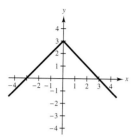

15. Reflected about the x-axis and shifted two units to the right and four units up

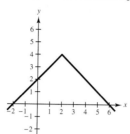

17. Shifted two units to the right

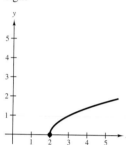

19. Shifted three units to the right and one unit up

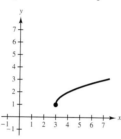

21. Reflected about the x-axis and shifted four units to the right and two units up

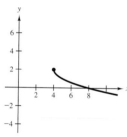

23. Reflected about the y-axis and shifted one unit up

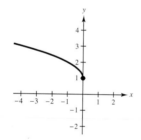

25.

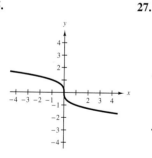

27.

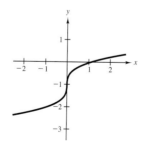

29.

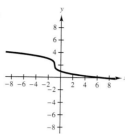

31.

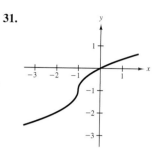

33.

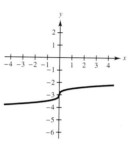

35. Common function: $y = x^3$
Transformation: shifted two units to the right
Equation: $y = (x - 2)^3$

37. Common function: $y = x^2$
Transformation: reflected about the x-axis
Equation: $y = -x^2$

39. Common function: $y = \sqrt{x}$
Transformation: reflected about the x-axis and shifted one unit up
Equation: $y = -\sqrt{x} + 1$

41. (a) Vertical shift of two units

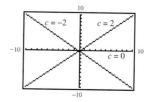

(b) Horizontal shift of two units

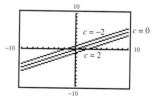

(c) Slope of the function changes

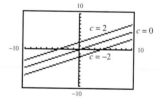

43. (a) $g(x) = (x - 1)^2 + 1$

(b) $g(x) = -(x + 1)^2$

45. (a)

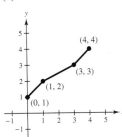

(b)

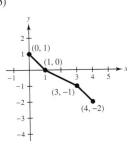

(c)

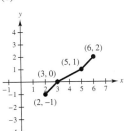

(d)

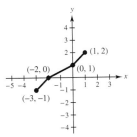

(e)

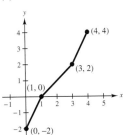

(f)

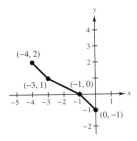

47. (a)

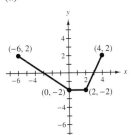

(b)

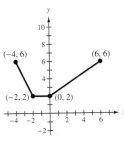

(c)

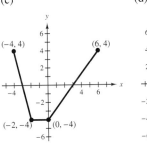

(d)

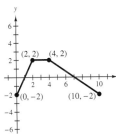

(e)

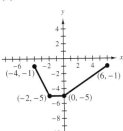

(f)

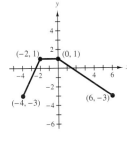

49. $g(x) = |x - 2| + 1$ **51.** $g(x) = -3|x|$

53. $h(x) = \sqrt{x + 3} + 2$ **55.** $h(x) = -4\sqrt{x}$

57. $g(x) = -x^3 + 3x^2 + 1$

59. Shifted one unit to the right and two units down

$g(x) = (x - 1)^2 - 2$

61. (a)

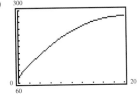

(b) $P(x) = 80 + 20x - 0.5x^2 - 25$

$P(x) = 55 + 20x - 0.5x^2, \quad 0 \le x \le 20$

Shifted 25 units down

(c) $P\left(\dfrac{x}{100}\right) = 80 + \dfrac{x}{5} - 0.00005x^2$

Vertical shrink

63.

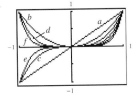

(a), (c), and (e) are odd functions; (b), (d), and (f) are even.
Also, (a), (c), and (e) are increasing for all real numbers;
(b), (d), and (f) are decreasing for all $x < 0$ and increasing
for all $x > 0$.

Section 1.7 *(page 80)*

Warm Up *(page 80)*

1. $\dfrac{1}{x(1-x)}$ **2.** $-\dfrac{12}{(x+3)(x-3)}$ **3.** $\dfrac{3x-2}{x(x-2)}$

4. $\dfrac{4x-5}{3(x-5)}$ **5.** $\dfrac{\sqrt{x^2-1}}{x+1},\ x \neq 1$

6. $\dfrac{x+1}{x(x+2)},\ x \neq 2$ **7.** $5(x-2)$

8. $\dfrac{x+1}{(x-2)(x+3)},\ x \neq -5, -1, 3$

9. $\dfrac{1+5x}{3x-1},\ x \neq 0$ **10.** $\dfrac{x+4}{4x},\ x \neq 4$

1.
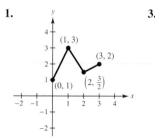

3.
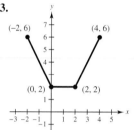

5. (a) $2x$ (b) 2 (c) x^2-1

(d) $\dfrac{x+1}{x-1}$; Domain: $(-\infty, 1) \cup (1, \infty)$

7. (a) x^2-x+1 (b) x^2+x-1 (c) x^2-x^3

(d) $\dfrac{x^2}{1-x}$; Domain: $(-\infty, 1) \cup (1, \infty)$

9. (a) $x^2+\sqrt{1-x}+5$ (b) $x^2-\sqrt{1-x}+5$

(c) $x^2\sqrt{1-x}+5\sqrt{1-x}$

(d) $\dfrac{x^2+5}{\sqrt{1-x}}$; Domain: $(-\infty, 1)$

11. (a) $\dfrac{x+1}{x^2}$ (b) $\dfrac{x-1}{x^2}$ (c) $\dfrac{1}{x^3}$

(d) $x, x \neq 0$; Domain: $(-\infty, 0) \cup (0, \infty)$

13. 9 **15.** $4t^2-2t+5$ **17.** -30

19. 26 **21.** 5 **23.** $\frac{3}{5}$

25. (a) x^2-2x+1 (b) x^2-1 (c) x^4

27. (a) $20-3x$ (b) $-3x$ (c) $9x+20$

29. (a) $\sqrt{x^2+4}$ (b) $x+4, x \geq -4$

31. (a) $x-\frac{8}{3}$ (b) $x-8$ **33.** (a) $\sqrt[4]{x}$ (b) $\sqrt[4]{x}$

35. (a) $|x+6|$ (b) $|x|+6$

37. (a) $x \geq 0$, or $[0, \infty)$

(b) All real numbers, or $(-\infty, \infty)$

(c) All real numbers, or $(-\infty, \infty)$

39. (a) All real numbers except $x = \pm 1$, or
$(-\infty, -1) \cup (-1, 1) \cup (1, \infty)$

(b) All real numbers, or $(-\infty, \infty)$

(c) All real numbers except $x = -2$ and $x = 0$, or
$(-\infty, -2) \cup (-2, 0) \cup (0, \infty)$

41. (a) 3 (b) 0 **43.** (a) 0 (b) 4

45. Answers will vary.

Sample answer: $f(x) = x^2, g(x) = 2x+1$

47. Answers will vary.

Sample answer: $f(x) = \sqrt[3]{x}, g(x) = x^2-4$

49. Answers will vary.

Sample answer: $f(x) = \dfrac{1}{x}, g(x) = x+2$

51. Answers will vary.

Sample answer: $f(x) = x^2+2x, g(x) = x+4$

53. $T = \frac{3}{4}x + \frac{1}{15}x^2$

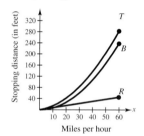

55. $(C \circ x)(t) = 2800t + 375$

$C \circ x$ represents the cost of production for t hours.

57. $R_1 + R_2 = -0.75t^2 + 0.67t + 742.8$,
$t = 0, 1, 2, 3, 4, 5$

Total sales have been decreasing.

59. (a) $y_1 = 4.91t + 4.7, y_2 = 1.95t + 12.5$

(b)

$y_3 = 2.96t - 7.8$;

y_3 represents the profit for 1994 through 2004; $42,520$

(c)

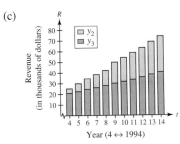

The heights of the bars represent the revenues for 1994 through 2004.

61. (a) $N(T(t)) = 100t^2 + 275$

(b) Approximately 2.18 hours

63.

Year	P	E	P/E
1992	$11.05	$0.65	17
1993	$13.07	$0.73	17.9
1994	$14.36	$0.84	17.1
1995	$18.81	$0.99	19
1996	$23.87	$1.11	21.5
1997	$24.15	$1.15	21
1998	$31.12	$1.26	24.7
1999	$42.26	$1.39	30.4
2000	$33.43	$1.46	22.9
2001	$28.42	$1.36	20.9
2002	$24.16	$1.32	18.3

65. Domain of $(f/g)(x)$: all real numbers $0 \le x < 3$, or $[0, 3)$
Domain of $(g/f)(x)$: all real numbers $0 < x \le 3$, or $(0, 3]$
The two domains differ because if $x = 3$, $(f/g)(x)$ is undefined (division by zero), and if $x = 0$, $(g/f)(x)$ is undefined (division by zero).

Section 1.8   (page 91)

Warm Up   (page 91)

1. All real numbers   **2.** $[-1, \infty)$

3. All real numbers except $x = 0, 2$

4. All real numbers except $x = -\frac{5}{3}$ **5.** x

6. x **7.** x **8.** x **9.** $x = \frac{3}{2}y + 3$

10. $x = \dfrac{y^3}{2} + 2$

1. $f^{-1}(x) = \frac{1}{2}x$ **3.** $f^{-1}(x) = x + 5$

5. (a) $5\left(\dfrac{x-1}{5}\right) + 1 = x$; $\dfrac{(5x+1)-1}{5} = x$

(b)

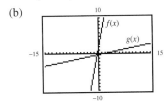

7. (a) $\left(\sqrt[3]{x}\right)^3 = x$; $\sqrt[3]{x^3} = x$

(b)

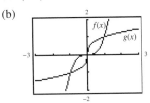

9. (a) $\sqrt{(x^2+4)-4} = x$; $\left(\sqrt{x-4}\right)^2 + 4 = x$

(b)

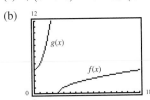

11. (a) $1 - \left(\sqrt[3]{1-x}\right)^3 = x$; $\sqrt[3]{1-(1-x^3)} = x$

(b)

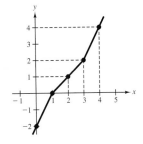

13.

x	0	1	2	3	4
$f^{-1}(x)$	-2	0	1	2	4

15.

x	-2	0	2	3
$f^{-1}(x)$	-4	-3	3	4

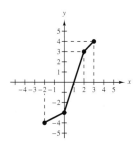

17. f doesn't have an inverse.

19. $g^{-1}(x) = 8x$ **21.** p doesn't have an inverse.

23. $f^{-1}(x) = \sqrt{x} - 3, x \geq 0$ **25.** $h^{-1}(x) = \dfrac{1}{x}$

27. $f^{-1}(x) = \dfrac{x^2 - 3}{2}, x \geq 0$

29. g doesn't have an inverse.

31. $f^{-1}(x) = -\sqrt{25 - x}, x \leq 25$

33. Error: f^{-1} does not mean to take the reciprocal of $f(x)$.

35. $f^{-1}(x) = \dfrac{x + 3}{2}$ **37.** $f^{-1}(x) = \sqrt[5]{x}$

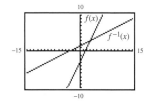   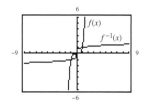

39. $f^{-1}(x) = x^2, x \geq 0$ **41.** $f^{-1}(x) = \sqrt{4 - x^2}$,
$\qquad 0 \leq x \leq 2$

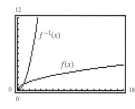   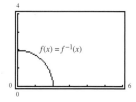

43. $f^{-1}(x) = x^3 + 1$

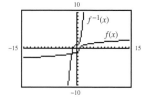

45. Because f is one-to-one, f has an inverse.

47. f doesn't have an inverse because two x-values share the same y-value.

49. g has an inverse.   **51.** h doesn't have an inverse.

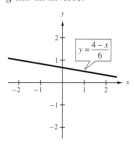   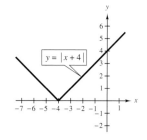

53. f doesn't have an inverse.

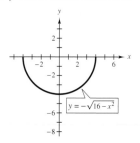

55. 32   **57.** 600

59. $(g^{-1} \circ f^{-1})(x) = \dfrac{x + 1}{2}$ **61.** $(f \circ g)^{-1}(x) = \dfrac{x + 1}{2}$

63. $C^{-1}(x) = \dfrac{x - 1200}{6.5}$; C^{-1} computes the number of pizzas

that can be made with a cost of x.

Domain of C: $[0, \infty)$

Domain of C^{-1}: $[1200, \infty)$

65. (a)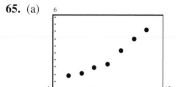

(b) $y = 0.227t + 3.07$

(c) $y^{-1} = \dfrac{t - 3.07}{0.227}$; y^{-1} represents the year in which the

average admission price is t dollars.

(d) 2007

67. After graphing $f(x) = x^2, x \geq 0$, and $f^{-1}(x) = \sqrt{x}$, it is observed that $f(x)$ and $f^{-1}(x)$ are reflections of each other about the line $y = x$. Because of this reflection, interchanging the roles of x and y seems reasonable.

69. $f^{-1}(x) = \dfrac{x^2 + 0.023}{0.054}$; $36.72

Review Exercises *(page 96)*

1. (a)

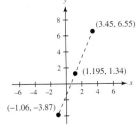

(b) $\sqrt{85}$

(c) $\left(0, -\frac{3}{2}\right)$

3. (a)

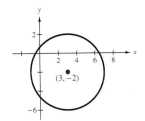

(b) ≈ 11.35

(c) $(1.195, 1.34)$

5. $x = -10$ or 30 **7.** (a) No   (b) No

9.

 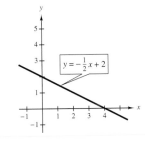

x	y
-2	3
0	2
2	1
3	$\frac{1}{2}$
4	0

$y = -\frac{1}{2}x + 2$

11.

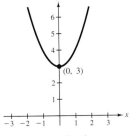

y-intercept: $(0, 3)$

Symmetry: y-axis

13.

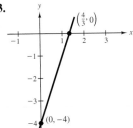

x-intercept: $\left(\frac{4}{3}, 0\right)$

y-intercept: $(0, -4)$

Symmetry: none

15.

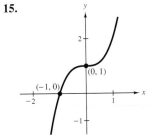

x-intercept: $(-1, 0)$

y-intercept: $(0, 1)$

Symmetry: none

17. $(x - 1)^2 + (y - 3)^2 = 25$

19. $(x - 3)^2 + (y + 2)^2 = 16$

21.

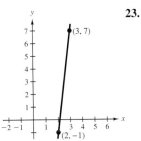

$m = 8$

23.

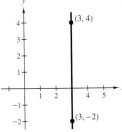

m is undefined.

25. $3x - 2y - 10 = 0$ **27.** $2x + 3y - 6 = 0$

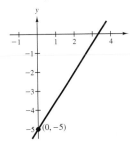 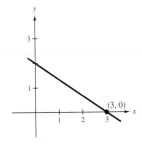

29. Slope: $\frac{5}{4}$; y-intercept: $\left(0, \frac{11}{4}\right)$

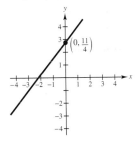

31. Parallel   **33.** Neither

35. (a) $5x - 4y - 23 = 0$ (b) $4x + 5y - 2 = 0$

37. $y = \frac{7}{3}x$ **39.** $y = 348x$

41. $y = 0.0405x$; $2632.50   **43.** $210,000

45. $V = -2750t + 30,000$, $0 \le t \le 10$

47. y is a function of x.   **49.** y is a function of x.

51. Function; Every element of A is assigned to an element of B.

53. (a) 1 (b) 0 (c) -1 (d) $\sqrt{x + 11} - 3$

55. All real numbers   **57.** $[-9, \infty)$

59. $[3, 5) \cup (5, \infty)$

61. The domain of $h(x)$ is all real numbers except $x = 0$, since division by zero is undefined. The domain of $k(x)$ is all real numbers except $x = -2$ and $x = 2$, since if $x = 2$ or $x = -2$, then $x^2 - 4$ equals zero, and division by zero is undefined. Using a graphing utility and the table feature, $h(0)$ results in an error and $k(-2)$ or $k(2)$ also results in an error.

63. (a) $B = 5000\left(1 + \dfrac{0.0625}{4}\right)^{4t}$ (b) $t \geq 0$

65. (a) Domain: all real numbers
      Range: $[1, \infty)$

    (b) Decreasing: $(-\infty, 0)$

       Increasing: $(0, \infty)$

    (c) Even

    (d) Minimum: $(0, 1)$

67. (a) Domain: all real numbers
      Range: all real numbers

    (b) Decreasing: $\left(0, \frac{8}{3}\right)$

       Increasing: $(-\infty, 0) \cup \left(\frac{8}{3}, \infty\right)$

    (c) Neither

    (d) Relative minimum: $\left(\frac{8}{3}, -\frac{256}{27}\right)$

       Relative maximum: $(0, 0)$

69. y is a function of x.   **71.** y is not a function of x.

73. y is a function of x.

75.

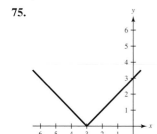

77.

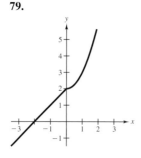

79.

81.

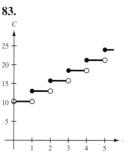

83.

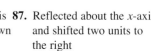

85. Reflected about the x-axis and shifted two units down and one unit to the right

87. Reflected about the x-axis and shifted two units to the right

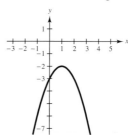

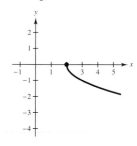

89. Shifted two units to the left

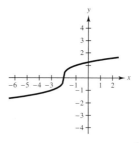

91. Common function: $y = \sqrt{x}$
     Transformation: reflected about the y-axis and shifted three units to the right
     Equation: $y = \sqrt{3 - x}$ or $y = \sqrt{-(x - 3)}$

93. $(f + g)(x) = \sqrt{x^2 - 4} + 3$
     $(f - g)(x) = \sqrt{x^2 - 4} - 3$
     $(fg)(x) = 3\sqrt{x^2 - 4}$
     $(f/g)(x) = \dfrac{\sqrt{x^2 - 4}}{3}$
     Domain of f/g: $x \leq -2, x \geq 2$, or $(-\infty, -2] \cup [2, \infty)$

95. 7   **97.** 40

99. (a) $x^2 + 6x + 9$
      Domain: All real numbers

    (b) $x^2 + 3$
      Domain: All real numbers

101. (a) $\dfrac{1}{3x + x^2}$

Domain: $x \neq 0, -3$

(b) $\dfrac{3}{x} + \dfrac{1}{x^2}$, or $\dfrac{3x + 1}{x^2}$

Domain: $x \neq 0$

103. Answers will vary.

Sample answer: $f(x) = x^2$, $g(x) = 6x - 5$

105. Answers will vary.

Sample answer: $f(x) = \dfrac{1}{x^2}$, $g(x) = x - 1$

107. $R = R_1 + R_2$

$R = 600.52 + 0.22t - 0.3t^2$

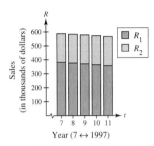

Year $(7 \leftrightarrow 1997)$

Total sales are decreasing.

109. $g(f(x))$; The bonus is based on sales over \$500,000, and is calculated by multiplying $x - 500,000$ by 0.03.

111. $f(g(x)) = x = g(f(x))$,

$f(g(x)) = 3\left(\dfrac{x - 5}{3}\right) + 5 = x$

$g(f(x)) = \dfrac{3x + 5 - 5}{3} = x$

so f and g are inverse functions of each other.

113. $f(x)$ does not have an inverse.

115. $f^{-1}(x) = \dfrac{1}{x}$

(f is its own inverse.)

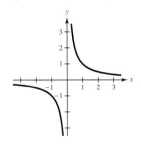

117. (a) $f^{-1}(x) = 2(x + 3) = 2x + 6$

(b)

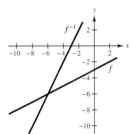

(c) $f^{-1}(f(x)) = 2\left(\tfrac{1}{2}x - 3\right) + 6$

$= x$

$f(f^{-1}(x)) = \tfrac{1}{2}(2x + 6) - 3$

$= x$

119. (a) $f^{-1}(x) = \sqrt{x}$, $x \geq 0$

(b)

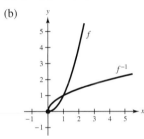

(c) $f^{-1}(f(x)) = \sqrt{x^2} = x$, $x \geq 0$

$f(f^{-1}(x)) = \left(\sqrt{x}\right)^2 = x$, $x \geq 0$

121. (a)

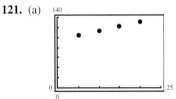

(b) $y = 1.791t + 94.73$

(c) $y^{-1} = \dfrac{t - 94.73}{1.791}$; 2010

Chapter Test (page 101)

1. Distance: $4\sqrt{5}$ **2.** Distance: 7.81

Midpoint: $(1, 0)$ Midpoint: $(0.44, 4.34)$

3. x-intercepts: $(-5, 0), (3, 0)$

y-intercept: $(0, -15)$

4. Symmetric with respect to the origin

5. $3x - 4y - 24 = 0$

6. $(x - 2)^2 + (y - 1)^2 = 9$

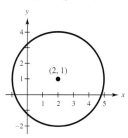

(2, 1)

7. True. To each value of x there corresponds exactly one value of y.

8. False. The element -9 is not included in set B.

9. (a) Domain: All real numbers

     Range: $(-\infty, 2]$

   (b) Decreasing: $(0, \infty)$

     Increasing: $(-\infty, 0)$

   (c) Even

   (d) Maximum: $(0, 2)$

10. (a) Domain: $(-\infty, -2] \cup [2, \infty)$

     Range: $[0, \infty)$

   (b) Decreasing: $(-\infty, -2)$

     Increasing: $(2, \infty)$

   (c) Even   (d) Minima: $(-2, 0), (2, 0)$

11.

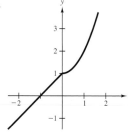

12.

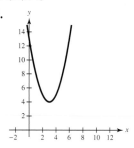

13. $(f - g)(x) = x^2 - 2x + 3$

14. $(fg)(x) = 2x^3 - x^2 + 4x - 2$

15. $(f \circ g)(x) = 4x^2 - 4x + 3$

16. $f(x)$ does not have an inverse.

17. $V = 25{,}000 - 4000t$

18.

$S = 3.74t + 46.1$

Chapter 2

Section 2.1   *(page 112)*

Warm Up   *(page 112)*

1. $\frac{1}{2}, -6$ **2.** $-\frac{3}{5}, 3$ **3.** $\frac{3}{2}, -1$ **4.** -10

5. $3 \pm \sqrt{5}$ **6.** $-2 \pm \sqrt{3}$ **7.** $4 \pm \dfrac{\sqrt{14}}{2}$

8. $-5 \pm \dfrac{\sqrt{3}}{3}$ **9.** $-\dfrac{3}{2} \pm \dfrac{\sqrt{5}}{2}$ **10.** $-\dfrac{3}{2} \pm \dfrac{\sqrt{21}}{2}$

1. g   **2.** e   **3.** c   **4.** f   **5.** b   **6.** a

7. h   **8.** d   **9.** $y = -(x + 2)^2$

11. $y = (x - 3)^2 - 9$ **13.** $y = -2(x + 3)^2 + 3$

15. Intercept: $(0, 0)$ **17.** Intercepts: $(\pm 4, 0), (0, 16)$

     Vertex: $(0, 0)$ Vertex: $(0, 16)$

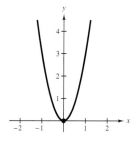 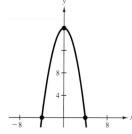

19. Intercepts: **21.** Intercepts:

     $\left(-5 \pm \sqrt{6}, 0\right), (0, 19)$ $(-1, 0), (0, 1)$

     Vertex: $(-5, -6)$ Vertex: $(-1, 0)$

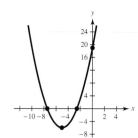 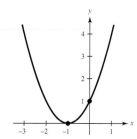

23. Intercepts:
$(1, 0), (-3, 0), (0, 3)$
Vertex: $(-1, 4)$

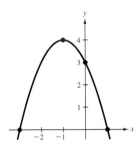

25. Intercept: $(0, \frac{5}{4})$
Vertex: $(\frac{1}{2}, 1)$

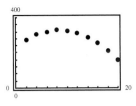

27. Intercepts:
$(1 \pm \sqrt{6}, 0), (0, 5)$
Vertex: $(1, 6)$

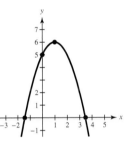

29. Intercept: $(0, 21)$
Vertex: $(\frac{1}{2}, 20)$

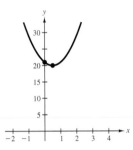

31. Intercepts: $(8 \pm 4\sqrt{2}, 0), (0, 8)$
Vertex: $(8, -8)$

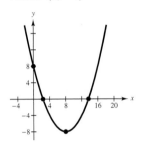

33. $y = -\frac{1}{2}(x - 2)^2 - 1$ **35.** $y = \frac{3}{4}(x - 5)^2 + 12$

37. Answers will vary.
Sample answer:
$f(x) = x^2 - x - 2$
$g(x) = -x^2 + x + 2$

39. Answers will vary.
Sample answer:
$f(x) = x^2 - 10x$
$g(x) = -x^2 + 10x$

41. Answers will vary.
Sample answer:
$f(x) = 2x^2 + 7x + 3$
$g(x) = -2x^2 - 7x - 3$

43. $A = 50x - x^2$; $(25, 625)$; The rectangle has the greatest area ($A = 625$ square feet) when its width is 25 feet.

45. $x = 25$ feet, $y = 33\frac{1}{3}$ feet; 50 feet $\times$ 33$\frac{1}{3}$ feet

47. 45,000 units   **49.** 20 fixtures   **51.** 14 feet

53. $y = 0.0285t^2 - 0.302t + 1.48$

55. (a)

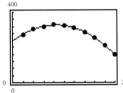

(b) $p(x) = -1.276x^2 + 21.63x + 236.5$

(c)

(d) $\approx (8.5, 328.2)$; Approximately 85,000 units returns the maximum profit, about \$32,820,000.

(e) Production costs may be growing faster than revenue, so profit decreases.

Math Matters   *(page 115)*

Both regions have the same area, $9\pi r^2$.

Section 2.2   *(page 123)*

Warm Up   *(page 123)*

1. $(3x - 2)(4x + 5)$ **2.** $x(5x - 6)^2$
3. $z^2(12z + 5)(z + 1)$ **4.** $(y + 5)(y^2 - 5y + 25)$
5. $(x + 3)(x + 2)(x - 2)$ **6.** $(x + 2)(x^2 + 3)$
7. No real solution   **8.** $3 \pm \sqrt{5}$
9. $-\frac{1}{2} \pm \sqrt{3}$ **10.** ± 3

1. e   **2.** c   **3.** g   **4.** d   **5.** f   **6.** h
7. a   **8.** b
9.

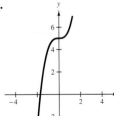

11.

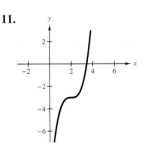

13.

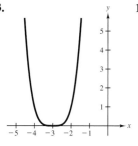

15.

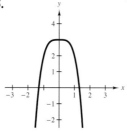

17. Rises to the left
Falls to the right

19. Rises to the left
Falls to the right

21. Rises to the left
Rises to the right

23. Rises to the left
Rises to the right

25. Falls to the left
Falls to the right

27. ± 4 **29.** -4 **31.** $1, -2$ **33.** No real zeros

35. $2, 0$ **37.** ± 1 **39.** $\pm \sqrt{5}$ **41.** 3

43.

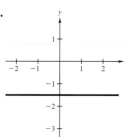

45.

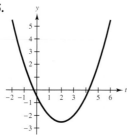

47.

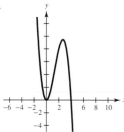

49.

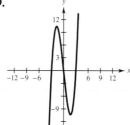

51.

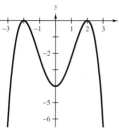

53.

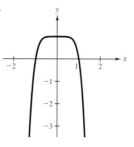

55. Answers will vary. Sample answers:

$a_n < 0$ $a_n > 0$

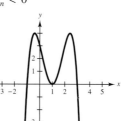   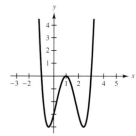

57. $f(x) = \frac{1}{2}(x^4 - 11x^3 + 28x^2)$

59. (a)

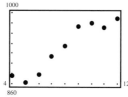

(b) Quadratic: $G(t) = -0.64t^2 + 27.5t + 756$

Quartic: $G(t) = 0.2893t^4 - 10.108t^3 + 126.10t^2$
$- 642.1t + 2001$

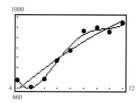

(c)

Year	$G(t)$	Quadratic	Quartic
4	876	855.8	877.3
5	862	877.5	860.3
6	879	898.0	879.6
7	914	917.1	912.8
8	935	935.0	944.3
9	972	951.7	965.6
10	980	967.0	975.0
11	970	981.1	977.9
12	989	993.8	986.5

(d) The quartic model represents the data more accurately
because the values are closer to the actual values, $G(t)$.

61. $(200, 320)$

63. The functions have a common shape because their degrees
are even, but their graphs are not identical because they
have different degrees.

Section 2.3   *(page 133)*

Warm Up   *(page 133)*

1. $x^3 - x^2 + 2x + 3$ **2.** $2x^3 + 4x^2 - 6x - 4$

3. $x^4 - 2x^3 + 4x^2 - 2x - 7$

4. $2x^4 + 12x^3 - 3x^2 - 18x - 5$

5. $(x - 3)(x - 1)$ **6.** $8(x + 2)(x - 5)$

7. $(3x + 5)(x - 1)$ **8.** $(3x - 4)^2$

9. $2x(x - 1)(2x - 3)$ **10.** $x(3x + 2)(2x + 1)$

1. $3x - 4$ **3.** $2x + 4$ **5.** $x + 3$

7. $x^3 + 3x^2 - 1$ **9.** $7 - \dfrac{25}{x + 4}$

11. $3x + 5 - \dfrac{2x - 3}{2x^2 + 1}$ **13.** $x + \dfrac{x - 27}{x^2 - 1}$

15. $x^2 - 6x + 17 - \dfrac{36}{x + 2}$

17. $2x^3 + 4x^2 - 2x - 8 - \dfrac{10x - 7}{x^2 - 2x + 1}$

19. $2x^2 - 3x + 5$ **21.** $4x^2 - 9$

23. $-x^2 + 10x - 25$ **25.** $5x^2 + 14x + 56 + \dfrac{232}{x - 4}$

27. $10x^3 + 10x^2 + 60x + 360 + \dfrac{1360}{x - 6}$

29. $2x^4 + 8x^3 + 2x^2 + 8x - 5 - \dfrac{7}{x - 4}$

31. $-3x^3 - 6x^2 - 12x - 24 - \dfrac{48}{x - 2}$

33. $-x^2 + 3x - 6 + \dfrac{11}{x + 1}$ **35.** $4x^2 + 14x - 30$

37. $f(x) = (x - 4)(x^2 + 3x - 2) + 3; \ f(4) = 3$

39. $f(x) = \left(x - \frac{1}{2}\right)(2x^2 - 8x + 6) - 6; \ f\left(\frac{1}{2}\right) = -6$

41. $f(x) = \left(x - \sqrt{2}\right)\left[x^2 + \left(3 + \sqrt{2}\right)x + 3\sqrt{2}\right] - 8;$
   $f\left(\sqrt{2}\right) = -8$

43. $f(x) = \left(x - 2 - \sqrt{2}\right)\left[-3x^2 + \left(2 - 3\sqrt{2}\right)x\right.$
   $\left. + 8 - 4\sqrt{2}\right]; f\left(2 + \sqrt{2}\right) = 0$

45. (a) -69 (b) -2081 (c) -6 (d) 446

47. (a) 1 (b) -267 (c) $-\frac{11}{3}$ (d) -3.8

49. (a) 72 (b) -68 (c) 37.648 (d) 30

51. (a) Proof   (b) $x + 2$

   (c) $f(x) = (x - 4)(x + 2)^2$

   (d) $4, -2$

   (e)

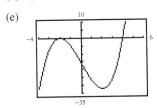

53. (a) Proof   (b) $x + 4$

   (c) $f(x) = (2x - 3)(x - 1)(x + 4)$

   (d) $\frac{3}{2}, 1, -4$

   (e)

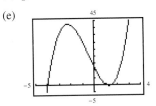

55. (a) Proof   (b) $x + \sqrt{3}$

   (c) $f(x) = \left(x - \sqrt{3}\right)(x + 2)\left(x + \sqrt{3}\right)$

   (d) $\sqrt{3}, -2, -\sqrt{3}$

   (e)

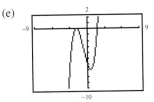

57. The second polynomial is a factor of the first polynomial.

59. b; 3, $\dfrac{-1 \pm \sqrt{17}}{2}$ **60.** d; $-2, \dfrac{3 \pm \sqrt{5}}{2}$

61. a; $-1, -2 \pm \sqrt{2}$ **62.** c; 3, $1 \pm \sqrt{5}$

63. Answers will vary. Sample answer:

   $f(x) = 3x^3 - 13x^2 + 4x + 20$

   $f(x) = -3x^3 + 13x^2 - 4x - 20$

   Infinitely many polynomial functions

65. $x^2 + x - 2$ **67.** $3x^2 + 5x + 7$ **69.** $x^2 + 3x$

71. $x^2 + 10x + 24$ square feet

73. (a) $\$192,116$ (b) Proof

Section 2.4   *(page 145)*

Warm Up   *(page 145)*

1. $f(x) = 3x^3 - 8x^2 - 5x + 6$

2. $f(x) = 4x^4 - 3x^3 - 16x^2 + 12x$

3. $x^4 - 3x^3 + 5 + \dfrac{3}{x + 3}$

4. $3x^3 + 15x^2 - 9 - \dfrac{2}{x + (2/3)}$

5. $\frac{1}{2}, -3 \pm \sqrt{5}$ **6.** $10, -\frac{2}{3}, -\frac{3}{2}$ **7.** $-\frac{3}{4}, 2 \pm \sqrt{2}$

8. $\frac{2}{5}, -\frac{7}{2}, -2$ **9.** $\pm\sqrt{2}, \pm 1$ **10.** $\pm 2, \pm\sqrt{3}$

1. Possible: $\pm 1, \pm 2, \pm 4$ **3.** $-3, \frac{1}{2}, 4$

 Actual: $-1, \pm 2$

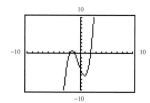

5. $\pm\frac{1}{2}, \pm 2$ **7.** $1, 2, 3$ **9.** $-1, -10$ **11.** $\frac{1}{2}, -1$

13. $\pm 3, \pm\sqrt{2}$ **15.** $-1, 2$ **17.** $-6, \frac{1}{2}, 1$

19. $-2, 0, 1$

21. (a) $\pm 1, \pm 3, \pm\frac{1}{2}, \pm\frac{3}{2}, \pm\frac{1}{4}, \pm\frac{3}{4}, \pm\frac{1}{8}, \pm\frac{3}{8}, \pm\frac{1}{16}, \pm\frac{3}{16}, \pm\frac{1}{32}, \pm\frac{3}{32}$

  (b)   (c) $1, \frac{3}{4}, -\frac{1}{8}$

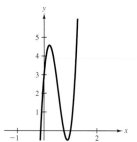

23. Real zero ≈ 0.7 **25.** Real zero ≈ 3.3

27. e; -1.769 **28.** c; 0.755 **29.** d; 0.206

30. a; $0.266, 1.175, 2.559$ **31.** f; 2.769

32. b; $-1.675, -0.539, 2.214$ **33.** $-1.164, 1.453$

35. $0.900, 1.100, 1.900$ **37.** $-1.453, 1.164$

39. $-2.177, 1.563$ **41.** d

42. a   **43.** b   **44.** c

45. (a) $V = x(12 - 2x)(10 - 2x)$

  Domain: $0 > x > 5$

  (b)

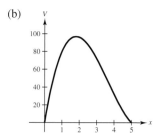

  Approximate measurements:
  2 inches $\times$ 8 inches $\times$ 6 inches or
  1.81 inches $\times$ 8.38 inches $\times$ 6.38 inches

  (c) $x \approx 1.628, 2, \approx 7.372$

  $x \approx 7.372$ inches is physically impossible because
  $10 - 2(7.372)$ and $12 - 2(7.372)$ would yield a negative length and width.

47. 18 inches $\times$ 18 inches $\times$ 36 inches   **49.** 4.5 hours

51. (a)

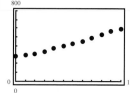

  (b) Linear: $R = 29.37t + 260.0$

   Quadratic: $R = 0.866t^2 + 19.85t + 275.9$

   Cubic: $R = -0.0961t^3 + 2.451t^2 + 13.17t + 280.6$

   Quartic: $R = 0.02320t^4 - 0.6059t^3 + 5.963t^2$
   $+ 5.38t + 283.3$

  (c) Linear   Quadratic

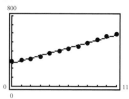   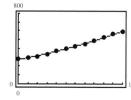

   Cubic   Quartic

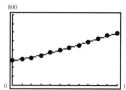   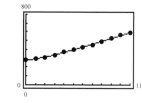

  (d) Answers will vary. Sample answer: The quartic model
   fits the data best because it follows the scatter plot most
   accurately. According to the model, recreation spending
   will reach $700 billion in 2003.

53. (a)

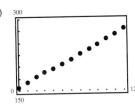

(b) Linear: $y = 10.19t + 156.5$

Quadratic: $y = 0.055t^2 + 9.53t + 157.7$

Cubic: $y = 0.0212t^3 - 0.326t^2 + 11.29t + 156.3$

Quartic: $y = -0.00297t^4 + 0.0925t^3 - 0.863t^2$
$$+ 12.59t + 155.8$$

(c) Linear   Quadratic

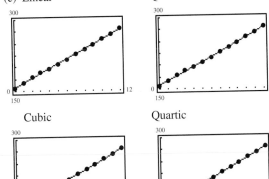

Cubic   Quartic

Each model fits the data well.

(d) 2005: Linear: 309.4   Quadratic: 313.0

Cubic: 323.9   Quartic: 312.1

2007: Linear: 329.7   Quadratic: 335.7

Cubic: 358.2   Quartic: 326.8

Answers will vary. Sample answer: The estimates from each of the four models seem reasonable for the years 2005 and 2007. However, for years after 2010 the cubic model yields values that are too high, and the quartic model yields values that are too low.

55. $384,356

57. (a) 30 and ≈ 38.91

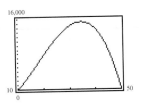

(b) You can solve $-x^3 + 54x^2 - 140x - 3000 = 14,400$ by rewriting the equation as $y = -x^3 + 54x^2 - 140x - 17,400$. Using the *table* feature of a graphing utility, you can approximate the solutions to be $x \approx -14.91$, $x = 30$, and $x \approx 38.91$. The company should charge $38.91.

59. (a) $x = -2, 1, 4$

(b) It touches the x-axis at $x = 1$ but does not cross the x-axis.

(c) f is of fourth degree or higher. The degree of f cannot be odd because its left-hand behavior matches its right-hand behavior.

(d) It's positive because f increases to the right.

(e) Answers will vary.

Sample answer: $f(x) = (x + 2)(x - 1)^2(x - 4)$

(f)

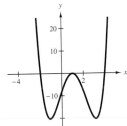

Mid-Chapter Quiz *(page 150)*

1. Vertex: $(-1, -2)$

Intercepts: $(-1 \pm \sqrt{2}, 0), (0, -1)$

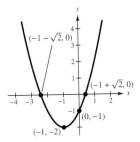

2. Vertex: $(0, 25)$

Intercepts: $(\pm 5, 0), (0, 25)$

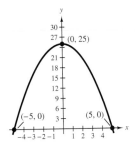

3. Falls to the left
Rises to the right

4. Falls to the left
Falls to the right

5. 13

6. $f(x) = (x - 1)(x^3 + x^2 - 4x - 4) + 0; f(1) = 0$

7. $f(x) = (x + 3)(x^2 + 2x - 8) + 0; f(-3) = 0$

8. $2x^2 + 5x - 12$ **9.** $\pm\sqrt{5}, -\frac{7}{2}$ **10.** $\pm 3, \pm\frac{1}{2}$

11. $1, -\frac{4}{3}$ **12.** $\frac{3}{2}$ **13.** $P = \$2,534,375; \$337,600$

14. (a)

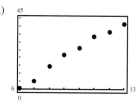

(b) Linear: $A = 6.00t - 34.7$

Quadratic: $A = -0.271t^2 + 11.14t - 57.7$

Cubic: $A = -0.0376t^3 + 0.802t^2 + 1.31t - 28.7$

Quartic: $A = 0.03381t^4 - 1.3223t^3 + 18.676t^2$
$- 106.42t + 208.3$

(c) Linear   Quadratic

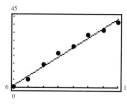   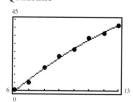

Cubic   Quartic

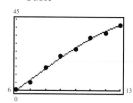   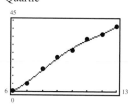

Each model fits the data well.

(d) 2005: Linear: 55.3   Quadratic: 48.4

       Cubic: 44.5   Quartic: 63.0

  2007: Linear: 67.3   Quadratic: 53.4

       Cubic: 40.6   Quartic: 123.9

Answers will vary. Sample answer: The cubic and quartic models are not valid after 2003 because they do not follow the path indicated by the scatter plot. Either the linear or the quadratic model could be accurate, depending on future trends.

Section 2.5 *(page 159)*

Warm Up *(page 159)*

1. $2\sqrt{3}$ **2.** $10\sqrt{5}$ **3.** $\sqrt{5}$ **4.** $-6\sqrt{3}$

5. 12   **6.** 48   **7.** $\frac{\sqrt{3}}{3}$ **8.** $\sqrt{2}$

9. $-\frac{1}{2} \pm \frac{\sqrt{5}}{2}$ **10.** $-1 \pm \sqrt{2}$

1. $i, -1, -i, 1, i, -1, -i, 1, i, -1, -i, 1, i, -1, -i, 1$
$i^{4n} = 1, i^{4n+1} = i, i^{4n+2} = -1, i^{4n+3} = -i,$
n is an integer.

3. $a = 7, b = 12$ **5.** $a = 4, b = -3$

7. $9 + 4i; 9 - 4i$ **9.** $-3 - 2\sqrt{3}i; -3 + 2\sqrt{3}i$

11. $-21; -21$ **13.** $-1 - 6i; -1 + 6i$ **15.** $-5i; 5i$

17. $-3; -3$ **19.** $2 + i$ **21.** $5 + 6i$

23. $3 - 3\sqrt{2}i$ **25.** $\frac{1}{6} + \frac{7}{6}i$ **27.** -10 **29.** $-2\sqrt{6}$

31. -10 **33.** $5 + i$ **35.** 41 **37.** $21 + 42i$

39. $-9 + 40i$ **41.** 8

43. $\left(16 + 4\sqrt{3}\right) + \left(-16\sqrt{2} + 2\sqrt{6}\right)i$

45. $\frac{3}{5} + \frac{4}{5}i$ **47.** $1 + \frac{1}{2}i$ **49.** $-7 - 6i$ **51.** $\frac{1}{8}i$

53. $-\frac{5}{4} - \frac{5}{4}i$ **55.** $\frac{35}{29} + \frac{595}{29}i$

57. Error: $(3 - 2i)(3 + 2i) = 9 - 4i^2 = 9 + 4 = 13$
(not $9 - 4 = 5$)

59. $1 \pm i$ **61.** $-2 \pm \frac{1}{2}i$ **63.** $-\frac{3}{2}, -\frac{5}{2}$

65. $\frac{1}{8} \pm \frac{\sqrt{11}}{8}i$

67. $x = -\frac{1}{2} \pm \frac{\sqrt{7}}{2}i$

The minimum point, the vertex of the parabola, occurs at $\left(-\frac{1}{2}, \frac{7}{4}\right) = (a, b^2)$.

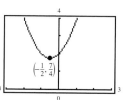

69. $x = -\frac{3}{2} \pm \frac{\sqrt{29}}{2}$

The x-intercepts of the graph correspond to the zeros of the function. The minimum point occurs at $\left(-\frac{3}{2}, -\frac{29}{4}\right) = (a, -b^2)$.

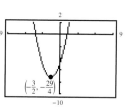

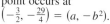

71.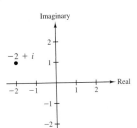

73.

75.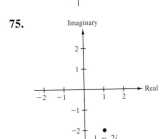

77. The complex number 0 is in the Mandelbrot Set because, for $c = 0$, the corresponding Mandelbrot sequence is 0, 0, 0, 0, 0, 0, which is bounded.

79. The complex number 1 is not in the Mandelbrot Set because, for $c = 1$, the corresponding Mandelbrot sequence is 1, 2, 5, 26, 677, 458,330, . . . , which is unbounded.

81. The complex number $\frac{1}{2}i$ is in the Mandelbrot Set because, for $c = \frac{1}{2}i$, the corresponding Mandelbrot sequence is $\frac{1}{2}i$, $-\frac{1}{4} + \frac{1}{2}i$, $-\frac{3}{16} + \frac{1}{4}i$, $-\frac{7}{256} + \frac{13}{32}i$, $-\frac{10,767}{65,636} + \frac{1957}{4096}i$, $-\frac{864,513,055}{4,294,967,296} + \frac{46,037,845}{134,217,728}i$, which is bounded.

Section 2.6   *(page 167)*

Warm Up   *(page 167)*

1. $4 - \sqrt{29}i, 4 + \sqrt{29}i$ **2.** $-5 - 12i, -5 + 12i$

3. $-1 + 4\sqrt{2}i, -1 - 4\sqrt{2}i$ **4.** $6 + \frac{1}{2}i, 6 - \frac{1}{2}i$

5. $-13 + 9i$ **6.** $12 + 16i$ **7.** $26 + 22i$

8. 29   **9.** i **10.** $-9 + 46i$

1. $\pm 4i; (x + 4i)(x - 4i)$

3. $\dfrac{5 \pm \sqrt{5}}{2}; \left(x - \dfrac{5 + \sqrt{5}}{2}\right)\left(x - \dfrac{5 - \sqrt{5}}{2}\right)$

5. $\pm 4, \pm 4i; (x - 4)(x + 4)(x - 4i)(x + 4i)$

7. $0, \pm\sqrt{5}i; x(x - \sqrt{5}i)(x + \sqrt{5}i)$

9. $-5, 8 \pm i; (x + 5)(x - 8 + i)(x - 8 - i)$

11. $2, 2 \pm i; (x - 2)(x - 2 + i)(x - 2 - i)$

13. $-5, 4 \pm 3i; (t + 5)(t - 4 + 3i)(t - 4 - 3i)$

15. $-10, -7 \pm 5i; (x + 10)(x + 7 - 5i)(x + 7 + 5i)$

17. $-5, -2 \pm \sqrt{3}i; (x + 5)(x + 2 - \sqrt{3}i)(x + 2 + \sqrt{3}i)$

19. $-\frac{3}{4}, 1 \pm \frac{1}{2}i; (4x + 3)(2x - 2 + i)(2x - 2 - i)$

21. $-\frac{1}{5}, 1 \pm \sqrt{5}i; (5x + 1)(x - 1 + \sqrt{5}i)(x - 1 - \sqrt{5}i)$

23. $2, \pm 2i; (x - 2)^2(x + 2i)(x - 2i)$

25. $\pm i, \pm 3i; (x + i)(x - i)(x + 3i)(x - 3i)$

27. $-4, 3, \pm i; (t + 4)^2(t - 3)(t + i)(t - i)$

29. Answers will vary.
Sample answer: $x^3 + 3x^2 + 36x + 108$

31. Answers will vary. Sample answer: $x^3 - 5x^2 + 9x - 5$

33. Answers will vary.
Sample answer: $x^5 + 4x^4 + 13x^3 + 52x^2 + 36x + 144$

35. Answers will vary.
Sample answer: $x^4 + 8x^3 + 9x^2 - 10x + 100$

37. Answers will vary.
Sample answer: $3x^4 - 17x^3 + 25x^2 + 23x - 22$

39. (a) $(x^2 - 8)(x^2 + 1)$

(b) $\left(x - 2\sqrt{2}\right)\left(x + 2\sqrt{2}\right)(x^2 + 1)$

(c) $\left(x - 2\sqrt{2}\right)\left(x + 2\sqrt{2}\right)(x - i)(x + i)$

41. (a) $(x^2 - 2x - 2)(x^2 - 2x + 3)$

(b) $\left(x - 1 + \sqrt{3}\right)\left(x - 1 - \sqrt{3}\right)(x^2 - 2x + 3)$

(c) $\left(x - 1 + \sqrt{3}\right)\left(x - 1 - \sqrt{3}\right) \cdot$
$\left(x - 1 + \sqrt{2}i\right)\left(x - 1 - \sqrt{2}i\right)$

43. $\pm 4i, \frac{5}{3}$ **45.** $\pm 6i, 1$ **47.** $-3 \pm i, \frac{1}{4}$

49. $1, 2, -3 \pm \sqrt{2}i$ **51.** $\frac{3}{4}, \frac{1}{2} \pm \frac{\sqrt{5}}{2}i$

53. $\pm 1, \pm 2.$ The x-axis intercepts occur at the solutions of the equation.

55. Answers will vary. Sample answer:
$f(x) = x^4 - 7x^3 + 17x^2 - 17x + 6$
Zeros: 1, 1, 2, 3
$g(x) = x^4 - 3x^3 + 3x^2 - 3x + 2$
Zeros: 1, 2, $\pm i$
$h(x) = x^4 + 5x^2 + 4$
Zeros: $\pm i, \pm 2i$

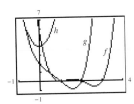

Similarities: f, g, and h all rise to the left and rise to the right, and are similar in shape.

Differences: f, g, and h have differing numbers of x-intercepts.

57. Solving the equation for $P = 9,000,000$ and graphing shows no real solutions.

59. No. The conjugate pair statement specifies polynomials with *real* coefficients. $f(x)$ has imaginary coefficients.

61. f's imaginary roots can only occur in pairs. f has only one unknown root and no unpaired complex roots, so the unknown root must be real.

63. f's imaginary roots can only occur in pairs. Because f has three roots, one or all of them must be real numbers.

65. Polynomials of odd degree eventually rise to one side and fall to the other side. Thus, f must cross the x-axis, so f must have a real root. You can show this by graphing several third-degree polynomials.

Section 2.7   *(page 177)*

Warm Up   *(page 177)*

1. $x(x - 4)$ **2.** $2x(x^2 - 3)$ **3.** $(x - 5)(x + 2)$

4. $(x - 5)(x - 2)$ **5.** $x(x + 1)(x + 3)$

6. $(x^2 - 2)(x - 4)$

7.    **8.**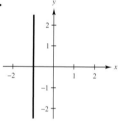

9. 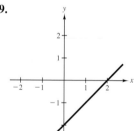   **10.**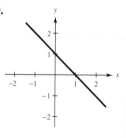

1. (a) All $x \neq -4$ (b) $y = 0$

3. (a) All $x \neq 5$ (b) $y = -1$

5. (a) All reals   (b) $y = 3$

7. (a) All reals   (b) None

9. (a) None
(b) None
(c) None

11. (a) $x = -1$
(b) None
(c) $y = x - 1$

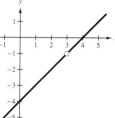

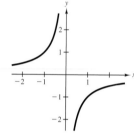

13. f   **14.** e   **15.** a   **16.** b   **17.** c   **18.** d

19. $g(x)$ shifts down two units. **21.** $g(x)$ is a reflection about the x-axis.

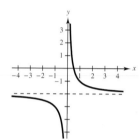

23. $g(x)$ shifts up three units.   **25.** $g(x)$ is a reflection about the x-axis.

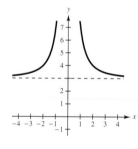   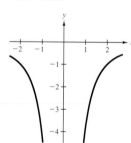

27. $g(x)$ shifts up five units.   **29.** $g(x)$ is a reflection about the x-axis.

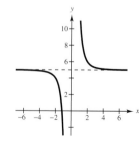   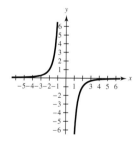

31.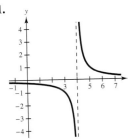

33.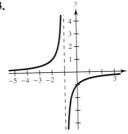

35.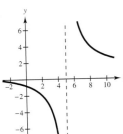

37.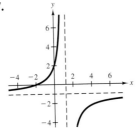

39.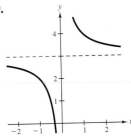

41.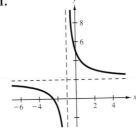

43.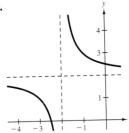

45.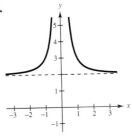

47.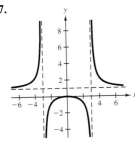

49.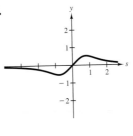

51.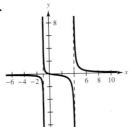

53.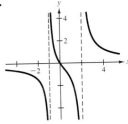

55. Answers will vary.
Sample answer:
$$f(x) = \frac{1}{x^2 + 1}$$

57. Answers will vary.
Sample answer:
$$f(x) = \frac{x}{x^2 - 2x - 3}$$

59. No. Given
$$f(x) = \frac{a_n x^n + \cdots + a_0}{b_m x^m + \cdots + b_0}$$
if $n > m$, there is no horizontal asymptote and n must be greater than m for a slant asymptote to occur.

61. (a) $176 million; $528 million; $1584 million
(b) No. The model would generate a zero in the denominator.

63. (a) 167 deer; 250 deer; 400 deer   (b) 750 deer

65. (a)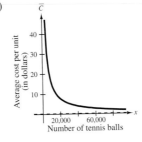

(b)

x	1000	10,000	100,000
$\overline{C}$	$150.25	$15.25	$1.75

Eventually, the average cost per tennis ball will approach the horizontal asymptote of $0.25.

67. (a)

n	1	2	3	4	5
P	0.50	0.74	0.82	0.86	0.89

n	6	7	8	9	10
P	0.91	0.92	0.93	0.94	0.95

(b) 100%

69. (a)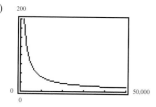

(b)

x	1000	10,000	100,000
$\overline{C}$	\$355	\$40	\$8.5

Eventually, the average recycling cost per pound will approach the horizontal asymptote of \$5.

71. (a) $S = \dfrac{2384.62t^2 + 36{,}846.2t + 728{,}154}{3.1981t + 250.452}$

(b) 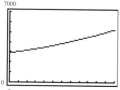   (c) 8.16 million

73. (a)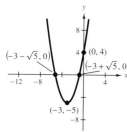

(b) 0.813 minutes $\approx$ 48.78 seconds

(c) This model does not have a horizontal asymptote. A model with a horizontal asymptote would be reasonable for this data because the winning times should continue to improve, but it is not humanly possible for the winning times to decrease without bound.

Review Exercises   *(page 183)*

1.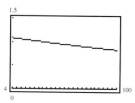

Vertex: $(-3, -5)$
Intercepts:

$(0, 4), \left(-3 \pm \sqrt{5}, 0\right)$

3.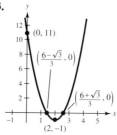

Vertex: $(2, -1)$
Intercepts:

$(0, 11), \left(\dfrac{6 \pm \sqrt{3}}{3}, 0\right)$

5. $f(x) = \frac{7}{9}(x + 5)^2 - 1$

7. $A(x) = -(x - 50)^2 + 2500$; 50 feet $\times$ 50 feet

9. (a)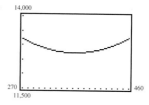

(b) About 363 or 364 units

(c) Calculate the vertex from the equation:

$$\left(\frac{4000}{11}, \frac{1{,}540{,}000}{121}\right)$$

The cost is minimum when $x = \frac{4000}{11} \approx 364$ units.

11. (a)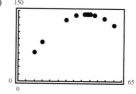

(b) $d(x) = -0.077x^2 + 6.59x + 2.4$

(c)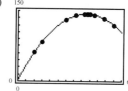

(d) $42.8°$

13. Falls to the left
Rises to the right

15. Rises to the left
Falls to the right

17. ± 4 **19.** $0, 2, 5$ **21.** $x^2 - 3x + 1 - \dfrac{1}{2x + 1}$

23. $x^2 + 11x + 24$ **25.** $x^2 - 3x + 3$

27. (a) -10 (b) 11

29. (a) Answers will vary.   (b) $x - 2$

(c) $(x - 5)(x + 3)(x - 2)$ (d) $-3, 2, 5$

(e)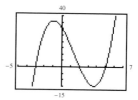

31. $x^2 + 11x + 28$ square feet

33. $\pm 1, \pm 3, \pm 5, \pm 15, \pm\frac{1}{2}, \pm\frac{3}{2}, \pm\frac{5}{2}, \pm\frac{15}{2}, \pm\frac{1}{4}, \pm\frac{3}{4}, \pm\frac{5}{4}, \pm\frac{15}{4}$

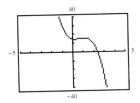

From the graph: $x \approx 2.357$

35. $-3, -1, 2$ **37.** $\pm 2, \pm\sqrt{5}$ **39.** $-2, 1, \pm\dfrac{\sqrt{3}}{3}$

41. $x \approx -2.3$ **43.** $-1.321, -0.283, 1.604$

45. $212{,}000$ **47.** $4\sqrt{2}i; -4\sqrt{2}i$

49. $-3 + 4i; -3 - 4i$ **51.** $5 + i$ **53.** $11 + 9\sqrt{3}i$

55. 89 **57.** $-10 - 8i$ **59.** $-7 + 24i$ **61.** 10

63. $3 - 2i$ **65.** $-3 - 4i$

67. $\dfrac{1 \pm \sqrt{23}i}{4}$ **69.** $\dfrac{-11 \pm \sqrt{73}}{8}$

71.

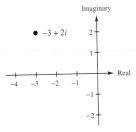

73. $\pm 3, \pm 3i; (x - 3)(x + 3)(x - 3i)(x + 3i)$

75. $-5, \pm\sqrt{3}i; (t + 5)\left(t - \sqrt{3}i\right)\left(t + \sqrt{3}i\right)$

77. $2, \pm\dfrac{3i}{2}; (x - 2)(2x - 3i)(2x + 3i)$

79. Answers will vary. Sample answer: $x^3 - 3x^2 + 16x - 48$

81. (a) $(x^2 + 8)(x^2 - 3)$
   (b) $(x^2 + 8)\left(x - \sqrt{3}\right)\left(x + \sqrt{3}\right)$
   (c) $\left(x + 2\sqrt{2}i\right)\left(x - 2\sqrt{2}i\right)\left(x - \sqrt{3}\right)\left(x + \sqrt{3}\right)$

83. $\pm 4i, \frac{1}{4}$ **85.** $-1 \pm 3i, -1, -4$

87. Domain: All $x \neq 6$
   Vertical asymptote: $x = 6$
   Horizontal asymptote: $y = 0$

89. Domain: All $x \neq \pm 2$
   Vertical asymptotes: $x = -2, x = 2$
   Horizontal asymptote: $y = 1$

91.

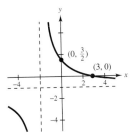

93.

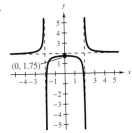

95. $x = \pm 1, y = x$

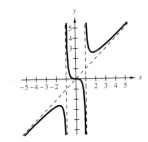

97. (a)

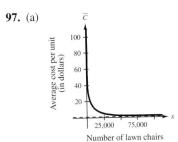

(b)

x	1000	10,000	100,000
$\overline{C}$	\$100.90	\$10.90	\$1.90

Eventually, the average cost per lawn chair will approach the horizontal asymptote of \$0.90.

99. (a) 304,000 fish; 453,330 fish; 702,220 fish
   (b) 1,200,000 fish

101. (a) $\approx \$31{,}666.67$ (b) \$142,500   (c) \$9,405,000
   (d) No. The model would generate a zero in the denominator at $p = 100$.

Chapter Test *(page 187)*

1. Vertex: $(-1, 2)$
   Intercepts: $(-1 \pm \sqrt{2}, 0), (0, 1)$

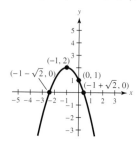

2. (a) Falls to the left   (b) Rises to the left
        Rises to the right   Rises to the right

3. $x^2 + 7x + 12$

4. $\pm 1, \pm 2, \pm 3, \pm 4, \pm 5, \pm 6, \pm 10, \pm 12, \pm 15, \pm 20, \pm 30,$
   $\pm 60, \pm \frac{1}{2}, \pm \frac{3}{2}, \pm \frac{5}{2}, \pm \frac{15}{2}$
   $f\left(\frac{5}{2}\right) = 0$
   $f(x) = \left(x - \frac{5}{2}\right)(2x - 6)(x + 4)$

5. $16 - 3i$ **6.** $7 - 9i$ **7.** $27 - 4\sqrt{3}\,i$

8. $23 - 14i$ **9.** i

10. $x = \dfrac{-5 \pm \sqrt{3}\,i}{2}$ **11.** $x = \dfrac{5 \pm 3\sqrt{7}\,i}{4}$

12. $x^4 - 7x^3 + 19x^2 - 63x + 90$ **13.** $\pm \sqrt{5}\,i, -2$

14.

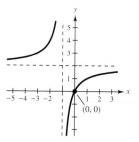

Intercept: $(0, 0)$
Vertical asymptote:
   $x = -1$
Horizontal asymptote:
   $y = 2$
Domain: All $x \neq -1$

15. \$200,000 ($x = 20$) and \$777,250 ($x = 77.725$) both result in profits of \$222,000. The interior decorating company should spend \$200,000, which is the smaller amount.

16.

x	10,000	100,000	1,000,000
$\overline{C}$	\$50	\$9.50	\$5.45

Conclusion: High volume yields lower average cost of recycling.

Chapter 3

Section 3.1 *(page 198)*

Warm Up *(page 198)*

1. 5^x **2.** 3^{2x} **3.** 4^{3x} **4.** 10^x **5.** 4^{2x}

6. 4^{10x} **7.** $\left(\frac{3}{2}\right)^x$ **8.** 4^{3x} **9.** 2^{-x} **10.** 2^x

1. 3.463   **3.** 95.946   **5.** 0.079   **7.** 54.598

9. 1.948   **11.** g   **12.** e   **13.** b   **14.** h   **15.** d

16. a   **17.** f   **18.** c

19.

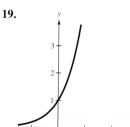

21.

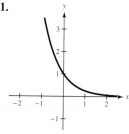

23.

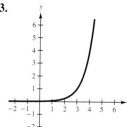

25.

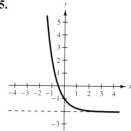

27.

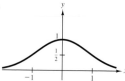

29.

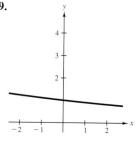

31.

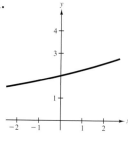

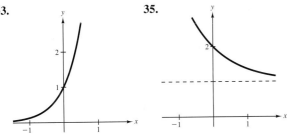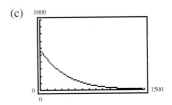

33.

35.

51. (a) $32.92

(b) $7.39

(c)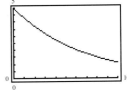

(d) No. The graph shows the demand curve, but not the cost curve.

53. (a) 100   (b) 300   (c) 900

55. 198.68 kilograms

37.

n	1	2	4
A	$7346.64	$7401.22	$7429.74

n	12	365	Continuous
A	$7449.23	$7458.80	$7459.12

39.

n	1	2	4
A	$24,115.73	$25,714.29	$26,602.23

n	12	365	Continuous
A	$27,231.38	$27,547.07	$27,557.94

57. (a)

(b) 1.11 pounds

(c) On the graph, when $P = 2.5$, $t \approx 4.6$ months.

59. (a)

Year	1994	1997	2000
Estimated shares (in millions)	77,000	140,000	250,000

(b)

Year	1994	1997	2000
Shares (in millions)	77,025	138,910	251,706

41.

t	1	10	20
P	$91,393.12	$40,656.97	$16,529.89

t	30	40	50
P	$6720.55	$2732.37	$1110.90

43.

t	1	10	20
P	$90,521.24	$36,940.70	$13,646.15

t	30	40	50
P	$5040.98	$1862.17	$687.90

61. (a)

Year	1980	1990	2002
Estimated age (in years)	22	24	25.3

(b)

Year	1980	1990	2002
Age (in years)	22.1	24.0	25.3

45. The account receiving 5% compounded quarterly earns more money. Even though the interest is compounded less frequently, the higher interest rate yields a higher return.

47. You should choose the account with free online access. The account receiving 5.75% compounded monthly yields a higher return, but the $5 monthly access fee reduces the total to less than the other account's return.

49. (a) $19,691.17   (b) $98,455.84   (c) $196,911.68

63. Women tend to marry about 2 years younger than men do. The median ages of both have been rising, and the age difference is decreasing.

Section 3.2   *(page 209)*

Warm Up   *(page 209)*

1. 3   **2.** 0   **3.** −1   **4.** 1   **5.** 7.389

6. 0.368   **7.** Graph is shifted two units to the left.

8. Graph is reflected about the *x*-axis.

9. Graph is shifted downward one unit.

10. Graph is reflected about the *y*-axis.

1. $\log_4 256 = 4$ **3.** $\log_{81} 3 = \frac{1}{4}$ **5.** $\log_6 \frac{1}{36} = -2$

7. $\ln e = 1$ **9.** $\ln 4 = x$ **11.** $4^2 = 16$

13. $2^{-1} = \frac{1}{2}$ **15.** $e^1 = e$ **17.** $5^{-1} = 0.2$

19. $27^{1/3} = 3$ **21.** 2   **23.** −4   **25.** $\frac{1}{3}$ **27.** 1

29. −4   **31.** 1   **33.** −4   **35.** 5   **37.** 2.538

39. −0.097   **41.** 0.452   **43.** 1.946   **45.** 2.913

47. 0.549

49.

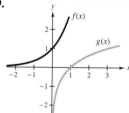

51.

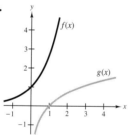

53. d   **54.** e   **55.** a   **56.** c   **57.** f   **58.** b

59. Domain: $(0, \infty)$ **61.** Domain: $(-4, \infty)$

Asymptote: $x = 0$ Asymptote: $x = -4$

x-intercept: $(1, 0)$ *x*-intercept: $(-3, 0)$

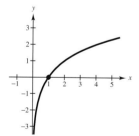   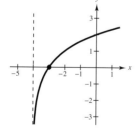

63. Domain: $(0, \infty)$ **65.** Domain: $(-\infty, 0)$

Asymptote: $x = 0$ Asymptote: $x = 0$

x-intercept: $(1, 0)$ *x*-intercept: $(-1, 0)$

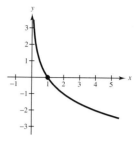

67. Domain: $(-1, \infty)$ **69.** $t \approx 23.7$ years

Asymptote: $x = -1$

x-intercept: $(0, 0)$

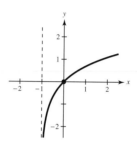

71. (a) 80.0   (b) 68.1   (c) In the third month $(t \approx 2.9)$

73. (a)

K	1	2	4	6	8	10	12
t	0	10.7	21.3	27.6	32.0	35.4	38.2

(b)

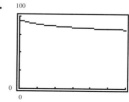

75.

The domain $0 \le t \le 12$ covers the period of 12 months or 1 year. The range $0 \le f(t) \le 95$ covers the possible scores.

77. (a) ≈ 29.7 years   (b) ≈ 33.1 years

(c) ≈ 5.2 years   (d) ≈ 8.7 years

Section 3.3   *(page 217)*

Warm Up   *(page 217)*

1. 2   **2.** -5 **3.** -2 **4.** -3 **5.** e^5

6. $\dfrac{1}{e}$ **7.** e^6 **8.** 1   **9.** x^{-2} **10.** $x^{1/2}$

1. $\dfrac{\log_{10} 8}{\log_{10} 5}$ **3.** $\dfrac{\log_{10} 30}{\log_{10} e}$ **5.** $\dfrac{\log_{10} n}{\log_{10} 3}$ **7.** $\dfrac{\log_{10} x}{\log_{10} \frac{1}{5}}$

9. $\dfrac{\log_{10} \frac{3}{10}}{\log_{10} x}$ **11.** $\dfrac{\log_{10} x}{\log_{10} 2.6}$ **13.** $\dfrac{\ln 8}{\ln 5}$ **15.** $\dfrac{\ln 5}{\ln 10}$

17. $\dfrac{\ln n}{\ln 3}$ **19.** $\dfrac{\ln x}{\ln \frac{1}{5}}$ **21.** $\dfrac{\ln \frac{3}{10}}{\ln x}$ **23.** $\dfrac{\ln x}{\ln 2.6}$

25. 2.585   **27.** 1.079   **29.** 2.633   **31.** -0.683

33. -1.661 **35.** 2.322   **37.** 1.1833   **39.** -0.2084

41. 1.0686   **43.** 0.1781   **45.** 1.8957   **47.** -2.7124

49. 0.9136   **51.** $\frac{1}{3}$ **53.** $-\frac{1}{2}$ **55.** -3

57. $\log_9 \frac{1}{2} - 1$ **59.** $\frac{1}{2} + \frac{1}{2}\log_7 10$ **61.** $-3 - \log_5 2$

63. $6 + \ln 5$ **65.** $\log_3 4 + \log_3 n$

67. $\log_5 x - \log_5 25$ or $\log_5 x - 2$ **69.** $4\log_2 x$

71. $\frac{1}{2}\ln z$ **73.** $\ln x + \ln y + \ln z$ **75.** $\frac{1}{2}\ln(a - 1)$

77. $2\ln(z - 1) - \ln z$ **79.** $2\log_6 x - 3\log_6 y$

81. $\frac{1}{3}(\ln x - \ln y)$ **83.** $\frac{3}{4}\ln x + \frac{1}{4}\ln(x^2 + 3)$

85. $\log_3 5x$ **87.** $\log_4\left(\dfrac{8}{x}\right)$ **89.** $\log_{10}(x + 4)^2$

91. $\ln\left(\dfrac{1}{216x}\right)$ **93.** $\ln\left(\dfrac{\sqrt[3]{5x}}{x + 1}\right)$ **95.** $\log_8\left(\dfrac{x - 2}{x + 2}\right)$

97. $\ln\dfrac{3^2}{\sqrt{x^2 + 1}}$ **99.** $\ln\left[\dfrac{x}{(x + 2)(x - 2)}\right]$

101. $\ln y = \frac{1}{4}\ln x$ **103.** $\ln y = -\frac{1}{4}\ln x + \ln\frac{5}{2}$

105. $\ln y = \frac{2}{3}\ln x + \ln 0.07$ **107.** ≈ 26 decibels

109. Graph the two functions in the same viewing window to show that their graphs coincide. Functions and graphs will vary.

111. Let $\log_a u = x$ and $\log_a v = y$.

$a^x = u$ and $a^y = v$

$\dfrac{u}{v} = \dfrac{a^x}{a^y} = a^{x-y}$

$\log_a a^{x-y} = x - y$

$\log_a \dfrac{u}{v} = \log_a u - \log_a v$

Mid-Chapter Quiz   *(page 220)*

1.
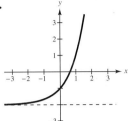

2.
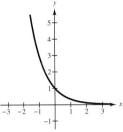

3.
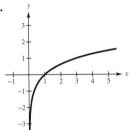

4.
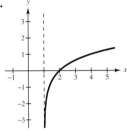

5. Monthly: $9560.92

Continuously: $9577.70

6. (a) 1997: 145.98 billion   (b) 2006: 811.61 billion

2002: 378.62 billion   2007: 982.05 billion

7. (a) 100   (b) 609   (c) 3713

8. $235.62   **9.** 2   **10.** 4   **11.** -2 **12.** 0

13.
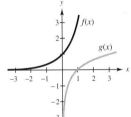

Domain of $f(x)$: $(-\infty, \infty)$

Domain of $g(x)$: $(0, \infty)$

Graphs are reflections about the line $y = x$.

14. $\frac{3}{2}$ **15.** $\frac{6}{5}$ **16.** $\frac{1}{3}(\log_{10} x + \log_{10} y - \log_{10} z)$

17. $\ln(x^2 + 3) - 3\ln x$ **18.** $\ln\left(\dfrac{xy}{3}\right)$ **19.** $\log_{10}\left(\dfrac{1}{64x^3}\right)$

20. $\ln y = \frac{1}{3}\ln x$

Section 3.4   *(page 228)*

Warm Up   *(page 228)*

1. $\dfrac{\ln 3}{\ln 2}$ **2.** $1 + \dfrac{2}{\ln 4}$ **3.** $\dfrac{e}{2}$ **4.** $2e$

5. $2 \pm i$ **6.** $\frac{1}{2}, 1$ **7.** x **8.** $2x$

9. $2x$ **10.** $-x^2$

1. 3   **3.** -2 **5.** 2   **7.** 64   **9.** $\frac{1}{10}$

11. x^2 **13.** $x^2 + 1$ **15.** $x^3 - 7$ **17.** $-8 + x^3$

19. $x + 5$ **21.** x^2 **23.** $\log_3 5 \approx 1.465$

25. $\ln 5 \approx 1.609$ **27.** $\ln 28 \approx 3.332$

29. $\frac{1}{2} \log_3 80 \approx 1.994$ **31.** 2

33. $\log_3 28 + 1 \approx 4.033$ **35.** $3 - \log_2 565 \approx -6.142$

37. $\frac{1}{3} \log_{10} \frac{3}{2} \approx 0.059$ **39.** $\log_5 7 + 1 \approx 2.209$

41. $\frac{1}{3} \ln 12 \approx 0.828$ **43.** $\ln\left(\frac{5}{3}\right) \approx 0.511$

45. $\ln \frac{1}{2} \approx -0.693$ **47.** $\frac{1}{3} + \frac{1}{3} \log_2 \frac{8}{3} \approx 0.805$

49. $\ln 5 \approx 1.609$ **51.** $\ln 4 \approx 1.386$

53. $2 \ln 75 \approx 8.635$ **55.** $\frac{1}{2} \ln 1498 \approx 3.656$

57. $\dfrac{\ln 4}{365 \ln\left(1 + \dfrac{0.065}{365}\right)} \approx 21.330$

59. $\dfrac{\ln 2}{12 \ln\left(1 + \dfrac{0.10}{12}\right)} \approx 6.960$

61. $e^{-3} \approx 0.050$ **63.** $\dfrac{e^{2.4}}{2} \approx 5.512$ **65.** 1,000,000

67. $\dfrac{3^{11/6}}{0.5} \approx 14.988$ **69.** $\frac{1}{5} e^{10/3} \approx 5.606$

71. $e^2 - 2 \approx 5.389$ **73.** $e^{-2/3} \approx 0.513$

75. $\dfrac{e^2}{1 - e^2} \approx -1.157$ **77.** $1 + \sqrt{1 + e} \approx 2.928$

79. No solution   **81.** 7   **83.** 1.562   **85.** 2

87. 180.384   **89.** $\log_2 7 \approx 2.807$ **91.** $e^3 \approx 20.086$

93. ≈ 8.15 years   **95.** ≈ 16.28 years   **97.** 26 months

99. (a) ≈ 184 model cars   (b) ≈ 240 model cars

(c)

101. (a) ≈ 29.3 years   (b) ≈ 39.8 years

103. (a)
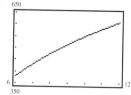

(b), (c) $1999 \, (t \approx 8.64)$

105. (a)

x	y
0.2	162.6
0.4	78.5
0.6	52.5
0.8	40.5
1.0	33.9

(b)
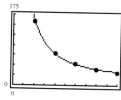

The model looks like a good fit for the data.

(c) 1.197 meters

(d) No. To reduce the g's to fewer than 23 requires a crumple zone of more than 2.27 meters, a length that exceeds the front width of most cars.

Section 3.5   *(page 239)*

Warm Up   *(page 239)*

1.   **2.**
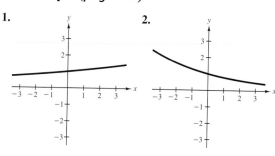

3.   **4.**
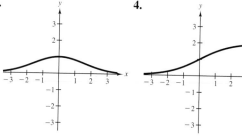

5.   **6.**
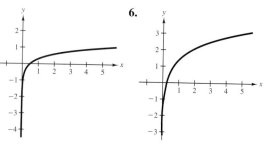

7. $\frac{1}{2} \ln \frac{7}{3} \approx 0.424$ **8.** $\frac{1}{5} e^{7/2} \approx 6.623$

9. $-5 \ln 0.001 \approx 34.539$ **10.** $\frac{1}{2} e^2 \approx 3.695$

	Initial Investment	Annual % Rate	Time to Double	Amount After 10 Years
1.	$5000	7%	9.90 yr	$10,068.76
3.	$500	6.93%	10 yr	$1,000.00
5.	$1000	8.25%	8.40 yr	$2281.88
7.	$6392.79	11%	6.30 yr	$19,205.00
9.	$5000	8%	8.66 yr	$11,127.70

	Isotope	Half-Life (Years)	Initial Quantity	Amount After 1000 Years
11.	^{226}Ra	1599	4 g	2.59 g
13.	^{14}C	5715	3.95 g	3.5 g
15.	^{239}Pu	24,100	1.65 g	1.6 g

17. Exponential growth   **19.** Exponential decay

21. $C = 1, k = \frac{1}{4} \ln 10 \approx 0.5756$

23. $C = 1, k = \frac{1}{4} \ln \frac{1}{4} \approx -0.3466$

25.

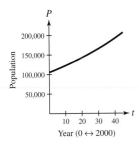

Year (0 ↔ 2000)

2024 ($t \approx 23.6$)

27. ≈ 6.84 hours to double

   ≈ 10.84 hours to triple

29. 12,180 years   **31.** $\approx 5.928 \times 10^{-13}$ **33.** $\approx 9.92\%$

35. (a) $N = 30(1 - e^{-0.05t})$ (b) 36 days

37. (a)

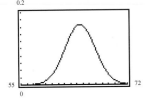

   (b) y changed from an exponential growth function to a decreasing linear function.

39.

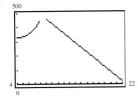

64.9 inches

41. (a)

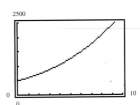

   (b) ≈ 1252 fish

   (c) ≈ 7.8 months

43. (a)

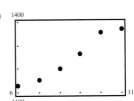

   (b) $S = 861.21e^{0.0431t}$

   (c)

   (d) 2002: $1445 million

   2003: $1508 million

45. (a)

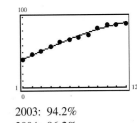

   (b) $P = 41.04e^{0.0744t}$

   (c) $P = \dfrac{105.294}{1 + 2.014e^{-0.2182t}}$

   (d) Exponential model   Logistic model

   2003: 108.0%   2003: 94.2%
   2004: 116.3%   2004: 96.2%

   (e) The logistic model is reasonable. The exponential model is not reasonable because its predictions exceed 100%.

47. (a) ≈ 7.906 (b) ≈ 7.684

49. (a) ≈ 20 decibels   (b) ≈ 70 decibels

51. $10^{-5.8} \approx 1.585 \times 10^{-6}$ **53.** $10^{4.5} \approx 31,623$

55. 3:00 A.M.

57.

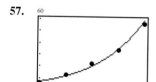

The data fit an exponential model.

Exponential model: $n = 3.9405e^{0.1086t}$

A logistic growth model would be more appropriate for this data because after the initial rapid growth in productivity, the worker's production rate will eventually level off.

Review Exercises   *(page 246)*

1. d   **2.** f   **3.** a   **4.** b

5. c   **6.** e   **7.** h   **8.** g

9.

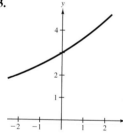

11.

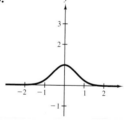

13.

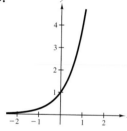

15.

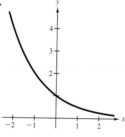

17.

n	1	2	4
A	\$9499.28	\$9738.91	\$9867.22

n	12	365	Continuous
A	\$9956.20	\$10,000.27	\$10,001.78

19.

t	1	10	20
P	\$184,623.27	\$89,865.79	\$40,379.30

t	30	40	50
P	\$18,143.59	\$8152.44	\$3663.13

21. \$10,144.60   **23.** $\log_4 64 = 3$

25. $\ln 7.3890\ldots = 2$ **27.** $3^4 = 81$ **29.** $e^0 = 1$

31. 5   **33.** 7   **35.** $-\frac{1}{2}$

37.

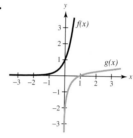

39. Domain: $(4, \infty)$

Vertical asymptote: $x = 4$

x-intercept: $(5, 0)$

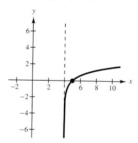

41. Domain: $(0, \infty)$

Vertical asymptote: $x = 0$

x-intercept: $(1, 0)$

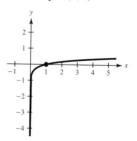

43. The average score decreased from 85 to about 75.

45. (a) 53.42 inches

(b)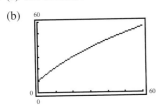

47. 2.096   **49.** 2.132   **51.** 0.9208   **53.** 0.2823

55. 2   **57.** 3.2   **59.** $\log_{10} x - \log_{10} y$

61. $\ln x + \frac{1}{2} \ln(x - 3)$ **63.** $3 \log_{10}(y - 5)$

65. $\log_4 6$ **67.** $\ln \sqrt{x}$ **69.** $\ln \dfrac{x}{(x - 3)(x + 1)}$

71. $\ln y = \frac{4}{3} \ln x$ **73.** $\ln 8 \approx 2.079$

75. $\dfrac{\log_{10} 146}{-3(\log_{10} 45)} \approx -0.436$ **77.** 1

79. $\frac{1}{3} e^{8.2} \approx 1213.650$ **81.** $\frac{1}{5} e^2 \approx 1.478$

83. $3e^2 \approx 22.167$ **85.** 7

87. (a) 197 desks   (b) 257 desks   **89.** 10.63 g

91. (a)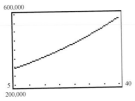

(b) 2009 $(t \approx 8.9)$

93. ≈ 4.54 hours; ≈ 7.19 hours

95. (a) $N = 50(1 - e^{-0.04838t})$ (b) 48 days

97. (a) $k \approx 0.3615$ (b) 302 deer   **99.** Yes

Chapter Test   *(page 250)*

1.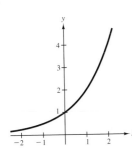

2.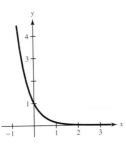

3.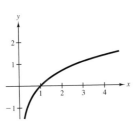

4.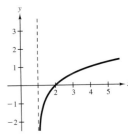

5. 87; 79.8; 76.5

6. After 12 months: 70.3
After 18 months: 67.8
Slowly over time, some (but not all) memory fades.

7. $2 \ln x + 3 \ln y - \ln z$

8. $\log_{10} 3 + \log_{10} x + \log_{10} y + 2 \log_{10} z$

9. $\log_2 x + \frac{1}{3} \log_2(x - 2)$ **10.** $\frac{1}{5} \log_8(x^2 + 1)$

11. $\ln \dfrac{yz^2}{x^3}$ **12.** $\log_{10} \sqrt[3]{x^2 y^2}$ **13.** ≈ 1.0981

14. $\ln 6 \approx 1.792$
$\ln 2 \approx 0.693$

15. 2   **16.** $e^6 - 2 \approx 401.429$ **17.** ≈ 16.6 years

18. About 2016 $(t \approx 15.5)$. Because the exponent on e is positive, the population of the city is growing.

19. ≈ 3.4 hours   **20.** ≈ 2.97 grams; ≈ 0.88 gram

21. Answers will vary.

Cumulative Test: Chapters 3–5   *(page 251)*

1. $x^2 + 3x - 4$ **2.** $x^2 - 3x + 6$

3. $3x^3 - 5x^2 + 3x - 5$ **4.** $\dfrac{x^2 + 1}{3x - 5}, x \neq \dfrac{5}{3}$

5. $9x^2 - 30x + 26$ **6.** $3x^2 - 2$

7.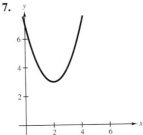

Domain: $(-\infty, \infty)$
Range: $[3, \infty)$

8.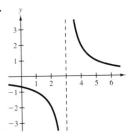

Domain: $(-\infty, 3) \cup (3, \infty)$
Range: $(-\infty, 0) \cup (0, \infty)$

9.

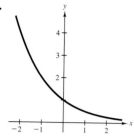

Domain: $(-\infty, \infty)$
Range: $(0, \infty)$

10.

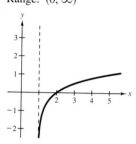

Domain: $(1, \infty)$
Range: $(-\infty, \infty)$

11.

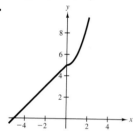

Domain: $(-\infty, \infty)$
Range: $(-\infty, \infty)$

12. 75,000 units   **13.** $38 - 34i$ **14.** $-9 + 40i$

15. $\dfrac{1 + 7i}{10}$ or $\dfrac{1}{10} + \dfrac{7}{10}i$ **16.** $\dfrac{5 \pm \sqrt{59}\, i}{6}$

17. $-3i, 3i, i, -i$. Because a given zero is $3i$ and $f(x)$ is a polynomial with real coefficients, then $-3i$ is also a zero. Using these two zeros, you can form the factors $(x + 3i)$ and $(x - 3i)$. Multiplying these two factors produces $x^2 + 9$. Using long division to divide $x^2 + 9$ into f produces $x^2 + 1$. Then, factoring $x^2 + 1$ gives the other two zeros of i and $-i$.

18. (a) $3x - 2 + \dfrac{2 - 3x}{2x^2 + 1}$

(b) $2x^3 - x^2 + 2x - 10 + \dfrac{25}{x + 2}$

19. $\ln 3 \approx 1.099$ **20.** $3 + e^{12} \approx 162{,}757.791$

$\ln 8 \approx 2.079$

21.

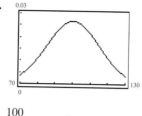
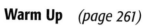

100

Chapter 4

Section 4.1   *(page 261)*

Warm Up   *(page 261)*

1.

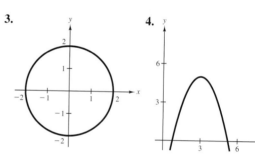

2.

3.

4.

5. x **6.** $-37v$ **7.** $2x^2 + 9$ **8.** -1

9. $x = 6$ **10.** $y = 1$

1. (a) No (b) Yes   **3.** (a) No (b) No   **5.** $(2, -3)$

7. $(-1, 2), (2, 5)$ **9.** $(-2, -2), (0, 0), (2, 2)$

11. $(0, 4), \left(\frac{12}{5}, -\frac{16}{5}\right)$ **13.** $(-1, 0), (1, 0)$ **15.** $(-1, 1)$

17. $\left(\frac{1}{2}, 3\right)$ **19.** $(10, 3)$ **21.** $(1.5, 0.3)$ **23.** $\left(\frac{20}{3}, \frac{40}{3}\right)$

25. No points of intersection   **27.** $\left(1 \pm \sqrt{2}, 2 \pm \sqrt{2}\right)$

29. $\left(\frac{29}{10}, \frac{21}{10}\right), (-2, 0)$ **31.** No points of intersection

33. $(0, 1), (\pm 1, 0)$ **35.** $(2, 1)$ **37.** $\left(\frac{1}{2}, \frac{3}{4}\right), (-3, -1)$

39. $(1, 4), (4, 7)$ **41.** $(0, 1)$

43. No points of intersection   **45.** 192 units

47. 233,333 units   **49.** 1200 CDs

51. 2000 ($t \approx 10.46$)   **53.** \$10,000 at 8%; \$15,000 at 14%

55. \$162,500

57. According to the model, ACT testing will not overtake SAT testing. The model for SAT participants may continue to be accurate, because it continues to increase. But the ACT model eventually decreases and becomes negative; it is unlikely that this model will continue to be accurate.

Section 4.2   *(page 272)*

Warm Up   *(page 272)*

1.

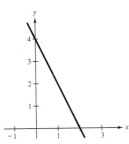

2.

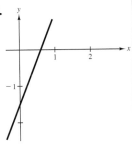

3. $x - y + 4 = 0$ **4.** $5x + 3y - 28 = 0$

5. $-\frac{1}{2}$ **6.** $\frac{7}{4}$ **7.** Perpendicular   **8.** Parallel

9. Neither parallel nor perpendicular

10. Perpendicular

---

1. $(2, 2)$ **3.** $(2, 0)$ **5.** Inconsistent

7. $(2a, 3a - 3)$, where a is any real number

9. $\left(-\frac{1}{3}, -\frac{2}{3}\right)$ **11.** $(-7, -13)$ **13.** $(4, -5)$

15. $(4, -1)$ **17.** $(40, 40)$ **19.** $\left(-\frac{6}{35}, \frac{43}{35}\right)$

21. Inconsistent   **23.** $\left(\frac{18}{5}, \frac{3}{5}\right)$ **25.** $\left(\frac{19}{7}, -\frac{2}{7}\right)$

27. $\left(a, \frac{5}{6}a - \frac{1}{2}\right)$, where a is any real number   **29.** $\left(\frac{90}{31}, -\frac{67}{31}\right)$

31. No. The lines intersect at $(79{,}400, 398)$.

33. 550 miles per hour; 50 miles per hour

35. $6\frac{2}{3}$ gallons at 20%   **37.** \$8000 at 12%
    $3\frac{1}{3}$ gallons at 50%   \$4000 at 10.5%

39. Yes; Let the number of adult tickets $= a$ and the number of children's tickets $= c$. By solving the system of equations $a + c = 740$, $4688 = 8.5a + 4c$, you obtain $a = 384$, $c = 356$.

41. $x = 309{,}091$ units; $p = \$25.09$

43. $x = 2{,}000{,}000$ units; $p = \$100.00$

45. (a) Fast-food   Full-service

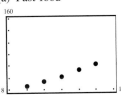   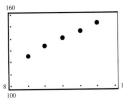

$y = 4.48x + 62.4$ $y = 6.79x + 65.1$

$x = 9$ corresponds to 1999.

(b) No

47. 200 units at \$130.00   **49.** $y = 0.97x + 2.1$

51. $y = 0.318x + 4.061$

53. (a) $y = 2.407t + 39.53$, $t = 0$ corresponds to 1998.

(b) \$53.97

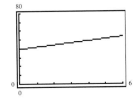

(c) Same model: $y = 2.407t + 39.53$, $t = 0$ corresponds to 1998.

(d) Same estimate: \$53.97 in 2004.

55. The equations form an inconsistent system.

57. Answers will vary. Sample answers:

(a) $x + 2y = 4$ (b) $3x - y = 6$
    $x - y = 1$ $0.3x - 0.1y = 0.6$
    $(2, 1)$ $(a, 3a - 6)$, a is any real number

(c) $3x - 2y = 12$
    $1.5x - y = 5$
    No solution

Section 4.3   *(page 284)*

Warm Up   *(page 284)*

1. $(15, 10)$ **2.** $\left(-2, -\frac{8}{3}\right)$ **3.** $(28, 4)$

4. $(4, 3)$ **5.** Not a solution   **6.** Not a solution

7. Solution   **8.** Solution   **9.** $5a + 2$

10. $a + 13$

---

1. $(4, -2, -2)$ **3.** $(2, -3, -2)$ **5.** $(-1, -6, 8)$

7. Inconsistent   **9.** $\left(1, -\frac{3}{2}, \frac{1}{2}\right)$

11. $(-3a + 10, 5a - 7, a)$

13. $\left(-4a + 13, -\frac{15}{2}a + \frac{45}{2}, a\right)$ **15.** Inconsistent

17. $(-3, 4, 2)$ **19.** $\left(-\frac{3}{5}a, \frac{4}{5}a, a\right)$

21. $\left(\frac{3}{4}a, -2a, a\right)$ **23.** $(-5a + 3, -a - 5, a)$

25. $\left(-\frac{3}{8}a - \frac{1}{4}, -\frac{3}{4}a + \frac{5}{2}, a\right)$ **27.** $(1, 0, 3, 2)$

29. Answers will vary.   **31.** Answers will vary.
Sample answers: Sample answers:

$a = 3$: $(3, -2, 3)$ $a = 2$: $(1, 6, 5)$

$a = 6$: $(6, 1, 5)$ $a = 4$: $(2, 12, 5)$

$a = -3$: $(-3, -8, -1)$ $a = 0$: $(0, 0, 5)$

33. $y = 2x^2 + 3x - 4$ **35.** $x^2 + y^2 - 4x = 0$

37. $300,000 at 8%

$400,000 at 9%

$75,000 at 10%

39. 25 pounds of grade A paper   **41.** 10 gallons of spray X

10 pounds of grade B paper   5 gallons of spray Y

15 pounds of grade C paper   12 gallons of spray Z

43. Invest $33,333.33 + 0.8a$ in certificates of deposit, $341,666.67 - 0.8a$ in municipal bonds, $125,000.00 - a$ in blue-chip stocks, and a in growth or speculative stocks, where $0 \le a \le 125,000$.

45. $y = 0.079x^2 + 0.63x + 2.9$

47. $y = -0.207x^2 - 0.89x + 5.1$

49. (a) $y = 3.407x^2 - 12.65x + 90.8$ (b) Same model

51. $y = 106.375x^2 - 567.55x + 768.3$; $4276.725 billion

53. Yes. See page 279 for a graphical explanation.

55. (a) $y = 0.125x^2 - 2.55x + 18$

(b) $y = 0.125x^2 - 2.55x + 18$ (c) No

Mid-Chapter Quiz   (page 289)

1. $(1, 3)$ **2.** $\left(-\frac{2}{5} \pm \frac{2\sqrt{11}}{5}, \frac{1}{5} \pm \frac{4\sqrt{11}}{5}\right)$

3. ≈ 1550 units   **4.** 500,000 units   **5.** $(2, -1)$

6. $\left(1, \frac{3}{2}\right)$ **7.** $x = 18,333$ units   **8.** $y = 3.32x + 53.3$

$p = 26.67

9. $(1, -2, 3)$

10. Answers will vary.

Sample answer: $(a + 6, a + 6, a)$, a is any real number.

11. Inconsistent   **12.** $y = 0.50x^2 - 6.9x + 161$

Section 4.4   (page 297)

Warm Up   (page 297)

1. Line   **2.** Parabola   **3.** Circle   **4.** Parabola

5. Line   **6.** Circle   **7.** $(1, 1)$ **8.** $(2, 0)$

9. $(2, 1), \left(-\frac{5}{2}, -\frac{5}{4}\right)$ **10.** $(2, 3), (3, 2)$

1. d   **2.** b   **3.** a   **4.** c   **5.** f   **6.** e

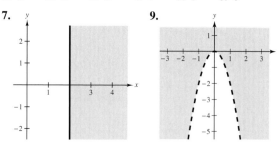

7.   **9.**

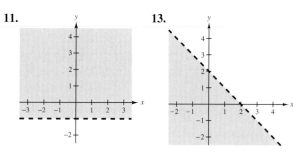

11.   **13.**

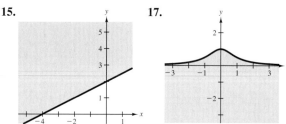

15.   **17.**

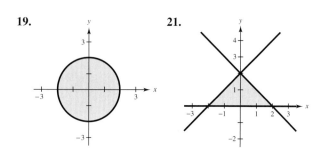

19.   **21.**

23.

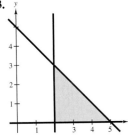

25.

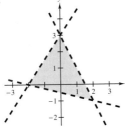

47. $2x + \frac{3}{2}y \le 18, \frac{3}{2}x + \frac{3}{2}y \le 15, x \ge 0, y \ge 0$

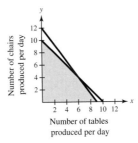

27.

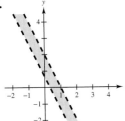

29.

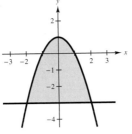

49. Consumer surplus: \$4,777,001.41

Producer surplus: \$477,545.60

51. Consumer surplus: \$40,000,000

Producer surplus: \$20,000,000

53. (a) $x + y \le 30,000$

$x \ge 6000$

$y \ge 6000$

$x - 2y \ge 0$

(b)

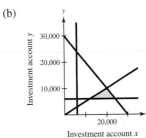

31.

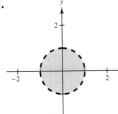

33.

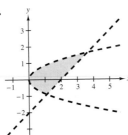

35.

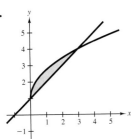

37.

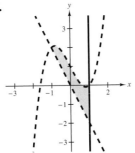

55. (a) $20x + 15y \ge 400$

$10x + 20y \ge 250$

$15x + 20y \ge 220$

(b)

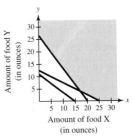

(c) Answers will vary. Sample answer: The nutritionist could give a ounces of food X and $-\frac{4}{3}a + 26.7$ ounces of food Y, where $0 < a \le 18$.

39.

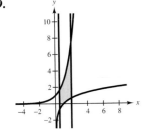

41. $1 \le x \le 8, 1 \le y \le 5$

43. $y \le \frac{5}{3}x, y \ge 0, y \le -\frac{5}{4}x + \frac{35}{4}$

45. $x^2 + y^2 \le 16, x \ge 0, y \ge 0$

57. (a)

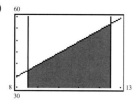

(b) $187.98 billion

Section 4.5   *(page 306)*

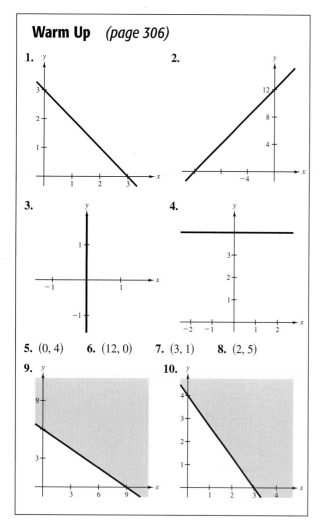

Warm Up   *(page 306)*

1.

2.

3.

4.

5. $(0, 4)$ **6.** $(12, 0)$ **7.** $(3, 1)$ **8.** $(2, 5)$

9.

10.

1. Minimum value at $(0, 0)$: 0
   Maximum value at $(6, 0)$: 36

3. Minimum value at $(0, 0)$: 0
   Maximum value at $(6, 0)$: 48

5. Minimum value at $(0, 0)$: 0
   Maximum value at $(3, 4)$: 17

7. Minimum value at $(0, 0)$: 0
   Maximum value at $(4, 0)$: 20

9. Minimum value at $(0, 0)$: 0
   Maximum value at $(5, 0)$ or $(0, 3)$: 30

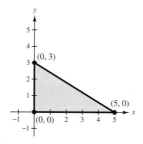

11. Minimum value at $(0, 0)$: 0
   Maximum value at $(5, 0)$: 45
   Same graph as in Answer 9

13. Minimum value at $(5, 3)$: 35
   No maximum value

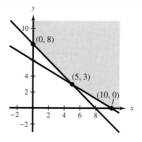

15. Minimum value at $(10, 0)$: 20
   No maximum value
   Same graph as in Answer 13

17. Minimum value at $(24, 8)$: 136
   Maximum value at $(40, 0)$: 200

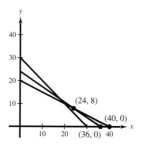

19. Minimum value at $(36, 0)$: 72
   Maximum value at $(24, 8)$: 88
   Same graph as in Answer 17

21. Maximum value at $(3, 6)$: 12

23. Maximum value at $(0, 10)$: 10

25. Maximum value at $(4, 4)$: 28

27. Maximum value at $(7, 0)$: 84

29. Answers will vary.
Sample answer:
$z = 2x + 11y$

31. Answers will vary.
Sample answer:
$z = 5x + y$

33. Crop A: 60 acres
Crop B: 90 acres
$33,150

35. Brand X: 3 bags
Brand Y: 6 bags
$240

37. Model A: 1600 bicycles
Model B: 0 bicycles
$120,000

39. 12 audits and 0 tax returns

41. Television: None
Newspaper: $1,000,000
250 million people

43. Type A: $62,500
Type B: $187,500
$26,875

45. Model A: 929 units
Model B: 77 units
$99,445

47. z is maximum at any point on the line segment connecting the vertices $(2, 0)$ and $\left(\frac{20}{19}, \frac{45}{19}\right)$.

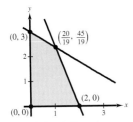

49. The constraint $x \leq 10$ is extraneous. The maximum value of z occurs at $(0, 7)$.

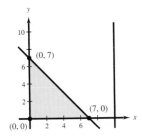

51. The constraint $2x + y \leq 4$ is extraneous. The maximum value of z occurs at $(0, 1)$.

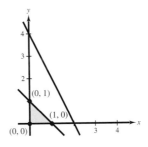

Review Exercises   *(page 312)*

1. $(-2, 4)$ **3.** $(8, -10)$ **5.** $(8, 6), (0, 10)$

7. $(1, 9), (1.5, 8.75)$

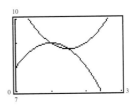

9. 5405 baskets

11. More than $200,000

13. Yes; During the fifth month of the new format

15. $(3, -5)$

17. Infinitely many solutions of the form $\left(a, \frac{5}{8}a - \frac{7}{4}\right)$

19. $(8, 9)$

21. Graph is a point. Solution: $(-1, 1)$

23. (a) $0.1x + 0.5y = 0.25(12)$
$ x + y = 12$

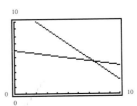

(b) 7.5 gallons of 10% solution
4.5 gallons of 50% solution

25. $x = 71,429$ units   **27.** $(3, 5, 2)$
$p = \$22.71$

29. $(2, -1, 3)$ **31.** $\left(\frac{1}{5}a + \frac{8}{5}, -\frac{6}{5}a + \frac{42}{5}, a\right)$

33. Inconsistent   **35.** $y = 2x^2 + x - 6$

37. $x^2 + y^2 - 4x + 2y - 4 = 0$

39. $200,000 in certificates of deposit

$100,000 in municipal bonds

$15,000 in blue-chip stocks

$185,000 in growth or speculative stocks

41. $25,000 at 6%

$26,000 at 8%

$50,000 at 10%

43. $y = 0.114x^2 + 0.8x + 1.59$

45.

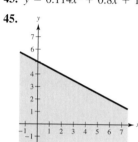

47.

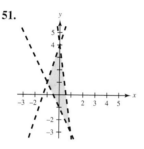

49.

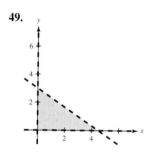

51.

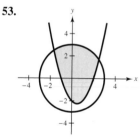

53.

55. $5x - 2y \geq \quad 3$

$5x - 2y \leq \quad 31$

$x - 6y \leq \quad -5$

$x - 6y \geq -33$

57. Consumer surplus: $4,500,000

Producer surplus: $9,000,000

59. $x \geq 2y$

$200x + 300y \leq 4000$

$x \geq 4$

$y \geq 2$

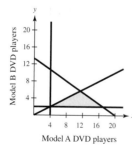

61. Minimum value at $(0, 0)$: 0

Maximum value at $(0, 8)$: 48

63. Minimum value at $(0, 0)$: 0

Maximum value at $(50, 15)$: 445

65. Minimum value at $(0, 0)$: 0

Maximum value at $(4, 3)$: 48

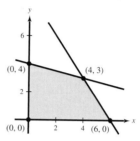

67. Minimum value at $(0, 0)$: 0

Maximum value at $(4, 5)$: 47

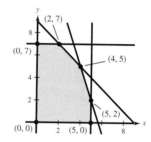

69. $525 model: 0 units

$675 model: 305 units

$38,125

71. 667 of the basic model

333 of the deluxe model

$133,330

73. Three bags of brand X

Two bags of brand Y

$105

75. 8 audits, 0 tax returns

$20,000

Chapter Test   *(page 317)*

1. $(2, 4)$ **2.** $(-2, 5), (3, 0)$

3. $(1.68, 2.38), (-4.18, -26.88)$ **4.** $(3.36, -1.32)$

5. $(2, -1, 3)$ **6.** $(2, -3, 4)$

7. $20,000 at 8%   **8.** $(50,000, 30)$

$30,000 at 8.5%

9. $y = -0.229t^2 - 1.18t + 7.4$; 1.8%

10.

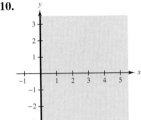

11.

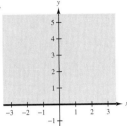

12.

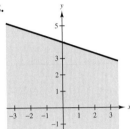

13.

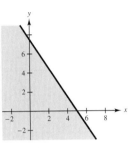

14.

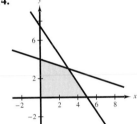

15. Maximum value at $(3, 3)$: 39

Minimum value at $(0, 0)$: 0

16. Model A: 297; Model B: 569

$215,875

Chapter 5

Section 5.1   *(page 330)*

Warm Up   *(page 330)*

1. -3 **2.** 30   **3.** 6   **4.** $-\frac{1}{9}$ **5.** Solution

6. Not a solution   **7.** $(5, 2)$ **8.** $\left(\frac{12}{5}, -3\right)$

9. $(40, 14, 2)$ **10.** $\left(\frac{15}{2}, 4, 1\right)$

---

1. 2×3 **3.** 4×3 **5.** 4×2 **7.** 2×4

9. Add 5 times R_2 to R_1.   **11.** Interchange R_1 and R_2.

Add 4 times R_1 to R_3.

13. Reduced row-echelon form

15. Not in row-echelon form

17. Not in row-echelon form

19. (a) $\begin{bmatrix} 1 & 1 & 2 \\ 0 & 1 & -9 \\ 2 & -1 & 1 \end{bmatrix}$ (b) $\begin{bmatrix} 1 & 1 & 2 \\ 0 & 1 & -9 \\ 0 & -3 & -3 \end{bmatrix}$

(c) $\begin{bmatrix} 1 & 1 & 2 \\ 0 & 1 & -9 \\ 0 & 0 & -30 \end{bmatrix}$ (d) $\begin{bmatrix} 1 & 1 & 2 \\ 0 & 1 & -9 \\ 0 & 0 & 1 \end{bmatrix}$

21. $\begin{bmatrix} 1 & 2 & -1 & 5 \\ 0 & 1 & -1 & 1 \\ 0 & 0 & 1 & -2 \end{bmatrix}$ **23.** $\begin{bmatrix} 1 & -1 & -1 & 1 \\ 0 & 1 & 6 & 3 \\ 0 & 0 & 0 & 0 \end{bmatrix}$

25. $\begin{bmatrix} 1 & 0 & 0 \\ 0 & 1 & 0 \\ 0 & 0 & 1 \end{bmatrix}$ **27.** $\begin{bmatrix} 1 & 2 & 0 & 0 \\ 0 & 0 & 1 & 0 \\ 0 & 0 & 0 & 1 \\ 0 & 0 & 0 & 0 \end{bmatrix}$

29. $\begin{cases} 2x + 4y = 6 \\ -x + 3y = -8 \end{cases}$ **31.** $\begin{cases} x \quad\quad + 2z = -10 \\ \quad 3y - z = 5 \\ 4x + 2y \quad\quad = 3 \end{cases}$

33. $\left[\begin{array}{cc:c} 2 & -1 & 3 \\ 5 & 7 & 12 \end{array}\right]$ **35.** $\left[\begin{array}{ccc:c} 1 & 10 & -3 & 2 \\ 5 & -3 & 4 & 0 \\ 2 & 4 & 0 & 6 \end{array}\right]$

37. $\left[\begin{array}{cccc:c} 9 & -3 & 20 & 1 & 13 \\ 12 & 0 & -8 & 0 & 5 \\ 1 & 2 & 3 & -4 & -2 \\ -1 & -1 & 1 & 1 & 1 \end{array}\right]$

39. $\begin{cases} x - 5y = 6 \\ \quad\quad y = -2 \end{cases}$

$(-4, -2)$

41. $\begin{cases} x + 3y - z = 15 \\ \quad y + 4z = -12 \\ \quad\quad\quad z = -5 \end{cases}$

$(-14, 8, -5)$

43. $(-4, 6)$ **45.** $(-4, -8, 2)$

47. $(3, 2)$ **49.** $(4, -2)$ **51.** $\left(\frac{1}{2}, -\frac{3}{4}\right)$

53. Inconsistent   **55.** $(-6, 8, 2)$

57. $\left(-\frac{3}{2}a + \frac{3}{2}, \frac{1}{3}a + \frac{1}{3}, a\right)$ **59.** Inconsistent

61. $(5a + 4, -3a + 2, a)$

63. $(0, 2 - 4a, a)$ **65.** $\left(\frac{435}{73}, -\frac{230}{73}, \frac{1184}{73}, -\frac{312}{73}\right)$

67. $(0, 0)$ **69.** $(-2a, a, a)$

71. $200,000 was borrowed at 8%, $500,000 was borrowed at 9%, and $100,000 was borrowed at 12%.

73. Both are correct. Because there are infinitely many ordered triples that are solutions to this system, a solution can be written in many different ways. If $a = 3$, the ordered triple is $(3, -3, 5)$. You obtain the same triple when $b = 5$.

75. $y = 1.091t + 17.69$

The total energy imports in 2006 will be about 35.146 quadrillion Btu. Because the data values increased in a linear pattern, this estimate seems reasonable.

77. If you solved Exercise 55 using Gauss-Jordan elimination, then the matrices are the same and produce the same solution, because the reduced row-echelon form of a matrix is unique. If you solved Exercise 55 using Gaussian elimination with back-substitution, then the matrices are different but produce the same solution.

Section 5.2 (page 344)

Warm Up (page 344)

1. -5　　**2.** -7　　**3.** Not in reduced row-echelon form

4. Not in reduced row-echelon form

5. $\begin{bmatrix} -5 & 10 & \vdots & 12 \\ 7 & -3 & \vdots & 0 \end{bmatrix}$

6. $\begin{bmatrix} 10 & 15 & -9 & \vdots & 42 \\ 6 & -5 & 0 & \vdots & 0 \end{bmatrix}$

7. $(0, 2)$　　**8.** $(2 + a, 3 - a, a)$

9. $(1 - 2a, a, -1)$　　**10.** $(2, -1, -1)$

---

1. $x = -3, y = 2$　　**3.** $x = -2, y = 5$

5. (a) $\begin{bmatrix} 8 & -1 \\ 1 & 7 \end{bmatrix}$　(b) $\begin{bmatrix} 2 & -3 \\ 5 & -5 \end{bmatrix}$

(c) $\begin{bmatrix} 15 & -6 \\ 9 & 3 \end{bmatrix}$　(d) $\begin{bmatrix} 9 & -8 \\ 13 & -9 \end{bmatrix}$

7. (a) $\begin{bmatrix} 7 & 3 \\ 1 & 9 \\ -2 & 15 \end{bmatrix}$　(b) $\begin{bmatrix} 5 & -5 \\ 3 & -1 \\ -4 & -5 \end{bmatrix}$

(c) $\begin{bmatrix} 18 & -3 \\ 6 & 12 \\ -9 & 15 \end{bmatrix}$　(d) $\begin{bmatrix} 16 & -11 \\ 8 & 2 \\ -11 & -5 \end{bmatrix}$

9. (a) $\begin{bmatrix} 3 & 3 & -2 \\ -2 & 5 & 7 \\ 1 & -8 & 11 \end{bmatrix}$　(b) $\begin{bmatrix} 1 & 1 & 0 \\ 4 & -3 & -11 \\ 1 & 6 & -5 \end{bmatrix}$

(c) $\begin{bmatrix} 6 & 6 & -3 \\ 3 & 3 & -6 \\ 3 & -3 & 9 \end{bmatrix}$　(d) $\begin{bmatrix} 4 & 4 & -1 \\ 9 & -5 & -24 \\ 3 & 11 & -7 \end{bmatrix}$

11. $\begin{bmatrix} -8 & -7 \\ 15 & -1 \end{bmatrix}$　　**13.** $\begin{bmatrix} -24 & -4 & 12 \\ -12 & 32 & 12 \end{bmatrix}$

15. $\begin{bmatrix} 10 & 8 \\ -59 & 9 \end{bmatrix}$　　**17.** $\begin{bmatrix} -6 & -9 \\ -1 & 0 \\ 17 & -10 \end{bmatrix}$　　**19.** $\begin{bmatrix} 3 & 3 \\ -\frac{1}{2} & 0 \\ -\frac{13}{2} & \frac{11}{2} \end{bmatrix}$

21. $\begin{bmatrix} -7 & 10 & -12 & 17 \\ 6 & 21 & 7 & 15 \\ -3 & 3 & -5 & 6 \end{bmatrix}$　　**23.** $\begin{bmatrix} -1 & 19 \\ 4 & -27 \\ 0 & 14 \end{bmatrix}$

25. $\begin{bmatrix} 1 & 0 & 0 \\ 0 & 1 & 0 \\ 0 & 0 & \frac{7}{2} \end{bmatrix}$　　**27.** $\begin{bmatrix} 60 & 72 \\ -20 & -24 \\ 10 & 12 \\ 60 & 72 \end{bmatrix}$

29. (a) $\begin{bmatrix} 0 & 15 \\ 6 & 12 \end{bmatrix}$　(b) $\begin{bmatrix} -2 & 2 \\ 31 & 14 \end{bmatrix}$　(c) $\begin{bmatrix} 9 & 6 \\ 12 & 12 \end{bmatrix}$

31. (a) $\begin{bmatrix} 3 & 5 \\ 1 & 6 \end{bmatrix}$　(b) $\begin{bmatrix} 11 & 5 & 0 \\ -7 & -4 & -1 \\ 14 & 8 & 2 \end{bmatrix}$　(c) Not possible

33. (a) $[11]$　(b) $\begin{bmatrix} -4 & 2 & 3 \\ 0 & 0 & 0 \\ -20 & 10 & 15 \end{bmatrix}$　(c) Not possible

35. (a) $\begin{bmatrix} -1 & 1 \\ -2 & 1 \end{bmatrix} \begin{bmatrix} x \\ y \end{bmatrix} = \begin{bmatrix} 4 \\ 0 \end{bmatrix}$　(b) $\begin{bmatrix} 4 \\ 8 \end{bmatrix}$

37. (a) $\begin{bmatrix} 1 & 2 \\ 3 & -1 \end{bmatrix} \begin{bmatrix} x \\ y \end{bmatrix} = \begin{bmatrix} 3 \\ 2 \end{bmatrix}$　(b) $\begin{bmatrix} 1 \\ 1 \end{bmatrix}$

39. (a) $\begin{bmatrix} 1 & -2 & 3 \\ -1 & 3 & -1 \\ 2 & -5 & 5 \end{bmatrix} \begin{bmatrix} x \\ y \\ z \end{bmatrix} = \begin{bmatrix} 9 \\ -6 \\ 17 \end{bmatrix}$　(b) $\begin{bmatrix} 1 \\ -1 \\ 2 \end{bmatrix}$

41. $\begin{bmatrix} 115 & 138 & 69 & 46 \\ 161 & 184 & 230 & 92 \end{bmatrix}$

43.

	Hotel x	Hotel y	Hotel z	
	97.75	105.80	126.50	Single
	115.00	138.00	149.50	Double
	126.50	149.50	161.00	Triple
	126.50	161.00	178.25	Quadruple

Occupancy

45. (a) $\$19{,}550$　(b) $\$21{,}450$

(c)

$$ST = \begin{bmatrix} \$15{,}850 & \$19{,}550 \\ \$26{,}350 & \$30{,}975 \\ \$21{,}450 & \$25{,}850 \end{bmatrix} \begin{matrix} 1 \\ 2 \\ 3 \end{matrix} \} \text{Outlet}$$

with column headers Wholesale　Retail

ST represents the wholesale and retail values of the computer inventory at each outlet.

47. Cannot perform operation.

49. Cannot perform operation.　　**51.** 2×2

53. 2×3　　**55.** $\begin{bmatrix} \$16{,}727.12 \\ \$9000.40 \\ \$12{,}399.58 \end{bmatrix}$

57. (a)

$$C = \begin{bmatrix} 0.05 & 0.085 \\ 0.12 & 0.10 \\ 0.30 & 0.25 \end{bmatrix} \begin{matrix} \text{In-state} \\ \text{State-to-state} \\ \text{International} \end{matrix}$$

with column headers A　B

Each entry c_{ij} represents the cost per minute charged by each company j for each type of call i.

(b)

$$T = \begin{bmatrix} \overset{\text{In-}}{\underset{\text{state}}{20}} & \overset{\text{State-}}{\underset{\text{to-state}}{60}} & \overset{\text{Inter-}}{\underset{\text{national}}{30}} \end{bmatrix}$$

Each entry t_{ij} represents the time (in minutes) spent on the phone for each type of call j.

(c)

$$TC = \begin{bmatrix} \overset{A}{17.20} & \overset{B}{15.20} \end{bmatrix}$$

Each entry represents the total monthly charges for each company.

(d) You should choose company B because the monthly charges are \$2 less than the monthly charges for company A.

59. $\begin{bmatrix} 0.40 & 0.15 & 0.15 \\ 0.28 & 0.53 & 0.17 \\ 0.32 & 0.32 & 0.68 \end{bmatrix}$ **61.** $\begin{bmatrix} -2 & 3 \\ 1 & -1 \end{bmatrix}$

Matrix is unique.

63. $AC = \begin{bmatrix} 12 & -6 & 9 \\ 16 & -8 & 12 \\ 4 & -2 & 3 \end{bmatrix} = BC$, but $A \neq B$.

65. (a) Gold subcribers: 28,750

Galaxy subscribers: 35,750

Nonsubscribers: 35,500

Multiply the original matrix by the 3×1 matrix

$$\begin{bmatrix} 25,000 \\ 30,000 \\ 45,000 \end{bmatrix}$$

which represents the current numbers of subscribers for each company and the number of nonsubscribers.

(b) Gold subscribers: 30,813

Galaxy subscribers: 39,675

Nonsubscribers: 29,513

Multiply the original matrix by the 3×1 matrix

$$\begin{bmatrix} 28,750 \\ 35,750 \\ 35,500 \end{bmatrix}$$

which represents the numbers of subscribers for each company and the number of nonsubscribers after 1 year.

(c) Gold subscribers: 31,947

Galaxy subscribers: 42,330

Nonsubscribers: 25,724

Multiply the original matrix by the 3×1 matrix

$$\begin{bmatrix} 30,813 \\ 39,675 \\ 29,513 \end{bmatrix}$$

which represents the numbers of subscribers for each company and the number of nonsubscribers after 2 years.

(d) The number of subscribers to each company is increasing each year. The number of nonsubscribers is decreasing each year.

Section 5.3 *(page 355)*

Warm Up *(page 355)*

1. $\begin{bmatrix} 4 & 24 \\ 0 & -16 \\ 48 & 8 \end{bmatrix}$ **2.** $\begin{bmatrix} \frac{11}{2} & 5 & 24 \\ \frac{1}{2} & 0 & 8 \\ 0 & 1 & 4 \end{bmatrix}$

3. $\begin{bmatrix} -5 & -2 & -13 \\ 4 & -13 & -2 \end{bmatrix}$ **4.** $\begin{bmatrix} -13 & 11 \\ -19 & 21 \end{bmatrix}$

5. $\begin{bmatrix} 1 & 0 \\ 0 & 1 \end{bmatrix}$ **6.** $\begin{bmatrix} 6 & 5 \\ 3 & -2 \end{bmatrix}$ **7.** $\begin{bmatrix} 1 & 0 & 0 \\ 0 & 1 & 0 \\ 0 & 0 & 1 \end{bmatrix}$

8. $\begin{bmatrix} 1 & 0 & 0 \\ 0 & 1 & 0 \\ 0 & 0 & 1 \end{bmatrix}$ **9.** $\left[\begin{array}{cc:cc} 1 & 0 & 3 & -2 \\ 0 & 1 & 4 & -3 \end{array}\right]$

10. $\left[\begin{array}{ccc:ccc} 1 & 0 & 0 & -6 & -4 & 3 \\ 0 & 1 & 0 & 11 & 6 & -5 \\ 0 & 0 & 1 & -2 & -1 & 1 \end{array}\right]$

1–9. Answers will vary.   **11.** $\begin{bmatrix} \frac{1}{4} & \frac{1}{2} \\ -\frac{1}{4} & -1 \end{bmatrix}$

13. Does not exist   **15.** $\begin{bmatrix} 1 & -1 \\ 2 & -1 \end{bmatrix}$ **17.** $\begin{bmatrix} \frac{1}{2} & -\frac{1}{3} \\ \frac{1}{4} & 0 \end{bmatrix}$

19. Does not exist   **21.** $\begin{bmatrix} 1 & 1 & -1 \\ -3 & 2 & -1 \\ 3 & -3 & 2 \end{bmatrix}$

23. $\begin{bmatrix} -175 & 37 & -13 \\ 95 & -20 & 7 \\ 14 & -3 & 1 \end{bmatrix}$ **25.** $\frac{1}{2}\begin{bmatrix} -3 & 3 & 2 \\ 9 & -7 & -6 \\ -2 & 2 & 2 \end{bmatrix}$

27. $\begin{bmatrix} \frac{1}{3} & 0 & 0 \\ 0 & -\frac{1}{2} & 0 \\ 0 & 0 & \frac{1}{4} \end{bmatrix}$ **29.** $\begin{bmatrix} 1 & 0 & 0 \\ -\frac{3}{4} & \frac{1}{4} & 0 \\ \frac{7}{20} & -\frac{1}{4} & \frac{1}{5} \end{bmatrix}$

31. Does not exist

33. $\begin{bmatrix} -\frac{1}{8} & 0 & 0 & 0 \\ 0 & 1 & 0 & 0 \\ 0 & 0 & \frac{1}{4} & 0 \\ 0 & 0 & 0 & -\frac{1}{5} \end{bmatrix}$ **35.** $\begin{bmatrix} -24 & 7 & 1 & -2 \\ -10 & 3 & 0 & -1 \\ -29 & 7 & 3 & -2 \\ 12 & -3 & -1 & 1 \end{bmatrix}$

37. $\frac{1}{19}\begin{bmatrix} 3 & 2 \\ -2 & 5 \end{bmatrix}$ **39.** Does not exist

41. $\frac{1}{59}\begin{bmatrix} 16 & 15 \\ -4 & 70 \end{bmatrix}$ **43.** (4, 8)   **45.** (10, 30)

47. $(-2, 3)$ **49.** $(2, 0)$ **51.** $(2, 1, 0, 0)$

53. $(2, -2)$ **55.** Inconsistent

57. $(-4, -8)$ **59.** $(-1, 3, 2)$

61. $\begin{cases} 2x + y + 3z = 16 \\ 4x - 2z = -2 \\ 3y + 2z = 1 \end{cases}$

63. AAA bonds: $20,000
   A bonds: $5000
   B bonds: $10,000

65. AAA bonds: $21,000
   A bonds: $5000
   B bonds: $10,000

67. 27 computer chips
   8 resistors
   6 transistors

69. 36 computer chips
   44 resistors
   28 transistors

71. 0 muffins
   300 cookies
   200 brownies

73. 100 muffins
   300 cookies
   150 brownies

75. (a) $\begin{bmatrix} \frac{949}{56} & -\frac{265}{56} & \frac{17}{56} \\ -\frac{265}{56} & \frac{229}{168} & -\frac{5}{56} \\ \frac{17}{56} & -\frac{5}{56} & \frac{1}{168} \end{bmatrix}$

$y = 0.049t^2 + 0.62t + 6.4$

(b) $20.9 billion

(c) The estimates are close and both seem reasonable.

77. $AB = \begin{bmatrix} 13 & 4 \\ 1 & 8 \end{bmatrix}$, $BA = \begin{bmatrix} 8 & 1 \\ 4 & 13 \end{bmatrix}$

Row 1 of AB is Row 2 of BA with reversed entries. Row 2 of AB is Row 1 of BA with reversed entries.

79. Answers will vary. Sample answers:
If $k = 3$, then

$\begin{bmatrix} 4 & 3 \\ -2 & 3 \end{bmatrix}^{-1} = \begin{bmatrix} \frac{1}{6} & -\frac{1}{6} \\ \frac{1}{9} & \frac{2}{9} \end{bmatrix}$.

If $k = -\frac{3}{2}$, then the matrix is singular.

81. (a) Answers will vary.
Sample answers:

$A_2 = \begin{bmatrix} 2 & 0 \\ 0 & 3 \end{bmatrix}$, $A_2^{-1} = \begin{bmatrix} \frac{1}{2} & 0 \\ 0 & \frac{1}{3} \end{bmatrix}$

$A_3 = \begin{bmatrix} 2 & 0 & 0 \\ 0 & 3 & 0 \\ 0 & 0 & 4 \end{bmatrix}$, $A_3^{-1} = \begin{bmatrix} \frac{1}{2} & 0 & 0 \\ 0 & \frac{1}{3} & 0 \\ 0 & 0 & \frac{1}{4} \end{bmatrix}$

(b) $A_n^{-1} = \begin{bmatrix} \frac{1}{a_{11}} & 0 & 0 & 0 & \cdots & 0 \\ 0 & \frac{1}{a_{22}} & 0 & 0 & \cdots & 0 \\ 0 & 0 & \frac{1}{a_{33}} & 0 & \cdots & 0 \\ \vdots & \vdots & \vdots & \vdots & & \vdots \\ 0 & 0 & 0 & 0 & \cdots & \frac{1}{a_{nn}} \end{bmatrix}$

Mid-Chapter Quiz *(page 359)*

1. Any matrix with three rows and four columns

2. Any matrix with one row and three columns

3. $\begin{bmatrix} 3 & 2 & \vdots & -2 \\ 5 & -1 & \vdots & 19 \end{bmatrix}$ **4.** $\begin{bmatrix} 1 & 0 & 3 & \vdots & -5 \\ 1 & 2 & -1 & \vdots & 3 \\ 3 & 0 & 4 & \vdots & 0 \end{bmatrix}$

5. $(2.769, -5.154)$ **6.** $(4, -2, -3)$ **7.** $\begin{bmatrix} 2 & -10 \\ 15 & 11 \end{bmatrix}$

8. $\begin{bmatrix} -5 & 2 & -13 \\ 5 & 6 & 11 \end{bmatrix}$ **9.** $\begin{bmatrix} 1 & 2 \\ -3 & 2 \end{bmatrix}$

10. $\begin{bmatrix} -6 & -2 \\ 3 & -5 \end{bmatrix}$ **11.** $\begin{bmatrix} \frac{2}{5} & \frac{1}{5} \\ -\frac{3}{10} & \frac{1}{10} \end{bmatrix}$ **12.** Not possible

13. $\begin{bmatrix} 3 & -2 \\ 3 & 10 \end{bmatrix}$ **14.** $\begin{bmatrix} 1 & 2 \\ -3 & 2 \end{bmatrix}$

15. $A = \begin{bmatrix} 1 & -3 \\ -2 & 1 \end{bmatrix}$, $X = \begin{bmatrix} x \\ y \end{bmatrix}$, $B = \begin{bmatrix} 10 \\ -10 \end{bmatrix}$; $X = \begin{bmatrix} 4 \\ -2 \end{bmatrix}$

16. $A = \begin{bmatrix} 2 & -1 & 1 \\ 3 & 0 & -1 \\ 0 & 4 & 3 \end{bmatrix}$, $X = \begin{bmatrix} x \\ y \\ z \end{bmatrix}$, $B = \begin{bmatrix} 3 \\ 15 \\ -1 \end{bmatrix}$; $X = \begin{bmatrix} 4 \\ 2 \\ -3 \end{bmatrix}$

17. $39.80   **18.** $84.40   **19.** $116.10

20.

$$LW = \begin{bmatrix} \$21.20 & \$18.60 \\ \$45.40 & \$39.00 \\ \$61.50 & \$54.60 \end{bmatrix} \begin{matrix} A \\ B \\ C \end{matrix} \} \text{ Model}$$

Plant 1   Plant 2

LW lists the total labor costs for each model at each plant.

Section 5.4 *(page 367)*

Warm Up *(page 367)*

1. $\begin{bmatrix} 3 & 5 \\ 4 & 0 \end{bmatrix}$ **2.** $\begin{bmatrix} -2 & 8 \\ 2 & -4 \end{bmatrix}$

3. $\begin{bmatrix} 9 & -12 & 6 \\ 3 & 0 & -3 \\ 0 & 3 & -6 \end{bmatrix}$ **4.** $\begin{bmatrix} 0 & 8 & 12 \\ -4 & 8 & 12 \\ -8 & 4 & -8 \end{bmatrix}$

5. -22 **6.** 35 **7.** -15 **8.** $-\frac{1}{8}$

9. -45 **10.** -16

1. -5 **3.** 1   **5.** 3   **7.** 0

9. 5   **11.** $\frac{11}{6}$ **13.** 4

15. (a) $M_{11} = -5, M_{12} = 2, M_{21} = 4, M_{22} = 3$

 (b) $C_{11} = -5, C_{12} = -2, C_{21} = -4, C_{22} = 3$

17. (a) $M_{11} = 30, M_{12} = 12, M_{13} = 11, M_{21} = -36,$

   $M_{22} = 26, M_{23} = 7, M_{31} = -4,$

   $M_{32} = -42, M_{33} = 12$

 (b) $C_{11} = 30, C_{12} = -12, C_{13} = 11, C_{21} = 36,$

   $C_{22} = 26, C_{23} = -7, C_{31} = -4,$

   $C_{32} = 42, C_{33} = 12$

19. -99 **21.** -145 **23.** 170   **25.** -58

27. -30 **29.** 0   **31.** -0.002 **33.** 0   **35.** 0

37. -108 **39.** 0   **41.** 412   **43.** -6

45. -16 **47.** 120   **49.** 40   **51.** -168

53. -20; Answers will vary.

55. -18; Answers will vary.

57. Matrices will vary. The determinant of each matrix is the product of the entries on the main diagonal, which in this case equals -18.

59. Matrices will vary. The determinant of each matrix is the product of the entries on the main diagonal, which in this case equals 28.

61. Rows 2 and 4 are identical.

63. Row 4 is a multiple of Row 2.

Math Matters   *(page 369)*

Answers will vary. There are $2^4 - 1 = 15$ possible ways in which the stack of four cards can be arranged according to the restrictions given. The numbers and cut-outs are placed on each card so that each choice between 1 and 15 results in a unique ordering of the stack of cards and so that the correct number from the back of the fourth card is displayed when the stack is overturned.

Section 5.5   *(page 376)*

Warm Up   *(page 376)*

1. 1   **2.** 0   **3.** -8 **4.** x^2 **5.** 8   **6.** 60

7. $\begin{bmatrix} 7 & -3 \\ -2 & 1 \end{bmatrix}$ **8.** $\begin{bmatrix} 1 & 2 & -1 \\ -1 & -2 & 2 \\ 2 & 5 & 0 \end{bmatrix}$

9. $\begin{bmatrix} 0.14 \\ 0.31 \end{bmatrix}$ **10.** $[7 \ 14]$

---

1. 11   **3.** 28   **5.** $\frac{33}{8}$ **7.** $\frac{5}{2}$ **9.** 28

11. $y = \frac{16}{5}$ or $y = 0$ **13.** $y = -3$ or $y = -11$

15. 250 square miles   **17.** Collinear

19. Not collinear   **21.** Not collinear   **23.** $y = -4$

25. $x - 6y + 13 = 0$ **27.** $x + 3y - 5 = 0$

29. $2x + 3y - 8 = 0$

31. $[3 \ 15][13 \ 5][0 \ 8][15 \ 13][5 \ 0][19 \ 15][15 \ 14]$

 48, 81, 28, 51, 24, 40, 54, 95, 5, 10, 64, 113, 57, 100

33. $[3 \ 1 \ 12][12 \ 0 \ 13][5 \ 0 \ 20][15 \ 13 \ 15]$

 $[18 \ 18 \ 15][23 \ 0 \ 0]$

 $-68, 21, 35, -66, 14, 39, -115, 35, 60, -62, 15, 32,$

 $-54, 12, 27, 23, -23, 0$

35. $1, -25, -65, 17, 15, -9, -12, -62, -119,$

 27, 51, 48, 43, 67, 48, 57, 111, 117

37. $-5, -41, -87, 91, 207, 257, 11, -5, -41,$

 40, 80, 84, 76, 177, 227

39. 34, 55, 43, 20, 35, 28, 19, 36, 33, 16, 24, 12, 56, 107, 111

41. HAPPY NEW YEAR   **43.** SOUND ALL CLEAR

45. SEND MORE MONEY

47. MEET ME TONIGHT RON

49. $-55, 3, 80, -20, 9, 8, -38, -67, 224, -35, -81, 168,$ $-61, 15, 64$

51. (a) Because $[45 \ -35]\begin{bmatrix} w & x \\ y & z \end{bmatrix} = [10 \ 15]$

   and $[38 \ -30]\begin{bmatrix} w & x \\ y & z \end{bmatrix} = [8 \ 14],$

   you can solve $\begin{bmatrix} 45 & -35 \\ 38 & -30 \end{bmatrix}\begin{bmatrix} w & x \\ y & z \end{bmatrix} = \begin{bmatrix} 10 & 15 \\ 8 & 14 \end{bmatrix}.$

 (b) JOHN RETURN TO BASE

Review Exercises   *(page 381)*

1. 2×4 **3.** $\begin{bmatrix} 1 & 3 & 0 & 2 \\ 0 & 1 & 1 & 2 \\ 0 & 0 & 1 & 2 \end{bmatrix}$

5. $\begin{bmatrix} 1 & 0 & 0 \\ 0 & 1 & 0 \\ 0 & 0 & 1 \end{bmatrix}$ **7.** $(3, -2)$ **9.** $(3, 2, -1)$

11. $(10 - 4a, a, a)$ **13.** No solution

15. $200,000 was borrowed at 8%, $600,000 was borrowed at 10%, and $100,000 was borrowed at 12%.

17. (a) $\begin{bmatrix} 3 & 7 \\ -4 & 4 \end{bmatrix}$ (b) $\begin{bmatrix} -5 & 3 \\ 8 & -2 \end{bmatrix}$

 (c) $\begin{bmatrix} -4 & 20 \\ 8 & 4 \end{bmatrix}$ (d) $\begin{bmatrix} -16 & 14 \\ 26 & -5 \end{bmatrix}$

19. (a) $\begin{bmatrix} 3 & 4 & 2 & 1 \\ 3 & -5 & 6 & 0 \end{bmatrix}$ (b) $\begin{bmatrix} -1 & 2 & -6 & 11 \\ -3 & 7 & 0 & 4 \end{bmatrix}$

(c) $\begin{bmatrix} 4 & 12 & -8 & 24 \\ 0 & 4 & 12 & 8 \end{bmatrix}$ (d) $\begin{bmatrix} -2 & 9 & -20 & 39 \\ -9 & 22 & 3 & 14 \end{bmatrix}$

21. $\begin{bmatrix} 8 \\ 5 \\ -6 \end{bmatrix}$ **23.** $\begin{bmatrix} 1 & 0 & 0 \\ 0 & 1 & 0 \\ 0 & 0 & 1 \end{bmatrix}$ **25.** Not possible

27. (a) $[4]$ (b) $\begin{bmatrix} 2 & -6 & 8 \\ -2 & 6 & -8 \\ -1 & 3 & -4 \end{bmatrix}$ (c) Not possible

29. $\begin{bmatrix} 4 & -11 \\ -3 & 1 \\ -1 & -3 \end{bmatrix}$ **31.** $\begin{bmatrix} 100 & 150 & 25 & 50 \\ 50 & 75 & 100 & 25 \\ 175 & 75 & 125 & 100 \end{bmatrix}$

33.

$$ST = \begin{bmatrix} \$5200 & \$8265 \\ \$5075 & \$7985 \\ \$5125 & \$8325 \end{bmatrix} \begin{matrix} 1 \\ 2 \\ 3 \end{matrix} \Big\} \text{Outlet}$$

Wholesale Retail

ST represents the wholesale and retail values of the car sound system inventory at each outlet.

35. Answers will vary.

37. $\begin{bmatrix} -1 & 0 & 0 \\ 0 & \frac{1}{2} & 0 \\ 0 & 0 & \frac{1}{4} \end{bmatrix}$ **39.** $\begin{bmatrix} -5 & 3 \\ 2 & -1 \end{bmatrix}$

41. $(3, 4)$ **43.** $\left(2, \frac{1}{2}, 3\right)$

45. $(-6, -1)$ **47.** $(2, -1, -2)$

49. 140 of model A

160 of model B

120 of model C

51. (a) $\frac{1}{70} \begin{bmatrix} 31,786 & -7173 & 395 \\ -7173 & 1627 & -90 \\ 395 & -90 & 5 \end{bmatrix}$

$y = -0.093t^2 + 3.54t + 23.8$

(b) 56.6%

53. 4 **55.** 0

57. (a) $M_{11} = 4, M_{12} = 7, M_{21} = -1, M_{22} = 2$

(b) $C_{11} = 4, C_{12} = -7, C_{21} = 1, C_{22} = 2$

59. (a) $M_{11} = 30, M_{12} = -12, M_{13} = -21,$

$M_{21} = 20, M_{22} = 19, M_{23} = 22,$

$M_{31} = 5, M_{32} = -2, M_{33} = 19$

(b) $C_{11} = 30, C_{12} = 12, C_{13} = -21,$

$C_{21} = -20, C_{22} = 19, C_{23} = -22,$

$C_{31} = 5, C_{32} = 2, C_{33} = 19$

61. 44; Answers will vary.

63. -12; Answers will vary.

65. -39; Answers will vary. **67.** 10

69. Not collinear **71.** Collinear

73. $x + 15y - 38 = 0$

75. $[20 \ 18][1 \ 14][19 \ 13][9 \ 20][0 \ 14][15 \ 23]$

94, 132, 44, 59, 77, 109, 78, 107, 42, 56, 99, 137

77. SEIZE THE DAY

Chapter Test (page 385)

1. $\begin{bmatrix} 3 & 1 & \vdots & 1 \\ 1 & -1 & \vdots & 7 \end{bmatrix}$ **2.** $\begin{bmatrix} 3 & 4 & 2 & \vdots & 4 \\ 2 & 3 & 0 & \vdots & -2 \\ 0 & 2 & -3 & \vdots & -13 \end{bmatrix}$

3. $\left(\frac{7}{3}a - \frac{10}{3}, -\frac{8}{3}a + \frac{29}{3}, a\right)$ **4.** $(-12, -16, -6)$

5. $(2, -3, 1)$ **6.** $\begin{bmatrix} 2 & 4 \\ 7 & 13 \end{bmatrix}$ **7.** $\begin{bmatrix} -4 & -8 \\ 13 & 29 \end{bmatrix}$

8. $\begin{bmatrix} 1 \\ 11 \end{bmatrix}$ **9.** $\begin{bmatrix} 7 & 15 \\ 10 & 22 \end{bmatrix}$ **10.** $\frac{1}{5}\begin{bmatrix} 4 & 1 \\ 3 & 2 \end{bmatrix}$

11. $\begin{bmatrix} 1 & 0 \\ 0 & 1 \end{bmatrix}$ **12.** $\frac{1}{5}\begin{bmatrix} -9 & 16 & -6 \\ 6 & -9 & 4 \\ 4 & -6 & 1 \end{bmatrix}$ **13.** 25

14. -23 **15.** -20 **16.** $(2, -2, 3)$

17. Matrices will vary. Sample answer:

$$\begin{bmatrix} 2 & -2 \\ -2 & 2 \end{bmatrix} \times \begin{bmatrix} 3 & 3 \\ 3 & 3 \end{bmatrix} = \begin{bmatrix} 0 & 0 \\ 0 & 0 \end{bmatrix}$$

18. $\frac{21}{2}$ square units **19.** Collinear

20. $BA = [260,000 \quad 514,000]$

BA represents the total value of each product at each warehouse.

Chapter 6

Section 6.1 (page 395)

Warm Up (page 395)

1. $\frac{4}{5}$ **2.** $\frac{1}{3}$ **3.** $(2n + 1)(2n - 1)$

4. $(2n - 1)(2n - 3)$ **5.** $(n - 1)(n - 2)$

6. $(n + 1)(n + 2)$ **7.** $\frac{1}{3}$ **8.** 24

9. $\frac{13}{24}$ **10.** $\frac{3}{4}$

---

1. $-5, -3, -1, 1, 3$ **3.** 2, 4, 8, 16, 32

5. $\frac{1}{2}, \frac{1}{4}, \frac{1}{8}, \frac{1}{16}, \frac{1}{32}$ **7.** $1, \frac{2}{3}, \frac{1}{2}, \frac{2}{5}, \frac{1}{3}$ **9.** $-1, \frac{1}{2}, -\frac{1}{3}, \frac{1}{4}, -\frac{1}{5}$

11. $\frac{5}{2}, \frac{11}{4}, \frac{23}{8}, \frac{47}{16}, \frac{95}{32}$ **13.** $3, \frac{9}{2}, \frac{9}{2}, \frac{27}{8}, \frac{81}{40}$ **15.** 0, 1, 2, 3, 4

17. $-1, \frac{1}{4}, -\frac{1}{9}, \frac{1}{16}, -\frac{1}{25}$ **19.** 3, 6, 9, 12, 15

21. $5, \frac{19}{4}, \frac{43}{9}, \frac{77}{16}, \frac{121}{25}$ **23.** -73

25. $\dfrac{4}{14{,}175}$ **27.** $\dfrac{44}{239}$ **29.** 56   **31.** 19,656

33. $n + 1; 1, 2$ **35.** $\dfrac{1}{2n(2n + 1)}$; undefined, $\dfrac{1}{6}$

37. $a_n = 2n - 1$ **39.** $a_n = (-1)^{n+1}(2n)$ **41.** $a_n = \dfrac{2}{n^2}$

43. $a_n = \dfrac{n + 1}{n + 2}$ **45.** $a_n = \dfrac{(-1)^{n+1}}{2^n}$ **47.** $a_n = 1 + \dfrac{1}{n}$

49. $a_n = \dfrac{1}{n!}$ **51.** 72   **53.** -84 **55.** 40   **57.** 30

59. $\dfrac{29}{40}$ **61.** 14   **63.** 4   **65.** $\dfrac{65}{24}$ **67.** $\dfrac{205{,}771}{33{,}264}$

69. $\displaystyle\sum_{i=1}^{9} \dfrac{1}{3i}$ **71.** $\displaystyle\sum_{i=1}^{8} \left[2\left(\dfrac{i}{8}\right) + 3 \right]$ **73.** $\displaystyle\sum_{i=1}^{6} (-1)^{i+1} 3^i$

75. $\displaystyle\sum_{i=1}^{20} \dfrac{(-1)^{i+1}}{i^2}$ **77.** $\displaystyle\sum_{i=1}^{5} \dfrac{2^i - 1}{2^{i+1}}$

79. $3.141587\ldots$; No

81. $\displaystyle\sum_{j=0}^{4} j^2 = 30 = \sum_{k=2}^{6} (k - 2)^2$; No

83. (a) $A_1 = \$10{,}212.50$

$A_2 = \$10{,}429.52$

$A_3 = \$10{,}651.14$

$A_4 = \$10{,}877.48$

$A_5 = \$11{,}108.63$

$A_6 = \$11{,}344.68$

$A_7 = \$11{,}585.76$

$A_8 = \$11{,}831.96$

(b) $\$23{,}189.04$

(c) No; $A_{80} = \$53{,}773.20$.

85. $a_0 = 974$

$a_1 = 998.56$

$a_2 = 1029.04$

$a_3 = 1065.44$

$a_4 = 1107.76$

$a_5 = 1156$

$a_6 = 1210.16$

2007: $\$1659.44$

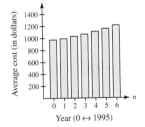

87. $\$152.26$ billion

Section 6.2   *(page 404)*

Warm Up   *(page 404)*

1. 36   **2.** 240   **3.** $\frac{11}{2}$ **4.** $\frac{10}{3}$ **5.** 18

6. 4   **7.** 143   **8.** 160   **9.** 430   **10.** 256

1. Arithmetic sequence; $d = 4$

3. Not an arithmetic sequence

5. Arithmetic sequence; $d = -\frac{1}{2}$

7. Not an arithmetic sequence

9. Arithmetic sequence; $d = 0.4$

11. 8, 11, 14, 17, 20; Arithmetic sequence, $d = 3$

13. 2, 8, 24, 64, 160; Not an arithmetic sequence

15. 1, 1, 1, 1, 1; Arithmetic sequence, $d = 0$

17. $\frac{1}{2}, \frac{1}{3}, \frac{1}{4}, \frac{1}{5}, \frac{1}{6}$; Not an arithmetic sequence

19. $-1, 1, -1, 1, -1$; Not an arithmetic sequence

21. $a_n = 5n - 1$ **23.** $a_n = -12n + 62$

25. $a_n = 5n + 2$ **27.** $a_n = 4n - 4$ **29.** $a_n = 6n - 4$

31. $10, 6, 2, -2, -6$ **33.** 5, 14, 23, 32, 41

35. 2, 6, 10, 14, 18   **37.** 54, 50, 46, 42, 38   **39.** b

40. d   **41.** c   **42.** a

43.

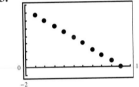

45.

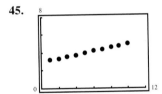

47. 84   **49.** 12   **51.** 23   **53.** 4000   **55.** 10,000

57. 1275   **59.** 22,770   **61.** 44,625   **63.** 520

65. 2725   **67.** (a) $\$40{,}000$ (b) $\$217{,}500$

69. 2625 seats

71. (a)   (b) $\$3016.2$ million

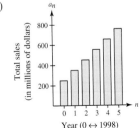

73. 1024 feet   **75.** 470 bricks

77. $\$425{,}000$; Answers will vary.

79. No. You must also know the number of terms involved.

81. (a)

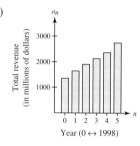

(b) $a_n = 265.63n + 1360$, $n = 0, \ldots, 5$

(c) $\displaystyle\sum_{n=0}^{5} 265.63n + 1360$

   Total revenue: \$12,144.45 million

Section 6.3   (*page 413*)

Warm Up   (*page 413*)

1. $\frac{64}{125}$ **2.** $\frac{9}{16}$ **3.** $\frac{1}{16}$ **4.** 324   **5.** $6n^3$

6. $27n^4$ **7.** $4n^3$ **8.** n^2 **9.** $\displaystyle\sum_{n=1}^{4} 2(3)^{n-1}$

10. $\displaystyle\sum_{n=1}^{4} 3(2)^{n-1}$

---

1. Geometric sequence; $r = 4$; $a_n = 4^{n-1}$

3. Geometric sequence; $r = -3$; $a_n = 4(-3)^{n-1}$

5. Geometric sequence; $r = -\frac{1}{2}$; $a_n = \left(-\frac{1}{2}\right)^{n-1}$

7. Not a geometric sequence

9. Not a geometric sequence   **11.** 4, 8, 16, 32, 64

13. $1, \frac{1}{3}, \frac{1}{9}, \frac{1}{27}, \frac{1}{81}$ **15.** $7, -\frac{7}{5}, \frac{7}{25}, -\frac{7}{125}, \frac{7}{625}$

17. $1, e, e^2, e^3, e^4$ **19.** $16\left(\frac{1}{4}\right)^4 = \frac{1}{16}$

21. $6\left(-\frac{1}{3}\right)^{11} = -\frac{2}{3^{10}}$ **23.** $100e^8$

25. $500(1.02)^{39} \approx 1082.37$ **27.** $16\left(\frac{3}{4}\right)^2 = 9$

29. $54\left(-\frac{1}{3}\right)^5 = -\frac{2}{9}$ **31.** a   **32.** c   **33.** b   **34.** d

35.

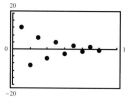

37.

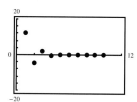

39.

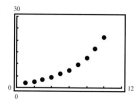

41. 16,368   **43.** $\dfrac{349,525}{32,768} \approx 10.67$ **45.** 511

47. $\approx 29,921.31$ **49.** 2   **51.** $\frac{2}{3}$

53. Sum does not exist; $|r| > 1$ **55.** $\frac{10}{3}$

57. (a) \$17,529.72; No   (b) \$961.96; Yes

59. \$18,416.57; \$18,444.46   **61.** \$2525.26 million

63. (a)

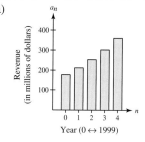

(b) \$1305.83 million

65. \$8186.54 million

67. \$5,368,709.11 for 29 days
   \$10,737,418.23 for 30 days
   \$21,474,836.47 for 31 days

   The job whose wage doubles daily pays more.

69. ≈ 91.12 square inches

Section 6.4   (*page 422*)

Warm Up   (*page 422*)

1. $5x^5 + 15x^2$ **2.** $x^3 + 5x^2 - 3x - 15$

3. $x^2 + 8x + 16$ **4.** $4x^2 - 12x + 9$ **5.** $\dfrac{3x^3}{y}$

6. $-32z^5$ **7.** 120   **8.** 336   **9.** 720   **10.** 20

---

1. 15   **3.** 1   **5.** 15,504   **7.** 4950   **9.** 4950

11. 56   **13.** 35   **15.** $x^4 + 4x^3 + 6x^2 + 4x + 1$

17. $x^3 + 6x^2 + 12x + 8$

19. $x^4 - 8x^3 + 24x^2 - 32x + 16$

21. $x^5 + 5x^4y + 10x^3y^2 + 10x^2y^3 + 5xy^4 + y^5$

23. $81s^4 + 216s^3 + 216s^2 + 96s + 16$

25. $64s^6 - 576s^5 + 2160s^4 - 4320s^3 + 4860s^2$
$- 2916s + 729$

27. $x^5 - 5x^4y + 10x^3y^2 - 10x^2y^3 + 5xy^4 - y^5$

29. $1 - 6x + 12x^2 - 8x^3$

31. $x^{10} - 5x^8y^2 + 10x^6y^4 - 10x^4y^6 + 5x^2y^8 - y^{10}$

33. $\frac{1}{x^5} + \frac{5y}{x^4} + \frac{10y^2}{x^3} + \frac{10y^3}{x^2} + \frac{5y^4}{x} + y^5$

35. The absolute values of corresponding numerical coefficients are the same. The terms of $(x - y)^n$ will alternate in sign.

37. -4 **39.** $-2035 + 828i$ **41.** 1

43. $32t^5 - 80t^4 + 80t^3 - 40t^2 + 10t - 1$

45. $81 + 216z + 216z^2 + 96z^3 + 16z^4$ **47.** $120x^7y^3$

49. $360x^3y^2$ **51.** $1,259,712x^2y^7$

53. $\frac{12!}{4!8!}(10x)^4(-3y)^8 \approx -3.248 \times 10^{10}x^4y^8$

55. $3360x^6$ **57.** $180x^8y^2$ **59.** $-326,592x^4y^5$

61. $70x^8$

63. $\frac{1}{128} + \frac{7}{128} + \frac{21}{128} + \frac{35}{128} + \frac{35}{128} + \frac{21}{128} + \frac{7}{128} + \frac{1}{128} = 1$

65. $0.07776 + 0.2592 + 0.3456 + 0.2304 + 0.0768$
$+ 0.01024 = 1$

67. Fibonacci sequence

69. (a) $g(t) = 0.015t^2 + 2.04t + 107.2$

(b)
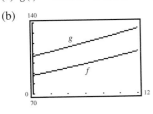

Mid-Chapter Quiz   *(page 424)*

1. 3, 5, 7, 9, 11   **2.** $\frac{1}{4}, \frac{2}{9}, \frac{3}{16}, \frac{4}{25}, \frac{5}{36}$ **3.** 20

4. 24,360   **5.** $\frac{2}{n^2}$ **6.** $\frac{1}{n + 1}$ **7.** $5n - 3$

8. $3n + 4$ **9.** $10(3)^{n-1}$ **10.** $4\left(-\frac{1}{2}\right)^{n-1}$

11. Arithmetic; Common difference is 3.

12. Neither; No common ratio, no common difference

13. Geometric; Common ratio is $\frac{1}{2}$.

14. Geometric; Common ratio is 2.

15. 35   **16.** 60

17. 1225   **18.** 126   **19.** $\frac{3}{4}$ **20.** $\frac{25}{4}$

21. (a) 126   (b) 1

22. (a) $x^4 - 20x^3 + 150x^2 - 500x + 625$
(b) $32x^5 + 80x^4 + 80x^3 + 40x^2 + 10x + 1$

23. $45x^4y^8$ **24.** (a) \$39,000   (b) \$208,500

25. \$59,294.72

Section 6.5   *(page 432)*

Warm Up   *(page 432)*

1. 6656   **2.** 291,600   **3.** 7920   **4.** 13,800

5. 792   **6.** 2300   **7.** $n(n - 1)(n - 2)(n - 3)$

8. $n(n - 1)(2n - 1)$ **9.** $n!$ **10.** $n!$

1. 6   **3.** 5   **5.** 3   **7.** 7   **9.** 30

11. 240

Permutations of
seating positions

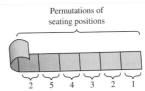

2   5   4   3   2   1

13. 6,760,000   **15.** 1024

17. (a) 900   (b) 648   (c) 180

19. A: 1800; B: 1575; A   **21.** 64,000   **23.** 720

25. (a) 40,320   (b) 384   **27.** 24   **29.** 336

31. 11,880   **33.** 9900   **35.** 720   **37.** 32,760

39. 142,506   **41.** 7,059,052   **43.** 30   **45.** 105

47. 2520   **49.** 3,921,225   **51.** 1800   **53.** 3744

55. Because order matters, use permutations.   **57.** 5

59. 20

61.

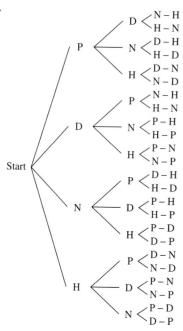

There are 24 possible arrangements.

63.

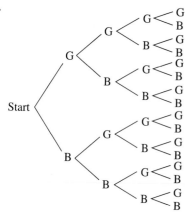

There are 16 possible arrangements.

65. No. $_nP_r = \,_nC_r$ if $r = 0$ or 1.

Section 6.6   *(page 443)*

Warm Up   *(page 443)*

1. $\frac{9}{16}$ **2.** $\frac{8}{15}$ **3.** $\frac{1}{6}$ **4.** $\frac{1}{80,730}$

5. $\frac{1}{804,722,688,000}$ **6.** $\frac{1}{24}$ **7.** $\frac{1}{12}$ **8.** $\frac{135}{323}$

9. 0.366   **10.** 0.997

1. $S = \{H1, T1, H2, T2, H3, T3, H4, T4, H5, T5, H6, T6\}$

3. $S = \{ABC, ACB, BAC, BCA, CAB, CBA\}$

5. $S = \{AB, AC, AD, AE, BC, BD, BE, CD, CE, DE\}$

7. $\frac{3}{8}$ **9.** $\frac{7}{8}$ **11.** $\frac{1}{9}$ **13.** $\frac{7}{12}$ **15.** $\frac{1}{3}$ **17.** $\frac{3}{13}$

19. $\frac{5}{13}$ **21.** $\frac{5}{9}$ **23.** $\frac{5}{12}$ **25.** 0.7   **27.** 0.9

29. 0.22   **31.** (a) $\frac{139}{198}$ (b) $\frac{59}{198}$ (c) $\frac{1}{33}$

33. (a) $\frac{1}{4}$ (b) $\frac{1}{2}$ (c) $\frac{9}{100}$ (d) $\frac{1}{30}$ **35.** $\frac{3}{28}$

37. (a) $\frac{5}{13}$ (b) $\frac{1}{2}$ (c) $\frac{4}{13}$ **39.** (a) $\frac{1}{120}$ (b) $\frac{1}{24}$ (c) $\frac{1}{6}$

41. (a) $\frac{3}{8}$ (b) $\frac{19}{30}$ **43.** $\frac{6}{4165}$

45. (a) ≈ 0.922 (b) ≈ 0.078

47. 0.82

49. (a) $\frac{1}{81}$ (b) $\frac{16}{81}$ (c) $\frac{65}{18}$ **51.** (a) $\frac{1}{64}$ (b) $\frac{1}{32}$ (c) $\frac{63}{64}$

53. $\frac{7}{32}$ **55.** $\frac{76,545}{524,288} \approx 0.1459980011$

57. Answers will vary.   **59.** 0.0729

61. $(0.73)^3 \approx 0.389$

Section 6.7   *(page 456)*

Warm Up   *(page 456)*

1. 24   **2.** 40   **3.** $\frac{77}{60}$ **4.** $\frac{7}{2}$ **5.** $\frac{2k + 5}{5}$

6. $\frac{3k + 1}{6}$ **7.** $8 \cdot 2^{2k} = 2^{2k+3}$ **8.** $\frac{1}{9}$

9. $\frac{1}{k}$ **10.** $\frac{4}{5}$

1. $\frac{5}{(k + 1)(k + 2)}$ **3.** $\frac{(k + 1)^2 \, k^2}{4}$

5–17. Answers will vary.   **19.** 210   **21.** 91

23. 979   **25.** 70   **27.** -3402

29. $S_n = n(2n - 1)$ **31.** $S_n = 10 - 10\left(\frac{9}{10}\right)^n$

33. $S_n = \frac{n}{2(n + 1)}$ **35–45.** Answers will vary.

47. Answers will vary. See page 448.

49. 0, 3, 6, 9, 12

First differences: 3, 3, 3, 3

Second differences: 0, 0, 0

Linear model: $a_n = 3n - 3$

51. 3, 1, -2, -6, -11

First differences: $-2, -3, -4, -5$

Second differences: $-1, -1, -1$

Quadratic model: $a_n = -\frac{1}{2}n^2 + \frac{1}{2}n + 4$

53. 0, 1, 3, 6, 10

First differences: 1, 2, 3, 4

Second differences: 1, 1, 1

Quadratic model: $a_n = \frac{1}{2}n^2 + \frac{1}{2}n$

55. 2, 4, 6, 8, 10

First differences: 2, 2, 2, 2

Second differences: 0, 0, 0

Linear model: $a_n = 2n$

57. 1, 2, 6, 15, 31

First differences: 1, 4, 9, 16

Second differences: 3, 5, 7

Neither linear nor quadratic

59. $a_n = n^2 - n + 3$ **61.** $a_n = n^2 + 5n - 6$

63. (a) First differences: 5, 4, 4, 4; Yes

(b) $a_n = 4n + 569$

(c) $a_n = 4.2n + 566.8$; Yes

Review Exercises   *(page 460)*

1. 5, 9, 13, 17, 21   **3.** $\frac{1}{2}, \frac{2}{3}, \frac{3}{4}, \frac{4}{5}, \frac{5}{6}$ **5.** $1, -\frac{4}{3}, 2, -\frac{16}{5}, \frac{16}{3}$

7. $a_n = 2^{n-1}$ **9.** $a_n = \frac{1}{n}$ **11.** $a_n = (-1)^{n+1}3^n$

13. 43,680   **15.** 51   **17.** 100

19. $\displaystyle\sum_{k=1}^{12} \frac{3}{1+k}$ **21.** $\displaystyle\sum_{k=1}^{13} (-1)^{k-1}2^k$

23. (a) $A_1 = \$4060.00, A_2 = \$4120.90, A_3 = \$4182.71,$
$A_4 = \$4245.45, A_5 = \$4309.14, A_6 = \$4373.77,$
$A_7 = \$4439.38, A_8 = \$4505.97$

(b) $A_{40} = \$7256.07$

(c) $A_{120} = \$23,877.29$

25. Arithmetic; $d = 5$ **27.** Arithmetic; $d = -\frac{1}{2}$

29. 9, 14, 19, 24, 29   **31.** 4, 7, 10, 13, 16

33. $a_n = 3n - 1$ **35.** $a_n = 107 - 7n$ **37.** 192

39. 765

41. (a) Company A: $40,500

Company B: $40,500

(b) $219,000

(c) $205,500

(d) For the short term, you should pick Company A, because at the end of 6 years Company A will have paid $13,500 more in salary. For the long term, you should pick Company B, because the annual raise is higher.

43. Geometric; $r = -3$; $a_n = (-3)^{n-1}$

45. Geometric; $r = \frac{1}{2}$; $a_n = 16\left(\frac{1}{2}\right)^{n-1}$

47. 3, 6, 12, 24, 48   **49.** $10, -2, \frac{2}{5}, -\frac{2}{25}, \frac{2}{125}$

51. $a_{40} = \frac{1}{2^{36}}$ **53.** $a_{10} = -\frac{25}{64}$ **55.** 8184   **57.** 1.5

59. The sum does not exist; $\left|\frac{5}{4}\right| > 1$

61. (a) $16,469.87

(b) No; it will be more because the interest is compounded.

63. (a)

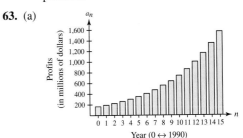

(b) $\approx \$10,237.83$ million

65. 35   **67.** 1   **69.** 792

71. $x^6 + 6x^5 + 15x^4 + 20x^3 + 15x^2 + 6x + 1$

73. $243x^5 + 405x^4y + 270x^3y^2 + 90x^2y^3 + 15xy^4 + y^5$

75. $x^8 - 16x^7y + 112x^6y^2 - 448x^5y^3 + 1120x^4y^4$
$- 1792x^3y^5 + 1792x^2y^6 - 1024xy^7 + 256y^8$

77. $90a^3b^2$ **79.** $25,344x^7$ **81.** 210   **83.** 144

85. (a) 9000   (b) 4536   (c) 4500 **87.** 504

89. 40,320   **91.** 5040

93. (a) 80,730   (b) 351   (c) 142,506

95. $S = \{0, 1, 2, 3, 4, 5, 6, 7, 8, 9\}$

97. $\frac{3}{13}$ **99.** (a) $\frac{5}{12}$ (b) $\frac{7}{12}$ (c) $\frac{5}{36}$

101. (a) $\frac{1}{120}$ (b) $\frac{1}{6}$ **103.** $\frac{396}{4165}$

105. (a) $\frac{1}{128}$ (b) $\frac{1}{64}$ (c) $\frac{127}{128}$ **107.** 0.43

109. Answers will vary.   **111.** Answers will vary.

113. 465   **115.** 1296   **117.** $S_n = \frac{5}{2}\left[1 - \left(\frac{3}{5}\right)^n\right]$

119. 2, 3, 5, 8, 12

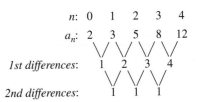

Quadratic model: $a_n = \frac{1}{2}n^2 + \frac{1}{2}n + 2$

Chapter Test   *(page 464)*

1. $3, 5, 7, 9, 11$ **2.** $-1, 4, -9, 16, -25$

3. $2, 6, 24, 120, 720$ **4.** $\frac{1}{2}, \frac{1}{4}, \frac{1}{8}, \frac{1}{16}, \frac{1}{32}$

5. Arithmetic; $d = 5$ **6.** Neither

7. Geometric; $r = 2$ **8.** Neither   **9.** 2600

10. $\approx 29{,}918.311$ **11.** $\frac{1}{2}$ **12.** $\$9096.98$

13. $32x^5 - 240x^4 + 720x^3 - 1080x^2 + 810x - 243$

14. $-192{,}456$ **15.** (a) 72 (b) $328{,}440$

16. (a) 330 (b) $8{,}936{,}928$ **17.** 120

18. $2^{10} = 1024$ **19.** $45{,}000$ **20.** $\frac{1}{16}$ **21.** 25%

22. Answers will vary.

Cumulative Test: Chapters 4–6
(page 465)

1. $x = 2, y = -3, z = 5$ **2.** $x = -2, y = 4, z = -5$

3.

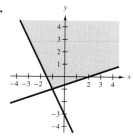

4.

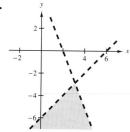

5. $\begin{bmatrix} 1 & 4 & 8 \\ -1 & 4 & 3 \\ 0 & 1 & 5 \end{bmatrix}$ **6.** $\begin{bmatrix} 16 & 9 \\ 8 & 7 \\ 7 & 2 \end{bmatrix}$ **7.** $\begin{bmatrix} 2 & 12 \\ -5 & -3 \\ 1 & 9 \end{bmatrix}$

8. $\begin{bmatrix} \frac{1}{2} & -\frac{1}{2} & 0 \\ \frac{1}{4} & \frac{1}{4} & -1 \\ -\frac{1}{2} & \frac{1}{2} & 1 \end{bmatrix}$

9. $|A| = 4$, $|B|$ does not exist, $|C| = 17$, $|D| = 6$

10. $x = 18{,}000$

$p = 66$

11. $BA = [170{,}000 \quad 230{,}000 \quad 237{,}000]$
$\$170{,}000$ is the value of the inventory in warehouse 1.
$\$230{,}000$ is the value of the inventory in warehouse 2.
$\$237{,}000$ is the value of the inventory in warehouse 3.

12. $7, 9, 11, 13, 15$ **13.** $2, \frac{2}{3}, \frac{2}{9}, \frac{2}{27}, \frac{2}{81}$ **14.** $\frac{130}{9}$

15. $64x^6 + 576x^5y + 2160x^4y^2 + 4320x^3y^3 + 4860x^2y^4$
$\quad + 2961xy^5 + 729y^6$

16. 720 **17.** ≈ 0.978 **18.** $\frac{1}{4}$

19. $2, 4, 7, 11, 16$

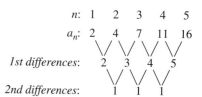

n:	1	2	3	4	5
a_n:	2	4	7	11	16

1st differences: 2   3   4   5

2nd differences: 1   1   1

Quadratic model

Appendix B

Section B.1   *(page A17)*

Warm Up   *(page A17)*

1. $9x^2 + 16y^2 = 144$ **2.** $x^2 + 4y^2 = 32$

3. $16x^2 - y^2 = 4$ **4.** $243x^2 + 4y^2 = 9$

5. $c = 2\sqrt{2}$ **6.** $c = \sqrt{13}$ **7.** $c = 2\sqrt{3}$

8. $c = \sqrt{5}$ **9.** 4 **10.** 2

1. a   **2.** c   **3.** d   **4.** h

5. g   **6.** f   **7.** e   **8.** b

9. Vertex: $(0, 0)$

Focus: $\left(0, \frac{1}{16}\right)$

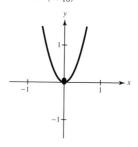

11. Vertex: $(0, 0)$

Focus: $\left(-\frac{3}{2}, 0\right)$

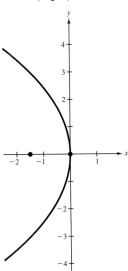

13. Vertex: $(0, 0)$

Focus: $(0, -2)$

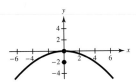

15. Vertex: $(0, 0)$

Focus: $(2, 0)$

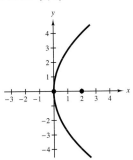

17. $x^2 = -6y$ **19.** $y^2 = -8x$ **21.** $x^2 = 4y$

23. $x^2 = -8y$ **25.** $y^2 = 9x$

27. Center: $(0, 0)$

Vertices: $(\pm 5, 0)$

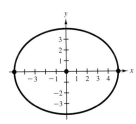

29. Center: $(0, 0)$

Vertices: $\left(\pm \frac{5}{3}, 0\right)$

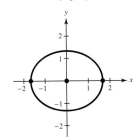

31. Center: $(0, 0)$

Vertices: $(\pm 3, 0)$

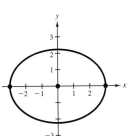

33. Center: $(0, 0)$

Vertices: $\left(0, \pm \sqrt{5}\right)$

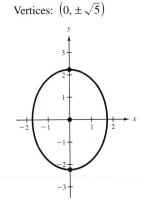

35. $\dfrac{x^2}{1} + \dfrac{y^2}{4} = 1$ **37.** $\dfrac{x^2}{25} + \dfrac{y^2}{21} = 1$

39. $\dfrac{x^2}{36} + \dfrac{y^2}{11} = 1$ **41.** $\dfrac{x^2}{400/21} + \dfrac{y^2}{25} = 1$

43. Center: $(0, 0)$

Vertices: $(\pm 1, 0)$

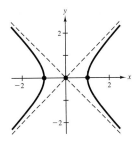

45. Center: $(0, 0)$

Vertices: $(0, \pm 1)$

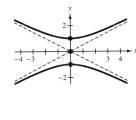

47. Center: $(0, 0)$

Vertices: $(0, \pm 5)$

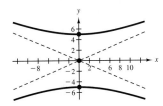

49. Center: $(0, 0)$

Vertices: $(\pm\sqrt{3}, 0)$

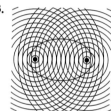

51. $\dfrac{y^2}{4} - \dfrac{x^2}{12} = 1$ **53.** $\dfrac{x^2}{1} - \dfrac{y^2}{9} = 1$

55. $\dfrac{y^2}{16/5} - \dfrac{x^2}{64/5} = 1$ **57.** $\dfrac{y^2}{9} - \dfrac{x^2}{9/4} = 1$

59. $x^2 = 12y$

61. Tacks should be placed at $\frac{3}{2}$ feet to the left and right of the center along the base. The string should be 5 feet long.

63.

65.

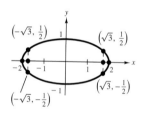

67.

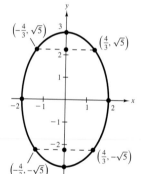

69. $x \approx 110.3$ miles

Section B.2   (page A26)

Warm Up   (page A26)

1. Hyperbola   2. Ellipse   3. Parabola
4. Hyperbola   5. Ellipse   6. Circle
7. Hyperbola   8. Parabola   9. Parabola
10. Ellipse

1. Shifted one unit up and two units to the left

3. Shifted three units down and one unit to the right

5. Shifted two units down and one unit to the right

7. Vertex: $(1, -2)$
Focus: $(1, -4)$
Directrix: $y = 0$

9. Vertex: $\left(5, -\frac{1}{2}\right)$
Focus: $\left(\frac{11}{2}, -\frac{1}{2}\right)$
Directrix: $x = \frac{9}{2}$

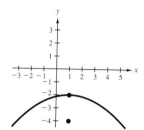

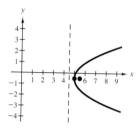

11. Vertex: $(1, 1)$
Focus: $(1, 2)$
Directrix: $y = 0$

13. Vertex: $(8, -1)$
Focus: $(9, -1)$
Directrix: $x = 7$

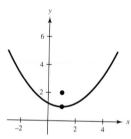

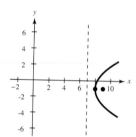

15. Vertex: $(-2, -3)$
Focus: $(-4, -3)$
Directrix: $x = 0$

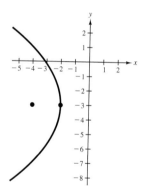

17. $(y - 2)^2 = -8(x - 3)$ **19.** $x^2 = 8(y - 4)$
21. $(y - 2)^2 = 8x$ **23.** $x^2 = -(y - 4)$

25. Center: $(1, 5)$

Foci: $(1, 9), (1, 1)$

Vertices: $(1, 10), (1, 0)$

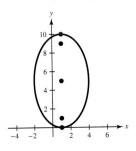

27. Center: $(-2, 3)$

Foci: $\left(-2, 3 - \sqrt{5}\right), \left(-2, 3 + \sqrt{5}\right)$

Vertices: $(-2, 0), (-2, 6)$

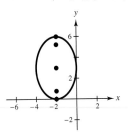

29. Center: $(1, -1)$

Foci: $\left(\frac{1}{4}, -1\right), \left(\frac{7}{4}, -1\right)$

Vertices: $\left(-\frac{1}{4}, -1\right), \left(\frac{9}{4}, -1\right)$

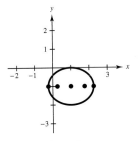

31. Center: $\left(\frac{1}{2}, -1\right)$

Foci: $\left(\frac{1}{2} - \sqrt{2}, -1\right), \left(\frac{1}{2} + \sqrt{2}, -1\right)$

Vertices: $\left(\frac{1}{2} - \sqrt{5}, -1\right), \left(\frac{1}{2} + \sqrt{5}, -1\right)$

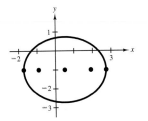

33. $\dfrac{(x - 2)^2}{4} + \dfrac{(y - 2)^2}{1} = 1$

35. $\dfrac{x^2}{48} + \dfrac{(y - 4)^2}{64} = 1$

37. $\dfrac{(x - 3)^2}{9} + \dfrac{(y - 5)^2}{16} = 1$

39. $\dfrac{x^2}{16} + \dfrac{(y - 4)^2}{12} = 1$

41. Center: $(1, -2)$

Vertices: $(-1, -2), (3, -2)$

Foci: $\left(1 - \sqrt{5}, -2\right), \left(1 + \sqrt{5}, -2\right)$

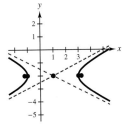

43. Center: $(2, -6)$

Vertices: $(2, -7), (2, -5)$

Foci: $\left(2, -6 - \sqrt{2}\right), \left(2, -6 + \sqrt{2}\right)$

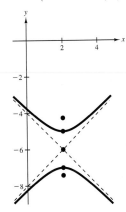

45. Center: $(2, -3)$

Vertices: $(3, -3), (1, -3)$

Foci: $\left(2 - \sqrt{10}, -3\right), \left(2 + \sqrt{10}, -3\right)$

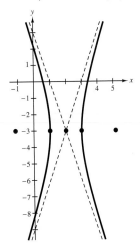

47. Center: $(1, -3)$

Vertices: $\left(1, -3 - \sqrt{2}\right), \left(1, -3 + \sqrt{2}\right)$

Foci: $\left(1, -3 - 2\sqrt{5}\right), \left(1, -3 + 2\sqrt{5}\right)$

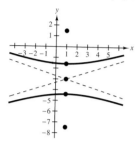

49. Center: $(-1, -3)$

Vertices: $\left(-1 - 3\sqrt{3}, -3\right), \left(-1 + 3\sqrt{3}, -3\right)$

Foci: $\left(-1 - \sqrt{30}, -3\right), \left(-1 + \sqrt{30}, -3\right)$

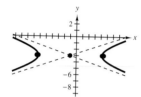

51. $\dfrac{(x-4)^2}{4} - \dfrac{y^2}{12} = 1$ **53.** $\dfrac{(y-5)^2}{16} - \dfrac{(x-4)^2}{9} = 1$

55. $\dfrac{y^2}{9} - \dfrac{(x-2)^2}{9/4} = 1$ **57.** $\dfrac{(x-3)^2}{9} - \dfrac{(y-2)^2}{4} = 1$

59. $\dfrac{(x-3)^2}{4} + \dfrac{(y+2)^2}{4} = 1$

Circle

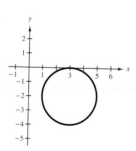

61. $\left(x - \frac{1}{2}\right)^2 - \dfrac{y^2}{4} = 1$

Hyperbola

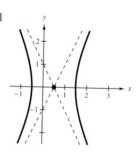

63. $\dfrac{(x+1)^2}{1/4} + \dfrac{(y-4)^2}{1/3} = 1$

Ellipse

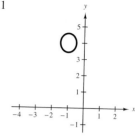

65. $\left(x - \frac{1}{5}\right)^2 = 8\left(y + \frac{3}{5}\right)$

Parabola

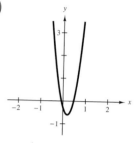

67. (a) $\approx 24{,}748.74$ miles per hour

(b) $x^2 = -16{,}400(y - 4100)$

69. $\dfrac{x^2}{25} + \dfrac{y^2}{16} = 1$

71. Perihelion: $\approx 9.138 \times 10^7$ miles

Aphelion: $\approx 9.454 \times 10^7$ miles

73. $e \approx 5.414 \times 10^{-2}$

75. $\dfrac{(x-h)^2}{a^2} + \dfrac{(y-k)^2}{b^2} = 1$

$\dfrac{(x-h)^2}{a^2} + \dfrac{(y-k)^2}{a^2 - c^2} = 1$

$\dfrac{(x-h)^2}{a^2} + \dfrac{(y-k)^2}{(a^2/a^2)(a^2 - c^2)} = 1$

$\dfrac{(x-h)^2}{a^2} + \dfrac{(y-k)^2}{a^2(1 - c^2/a^2)} = 1$

$\dfrac{(x-h)^2}{a^2} + \dfrac{(y-k)^2}{a^2[1 - (c/a)^2]} = 1$

$\dfrac{(x-h)^2}{a^2} + \dfrac{(y-k)^2}{a^2(1 - e^2)} = 1$

77. $\dfrac{x^2}{321.84} + \dfrac{y^2}{19.02} = 1$

Index of Applications

BIOLOGY AND LIFE SCIENCES

Ancient civilization, 240
Antler spread, 247
Average recycling cost, 179, 186, 187
Bacteria count, 79, 83
Bacteria growth, 199, 220, 240, 248, 249, 250
Bear population, 250
Body temperature, R137
Carbon dating, 234, 240
Deer population, 238
Diet supplement, 299
Endangered species, 241
Erosion, R38
Forest yield, 229
Galloping speed of animals, 218
Genders of children, 434, 445, 463
Height and forearm length, 34
Human height, R78, R137
Human memory model, 179, 186, 208, 210, 229, 247, 248, 250
Learning curve, 240, 249
Native prairie grasses, 229
Nitrogen dioxide emissions, 148
Nutrition, 296
Oxygen level, R61
Population of deer, 179
Population of elk, 179
Population of fish, 186
Radioactive decay, 197, 200, 239, 240, 248, 250
Recycling plan, 443
Renewable energy, 103
Skill retention model, 211
Smokestack emission, 175, 186
Spread of a virus, 236
Stocking a lake with fish, 241
Suburban wildlife, 229
Water pollution, 178
Wildlife management, 249

BUSINESS

Acquisition of raw materials, 357
Advertising cost, 148, 149, 185
Advertising expenses, 125
Annual operating cost, R136
Annual payroll, 397
Average cost, 52, 176, 179, 186
Break-even analysis, R135, R136, R152, 259, 262, 289, 312
Budget variance, R7, R9

Camping equipment sales, 241
Cassette shipments, 241
Cell phone market share, 319, 379
Compact disc sales, 313
Company profits, R146, R153
Company revenue, 63
Comparing product sales, 312
Comparing profits, 81
Comparing sales, 82, 100
Competing restaurants, 433
Computer inventory, 298
Computer models, 383
Concert ticket sales, 299, 315
Consumer and producer surplus, 295, 298, 315
Contract bonuses, 348
Cost, 81, 93, 100
Cost-benefit model, 175
Cost equation, R111
Cost, revenue, and profit, 50, 82
Daily sales, R136
Declining balances depreciation, R37
Defective units, 442, 445, 462, 463, 465
Demand function, 149, 199, 220, 229, 248
Digital music sales, 287
Dollar value, 97
DVD player inventory, 315
Earnings per share
   Commerce Bancorp, Inc., 114
   Home Depot Stores, 15
   Microsoft Corporation, 15
Earnings-dividend ratio
   Wal-Mart Stores, 93
e-retail, R150
Factory production, 346, 381
Fast food sales, R103
Fourth quarter sales, 26, 97
Full-length cassette sales, 180
Furniture production, 298
Gold prices, 14
Hotel pricing, 346
Hotel room rates, 358
Increasing profit, R143
Inventory, 465
Inventory levels, 346, 381, 385
Job applicants, 432, 434, 462
Labor/wage requirements, 346, 359, 382
Lottery sales, 26
Making a sale, 443, 445, 446
Market research, R122, R125, R152
Market share, 442

Maximum profit, 64, 316
Maximum revenue, 316
Media selection, 308
Minimum cost, 316
Mobile homes manufactured, 51
Monthly cost, R103
Monthly production costs, R114
Monthly profit, R88, R154
Multiplier effect, 387, 458
National defense budget, 51
New truck sales, R124
New York Stock Exchange, 200, 220, 227
Non-farm employees, 423
Operating revenue for airline industry, 38
Optimal cost, 113, 183, 305, 308
Optimal profit, 114, 183, 304, 307, 308, 309, 316, 317
Optimal revenue, 113, 183, 308
Outerwear sales trends in Arizona, 283
Patents issued, 46
Payroll mix-up, 445, 463
Prescription drug expenditures, R113
Price of a product, R153
Price-earnings ratio
   McDonald's Corporation, 83
   Walt Disney Company, 83
Production cost, R102
Production limit, R91
Profit, R144, R154, R155, 73, 93, 115, 135, 168, 184, 251, 406, 461
   Biomet, Inc., 415
   Harley-Davidson, Inc., 414
   Home Depot, Inc., 415
Profit and advertising expenses, 144, 150, 187
Projected expenses, R90
Projected revenue, R90, R151
Projected science spending, R113
Property tax, 36, 97
Quality control, R9
Raw materials, 357
Real estate rentals, 39
Reimbursed expenses, 37
Research and development expenditures, 242
Restaurant sales, 274
Retail trade sales, 230
Revenue, 26, 99, 168
   American Express, 26
   California Pizza Kitchen, Inc., 415

Krispy Kreme Doughnuts, Inc., 415
McDonald's Corporation, 314, 406
Outback Steakhouse, Inc., 406
United Parcel Service, Inc., 397
Revolving credit, 275
Sales commission, 199
Sales price and list price, 37
Sales, 19, 53, 241, 403, 406, 461
  Biomet, Inc., 415
  The Cheesecake Factory, Inc., 405
  for electronics and appliance stores, 101
  Microsoft Corporation, 39
  of nondurable toys and sports supplies, 289
  Starbucks Corporation, 405
  tortilla industry, 15
  Yankee Candle, 201
Shoe sales, 274
Sporting goods sales, R88
State income tax, 29
State sales tax, 36
Straight-line depreciation, 32, 37, 97, 101
Supply and demand, 271, 274, 289, 313, 317, 465
Tax liability, 132
Textbook sales, 241
Ticket sales, 274
  for Broadway shows, 249
Total cost, R87
Total revenue, R87, R102, R113, R114, R151, R154
Transportation cost, 147
U.S. car sales, R89
Value of cassettes shipped, 110
Worker's productivity, 243
Years of service for employees, 440

CHEMISTRY AND PHYSICS

Acid mixture, 273, 285, 313
Acid solution, R87
Automobile aerodynamics, 73
Automobiles, 230
Charge of an electron, R27
Chemical reaction, 243
Comet orbit, A25, A28
Daily high temperature, 23
Diesel mechanics, 93
Distance between sun and planets, 215
Drug concentration, 179
Earthquake magnitudes, 98, 237, 243
Escape velocity, R35, R38
Estimating the time of death, 243

Falling object, R97, R101, R112, R114, 406
Fluid flow, A28
Fuel mixture, 273
Hot air balloon, R112
Hyperbolic mirror, A19
Intensity of sound, 243
Land area of Earth, R27
Light year, R27
Mass of an electron, R23
Maximum height of a baseball, 109
Maximum height of a diver, 114
Maximum height of a shot put throw, 114
Medicine, 147
Mixture, R91, R151
Molecules in a drop of water, R23
Newton's Law of Cooling, 189, 244
Number of air sacs in lungs, R27
Parabolic reflector, A11
Path of a ball, 51
Path of a baseball, 45
Period of a pendulum, R38
pH levels, 237, 243
Planetary motion, A28
Power line, R125
Pressure change, R92
Radio waves, R90
Refrigeration, R61
Satellite antenna, A18
Satellite orbit, A27, A28
Sound intensity, 216, 218
Speed of atomic particles, R28
Speed of a baseball, 52
Speed of light, R23
Temperature, 25
  in Bismarck, North Dakota, R9
  in Chicago, Illinois, R9
  of core of sun, R27
Temperature and hour of the day, 40
Thawing a package of steaks, 243, 249
Thickness of a soap bubble, R27
Throwing an object on the moon, R109, R152
Vertical motion, 98
Water area of Earth, R27
Width of human hair, R27
Work, 210

CONSTRUCTION

Architecture, A18
Brick pattern, 406
Mountain tunnel, A19
Nail length, 218

Suspension bridge, A18
Wheelchair ramp, 26

CONSUMER

Annual salary, R79, R150, 26
Apartment rent, 26
Cable television subscribers, 348
Cell phone bill, 275
Charitable contributions, 111, 122
Charter bus fares, 52
Child care demand, 11
Choice of two jobs, 263, 312
College costs, R102
College textbooks expenditures, 51
Comparative shopping, R132, R135
Consumer spending, R112
Cost of books and supplies, 53
Cost of dental care, 148
Cost of higher education, R147
Cost of housing, 51
Cost of overnight delivery, 64, 99
Cost of renting, 38
Discount, R87, R91
Discount rate, R89, R150
Food budget, R89
Hourly wages, 37
Job offer, 405, 461
List price, R89, R150
Loan payments, R89
Long-distance phone plans, 260, 343, 347
Markup, R91
Monthly basic cable rates, R155
Original price, R80
Percent of a raise, R81
Percent of a salary, R81
Personal watercraft purchased, 286
Price of a telephone call, 58
Prices of homes, 288
Rebate and discount, 79
Recreation spending, 147
Reduced rates, R120
Renting a moving truck, 260
Repaying a loan, 37
Salary, 415, 424
Salary and bonus, 100
Salary increase, R80, R135
Sales commission, 37
Sales tax and selling price, 52
Sharing the cost, R124, R152
Sound recordings purchased, 317
Trade in value of a car, 11
Vacation packages, 346
Value of the dollar, 60
Weekly paycheck, R88
Weekly salary, R89

GEOMETRY

Accuracy of a measurement, R133, R137, R152
Area of a circle, 40
Area and circumference of a circle, 53
Area of a forest region, 377
Area of a foundation, 100
Area of a sector of a circle, R92
Area of a shaded region, R47
Area of a square, R136, R137
Area of a tract of land, 377
Area of a trapezoid, R92
Area of a triangle, R87, R91
Base and height of a sign, R101
Centimeters and inches, 36
Circles and pi, 115
Cutting across the lawn, R98
Depth of a submarine, R102
Depth of an underwater cable, R151
Diagonals of a polygon, 434
Dimensions of a billboard, R101, R151
Dimensions of a box, 146, 147
Dimensions of a building, R101
Dimensions of a classroom, R37
Dimensions of a container, 45
Dimensions of a corral, R111
Dimensions of a cube, R37, R39
Dimensions of a house, 135
Dimensions of a lot, R101
Dimensions of a package, 147
Dimensions of a picture frame, R89
Dimensions of a room, R82, R89, R96, R150, 135, 184
Dimensions of the square base of a box, R111, R114
Dimensions of a volleyball court, R150
English and metric systems, 30
Feet and meters, 97
Floor plan, R47
Formula for pi, 396
Fractals, 158
Geometric modeling, R54
Height of a can, R86
Lateral surface area of a cylinder, R92
Length of a playing field, R146
Length of a room, R54, R146
Length of the sides of an isosceles right triangle, R102
Length of the sides of an equilateral triangle, R102
Length of a tank, R91
Liters and gallons, 36
Number of meters in one foot, R64
Number of sides and diagonals of regular polygons, 280

Optimal area, 113, 183
Perimeter of a rectangle, R87, R91
Pythagorean Theorem, 27
Regular polygons and regular polyhedra, 288
Ripples in a pond, 52, 83
Salvage depth, R102
Sierpinski Triangle, 158
Snowflake curve, 416
Square pattern, 416
Surface area of a cone, R92
Volume of a box, R45, R47, R111, R152, 50, 98
Volume of M&Ms, R92
Volume of a package, 50
Volume of a rectangular prism, R91
Volume of a right circular cylinder, R91
Volume of a sphere, 44
Volume and surface area of a sphere, R22
Water depth, R91
Weather balloon, R22

INTEREST RATE

Account balance, 238
Balance in an account, R25, R28, R39, R65, R67, 98
Bond investment, 357
Borrowing money, R124, 332, 333, 381
Cash advance, R124, R152
Cash settlement, 199
Comparing investment returns, R91
Compound interest, R26, R47, R92, R121, R122, R124, R146, R153, 195, 196, 199, 229, 239, 246, 250, 397, 412, 414, 424, 460, 461, 464
Doubling and tripling an investment, 226
Investment, 285, 298, 313, 317
Investment groups, 347
Investment mix, R90
Investment plan, 246
Investment portfolio, 273, 282, 286, 312, 314
Investment time, 210, 247
Investments, 309
Monthly payment, R61, 211
Rate of inflation, R26
Savings plan, R44, R47
Simple interest, R84, R89, R91, R135, R150, 26, 36, 40, 256, 263

MISCELLANEOUS

100-meter freestyle winning times, 180
Aircraft boarding, 462

Airline passengers, R124
Alternative fuel, 263
Alumni association, 444
Archimedes and the gold crown, 219
Arranging flowers, 431
Athletes and jersey numbers, 40
Australian football, A28
Baseball, 446
Baseball salaries, R69, R148
Batting average, R125
Bike race, 433
Birthday months, 41
Birthday problem, 442
Car insurance, 47
Cash scholarship, 445
Choosing officers, 433
Coffee maker types, 313
College bound, 444
College enrollment, R19, 26
Combination lock, 433
Community service, R125
Computer systems, 432, 462
Concert seats, 433
Concert tickets, 434
Cost of a postage stamp, R26
Counting card hands, 431
Course grade, R89
Course schedule, 432
Crop spraying, 285
Cryptography, 373, 374, 375, 377, 378, 384
Cutting a piece of yarn, 412
Digital camera depreciation, 37
Discus throw, 38
Dominoes, 448
Drawing a card, 436, 439, 443, 444, 462
Drawing marbles, 443
Drawing pieces of paper, 425
Education, 463
Emails received, 446
Exercise program, R132
Federal income tax and adjusted gross, 47
Fitness, R150
"Flying Elvi," R101
Forming a committee, 434
Forming an experimental group, 433
Fuel-efficient car, 36
Game show, 444, 445, 463, 465
Grades and happiness, 47
Grades of paper, 285
Grand Canyon, R151
Guessing game, 369
Hair products, 285
Heads or tails?, 435, 436, 443, 462, 464

High school enrollment, 38
Home theater system, 462
Horse race finishes, 428
Humidity control, R137
IQ scores, R136, 251
Land size of the U.S., 60
Letter mix-up, 445
Letter pairs, 426
License plate numbers, 432
Long-distance telephone calls, 47
Lottery, 433, 438
Man and mouse, R137
Movie sequels, R88
Musical notes and frequencies, R1, R62
New York City Marathon, R91
News cast competition, 312
Notes on a musical scale, R38
Olympic diver, R101
Olympic swimming, 35
Owl and the mouse, R101
Pin numbers, 431
Pizza toppings, 434
Poker hand, 434, 445, 463
Population, 239, 240, 246, 248, 250
Population growth, 200, 210
Posing for a photograph, 433
Preparing for a test, 444, 463
Prize money won at the Indy 500, 34
Random number generator, 441, 444
Ratio of day to year, R65
Research habits, 446
Riding in a car, 433
Roller coaster ride, 462
Royal Gorge Bridge, R101
Sailboat stays, R125
Salary and amount in savings, 52
Scuba diving, 26
Seating capacity, R111, 402, 405, 461
Seizure of illegal drugs, 178
Single file, 433
Snow removal, 247
Sound system, 464
Space shuttle meals, 431
Spam, 446
Taste testing, 443
Telephone numbers, 462, 426
Test questions, 433
Test scores, 52, 249
Theater tour, 434
Time study, R137
Toboggan ride, 432
Tossing a coin, 445
Tossing a coin and a die, 443
Tossing a die, 443
Tossing two dice, 437

True-false exam, 432, 464
Voting preference, 348
Weather report, 464
Weight loss program, R135
Weightlifting, R136
Winning an election, 444
Work rate, R125

TIME AND DISTANCE

Airplane speed, 270, 273
Average speed, R90
Catch-up time, R90
Distance, R82
    of a baseball, 115, 183
    between sun and Jupiter, R65
    from a dock, R112
    to a star, R90
    traveled, R87
    traveled by a car, 37
    traveled on a bicycle, 40
Flying distance, R102, R112
Flying speed, R113
Height
    of a balloon, 50
    of a building, R90
    of a flare, R153
    of a mountain climber, 31
    of a parachutist, 37
    of Petronas Tower 1, R83
    of a projectile, R145, R153
    of a tree, R90
Navigation, A19
Stopping distance, 81, 288
Travel, R9
Travel time, R87, R90, R150

U.S. DEMOGRAPHICS

Adult amateur softball teams, 289
Age at first marriage, 201
Aircraft departures, 33
Alternative fuel consumption, 125
Alternative fuel vehicles in use, R108,
    R111
Amount spent on athletic footwear, 26
Average grade of college freshmen,
    R154
Average heights of males and females in
    the U.S., 230
Average miles per gallon for passenger
    cars, 64
Cable television advertising, R78
Cable television subscribers, R136
Car dealerships, R147
Cell phone subscribers, R66, R99

Cell sites, R66
Child support collections, 358
College applications, 383
College graduates, R147
College tuition, R28
Colleges and universities, 438
Colleges and universities expenditures,
    R67
Comparing populations
    Alaska and North Dakota, 312
    Oklahoma and Oregon, 262
    Wisconsin and Arizona, 263
Daily fee golf facilities, A6
Domestic travelers, 148
Employment, 35
Energy imports, 333
Federal government expenses, R19
Federal Pell Grants, R47
Federal student aid, 299
Female labor force, 82
Fossil fuel consumption, 333
Fuel consumption, R124
Fuel efficiency, 14
Full-length CDs, 242
Genetically modified cotton crops
    planted, 287
Grade level salaries for federal
    employees, 64
Head Start enrollment, R103
Health care expenditures, 39
Health care spending, 180
High school graduates, 242
Home prices, R65, R66
Hospital costs, 397
Hours spent on the Internet, 28
Ice cream produced, 124
Land area of genetically modified crops,
    185
Land area of genetically modified
    soybean crops, 150
Life expectancy, R124, 15
Life insurance per household, 423
Mail-order prescriptions, R155
Median age of U.S. population, 211
Men's heights, 241
Minimum wage, R28
Movie theater admission prices, 93
Newspapers, 253, 310
Number of cable TV systems, 64
Number of shopping centers, 26
Occupied housing units, 97
PC and Internet users, 274
People per household, 1, 94
Per capita land area, 176
Personal income, R153

Political makeup of the U.S. Senate, 78
Population
    of California, R64
    of Connecticut, 63
    of North Dakota, 63
    of Orlando, Florida, 32
    of the U.S., R103, 60, 394, 441
    of Vermont, 457
Preparing weekday meals, R19
Projected energy consumption, 100
Public college enrollment, R136

Radio stations, 63
Ratio of males to females, 397
Registered vehicles, R153
Research and development expenditures, 64
Revolving credit, 93
SAT or ACT?, 263
SAT scores, 235
Snowboarders, 299
Social Security benefits, R78
Software piracy, 183

Studying social science, 384
Sunday newspapers in circulation, 39
Supreme Court cases, R78
Television owners, R88
U.S. currency, 230
Voting-age population, 445
Weekly earnings, R75
Women's heights, 241
World Internet users, R88
World population, R147
World population increase, 232

Index

A

Absolute value, R5
  function, graph of, 60
  inequality, solving, R131
  properties of, R6
Adding fractions
  with like denominators, R14
  with unlike denominators, R14
Addition, R11
  of complex numbers, 152
  of a constant, R128
  of inequalities, R128
  of matrices, 335
    properties of, 337
  of polynomials, R41
Additive identity
  for complex numbers, 152
  for matrices, 338
  for real numbers, R12
Additive inverse
  for complex numbers, 152
  for real numbers, R12
Adjoining matrices, 351
Algebraic equation, R79
Algebraic expression, R10
  constant, R10
  evaluating, R11
  terms of, R10
  variable, R10
Algebraic function, 190
Approximately equal to, R3
Area
  common formulas for, R85
  of a triangle, 370
Arithmetic sequence, 398
  common difference of, 398
  finite, sum of, 401
  nth partial sum, 402
  nth term of, 399
Associative Property of Addition
  for complex numbers, 153
  for matrices, 337
  for real numbers, R12
Associative Property of Multiplication
  for complex numbers, 153
  for matrices, 341
  for real numbers, R12
Associative Property of Scalar
    Multiplication, 337
Asymptote
  horizontal, 170

  of a hyperbola, A15
  oblique, 174
  slant, 174
  vertical, 170
Augmented matrix, 321
Average cost per unit, 176
Average value for a population, 235
Axis
  imaginary, 156
  of a parabola, 105, A9
  real, 156
  of symmetry, 105

B

Back-substitution, 255
Balance in an account, R25
Base, R20
  of an exponential function, 190
  natural, 194
Basic rules of algebra, R12
Bell-shaped curve, 235
Binomial, R40
  cube of, R43
  expanding, 420
  square of, R43
Binomial coefficient, 417
Binomial Theorem, 417
Bounded, 156
  interval(s), R126
    on the real number line, R126
Branches of a hyperbola, A14
Break-even point, 259

C

Cartesian plane, 2
Center
  of a circle, 10
  of an ellipse, A11
  of a hyperbola, A14
Certain event, 436
Change-of-base formula, 204, 212
Characteristics of exponential functions,
  192
Characteristics of logarithmic functions,
  205
Checking a solution, R72
Circle, 9
  center of, 10
  general form of the equation of, 10
  radius of, 10
  standard form of the equation of, 9,
    10, A20

Closed interval, R126
Coded row matrices, 373
Coefficient
  binomial, 417
  leading, R40
  of a term, R40
  of a variable term, R10
Coefficient matrix, 321
Cofactors of a square matrix, 362
  expanding by, 363
Collinear points, test for, 371
Column matrix, 320
Column of a matrix, 320
Combinations of n elements taken r at a
  time, 430
Common difference, 398
Common formulas
  for area, R85
  miscellaneous, R85
  for perimeter, R85
  for volume, R85
Common graphs of functions, 60
Common logarithmic function, 203
Common ratio, 407
Commutative Property of Addition
  for complex numbers, 153
  for real numbers, R12
Commutative Property of Multiplication
  for complex numbers, 153
  for real numbers, R12
Complement of an event, 442
  probability of, 442
Completely factored, R48
Completing the square, R104, 10
Complex conjugate, 154
Complex fraction, R59
Complex number, 151
  addition of, 152
  additive identity, 152
  additive inverse, 152
  Associative Property of Addition, 153
  Associative Property of
    Multiplication, 153
  Commutative Property of Addition,
    153
  Commutative Property of
    Multiplication, 153
  Distributive Property, 153
  equality of, 151
  standard form of, 151
  subtraction of, 152

Complex plane, 156
    imaginary axis, 156
    real axis, 156
Complex zeros occur in conjugate pairs, 164
Composite number, R15
Composition of two functions, 76
Compound interest, 195
    formulas for, 196
Condense a logarithmic expression, 214
Conditional equation, R70
Conic (conic section), A8
    degenerate, A8
    ellipse, A11
    hyperbola, 170, A13, A14
    parabola, 104, A9
    standard forms of equations of, A20, A21
    standard position, A20
Conjugate, 164
    complex, 154
    of a radical expression, R32
Conjugate axis of a hyperbola, A15
Consistent system, 268
Constant, R10
    addition of, R128
    function, 56, 60, 104
    multiplication of, R128
    of proportionality, 29
    term, R10
        of a polynomial, R40
    of variation, 29
Constraints, 300
Consumer surplus, 295
Continuous, 116
Continuous compounding, 195
Continuously compounded interest, 195
Coordinate(s), 2
    axes, reflection in, 67
    of a point on the real number line, R4
    x-coordinate, 2
    y-coordinate, 2
Critical numbers
    of a polynomial inequality, R138
    of a rational inequality, R142
Cross-multiplying, R74
Cryptogram, 373
    decode, 373
    encode, 373
Cube(s)
    of a binomial, R43
    difference of two, R49
    perfect, R30
    root, R29
    sum of two, R49

Cubing function, graph of, 60
Curve
    bell-shaped, 235
    logistic, 236
    sigmoidal, 236

D

Decode, 373
Decreasing function, 56
Defined function, 47
Degenerate conic, A8
Degree
    of a polynomial, R40, R41
    of a term, R40, R41
Denominator, R12
    least common, R15
    rationalizing the, R31, R32
Dependent system, 268
Dependent variable, 41, 47
Determinant
    of a 2×2 matrix, 360
    area of a triangle, 370
    of a matrix, 353
    of a square matrix, 360, 363
    test for collinear points, 371
    two-point form of the equation of a line, 372
Diagonal matrix, 366
Difference
    of two cubes, factoring, R49
    of two functions, 74
    of two squares, factoring, R49
Differences
    first, 454
    second, 454
Diminishing returns, point of, 125
Direct variation, 28, 29
Directed distance, A23
Directly proportional, 29
Directrix of a parabola, A9
Discriminant, R105
Distance
    between two numbers, R6
    directed, A23
    traveled formula, R85
Distance Formula, 3
Distinguishable permutations, 429
Distributive Property
    for complex numbers, 153
    for matrices, 341
    for real numbers, R12
Divide evenly, 127
Dividing fractions, R14
Division, R11
    Algorithm, 127

long, 126
synthetic, 129
    results of, 131
Divisors, R15
Domain
    of an expression, R55
    of a function, 40, 47
    implied, 44, 47
    of a rational function, 169
Double inequality, R5
Double solution, R94
Double subscript, 320

E

e, natural base, 194
Eccentricity of an ellipse, A28
Elementary row operations, 321
Elimination
    Gaussian, 277
        with back-substitution, 326
    Gauss-Jordan, 327
    method of, 264, 266
Ellipse, A11
    center of, A11
    eccentricity of, A28
    foci of, A11
    latus rectum of, A19
    major axis of, A11
    minor axis of, A11
    standard form of the equation of, A12, A21
    vertices of, A11
Encode, 373
Endpoints of an interval, R126
Entry of a matrix, 320
    main diagonal, 320
Equal matrices, 334
Equality
    of complex numbers, 151
    hidden, R79
    properties of, R16
Equation(s), R16, R70
    algebraic, R79
    checking a solution, R72
    circle, standard form, 9, 10, A20
    conditional, R70
    conics, standard form, A20, A21
    ellipse, standard form, A12, A21
    equivalent, R71
    general form, circle, 10
    graph of, 5
    hyperbola, standard form, A14, A21
    identity, R70
    linear, R71

of a line
  general form, 21
  horizontal, 21
  intercept form, 25
  point-slope form, 18, 21
  slope-intercept form, 20, 21
  summary of, 21
  two-point form, 19, 372
  vertical, 21
parabola, standard form, A9, A20
polynomial
  second-degree, R93
  solution of, 120
position, R97
quadratic, R93
quadratic type, R116
recognition, 9
solution(s) of, R70, 5
solve, R70
system of, 254
  solution of, 254
true, R70
Equation editor, A1
Equilibrium, point of, 271, 295
Equivalent
  equations, R71
    forming, R71
  expressions, R55
  fractions, R14
    generate, R14
  inequalities, R128
  systems of equations, 265, 277
    operations that produce, 277
Escape velocity, R35
Evaluate an algebraic expression, R11
Even function, 59
  test for, 59
Event(s), 435
  certain, 436
  complement of, 442
  impossible, 436
  independent, 440
  mutually exclusive, 439
  probability of, 436
Existence theorem, 161
Expand a logarithmic expression, 214
Expanding
  a binomial, 420
  by cofactors, 363
  by minors, 363
Experiment, 435
Exponent, R20
  properties of, R21, R33
  rational, R32, R33

Exponential
  decay model, 231
  equation
    inverse property, 221
    one-to-one property, 221
    solving, 221
    strategies for solving, 221
  form, R20
  function
    with base a, 190
    characteristics of, 192
    natural, 194
  growth, 197
    model, 231
  notation, R20
Exponentiating, 224
Expression
  algebraic, R10
  domain of, R55
  equivalent, R55
  fractional, R56
  radical, simplest form of, R31
  rational, R56
Extended principle of mathematical
    induction, 448
Extracting square roots, R95
Extraneous solution, R73, R117
Extrapolation, linear, 19

F

f of x, 42
Factor(s), R15
  of a polynomial, 120, 165
Factor Theorem, 130
Factorial, 390
Factoring, R48
  difference of two cubes, R49
  difference of two squares, R49
  by grouping, R52
  perfect square trinomial, R49
  special polynomial forms, R49
  sum of two cubes, R49
Family of functions, 66
Feasible solutions, 300
Finding a determinant
  expanding by cofactors, 363
  expanding by minors, 363
Finding a formula for the nth term of a
    sequence, 452
Finding inverse functions, 87
Finding an inverse matrix, 351
Finding test intervals for a polynomial,
    R138

Finding zeros of a polynomial
  Intermediate Value Theorem, 139
  Rational Zero Test, 136
  zoom-and-trace technique, 140
Finite
  sequence, 388
  series, 392
First differences, 454
Focus (foci)
  of an ellipse, A11
  of a hyperbola, A14
  of a parabola, A9
FOIL Method, R42
Forming equivalent equations, R71
Formula(s), R85
  change-of-base, 204, 212
  common
    area, R85
    miscellaneous, R85
    perimeter, R85
    volume, R85
  for compound interest, 196
  Distance, 3
  for distance traveled, R85
  Midpoint, 4
  Quadratic, R93, R104
  recursion, 400
  simple interest, R85
  temperature, R85
Fractal, 156
Fractal geometry, 156
Fraction, R3
  adding
    like denominators, R14
    unlike denominators, R14
  complex, R59
  denominator, R12
  dividing, R14
  equivalent, R14
    generate, R14
  least common denominator, R15
  multiplying, R14
  numerator, R12
  properties of, R14
  rules of signs, R14
  subtracting
    like denominators, R14
    unlike denominators, R14
Fractional expression, R56
Function(s), 40, 47
  absolute value, graph of, 60
  algebraic, 190
  common graphs of, 60
  composition of, 76

constant, 56, 60, 104
continuous, 116
cubing, graph of, 60
decreasing, 56
defined, 47
dependent variable, 41, 47
difference of, 74
domain of, 40, 47
even, 59
exponential, 190
    characteristics of, 192
    natural, 194
family of, 66
graph of, 54
greatest integer, 58
identity, graph of, 60
implied domain, 44, 47
increasing, 56
independent variable, 41, 47
inverse, 84, 85
    finding, 87
linear, 104
logarithmic, 202
    characteristics of, 205
    common, 203
    natural, 206
    properties of, 203, 213
name of, 47
notation, 42
objective, 300
odd, 59
piecewise-defined, 43
polynomial, 104
product of, 74
quadratic, 104
quotient of, 74
range of, 40, 47, 54
rational, 169
relative maximum of, 56, 57
relative minimum of, 56, 57
square root, graph of, 60
squaring, graph of, 60
step, 58
sum of, 74
terminology, summary of, 47
test for even and odd, 59
transcendental, 190
undefined, 47
value at x, 42, 47
Vertical Line Test for, 55
zero of, 54, 120
Fundamental Counting Principle, 426
Fundamental Theorem of Algebra, 161
Fundamental Theorem of Arithmetic, R15

G

Gaussian elimination for a system of
    linear equations, 277
    with back-substitution, 326
Gaussian model, 231
Gauss-Jordan elimination, 327
General form of the equation
    of a circle, 10
    of a line, 21
General problem-solving strategy, R81
Generate equivalent fractions, R14
Geometric sequence, 407
    common ratio of, 407
    finite, sum of, 410
    nth term of, 408
Geometric series, 411
    sum of, 411
Graph
    of absolute value function, 60
    of cubing function, 60
    of an equation, 5
        intercept of, 7
        symmetry of, 8
    of a function, 54
    horizontal shift of, 65
    of identity function, 60
    of an inequality, R126
        in two variables, 290
    nonrigid transformation of, 69
    of a rational function, 172
    reflection of, 67
    rigid transformation of, 69
    of square root function, 60
    of squaring function, 60
    turning points of, 120
    vertical shift of, 65
    vertical shrink of, 69
    vertical stretch of, 69
Graphical interpretation of solutions of
        systems of linear equations, 268
Graphing utility
    equation editor, A1
    features
        intersect, A5
        maximum, A7
        minimum, A7
        regression, A6
        root, A4
        table, A2
        trace, A3
        zero, A4
        zoom, A3
    uses of, A1

    viewing window, A2
        square setting, A3
Graphing, point-plotting method, 5
Greater than, R4
    or equal to, R4
Greatest integer function, 58
Guidelines for graphing rational
    functions, 172

H

Half-life, 197
Hidden
    equality, R79
    product, R84
Horizontal asymptote, 170
    of a rational function, 171
Horizontal line, 21
Horizontal Line Test, 89
Horizontal shift, 65
Human memory model, 208
Hyperbola, 170, A13, A14
    asymptotes of, A15
    branches of, A14
    center of, A14
    conjugate axis of, A15
    foci of, A14
    standard form of the equation of, A14,
        A21
    transverse axis of, A14
    vertices of, A14

I

Identity, R70
    function, graph of, 60
    matrix, 341
Imaginary axis, 156
Imaginary number, 151
    pure, 151
Imaginary unit i, 151
Implied domain, 44, 47
Impossible event, 436
Improper rational expression, 127
Inclusive or, R14
Inconsistent system, 268
Increasing function, 56
Independent events, 440
    probability of, 441
Independent system, 268
Independent variable, 41, 47
Index
    of a radical, R29
    of summation, 392
Induction, mathematical, 447

Inductive, 363
Inequality
  absolute value, solving, R131
  double, R5
  equivalent, R128
  graph of, R126
  linear, R129
    in two variables, 291
  properties of, R128
  satisfy, R126
  solution set of, R126
  solutions of, R126
  solve, R126
  symbols, R5
  in two variables
    graph of, 290
    sketching the graph of, 290
    solution of, 290
Infinite
  geometric series, 411
    sum of, 411
  sequence, 388
  series, 392
  wedge, 294
Infinity
  negative, R127
  positive, R127
Integer, R2
  irreducible over, R48
  negative, R2
  positive, R2
Intercept
  of a graph, 7
  x-intercept, 7
  y-intercept, 7
Intercept form of the equation of a line, 25
Interest
  compound, 195
  simple, R85
Intermediate Value Theorem, 139
Interpolation, linear, 19
Intersect feature, A5
Intersection
  of A and B, 439
  point of, 258
Interval
  bounded, R126
  closed, R126
  endpoints of, R126
  open, R126
  unbounded, R127
Inverse
  for a real number
    additive, R12
    multiplicative, R12

of a square matrix, 349
  finding, 351
Inverse function, 84, 85
  finding, 87
  Horizontal Line Test for, 89
Inverse property
  of exponential equations, 221
  of logarithmic equations, 221
Invertible matrix, 350
Irrational, R3
Irreducible
  over the integers, R48
  over the rationals, 165
  over the reals, 165

L

Latus rectum of an ellipse, A19
Law of Trichotomy, R5
Leading coefficient of a polynomial, R40
Leading Coefficient Test, 118
Leading one, 324
Least common denominator, R15
Less than, R4
  or equal to, R4
Like radicals, R34
Like terms of a polynomial, R41
Line(s)
  equation of
    general form, 21
    intercept form, 25
    point-slope form, 18, 21
    slope-intercept form, 20, 21
    summary of, 21
    two-point form, 19, 372
  horizontal, 21
  parallel, 22
  perpendicular, 22, 23
  slope of, 16, 18
  vertical, 21
Linear equation, R71
Linear extrapolation, 19
Linear Factorization Theorem, 161
Linear function, 104
Linear growth, 197
Linear inequality, R129
  in two variables, 291
    system of, 292
Linear interpolation, 19
Linear programming, 300
Linear programming problem
  optimal solution of, 300
  solving, 302
  unbounded, 303
Linear regression, 28, 33
Logarithmic equation
  inverse property, 221

one-to-one property, 221
  solving, 221
    exponentiating, 224
    strategies for solving, 221
Logarithmic expression
  condense, 214
  expand, 214
Logarithmic function
  with base a, 202
  characteristics of, 205
  common, 203
  natural, 206
    properties of, 206, 213
  properties of, 203, 213
Logarithmic model, 231
Logistic curve, 236
Logistic growth model, 231
Long division of polynomials, 126
Lower limit of summation, 392
Lower triangular matrix, 366

M

Main diagonal entries, 320
Major axis of an ellipse, A11
Mandelbrot Set, 156
Mathematical induction, 447
Mathematical model, R79, 28
Mathematical modeling, R79
  translating key words and phrases, R80
Matrix (matrices), 320
  addition, 335
    properties of, 337
  additive identity, 338
  adjoining, 351
  augmented, 321
  coefficient, 321
  column, 320
  column of, 320
  determinant of, 353
  diagonal, 366
  Distributive Property, 341
  elementary row operations, 321
  entry of, 320
    main diagonal, 320
  equal, 334
  finding an inverse, 351
  identity, 341
  invertible, 350
  multiplication, 339
    properties of, 341
  nonsingular, 350
  order of, 320
  product of two, 339
  row, 320
    coded and uncoded, 373

row-echelon form, 324
    reduced, 324
row-equivalent, 321
row of, 320
scalar, 335
scalar multiple, 335
scalar multiplication, 335
    properties of, 337
singular, 350
square, 320
    cofactors of, 362
    determinant of, 360, 363
    inverse of, 349
    minors of, 362
stochastic, 348
sum of, 335
triangular, 366
    lower and upper, 366
zero, 338
Maximum feature, A7
Method of elimination, 264, 266
Method of substitution, 254
Midpoint Formula, 4
Midpoint of a line segment, 4
Minimum feature, A7
Minor axis of an ellipse, A11
Minors of a square matrix, 362
    expanding by, 363
Miscellaneous common formulas, R85
Mixture problem, R84
Model
    mathematical, R79, 28
    verbal, R79
Monomial, R40
Mount, A19
Multiplication, R11
    of a constant, R128
    matrix, 339
    scalar, 335
Multiplicative Identity Property for real
    numbers, R12
Multiplicative Inverse Property for real
    numbers, R12
Multiplicity, 121
Multiplying fractions, R14
Mutually exclusive events, 439
    probability of, 439

N

n factorial, 390
Name of a function, 47
Natural base, 194
Natural exponential function, 194
Natural logarithmic function, 206
    properties of, 206, 213

Natural number, R2
Negation, properties of, R13
Negative
    infinity, R127
    integers, R2
    number, R4
Nonnegative real numbers, R4
Nonrigid transformation of a graph, 69
Nonsingular matrix, 350
Nonsquare system, 281
Normally distributed population, 235
Not equal to, R3
Notation
    exponential, R20
    function, 42
    scientific, R23
    sigma, 392
    summation, 392
nth partial sum, 392
    of an arithmetic sequence, 402
nth root of a number, R29
    principal, R29
nth term
    of an arithmetic sequence, 399
    of a geometric sequence, 408
Number(s)
    of combinations of n elements taken r
        at a time, 430
    complex, 151
    composite, R15
    imaginary, 151
        pure, 151
    irrational, R3
    natural, R2
    negative, R4
    nth root of, R29
    of permutations of n elements, 427
        taken r at a time, 428
    positive, R4
    prime, R15
    principal nth root of, R29
    rational, R3
    real, R2
        nonnegative, R4
    of solutions of a linear system, 279
Numerator, R12

O

Objective function, 300
Oblique asymptote, 174
Odd function, 59
    test for, 59
One-to-one property
    of exponential equations, 221
    of logarithmic equations, 221

Open interval, R126
Operations that produce equivalent
    systems, 277
Optimal solution of a linear program-
    ming problem, 300
Optimization, 300
Order of a matrix, 320
Order on the real number line, R4
Ordered pair, 2
Ordered triple, 276
Origin
    of the real number line, R4
    of the rectangular coordinate system,
        2
    symmetry, 8
Outcomes, 435

P

Parabola, 104, A9
    axis of, 105
    axis of symmetry, 105
    directrix of, A9
    focus of, A9
    standard form of the equation of, A9,
        A20
    vertex of, 105, A9
Parallel lines, 22
Pascal's Triangle, 419
Perfect
    cubes, R30
    square trinomial, factoring, R49
    squares, R30
Perimeter, common formulas for, R85
Permutation, 427
    distinguishable, 429
    of n elements, 427
        taken r at a time, 428
Perpendicular lines, 22, 23
Piecewise-defined function, 43
Point of diminishing returns, 125
Point of equilibrium, 271, 295
Point of intersection, 258
Point-plotting method of graphing, 5
Point-slope form of the equation of a
    line, 18, 21
Polynomial, R40
    addition of, R41
    constant term, R40
    degree of, R40, R41
    division
        long, 126
        synthetic, 129
    equation
        second-degree, R93
        solution of, 120

factor(s) of, 120, 165
finding test intervals for, R138
finding zeros of
    Intermediate Value Theorem, 139
    Rational Zero Test, 136
    zoom-and-trace technique, 140
function, 104
    real zeros of, 120
    of x with degree n, 104
inequality
    critical numbers of, R138
    test intervals of, R138
leading coefficient of, R40
like terms of, R41
prime, R48
product of, R42
special products of, R43
standard form of, R40
subtraction of, R41
term
    coefficient of, R40
    degree of, R40, R41
in x, R40
x-intercept of, 120
zero, R41
zeros of, R138
Position equation, R97
Positive
    infinity, R127
    integers, R2
    number, R4
Power, R20
    reciprocal, R117
Powers of integers, sums of, 451
Prime
    factorization, R15
    number, R15
    polynomial, R48
Principal nth root of a number, R29
Principal square root, 155
Principle of Mathematical Induction, 448
    extended, 448
Probability, 435
    of a complement, 442
    of an event, 436
        properties of, 436
    of independent events, 441
    of mutually exclusive events, 439
    of the union of two events, 439
Producer surplus, 295
Product
    of polynomials, R42
    of two functions, 74
    of two matrices, 339
    ˙ rational expression, 127

Properties
    of absolute value, R6
    of Equality, R16
        Reflexive, R16
        Symmetric, R16
        Transitive, R16
    of exponents, R21, R33
    of fractions, R14
    of inequalities, R128
    inverse, 221
    of logarithms, 203, 213
    of matrix addition, 337
    of matrix multiplication, 341
    of natural logarithms, 206, 213
    of negation, R13
    one-to-one, 221
    of probability of an event, 436
    of radicals, R30
    of scalar multiplication, 337
    of sums, 393
    of zero, R14
    Zero-Factor, R93
Proportion, R83
Proportionality, constant of, 29
Pure imaginary number, 151
Pythagorean Theorem, R98
Pythagorean triple, R99

Q

Quadrants, 2
Quadratic equation, R93
    solutions of, R105
    solving
        by extracting square roots, R95
        by factoring, R93
        using Quadratic Formula, R104
Quadratic Formula, R93, R104
Quadratic function, 104
    standard form of, 107
Quadratic type, R116
Quotient of two functions, 74

R

Radical
    conjugate, R32
    index of, R29
    like, R34
    properties of, R30
    simplest form, R31
    symbol, R29
Radicand, R29
Radius of a circle, 10
Range of a function, 40, 47, 54
Rate, 16

Rate of change, 16, 28, 31
Ratio, R83, 16
Rational
    exponent, R32, R33
        properties of, R33
    expression, R56
        improper, 127
        proper, 127
        zeros of, R142
    function, 169
        domain of, 169
        guidelines for graphing, 172
        horizontal asymptote of, 170, 171
        slant asymptote of, 174
        vertical asymptote of, 170
    inequality, critical numbers of, R142
    number, R3
        irreducible over, 165
Rational Zero Test, 136
Rationalizing the denominator, R31, R32
Real axis, 156
Real number, R2
    Additive Identity Property, R12
    Additive Inverse Property, R12
    Associative Property of Addition, R12
    Associative Property of
        Multiplication, R12
    Commutative Property of Addition,
        R12
    Commutative Property of
        Multiplication, R12
    Distributive Property, R12
    irreducible over, 165
    Multiplicative Identity Property, R12
    Multiplicative Inverse Property, R12
    nonnegative, R4
Real number line, R4
    bounded intervals on, R126
    coordinate, R4
    order on, R4
    origin of, R4
    unbounded intervals on, R127
Real zeros of polynomial functions, 120
Reciprocal power, R117
Rectangular coordinate system, 2
    origin of, 2
    quadrants, 2
    x-axis, 2
    y-axis, 2
Recursion formula, 400
Red herring, R92
Reduced row-echelon form, 324
Reflection, 67
    in the x-axis, 67
    in the y-axis, 67